ANALYSIS, DESIGN AND EVALUATION OF MAN-MACHINE SYSTEMS 1998
(MMS'98)

A Proceedings volume from the 7th IFAC/IFIP/IFORS/IEA Symposium, Kyoto, Japan, 16 - 18 September 1998

Edited by

S. NISHIDA
Department of Systems and Human Science, Graduate School of Engineering Science, Osaka University, Osaka, Japan

and

K. INOUE
Department of Aeronautics and Astronautics, Graduate School of Engineering, Kyoto University, Kyoto, Japan

Published for the

INTERNATIONAL FEDERATION OF AUTOMATIC CONTROL

by

PERGAMON
An Imprint of Elsevier Science

UK Elsevier Science Ltd, The Boulevard, Langford Lane, Kidlington, Oxford, OX5 1GB, UK
USA Elsevier Science Inc., 660 White Plains Road, Tarrytown, New York 10591-5153, USA
JAPAN Elsevier Science Japan, Tsunashima Building Annex, 3-20-12 Yushima, Bunkyo-ku, Tokyo 113, Japan

First edition 1999

Library of Congress Cataloging in Publication Data

A catalogue record for this book is available from the Library of Congress

British Library Cataloguing in Publication Data

A catalogue record for this book is available from the British Library

ISBN: 9780080430324

Printed and bound in Great Britain by
CPI Antony Rowe, Chippenham and Eastbourne

IFAC SYMPOSIUM ON ANALYSIS, DESIGN AND EVALUATION OF MAN-MACHINE SYSTEMS 1998

Sponsored by
International Federation of Automatic Control (IFAC)
Technical Committees on
- Man-machine Systems
- Social Impact of Automation
- Fault Detection, Supervision and Safety of Technical Processes
- Aerospace
- Automotive Control
- Marine Systems
- Air Traffic Control Automation
- Transportation Systems

Co-sponsored by
International Federation for Information Processing (IFIP)
International Federation of Operational Research Societies (IFORS)
International Ergonomics Association (IEA)

Organized by
Technical Committee for Human Interface
Society of Instrument and Control Engineers (SICE)

PREFACE

Automatic machines or smart robots are interconnected with distributed knowledge bases by nation- or world wide digital networks. They are supporting human life in various places, at the same time saving energy and resources. Technology and economic development are effecting rapid changes depending on the fields, regions and culture to the way people interact with the systems, while sometimes keeping, otherwise forsaking their own traditions. Man machine systems have to create new interaction styles between human and machine intelligence, support cooperation among different organizations and enhance situation understanding for the long and short term, or remote and local activities regarding performance, safety, security and satisfaction.

The seventh IFAC/IFIP/IFORS/IEA Symposium on Analysis, Design and Evaluation of Man-Machine Systems was held in Kyoto, Japan on September 16-18, 1998, reflecting the above-mentioned situations. At the symposium, 99 papers were presented including four plenary papers among 131 submissions. The International Program Committee (IPC) selected 87 excellent papers for the Proceedings volume. Furthermore, we added three special reports: one is the overview of the symposium by the IPC Chairman and the others are the reports on workshops held at the symposium. We believe these papers will be beneficial to the engineers and scientists in the field of Man-Machine Systems.

Shogo Nishida and Koichi Inoue
Osaka University Kyoto University

CONTENTS

FACTORY AUTOMATION II

INFORMATION SYSTEMS I

INFORMATION SYSTEMS II

FUZZY AND NEURAL NETWORK

RELIABILITY AND FAULT DIAGNOSIS

PROCESS CONTROL I

PROCESS CONTROL II

TRAFFIC CONTROL AND CAR DRIVER MODEL

SUPERVISORY CONTROL

NAVIGATION DEVICES AND SYSTEMS

FLIGHT MANAGEMENT AND SYSTEM CONTROL

VR AND TELEOPERATION

PHYSIOLOGICAL MEASUREMENT

NUCLEAR I

NUCLEAR II

INTERFACE DESIGN I

INTERFACE DESIGN II

HUMAN BEHAVIOR MODEL I

HUMAN BEHAVIOR MODEL II

INTERFACE DEVICES

CSCW

OVERVIEWS of MMS' 98 and PROPOSALS for 21st Century

Hiroshi TAMURA
Chairman International Program Committee

Scope of the Symposium

Automatic machines or smart robots are now interconnected with distributed knowledge bases by nation- or world-wide digital networks. They are supporting human life in various places, at the same time saving energy and resources. Technology and economic development are effecting rapid changes depending on the fields, regions and culture to the ways people interact with the systems, while sometimes keeping, otherwise forsaking their own traditions. Man-machine systems have to create new interaction styles between human and machine intelligence, support cooperation among different organizations and enhance situation understanding for the long and short term, or the remote and local activities regarding performance, safety, security, and satisfactions.

In the symposium, 98 qualified papers were presented in the technical sessions. They are carefully selected by reviewing 131 papers, proposed by 269 authors from 22 countries. In the opening speech IPC chairman mentioned, that while Japanese accepted new technology and automation most positively, at the same time interested in preserving the traditional culture and human based technology. Especially in Kyoto they devote more effort in maintaining the tradition than in pursuing the new. Above all, tea service and flower arrangement are the disciplines important in Japanese life.

'Tea Ceremony and Fine Technology'

Special talk titled above by **Soushu SEN**(Grand Master of Mushanokoji Senke School) was arranged together with experiencing session in the symposium. The formal tea ceremonies were explained by Grand master using well prepared slides. Tea ceremony was established by Rikyu Sen in 16th Century, and the spirits of the serving tea are summarized as below.

The seven Precepts of RIKYU:

1. Flower arranged as natural in the field.
2. Charcoal placed to let the water boil.
3. Feel cool in summer.
4. So do cozy in winter.
5. Everything ready in advance.
6. Preparc for rain, for just in casc.
7. Harmony in encounter of guests.

When Rikyu Sen established the ways of tea ceremony, it was war time in Japan. Merchants and Samurai (military class), which were newly growing by that time, used the tea services as the art of communication among and between these completely different social powers. Fine art and technology have been cultivated along with the tea ceremony. Tea ceremony has been appreciated in Japan as the ways of fostering communication mind and making moment of truth.

Discussions were made among audiences and grand master, how the formal tea ceremony changed adapting to the change of social lives of participants. Communication environment is changing by recent introduction of multimedia and other new technology. It is worthwhile to consider the essence of communication in new media age.

From Machine Age to Human Age

This was the last MMS symposium in the 20th century. Thus IPC chairman asked participants to write proposals for the man-machine systems in 21st century, - from machine age to human age. Proposals were reviewed in the closing session and transferred to the next symposium chairman. Below is the proposal of this author.

Automatic systems will be filling up the factory, office, shop, street & square, and even the home. Those systems should work as a part of nature, and look natural beyond nature.

So far, machines are designed as tools for tasks. And tools are manipulated by operators. They are required to be efficient, exact, safe and reliable.

Automatic systems in 21st century are no longer simple tools for limited users. They should be designed as essential parts of shared environment among **users and non-users**, as well as among **users in use and not in use**.

They should be friendly not only to users but also for nonusers, lovable even when not in use. **Environment should not be manipulated by limited users.** This is most essential for barrier free design.

Automatic systems should be designed thoughtfully and carefully prepared so that everybody might be encouraged to use, but never enforce one to use. Leave enough freedom for people to create new relations and functionalities.

PERFORMING ARTS' IMPACT ON ENGINEERING WORKSHOP REPORT

James L. Alty* and Gunnar Johannsen**

* *LUTCHI Research Centre, Department of Computer Science*
Loughborough University
Loughborough LE11 3TU, UK
(phone +44 1509 222648; e-mail j.l.alty@lboro.ac.uk)

** *IMAT-Lab. Systems Engineering & Human-Machine Systems*
University of Kassel
D-34109 Kassel, Germany
(phone ++49-561-804-2658; e-mail joh@imat.maschinenbau.uni-kassel.de)

Organizer: G. Johannsen (Germany)

Panelists: J.L.Alty (UK), T. Calvert (Canada), Y. Nagashima (Japan), S. Nishida (Japan), Y. S. Sagata (Japan), Y. Shibuya (Japan), H. Tamura (Japan)

Other Speakers: K. Kuutti (Finland), C. Skourup (Norway), and some others

1. PRESENTATIONS OF PANELISTS

By introducing the workshop, Johannsen stated again that we possibly can learn a lot in the future from some of the entertainment sectors, particularly the performing arts, for the field of human-machine systems (HMS) and human-machine interfaces (HMI) in industrial, transportation and service sectors. Some objectives for new and improved HMS/HMI may be more efficient and safer interaction, less routine and more creativity, as well as healthier and more enjoyable interaction.

Possible directions of R&D include the use of motion (dance, musical instruments, free conver-sation), gesture (display and) control, acoustic/sound displays (particularly, more musical ones), multi-media (when to use which media? - in which combination?), and expressiveness.

An exciting new input technique was presented by Nagashima. It consisted of a thin flexible piece of wood which could be attached to the back of the hand. It was programmed so that various actions would occur when it was bent (given a signal 0 -> 120). Various programmes were shown

1. play random notes on bend <80
2. Play complete set of arpeggios on bending
3. Alter rhythm when bent
4. Etc....

The output could also control video using MIDI.

The Ease of Use concept was referred to. Nagashima pointed out that in the arts only novices wanted easy-to-use interfaces. Professionals wanted hard-to-use interfaces which would yield interesting output with improved skill.

Johannsen wondered if the important aspect contributed by the arts was "enjoyment". Should HMS goals be changed to include enjoyment?

Alty pointed out that we needed to exploit techniques used in the arts. However, we must remember that the goals of artists are quite distinct from those of engineers. We therefore needed to "hone" the techniques used by artists to make them applicable in engineering. He pointed out that there were two quite different directions

1. Using Artistic techniques in Engineering
2. Using Technology in the Arts.

The raw results from artistic endeavour were not immediately applicable to Engineering.

Take Music: The theory was well developed (harmony, orchestration, tunes) but was too vague to be of direct use. One needed to do experiments to determine what the average person could distinguish in tone difference or timbre, and from this define design specifications.

Sagata described a system for assisting children in setting up animated computer stories. The computers enabled the children to express themselves. The contrast between their screens and the boring ones those used by adults was striking.

Shibuya wondered if entertainment might help in human-human interaction. Enjoyment was also important.

Calvert suggested that computers freed the artist from physical constraints, giving new insights which were different for everyone. He suggested that having an Artist on the Designer team would help in providing better interfaces.

Enjoyability was again stressed by Nishida. Performing arts might give new insights into research of enjoyability resulting in better interfaces for industry.

Tamura pointed out that an Electric Supply company gave the same service to customers with different needs. For example, a PC would respond badly to a power interrupt of 20 msec, whereas Air Conditioning would not be affected with a break of 20 minutes. They should not be aimed to the same market.

2. DISCUSSION

Many artistic endeavours were becoming big industry - Broadcasting, Theme Parks. Will there be whole new industries developed from the Arts?

A comment was made that it was easier to train an Engineer than an Artist. For example, PhD degrees do not guarantee creativity.

Is enjoyment the new goal that we are seeking? Will artists make things more usable?

It was pointed out that artists, in general, did not care about usability. They expected their audiences to catch up with their endeavours. They did not "dumb-down" to them.

Enjoyability is actually part of two different things:

1. The enjoyment of creativity
2. The enjoyment of using

Artists tend to give new viewpoints to Engineers. Keep their approach fresh. Another contribution to interface design was from users, often unintentionally. We could tap more into this ignorance.

Kuutti pointed out that the performing arts were very demanding and that the spin-off would always be useful. He mentioned the Lightfoot system and the Conductors Baton. The Lightfoot system had many sensors to determine foot movement, but had been found really useful for all sorts of input tasks, not envisaged at the outset. This was also true of the baton.

He also commented on the issue of "Engagement" where artists and performers are caught up in the flow of the performance. He wondered if one more arrow could be added to Alty's two arrows, that using technology to make the arts more accessible.

Some people pointed out that many tasks were quite different, and it would not be easy to make all tasks more enjoyable.

Finally, it was suggested that the goal of Enjoyment was unattainable. Once something becomes enjoyable it ceases to be so.

Enjoyment only results from the application of constraints. Total freedom is not enjoyable.

COORDINATION QUALITY IN AUTOMATION

Tetsuo Sawaragi * **Akira Ishibashi** ** **Yoshio Nakatani** ***
Kensuke Kawai **** and **Osamu Katai** †

* *Dept. of Precision Engineering, Graduate School of Engineering, Kyoto University*
Yoshida Honmachi, Sakyo, Kyoto 606-8501, Japan.
** *Dept. of Human Health Sciences, Faculty of Human Sciences, Waseda University*
2-579-15 Mikashima, Tokorozawa, Saitama 359-1192, Japan.
Also Senior Captain of All Nippon Airways Co., LTD.
*** *Dosys Co., LTD.,*
10-4 Haccyobori 4 chome, Chuo, Tokyo 104-0032, Japan.
**** *Toshiba Corporation*
1-1 Shibaura 1-chome, Minato, Tokyo 105-8001, Japan.
† *Dept. of Systems Science, Graduate School of Imformatics, Kyoto University*
Yoshida Honmachi, Sakyo, Kyoto 606-8501, Japan.

Abstract. This paper summarizes the topics and discussions made at the workshop on coordination quality in automation during the symposium. The goal of this workshop was to identify the basic problems and issues of the human's interaction with automated systems in manufacturing systems, vehicles, aviation, computer-human interaction, and process control, etc. The introduction of complex and powerful automation to a variety of high-tempo high-risk domains has led to unexpected difficulties which are the result of an increased need for, but lack of support of, human-machine communication and coordination. After reviewing the status quo of human and automated system coordination from three different practical fields of aviation, manufacturing and process control, we discuss about the new styles of coordination among human and automated agents from perspectives of social dimensions of the human-machine cooperation.

Keywords. Man-machine systems, interfaces, human-centered design, aviation, CAD, plant engineering.

1. INTRODUCTION

The goal of this workshop was to identify the basic problems and issues of the human's interaction with automated systems in manufacturing systems, vehicles, aviation, computer-human interaction, and process control, etc. The introduction of complex and powerful automation to a variety of high-tempo high-risk domains has led to unexpected difficulties which are the result of an increased need for, but lack of support of, human-machine communication and coordination. That is, new computerized and automated devices create new burdens and complexities for the individuals and teams of practitioners responsible for operating and managing high-consequence systems. Wherein, cognitive behaviors and strong affective elements come into play, and unanticipated interactions between the automated system, the human operator and other system in the workplace begin to emerge, causing serious deficiencies that can become apparent after system is delivered and is put to work.

In this workshop, we do not focus into the automation *per se*, but to the way the automated device is implemented in practice. Following the position statements by the practioners from three different domain fields, we went into the discussions on general principles for coordination design that are applicable across a variety of domains. The discussion were extended towards system development methodologies for facilitating user acceptance by allowing more extensive user participation, which is a key to success in both innovation promotion and the integration of human factors. The focused topics to be discussed at this workshop are; automation-induced tradeoffs between human workload and situation awareness, implications of automation for effective teamwork and communication, how to deal with the presence of "electric" team member (i.e., "agents"), and social/cultural dimensions on user acceptance of automation. In each of the following sections of this paper, we present the brief summaries of the position statements provided by three of the authors followed by the summary of the discussions made with the audience of the workshop.

2. TOPICS ON AUTOMATION AND HUMAN: FROM A PILOT'S VIEWPOINT

2.1 *Summary of Talk Presented by A. Ishibashi*

In the last decade, the advantages and disadvantages of highly automated advanced technology systems in aviation have been pointed out by many engineering researchers. The automation in avionics has indeed contributed much to the improving safety, reliability, economy and comfort of the flight, but at the same time it has introduced a number of "automation costs" such as the increase of task complexity caused by the introduction of FMS/CDU (Flight Management System/Control Display Unit), the lack of opacity due to strong and silent character of machine agent and to lack of feedback of the operations, etc. An advanced technology airplane shows a drastic change of the quality of the pilot's workload from "think ahead" to "plan ahead". When pilots fly the airplane manually, they have to think a little bit ahead of the current situation and to prepare for the next action. In the current glass cockpit they have to plan next flight profile data and to prepare to input them into the FMS/CDU. The pilot has to execute his tasks only based on the information appearing on the panel (e.g., the CDU displays), which causes the pilot's difficulty in keeping situation awareness.

Consequently we have had more latent factors of human errors induced by the complex highly automated system. FAA's Human Factors team pointed out some accidents specific to such a new type of airplanes in their report in 1996. Taking those accidents and other ones induced by the difficulties in interactions between flight crews and cockpit automation serious, the FAA has launched a study to evaluate the flight crew/cockpit automation interfaces of current generation transport category airplanes. We have learned from this report that the most important point these accidents imply is that the common cause of those accidents is a "failure of situation awareness". Pilots could not be aware of the situation that the airplane encounters due to complex automated systems and lack of their feedback, which led to CFIT (Controlled Flight Into Terrain) accidents.

Thus, new advanced automation systems begin to create new human factors problems. Now we have to make an attempt to prevent such human errors systematically; changing our ways of operation by revising operation policy manuals to catch up with new type of airplanes. Actually all airlines have specified new "operation policies" for the current generation transport category airplanes. In addition to them, "operation concept" is clarified concerning with task sharing, crew coordination, crew communication, monitor and crosscheck and workload management, etc. These operation policies and operation concepts are specified by the operating organization, but how to put them into practice is more important at the individual level. Pilot training is the most important method. The Crew Resource Management (CRM) training concept has been developed by NASA and US major airlines in the 1980s. And in recent years worldwide airlines are using this training concept for their pilot training. Many airlines have modified CRM to adapt it to their own habit and cultural background. Most of the airlines provide a CRM program for all pilots during annual current training which is mandatory. Usually they are a set of ground school and LOFT (Line Oriented Flight Training) simulator training. The CRM program aims not only at improvement of pilot's individual competence but at also improvement of team performance as the crew.

The highly developed automation system creates many unexpected "automation-induced surprises" and some other human factors problems which lead to some kinds of incidents and accidents. This is caused by problems of coordination or communications between the designers and users of the automated airplanes. The designers want to design airplanes in the manner of "technical-centered" automation, and the users hope aircraft to be designed according to "human-centered" automation. In the design process of the new automated cockpit the designers should hear users' opinions from their practitioners' viewpoints and users should participate design discussions to offer information learned from their abundant operation experiences. Actually Boeing Co. Ltd. has initiated a "Working Together" concept during the design of B777 and the collaboration between the users including ANA Co. Ltd. and the designers was quite successful. Cockpit automation must provide the crew with appropriate information about its intended course of action. The system must support the crew's ability to maintain a higher level of situation awareness about the automation status, behavior, intention and limitations as well as the airplane status in order to allow crews to reliably and efficiently coordinate with the system.

2.2 *Discussions*

Following the position statements summarized in the previous subsection, an active discussion was made, which can be summarized into the following three issues; Crew Resource Management, soft protection vs. hard protection, and impacts of automation on communication styles.

Fisrt, a discussion on the CRM was made. In the past, most of the training was mainly focused on the purpose of keeping the pilot's technical skills of correctly executing the prescribed procedures including normal procedures and abnormal procedures. However, these are restricted to representative and frequently occurring ones and training for a rare abnormal situation is not sufficient. To supplement this, the CRM training was introduced, the main purpose of which is to keep the pilot's problem solving skills (i.e., a capability of inventing new solutions) through coordination among crew members. Concerning with this, customization of the original CRM programs so that they can fit to their own "cultures" (i.e., nationalities and company policies) is of importance. The current statistics shows that the rate of the accidents has been decreasing in United States, Europe and Japan, but vise versa in the developing countries. This may be related to the quality and quantity of the efforts for establishing appropriate styles of coordination between the human and the artifacts including the automated systems as well as the CRM programs.

Second, concerning with the involvement of the human pilot in the control loop, pilots would like to stay in the loop. We often discriminate concepts of "hard" protection and "soft" protection; along the former concept the machine system should override the human any time, while the latter insists that the machine's task must be restricted to presenting warnings and suggestions, and the final decisions should be left to the human. This corresponds to the well-known difference of policies on the design of human-machine systems between Boeing and Airbus companies. The pilot's preference to either of soft or hard protection may depend on the degrees of trust to his own skills and other factors including the quality of the training, the geographic conditions of the airports, etc., but anyway the main reason why they want to stay in the loop is that they want to keep their active mindedness during the flight, and this is the most important factor to keep the pilot's high vigilance and good situation awareness as well as to prevent their skills from degrading. Actually during the flight the cross-checking among the crews is made and this is very effective to keep their vigilance.

Finally, the discussion was extended to the impact of automation on the shift of communication styles among the humans mediated by kinds of automation tools. Especially in the field of aviation, the users of the automated systems are pilots as well as the air traffic controllers on the ground. The introduction of automated systems to the aircraft may have effects on the styles of coordination between the controllers and the pilots. For instance, in monitoring on the ground, the controllers cannot understand the behavior of the aircraft operated by the automated systems if the pilot is unaware of the operations made by the automated system and fails to report to the ground. Currently, a new application of automation to communications is emerging, which is called "Datalink". This is designed to replace the traditional audio-voice link between pilot and controller with an automated transfer of digital information. Implementation of datalink systems has a number of generic implications. This may increase the pilot's and controller's "head down" time for interpreting the displayed digital information, and automatically transmitted data cannot be always shared in awareness by the both parties, which may cause contradiction of situation awareness between the air and the ground. As this example shows, the implications of automation with respect to the coordination quality between human and the machine as well as inter-human communication must be discussed from a variety of viewpoints.

3. COORDINATION IN ENGINEERING DESIGN

3.1 *Summary of Talk by Y. Nakatani*

In a field of manufacturing, an automation tool have contributed much with respect to a design phase. Most representative one is a CAD (Computer Aided Design) tool automating drawing tasks and parametric design tasks. The next generation CAD will come to contain product data in addition to drawing data so that they could be shared in realtime among distributed design group members as well as among different sections (i.e., designers, production engineers, service engineers dealing with the maintenance, etc.).

Advanced CAD tools should have a capability of supporting a designer's creative task in a coherent way with the human expert designer does, and it has to enable not only design data but also design knowledge to be transferred among the design participants. One of the characteristics of the expert designers' ways of thinking is a case-based reasoning (CBR). That is, an appropriate prior design precedents is retrieved and is flexibly utilized in a new design problem. Wherein, designers must be able to understand the intention of the original design precedents to successfully use them, but current design automation tools are insufficient for supporting such a task since they do not contain any design intention, decisional process information nor design knowledge. They are to be reconstructed by the human designer using a diagrammatic reasoning from the digitized data of CAD, which requires much design expertise. Design modification with lack of such expertise and with mis-understanding of the design precedents would lead to the design causing a malfunction of a product, which incurs a lot of time and expense cost.

In Japan such deep knowledge of design expertise have been transferred from expert designers to novice designers on the job training (OJT) directly through a man-to-man communication. However, current severe economic recession causes frequent re-assignment of expert designers to different work sections, which makes the transfer of knowledge quite difficult. We are now seeking for alternative styles of knowledge transfer such as schooling education and formal documentation of design expertise, both of which are not going well as expected. The main reason why the experts are unwilling to transfer their knowledge is very simple; it is impossible to teach or formally describe design expertise by isolating that from the practice of design activity. A design support and design expertise transfer using a case-based approach is very useful and practical as an alternative to such a formal documentation. In order to offer an effective support to a new design problem, design precedents must be stored together with implicit problem solving know-how used and realized in them. Moreover, the efficient and effective retrieval of prior cases may depend upon how clearly a designer identifies the problems in current design situation. This implies the importance of a situation awareness and/or situation assessment in an engineering design field.

Based upon the above investigation, we are now developing a new design support tool based on a concept of a "Post-It" function. Since it is not practical to ask the design expert to make a formal documentation, more informal transfer of expertise using simple memos is now under investigation. Examples of such memos are like "Do not change the length here!", "Warning! This is a special device", and other list of cautions, and these memos are put electrically on the corresponding part of the CAD data. To guarantee the usability of such a function, we are now designing interfaces using a fixed memu-type memos as well as a free formatted memos.

3.2 *Discussions*

The idea of a "Post-It" function implies a number of important issues with respect to human-automation and/or human-human communication. This can be discussed from the perspectives both of the sender and the receiver of the messages (i.e., memos). From the sender's perspectives, its easiness of just putting memos is quite attractive and can save much efforts needed for communication in comparison with the formal documentation of design expertise. Moreover, the memo attached on CAD data may function as an useful index for his own memory retrieval of the design precedent. From the receiver's perspectives, it is very useful because the memos highlight what and where is important within a total design data, thus he/she can focus the attention without difficulty. Besides that point, such incompleteness of the exchanged message via memos would contribute much to promoting the receiver's proactive efforts to interpret the meaning of the message and the reason why the original designer has put that there.

The contents to be carried by memos must be investigated with respect to their potentials for activating such knowledge sharing and transfer without any explicated communication protocols. For instance, two kinds of messages would be possible such as a memo saying "I designed this portion in this way because ..." and a memo just saying "do not change here". Such a difference may cause the subsequent ways of knowledge sharing among the sender and the receiver as well as the following communication styles. Another issue pointed out by the audience of the workshop was concerning with its applicability toward a knowledge transfer tool for an cross-section design team such as a knowledge transfer from a manufacturing designer to a production engineer. In this case, the words and ontologies are not common and this may cause communication problems even when they exchange a simple memo. Moreover, the transfer must be made in a bi-direction with frequent exchanges of mutual feedback rather than in a uni-direction.

Finally, the problem of knowledge transfer may be caused by the frequent re-assignments of the experts designers. But on the other hand, fluid participation in a variety of tasks or regular turnover of the team members from experts to novices do contribute much to enriching experts' adaptability and to keep the level of organizational knowledge that is learned by the organization itself rather than within the individual members. The most big problem causing the transfer of expertise is individual designer's "closed" attitude keeping him/her from learning new things and/or new ways of thinking (i.e., adherence to what they are getting accustomed). Such an attitudes is often observed in a medical field, and a proposed "Post-It" function should be investigated concerning with how it can get rid of such barriers and make individuals open to new ideas.

4. AUTOMATION IN PLANT ENGINEERING

4.1 *Summary of Talk by K. Kawai*

Design of automation systems in power generating plants is quite different among United States, Europe, and Japan. In Japan, during the last 20-25 years, the efficient and effective usage of process computers have been sought for, and we established a one-man operation. We introduced a voice announcement system, which is connected to the automation server machine and provides voice messages to the human operator. The purpose of this is to help the operator identify what kind of action is executed by the automation system, which was designed so that it can tell the operating stuff by giving them pre-messages before it initiates the automated actions. This automation system has a number of different automation modes such as normal startup and shutdown modes and an emergency mode, and the automated recovery from the anomalous states is designed respectively. In this system, the human operator can maintain a high awareness of the plant situation as well as be able to confirm the automation system's intention by just looking into the CRT and by selecting the operation modes. The design policy herein does not concern with elimination or reduction of the operating stuff. Rather, our focus is how to design the work environment so that even just one operator can handle and manage his tasks without heavy overload. Different from Japan, in European countries a function group control (FGC) concept is prevailing, and the basically sequentially automated systems using a micro processors are well developed, and in the United States sequential control as well as modulating control are of major concern. The design specifications demanded by India, China and Middle East countries are quite different, and we plant engineers repeat the discussions with the customers and propose ideas that are customized to their respective cultures and peculiarities of their operating environments.

As a more advanced plant automation system, we designed a turbine-generator vibration diagnosis expert system. The system configuration consists of monitoring and diagnosis subsystems, which utilize two kinds of reasonings of forward and backward reasonings. Operating stuff is forced to input data that are not measured by sensors. To prevent the delay and mistakes of their inputs, the system installs both automated and semi-automated modes to supersede the human. Moreover, certifying the operator's situation assessment is very important, thus we designed a special window and a way of utilizing the plant trend history data for this purpose. The trend of developing such a highly advanced automation system is currently refrained. This is because of the enormous R&D costs needed to tackle with the complexity of the problems to be solved. In order to manage this complexity, we are now developing a new design architecture in which emerging agent technologies are applied to concentrated monitoring of multi units. The design principle therein is to divide the system appropriately into a number of subsystems, to apply the most suitable algorithm to the individuals, and to let them coordinate with each other using agent-related techniques.

Concerning with the issue of coordination, communication design is of importance, and we are now investigating into a typical man-to-man coordinating activity of "negotiations". Negotiation is a formal communication opportunity as well as a complex problem solving task, and an actual business negotiation must be carefully designed to establish a "win-win relationship" among the two parties through debates and discussions. The critical factors for successful negotiations are summarized as follows:

- Do a homework (i.e., much preparation is needed)
- Share agenda
- Ask suitable questions
- Make efforts to validate hypotheses
- Be aware of the danger of last hours

We are now developing a new computer-operator communication design based on the above characteristics of negotiation.

4.2 *Discussions*

An idea of an ideal future human-automation relationship using a metaphor of negotiation is quite profound in the following senses. First, a "win-win relationship" symbolizes regarding the human and the automation system as the equivalent partners, rather than regarding one of them superior to the other nor regarding the roles assigned to the human and the automation as fixed. It is still usual that there happens a contradictory situation where the designer and the user of the automation system meet conflicts. If they keep insisting on what they believe and do not change their views, it may lead to a catastrophe. To avoid that they have to cooperate with each other to change a current conflictive situation into a new coordinated one that can lead to a safer plant status and the better operations. Human and computer should be looked as two parties coming from the same positions and having different expertise. For instance, computers can know about the feasibility of certain solutions in a physical sense and it can search a huge state space, whereas the human comes to the negotiation table with ideas about what to prefer in terms of criteria. Thus, negotiation should be done between them by exchanging those expertise defined on the feasibility space and the preference space back and forth so that they can reach to the satisfactory solution.

Related to this, we have to consider about what the ideal commutation style is and how the relationship between the human and the computer should be especially in a time-critical situation. Unfortunately, we have no theory on how to handle such a situation properly. Wherein, a conventional formal communication is not expected to function well, but some higher level of communication must emerge among the two parties. We have to extend our views on a human-computer relationship toward a social interaction relationship between the human and the computer. How to establish a consensual domain between the computer and the human should be discussed in the future interface design (Winograd and Flores, 1986).

Another interesting point clarified during the talk is the need of customization of the automation system according to the difference of the customers. The automation system is exported together with its training system using a full-scale simulator. At that time, not only the automated system but also the training program should be customized. Automation cannot used in a stand alone way but its function is determined by how it is used by the human operators. Therefore, some programs corresponding to the CRM programs in aviation should be developed also in the plant engineering.

From a more broader point of view, the future automation systems should be designed based on its life-cycle perspectives. That is, it is very important to design the automation depending upon how it is maintained by the customers after the operation has set in. Automation will bring about a drastic change in the styles and quality of maintenance. For instance, conventional corrective and regularly scheduled preventive maintenace may be abandoned due to the improvement of reliability of sensing and self-inspecting capabilities of the automation systems. The advanced plant automation system based on an expert system architecture may suffer from a troublesome problem concerning with the maintenance of a huge knowledge base. It is expected for the automation system to have a capability of self-recovering when it meets any malfunctions caused by the inappropriate hardware, software and knowledge installation.

5. CONCLUSIONS

In concluding the discussions of this workshop, we would like to highlight a number of common problems that are clarified by the experienced user and system designers. The discussions made after the individual position statements from three different domains commonly contain a number of fundamental issues that the designers of human-machine systems have to think of when a new automation is introduced to a particular task domain; how the communication between the human and the automated system should be designed, and how the automation effects on the conventional communication styles of the humans. And these issues can be translated into the following more general query on "Who is responsible for the design of coordination; a human user, a system designer or an artifact of automated agent ?"

In the conventional design of human-machine systems, the designer of the system has taken an external position lying outside of the interactions made between the human user and the automated system. He has been an external observer and his task has been on how to design the optimized interaction between them based on the technology-centered idea. In the past, a human-centered design concept was proposed (Billings, 1997). However, this is still a mirage and its practical realization is thought to be suspicious (Sheridan, 1997). As far as a human-centered design concerns, this is no more than a conventional design principle; how to optimize the predicted interactions between the human and the machine. As all three speakers commonly pointed out in their position statements, the interactive domain formed by the human and the automated system as well as by the humans mediated by the automated system is too complex to be predicted exhaustively and to be optimized at the design stage. Wherein, a number of novel, unknown relations do emerge among them, some of which may indeed lead to the bottom-up design of coordination, but some others bring about a catastrophe (i.e., a disorder). Depending upon how they are used by the human, the artifacts do not always function as the original designer expected, rather ironically they may contribute to enlarging the discrepancy between the human and the machine. It is often said that we are killed by the kindness of the current interface design, and this may be true in some sense.

The conventional design principle of human-machine systems has been based on the conception that a human user is a likely source of significant variance in system performance, then it is better to control this source of variance early and/or to minimize such human factors in system development. However, a human itself is too complex to be controlled, and the human is at the same time a source of creativity that sometimes exceeds out of the designer's restricted views.

Another point commonly depicted by the speakers is to establish "mutual knowledge of intent" among those three parties (i.e., a user, a designer and an automated system). This is not novel idea per se, but what is to be emphasized here is that "mutual knowledge of intent" is not any product to be obtained and shared, but is a continuous process of getting more and more aware of their partners (i.e., an emerging property). The pilots' preference to "soft protection" and the ideas of "Post-It" function as well as of a metaphor of "negotiation" provided by the speakers are all imply that extensive and proactive user participation is a key to success in both innovation promotion and the integration of human factors. Moreover, as a true partner of the user an automated system also has to have an analogous capability of "sociality" to be embedded within the interactions with the human user and to form a creative interactive domain together with him. This may lead to the bottom-up emergence of coordination, rather than the a priori designed coordination.

Finally, we would like to express our thanks to more than 40 participants for their coming to the workshop in spite of the late evening schedule. Especially we would like to appreciate Prof. Thomas B. Sheridan (MIT, US), Prof. Hendrik G. Stassen (Delft University, the Netherlands), Prof. Ernst A. Hartmann (Germany) and Prof. T. Buro (Germany) for their active contribution to the workshop. Our appreciation is extended to other active participants including Prof. P.A. Wieringa (the Netherlands), Prof. B. Riera (France), Prof. R. Beyer (Germany), Prof. M. Kitamura (Japan), Prof. T. Inagaki (Japan) and Prof. M. Takahashi (Japan) for their useful comments provided during the workshop.

REFERENCES

Billings, C.E. (1997). Aviation Automation: The Search for a Human Centered Approach, Lawrence Erlbaum Ass. Pub., Mahwah, N.J..

Sheridan, T.B. (1997). Human-Centered Automation: Oxymoron or Common Sense?, *Proc, of IEEE Int. Conf. on System, Man and Cybernetics*, Vancouver, Canada.

Winograd, T. and Flores, F. (1986). *Understanding Computers and Cognition: A New Foundation for Design*, Ablex Pub. Corp., Norwood, N.J..

KANSEI SCIENCE AND MAN-MACHINE INTERACTION IN PERFORMING ART

Tsutomu OOHASHI *
Emi NISHINA* *

* Chiba Institute of Technology,
* ATR Human Information Processing Laboratories
** National Institute of Multimedia Education

Abstract : The drastic change of the direction circumstance of performing arts and considerable progress of Kansei science, we are confronted some complex and sensitive issued in man-machine interaction in performing arts that have not been considered before. Applying Kansei science, we have developed a new approach named Expression engineering to overcome this situation. And we would like to introduce a large scale of live performing art in Expo that is one of the good example of highly systematized man-machine interaction.

Keywords : Kansei science, Expression engineering, man-machine interaction, performing art

1. INTRODUCTION

Considerable progress in human science, especially brain science, have offered new observation on the characteristics about information processing in humans. Thus, we are confronted some complex and sensitive issues in man-machine interaction in performing art that have not been considered before.

In dealing with the issue on man-machine interaction in performing art, we have laid stress on "Kansei", which is a concept developed from the inherent Japanese way of thinking. In the way of traditional Japanese performing arts, such as Noh and Bunraku, there are a variety of elements that emphasize interaction between different senses, attach importance to information that is difficult to symbolize or segment, possesses a heterogeneous and/or non-linear spatio-temporal structure, and the expression to approach the border between the area of conscious perception and that of non-conscious perception, etc.. We cannot ignore these achievements of traditional Japanese performing arts. In this sense, "Kansei" is valuable for our research nevertheless, there is not a scientifically adequate definition by which to establish the traditional concept of Kansei. It is, therefore, difficult to effectively use this concept in the domain of science.

For Kansei to be a useful concept in scientific fields, we must first define several terms related to Kansei : Kansei, a general term for the function of the brain which contains a positive emotion such as beauty, amenity, interest, pleasure as an essential element : Kansei information, information that induces a Kansei reaction : and, Kansei information processing, a procedure to generate, synthesize, arrange, store and transmit Kansei information. These definitions are based on an examination of functions of the brain, which administers Kansei, applying the latest research in brain science. We first developed a method to measure Kansei reactions in relation on brain function, and then examined reactions to various kinds of Kansei information, both consciously perceived and non-consciously, synthesizing the results of physiological and psychological experiments [1].

2. THE DRASTIC CHANGE IN THE DIRECTION CIRCUMSTANCE OF PERFORMING ARTS

2.1 Change of users' needs

Let's pay attention to a subject that actually influences man-machine interaction but is not easily recognized from the view point of technology. First, the culture of uses, that is, audience or spectators, has recently drastically changed. We should review this fundamental change of needs in performing art. In our circumstance, a traditional "cultural code"

collapsed and the diversified cultural contexts intermingle in different fashion without a common code. In other words, this is a collapse of genre, a structure without a common denominator.
Under this circumstance, the technical field concerning man-machine interaction is no longer allowed to rely on just past promises. Actually, we are required to apply a wide variety of materials originating from different cultures and different systems to our work. Nevertheless, with sound reinforcement in theaters, for example, there are very few operators who can handle all the heavy metal music, Japanese Jiuta and Bulgarian chorus by him or herself without difficulty.

2.2. Subsidence of cultural code, the West
Westernized idea or philosophy is losing its power it once enjoyed and the Pan-global idea or philosophy has begun to hold the controlling power. Also in the field of man-machine interaction, relative culture has begun prevailing. In this circumstance an idea or philosophy from "follow the West" that persists in us raises friction here and there.
The western culture that exported the direction paradigm of performing art actually has a history no longer than 500 years since the Renaissance. On the other hand, Georgia or Bulgaria, for instance, has a much longer history since B.C., and has enjoyed the blossom of remarkable culture in the 12th-13th century, more than 200 years before the Renaissance. The cultural sphere of Hinduism also preceded the western culture over 30 centuries. These historical point of view is very important to evaluate the maturity of culture. We should know that these problems lie not only in directing software, but also technological aspects of performing art.

2.3 From language-depending expression to Kansei-appealing expression
Internationalization have made the opportunity to gather a lot of people from different countries and societies together to enjoy the same performing arts. They grow up in their own culture, and generally understand only their own language. So, it is difficult for us to overlook the movement from language-depending expression to Kansei- appealing expression in today's stage performance.
For example, in Shakespeare's "Hamlet" translated by Kouson Fukuda, a famous translator, lines of 188 pages with 697 characters per page were uttered in the three-hour performance. Against this, in the Landscape Opera "Gaia," which we would mention later, the words used were less than one hundredth of that. In addition, since most of the words were composed within the songs, so pure lines would be less than one page, well constructed to those of Hamlet. The conceptual effect of the story or literature has a less impact today. A sensitive and physiological message that stimulates audience directly via their eyes and ears, provides a Kansei reaction more deeply and easily. When we encounter a foreign culture, we can enjoy much more if there is no language barrier. Therefore, as the program of recent performing arts becomes new and large scale, the restraint of language is more significant. Instead, sound and lighting become more important for direction. A large-scale outdoor event involved with international enterprise is a typical example. This tendency increases the needs for new Kansei direction technology including stage mechanism, lighting, acoustics and special effects. No doubt, it is dangerous to seat a conventional type of director at the head of a new, Kansei-oriented direction system if he has poor knowledge of technology and is just armed with literary or psychological translation. To avoid the risk, it is important to select and educate some staff members who are qualified and strong in the technological world, including sound and light. We should entrust the nucleus of Kansei-oriented direction to him or her. It may sound rough, but it may be the most safety way.

3. KANSEI SCIENCE AND CHANGE IN EVALUATION STRUCTURE

Declining of words and literature in performing arts should have made a great impact on the evaluation system for the acceptance of the work by society. "Criticism," the main system of evaluation, the matrix of which is literature and aesthetics depending on words, cannot do much for the Kansei information. Thus, in recent years, more realistic methods like questionnaires or statistics have often been used. Moreover, a completely new method for reliable evaluation on Kansei reaction has been tried by extracting the physiological reaction for obtaining evaluating material. This has strong persuasive power completely different from the critic's subjectivity or from the questionnaire survey. We believe it will have great influence on society.
For example in our studies, which were conducted by our newly developed Kansei measurement methods, we discovered that the Gamelan music of Bali, which is rich in non-stationary high frequency components above the audible range can significantly increase the α-EEG activity of listeners relative to otherwise identical sounds from which the high frequency components had been removed. Although no subject would recognize the high frequency components as sound in a listening experiment, they nevertheless perceived, with

statistical significance, the sounds containing such frequency components as more pleasant. We use the term "hypersonic effect" to describe this phenomenon [2]. In our latest research, we have begun to probe the anatomical and neuronal mechanisms underlying those phenomena using positron emission tomography (PET). We found that when the recordings contained high frequency components above the audible range, regional cerebral blood flows in some deep brain structure increased dramatically, with a statistical significance as compared to the flow in these regions when they listened to the same sound but lacking frequencies above 22kHz [3].

We believe that such clear results on hypersonic effect, which is a phenomenon at the delicate location between the perception and non-perception, would be brought out by our newly developed approach based on the scientific concept Kansei. This suggest that the Kansei could be a useful scientific conceptual apparatus for man-machine interaction.

4. PROFILE OF "EXPRESSION-ENGINEERING"

Now the situation of the performing arts shows that introducing, uniting, and orchestrating various separated fields with each own special discipline breaks the current situation and leads to a broader perspective. We propose a new interdisciplinary approach named "Expression engineering", that is designed for the innovative intellectual system in order to play this role as a functional core of direction of performing arts. It is an integrated system of science, art and technology. It is not a summation set of multiple fields but it gives an all-around, intellectual and Kansei function in one person.

What elements of system constitute our Expression engineering for direction of performing arts? The way of developing its outline is sketched by quoting the elemental fields as follows:

First, the main task of Expression engineering is to grasp the situation of media technology under explosive development and to extract its full potential. To achieve this, gather all information on many kinds of man-machine interface, sound, lighting, special effects and stage setting as well as computer technology to control their functions. Then, develop, evaluate and systematize these. Scientifically speaking, the application technique of these elements is essential, including electronic engineering, mechanical engineering, acoustic engineering, visual information science, computer science, system and control science and audio-visual techniques. Moreover extend the intellectual network to less conspicuous fields such as the strong electrolyte techniques involved with the power supply, inorganic chemistry, basic physics, etc. that are directly related to special effects.

Secondly, we are dealing with an urgent and important task to build a supporting system for new production method to enhance Kansei dimension. For this purpose, it is controlled by the physiological aspect. Most of all, stage logistics based on past literature and aesthetics do not work. An approach to integrate human science and information science, for example, biological information science, brain science and ethology, has been developed. It created some useful information for direction of performing arts as we wrote before.

To handle the diversity of many cultural codes from the point of global view, a circuit to input knowledge on cultural anthropology is developed. This allows us to effectively fill the great void that had not been touched. In order to help the development of innovative direction style, direction device and work, their evaluation system, we developed a method to determine the audience's Kansei reaction objectively by direct measurement of the physiological reaction of the brain's activation in particular. It is beginning to show its powerful effect as we mentioned before.

These elements of Expression engineering are not used independently or separately, but they are used in unit after highly organic integration and are making their functions effective. Here, systematization and control is a key point. It is not enough to organize qualified experts - at least one person, and much more is desirable, who can integrate the whole system, is required to reach the goal. The person has to control both the creating field and the technical field. Improving the efficiency of joint possession of information is one reason. Another reason is that creation of an unknown image in information space, which is a subject of performing arts direction, can be completed only in a single person.

As a prototype from the method of "Integration with intellectual interface" is now in the practical phase (Fig. 1) [4]. With this, we are facing a preposterous idea quite different from the conventional concept, that is, one person always has to master many specialty fields that seem impossible even for one lifetime. However, it is not impossible to attain the technical knowledge over the multiple fields required for the solution by application of a new approach. The source of activity in Expression engineering comes from a point where all information related to man-machine interaction is eventually centralized in one person.

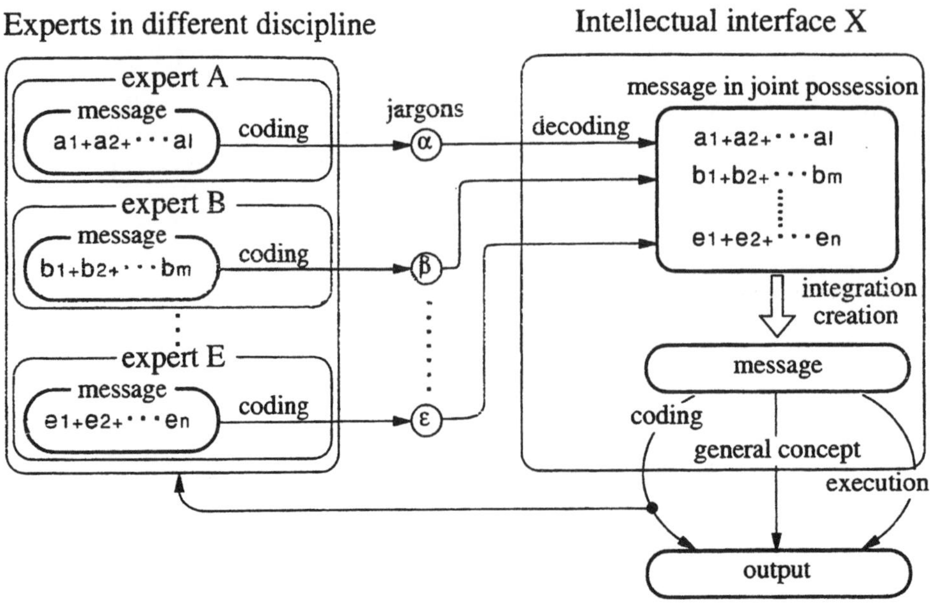

Fig.1 Advanced information processing with intellectual interface

5. EXAMPLE OF PERFORMING ART APPLYING EXPRESSION ENGINEERING

Finally, we would like to introduce an example developing a large scale of the performing art by applying our Expression engineering : Landscape Opera "Gaia" presented in the International Garden and Greenery Exposition, Osaka, Japan 1990.

5.1 Concept of the performance

Landscape Opera "Gaia" was one of the largest outside event held by the Association of Expo90 on May 19 and 20, 1990 to embody the idea of the Expo. The executive producer was Mr. Sakyo Komatsu who is a famous SF novelist. The production, performance direction and music composition were conducted by Shoji Yamashiro which is T.Oohashi' s artist name. By fifteen hundreds of performers combining with the latest technology for performance, Landscape Opera "Gaia" described the earth and its ecosystem from their birth to the near future.

To create this large scale performing art, a completely new direction system was constructed by integrating multimedia and breaking down the genre of performance. We selected suitable direction techniques from ordinary method, developed hardware and software to overcome their insufficient function, and reconstructed a new style of integrated functions around key persons of this project. In this international Expo, meta-cultural Kansei-appealing expression was required to overcome the difference in nationality, language, generation and gender of the huge audience from all over the world. The structure and function of the space for this performance was substantially different from the ordinary style, as we mentioned later. The knowledge from Kansei science were used to control this space. We took the audio-visual technologies more seriously to induce many kinds of Kansei reactions directly to humans. Since there was a strong suspicion that ordinary system for direction of performance was limited to this huge scale of performance, the direction system was mere extensively and carefully considered than the content of the performance itself. Thus a real time performance control system and its network was developed.

5.2 Space for the performance

This Expo sent a message throughout the world that human and nature can co-exist in a healthy, symbiotic relationship using high technology as basis. To symbolize the message, a lake of forty thousand square meters, named "Sea of Life", was constructed in the central area of the Expo. There was an outside pavilion named "Aleph" designed by Oohashi using this lake. "Aleph" consisted 141 fountain elements of 9 groups and a water split system as hardware. These fountains generated sound containing high frequency components above the audible range with natural fluctuation structure, and high density visual information with fluctuation structure. These audio-visual information with many of flowers and greenery planted around the lake activate our Kansei brain. In the "Gaia", this complex landscape was used as the performance space. Several stages with different structure and function were constructed around the lake which could be seen from all its bearings. Temporal stage-passage was built across the lake for the performance with parade (Fig.2).

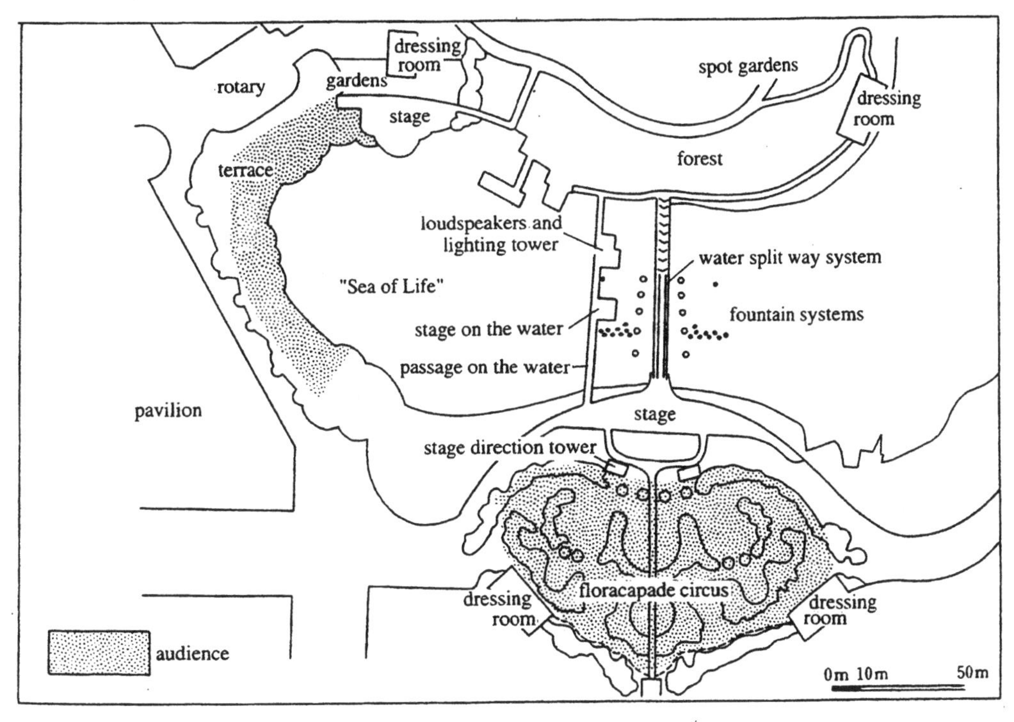

Fig.2 Performance space for Landscape Opera "Gaia" in Expo'90

As a result, this large space with a complex structure consisted a unique stage without comparable example before.
For performing arts played outside, it would be one of the ultimate target to find out a pleasant environmental information for humans. Through "Gaia", we were engrossed in this difficult but important target, and realized the importance of highly harmonized space with natural environment. In addition, the result of the field works of various kind of festival in different culture over the world.

5.3 Human interface system for expression
In the performance space, many kinds of systems for expression were built in as man-machine interface. One of the main system was water expression system. As we mentioned before, the fountains and the water split system were controlled by a computer combined with originally created contemporary music. Applying the conception of the Eastern aesthetics, these computer-controlled music fountain systems interacted with the wind, and changed their figure. For the water split system, a large water tank of six hundred tonnage was built under the lake to appear the passage on the bottom of the lake dramatically. To control these systems by a computer simultaneously with sound and lighting system, the development of the hardware, the composition of computer music and the lighting program should be completely well organized and developed by a person, T.Oohashi, who was familiar with all equipment as well as the software. The same style of inter-disciplinary direction was employed for "Gaia". As a result, water expression system was controlled by the MIDI code of the music, and lighting and all the direction were controlled by the SMPTE time code on computer system. This control system using computers was a record-breaking success at that time (Fig.3).
The sound reproduction system and lighting system, which had been prepared for "Aleph", were improved. Special system using fireworks and laser lighting were also prepared in this space.

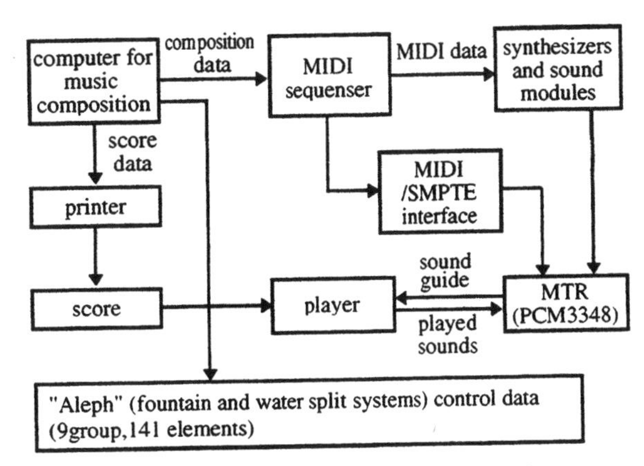

Fig.3 Sound and water control system of "Gaia"

5.4 Construction of the performance
As the audience of the Expo came from all over the world, Kansei appealing expression should be effective independently from their cultural basis. Applying Kansei science, we selected

Kansei-appealing performance and music from various cultures as element of the performance, for example, polyphony of Mubuti Pygmy, Balinese Gamelan and Cak, Japanese drum performances, Noh, western classic music, jazz, disco music etc.. By using these elements based on different cultural origin, a new contemporary music "Symphonic Suite Gaia" was composed by Shoji Yamashiro. The pre-recorded music material and live music played by the musicians participating "Gaia" were mixed at the performance space. To realize effective Kansei-appealing expression, lighting, special effects and timbre of sound should be controlled exactly. For this purpose, a person who was well versed in sound and lighting technology was appointed to the general director, and the hardware suitable for the performance were prepared. A lighting designer, sound director and a water performance director worked under the general director's instruction.

5.5 Human interface system for direction

In such a large scale of outside event, there are many impediments to the direction. Moreover, it was difficult to repeat rehearsal many times with fifteen hundreds of performers. Therefore we took a new method that developed and relied on a structure for real-time expression inducement and feedback control during the performance. For this purpose, we built a tower for stage direction of six meters height, so that the general director could look around the whole performing space and system. The general director and the other directors of each technology stayed in this tower and gave instruction to the operators and floor directors (Fig.4). This real-time network made possible the direction of dynamic live performance "Gaia".

The Landscape Opera "Gaia" engrossed eighty thousand audience and achieved its purpose. We think that this live performing art is also one of the good example of highly systematized man-machine interaction.

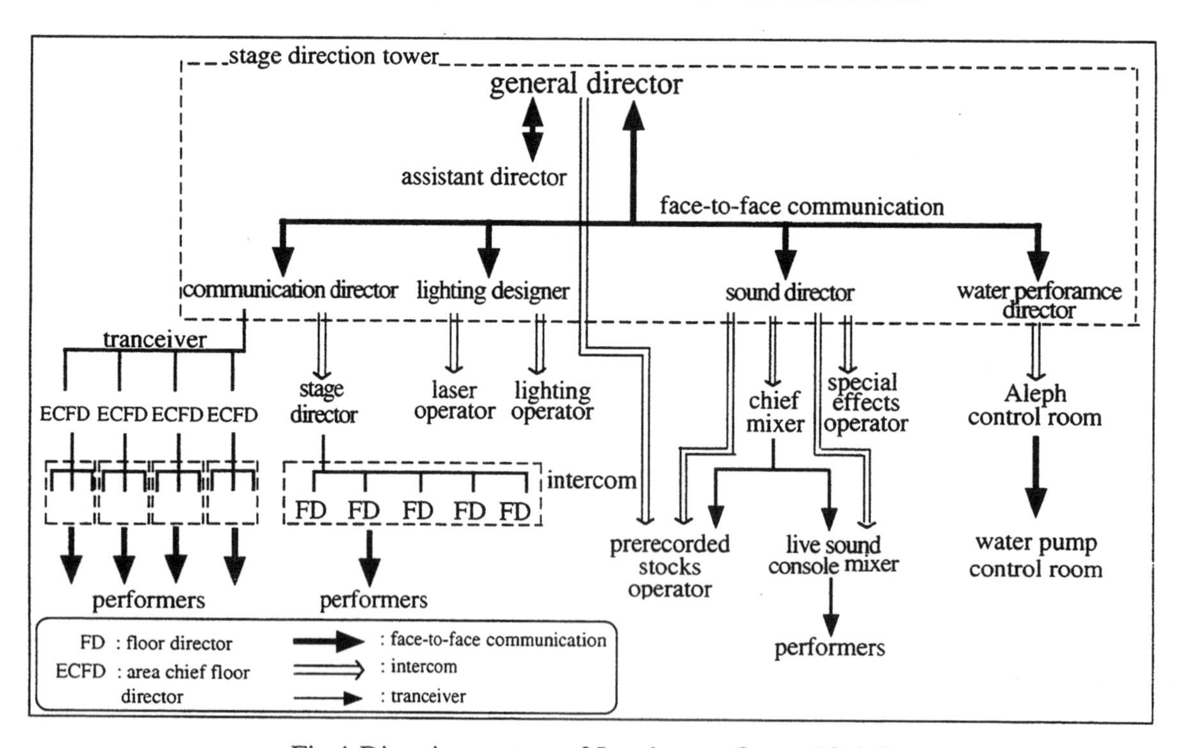

Fig.4 Direction system of Landscape Opera "Gaia"

REFERENCES

[1] Oohashi,T. et al. (1997) Physiological and psychological effects of high frequency components above the audible range - An approach to KANSEI information, AIMI International Workshop for KANSEI, Processings, 110-115, Genova.

[2] Oohashi, T. et al. (1991) High frequency components above the audible range affects brain electric activity and sound perception, Audio Engineering Society 91st Convention, preprint 3207, 1-25, New York.

[3] Oohashi, T. et al. (1995) Activation of deep brain structure by sound frequencies above the audible range in human, IBRO World Congress of Neuroscience Abstract, 478, Kyoto.

[4] Oohashi, T. (1989) *Science for Information Environment*, Asakurasyotenn, Tokyo.

MAN-MACHINE ASPECTS OF MINIMALLY INVASIVE SURGERY [1]

Henk G. Stassen * Jenny Dankelman *
Kees A. Grimbergen *, Dirk W. Meijer ****

** Man-Machine Systems and Control Group,*
Faculty of Design, Construction and Production,
Delft University of Technology, Mekelweg 2, 2628 CD Delft
*** Academic Medical Center, Meibergdreef 9, 1105 AZ*
Amsterdam, the Netherlands

Abstract: Minimally invasive surgery is at this moment one of the outstanding developments in surgery. In this type of surgery the actual operation is performed through a number of small incisions in the skin. In the operations special instruments are inserted via trocars, i.e. tubes which allow the surgeon to bring instruments or sensors into the body. The view at the operating field comes from a laparoscope, a camera presenting a two-dimensional image on a monitor. The minimally invasive surgical technique has many potentional benefits for the patients. However, compared to open surgery there are severe disadvantages for the surgeon, such as the loss of three-dimensional visual feed back and proprioception, the disturbed eye-hand coordination, and the poor ergonomic design of the surgical instrumentation, and of the working place. At this moment the differences beween open and minimally invasive surgery can mainly be ascribed to differences in the manual control task. In this paper, the man-machine aspects of the traditional open operation process will be compared with those of the minimally invasive surgery process. Especially the consequences of the restricted perception in minimally invasive surgery will be discussed. Some future developments will be discussed. *Copyright © 1998 IFAC*

Keywords: Minimally invasive surgery, man-machine systems, eye-hand coordination, manual control

1. INTRODUCTION

In conventional, open surgery access to the internal body is provided via one large incision. Via this large incision the surgeon is able to use the hands to palpate and to manipulate the tissue. In addition, simple instruments are used. In this way, the surgeon directly observes the intestines of the patient, the hands and instruments, Fig. 1. So, his eye-hand coordination is not disturbed.

In minimally invasive surgery the access to the internal body is achieved by creating a working space by means of abdominal insufflation with CO_2 gas. Subsequently several cylindrical trocars or cannulas, inserted in for instance the abdominal wall of a patient. This method of access allows the introduction of a laparoscope and of thin instruments to treat the internal tissue of the patient, see Fig. 2 (Cuschieri *et al.*, 1992). To observe the interior of the body, i.e. the abdomen, a laparoscope with a camera system is used, providing the

[1] This research is part of the Minimally Invasive Surgery and Interventional Techniques (MISIT) program of the Delft Interfaculty Research Center on Medical Engineering (DIOC-9)

Table 1. Taxonomy of abdominal surgical techniques (Sjoerdsma, 1998).

Visual observation / Manipulation	Direct	Indirect via the laparoscope
Direct	Open surgery (1)	Hand-assisted laparoscopic surgery (2)
Indirect via instruments	Small-incision surgery (3)	Laparoscopic surgery (4)

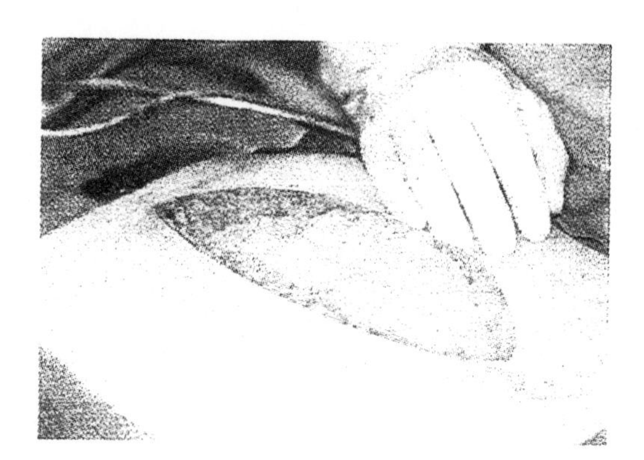

Fig. 1. The conventional open surgical process.

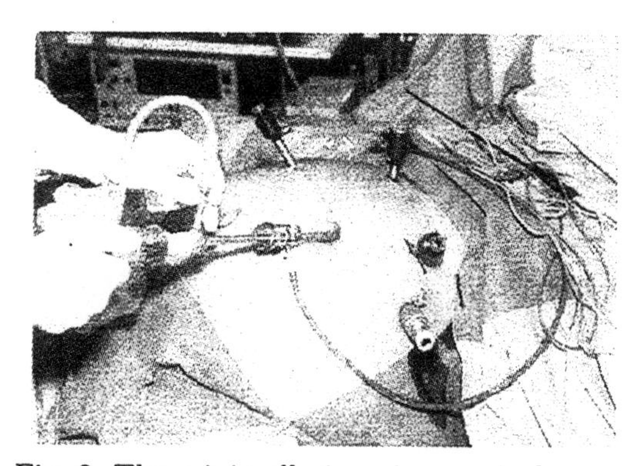

Fig. 2. The minimally invasive surgical process.

surgeon with a 2D-image of the anatomy and the instrument handling. The laparoscope is directed by the assistant. To manipulate and to treat the tissue, long instruments like shears and forceps are used (Melzer *et al.*, 1992; Jansen and Cuesta, 1993). Exposure of the working space inside the body, i.e. the abdominal cavity, is achieved by insufflation with carbondioxide gas.

The minimally invasive surgery has great benefits for the patients, however, it yields a series of disadvantages to the surgeon. Firstly, the eye-hand coordination is severely disturbed, since the images on the monitor are just 2D-images on a flat monitor screen in contrast to the 3D-visual feedback in open surgery. Moreover, the control of the images is performed by the assistant, so the surgeon is provided with a 2D-image taken from a point of view, different from his own eyes. Also the assistant will move the laparoscope, which is experienced as annoying, and which will not contribute to movement parallax. In addition, the laparoscopic instruments have certainly not the rich functionality of the human hand. The freedom of movement is limited and the perception of forces, velocities and positions is to a large extend decreased. Furthermore, the movement of the hands controlling the laparoscopic instruments, is in some directions just opposite to the movement of the tip of the laparoscopic instruments. In conclusion the short term benefits for the patients, such as less damage to the body, reduced risk for inflammations and shorter stay in the hospital, are accompanied by a more difficult way of operating for the surgeons. The reasons to still treat the patients by the difficult, laparoscopic surgery technique, and the role of the man-machine discipline to reduce the limitations and to improve the minimally invasive surgery process are very well described by Cuschieri. He stated: "*Indeed the essential attribute of the new surgical approach is the execution of established surgical operations in a manner which leads to the reduction of the trauma of access and thereby accelerates to recovery of the patient. In this respect it is far wiser that we should adapt and develop existing technology to enable the performance of well-tested and validated procedures than to embark on new, substitute and invalidated operations imposed by the restriction of the current technology*" (Cuschieri, 1992). Or to say it in the words of Johnson in the Lancet: "*The operation inside is identical with the open procedure, the only difference being the method of access*" (Johnson, 1997).

The trade-off between trauma of access and operation difficulty has led until now to several techniques for the treatment of abdominal pathologies. The way of access primarily determines the difficulty by restricting the direct hand contact, the eye-sight, or both. Sjoerdsma gives a taxonomy of the four different techniques which can be distinguished according to the tissue contact and visualisation, Table 1 (Sjoerdsma, 1998).

The open surgery (1) and the laparoscopic surgery (4) have just been introduced. The two newly mentioned techniques (2 and 3) are introduced to partly overcome the disadvantages of the minimally invasive procedure. The small-incision surgery (3), a technique to minimize the incision of open surgery, leads to a technique where the surgeon still has a direct view on the tissue, but is not able to manipulate the internal tissue by the hands or finger tips (Majeed *et al.*, 1996). The instruments used are the same as with the open surgery. The small-incision access or mini-laparotomy is mainly applied for operations where one small incision is sufficient to reach the entire operation field, such as gallbladder removal or a hernia repair. The last technique (4) is the hand-assisted laparoscopic procedure. This technique is a mixture of the laparoscopic and the open operation techniques (Bemelman *et al.*, 1996; O'Reilly *et al.*, 1996). The essential feature is that the surgeon introduces one hand in the abdominal cavity in the standard laparoscopic set-up; a plastic sleeve fitted to the abdominal wall and tightened around the arm of the surgeon prevents the leakage of gas. In this way the surgeon has direct contact with the tissue to be treated with one hand. The method is in particular used for those cases where a larger incision is needed to remove a resected bowel or spleen.

In order to elucidate the man man-machine system challenges, it is fruitful to summarize the advantages and disadvantages for patient and surgeon. Table 2 indicates the consequences and effects of the different operation procedures for patient and surgeon for a number of aspects. Here it should be mentioned that some aspects are of vital importance to the patient, whereas others are vital for the surgeon. For example, for the patients the consequences of the minimally invasive surgery (4) with reference to damage to the body, hospital stay and recovery time after the operation seem less than those for open surgery. However, the complexity of the operation procedures and the handling of possible disturbances are experienced by the surgeon as far more difficult in minimally invasive surgery. Other important issues are possible wound infection, the number of persons in the operation theatre, and the training of the surgeons. An additional feature of minimally invasive surgery is the easy access to on-line teleconsulting. Finally, it is interesting to see that although the actual minimally invasive operation process is more expensive, the total cost of the overall medical treatment can be substantially lower.

For the Man-machine disciplines the aspects: operation complexity, handling of disturbances, number of persons in the operation theatre (including the logistics and work organisation), the training of the surgeons, and the possibility of teleconsulting are of direct concern. In order to understand the different operation techniques, the operation procedures will be presented in the form of block diagrams, elucidating the interactions between patient and surgeon.

2. SURGEON-PATIENT INTERACTION

In open surgery, the surgeon has two possibilities to manipulate the tissue in the operating area,

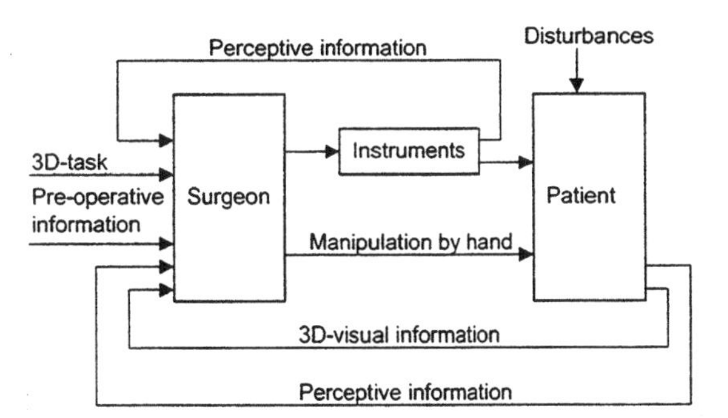

Fig. 3. Block diagram of the open surgical process (1).

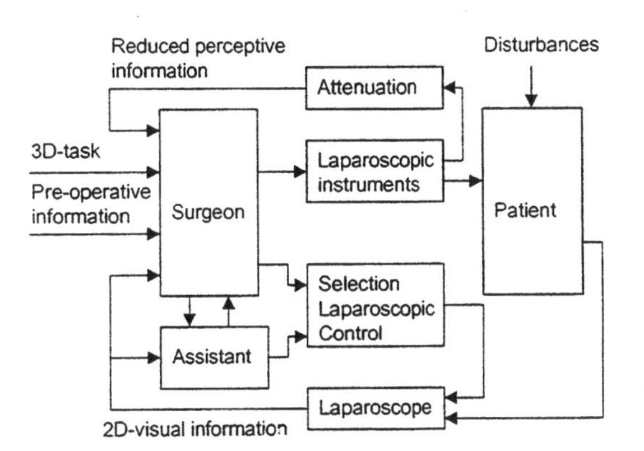

Fig. 4. Block diagram of the minimally invasive surgical process (4).

Fig 3, i.e. by the hands and by the surgical instruments; both the activities provide the surgeon with direct feedback (Stassen *et al.*, 1998). In addition 3D-visual cues inform the surgeon about the actual state of the surgical process. The actions are initiated and based on the 3D-task to be executed, the pre-operative information, such as CT-scans, MRI- and Röntgen images, and the on-line information that is fed back. In this case the eye-hand coordination is normal; disturbances or large variation in the patient's anatomy are accurately detectable by direct vision and direct palpation.

In minimally invasive surgery the surgeon manipulates the tissue via the laparoscopic instruments, Fig 4. Due to friction and, in general, the poor ergonomic design of the instruments, the feedback of perceptive information is disturbed, so only reduced perceptive information will reach the surgeon (Sjoerdsma *et al.*, 1997). In addition, no 3D-visual information is available; instead only 2D-visual information originating from the laparoscope is fed back. Moreover, most of the time the laparoscope is controlled by the assistant, and is providing the surgeon with an image from a point of view different from his eyes. Hence the eye-hand coordination is totally disturbed.

In order to overcome the difficulties introduced by the minimally invasive procedure, see Table 1, two

Table 2. Consequences of the operation procedures - open surgery (1), hand-assisted laparoscopic surgery (2), small-incision surgery (3) and laparoscopic surgery (4) - for patient (pat.) ans surgeon (surg.). - negative consequence, + positive consequence, o - no negative/no positive consequence.

Operation technique	open (1)		small-inc.(2)		hand-assist (3)		laparosc. (4)	
Aspects patient/surgeon	pat.	surg.	pat.	surg.	pat.	surg.	pat.	surg.
Damage to the body	-		o		o		+	
Hospital stay	-		+		+		+	
Recovery time, before going to work	-		+		+		+	
Operation complexity		+		o		o		-
• observation		+		-		o		-
• handling		+		o		o		-
• operation time	+	+	o	o	o	o	-	-
Disturbances		+		+		-		-
Wound infection	-	-	o	o	o	o	+	+
Number of persons in operation theatre		+		+		-		-
Training surgeons		+		-		+		-
On-line teleconsulting		o		o		+		+
Medical cost of surgery	+	+	+	+	-	-	-	-
Overal cost of treatment	-	-	o	o	o	o	+	+

other procedures were mentioned, i.e. the small incision surgery and the hand-assisted laparoscopic surgery. The small incision procedure, Fig 5, is in fact a variant of the open surgery, however, the incision is that small that the hand can not be brought into the cavity, so the hand perception and manipulation are missing, but the eye-hand coordination as well as the 3D-vision are in tact. In the hand-assisted procedure, Fig 6, one tries to compensate for the missing hand perception. Therefore a small incision is made through which one hand can be brought into the body; so the surgeon is directly connected via a sleeve with the patient. In this way the hand perception and manipulation are partly restored.

Often a combination of open and laparoscopic surgery is achieved. For example, in removing a part of the small and/or large bowel, the dissection of the bowel is minimally invasively performed, whereas the removal of the sick part of the bowel is done by the open surgery procedure.

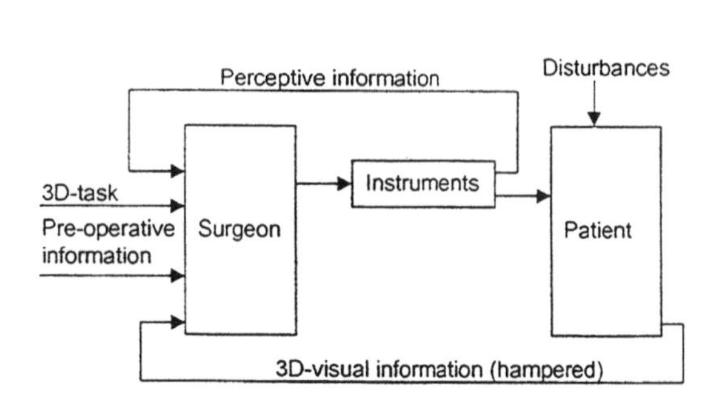

Fig. 5. Block diagram of the small-incision surgical process (3).

3. MODEL OF THE SURGEON'S COGNITIVE BEHAVIOR

Many human control behavior models have been reported in literature (Stassen *et al.*, 1990). In the excellent review paper on Human Performance Models, Pew stated that "*Modelling: builds from the task-analytic output to produce a formal, often quantitative description of the behavior of one or more people in interaction with equipment. A model of human performance requires first a model or representation of the system and environment with which the people are to function*" (Pew and Baron, 1982). This statement makes two important points. Firstly it says that a human behavior model is always related to a particular technical or medical environment, and secondly it tells us that the human operator should have an internal representation of the system and environment with which he is interacting. In the process described in this paper the interacting

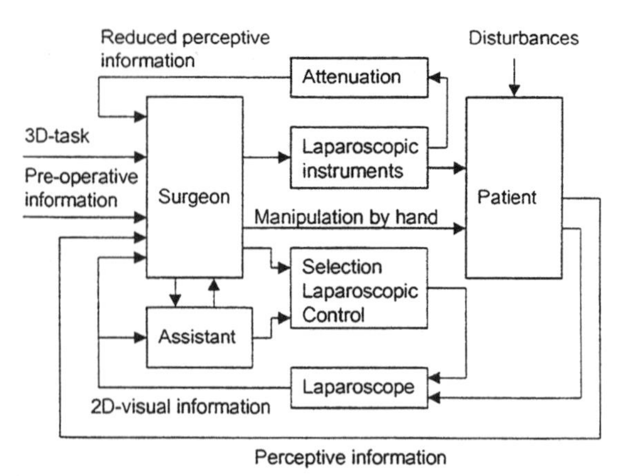

Fig. 6. Block diagram of the hand-assisted laparoscopic surgical process.

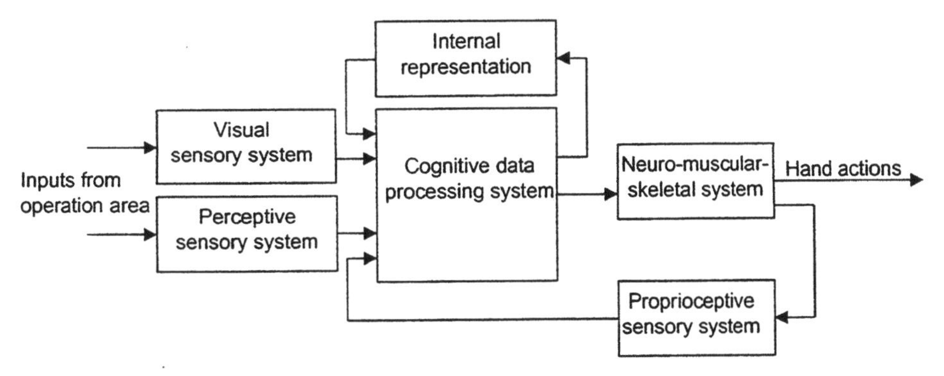

Fig. 7. Block diagram of the surgeon's cognitive functions.

environment and system is the operation theatre, the operation protocol and the patient. The internal representation includes the anatomy and pathology of the patient (Stassen *et al.*, 1990).

There are many choices in deciding which modelling approach will be used, i.e. one can select from a behavioral, physiological or system engineering approach. Here we will describe the surgeon's cognitive behavior from the view point of the functional, system engineering methodology, taking into account the physiological basic functions, Fig. 7. So, the surgeon's behavior is to be described as a cognitive data processing system, that receives the inputs of the environment from visual and perceptive cues via the visual and perceptive sensory systems. The cognitive data processing is considered from the visual and perceptive sensory system inputs to build an internal representation, stored in a long term memory, and to generate control signals to the neuro-muscular-skeletal system, in order to achieve an operation action. Via the proprioceptive sensory system the actual state of the neuro-muscular actions will be fed back to the cognitive data processing system. With this very simplified human behavior model the basic man-machine problems in minimally invasive surgery can be elucidated. Finally one important aspect should be mentioned, in particular with reference to future developments. Rasmussen states that human behavior can be distinguished at three levels (Rasmussen, 1983): "*Skill-Based Behavior represents sensory-motor performance during acts or activities which, following a statement of an intention, take place without conscious control as smooth, automated, and highly integrated patterns of behavior ...... At the next level of Rule-Based Behavior, the composition of such a sequence of subroutines in a familiar work situation is typically controlled by a stored rule or procedure which may have been derived empirically during previous occasions, communicated from other persons know-how as instruction or a cookbook recipe, or it may be prepared on occasion by conscious problem solving and planning ...... During unfamiliar situations, faced with an environment for which no know-how or rules for control are available from previous encounters, the control of performance must move to a higher conceptual level, in which performance is goal-controlled and Knowledge-Based.*" So, at the Skill-Based Behavior level the inputs are signals, at the Rule Based Behavior signs and at the Knowledge Based Behavior symbols or symptoms, Fig. 8. All three levels occur in minimally invasive surgery, i.e. motor skills are needed at the Skill-Based Behavior level, the operation protocol determines the Rule-Based Behavior and finally the interpretation of the 2D-images, the variations in the patient's anatomy and pathology and possible disturbances during the operation may require creativity and decision power at the Knowledge-Based Behavior level.

4. BASIC MAN-MACHINE PROBLEMS IN MINIMALLY INVASIVE SURGERY

In the introduction it has been stated that the patient's gain from the minimally invasive techniques will fire back to the surgeon's performance difficulties. Following Rasmussen's taxonomy we will discuss the basic man-machine problems (Breedveld, 1998a) at the three levels.

Skill-Based Behavior: The skills needed to perform the different minimally invasive operations are seriously hampered for two reasons. First, the perception of forces, velocities and displacements of the tissue treated can hardly be perceived, because of the poor dynamics of the instruments used (Sjoerdma *et al.* 1997). Hence the extended proprioception (Simpson, 1969), needed in order to achieve the motor skills, is missing. Secondly the number of Degrees of Freedom (DOF) of the instruments used by the surgeon is far below those of the hand. Mostly there is one DOF in the grasping device, whereas there are only four DOF's in the position of the instrument via the trocar, Fig. 9. So, the possibilities to handle the instruments are hampered by the limitations of the surgical instruments. Therefore intensive training

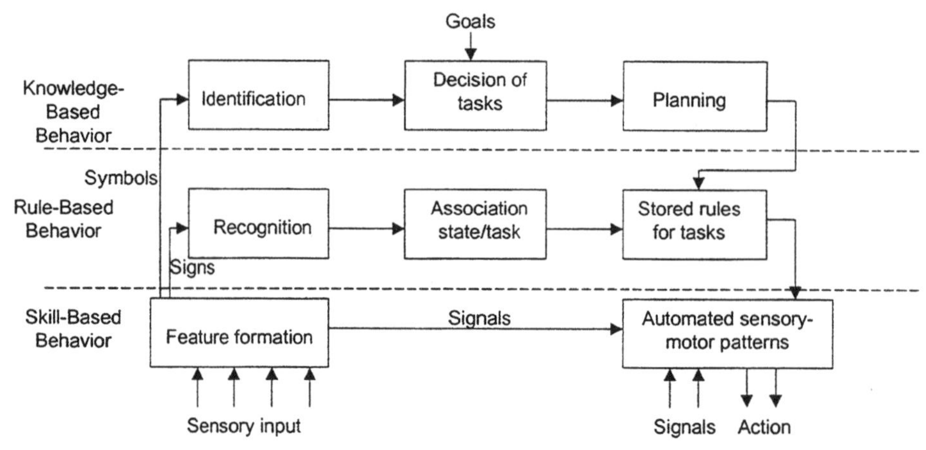

Fig. 8. The three-level behavior model (Rasmussen, 1983).

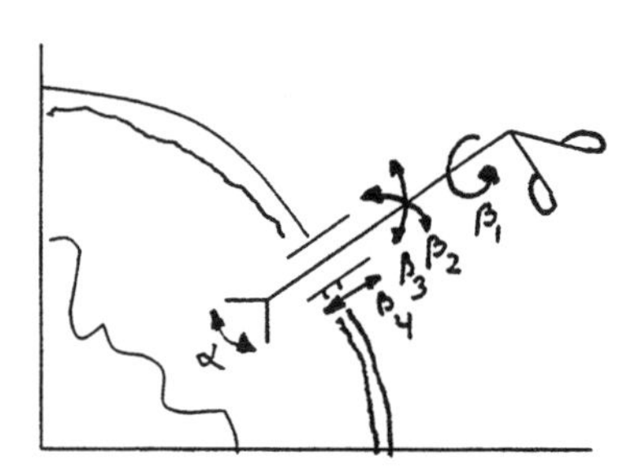
Fig. 9. The five DOF for the manipulation of the grasping device via the trocar. α: opening and closing; $\beta_1 \cdots \beta_4$: positioning.

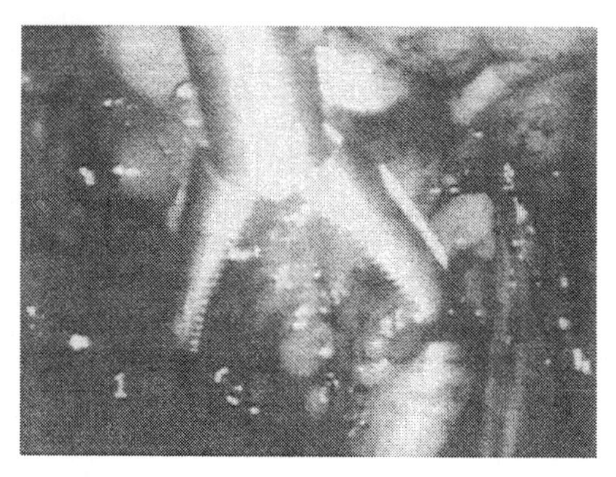
Fig. 10. Laparoscopic camera picture during dissection of the large intestines. The picture shows that the surgeon is only manipulating the tissue which is fully visible.

sessions are required in order to achieve the experiences needed to perform a successful operation. In addition to this fact the visual information of the operating area is a 2D-laparoscopic camera picture with no shadow, no stereovision and no movement parallax, whereas there is also a misfit of accomodation and convergence. Hence, it is extremely difficult to interpret the visual information directly. So, the expected Skill-Based Behavior can hardly be achieved due to the missing eye-hand coordination.

Rule-Based Behavior: In fact the operation protocol is to a large extent determining the rules to be followed during an operation. The rules may be disturbed by the uncertainty of the 2D-camera picture, by the variations in the anatomy and pathology of the patient, and by non-expected events that may occur. Here it is almost needless to say that the surgeon has to execute a 3D-task. Examples of unexpected events are:

- Internal bleedings or damage to tissue.
- Variations in anatomy and pathology not detected on CT-scans or MRI-images.
- Missing surgical instruments due to failing logistic procedures and instruments that break during the operation process (Sjoerdsma, 1998).

All those events ask for rules and procedures to overcome the problems, for better logistics and for an integration of the pre- and peroperative information. A good example of Rule-Based Behavior is the actions to be taken during the dissection of the colon. After removing a sick part of the intestines one has to suture both the ends of the intestines to each other, so one shortens the length of the bowel. Therefore, one has to dissect the intestines over a large area from the abdominal wall. The only information available to the surgeon is the 2D-laparoscopic camera picture on which he sees the tissue to be treated and the tip of the laparoscopic instrument, Fig. 10. A standard rule now is that the surgeon only will dissect that part of the bowel which together with the instrument is fully visible. He never will try to dissect the tissue at a not clearly exposed area with the risk of damaging intestines, nerves or arteries and venes. So many actions to be performed at the Rule-Based Behavior level are the consequence of the difficult operation technique.

Knowledge-Based Behavior: To build a 3D internal representation of the operation area on the basis of the 2D laparoscopic camera pictures, and the CT-scans and MRI-images, is a very complex but essential problem. Additional complications are the complex mental processing of the different coordinate system of the laparoscope and the instruments, as well as the mixing of the peroperative, laparoscopic information with the preoperative CT- and MRI-images, since these visual information sources are obtained under different circumstances. Firstly, in minimally invasive surgery the abdomen is insufflated with carbondioxide gas under pressure in order to create an operating space, and secondly the position of the body with reference to the gravitational force, i.e. the per- and pre-operative visual information is principally different. Taking into account these facts, it is clear why it is so difficult to establish a fine-tuned eye-hand coordination. These mental processes of the surgeon will be executed at the knowledge-based level. Moreover, the handling of suddenly unexpected events will be achieved at the highest cognitive level.

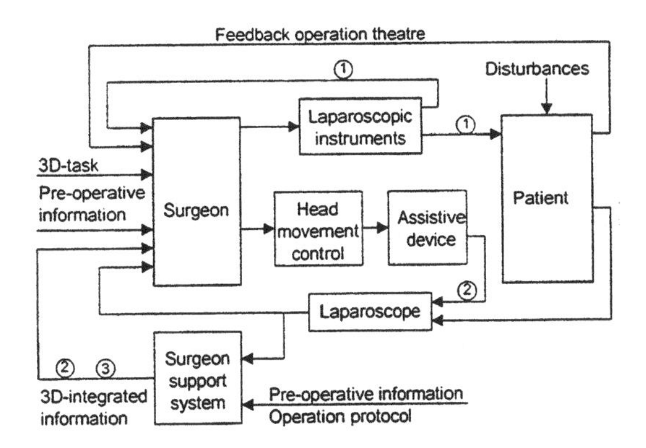

Fig. 11. Block diagram of the minimally invasive surgery process in the near future: Manual control mode. The meaning of the symbols ⊙ are explained in the text.

5. RESEARCH APPROACHES

There are two principally different ways to approach clinical problems in minimally invasive surgery (Stassen, 1997). On request of a medical professional or on the base that an engineer gets a bright idea, a new instrument or system is developed and designed. In this case engineers are showing their medical counterparts what technologically is possible, and how inventive they are. The result of such an approach is often a hi-tech instrument or system that does not fulfil the needs and demands of the practicing medical doctor, and then the system can be added to the collection of unused instruments and systems. We will call this approach the *technologically-driven approach.*

Another approach is the *clinically-driven approach.* Here we observe the surgeon in his work environment; we perform a task analysis and discuss his activities during and after the actual operation in order to detect fundamental problems and limitations occuring during the operation process. In this way, as a joined enterprise, the functional specifications for an instrument or system can be defined. This is a very difficult process, since the (para) medical professionals and the engineers talk different languages, have different cultures and do not know each other field. However, it is the best possibility to come to real, applicable systems. During the entire process of defining functional specifications, the development and design of a prototype, the technical evaluation, the functional laboratory evaluation, the clinical animal experiments by the experimental surgeon, the development of a series of clinical prototypes, the clinical evaluation, and finally the design of the prototype, the medical input should be garanteed. In the field of industrial man-machine systems research we call this user-participation (Johanssen *et al.*, 1994). Although this approach is time consuming for both the parties, it is our experiences that this approach is the only one to guarantee a minimum number of risks to develop unusable systems.

In particular the clinically-driven approach is the most suitable to reach short-term successes. The technologically-driven approach may in the long term also lead to medical innovations in a more inefficient process of adaptations.

6. SHORT-TERM RESEARCH ON MINIMALLY INVASIVE SURGERY

The Figs 3, 4, 5 and 6 showed that in all four operation procedures the surgeon can be considered to be a manual controller; the comparison of Fig. 3 and Fig. 4 showed the typical problems in minimally invasive surgery. Short-term research now is dedicated towards the elimination of the limits of the presently used minimally invasive operation protocols in order to improve the manual control tasks of the surgeon, Fig. 11. The authors believe that this only can be achieved at short notice by following the clinically-driven approach, since the solutions to the problems which will become available are clinically recognized, and are well supported by the medical profession. Here it should be noted that medical doctors are individually educated and trained in their practical skills; as a consequence in general they are hesitant to introduce new concepts. Moreover, the language and culture of the disciplines are very different.

Skill-Based Behavior: The surgical instruments used now a days are in general ergonomically poor designs. Many very basic characteristics are not known, such as the damage and trauma due to grasping and manipulating tissue, and the DOF's needed to perform an operation in a posture acceptable for the surgeon during long operation procedures. The consequences of the loss of proprioceptive and perception feedback of forces, velocities and position of the grasping devices, due to friction in the surgical instruments are also unknown. To overcome this problem, in principle two ways can be followed: active and passive feedback.

• Active feedback: One accepts the poor mechanical design, and fits sensors at the tip of the surgical instruments. The output signals of these sensors are used to activate small actuators in the handgrip of the surgical instruments. In this way a feedback is realised which easily can be adjusted to the sensitivity characteristics of the hand; it can be optimized in terms of man-machine systems (Sheridan and Ferrell, 1974). The problem is that the instrument becomes complex, fragile, expensive and difficult to sterilize. From the viewpoint of control engineering it is the wrong way. One should always try to design the best open loop system, so that closed loop system is not needed.

• Passive feedback: The best approach is to take away the friction and hysteresis in the open loop, so that a direct feedback is guaranteed by the surgical instrument. An example is the low-friction grasper (Herder *et al.*, 1997) based on the rolling link mechanism (Kuntz, 1995). In this way, instruments are obtained that are simple in construction, relatively cheap and sterilizable. In fact one has achieved extended proprioception (Simpson, 1969).

An other important aspect of the minimally invasive instrument is the decrease in DOF's, which limits the surgeon in manipulating the tissue. As far as known to the authors, there are no reports on the number of DOF's needed to perform a minimally invasive operation (Cuschieri, 1992). New tools, very different from normal grasping assistive devices are based on the cosmetic hand prothesis with individually controlled fingers (Cool and Hooreweder, 1971). Summarizing, this part of the research is focussed on the restoration of the functions available in open surgery, by improving the laparoscopic instruments. This aspect is indicated by the symbol (I) in Fig. 11.

Rule-Based Behavior: In order to execute 3D-tasks at a rule-based behavior level one should be able to generate on-line 3D-information of the operation area. Unfortunately, only 2D-laparoscopic video images recorded by an assistant from a different point of view, are available. The assistant collects visual information by controlling the laparoscope according to the instructions of the surgeon and to a set of rules. Common rules are (Danis, 1996) "*Tip of the moving instrument should stay in the middle of the picture.*" and "*the abdominal wall should stay at the top of the picture.*" So the three basic problems - 2D-video images, the control of the laparoscope by the assistant and the different points of view - are disturbing the depth perception. There are many ways to try to restore the loss of depth perception. An excellent and complete review on the state of the art is reported by Breedveld; a selection of the most promising approaches will be discussed (Breedveld 1998a). Basically three depth information sources exist, i.e. pictorial information, parallax and visuomotor cues (Sheridan, 1996). Because the laparoscope is carrying its own light source at the tip, no pictorial information such as shadows will appear in the laparoscopic pictures. Hence, it will be difficult to spatially position the laparoscopic instruments accurately. By introducing several light sources via additional trocars, called illumination cannulas (Schurr *et al.*, 1996) - thus not connected to the laparoscope - shadows are introduced. However, the first rather primitive experiments were not very successful (Voorhorst, 1998). In addition, there are at least two more incisions in the skin of the patient to be made. Two types of parallax can be recognized: stereovision and movement parallax. Stereovision can not be used, because all standard laparoscopes are monocular. However, with a monocular laparoscope the possibility exists to introduce movement parallax which concerns shifts in the picture seen by one eye introduced by head movements. So, by moving the head the relative position of visible objects changes, creating spatial information. In literature this is often incorrectly called motion parallax. Motion parallax is caused by moving the objects externally. Both the forms of parallax introduce shifts in the retinal images. Experiments (Smets *et al.*1987; Stappers, 1992) have shown that motion parallax and movement parallax have important contributions to depth perception (Stassen and Smets, 1997). In particular, Voorhorst showed the importance of the perception - action coupling in the case that the surgeon is controlling the laparoscope by the head (Voorhorst, 1998). Finally the two visuomotor cues, accommodation and convergence, are of no direct use in laparoscopic surgery, since the surgeon's eye cannot control the lens of the laparoscope.

Many supporting aids for depth perception have been reported (Breedveld, 1998a). The majority of these supporting aids is based on the disparity of both the eyes. In particular stereo-endoscopes with active or passive eyewear shutter-glass systems are discussed (Cuschieri, 1996; 1998). A detailed review on stereo-television systems can be found in Motoki et al. (1995). Here it should be noted that in the medical literature the emphasis often is put on stereovision, and that the movement parallax and motion parallax are considered to be less important additional cues. Although no evidence for this statement can be found in literature, the authors believe that it is just the other way around, and the motion parallax and the movement parallax might be the most important depth perception cues. Different movement parallax systems have been reported; they can be categorized in two groups: independent movement parallax systems and head-coupled movement parallax systems. The first category supports the surgeon with motion parallax. In general those systems are very complex and expensive, and therefore they are hardly used in practice. One of the very few commercially available systems is the Varifocal Mirror Display (Enderly, 1987). This system projects the points of a set of 2D-images on a mirror mounted on a loudspeaker, excitated by a 30 Hz signal. The forward/backward movement of the mirror generates some kind of a third dimension. A great advantage of this system is that movement parallax, stereovision, accommodation and convergence are all in harmony with each other, and so the observer thinks he really observes a 3D-object. Head coupled movement parallax systems are all controlled by the surgeon himself. This is realised in such a way that the picture on the monitor moves in the direction opposite to the surgeon's head. In this way it gives the surgeon the impression of looking through the monitor (Breedveld, 1998a). Note that the surgeon experiences movement parallax, whereas the rest of the operation team in the operation theatre experiences motion parallax. Head movement parallax systems are relatively simple and cheap, and therefore they are used in many applications. Several head - coupled movement parallax systems for laparoscopic application are described in literature (Cuschieri, 1994; 1995). An advantage of this system is that the surgeon himself is able to move the laparoscope intuitively in the direction he wants. The disadvantage is that he is limited by the constraints of the restricted number of DOF's of the laparoscope. Therefore there is a strong need to develop flexible laparoscopes (Treat, 1996) or probably endoscopes with a 90° flexion (Breedveld, 1998b).

A system, widely used in industry and in particular in aviation, is the Head Mounted Display, HMD. Such a display generates the pictures on two small television screens mounted in a helmet in front of the eyes. The surgeon's head movements are controlling the pictures in order to create movement parallax. Disadvantages of these systems are the large computing time delays between the surgeon's head movement and the movement of the pictures, hence the perception-action coupling is disturbed (Voorhorst, 1998). An additional problem is that the HMD isolates the surgeon from the medical team and the patient (Cuschieri, 1994; 1995). However, by projecting the picture on the patient this problem can be prevented (Bajura *et al.*, 1996). Laparoscopic applications with different kinds of HMD have shown some interesting results (Geis, 1996).

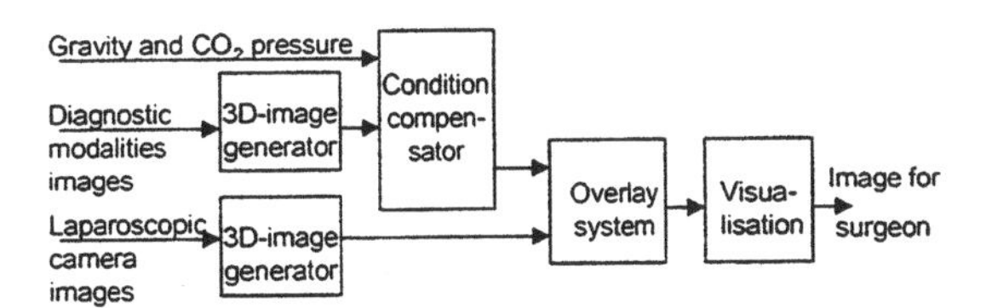

Fig. 12. The Surgeon Support System

Looking back to the original block diagram of Fig. 4, one sees that the introduction of the head control of the laparoscope changed the block diagram remarkably, Fig. 11. The symbols ② in this figure elucidates where the changes occur.

Knowledge-Based Behavior: As said before, the major Knowledge-Based Behavior activities are:

• The transformation of the coordinate system of the laparoscope and the laparoscopic instruments.
• The 3D-reconstruction and the remembrance of the 2D-CT and MRI pictures.
• The interpretation of the pre-operative images in relation to the laparoscopic camera pictures.
• The interventions needed when unexpected events occur.

These kind of activities are particularly human, they request creativity of the surgeon; therefore they cannot be automated. However, it is certainly possible to simplify these tasks by supporting the surgeon with High-Tech assistive systems. With the increasing progress on the information technology and computing speed and the computing memory, a lot of research has been executed in the field of 3D-computer vision (Faugeras, 1993; IEEE, 1996). One of the major contributions is in medical diagnostics, since for this application off-line image processing is possible. Applying 3D-computer vision in surgery requires a level of real time image processing, which is now a days hardly available. However, with the expected progress in computing speed, the authors believe that within 4 to 6 years a high-tech Surgeon Support System, SSS, can be built, in order to provide real time 3D-feedback to the surgeon. Such a SSS will contain the following subsystems (Fig. 12):

• A 3D-image generator that builds from the diagnostic pictures from various modalities a 3D-picture of the anatomy and pathology of the patient.
• A dynamic module that describes the influence of gravity and the carbon-dioxide pressure on the position of the internal organs in order to compensate for the different conditions under which the patient is diagnosed, and will be operated. With the help of such a module (the condition

compensator) a compensation for the different conditions can be achieved.

• A subsystem to generate the 3D-laparoscopic camera picture.

• A module to overlay the laparoscopic pictures over the pictures generated by the condition compensator.

• A visualisation unit that provides the 3D-pictures for the surgeon.

Although a lot of work has to be done, the authors feel that this approach is feasible. Such a SSS will take away a lot of mental burden from the surgeon, leaving him the cognitive data processing capacity for his real task: To perform a successful operation and to react adequately to unpredicted events and disturbances. As before, the changes from Fig. 4 to Fig. 11 are depicted in Fig. 11 by the symbol ③.

7. LONG-TERM RESEARCH ON MINIMALLY INVASIVE SURGERY

Long-term research on minimally invasive surgery is without any doubt a challenge for robotic and automation engineers. It is their dream to automate the entire process, leaving the operator only the supervision of the automated plant. This move from manual control to supervisor (Johannsen *et al.*, 1994; Sheridan, 1992), is to a great extend realized in industry and transport the last 50 years. However, the authors claim that it is an illusion to think that this approach will be succesful in medicine. It is remarkable to see that the US government, the NATO, and the EC are willing to spend millions of dollars or ecu's to this approach. Early projects, such as the SRI project on Telepresence Surgery (Hill, 1994; Green, 1995) are in fact advanced manual control configurations, where based on the concept of a Master-Slave system, a remote telemanipulator is controlled by the surgeon. Green states in his article (1995)" *Telepresence laparoscopic surgery will result in significant lower health costs. OR time for complex procedures such as colon resections and other bowel surgery, Nissen fundoplication and laparoscopically assisted vaginal hysterectomy will be only marginally higher than those of an open procedure - much less than for laparoscopic surgery as currently practiced. Costs associated with the hospital stay and return to normal function will be the same as for current laparoscopic procedures - much lower than for open surgery*". No arguments are given, and there are severe doubts whether this statement is correct for civil medicine; in any way the study of Sjoerdsma (1998) shows just the opposite. Another much too optimistic view is given by the following phrase (Green, 1995): " *Technologically, there seems to be no fundamental barrier to implementing telesurgery, although achieving fail-safe performance of both the sophisticated equipment and the communication link is a significant task, and propagation delays may present a challenge at distances greater than a few hundred kilometers*". Here it should be noted that Master-Slave systems are complex and sensitive for disturbances and contact instability (Sheridan, 1996; Lazeroms *et al.* , 1997). Furthermore, the mechatronical instruments are fragile and difficult to sterilize. The authors feel that this technologically-driven approach will not lead to direct benefits to the patient and surgeon, that it will not directly be accepted by the society, that it may lead to serious problems in terms of the medical responsibility - ethical as well as juridical - and that it is questionable whether it will ever become commercially feasible. Technologically seen, it is a very interesting approach of which, however, the spin-offs are difficult to predict.

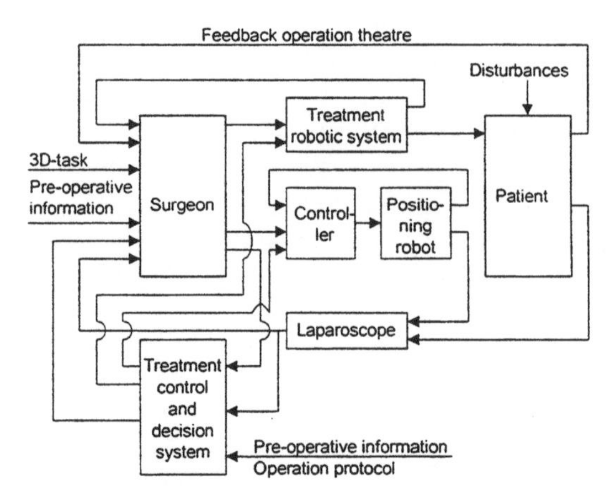

Fig. 13. Block diagram of the minimally invasive surgery process in the far future: Supervisory control mode.

What will be necessary to design a control mode where the surgeon is just a human supervisor, which can fulfil the six basic supervisory tasks, i.e. monitoring, task interpretation, planning, fault management, setpoint control or fine process tuning and intervention?, Fig. 13. At the actuator side one needs a treatment robotic system that is able to manipulate (laparoscopic) instruments with 6 DOF's. At the sensor side one needs a laparoscope or TV-camera that observes the operation area; laparoscope and/or camera also should be manipulated in 6 DOF's by a positioning robot. The decisions on the control actions should be generated by the " *brains*" of the system, the Treatment Control and Decision System, TCDS. In order to be able to build such a TCDS an accurate, dynamic, numeric, 3D-model of the entire operation area of the individual patient should be available, i.e. of his anatomy and pathology. Moreover, the task, i.e. the operation protocol should be numerically available. Because in general these models do not exist, a TCDS can hardly be developed. Even in the normal supervisory control mode a reliable system cannot be designed. Needless to say, that when unpredicted events may occur, so in case of malfunctioning of the system, the surgeon as supervisor is completely lost. In this case, only skills, flexibility and creativity of the surgeon may help to overcome the unforeseen situation.

In some specific minimally invasive procedures, however, the risks on unpredicted events and the anatomical variations among the patients is relatively small. In those cases operation protocols can be automated. The surgeon then acts as a supervisory controller, tunes the robotic system and monitors the operation process. He is able to intervene when an unpredicted event occurs. He then stops the robot, and takes over the task in a manual control mode. An example of a minimaly invasive protocol that is becoming more or less clinically accepted is the transuretral prostatectomy: a removal of a part of the prostate via the urethra (Davies *et al.*, 1996; Harris *et al.*, 1997; Rovetta *et al.*, 1997). So in general, telemanipulation and supervisory control surgery exist only in the mind of research engineers, much less in that of the clinical professionals.

8. CONCLUDING REMARKS

Minimally invasive surgery is an extremely interesting field for the man-machine discipline to be involved. It covers the entire area of the early manual control and the present attention to supervisory control. It requests the insights obtained from studies on process control, robotics, manipulators and logistic systems. It is a challenge to contribute to this medical field.

The authors strongly believe that the real progress in minimally invasive surgery, that is the progress for the benefits of the patient and surgeon can optimally be achieved by improving the manual control tasks of the surgeon based on the clinically-driven approach. Task analysis, observations of the surgeon in the operation theatre, discussion with the medical staff will learn where the actual problems are. Improvements of the laparoscope, the laparoscopic instruments, the visual feedback and the logistics and organisation of the operation protocols are the essential elements for research. Restoring of the lost eye-hand coordination is essential; it is the central theme. The clinical value of telediagnostics should also be taken into account; it may be a real contribution. Another important issue will be the commercial aspects.

Studies on telemanipulation (Green, 1995; Kühnapfel, 1994) in minimally invasive surgery are from the engineering point of view very interesting, but this technologically-driven research will in our view not lead to direct civil, medical applications. It is therefore from the viewpoint of the patient, the medical staff and the society in general not very attractive.

9. ACKNOWLEDGEMENT

The paper is the direct result of a program on Minimally Invasive Surgery and Intervention Techniques, MISIT, which is financially sponsored by the Delft University of Technology, DUT, The Netherlands. One of the aims of the projects is to create a clinically-driven program on Medical Engineering where the engineers of six different faculties, i.e. Mechanical Engineering, Industrial Design, Applied Physics, Electronical Engineering, Informatics and Aeronautical Engineering of the DUT closely cooperate with the medical staff of the Academic Medical Center of the University of Amsterdam, i.e. the departments of gynaecology, internal medicine, experimental surgery, general surgery, cardiology and others. Also some local hospitals contribute to the program.

10. REFERENCES

Bajura, M., H. Fuchs and R. Ohbuchi (1996). Merging virtual objects with the real world: Seeing ultrasound imagery within the patient. *Computer Intergrated Surgery - Technology and Clinical Applications* (R.H. Taylor, S. Lavallee, G.C. Birdea, R. Mösges ((Eds)), pp.245-254, The MIT Press, ISBN 0-262-20097-X.

Bemelman, W.A., J. Ringers, D.W. Meijer, C.W. de Wit and J.J.G. Bannenberg (1996) Laparoscopic-assisted colectomy with the dexterity pneumo sleeve. *Diseases of the Colon and Rectum* **39**, pp 59-91.

Breedveld, P. (1998a). *Observation, Manipulation and Eye-Hand Coordination in Minimally Invasive Surgery. Overview of Negative Effects, Experiments and Supporting Aids.* Technical Report N-510,45p, DUT, Delft.

Breedveld, P. (1998b). Improvement of depth perception and eye-hand coordination in laparoscopic surgery. *Proc. of IFAC Symposium on Analysis, Design and Evaluation of Man-Machine Systems*, 6p, Kyoto, Japan

Cool J.C. and G.J.O. Hooreweder (1971). Hand prosthesis with adaptive internally powered fingers. *Medical and Biological Engineering* **9**, pp33-36.

Cuschieri, A., G. Buess and J. Perissat (1992) *Operative Manual of Endoscopic Surgery*, 353p, Springer Verlag, Berlin.

Cuschieri, A. (1994) Shape of things to come: expectations and realism. *Surgical Endoscopy* **8**, pp 83-85.

Cuschieri, A. (1995) Whither minimal access surgery: tribulations and expectations. *The American Journal of Surgery* **169**, pp 9-19.

Cuschieri, A. (1996) Visual display technology for endoscopic surgery. *Minimally Invasive Therapy and Allied Technologies* **5**, pp 427-434.

Cuschieri, A. (1998) Randomised study of influence of two-dimensional versus three-dimensional imaging on performance of laparoscopic cholecystectomy. *The Lancet* **351**, pp. 248-251.

Danis, J. (1996) Theoretical basis for camera control in teleoperating. *Surgical Endoscopy* **10**, pp 804-808.

Davies, B.L., R.D. Hibberd, A.G. Timony and J.E.A. Wickham (1996) A clinically applied robot for prostatectomies. In *Computer Intergrated Surgery - Technology and Clinical Applications* (R.H. Taylor, S. Lavallee, G.C. Birdea, R. Mösges ((Eds)), pp.593-601, The MIT Press, ISBN 0-262-20097-X.

Enderle, G. (1987) Schwingspiegel-display (In German). *Informatik Spektrum* **10**, pp. 43-45.

Faugeras, O. (1993) *Three-Dimensional Computer Vision: A Geometric Viewpoint.* 663p., The MIT Press, Cambridge, Mass, ISBN 0-262-06158-9

IEEE. (1996) Computers in surgery and therapeutic procedures. *IEEE Computer magazine* **29**, pp. 20-72.

Finlay, P.A., and M.H. Ornstein (1995) Controlling the movement of a surgical laparoscope. *IEEE Engineering in Medicine and Biology Magazine*, pp 289-291.

Geis, W.P. (1996) Head-mounted video monitor for global visual acces in mini-invasive surgery: an initial report. *Surgical Endoscopy* **10**, pp 768-770.

Green, P.S., J.W. Hill, J.F. Jensen and A.S. Shah (1995). Telepresence surgery. *IEEE Engineering in Medicine and Biology*, pp 324-329.

Harris S.J., O. Mei, R.D. Hibberd, B.L. Davies (1997). Experiences using a special purpose robot for prostate resection. *Proc 8th IEEE Int. Conf. on Advanced Robotics (ICAR '97)*, pp 161-166, Hyatt Regency, Monterey, CA.

Herder J.C., M.J. Horward and W. Sjoerdsma (1997) A laparoscopic grasper with force perception. *Minimally Invasive Therapy and Allied Technologies* **6**, pp. 279-286.

Hill, J.W., P.S. Green, J.F. Jensen, Y. Gorfu and A.S. Shah (1994) Telepresence surgery demonstration system. *Proc. IEEE Int. Conf. on Robotics and Automation*, San Diego, Ca, pp 2302-2307.

Jansen, A. and M.A Cuesta (1993). Basic and advanced instruments needed for developments in minimally invasive surgery. In: *Minimally Invasive Surgery in Gastrointestinal Cancer* (M.A. Cuesta and A.G. Nagy (Eds)), pp 15-25, Churchill Livingstone, Edinburgh

Johannsen, G., A.H. Levis; H.G. Stassen (1994). Theoretical problems in man-machine systems and their experimental validation. *Automatica* **30**, pp. 217-231.

Johnson, A. (1997). Laparoscopic surgery. *Lancet* **349**, pp 631-635.

Kuntz, J.P. (1995) *Rolling Link Mechanisms.* 144p, PhD-Thesis, TUD, Delft.

Lazeroms, M., W. Jongkind, G. Honderd (1997) Telemanipulation design for minimally invasive surgery. *Proc. Am. Control Conf.* pp 2982-2986, Albequerque, New Mexico.

Majeed, A.W., G. Trou, J.P. Nicholl, A. Smythe, M.W.R. Reed, C.J. Stoddard, J. Peacock and A.G. Johnson (1996). Randomised, prospective, single-blind comparison of laparoscopic versus small-incision cholecystectomy. *Lancet* **347**, pp. 989-994.

Melzer, A., G. Buess and J. Perissat (1992). Instruments for endoscopic surgery. In: *Operative Manual of Endoscopic Surgery* (A. Cuschieri, G.Buess and J. Perissat (Eds)), pp 14-36, Springer Verlag, Berlin.

Motoki, T., H. Isono and J. Yuyama (1995) Present status of three dimensional television research. *Proc. of the IEEE* **83**, pp 1009-1021.

O'Reilly M.J., W.B. Sage, S.G. Mullins, S.E. Pinto and P.T. Falkner (1996) Technique of hand-assisted laparoscopic surgery. *Journal of Laparoscopic Surgery* **6**, pp. 239-244.

Pew, R.W. and S. Baron (1982). Perspectives on human performance modeling. *Proc. of IFAC Conf. on Analysis, Design and Evaluation of Man-Machine Systems*, Baden-baden, FRG, pp. 1-13.

Rasmussen, R (1983) Skills, rules and knowledge; signals, signs and symbols, and other distinctions in human performance models. *IEEE Transactions on Systems, Man and Cybernetics* **SMC-13**, pp. 257-266.

Rovetta, A., R. Sala, M. Bressanelli and L.M. Tossati (1997) A robotic system for the execution of a transurethral laser prostatectomy (TULP). *Proc. 8th IEEE Int. Conf. on Advanced Robotics (ICAR '97)*,pp. 167-172, Hyatt Regency, Monterey, CA.

Schippers, E. and V. Schumpelick (1996). Requirements and possiblities of computer-assisted endoscopic surgery. In: *Computer Integrated Surgery - Technology and Clinical Implications.* (R.H. Taylor, S. Lavallee, G.C. Burdea, R. Mösges (Eds)), pp 561-565, The MIT Press, ISBN 0-262-20097-X.

Sheridan, T.B. (1996) Human factors in telesurgery. In: *Computer Integrated Surgery - Technology and Clinical Implications.* (R.H. Taylor, S. Lavallee, G.C. Burdea, R. Mösges (Eds)), pp 223-229, The MIT Press, ISBN 0-262-20097-X.

Sheridan, T.B. and W. R. Ferrel (1974) *Man-Machine Systems. Information, Control and Decision Models of Human Performance.* 452p., The MIT Press, Cambridge, U.S.A.

Sheridan, T.B. (1992) Telerobotics, automation and human supervisory control. 393 p, The MIT Press, Cambridge, Mass., ISBN 0-262-19316-7.

Simpson, D.C. (1969). An externally powered prosthesis for the complete arm. *Biomedical Engineering* **4**, pp. 106-119.

Sjoerdsma, W. (1998) *Surgeons at Work. Time and Action Analysis of the Laparoscopic Surgical Process.* PhD-thesis, Delft, 103p.

Sjoerdsma, W., J.L. Herder, M.J. Horward and A. Jansen (1997). Force transmission of laparoscopic grasping instruments. *Minimally Invasive Terapy and Allied Technologies* **6**, pp 274-278.

Stappers, P.J. (1992) *Scaling the Visual Consequences of Active Head Movements. A Study of Acitve Perceivers and Spatial Technology.* 248p, PhD-Thesis, DUT, Delft. ISBN 90-9005605-X.

Smets, G.J.F., C.J. Overbeeke, M.H. Sratmann (1987) Depth on a flat screen. *Perceptual and Motor Skills* **64**, pp 1023-1034.

Stassen, H.G. (1997) Technical assesment of new technology. *Proc. New Interventional Technology in an era of Evidence Based Medicine*, AMC, Amsterdam, pp. 18-19.

Stassen, H.G., G. Johannsen and N. Moray (1990) Internal representation, internal model, human performance model and mental workload. *Automatica* **26**, pp. 811-820.

Stassen, H.G. and G.J.F. Smets (1997). Telemanipulation and telepresence. *IFAC Control Engineering Practice* **5**, pp. 363-374.

Stassen, H.G., J. Dankelman and C.A. Grimbergen (1998). Developments in minimally invasive surgery and interventional techniques (MISIT). *Proc. of the 16th EAC on HDM and MC*, 7p., Kassel, Germany.

Treat, M.R. (1996) A surgeon's perspective on the difficulties of laparoscopic surgery. In *Computer Intergrated Surgery - Technology and Clinical Applications* (R.H. Taylor, S. Lavallee, G.C. Birdea, R. Mösges ((Eds)), pp.559-560, The MIT Press, ISBN 0-262-20097-X.

Voorhorst, F.A. (1998) *Affording action. Implementing perception-action coupling for endoscopy.* PhD-thesis, DUT, Delft, 170p.

RUMINATION ON AUTOMATION, 1998

Thomas B. Sheridan

Massachusetts Institute of Technology

Abstract: In recent years there has been great interest in the evidence that with experience human decision-making becomes "automatic", "recognition-primed", "naturalistic", "situated", and "ecological". Advocates of these approaches have rejected classical decision theory as a description of decision-making in real tasks. This rejection poses a challenge to better understand human automaticity by extending normative decision modeling. Two approaches, both based on state-space reduction, are discussed. There is also the challenge of relating these models to what is being learned in neurobiology, and again a model is presented which potentially explains how behavior can become automatic. Implications for computer-based automation of decision aiding and control are also proposed, including a revised four-stage format for scaling levels of automation. *Copyright ©1998 IFAC*

Keywords: Decision, models, memory, criteria, neurobiology, automation, control, safety.

1. INTRODUCTION

This paper deals with two kinds of automatic response. The first is that form of human behavior which occurs naturally with habituation in a particular kind of task. With experience a person tends to become automatic: going from sensing and perception directly to action without appearing to consider alternative decisions, or to contemplate the decision criteria, or to weigh the alternatives against the criteria, or to think consciously. Experienced decision-makers show little evidence of having to develop the whole tree of possibilities and do all the corresponding probability-utility multiplications to determine the decision alternative with the "best" outcome, as decision theory would seem to require (Keeney and Raiffa, 1976). Such automaticity can also be forced by the urgency of a deadline, in which case the decision-maker does what seems most evident, best known, or most convenient.

The second kind of automatic response is mechanized, or artificial automation, what we usually mean by the term automation. It is accomplished by some combination of artificial sensors, computer-based processing of signals from these sensors and estimation of states of the sensed environment, computer-based decision, and mechanized control of actions to implement the decision. Human automatic response and automation are not independent of one another. The latter is likely to be under supervisory control, where the human gives goals and constraints to the machine and the machine seeks to achieve those goals while the human monitors. However the relation between automatic human behavior and automated machine behavior remains to be clarified.

Naturally we seek to use machine automation so as to achieve a best combination of human-plus-machine. But there is doubt that the allocation of tasks to humans or machines can be accomplished on a very rational basis (Sheridan, 1998), in spite of the long-standing Fitts List (Fitts, 1951) and other assertions of how people are superior to machines and vice-versa. A later section the paper suggests some tentative connections between human automaticity and mechanized automation — when to use how much of the latter for what part of the task as a function of how much experience the human has and how automatic the human's behavior has become.

2. THE NATURE OF HUMAN DECISION-MAKING

In recent years there has been growing appreciation of ways in which human behavior is "automatic". The idea is certainly not new, for Pavlov showed how autonomic responses could be conditioned on certain stimuli which are paired in time and place with other stimuli which originally elicited that response, as salivation became conditioned on the sound of a bell when the bell was paired with food. Later Skinner (1953) showed that "free operant" responses, starting as essentially random behavior, could be shaped (operant conditioning) by rewards which on successive trials demanded more and more refinement, and by withholding rewards when responses were outside the category of what was acceptable. Chomsky (1968) theorized how the elements of linguistic behavior are built into the brain from birth, and won out over Skinner's conflicting assertion that all of language learning is by operant conditioning. However the great majority of Skinner's empirical findings remain pertinent today, this in spite of the fall from grace of "behaviorism" in the contemporary beauty contest in which "cognitive science" and computation are the popular metaphors for thinking. (It will later be argued that thinking is not computation, but there is nothing wrong with comparing human thinking to the behavior of normative computational models.)

The perceptual psychologist Gibson (1979) postulated visual mechanisms which are *ecological*, in the sense that they have evolved to best perceive those patterns in the natural environment which the organism needs to cope. Thereby he set into motion a bevy of followers as well as critics. Dennet ('98), in the latter camp, asserts that Gibson "viewed all cognitivistic, information-processing models of vision as hopelessly entangled in unnecessarily complex Rube Goldberg mechanisms posited because the theorists had failed to see that a fundamental reparsing of the inputs and outputs was required. Once we get the right way of characterizing what vision receives from the light, he thought, and what it must yield ('affordances') the theory of vision would be a snap".

Gibson's followers, however, are having significant impact, and have applied his ideas to the man-made world as well as the natural world. Rasmussen (1983) has made a large impact on the human-machine systems engineering community with his well-known trinity of behavioral levels: skill-based, rule-based and knowledge-based, organized in a hierarchy as shown in Figure 1.

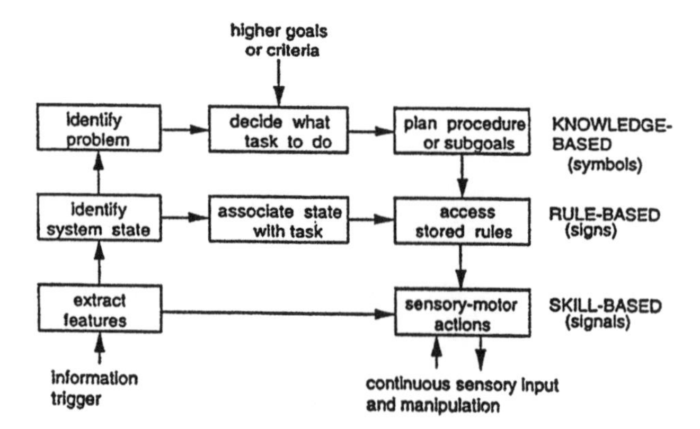

Figure 1. Rasmussen behavioral trinity

As learning progresses behavior moves naturally to a lower level, becoming more automatic. Vicente and Rasmussen have called for human-machine interface design to be ecological in the Gibson sense, meaning that "to properly control the process, the human-machine system must take account, or embody, the constraints inherent in the work domain." This, of course, is something all control engineers appreciate in designing a controller to accommodate a dynamic process, but Vicente and Rasmussen saw that the implications are much broader, and that engineering interface design benefits from seeing the world in terms of a hierarchy of levels of abstraction (from purpose at the top to physical details at the bottom). They argue that the interface design should be isomorphic with the ways humans think and operate — simultaneously at all three levels of behavior as well as at the higher abstraction levels.

Klein, in a retrospective interview study of fire fighting commanders faced with dynamic, high pressure decision situations, asserted that they showed no signs of adherence to the "canonical procedures" of classical decision theory, namely identification of feasible options, identification and weighting of evaluative dimensions, and rating of each option on each evaluation criterion. It was on this basis that he postulated recognition-primed or naturalistic decision-making (see Szambok and Klein, 1997 for a riview).

There is also much evidence from controlled laboratory experiments on how people deviate from normative decision theory. Edwards (1968) has shown how human decision-makers are more conservative than Bayes' theorem would dictate, and Tversky and Kahneman (1971) have shown other ways in which humans are not fully rational in estimating probabilities.

Suchman videotaped Xerox employees trying to use a Xerox copier and discovered that they had many kinds of difficulties that were almost impossible to predict. She characterized such behavior as *situated*, or highly context-dependent, and her studies have precipitated many similar observational studies which confirm her assertions. See Landauer (1995) for a review and commentary.

Within the artificial intelligence subcommunity of computer science there is a related move away from seeing behavior and intelligence in terms of *ab initio* problem solving, and more toward appropriate adaptation of bodily response to current environmental and social (external "agent") context. In place of the classical Turing test, wherein a computer is judged intelligent if a human observer cannot tell whether responses to typed questions are being generated by human or computer, some AI professionals want to test intelligence by whether a computer's immediate physical interaction with a human is judged "natural." This has been called *embodied AI* by Brooks (1998) who heads a team at MIT to construct a robot to demonstrate these ideas.

The author completely agrees with the above-discussed propositions that real decision making tends to be recognition-primed and situated and that the human intelligence that matters is embodied. He once stood in front of a group of Airbus test pilots and asserted that pilots do not make decisions (in the classical sense of multiplying utilities and probabilities), that they are mostly automatic (recognition-primed). The thought was not well received. Perhaps we are loathe to admit to acting like automatons. (However, when no one is looking, operators seem to prefer the automatic or skill-based decision mode to the more thoughtful knowledge-based mode.) The social psychologist Langer (1989) and others exhort us to be less automatic and more consciously thoughtful in our social interaction (her term is *mindful).* Our natural tendencies are otherwise.

3. USE AND ABUSE OF NORMATIVE MODELS

In the behavioral and social sciences to a much greater degree than in engineering there is a tendency to reject old theories when new ones are proposed in a repeated cycle of thesis-antithesis. Intellectual bandwagons and the heralding of paradigm shifts seem to be *de rigeur.* In engineering and the physical sciences theoretical and experimental contributions are more incremental, with newer contributions accreting to form more and more codified structures. Obsolete structures eventually collapse of their own weight and are ignored; or the older methods of analysis are simply replaced with the newer methods. There is little call or need for decrying the older ideas; they simply cease being useful.

The reason for this difference probably has to do with the fact that the engineering and physical sciences are more mature and have become quantified, whereas the behavioral and social sciences have not. Theories in the latter arena are mostly verbal, supported by experiments which are much harder to control. In the behavioral/social sciences the difference between the laboratory and the real world is vast, and the degree to which an independent variable can be controlled is more or less the degree to which experimental results can be challenged as being unrealistic. Linearity and the superposition of independent effects is always a stretch. In the still largely Newtonian world of the engineering/physical sciences most artifacts have been engineered on the basis of linear assumptions. Even so, non-linear analysis, when it is employed in the engineering world, is far better codified with much greater predictive clout than non-linear analysis in the behavioral/social world.

Systems models, i.e., models which predict the behavior of some configuration of elements from known behavior of the component elements, are usually normative. Assumptions are made about the components, and the system behavior follows logically. This is true of decision theoretic models as well as those for information, communication and control, indeed all of applied mathematics. In this sense a normative model is immune from criticism on the basis that it does not fit the real world. It can only be criticized on the basis that it is not rational or that it requires too many assumptions to ever be useful (by "Occam's razor" the most powerful models are those with the fewest assumptions) or is too complex or too obscure.

For these reasons criticism of classical decision theory by proponents of recognition-primed or naturalistic decision-making (asserting that human operators cannot follow the prescriptions of normative decision theory) is misguided. Classical decision theory states what is, or what is best logically, based on axioms of probability, utility and methods for combining probability and utility to rate relative desirability of decision alternatives. By itself it has no credentials for descriptive modeling, i.e. for generalizing on what is true in the real world, especially what is true for humans.

Normative models, of course, are often used as baselines for comparison to experimental results. In that sense the following question is posed: can experimental results (dependent-variables such as time, errors. etc.) for different experimental conditions (independent variables or parameters) be matched to the model results when some parameter(s) are adjusted? If that is the case the normative parameter variation begins to model the variation in experimental condition, the model otherwise being constant.

The fact that recognition-primed decision behavior or naturalistic decision-making does not seem to correspond to classical decision theory, in the sense that the latter implies working through the tree of alternatives using probability and utility products to maximize subjectively expected utility, surely begs the question of what normative model would be better. That is, what axioms and best procedure for those given axioms, will yield results which come closer to matching experimental results and finding correspondences between experimental conditions and model parameters? In Section 5 two examples are offered which point out why, and perhaps allude to how, a person implements recognition-primed response, or skill-based behavior in the Rasmussen paradigm.

There is also the question of how such a normative model corresponds to what we know of neurobiological mechanisms, which until recently have eluded science. A crude qualitative model of this process is suggested in Section 5.

4. ERGONOMICS OF PROBLEM STATE SPACE REDUCTION: A NORMATIVE BASIS FOR RECOGNITION-PRIMED DECISION

The consideration of many decision alternatives and paths, the very property which an experienced human seems not to bother with, or treats so quickly and easily as to leave no trace, is the undoing of a perfectly rational but brute-force computer decision-maker.

What follows are two examples of how generic brute-force decision-making can be made many orders of magnitude simpler. The implication is that humans, through experience and the ability to see similarities in problem structures, perform such simplifications on the "problem spaces" they encounter, so that their decision-making can be simplified from knowledge-based to rule-based and eventually to skill-based behavior — in other words made more recognition-primed or situated, and appear more automatic.

4.1 Partitioning the state space for multi-step multi-object manipulations

In general the state space, the vector of all possible states a system can take, is the product of the number of possible states for each salient variable. In an analog world where continuous variables are approximated by discrete quantization for even a small number of variables one very quickly runs into computer memory limitations: the so-called curse of dimensionality.

For example, Figure 2 (a) shows (on a plane to make matters simple) a "hand" (two "fingers") which can open or close to 100 levels, rotate to 100 different angular orientations, and translate to any point within a 100x100 physical space. Let us suppose this hand must pick up a magnetic "tool" which can also be rotated to 100 orientations and maneuvered to within the 100x100 translation space. The magnetic tool must be aligned to make proper contact with an "object" which itself can be positioned to one of 100 orientations and within the 100x100 space. The object must then be maneuvered into a given final destination position. Thus, considering the generic state space of all 10^8 combinations of the hand and 10^6 combinations of the tool and 10^6 of the object, there are 10^{20} possible states (X)! Add to this the fact that in planning a robot or human handling operation (e.g., grabbing a screw driver and using it on a screw) one must (in general) consider a number of possible state transitions, X factorial, and a much larger number of sequences, each consisting of many such transitions. Quite clearly solving any practical problem of several variables in a continuous physical space by evaluating all possible alternatives is doomed to computational failure from the beginning. There are many state space searching algorithms (such as dynamic programming) and heuristics (rules of thumb) which must be used to simplify complex problems for computer solution.

What does a human do to simplify such a decision problem? The human observes that there is a 3-step sequential order to the task, namely that first the hand must move to and grasp the tool, and next the hand-plus-tool (as a single entity) must align with the object, and finally the hand-plus-tool-plus-object must maneuver the object into the goal position. The human also observes that each of these operations can be done within its own limited state space. Using this simple and crude partitioning of the task, we first have translation and grasping in the lower left quadrant of Figure 2(b) with no rotation required, a 50x50 = a 2500 element state space. Then there is an operation of aligning the hand-plus-tool, a 50x50 translations x100 orientations = a 250000 element

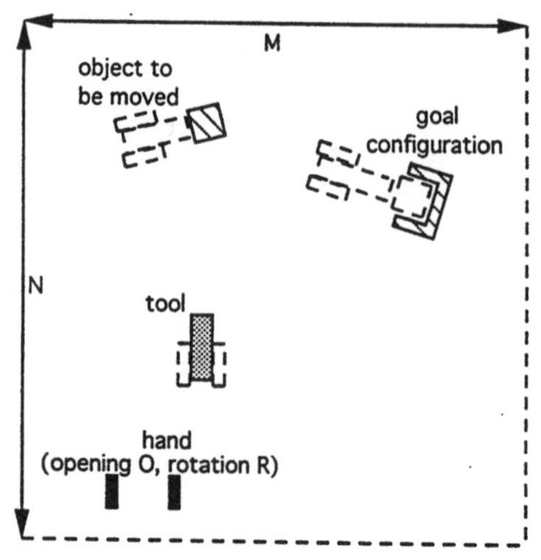

Search state space = (MNOR) (MNR) (MN)

Assume M = N = R = O = 100

Search state space = 100000000000000000000

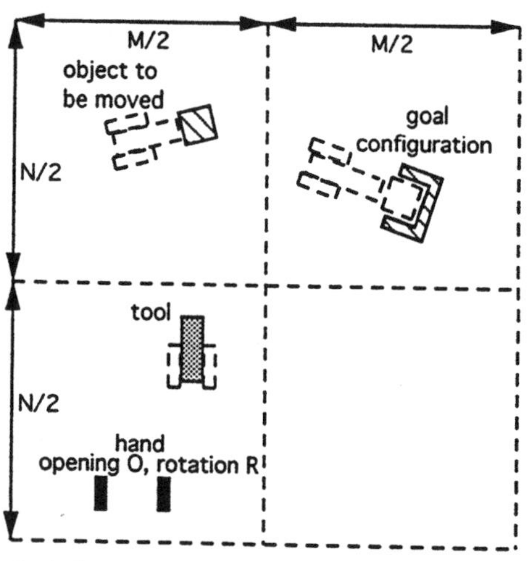

Revised search space:
(M/2) (N/2) O + (M/2) (N/2) R + (M/2) (N/2) R =
1000000/4 + 100000/4 + 100000/4

a reduction by more than 10000000 times!

Figure 2. Sample problem of state space partitioning. (a) Above, the initial given problem and full state space. (b) Below, the partitioned state space

state space. Finally there is the maneuvering operation of hand-plus-tool-plus-object, perhaps a 50 tall by 100 wide by 100 orientations = a 500000 element state space, the sum of which is 752500 states. This is reduction in the state space by a factor of over 10^{14}. People seem to be very good at such state space reductions by task partitioning.

State space partitioning which is good for one particular task is likely to be generalizeable to many other tasks which have similar properties (in this example, grasping a tool and using the tool to manipulate another object in the environment).

4.2 Finding and fitting the Pareto frontier in multi-attribute utility space

The previous example dealt with a state space of manipulations in physical space. Another kind of state space to be considered is a multi-attribute utility space, where each system state is an alternative entity which has multiple attributes, each having a utility or good-bad value. Here the decision is not how best to move an object, but rather how to make a best trade-off among available choices that have multiple-objectives.

Consider the problem of deciding how fast to drive. In the simplest case we have a decision space (Figure 3) of two evaluative attributes, trip time plotted on the X axis, and probability of death or serious injury plotted on the Y axis. (Both attributes are scaled to be bad in this example. Had we used their inverses they both would be good.) In general there are many more attributes in the state space, for example pleasure of the trip, wear on the vehicle, and so on. However, we need only two attributes to make the point.

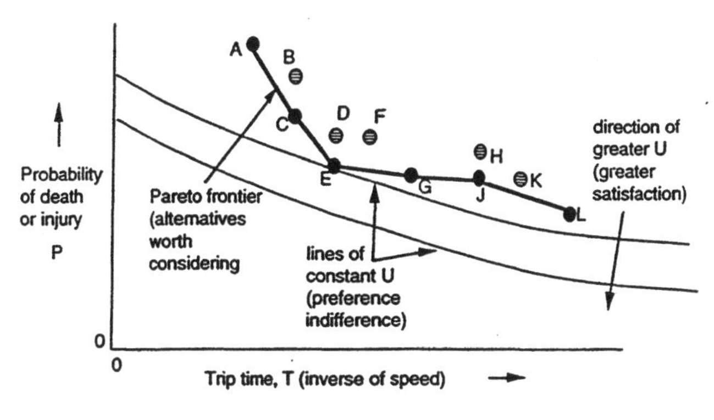

Figure 3. Pareto frontier showing dominated points and utility curves.

There are two kinds of data which are mapped into this space. The first kind of data is the set of feasible decision alternatives, shown as circles. One of these must be chosen, and the question is: which one? There may be infinitely many alternative choices or they may be finite in number as shown. Notice that the decision alternatives do not cover the entire state space of attributes. There are physical or economic constraints which are independent of the human decision-maker which constrain the set. For example they may dictate that there are no options for a trip which takes very little time which at the same time poses a very small probability of death or serious injury. Further, many (happily, in fact, most) alternatives are *dominated,* meaning that while they are at the same level of one attribute, one alternative is better with respect to the other attribute. All the non-dominated alternatives form a line (in hyperspace a surface) called the Pareto frontier (after an Italian economist), which itself is the tradeoff imposed by the physical and economic factors outside the control of the human decision-maker.

The second kind of data mapped into the multi-attribute state space is the preference trade-off function, the decision-maker's constant utility curve, shown on the figure as a set of continuous lines. Each curve plots all hypothetical decision alternatives which, if they were available (which they are not!) would be equally preferable. In other words the decision-maker would be indifferent to any point on any single curve. In general there is a hypothetical family of such curves, or still more generally a function of all attribute variables which yields a measure of subjective value, called *utility* by economists. Theoretically this function exists independently of what is available.

The elicitation of utility functions from decision-makers is well established normatively (Von Neumann and Morgenstern, 1944) , but in practice is not easy, and requires some training, especially as the number of attributes exceeds two or three. March and Simon (1958) have suggested approximation techniques called *satisficing*. Charney and Sheridan (1989) and others have implemented some of these on a computer-interactive basis.

The normative optimum is where the best (least bad in the example) utility line just intersects the Pareto frontier. In Figure 3 this is point E. This is the best match of what is preferable with what is available.

4.3. From normative simplification to naturalistic decision-making

Both 4.1 and 4.2 present normative models. Normative models are generated by applied mathematicians in hopes that they lend some rationality to different real problems. Empirical science is a matter of fitting descriptive models to experimental data — means of generalizing the data by the simplest descriptions which are at the same time acceptable as to fit. Both engineering design and everyday decision-making can be said to be a matter of applying rational simplification (as normatively modeled) to real observations (as descriptively modeled) in order to predict. Rationality becomes art with the simplification of experience. When observing art no rationality is evident!

Both manipulation-state-space-parsing and Pareto-frontier-utility-fitting are rational procedures by which humans can simplify decisions and render them more automatic and naturalistic. Certainly these are not the only such ergonomic simplifications, but they clearly indicate why a human decision maker, with a little experience and memory, need not go through the whole formal procedure suggested by the axioms of classical decision theory.

5. HOW IT SURELY DOESN'T WORK, AND HOW IT IS MORE LIKELY TO WORK

Neither the applied mathematicians espousing normative decision theory nor the cognitive psychologists espousing naturalistic decision-making have purported to explain the neural mechanisms by which real naturalistic decision-making occurs. Observing from the outside it certainly does appear largely automatic. Are there observable neurobiological correlates of "automatic" or "recognition-primed" decision-making?

Computers are attractive as engines for implementing normative algorithms as well as heuristic procedures, but there is little reason to expect the internal neurophysiological and biochemical mechanisms of real cognition work like computers.

It would seem there must be automatic or recognition-primed mechanisms inside the organism, though the "wet" science here is yet at a very early stage. What has been shown clearly at a biochemical level is the astounding ability of exceedingly complex protein molecules to "automatically" (by chemical bonding) recognize one another as friend (to pair, reproduce, etc.) or foe (to destroy). It is notable that two Nobel laureates in the field (who both, it might be noted, are employed at the same institution but work quite independently of one another) have written books to popularize what is being learned about how real neural nets (which are different in key ways from artificial ones) form internal representations of external events (see for example Edelman, 1992, or Crick, 1994).

Normative network models which share some properties with the neurobiology allow for interesting experiments. The neuronal network learning models put forth by Hebb (1949) are classic proposals in this area. Yufik and Sheridan (1996) describe a "virtual network" model (Figure 4) which demonstrates clustering or "chunking" of abstract elements, corresponding to the Miller (1956) use of the term, with repeated use of certain pathways, presumably triggered by external stimuli.

Whenever any two nodes (small circles) are exercised together (by whatever criterion one chooses) the weight W of connection between those two nodes strengthens, meaning that in the future when either node is excited by external stimuli the other tends to be also. When the sum of weights within any combination of nodes sufficiently exceeds the sum of weights connecting those nodes to nodes outside that combination, that combination becomes an association packet or "chunk". The ratio σ of the internally binding sum to the externally pulling sum of weights is continually changing because of activity. A packet adhesion (resistance to breakdown) barrier U (indicated by the height of the black bars) is determined by σ minus a product of a general activity or ambient "temperature" variable T and the sensitivity of σ to T. The latter term, in other words, is tending to break down the packet. This model has been implemented in a computer with some very efficient algorithms and shows interesting packet formation and decay properties as a function of experience, T and other parameters.

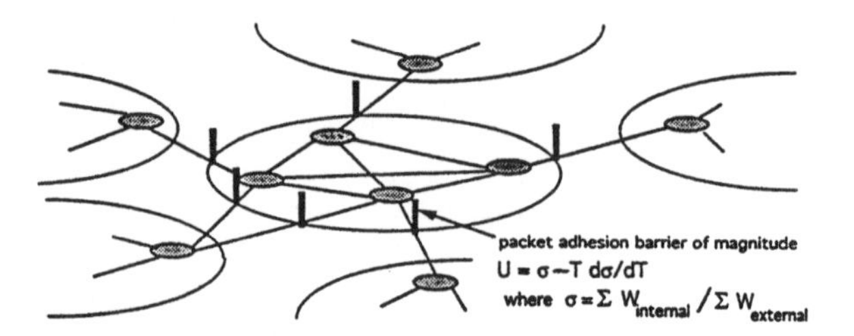

Figure 4. The virtual network model.

How does this "virtual network" model effect a reduction in the state space (as described in the previous section) as it gains experience with a task? The nodes may be regarded as states in the state space of all decisions or actions. With experience, as many nodes are combined into a packet, those nodes tend to act together as one, effectively becoming a single state, while still embracing all the functionality that was present before. In this way the irrelevant state combinations are cast aside, since for the given habituated task they need not be considered. The state space is thus reduced to any degree consistent with similarities in tasks (stimuli) encountered in the environment.

The virtual network model is presently being applied to air traffic controllers' conceptual organization of aircraft into pairings which may be in danger of violating aircraft separation rules. Such network clustering mechanisms are not computational in the usual sense of that term, but more represent the kind of operant conditioning which Skinner (1953) theorized. They may come closer to what happens within the body, but still beg many questions about how the nervous system remembers, associates, perceives patterns, and automatically responds.

6. HOW SHOULD AUTOMATION OF DECISION AIDS AND CONTROL RELATE TO THE DEGREE TO WHICH HUMAN RESPONSE IS AUTOMATIC?

Many tasks must be done by a human because it is not known how to automate them. Other tasks are done by humans because humans like to do them. Still others are done by humans because the tasks are trivial, or are embedded in either of the other two types of task, and therefore it is inconvenient to set some automation to do them. To the extent to which a task is easily done by a human but is boring, fatiguing or hazardous, it should be automated if feasible and convenient to do so. When a task can be done by either human or machine it is worth considering whether human and machine might work together and complement one another.

Decision aiding seems obvious for a novice or an unfamiliar or unexpected situation where response is not automatic or recognition-primed. Taking time for knowledge-based contemplation would seem to be in order. Unfortunately unfamiliar or unexpected situations are often the very circumstances where fast response is essential, and operators are therefore forced to employ whatever recognition-primed behavior can be brought to bear. These are also the situations where decision-aiding and automation might seem appropriate. But upon a little reflection we realize that decision-aids and automation are least reliable when situations cannot be predicted and modeled. Further, the burdens of using decision aids under stress can outweigh the benefits, even if they work (Kirlik, 1993). The designer is left in a "catch 22". To automate or not?

These kinds of considerations require that one take a hard look at the elements of each particular task, and break it into task elements to determine what parts cannot be automated because we don't know how, what parts do humans like to do, what parts are trivial for humans and inconvenient to automate, what parts are boring, fatiguing or hazardous and should be automated if they can be. Also, what degree of unfamiliarity can be coped with by humans with their best recognition-primed behavior, and what degree of unfamiliarity can be accommodated by automation using the best available predictive models? Surely automation need not apply uniformly to entire tasks, but rather can be of benefit to some parts more than other parts. But what elemental breakdown?

What seems an obvious breakdown, a taxonomy that applies to most complex human-machine systems, is the sequence of operations shown in Figure 5: acquire information; analyze and display; decide action; and implement action. Surely human response is automatic (or recognition-primed) to a different degree at these different stages, depending of course on the task context. Below each box in Figure 5 are noted some typical variables relevant to that stage.

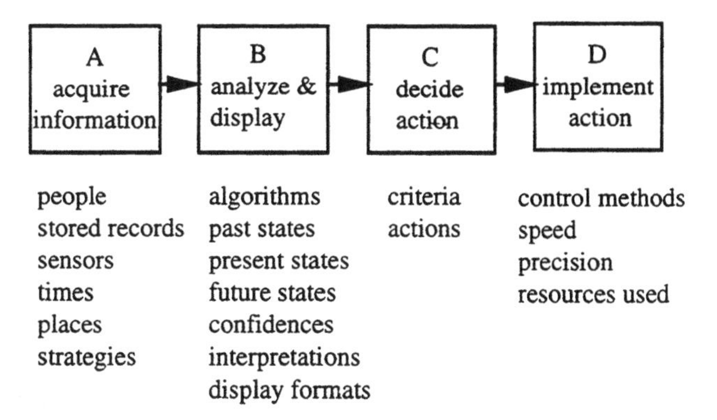

Figure 5. Four stages of a complex human-machine task.

Human errors surely occur differently at these different stages. Reason (1990) has already suggested that errors of recognition in the face of noise (occurring at the analysis stage) should be distinguished from errors of taking risky decisions where payoffs are probabilistic. One can also distinguish errors in attention (further upstream, at the data acquisition stage) since recognition can only occur if attention has been properly allocated. And action slips, of course, are further downstream at the implementation stage.

7. LEVELS OF AUTOMATION: A FOUR - STAGE APPROACH TO DESIGN

Twenty years ago Sheridan and Verplank (1978) proposed a ten-level scale of degrees-of-automation, which folded together several relevant dimensions such as:

a. the degree of specificity required of the human for inputting requests to the machine;

b. the degree of specificity with which the machine communicates decision alternatives or action recommendations to the human;

c. the degree to which the human has responsibility for initiating implementation of action;

d. the timing and detail of feedback to the human after machine action is taken, and so on.

Surprisingly, the level descriptors as published have been taken more seriously than were expected, since it was evident then (and remains so) that these many dimensions of automation can be ordered in various different ways on multi-dimensional as well as one-dimensional scales. It is the same as with mental workload and other complex properties of human-machine systems: both one and several dimensional scales are appropriate. When combining dimensions to fit a single scale questions arise as to whether there is rigorous justification for claiming the scale to be ordered as to "degree of automation". However when attempting to design a multi-dimensional scale (ordering being easier and more easily justified along any one dimension) one quickly realizes the impossibility of claiming that one has been comprehensive and included all dimensions relevant to a broad class of tasks. The main point, for a system designer, is to have some way of considering many options which are more or less ordered, and then to make a satisficing (as mentioned above) decision as to the degree of automation.

What is easier to justify is that for different stages of any relatively complex task the appropriate level of automation is likely to be quite different. Therefore, at least initially, it is improper and misleading to consider any one level of automation to fit the entire task. Accordingly a somewhat simplified scale (eight levels would seem to serve the purpose, but the reader is encouraged to make his own modifications) is proposed — to use to consider appropriate degree of automation at each stage of the given task. Such a scale is shown in Figure 6.

1. The computer offers no assistance: the human must do it all.

2. The computer suggests alternative ways to do the task.

3. The computer selects one way to do the task, and

4. .... executes that suggestion if the human approves, or

5. .... allows the human a restricted time to veto before automatic execution, or

6. .... executes automatically, then necessarily informs the human, or

7. .....executes automatically, then informs the human only if asked.

8. The computer selects, executes, and ignores the human.

Figure 6. Simplified scale of degrees of automation of each of: (A) acquire information, (B) analyze and display, (C) decide action, (D) implement action

Figure 7 illustrates how different kinds of tasks call for different levels of automation at different stages. The numbers correspond to the eight-point scale, 1 being the least and 8 the most automation. The circles might represent, for example, the task of holding an election of officers for an organization. At the ACQUIRE stage data are collected manually from eligible individuals, though email (level 2 automation) might have been used to make suggestions to solicitors on how to operate. The results are then ANALYZEd by computer and the winners are DECIDEd automatically. The transfer of power is then executed with only a modest amount of computer-stored procedural advice.

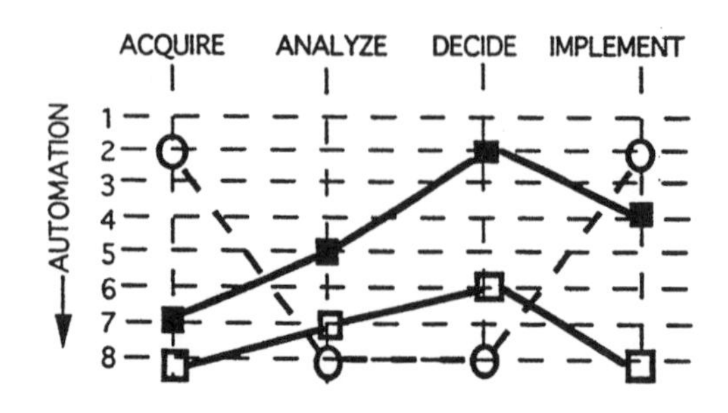

Figure 7. Different levels of automation at different task stages.

The black squares represent advice recently given by a U.S. National Research Council (1998) committee on which the writer served. Its advice pertained to the proper amount of automation for new civilian air traffic control systems. After much debate the committee decided that acquisition and analysis (these two they lumped together) could and should be highly automated — in fact they already are (radar, weather, schedule information, etc.) However decision making, except for certain decision aids now under development, should be done by human air traffic controllers. Implementation is in the hands of the pilots, which in turn is largely turned over to autopilots and the other parts of the flight management system.

The open squares represent a typical manufacturing robotics task. A computer vision system acquires all the data. This is analyzed in a computer where the analysis results are available to a human supervisor should he need to check the data. The analysis is passed on to a computer decision algorithm and the decision results are displayed for the operator. The decision is implemented by a robot in fully automatic fashion.

8. CONCLUSION

Two kinds of automatic behavior have been discussed: automatic human response which occurs as tasks become well learned; and machine automation. There is keen interest these days in both kinds. Experimental researches of automatic or recognition-primed human response have rejected classical decision theory as a basis for insight. This seems to stem from a degree of misunderstanding of what classical decision theory is: mathematical reasoning which does not purport to describe reality.

For a computer to become effectively automatic through experience it must reduce the problem state space by orders of magnitude. Two ways this might be modeled in a normative way are proposed: (1) partitioning the state space to handle sequential object manipulations; and (2) finding and fitting the Pareto frontier in multi-attribute utility space. It is claimed here that these normative decision models are a means to understand what appears to be artful naturalistic decision-making. The mechanism by which the nervous system actually decides is still a mystery, is surely not like a computer, and is probably closer to network learning by reinforcement and formation of associative structures. A particular network model is discussed which demonstrates extensive state space reduction (and hence automaticity) with experience.

It is proposed that the proper degree of automation is best considered by separately evaluating human needs and machine capabilities at four different stages of a task: information acquisition, analysis and display, action decision, and action implementation. An eight-level scale of degrees of automation is proposed for doing so. Three examples are offered to suggest how different tasks might fare under such evaluation.

REFERENCES

Brooks, R.A., C. Brazeal, R. Irie, C.C. Kemp, M. Marjanovic, B. Scassellati and M. Williamson (1998). Alternate essences of intelligence. *Proc. AAAI 1998.*

Charney, L. and T.B. Sheridan (1989). Adaptive goal setting in tasks with multiple criteria. In *Proc. 1989 Intl. Conf. on Cybernetics and Society*, Cambridge, MA.

Chomsky, N. (1968). *Language and Mind.* New York: Harcourt, Brace & World.

Crick, F.H.C. (1994). *The Astonishing Hypothesis: the Scientific Search for the Soul.* New York: Scibners.

Dennett, D.C. (1998). *Brainchildren.* Cambridge, MA: MIT Press, p. 231.

Edelman, G.M. (1992). *Bright Air, Brilliant Fire: On the Matter of the Mind.* New York: Basic Books (Harper Collins).

Edwards, W. (1968). Conservatism in human information processing.. In Kleinmuntz, B. (Ed.), *Formal Representations of Human Judgment,* New York: Wiley.

Fitts, P.M., Editor (1951). *Human Engineering for an Effective Air Navigation and Traffic Control System.* Washington, D.C., National Researech Council.

Gibson, J.J. (1979). *The Ecological Approach to Visual Perception,* Boston, MA: Houghton Mifflin.

Hebb, D.O. (1949). *The Organization of Behavior, a Neuropsychological Theory.* New York: Wiley.

Keeney, R.L. and H. Raiffa (1976). *Decisions with Multiple Objectives: Preferences and Value Tradeoffs.* New York: Wiley.

Kirlik, A. (1993). Modeling strategic behavior in human-automation interaction: why an "aid" can (and should) go unused. *Human Factors,* 35 (2), 221-242.

Landauer, T.K. (1995). *The Trouble with Computers.* Cambridge, MA: MIT Press, pp. 246, 307, 389.

Langer, Ellen (1989). *Mindfulness.* Reading, MA: Addison Wesley.

March, J.G. and H.A. Simon (1958). *Organizations.* New York: Wiley.

Miller, G.A. (1956). The magical number seven, plus or minus two: some limits on our capacity for processing information. *Psych. Review,* 83, 81-97.

Rasmussen, J. (1983). Skills, rules and knowledge: signals, signs and symbols and other distinctions in human performance models. *IEEE Trans. on Systems, Man and Cybernetics,* Vol. SMC-13, pp.257-267.

Reason, J.T. (1990). *Human Error.* Cambridge University Press.

Sheridan, T.B. (1998). Allocating functions betwen humans and machines: any hope for rationality? *Ergonomics in Design,* in press.

Sheridan. T.B. and W.L. Verplank (1978). *Human and Computer Control of Undersea Teleoperators.* MIT Man-Machine Systems Laboratory Report. See also Sheridan, T.B. (1987). Supervisory control, in Salvendy, G. (Ed.), *Handbook of Human Factors,* NY: Wiley.

Skinner, B.F. (1953). *Science and Human Behavior.* New York: MacMillan.

Szambok, C.E. and G.A. Klein (1997). *Naturalistic Decision Making.* Mahwah, NJ: Lawrence Erlbaum Associates.

Tversky, A. and D. Kahneman (1971). The belief the law of small numbers. Psych. Bulletin 76, 105-110.

Vicente, K.J. and J. Rasmussen (1992). Ecological interface design: theoretical foundations. *IEEE Trans. on Systems, Man and Cybernetics,* Vol. SMC-22, No. 4, July/August, pp. 589-606.

Von Neumann, J. and O. Morgenstern (1944). *Theory of Games and Economic Behavior.* Princeton Univ. Press.

Wickins, C.D., A.S. Mavor, R. Parasuraman and J.P. McGee (Eds.) (1997). *The Future of Air Traffic Control.* Washington, DC: National Academy Press.

Yufik, Y. and T.B. Sheridan (1996). Virtual networks: new framework for operator modeling and interface optimization in complex supervisory control systems. *Annual Reviews in Control,* vol. 20 , pp. 179-195, Oxford, UK: Pergamon.

VIRTUAL REALITY TECHNOLOGY AND ITS INDUSTRIAL APPLICATIONS

Junji Nomura*, Kazuya Sawada**

**Systems Development Center, Matsushita Electric Works, Ltd.
1048, Kadoma, Osaka 571-8686, Japan
**Advanced Technology Research Laboratory, Matsushita Electric Works, Ltd.
1048, Kadoma, Osaka 571-8686, Japan*

Abstract: Virtual reality (VR) is well-known and is currently being investigated for practical use in various industrial fields. Using three-dimensional computer graphics, interactive devices, and a high-resolution display, a virtual world can be realized in which imaginary objects can be picked up as if they were in the physical world. Using VR technology, Matsushita Electric Works, Ltd. has been developing application systems for industrial use since 1990. This paper details four VR application systems: a relax/refresh system employing VR, a horseback riding therapy simulator with VR technology, a VR-supported kitchen layout design system, as well as an urban planning VR system. *Copyright© 1998 IFAC*

Keywords: Virtual reality, Human-centered design, Human-machine interface, Computer applications, Simulators, Computer-aided design

1. INTRODUCTION

Virtual Reality is well-known and is currently being investigated for practical use in the various industrial fields such as computer graphics, CAD, CAM, CIM, robotics, medical/health care, multi-media, games and so on. Many companies have released commercial products which can be used for developing VR application systems.

Since 1990, Matsushita Electric Works, Ltd. has developed several VR application systems for industrial use. In this paper, four VR systems are introduced.

First, a relax/refresh system employing virtual reality suitable for the health care field. This system employs a physiological feedback mechanism which controls stimulation by detecting the sequential data of the user's heart rate.

Second, a VR-based system that reproduces the effect of horseback riding therapy. This system may be enjoyed by healthy senior citizens, and has been specifically designed to rejuvenate both the mind and the body.

Third, a VR-supported kitchen layout design system. With this VR system, customers can have a pseudo-experience of their "virtual kitchen" and modify its design. This system helps our customers to make decisions when designing the layout of their new kitchen. The VR system has been in practical use in Matsushita's showrooms since 1994.

Finally, a distributed VR system for use in urban planning. The core of this VR system is a domed VR facility where virtual reality can be experienced. This VR system presents a virtual city, so that various specialists can cooperate with each other in city planning and plan evaluation, with citizens also enabled to experience and evaluate the planned city.

2. A RELAX/REFRESH SYSTEM EMPLOYING VR

In this modern day and age, people are continually subjected to stress in a variety of different forms due to the fluctuating nature of politics, economics, technology and so on. Such stress results in people becoming more interested in their physical and mental health, and consequently they go to special resorts, join fitness clubs or exercise in sports clubs where there are a variety of health care systems which are designed to remove stress or fatigue. However, use of those systems is determined by the user; there is no "autonomous" function that determines the system's own action based on its receiving information such as physiological data of the user. Thus, to develop an intelligent health care system, such "autonomy" has to be embodied in the system. Furthermore, to make the system human-friendly, a novel human interface has to be introduced, which realizes real-time "interaction" between the user and the system, and "presence" allowing the user to be situated in a comfortable virtual world (Zeltzer, 1992).

In this section, a novel intelligent health care system is proposed, which employs virtual reality technology realizing the autonomy, interaction and presence mentioned above. The purpose of this VR system is to allow people to relax and become refreshed in a short period of time. The system removes fatigue caused by continuous office work such as VDT jobs, and frees users from stress by lowering the activity of their body so that they feel relaxed. Moreover, since the user may feel drowsy after relaxation, the system refreshes the user by stimulation.

2.1 Autonomy, Interaction and Presence in the Proposed System

Figure 1 illustrates the overview of the developed VR relax/refresh system. From physiological information from the user, sequential data on the heart rate acquired by the sensor is utilized.

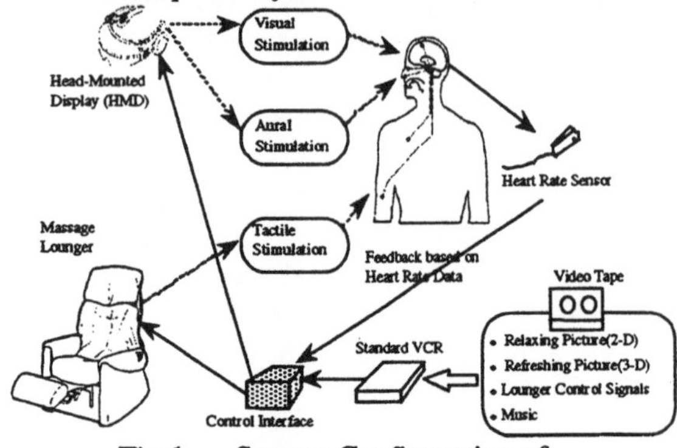

Fig.1 System Configuration of Relax/Refresh System.

When the user is fully relaxed, the heart rate data decreases (Schmidt, 1979). If this phenomenon is detected by the heart rate sensor, then the system reduces the three types of stimulation, i.e., visual, aural and tactile stimulation. This control can be regarded as the autonomy of the system.

While the user experiences the system, the heart rate is measured. According to the data obtained, the strength of stimulation is varied in real-time fashion. Thus, the unconscious interaction between the user and the system is realized. That is, it is not necessary for the user to actively input information to the system.

In addition, as mentioned above, the purpose of this system is to allow the user to relax and become refreshed in a short period of time. Thus, the strength of the three types of stimulation has to be controlled along with the desires of the user via specific scenarios. One example of a scenario could be a sun setting at a seashore at a resort, and rising after a moonlit night. While this kind of motion picture is given to the user accompanied by appropriate music, a massage lounger applies professional massage. This combination of three types of stimulation: visual, aural and tactile presents the user with an environment like a hotel room in a resort.

2.2 Overview of the Prototype System

The functions called autonomy, interaction and presence are realized in the prototype system as follows. As shown in Figure 1, the system comprises a massage lounger, a head-mounted-display (HMD), a standard VCR and a control interface which controls the three types of stimulation based on heart rate data. The HMD enables the user to directly experience the scene and music with all stimulation from the outside being shut out. An illustration of the proposed system is shown in Figure 2.

Fig.2 An Outlook of the Prototype System.

The scene and music are recorded onto a video tape. At the first relaxation stage, a simple two-dimensional picture is used to gently stimulate the visual sense of the user. At the next refreshment stage, a three-dimensional scene is used to stimulate the sense considerably. Signals controlling the massage lounger are also recorded on the video tape. This enables the rythm of the massage to be matched with the scenario and the music. That is, a soft massage is applied with the moderate scene and music during the initial relaxation stage, and a strong massage is applied with the stereographic scene and louder music during the refreshment stage. In the micro-computer mounted on the control interface, an algorithm is implemented, which detects the point in time when the user is fully relaxed by recognizing the decrease in heart rate.

2.3 *Evaluation of the Prototype System*

An overview of the experiment is as follows. Five subjects were required to perform a specified VDT job for 40 minutes, and to perform one of the following tasks for the next 10 minutes:

1) Experiencing the VR relax/refresh system,
2) Resting,
3) Continuing VDT work.

Here, resting means that the subject just sits on the chair on which he did the VDT job. After he does one of those tasks, he is required to perform the same VDT job again for another 40 minutes.

To determine how the proposed VR relax/refresh system is effective on a living body, fluctuations of the human heart are used to evaluate the activity of the autonomic nervous system. Figure 3 illustrates the changes of indexes α and β, which represent the activity level of the sympathetic nervous system and that of the parasympathetic nervous system respectively, during measurement. The values of these indexes are calculated as the average data of the 5 subjects. As shown in Figure 3, as subjects experience the VR relax/refresh system, i.e., from 40 to 50 minutes, α becomes lower while β becomes higher. This indicates that the VR relax/refresh system is effective in inducing relaxation. In contrast, index α is higher while β is lower during the final 40-minute VDT job. This indicates that the VR relax/refresh system is effective in giving refreshment.

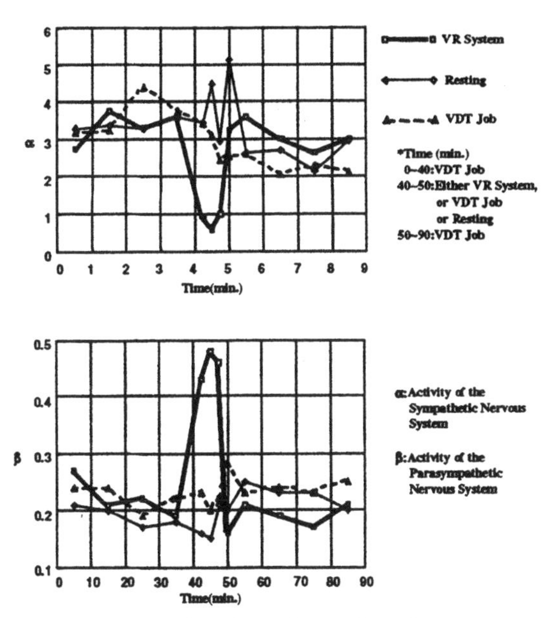

Fig.3 Experiment Results.

2.4 *Conclusion*

The prototype system was evaluated by experiment, and showed that the system is effective for relaxation and refreshment. Regarding future research, we are considering enhancing the resolution of the HMD used in the prototype system and to enrich the vibration/massage patterns set in the massage lounger. Also, it is necessary to evaluate further the physiological effects of the proposed VR relax/refresh system to fully understand the physiological phenomena of the user.

3. HORSEBACK RIDING THERAPY SIMULATOR WITH VR TECHNOLOGY

It has been said since bygone days that horseback riding gives both physical and psychological benefits to the rider, and consequently horseback riding has often been used to help treat disabled persons and for its curative effects in general, especially in Northern Europe. It is gaining attention in Japan as well. Horseback riders experience a magnificent "sensation of release" by riding a horse outdoors which also elevates the eye level. In addition, the "dialogue with a living animal" that takes place every time a rider swings up into a saddle, and the nurturing involved in caring for a horse, bring psychological benefits to the rider. The activities of maintaining or recovering one's balance, shifting one's posture, and shifting one's weight in response to a horse's movements are part of the process of giving physical stimulation and are considered as some of the positive aspects of the exercise (Ohta, 1997). However, there are few examples of reports that state that these balance maintaining functions, posture shifting functions, and possibilities for weight shifting through up-and-down movements give the same benefits to everyone (Kimura, 1987). We have developed a system that reproduces the effect of horseback riding therapy (Shinomiya, 1997). This training device may be enjoyed by healthy senior citizens, and it has been specifically designed to rejuvenate both the mind and the body. By conducting horseback riding therapy with a machine, it is possible to accurately control the distribution of loads, and the effects of bearing these loads on the human body can be evaluated.

There have been many reports of horseback riding simulations, but these simulators were not developed specifically to include applications to medical research. We have developed a simulator that accurately reproduces the movements of a saddle at pre-determined points in order to mimic actual horseback riding, while guaranteeing the safety of the rider.

3.1 Analysis of Actual Riding Data

Before designing the simulator, we collected and analyzed data on the movements of the saddle during actual horseback riding. Only then did we put together our design for the physical simulator. Data on the movements were collected by making a mark on a horse's saddle, filming the horse's and saddle's movements using high-speed photography, and then used a special method to reproduce the data 3-dimensionally. The locations of the mark are shown in Figure 4.

Fig.4 The location of the marks.

Since it is said that the walking gait of the horse is the most effective at producing the curative effects of horseback riding, data during normal walking motion was analyzed. Also, for providing enjoyment for the rider, data was analyzed to reproduce the trot and canter gaits as well. Normal walking positions are shown in Figure 5. Three-dimensional movement is represented by the X, Y, and Z axes, where the forward component of movement is represented along the X axis, the side-to-side component of movement is shown along the Y axis, and the up-down component is displayed along the Z axis. The amount of revolving motion with the X axis as the center is referred to as "roll," the amount of revolving motion with the Y axis as the center is referred to as "pitch," and the amount of revolving motion with the Z axis as the center is referred to as "yaw."

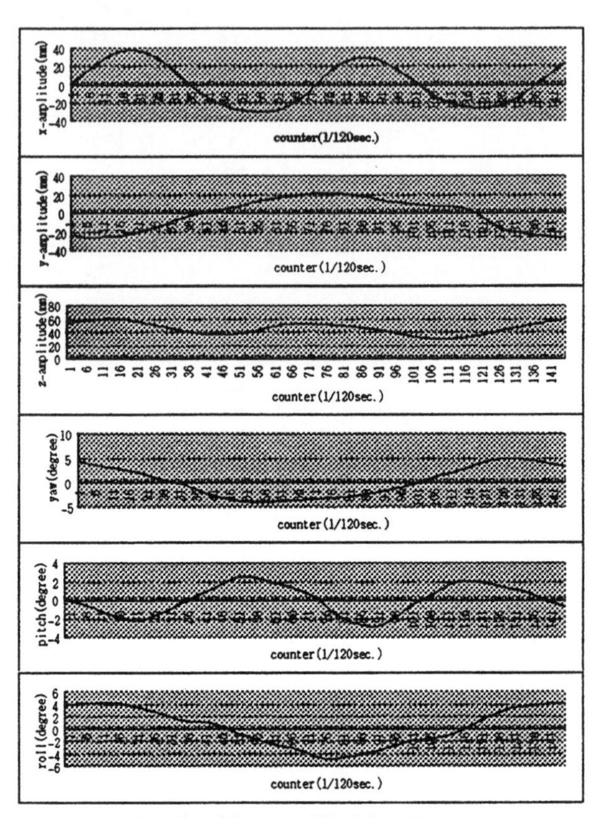

Fig.5 Normal Walking Data.

3.2 Overview of the System

System Configuration; The system configuration is shown in Figure 6. and can be divided into four sections: the drive section, the control section, the virtual reality section, and the VR operating section. Two computers are used; one for the control section and the other for the VR section.

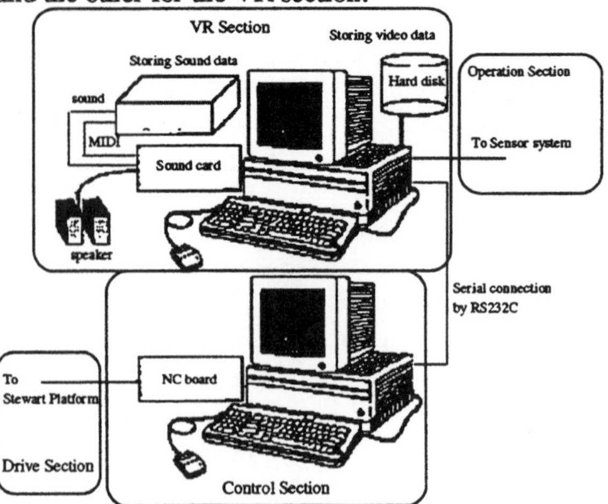

Fig.6 System Configuration of the Simulator.

Drive Section and Control Section; A small, simple, light-weight structure for the simulator drive section is required, and a Stewart platform-shaped parallel mechanism is employed. In the direct drive actuator, an AC servo motor drive is employed because it allows for precise positioning. The capacity required for the AC servo motor is 750 W.

Virtual Reality Section; In normal horseback riding, the horse is controlled by the rider through the reins, the legs, and by shifting of the hips to go forward or backward, increase and decrease speed, turn, stop, and to obtain different gaits, and so on. By simultaneously displaying motion pictures and sounds to the user of the simulator, a feeling of actually being on horseback is reproduced and enjoyed.

The following controls are possible:

1) The horse moves as desired.
- Changing speed: The speed is changed by changes in the acceleration and amplitude of the drive section.
- Changing direction: The direction is changed by increasing or decreasing the amount of possible yaw and pitch.
- Changes in walking style: changes among normal walking, trot and canter gait.

2) Visible scenery
- The scenery changes in accordance with the movements of the horse: Digitalized, subjective motion pictures are taken, stored in a hard disk and are played back to match the scenes. When the speed of the drive section increases, the frames of motion pictures are skipped, and when the speed decreases, the frames are repeated.
- A wide range of vision: A fish-eye lens was used for taking the pictures, and a large-screen television is used for the playback.

3) Auditory effects
- The sound of the hoofs changes along with the movements of the horse: The sounds of hoofs have been digitalized and are played back by an MIDI system.
- The sounds of the environment change to match the motion pictures.

4) Makes the movements displayed in the image actually occur.
- When a road climbs on screen, the accompanying effects are felt by the rider: The sensation of riding up a hill or round a curve is produced through the operation of the drive section. On an incline, the value of the pitch is changed, while on a curve, the value of the yaw is changed.

5) Designed to always be exciting.
- The rider can choose which way to go at a crossroad.: motion pictures and sounds for all directions are stored in the simulator.

VR Operation Section; The simulator has two kinds of control units; the rein control unit and the leg control unit. The rein control unit is for turning and slowing down. The leg control unit is for starting and speeding up. Sensors (shock sensors) are attached at the ends of the reins and beneath the saddle.

3.3 Evaluation and Conclusions

We have developed a horseback riding therapy simulator, which gives the sensation of riding a real horse through the motion of the drive system, motion pictures and sounds. An illustration of the simulator is shown in Figure 7.

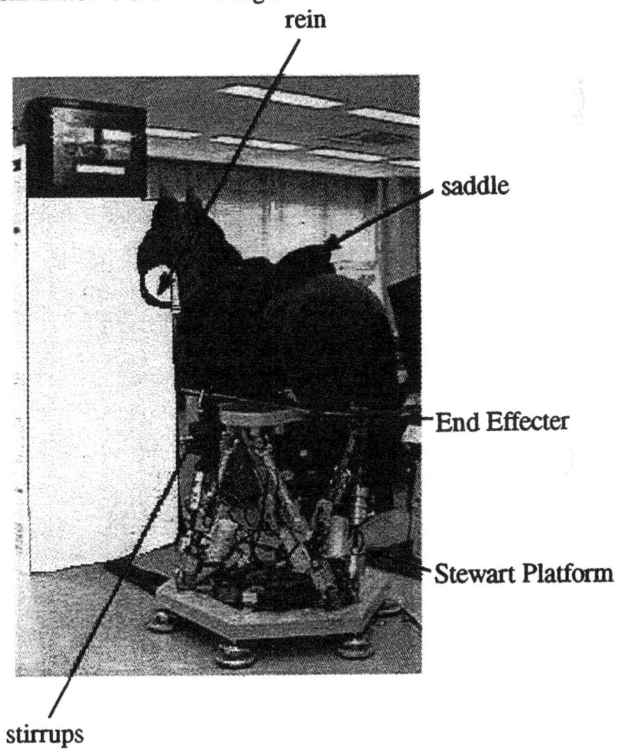

Fig.7 An Outlook of the Horseback Riding Therapy Simulator.

We had 5 horse riders ride the simulator to see how well the machine reproduced the sensations of riding an actual horse. All riders said that real horses have the same walking movement as that produced by the simulator. The evaluations of the canter and trot were inconclusive, however, as the acceleration in these modes was judged too slow to be realistic.

We will continue to research the influence of the movements involved in horseback riding on people, and carry out measurements of the muscular effects produced in the human body using electrodes.

4. VR-SUPPORTED KITCHEN LAYOUT DESIGN SYSTEM

The changing needs and values of today's consumer have had a significant impact on sales and manufacturing processes. A customer must get the necessary goods in the required quantity when needed. But when the goods desired by many users are diversified, the manufacturer has a difficult time coping with the increased workload. Changing only the production system cannot deal effectively with the situation; the corresponding sales system, including marketing, distribution, and information services, must also be improved.

Computer technology is advancing at a rapid pace. The development of a total production system incorporating CAD, CAM, and CAE is now possible. This technology also permits movement from mass production to the production of a variety of goods in small quantities. At present, however, most computer-aided manufacturing is geared towards mass production and is unable to handle one-of-a-kind products. The specifications of these products should be easily changed to accommodate individual customer's needs. To execute this concept, virtual reality (VR) technology can be employed.

VR allows customers to examine a design and make changes at the initial stage of the process (Frampton, 1995). Examples of VR applications are to be found in design, prototyping, and space layout planning, as well as in teleoperations, operator training, and entertainment.

4.1 Kitchen Layout Design and VR

One of Matsushita's strongest product lines is the "System Kitchen," a custom-planned and built kitchen combining our cabinets and appliance units. Customers can choose from over 30,000 kitchen unit products and an infinite number of possible kitchen layouts. They make many detailed and difficult decisions when specifying the layout of their new kitchen.

The kitchen planning process is detailed in Figure 8. When an interested customer comes to the showroom, a person we call the "kitchen planner" first explains the kitchen products' description using catalogues and displays. Using pencil and paper, the kitchen planner next draws a rough layout according to the customer's wishes. Then a floor plan, an elevation view, a perspective drawing, and a written estimate are created on a CAD system based on the rough sketch. At this early design stage, no VR walk-through capability is provided, hence there are often many discrepancies between the system kitchen actually manufactured and the customer's original idea.

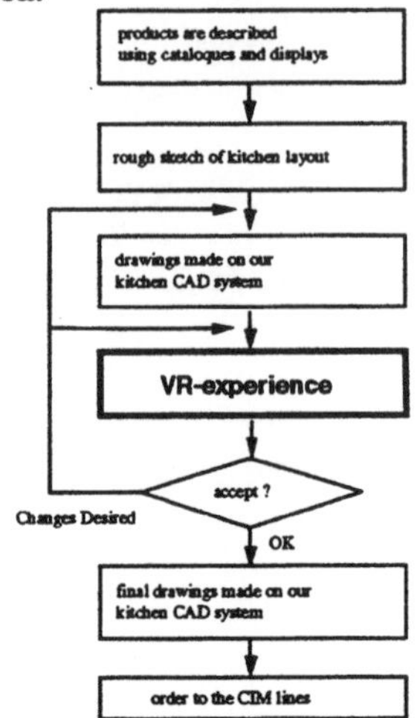

Fig.8 System Kitchen Planning Process.

Our VR system helps to eliminate these problems. The next time the customer comes to the showroom, they can experience the many aspects of their kitchen using our VR system. Customers can check their own kitchen and decide if it matches their idea of how the kitchen should be. Customers can also modify the design of the kitchen, if needed. Details of the VR experience are described below. Once the customer approves the kitchen layout design, final approval and appliance drawings are made and the order is sent to the Computer-Integrated Manufacturing (CIM) lines.

4.2 Overview of KiPS System

Our VR system named KiPS (Kitchen Planning-support System) was developed on personal computers (PC-AT clones) and replaces our previous VR system which was developed on the SGI(Silicon Graphics Inc.) machines (Yamamura, *et al.*, 1996).

The KiPS system has made it possible to achieve hardware cost reduction and a user-friendly interface. The hardware cost for the VR system has been cut from over US$500,000 (the previous system) to US$20,000 (KiPS system). User-friendly interface devices like joysticks and touch screens allow novice users to experience the VR environment more easily. Though the detail, rendering resolution, and frame rate under the KiPS system is not as fine as under the previous system, the KiPS system provides enough rendering performance for customers to have a good idea of their kitchen layout design.

Figure 9 shows the system configuration of our VR system, KiPS. A graphic accelerator card, a voice synthesis card and a network interface card are embedded in the PC. Sense8 Corporation's World Tool Kit was adopted as our virtual environment development software.

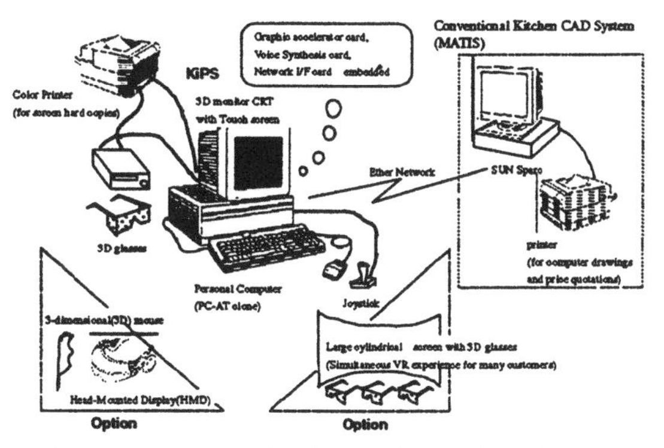

Fig.9 System Configuration of KiPS System.

As an option, a Head-Mounted Display (HMD) and a 3-dimensional mouse (3D-mouse) can be connected to the system for an immersive VR experience. And, to allow participation in the VR experience by several persons, a large cylindrical screen (4.4 m wide by 1.7 m high) can be introduced for simultaneous VR-experience on an optional basis. Two stereoscopic images with a rendering resolution of 1280x1024 are projected on the screen by two video projection systems. (This resolution is possible because SGI's Onyx RE2 is used for the rendering rather than a PC system.)

4.3 Presentation with KiPS

The following steps are typical of the presentation of a "virtual kitchen" using our VR system:

1) After drawing the plan on the SUN-based conventional CAD system (MATIS), the wireframe models in the CAD system are automatically translated into 3-dimensional VR models to be experienced in the KiPS system, and the plan data (VR data of each unit and information on unit arrangement) is transferred from MATIS to KiPS.

2) With our new VR system, customers can maneuver through their "virtual kitchen" using a joystick, and can open and close the doors of the kitchen cabinets to gain valuable insight regarding their potential investment. Figure 10 shows a sample of the "virtual kitchen." Figure 11 shows a presentation on a large cylindrical screen and HMD for simultaneous VR-experience. Customers, family members in many cases, can share the "virtual kitchen" with the HMD-wearer watching the "virtual kitchen" from the same viewpoint. A stereoscopic view is experienced through the use of polarizing glasses. The following items can be experienced in the "virtual kitchen":

- The arrangement of cabinets and appliances
- Colors and finishes
- The general feeling of available space
- Overall ergonomic design: The customer can open and close cabinet doors, turn on and off water faucets and stove burners, etc.

Fig.10 A Sample of Virtual Kitchen.

Fig.11 Presentation with a Large Screen and HMD.

3) Customers can replace the kitchen units and also change the color and texture of the cabinet doors using a touch screen to edit their own kitchen design through interaction with the "virtual kitchen." The following modifications can be made in the "virtual kitchen":

- The replacement, addition and deletion of kitchen units
- The change in color and texture of the cabinet doors, the walls, the floors and the ceilings
- The layout of decorative items like kitchen furniture, electrical appliances, human images, etc. for easy understanding of the available space and for verisimilitude

4) After all desired modifications have been made, the final plan can be transferred back to MATIS. Then, the plan data is automatically converted and a computer drawing of the kitchen and price quotation can be printed out.

4.4 *KiPS Results*

The VR system, KiPS, has been in practical use in Matsushita's four showrooms in Tokyo, Yokohama, Osaka and Hiroshima since 1994. More than three thousand customers have experienced our VR system. Table 1 shows a summary of customer feelings toward the KiPS system, suggesting that the VR system is a very useful design, sales, and promotional tool. The probability of acceptance of Matsushita's System Kitchen is 55.5% among the customers using the KiPS system in designing their kitchen, and 37% among the customers not using the KiPS system.

Table 1 A Summary of Customer Feelings toward KiPS.

(survey conducted: Oct.1994-Feb.1995, number of questionnaires: 322)

good points		poor points		Useful for selecting your kitchen components?	
easy to have a good idea of the kitchen	37.3%	not realistic	20.2%	very useful	18.0%
able to see 3D image	29.1%	low resolution (screen & HMD)	17.7%	useful	51.5%
realistic	21.7%	slow rendering	7.8%	moderately useful	13.4%
able to modify the kitchen	18.0%	not enough detail	7.8%	useless	0.3%
easy to understand colors and textures	3.7%	difficult to operate (3D-mouse)	5.0%	undecided	16.8%
		eye fatigue	4.3%		
		unable to see 3D image	3.7%		

Note: totals exceed 100% in cases where more than one item might be selected.

4.5 *Kitchen Layout Design System on the Internet*

In the future, we plan to put our VR-supported kitchen layout design system on the Internet. The VR system can be another initial design specification entry point. Customers can experience the virtual kitchen automatically designed by the VR system according to kitchen parameters such as shape, dimension, fixtures, etc. given by the customer. Thus customers can make a kitchen layout design at home and have clear design images before they come to the showrooms. Then they work on their designs in detail with the "kitchen planner," at the showroom.

4.6 *Prototype of the Kitchen Layout Design System on the Internet*

The meteoric development of the VRML (Virtual Reality Modeling Language) is one of the most important standards developments in graphics and internetworking (Don, 1996). It is an informal, intense, open, and collaborative design process; the VRML, its browser, and its converter are cross-platform supported and many of them are freeware. Accordingly, the VRML is placed as the basic language of our new VR system. The VRML describes not only 3-D worlds but also has many functions such as hyper-link and script. The following can be realized when using the VRML functions:

- Using the Anchor node, information such as size and price of component products can be added. We can make hyper links between a 3-D object and a text, an image, a movie, a sound, and other 3-D objects.
- Using the Inline node, the necessary 3-D objects can be read from other databases on the Internet.
- Using the Sensor or Interpolator node, the function of each component product can be explained effectively. For example, cabinet doors can be opened and closed.
- In the final process, the design environment is needed to coordinate kitchen design interactively. Using the Script node, a new palette for color or texture can be shown.

Using these functions of the VRML, we developed the new VR system that is Internet compatible (Fukuda, *et al.*, 1997). The following are the typical steps used in kitchen layout design using the prototype system:

- Customers select the kitchen cabinet products.
- If required, they can get more information about the products, such as function, price, size, and detailed appearance.
- The initial layout design, including the selected cabinets are presented, and customers can experience the virtual kitchen. Figure 12 shows an example of the virtual kitchen using VRML.
- If required, customers can modify the layout design with the palette on the screen. When they select a favorite product in the palette, the cabinet's data is loaded from a database on the network.
- After all desired modifications have been made, the final design can be transferred and stored at a Matsushita's showroom, so they can then design in detail with the "kitchen planner."

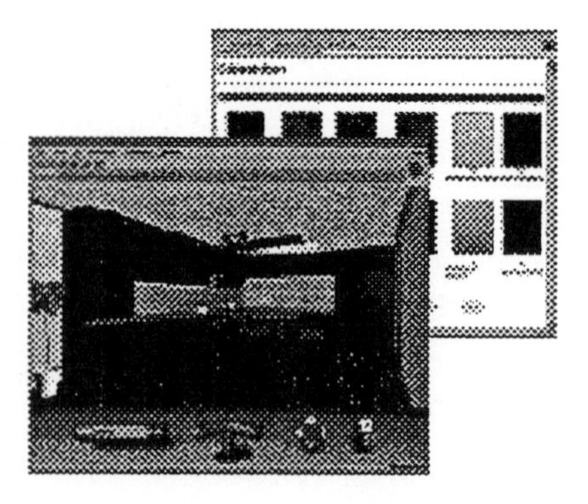

Fig.12 A Sample of Virtual Kitchen using VRML.

4.7 Conclusion

Our VR system, "KiPS," was developed on personal computers, which made it possible to achieve hardware cost reduction and a user-friendly interface. The system allows our customers to have a good idea of their potential purchase and to also take part in the design process of the kitchen layout. Positive customer reaction to our system has been reported, which leads us to the conclusion that the VR system is a very useful design, sales and promotional tool.

In the future, we plan to put KiPS on the Internet. Customers can then make a kitchen layout design at home. A prototype system has been developed. We are now developing a system employing a Genetic Algorithm (GA) which will make kitchen design candidates according to kitchen parameters given by the customer. Furthermore, our goal is to expand the scope of our system so that an entire home may be designed, instead of a kitchen only. Figure 13 shows a system configuration of the House Design Advisory System using the Internet.

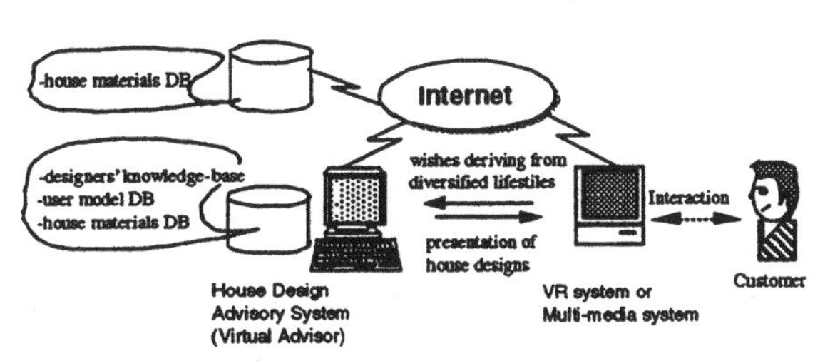

Fig.13 House Design Advisory System.

5. URBAN PLANNING VR SYSTEM

Planning of urban environment, whether for an entire city, a city block, or buildings and structures within a city, presents difficulties due to the following problems:

1) No technique has been established to support communication among specialists in the design stage;
2) It is impossible to produce urban environment on a test basis;
3) No technique for evaluation of amenities has been established.

To provide solutions to these problems, a system based on VR technology is under development which will allow creation of a virtual city, enabling many specialists to cooperate in city design and evaluation, simultaneously enabling citizens also to experience and evaluate the virtual city.

5.1 Virtual City Multimodal Presentation System

The virtual city multimodal presentation system is a distributed system whereby VR can be experienced. This system comprises VR terminals connected via Internet, with a computer system based on clustered workstations interconnected through a high-speed network. This system provides a framework whereby many specialists and citizens can participate in the process of developing a virtual city regardless of their physical location. VR experience facilities which can present a high-definition virtual city to many people should be installed at some terminals.

For this purpose, a domed VR experience facility has been produced on a test basis. This facility comprises a domed screen, six stereoscopic projectors, graphics computers, liquid-crystal shutter glasses, infrared ray emitters and magnetic sensors. Figures 14 and 15 are a side view and a photograph, respectively, of the domed VR experience facility. The domed screen is 6.8 m in diameter. This screen, in conjunction with the six forward-projection type projectors, presents a VR world of an angle of view (180 degrees in horizontal angle and 90 degrees in vertical angle) and scale matching that experienced in the real world. A high-resolution, stereoscopic picture of 3,086 pixels (horizontal) × 1,536 pixels (vertical) is projected over the entire screen. In the future, we plan to incorporate a stereophonic sound system, a voice input system, and oscillating apparatus etc. into this facility, to realize an advanced multimodal system providing a more natural-looking interface through which people can experience a virtual city of heightened realism.

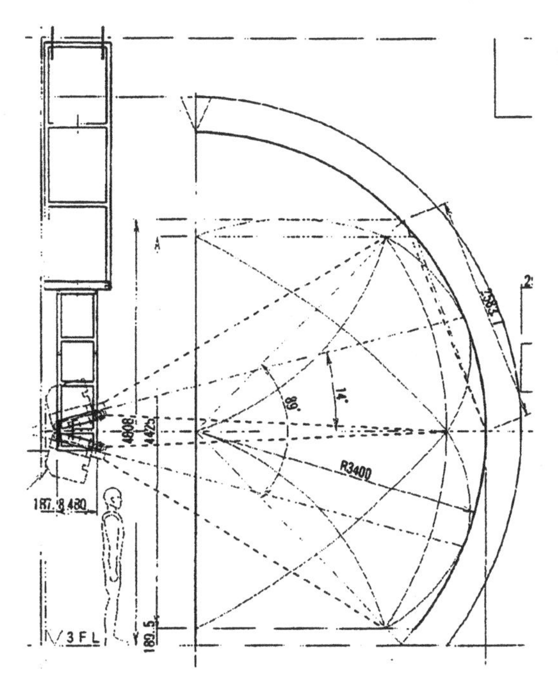

Fig. 14 Design (Side View) of Domed VR Experience Facility.

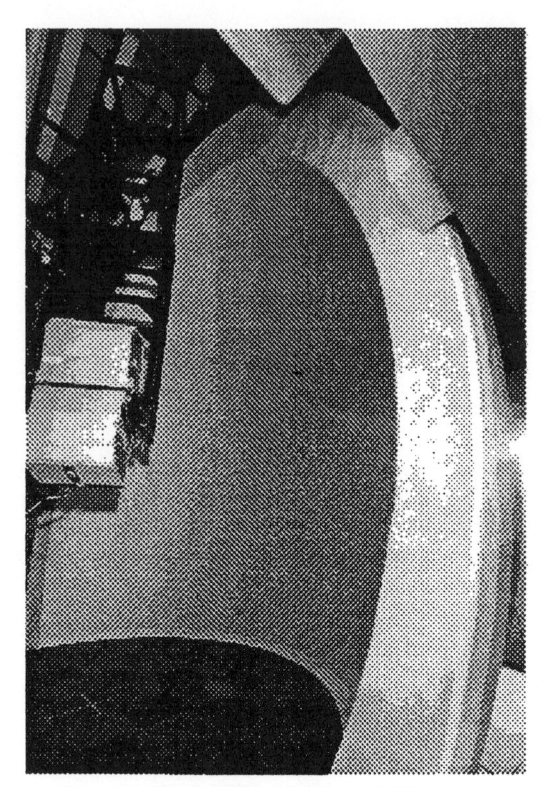

Fig. 15 Photograph of Domed VR Experience Facility.

5.2 *VR Simulation Application to Urban Planning*

Two examples of application of VR simulation to urban planning are presented.

Figure 16 shows an example of application to a mixed commercial facilities plan based on a station plaza redevelopment project. The major purpose of this application was to VR-simulate the client's vision of Spanish buildings and Spanish townscape in the design stage.

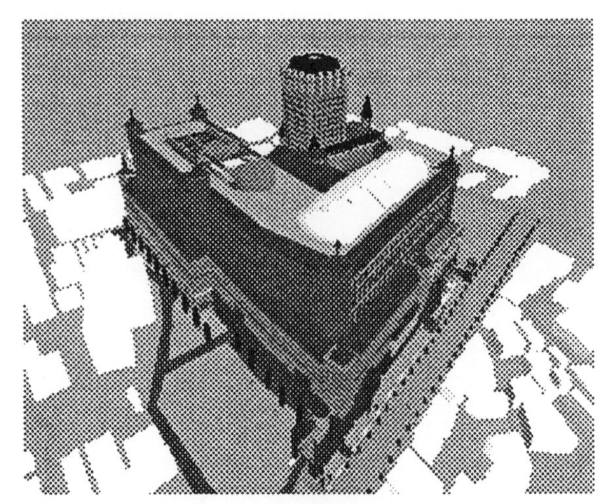

Fig. 16 VR Simulation Example (Mixed Commercial Facilities).

Figure 17 shows an example of application to a street renewal plan also based on a redevelopment project. The major purpose was to develop a consensus among regional residents as to the type and arrangement of roadside trees, the color of sidewalks etc., in the design stage.

Fig. 17 VR Simulation Example (Street Renewal).

These two examples have verified the effectiveness of VR simulation in urban planning. Based on shared experience of a virtual space, the client, designer and all other persons concerned can evaluate and discuss street width, building colors, and various other specification details in the design stage; without a VR system, these details could be evaluated only after completion of construction.

5.3 *Conclusion*

A distributed VR system based on a domed VR experience facility has been described. If this system is developed for practical use, persons concerned with urban planning, public space design etc. (constructors, residents/users, general contractors, subcontractors, and manufacturers) will be able to access a virtual design space and related necessary information, so that specialists can cooperate with each other in carrying out high-quality design evaluations, pooling their expertise.

(p.s.)
This study is being conducted by the Laboratories of Image Information Science and Technology as part of the "Human Media R & D" project, for the New Energy and Industrial Technology Development Organization, under the Ministry of International Trade and Industry's Industrial Technology Development and Organization System.

6. CONCLUSION

In this paper, four VR application systems for industrial use which have been developed in Matsushita Electric Works, Ltd. are described. As stated in section 1, VR technology has recently been investigated for industrial use, since the basic components for developing the VR application system are decreasing in cost. In addition, the research and development into high-performance VR systems which handle large 3-D models and provide extremely high-resolution 3-D graphics images is to be continued. Thus, the development of VR application systems can be divided into two: a low-cost VR system and a high-performance VR system. From this point of view, all four application systems are categorized into the low-cost VR systems. Accordingly, we intend to produce low-cost VR systems for practical use, while developing high-performance VR systems for research.

REFERENCES

Don, B. (1996). VRML: Prelude and Future. *SIGGRAPH'96, Panels Sessions.*

Frampton, R. (1995). VR: Multimedia and More - The Role of Virtual Reality in the Commercial and Industrial Society. *Proc. of VSMM'95 (International Conference on Virtual Systems and Multimedia '95)*, 225-229.

Fukuda, T. *et al.* (1997). Networked VR System: Kitchen Layout Design for Customers. *VRML'97.*

Kimura, T. (1987). The Development of an Intelligent Simulator of Horses. *Medical Treatment*, Vol. 40, No. 8, pp. 749-755.

Ohta, E. (1997). The Curative Effects of Horseback Riding. *Stock Breeding Research*, Vol. 51, No. 1, pp. 148-154.

Schmidt, R. (1979). *Fundamentals of Neurophysiology*, Kinpoudou.

Shinomiya,Y. *et al.* (1997). Horseback Riding Therapy Simulator with VR Technology. *Symposium on Virtual Reality Software and Technology 1997 - VRST'97*, 9-14, ACM press.

Yamamura, A. *et al.* (1996). Kitchen Layout Design in Virtual Environments. *Proc. of The 1996 ASME Design Engineering Technical Conferences and Computers in Engineering Conference.*

Zeltzer, D. (1992). Autonomy, Interaction, and Presence. *PRESENCE 1*, **1**, pp.127-132.

HUMAN-COMPUTER COMMUNICATION FOR DESIGN AND OPERATION IN PERFORMING ARTS AND INDUSTRIAL ENGINEERING

Gunnar Johannsen

University of Kassel, Lab. Systems Engineering & Human-Machine Systems
D-34109 Kassel, Germany
(phone ++49-561-804-2658; e-mail joh@imat.maschinenbau.uni-kassel.de)

Abstract: Activities in performing arts and industrial engineering are briefly compared. This gives an overview on several professions. The modes of communication in both domains, including the view on agent-agent communication with different forms of control and display, as well as dimensions of expression and cognition, are discussed. Then, examples of human-computer interfaces in both areas, the performing arts as well as engineering and the service sectors, are explained. The possibilities of transferring experience between both worlds are analysed.

Keywords: Human-machine interface, human-centered design, cognitive systems, communication environments, agents, user interfaces, performing arts, industrial production systems, multimedia, movement.

1. INTRODUCTION

The main idea of this paper is to learn as much as possible from some of the entertainment sectors, particularly the performing arts, for the field of human-machine systems in industrial, transportation and service sectors. Audio and music technologies, high-performance computer graphics, multimedia technologies, and virtual reality approaches seem to be some of the most important key areas which may be capable of bridging between the two seemingly different worlds.

2. ACTIVITIES IN ARTS AND ENGINEERING

An overview on several professions in performing arts and industrial engineering can be provided by introducing and comparing their characteristic activities. In both fields, these activities are classified into those belonging more to the areas of design and those belonging more to the areas of operation. As Table 1 indicates, the different categories of activities are mainly classified under task and job criteria.

The areas of design, as listed in Table 1, comprise creative activities with the main goal of production. The products are pieces of music, dance, play etc. in the performing arts, and interactive systems, machines, interfaces and work places in industrial engineering. Many disciplines in these areas have established their own repertoire of methodologies. This may be exemplified with „Design for Success“ by Rouse (1991) and with „Fundamentals of Musical Composition“ by Schoenberg (1967). Human-centered approaches seem to prevail in most cases, i.e., the designed products have to serve people. The principles of construction are sometimes even stronger in art than in engineering. In both domains, complex products are built from smaller subunits. Craftsmanships combined with creativity, knowledge, and improvisations are needed.

Table 1 Activities in performing arts and industrial engineering.

	Performing Arts	Industrial Enginering
Design	scene-painting composing graphics/video/audio design choreographing multimedia design play/game design	work design graphical user interface design dialogue design application/user/task modelling motion/audio cues design multimedia design
Operation	conducting playing musical instruments stage-managing play-/opera-acting singing/playing dancing	operating supervising maintaining consulting managing navigating

The areas of operation in Table 1 deal with the products of the design areas. Operation in the widest sense means usage of products and re-production, thereby also product evaluation. All types of operation in the performing arts are also regarded as highly creative. Conducting is even considered to be the most intellectual of the reproducing arts (Scherchen, 1953). In industrial engineering however, the operational activities are often assumed to be less creative than the design activities, particularly in direct sensory control of technical plants and vehicles (Johannsen, 1993). A distinction is made between the technical-scientific knowledge of functioning versus the practical knowledge of utilisation (De Montmollin and De Keyser, 1986). Considering safety-relevant fault management tasks in risky systems, the creativity of the operational staff should better be taken full advantage of. Maybe, engineering can learn from the arts in this respect. As Goodwin (1996) stated: „ ... whereby science can begin to connect with the arts: people being creative and playing. There's nothing trivial about play. Play is the most fundamental of all human activities, and culture can be seen as play."

3. MODES OF COMMUNICATION

3.1 Communication between Agents

All activities listed in Table 1 can be explained more or less with a general scheme of communication between different agents. As Figure 1 shows, an interface generally exists between two agents to allow communication between them. An agent can act and, thus, execute control on another agent while perceiving displayed information from that agent.

In case of one agent being in control, this agent has overall responsibility and the other agent(s) is (are) subordinate. This is shown in the upper part of Fig. 1. Examples are a musician playing a musical instrument (this instrument being the subordinate agent), a person who is conducting an orchestra, or is stage-managing an opera scene, as well as a pilot flying an aircraft, a shift supervisor managing a power plant or a financial expert advising a client during a consultation. The controlled agents of these examples are certainly of different nature and have different degrees of autonomy. Whereas the controlling agent is always a human, the controlled agents can either be also one or several humans, or they can be instruments, tools or machines. Thus, the traditional human-machine system also belongs into this category.

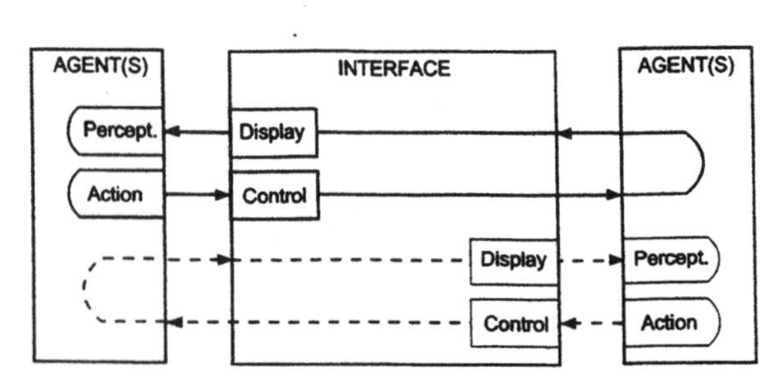

Fig. 1. Displays and controls in agent-agent communication

In case of more cooperation and distributed responsibility among the agents of both sides of the interface, the lower part of Fig. 1 needs to be considered in addition. The above example of the consultation between a client and a financial expert can already be of this type of communication between agents. Certainly, a team of players in an opera, the corps de ballet, pilots communicating with air traffic controllers, cooperating maintenance personnel in industrial plants or intelligent computer agents and computerised machine agents which are sharing initiative with human agents are examples for this type of agent-agent communication with distributed control responsibilities.

Different forms of control can be exerted, either direct (as with the conductor), via biosignals or via control devices. Speech and motion (displacements, forces, gestures) can be used directly or via control devices. Displayed information can also be either direct (views, sounds) or mediated via computer graphics as well as by means of video, audio and music technologies. Generally, visual, auditory, mimic (including face-to-face) and haptic informations as well as vibrations can be used as forms of displays. Modes of communication suitable for handicapped people have also to be considered. An example is the lightspot operated typewriter for physically disabled persons (Stassen, 1989).

3.2 Dimensions of Expression and Cognition

There are certainly fundamental differences between the different modes of agent-agent communication as explained with Fig. 1 for the different activities of Table 1. The dimensions of expression and cognition are regarded here for clarifying these differences.

First, the superficial argument that the arts relate more to emotions than engineering, needs to be rejected. Copland (1988) differentiates between three separate planes we all listen to in music: „(1) the sensuous plane, (2) the expressive plane, (3) the sheerly musical plane." The sensuous plane „is the plane on which we hear music without thinking, without considering it in any way." The second and the third planes are the dimensions of expression and cognition. Goodman (1976) rejects the popular opinion even more that a primary objective of art is to generate emotions. He strongly uses the term expression which a work of art conveys. Then, cognitive perception and cognitive processes are initiated in the human interaction with any such work of art. Thereby, emotions may not or may occur which, however, function cognitively in the aesthetic experience.

Emotions may also arise in engineering. Probably, an unexpected alarm of a severe safety-critical incident in a technical plant may generate strong fear and anxiety. As any blind emotions can be dangerous, they should be avoided or cognitively transformed. This necessity is independent of the degree of reality, virtual reality or imagination.

Thus, the dimension of cognition plays a general role in the arts and in engineering. It possesses controlling and evaluative power. This varies with the different modes of agent-agent communication in design and operation. Johannsen (1996, 1997b) introduced the Cognition Space Metaphor as a multidimensional concept. Two 3D-coordinate systems comprise the designating dimensions TAU (task, application, user) and the execution-related dimensions STI (space, time, interaction). Each of the dimensions contains, at least, two levels on which selections from a set of discrete choices determine different communication scenarios in design and operation. This is equally applicable in art and engineering which was exemplified by Johannsen (1996).

It would be interesting to compare the specifics of the performing arts and industrial engineering along all dimensions of the Cognition Space. The restrictions of this paper only allow a brief discussion of the time-dimension and of two subdimensions of the application-dimension. The time-dimension is generally important in industrial engineering and in all performing arts. Time relates to the different dynamics and temporal characteristics as well as to runtime conditions and historical periods (past, present, future). All these subdimensions determine such different professions as conductor, dancer, power plant operator and pilot, although with different parameters.

Combining time and space and, particularly, looking at movements in space and time, seems to be much more important in the performing arts than in engineering. Good examples are musicians and dancers. Movements are also used for or support expressions. Biomechanically, movements are necessary in offices and industrial workplaces, too. Tietze (1989) has criticised industrial ergonomics for suppressing appropriate movements. New designs, for real and virtual environments, need to consider this more strongly.

The two subdimensions of the application-dimension in the Cognition Space concern the structural aggregation/decomposition and the functional abstraction. More abstract representations are supported by systems of symbols in arts, science and engineering (Goodman, 1976). Different types of notation symbolise and preserve the syntax and the semantics of specific works. Examples of formalised notation systems are scores in music and the Laban-notation in choreography (Goodman, 1976), as well as, but less strictly, standards for style guides and for symbols or icons of graphical user interfaces in engineering.

Structural decomposition leads finally to the most fine-grained elements in each work of art or engineering. Even the elements have a certain power of expression in the arts. The basic elements are line (melody), body or solid (harmony), colour (tone colour), and spatial orientation (rhythm). Motion is used for expressions in dance, with the whole body and the extremities of legs, arms, hands and fingers. Gestures of the upper extremities convey speed, rhythm, and expression in conducting.

Expressiveness would also be required in engineering. However, it is almost completely missing. Or it is used in a very simple manner, as with most audio alarm signals (Sorkin, 1987). Expressive melodies for certain failures still need to be investigated. The fierceness during face-to-face communication may express a severe emergency or time pressure. Motions or gestures may be expressive inputs in agent-agent communication.

4. EXAMPLES OF HUMAN-COMPUTER INTERFACES

The main functionalities of human-computer interfaces for supporting design and operation will be exemplified in the following. Such computer-based interfaces for display and control have been developed over decades in both areas, the industrial, transportation and service sectors as well as the performing arts. These interfaces often do not only include displays and controls but also software modules for information handling and processing as well as for knowledge support (Johannsen, 1997b).

4.1 Examples in Performing Arts

The examples in performing arts are mainly from the areas of computer music and media arts. More than one hundred years ago, the first electronic music machine was patented in the United States (Manning, 1994). Particularly since the 1950s, electronic music studios and research projects strongly developed in many parts of the world (Winckel, 1955; Manning, 1994). Sound synthesisers and all kinds of musical instruments have been created. The control input parts of traditional instruments have been modified for being input devices for synthesisers. Examples are particularly the piano keyboard, percussions and the guitar, but also violin, cello and woodwind instruments (Paradiso, 1997). Thereby, a much wider control input spectrum exists than with the typical control devices in engineering.

Computer music mainly dates back to the early 1980s (Manning, 1994). It has been used for composition and for performance activities. The completely digital synthesiser was developed, and Musical Instrument Digital Interface, or MIDI, was created for interconnecting all kinds of equipment, particularly synthesisers and musical instruments.

Another use of computers has created digital audio and sound sampling. With samplers, audio wave files are generated which can be edited, today even online on personal computers. Thereby, a large number of sound effects can be created interactively. A complete overview on music technologies is given by Williams and Webster (1996). Samplers and synthesisers are the most common hardware devices in today's sound studios. Several powerful software packages exist which allow digital audio editing. Some of these software packages additionally include a score editor and related editors with support functions for music notation. Even full orchestra scores can be created with direct notation on the computer screen or via a MIDI-keyboard. Fig. 2 shows a screen set of the professional software Logic Audio for digital audio and MIDI sound production (without its score editor). An interface for the dance notation as an interactive tool for the support of choreographers is presented in this conference (Calvert, *et al.*, 1998).

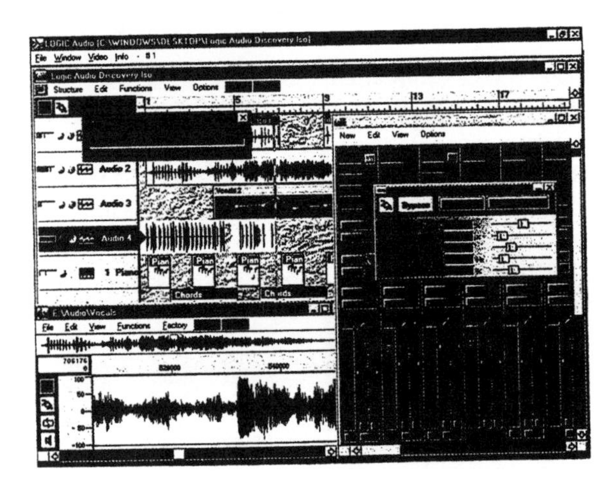

Fig. 2. Screen set for digital audio and MIDI sound production (reprinted with permission by Emagic GmbH, Rellingen, Germany).

Gestures and movements in playing electronic musical instruments are sometimes partially recorded. Capturing of gestures and movements is much more important in conducting and dancing. Movements of digital batons can be tracked and used as inputs for electronic or computer music production or interpretation (Williams and Webster, 1996, pp. 372-373; Paradiso, 1997). A direct sound generation with the movements of the pure hands, with holding nothing at all, was already developed in the early 1920s as the Theremin, called after its inventor Leon Theremin (Manning, 1994; Paradiso, 1997). This free-gesture controller has a pair of sensors. The player moves the hands in the air above these sensors, one hand controlling pitch, and the other amplitude.

Electrical fields and other sensing modes for non-contact detection of musical gestures and body movements in dancing are investigated in the MIT Media Lab and in several other research labs and commercial companies (Paradiso, 1997). An environment for interactive multimedia art has been developed by Katayose, *et al.* (1995) which has further been developed since then and is also presented at this conference (Nagashima, 1998). In these multimedia worlds, human performers interact in real time with computer graphics, computer music, and video. Another paper in this conference deals with the cooperation between the human performer and computer agents with the expression of feelings during musical ensemble play (Saiwaki, *et al.*, 1998).

4.2 Examples in Engineering and Service Sectors

The design process for cooperative graphical user interfaces and multimedia interfaces has been based on systems engineering development procedures (Johannsen, 1997a; Borys and Tiemann, 1997). Task orientation and user participation are particularly considered. Such systems life-cycle methodologies, combined with prototyping and usability testing, are more systematically applied for better human-centered interface designs in industrial engineering.

The example of the cement industry shows that even the traditional process control room allows multimedia-based operation (Johannsen, 1997a). Audio and video informations as well as voice and face-to-face communications are widely used in addition to interactive computer graphics. New multimedia technologies can be based on current experiences in order to improve the task-oriented communication and control processes further. Thereby, new forms of knowledge visualisation in functional human-machine interfaces have also to be introduced in industrial applications. An overview on such new interface technologies with advanced laboratory implementations for chemical and cement plants is discussed by Johannsen, *et al.* (1997); see also Ali (1998).

New ways of multimedia communication and control will be investigated with an experimental multimedia process control room which is currently under development (Borys and Johannsen, 1997). At present, it already consists of two powerful computer-graphics workstations capable of displaying stereo and holographic images, and with video and audio inputs/outputs, camera, active stereo glasses with liquid crystal shutters, space mouse and keyboards. The audio equipment will include audio editing facilities for developing acoustical displays as well as equipment for sound spatialisation. A vibration device is also under consideration. The main idea of this experimental multimedia process control room is to bring the operational staff closer back to the technical process. Examples from the management and operation of power plants show that human operators traditionally used a lot of sound and vibration information which is nowadays often replaced by abstract graphics visualisations. In making again better use of other sensory modalities, the redesign of industrial process control rooms should also consider experiences from computer interface developments in the performing arts. Similar extensions with the additional use of other sensory modalities will also be investigated in aircraft guidance with more "musical" acoustical displays as support tools in 3D-space, which will go beyond more recent developments of perspective flight path displays (Theunissen, 1995).

Other applications of multi-modality and multimedia approaches for cooperative interfaces are summarised by Johannsen (1997b) with the examples of design work in teams, distributed maintenance in industrial plants, and training of maintenance situations in virtual reality environments. A remote consultation system for advising clients of insurance companies is a good example for the application of multimedia technologies, including computer graphics, video, audio, and teleconferencing techniques, in the service sectors (Tanaka, *et al.*, 1997).

5. TRANSFER OF EXPERIENCE BETWEEN BOTH DOMAINS AND FUTURE PROSPECTS

Several of the above examples relate to more recent research from the author's laboratory as well as to his own experience in computer music. The discussion of the modes of communication and of the examples of human-computer interfaces has shown that there are many similarities in both domains of performing arts and industrial engineering. However, remarkable differences exist as well. Particularly, expressions and movements are much more involved in performing arts. That leads to much richer forms of control input also in the human-computer interfaces. It needs to be explored in further research whether at least some of these possibilities may be transferred into the industrial engineering domain. The question arises whether we can "conduct a power plant". Some ideas for using human gestures in device control and menu selection are presented at this symposium (Shibuya, *et al.*, 1998).

Too little use is generally made of audio information in industrial, transportation and service sectors. Recently, speech is more often used. But sound is applicd only in quite primitive ways, if at all, as compared to the use of graphics. More musical acoustic displays have to be investigated in the future. Other examples of how we can learn from different sectors of the performing arts for the interface design in industrial engineering are presented by Alty (1998). On the other hand, the performing arts can also take advantage of the developments in the industrial and service sectors. Particularly, real-time requirements as well as hardware and software developments can further support new interactive technologies in the performing arts.

REFERENCES

Ali, S. (1998). In: *this IFAC-MMS Symp.*

Alty, J.L. (1998). In: *this IFAC-MMS Symp.*

Borys, B.-B. and G. Johannsen (1997). An experimental multimedia process control room. In: *Advances in Multimedia and Simulation (Proc. Conf. Human Factors and Ergonomics).* (K.-P. Holzhausen, Ed.), pp. 276-289. Bochum.

Borys, B.-B. and M. Tiemann (1997). The DIADEM software development methodology extended to multimedia interfaces. In: *Advances in Multimedia and Simulation (Proc. Conf. Human Factors and Ergonomics).* (K.-P. Holzhausen, Ed.), pp. 117-125. Bochum.

Calvert, T., S. Mah and Z. Jetha (1998). In: *this IFAC-MMS Symp.*

Copland, A. (1988). *What to Listen for in Music.* Mentor ME 2735, Penguin Books, New York.

De Montmollin, M. and V. De Keyser (1986). Expert logic v. operator logic. In: *Analysis, Design and Evaluation of Man-Machine Systems* (G. Mancini, G. Johannsen, L. Mårtensson, Eds.), pp. 43-49. Pergamon Press, Oxford.

Goodman, N. (1976). *Languages of Art. An Approach to a Theory of Symbols* (2nd ed.). Hackett, Indianapolis. (Suhrkamp, Frankfurt, 1995).

Goodwin, B. (1996). Biology is just a dance. In: *The Third Culture* (J. Brockman, Ed.), pp. 96-110. Touchstone, Simon & Schuster, New York.

Johannsen, G. (1993). *Mensch-Maschine-Systeme* (Human-Machine Systems; in German). Springer, Berlin.

Johannsen, G. (1996). Cognition space metaphor for human-machine interaction. In: *Proc. Cognitive Systems Engineering in Process Control (CSEPC 96)*, pp. 253-260. Kyoto.

Johannsen, G. (1997a). Conceptual design of multi-human machine interfaces. *Control Engineering Practice*, **5**, 349-361.

Johannsen, G. (1997b). Cooperative human-machine interfaces for plant-wide control and communication. In: *Annual Reviews in Control* (J.J. Gertler, Ed.), Vol. 21, pp. 159-170. Pergamon, Oxford.

Johannsen, G., S. Ali and R. van Paassen (1997). Intelligent human-machine systems. In: *Methods and Applications of Intelligent Control* (S. G. Tzafestas, Ed.), pp. 329-356. Kluwer, Dordrecht.

Katayose, H., T. Kanamori, T. Sakaguchi and Y. Nagashima (1995). Virtual performer: An environment for interactive multimedia art. In: *Symbiosis of Human and Artifact (Proc. 6th Conf. Human-Computer Interaction, Yokohama).* (Y. Anzai, K. Ogawa, H. Mori, Eds.). Vol. 20A, pp. 107-112. Elsevier, Amsterdam.

Manning, P. (1994). *Electronic and Computer Music* (2nd ed.). Clarendon Press, Oxford.

Nagashima, Y. (1998). In: *this IFAC-MMS Symp.*

Paradiso, J.A. (1997). Electronic music: New ways to play. *IEEE Spectrum*, **34**, 12 (Dec.), 18-30.

Rouse, W.B. (1991). *Design for Success: A Human-Centered Approach to Succesful Products and Systems.* Wiley, New York.

Saiwaki, N., J. Kawabata and S. Nishida (1998). In: *this IFAC-MMS Symp.*

Scherchen, H. (1953). *Lehrbuch des Dirigierens.* Schott, Mainz.

Schoenberg, A. (1967). *Fundamentals of Musical Composition.* Faber and Faber, London.

Shibuya, Y., K. Takahashi and H. Tamura (1998). In: *this IFAC-MMS Symp.*

Sorkin, R.D. (1987). Design of auditory and tactile displays. In: *Handbook of Human Factors* (G. Salvendy, Ed.), pp. 549-576.Wiley, New York.

Stassen, H.G. (1989). The rehabilitation of severely disabled persons. In: *Advances in Man-Machine Systems Research* (W.B. Rouse, Ed.), Vol. 5, pp. 153-227. JAI Press, Greenwich, CT.

Tanaka, T., H. Mizuno, H. Tsuji, H. Kojima and H. Yajima (1997). Tele-consultation system supporting asymmetrical communications between customers and expert staff in a distributed environment. In: *Design of Computing Systems: Cognitive Considerations. (Proc. 7th Conf. Human-Computer Interaction, San Francisco).* (G. Salvendy, M.J. Smith, R.J. Koubek, Eds.). Vol. 21A, pp. 27-30. Elsevier, Amsterdam.

Theunissen, E. (1995). In-flight application of 3-D guidance displays. In: *Proc. 6th IFAC/IFIP/IFORS/IEA Symp. Analysis, Design and Evaluation of Man-Machine Systems,* MIT, Cambridge, MA. (Preprints), pp. 285-290.

Tietze, B. (1989). L'ufficio urla e chiede liberta. (The office cries and asks for freedom). Interview with B. Tietze by S. Carbonaro. *Ufficiostile,* **22,** 82-93.

Williams, D.B. and P.R. Webster (1996). *Experiencing Music Technology.* Schirmer Books, New York.

Winckel, F. (Ed.) (1955). *Klangstruktur der Musik.* Verlag für Radio-Foto-Kinotechnik, Berlin.

Cooperative Performance System based on Agents with Mental Model

Naoki Saiwaki, Jun'ichi Kawabata and Shogo Nishida

Graduate School of Engineering Science, Osaka University
1-3, Machikaneyama, Toyonaka, Osaka 560 JAPAN

Abstract: In this paper, we propose a cooperative performance system based on agents. To realize more intimate cooperation, it is necessary for the system to consider the feelings of a user which reflect his personal taste and mental state. We propose a model of cooperation of the agents considering those feelings. In the model, emotional factors are expressed by mental potential, which is estimated based upon the relation between the agent's behavior and the situation. The mental potential is used for cooperative decision making.

A prototype system of creating a musical ensemble was developed for evaluation and it was confirmed that the users were relatively satisfied with playing music together with the agents.

Keywords: cooperative performance, music ensemble, agent, mental model

1. INTRODUCTION

In recent years, it is requested that the computer supports users with regard for their personalities and individualities. Then researchers have become increasingly interested in the field of agents to make a system behave intellectually (Genersereth, *et al.*, 1994; Ishida, 1995; Maes, 1994). We aim more intimate cooperation between human and the agents in this paper. To realize such cooperation, it is necessary that the system considers not only the logical information, but also the feelings of a user which reflect his personal taste and mental state. Although a large number of studies have been made on the former, little attention has been given to the latter. Therefore we take notice of the latter, and propose the model of the cooperative agent considering those feelings. In this model, an agent has the parameter, which expresses the virtual emotional state, called "mental potential." It is estimated based on the relation between agent's behavior and the situation, and contributes to the cooperative decision making.

We selected a musical ensemble for an example of such cooperation, because a musical ensemble is an interesting example of such cooperation (Rowe, 1993; Wake, *et al.*, 1994). A prototype system of creating a musical ensemble was developed, and we verified the effectiveness of the cooperative agent.

2. MODEL OF COOPERATION

2.1 Modelling Cooperation

The information on feelings has the following features: ambiguity, polysemy and dependence on a situation. The most of the information with which the computer has usually dealt is for the logical information processing, and such information has been excluded.

To realize more intimate cooperation, however, it is necessary that the system considers not only the logical information, but also the information on feelings which reflects user's personal taste and mental state. However, it is difficult to get objective and rational results because of such attributes. Therefore we introduce the following ideas for constructing a model of the cooperative agent (Fig. 1).

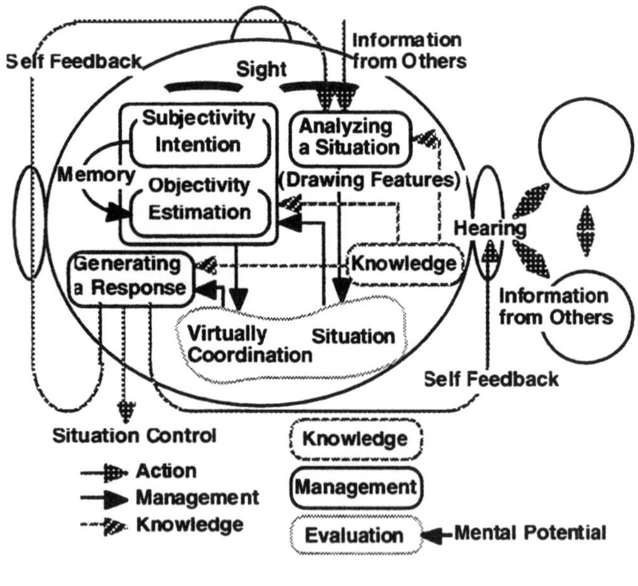

Fig. 1 Model of Cooperation

1) Parallel Recognition of Situation and Satisfaction
Each agent recognizes the situation and estimates its satisfaction level by utilizing its own criterion. The entire situation including all agents' reactions cannot be objectively evaluated by external rules.

2) Intention
Each agent has an intention for changing the situation. The individuality of each agent is realized by different intentions in the situation.

3) Operation of Behavior
Each agent operates the next performance in order to change the situation to its satisfaction.

The cooperation of such a model is realized when each agent intends to change the situation in order to balance its satisfaction level with another. Thus, conflicts are virtually avoided.

The system, which consists of agents, is not stable as a single stationary result. It tends to gravitate toward a partially stable situation. In other words, the cooperation is attained at strategic points that are significant for making the agents approximate "a human touch" in an ever-changing situation. That is, the above mentioned inadequate attributes disappear in real time interaction for a user.

2.2 Mental Potential

The mental state is virtually calculated in each agent. We propose "mental potential" which is estimated based on the relation between agent's behavior and the situation. It is used for the cooperative decision making (Fig. 2).

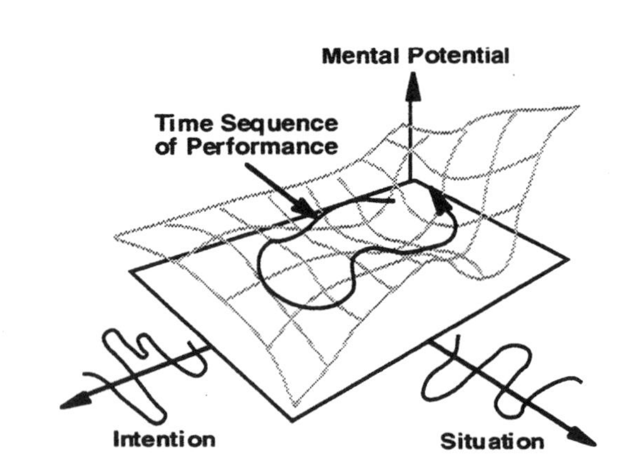

Fig. 2 Mental Potential

The mental potential has two aspects, "stress level" and "climax level." The stress level represents the level of satisfaction. If the situation is not as expected, the stress level is added. The reverse is also true. The climax level shows whether the situation is lively or not. If the situation is lively from the theoretical point, the climax level is added, and the reverse is also true. A mental state is expressed by the combination of the stress level and the climax level.

3. ARCHITECTURE OF AGENT

Each agent is composed of the following three parts. The architecture of the agent is shown in Fig. 3.

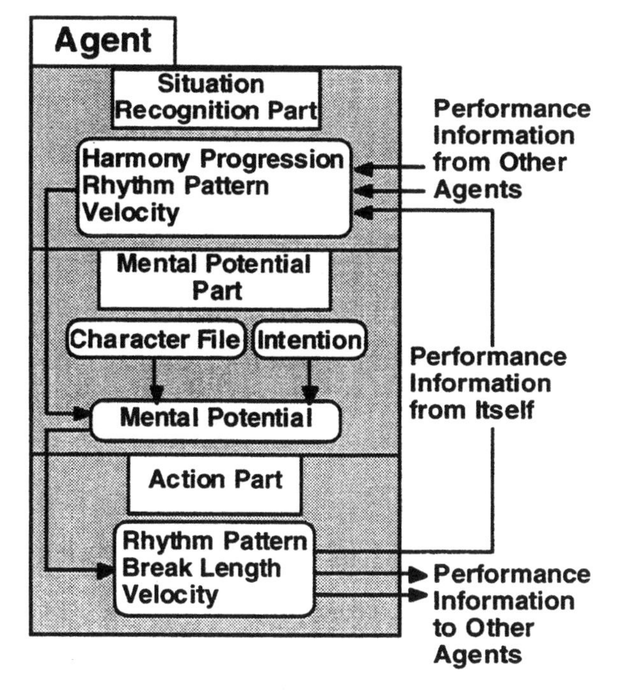

Fig. 3 Architecture of Agent

1)Situation recognition part
This part recognizes the situation of the whole environment by drawing the features of the incoming information. In the musical ensemble case, a situation means the concrete harmony progression and the rhythm pattern.

2)Mental potential part
In this part, the mental potential is estimated based on the situation obtained in the situation recognition part. The mental potential, for example, plays such a role as the feeling of human when he changes the rhythm in the musical ensemble.

Personality of the agent are represented as the difference in a situation's transition probability, called "character file", which is utilized to estimate mental potential. These are given in advance. Each agent has a different character file. If the character file is different, the reaction of the agent is different. That is to say, the personality of the agent is realized in this part.

3)Action part
By using the mental potential as a parameter, the next action pattern is decided in real time in this part. In the musical ensemble case, for example, the rhythm pattern and the volume are changed according to the mental potential.

A cooperative performance system is constructed by using the above mentioned agents.

4. ENSEMBLE SYSTEM

4.1 Outline of Prototype System

To verify the effectiveness of the agent, a prototype system of creating a musical ensemble was developed (Fig. 4). In this system, the session by three parts, the piano part, the drum part and the bass part, is held. A user plays a synthesizer and two agents play drum (agent 1), and bass (agent 2). It cooperatively produces and plays a musical performance pattern of the instrument using the performance information of all agents and the human player.

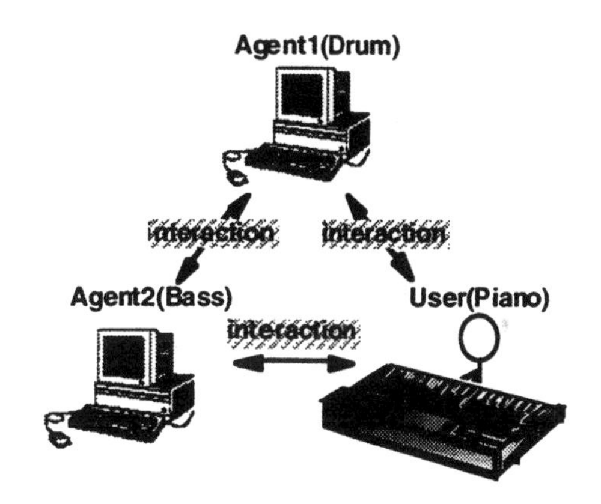

Fig. 4 Prototype System

4.2 Algorithm of Agent

In the system, a musical ensemble is selected for cooperation. Generally speaking, music assumes the form that four bars are in a mass. Therefore, an agent deals with four bars in a mass. The flow of agent's management in every four bars is shown in Fig. 5.

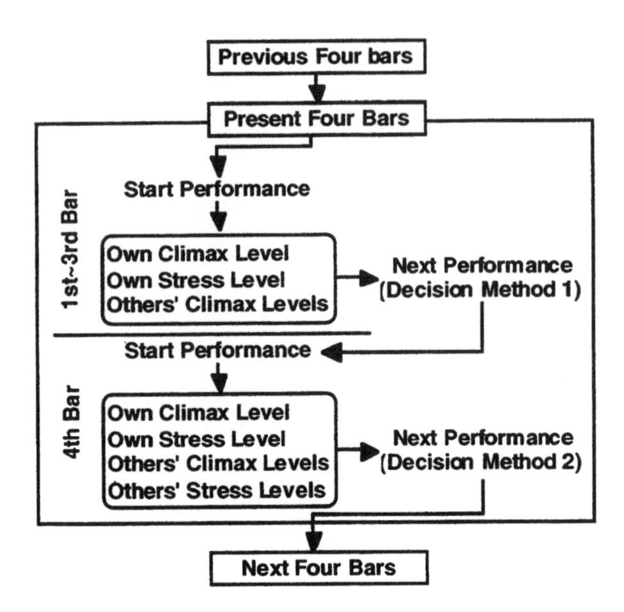

Fig. 5 Flow chart of Every Four Bars

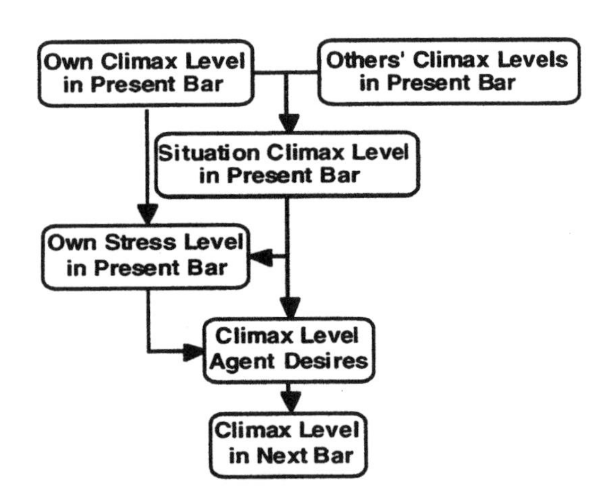

Fig. 6 Decision Making Algorithm 1

From first to third bar, the agent draws the features of its and others' performances as shown in Fig. 5 and calculates its own climax level, its own stress level and others' (a user and the other agent) climax levels (Fig.6). The climax level and the stress level are ranked from first to ninth. The first rank is the smallest and the ninth rank is the largest. Agent's own climax level and others' climax levels are directly obtained from their performances.

The flowchart of deciding the climax level in the next bar is shown in Fig. 7.

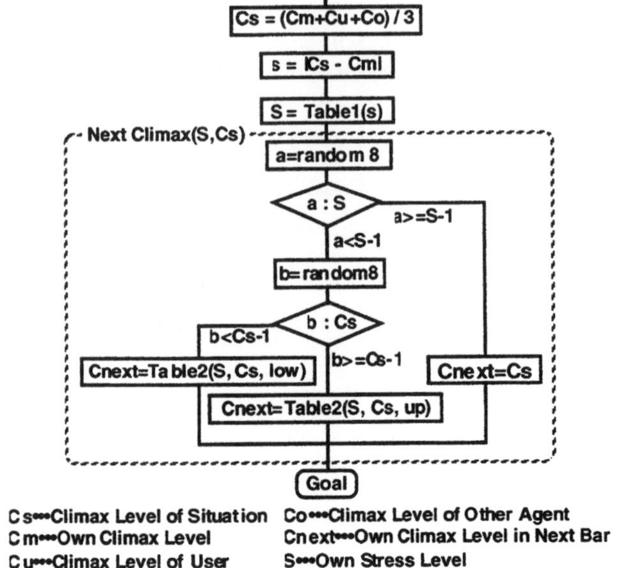

Fig. 7 Flowchart of D. M. A. 1

At first, the agent calculates the climax level of the situation in the present bar by its own climax level and others' climax levels directly obtained from their performances. The climax level of the situation is the averaged value of everybody's climax level. Then agent's own stress level is decided by the difference between its own climax level and the climax level of the situation. The concrete values of the stress level are shown in Table 1.

Table 1 Stress Level

Difference between Climax Level of Situation and Climax Level of Agent (columns); Stress Level of Agent, Small ↔ Large (rows)

	0	0 ~0.67	0.67 ~1.34	1.34 ~2	2 ~2.67	2.67 ~3.34	3.34 ~4	4 ~4.67	4.67~
9	-8	-7	-6	-5	-4	-3	-2	-1	0
8	-7	-6	-5	-4	-3	-2	-1	0	+1
7	-6	-5	-4	-3	-2	-1	0	+1	+2
6	-5	-4	-3	-2	-1	0	+1	+2	+3
5	-4	-3	-2	-1	0	+1	+2	+3	+4
4	-3	-2	-1	0	+1	+2	+3	+4	+5
3	-2	-1	0	+1	+2	+3	+4	+5	+6
2	-1	0	+1	+2	+3	+4	+5	+6	+7
1	0	+1	+2	+3	+4	+5	+6	+7	+8

The next step is to decide the climax level in the next bar which the agent desires. The desirable climax level is decided by agent's own stress level and the climax level of the situation. In this execution, the agent decides whether it keeps the climax level of the situation in the next bar or not by using its own stress level. The way of the calculation in this decision is shown in Fig. 7. When the agent decides that it changes the climax level of the situation, then it decides whether it changes the situation toward lively one or calmly one. In this decision making algorithm, the climax level of the situation is used, and the way of the calculation is shown in Fig. 7. The concrete values of the desirable climax level in the next bar are shown in Table 2.

As the above mentioned way, in decision method 1, the climax level which the agent desires in the next bar is decided by its stress level and the climax level of the situation in the present bar, and it will be its own climax level in the next bar. Then the musical performance is operated in the action part based on this value.

In fourth bar, the agent calculates others' stress levels in addition to its climax level, its stress level and others' climax levels from a musical performance. They are related as shown in Fig. 8, and the flowchart of deciding the climax level in the next bar is shown in Fig. 9. Agent's own climax level, others' climax levels and others' stress levels are obtained directly from their performances.

Table 2 Climax Level Agent Desires

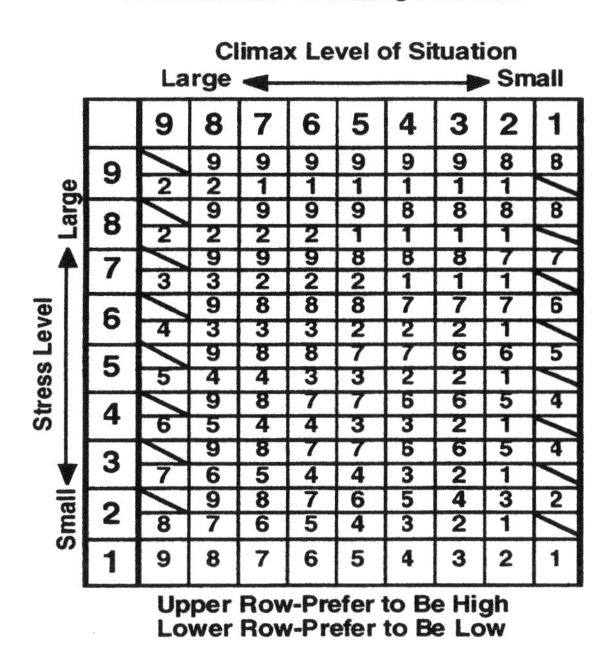

Climax Level of Situation: Large ← → Small; Stress Level: Large ↑ ↓ Small

Stress Level		9	8	7	6	5	4	3	2	1
9	Upper		9	9	9	9	9	9	8	8
	Lower	2	2	1	1	1	1	1	1	
8	Upper		9	9	9	9	8	8	8	8
	Lower	2	2	2	2	1	1	1	1	
7	Upper		9	9	9	8	8	8	7	7
	Lower	3	3	2	2	2	1	1	1	
6	Upper		9	8	8	8	7	7	7	6
	Lower	4	3	3	3	2	2	2	1	
5	Upper		9	8	8	7	7	6	6	5
	Lower	5	4	4	3	3	2	2	1	
4	Upper		9	8	7	7	6	6	5	4
	Lower	6	5	4	4	3	3	2	1	
3	Upper		9	8	7	7	6	6	5	4
	Lower	7	6	5	4	4	3	2	1	
2	Upper		9	8	7	6	5	4	3	2
	Lower	8	7	6	5	4	3	2	1	
1		9	8	7	6	5	4	3	2	1

Upper Row-Prefer to Be High
Lower Row-Prefer to Be Low

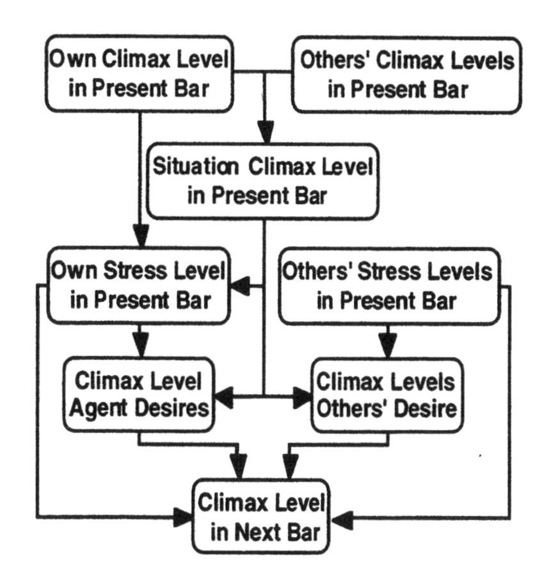

Fig. 8 Decision Making Algorithm 2

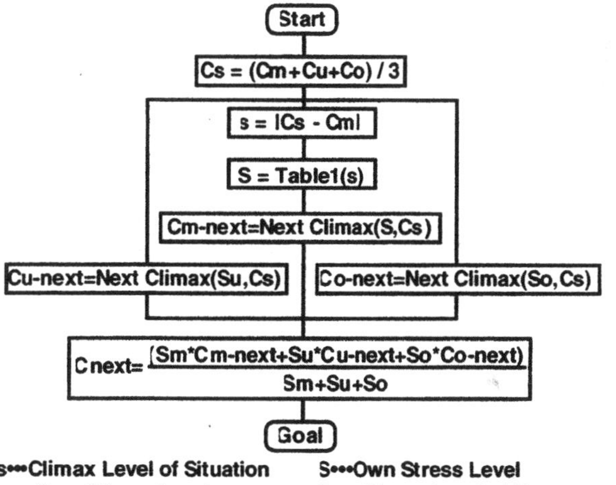

Cs•••Climax Level of Situation
S•••Own Stress Level
Cm•••Own Climax Level
Su•••Stress Level of User
Cu•••Climax Level of User
So•••Stress Level of Other Agent
Co•••Climax Level of Other Agent
Cm-next•••Own Climax Level in Next Bar(Desired Value)
Cu-next•••Climax Level of User in Next Bar(Expected Value)
Co-next•••Climax Level of Agent in Next Bar(Expected Value)
Cnext•••Own Climax Level in Next Bar(Coordinated Value)

Fig. 9 Flowchart of D. M. A. 2

In deciding the climax level in the next bar which the agent desires, the same way of the calculation is carried out. On the other hand, the agent guesses others' climax levels in the next bar by using others' stress levels directly obtained from their performances and the climax level of the situation. The same way to decide the desirable climax level based on agent's stress level and the climax level of the situation is also used. Then by four values, others' expected climax levels in the next bar, others' stress levels, agent's own desirable climax level and stress level, agent's own climax level in the next bar is decided. The way of the calculation in this decision is shown in Fig. 9. The agent considers the desire whose stress level is large by using this way of the calculation. Therefore everybody's stress level will be balanced.

To draw the climax level and the stress level actually from the performance, the information each pays attention to is shown in Fig. 10. The actual values of the climax level are decided based on the musical theory.

5. EXPERIMENT AND EVALUATION

The prototype system was experimented by a few players to evaluate its performance. The following results were obtained from the experiments.

1)The novice player enjoyed the musical ensemble which was cooperatively produced by agents in real time.

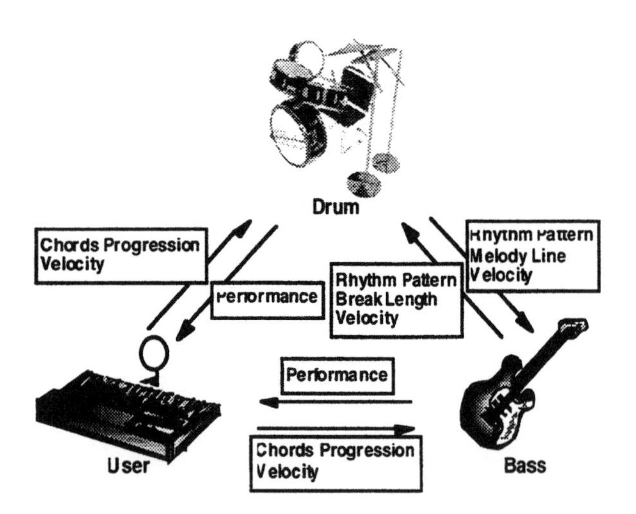

Fig. 10 Exchanged Information in Ensemble

2)The more advanced player complained about the musical performance level of agents.

3)The ignorant player of music could not enjoy the ensemble so much, because he did not know the effect of his performance.

4)It is more difficult to produce interesting performance patterns with the bass than the drum, as the bass has various musical aspects. The level of satisfaction from the drum part was higher than that of the bass for human players.

6. CONCLUSIONS

This paper proposed a cooperative performance system based on agents. A prototype system of creating a musical ensemble was developed, and it was confirmed that the novice users were adequately satisfied with the music made together with the agents.To realize more intimate cooperation, the following problems will need to be solved.

1)The rule of making a musical performance is the minimum requirement and very simple. Therefore, agent's musical performance is not enough to satisfy the advanced player. It is necessary to improve agent's performance and satisfy the advanced player.

2)Various parameters used in recognizing the music and deciding intentions are fixed values given in advance and cannot be changed during the period that the system is working. User's personal taste which differs variously may exceed the range which fixed values can express. To avoid this problem, they should be changed in real time through the feedback from the situation.

3)Catastrophe is applied to animal's actions, psychology, cultural sciences and social science as the mathematical method to describe the evolution of the form in nature. Induction of this theory to the agent will make the agent behave interestingly.

It is expected that the architecture of the proposed agents will not only be applied to music, but also human interface, which is required to consider the feelings of a user in general.

REFERENCES

Genersereth, M. R. And S. P. Ketchpel (1994). Software Agents. *Communication of the ACM*, **Vol.37, No.7**, pp48-53

Ishida, T. (1995). Discussion on Agents. *Journal of Japanese Society for Artificial Intelligence,* **Vol.10, No.5**, pp663-667.

Maes, P. (1994). Agents that Reduce Work and Information Overload. *Communication of the ACM*, **Vol.37, No.7**, pp31-40.

Rowe, R. (1993). *Interactive Music Systems,* The MIT Press, Cambridge

Wake,S. H. Kato, N. Saiwaki and S. Inokuchi (1994). Cooperative Musical Partner System Using Tension-Parameter:JASPER(Jam Session Partner). *Journal of Information Processing Society of Japan*, **Vol.35, No.7**, pp1469-1481

ELECTRONIC SOUND EVALUATION SYSTEM BASED ON KANSEI(HUMAN FEELING)

Hiroaki KOSAKA and Kajiro WATANABE

Yokogawa Electric Corp.
College of Engineering, Hosei University

Abstract: Electronic sound is used as one of the very common means of human-machine interface. Few investigations of how human feels to such the simple sounds noisy or comfortable, exist. The electronic sound must be designed by considering how or in what situation it is used. This research is aimed at investigating how human feels to a variety of electronic sound and describes the relation between human feeling and physical characteristics of sound. We found relations between the human auditory feeling and the physical characteristics of the sound. We showed a guideline of how to design a variety of electronic sound that satisfy the given human feeling specifications.

Keywords: Human-centered design, Human-machine interface, Evaluation, Neural networks, Human factors

1 INTRODUCTION

An opportunity for making use of human-machine interface like electronic sound is increased. Comfortable human interface is required. This paper deals with electronic sounds as one of the most common means of human-machine interface. Few investigations for how human feels to such the simple sound yield the situation that it is sometimes called sound pollution. The electronic sound must be designed by considering how or in what situation it is used. Our goal is developing a system evaluating electronic sounds to support designing electronic sound based on feeling evaluation. 6 words for evaluating electronic sounds are selected. To decrease the number of evaluation, the design of experiments is applied. Feeling evaluation experiments are carried out and the human auditory physiological feeling data about each 32 electronic sound are obtained. The experimental data obtained by evaluation tests were stored into artificial neural network. The network can automatically evaluate electronic sounds based on human feelings.

2 PROBLEM DESCRIPTIONS

2.1 Electronic Sound

Fig. 1 shows 1 cycle wave form of typical electronic sound. This sounds "Pi Pi Pi Pi...". Fig. 2 shows spectrum of the sound shown in Fig. 1. In most cases, electronic sounds in automobile can be designed using physical parameters in Fig. 3. In Table 1, we defined parameters necessary to form the electronic sound in Fig. 1.

We set 4 combinations of 8 physical parameters. Table 2 shows the values of the combinations. All existing electronic sound can be given by the 4 combinations. Fig. 4 shows the envelope of square, sin, triangle, sawtooth in Table 2.

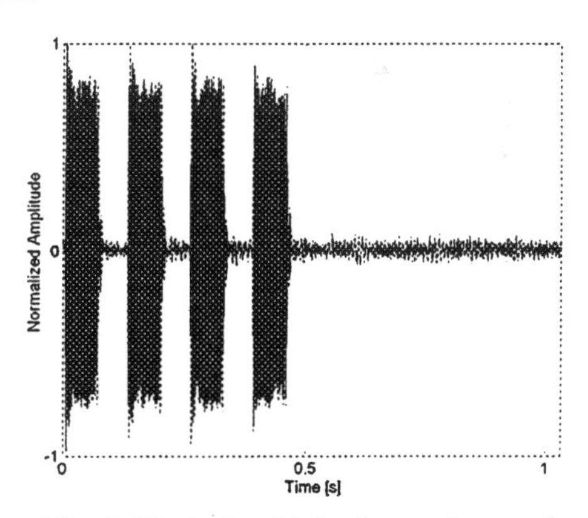

Fig. 1. Typical vehicle electronic sound

Table 1 Definition of physical sound parameters

Basic Frequency	$F = T_1^{-1}$ [kHz]
Outside Envelope Form	W_1 [type]
Inside Envelope Form	W_2 [type]
Outside Cycle	T_6 [s]
Duty Ratio(Outside)	$R_1 = T_2 / T_3$
Duty Ratio(Inside)	$R_2 = T_4 / T_5$
Number of Inside Wave	N [wave]
Volume	V [dB]

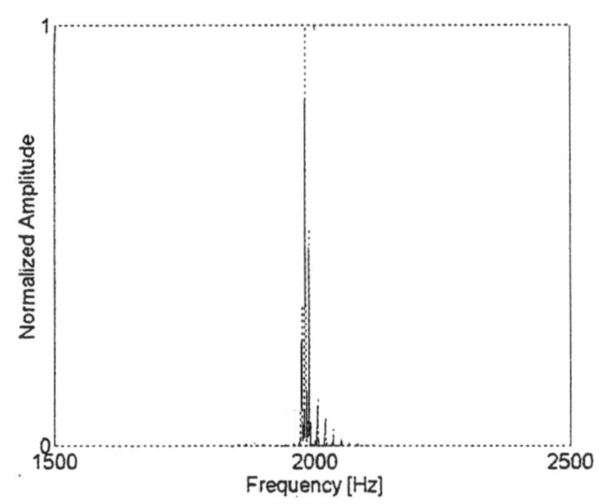

Fig. 2. Spectrum of typical vehicle electronic sound

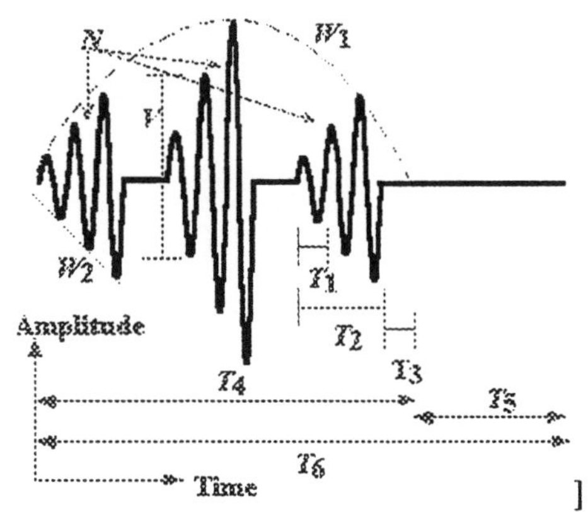

Fig. 3. Generalized form of electronic sound wave

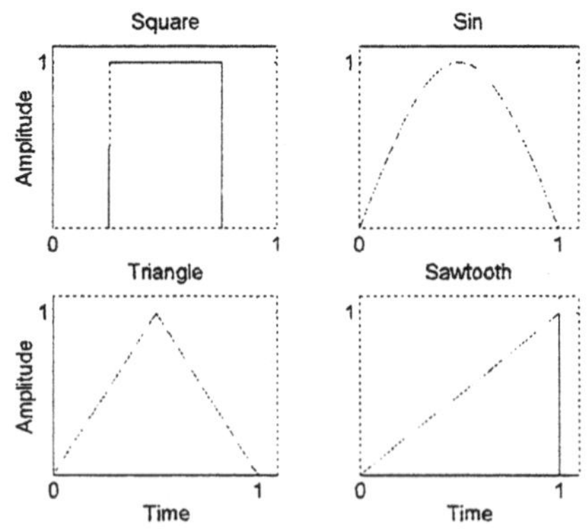

Fig. 4 Envelope forms (square, sin, triangle and sawtooth)

Table 2 Physical values of electronic sound

Case	1	2	3	4
F	0.5	1.0	1.5	2.0
W_1	Square	Sin	triangle	sawtooth
W_2	Square	Sin	triangle	sawtooth
T_6	0.5	0.8	1.1	1.4
R_1	0.286	0.444	0.556	0.714
R_2	0.286	0.444	0.556	0.714
N	1	2	3	4
V	60	64	68	72

2.2 *Problem Description*

We consider the following problems.

(P1) Select the words to express the human feeling.

(P2) Investigate the way to decrease the load of a subject.

(P3) Obtain the Kansei data by evaluation experiment about electronic sounds.

(P4) Develop the electronic sound evaluation system based on the data obtained (P3).

3 SELECTION TO EXPRESS KANSEI

Here we consider about problem (P1).

Few studies about the human feeling expressing words to evaluate electronic sound feeling exist. It is desirable that the electronic sound feeling is expressed or evaluated by few words as possible. The purpose of this section is to find the few words appropriate to evaluate and/or express the feeling of electronic sounds. First, we collected 83 words related with sound from dictionaries and related publication. Among those, we selected 14 from the 83 words removing the words which were not appropriate to evaluate or to express the human feeling of electronic sounds clearly. The 14 words are listed in Table 3.

Table 3 Words to evaluate the electronic sound human feeling

Bright	Clean	Emergency
Pleasant	Dim	Loud
High-pitched	Artificial	Harsh
Smooth	Clear	Conspicuous
Soft	Comfortable	

Questionnaire tests by the rating method about 14 words above were carried out by the subjects of 15 engineering students. For example, the electronic sound feeling for the word "Bright" was evaluated by the rating scale in which the degree of, "Bright" to "Not Bright" is plotted. The rating scale was 5 grades. The subjects evaluate electronic sounds by selecting one of the 5 grades. We let subjects understand the meanings of the words in Table 3 and let them heard the 8 sounds. The factor analysis to the averages of questionnaire grades was applied. The 14 points in Figure 5 show the results of the factor analysis. In order to cluster to several groups, cluster analysis was applied. The 6 groups in Figure 5 were clustered by the cluster analysis. We named 6 group from G_1 to G_6.

The 6 words on behalf of groups are shown in Table 4. After this, we investigated about the 6 words(groups) as evaluation words for electronic sound.

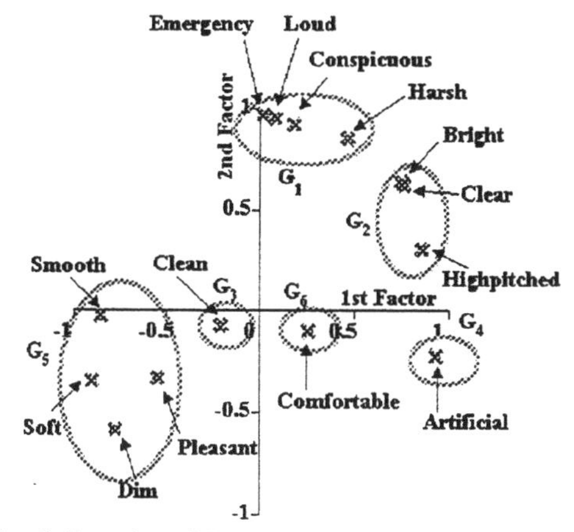

Fig. 5. Results of factor analysis and cluster analysis

Table 4 Selected words from experimental results

G_1	Emergency
G_2	Clear
G_3	Clean
G_4	Artificial
G_5	Soft
G_6	Comfortable

4 DECREASE LOAD OF SUBJECT IN EVALUATION TESTS

Here we consider about problem (P2).

We set 4 combinations of 8 physical parameters of electronic sound. We evaluated electric sounds by the 6words. If we were to evaluate electronic sound for all the combinations, we would have to evaluate $4^8 \times 6 = 393{,}216$ times, very many cases. Therefore, we needed a way to decrease the number of evaluation tests.

4.1 Reduction of the number of tests

We applied the design of experiments (Taguchi method) to problem (P2). The design of experiments assesses that varying factors have the most effect to a desired outcome. And we can decrease the number of the evaluation tests if we determine the combination of physical parameters of electronic sound that we must evaluate by the table of orthogonal arrays.
Thus, it seemed appropriate to use the Taguchi method.

4.2 Assignment

We designed the combination of physical parameters using the table of orthogonal arrays $L_{32}(2^{31})$. The individual factors in $L_{32}(2^{31})$ are varied between two values each. Since we designed 4 cases of 8 factors(physical parameters), we used two column per factor in designing. We can design 4 cases using two column per factor. We assigned F to the column no.8 and 16, W_1 to 9 and 19, W_2 to 10 and 20, T_6 to 11 and 23, R_1 to 12 and 17, R_2 to 13 and 18, N to 14 and 21 and V to 15 and 22 in $L_{32}(2^{31})$. The results of the assignment is shown in Table 5. The number of row of Table 5 is the number of sound we must evaluate. Therefore we only have to evaluate 32 electronic sound.

Here we focus on the first row in Table 5. The cases of sound No.1 are all 1. Then, values of physical parameters of sound no.1 are the values of case 1. Physical parameters of 32 sound are shown in Table 6. The graph of wave of Sound No.6 is shown in Fig. 6 as a example of designed sound.

Table 5 A part of orthogonal array

No.	F	W_1	W_2	T_6	R_1	R_2	N	V
1	1	1	1	1	1	1	1	1
2	2	2	2	2	2	2	2	2
3	3	3	3	3	3	3	3	3
4	4	4	4	4	4	4	4	4
5	1	1	2	2	3	3	4	4
6	2	2	1	1	4	4	3	3
7	3	3	4	4	1	1	2	2
8	4	4	3	3	2	2	1	1
9	1	2	3	4	1	2	3	4
10	2	1	4	3	2	1	4	3
11	3	4	1	2	3	4	1	2
12	4	3	2	1	4	3	2	1
13	1	2	4	3	3	4	2	1
14	2	1	3	4	4	3	1	2
15	3	4	2	1	1	2	4	3
16	4	3	1	2	2	1	3	4
17	1	4	1	4	2	3	2	3
18	2	3	2	3	1	4	1	4
19	3	2	3	2	4	1	4	1
20	4	1	4	1	3	2	3	2
21	1	4	2	3	4	1	3	2
22	2	3	1	4	3	2	4	1
23	3	2	4	1	2	3	1	4
24	4	1	3	2	1	4	2	3
25	1	3	3	1	2	4	4	2
26	2	4	4	2	1	3	3	1
27	3	1	1	3	4	2	2	4
28	4	2	2	4	3	1	1	3
29	1	3	4	2	4	2	1	3
30	2	4	3	1	3	1	2	4
31	3	1	2	4	2	4	3	1
32	4	2	1	3	1	3	4	2

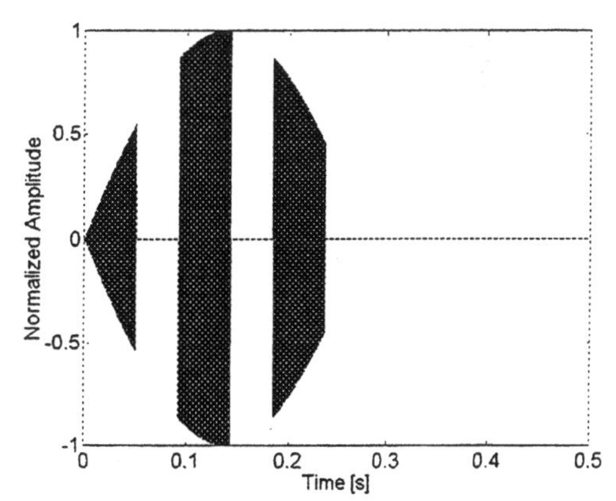

Fig. 6. A electronic sound designed by the design of experiments

5 PHYSICS OF ELECTROCNIC SOUND AND KANSEI

Here we consider about problem (P3).

5.1 Questionnaire

Questionnaire tests by the rating method of 6 feeling evaluation words were carried out by the subjects of 34 engineering students. They feeling-evaluated the 32 electronic sounds by selecting one of the 5 grades.

5.2 Dual Scaling

To apply the multiple regression analysis, the categorical data are quantified by the dual scaling[3]. The dual scales obtained are shown in Figure 7.

Table 6: Physical values

No.	F	W_1	W_2	T_6	R_1	R_2	N	V
1	0.5	square	square	0.5	0.286	0.286	1	60
2	1.0	sin	sin	0.8	0.714	0.714	2	64
3	1.5	sawtooth	sawtooth	1.1	0.444	0.444	3	68
4	2.0	triangle	triangle	1.4	0.556	0.556	4	72
5	0.5	square	square	0.8	0.444	0.444	4	72
6	1.0	sin	sin	0.5	0.556	0.556	3	68
7	1.5	sawtooth	sawtooth	1.4	0.286	0.286	2	64
8	2.0	triangle	triangle	1.1	0.714	0.714	1	60
9	0.5	square	sin	1.4	0.286	0.714	3	72
10	1.0	sin	square	1.1	0.714	0.286	4	68
11	1.5	sawtooth	triangle	0.8	0.444	0.556	1	64
12	2.0	triangle	sawtooth	0.5	0.556	0.444	2	60
13	0.5	square	sin	1.1	0.444	0.556	2	60
14	1.0	sin	square	1.4	0.556	0.444	1	64
15	1.5	sawtooth	triangle	0.5	0.286	0.714	4	68
16	2.0	triangle	sawtooth	0.8	0.714	0.286	3	72
17	0.5	square	triangle	1.4	0.714	0.444	2	68
18	1.0	sin	sawtooth	1.1	0.286	0.556	1	72
19	1.5	sawtooth	sin	0.8	0.556	0.286	4	60
20	2.0	triangle	square	0.5	0.444	0.714	3	64
21	0.5	square	triangle	1.1	0.556	0.286	3	64
22	1.0	sin	sawtooth	1.4	0.444	0.714	4	60
23	1.5	sawtooth	sin	0.5	0.714	0.444	1	72
24	2.0	triangle	square	0.8	0.286	0.556	2	68
25	0.5	square	sawtooth	0.5	0.714	0.556	4	64
26	1.0	sin	triangle	0.8	0.286	0.444	3	60
27	1.5	sawtooth	square	1.1	0.556	0.714	2	72
28	2.0	triangle	sin	1.4	0.444	0.286	1	68
29	0.5	square	sawtooth	0.8	0.556	0.714	1	68
30	1.0	sin	triangle	0.5	0.444	0.286	2	72
31	1.5	sawtooth	square	1.4	0.714	0.556	3	60
32	2.0	triangle	sin	1.1	0.286	0.444	4	64

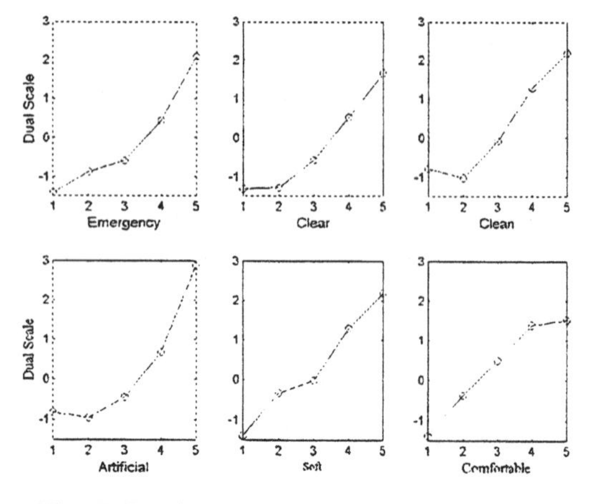

Fig. 7. Dual scales about 6 evaluation words

6 ELECTRONIC SOUND EVAULUATION SYSTEM VIA ARTIFICIAL NEURAL NETOWORK

Here we consider about problem (P4).

Here we developed a neural network system that evaluate electronic sounds. Three layer perceptron type neural network was employed to develop the electronic sound evaluation system.

A set of 7 physical values of a electronic sound are the inputs to the neural network. The values scaled by dual scaling about each word are the outputs. These values are normalized into the range of [0,1]. The number of middle layer was 12. The back-propagation algorithm was employed to train the network. The learning coefficient was within 0.1. The iteration number was 20662. The neural network is same as that in Fig. 8.

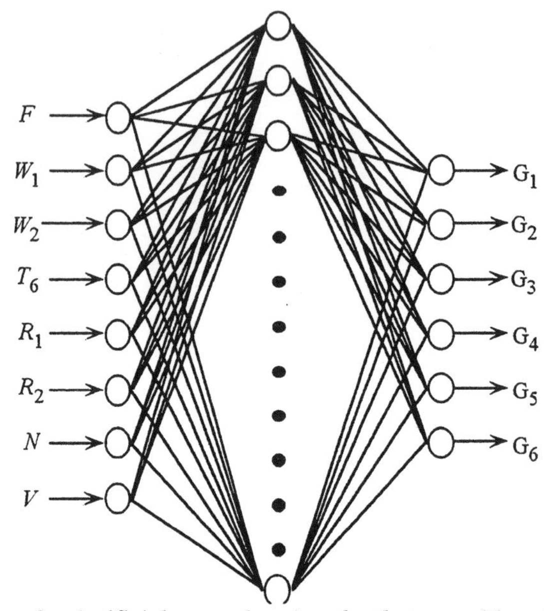

Fig. 8. Artificial neural network that provides the physical parameters of the desired a electronic sound

7 ELECTRONIC SOUND EVALUATION EXPERIMENT

In order to input unknown data to the neural network and investigate the outputs, we designed 18 electronic sounds that we feel {very, neither, little} about the given human feeling from G_1 to G_6. The parameters not influence to the feeling in Table 8 were selected randomly. The neural network do not learn about these 18 electronic sounds. A example of electronic sound wave form ("Very Emergency") is shown in Fig. 9.

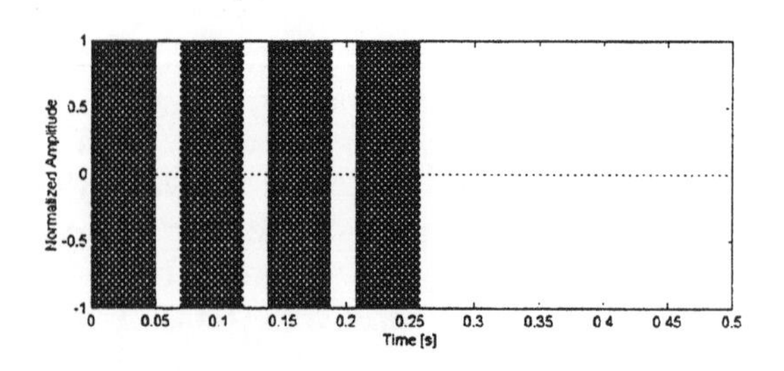

Fig. 9. Electronic sound which has the Kansei "Very Emergency"

7.1 Experimental Condition

Questionnaire tests by the rating method of 18 sounds with the 6 feeling words were carried out by subjects of 15 engineering students. The subjects evaluated electronic sound by selecting one of the 5 grades. The 15 subjects did not join to the questionnaire tests described in 5.1.

7.2 Results of the Experiment

We compared the results of the experiment in 7.1 and the output of the neural network when we inputted the results in 7.1.

For example, we compare between the dual-scaled average of the evaluation about the sounds {"Very Emergency", "Neither", "Little Emergency"} obtained in 7.1 and the dual-scaled output value of the G_1 ("Emergency") in Fig. 7 when we inputted the physical values of the sound "Very Emergency" to the neural network. If the error between the dual-scaled average of the evaluation about the sounds "Very Emergency" and the dual-scaled output value of the G_1 ("Emergency"), We will consider that the neural network outputs the correct answer. The results of the comparison mentioned above about 18 sounds are shown in Table 7. The percentage of correct answer was 67%.

Table 7 Results of the comparison

Group	Aim	Comparison
G_1	Very Emergency	O
	Neither	O
	Little Emergency	O
G_2	Very Clear	O
	Neither	O
	Little Clear	O
G_3	Very Clean	×
	Neither	×
	Little Clean	O
G_4	Very Artificial	×
	Neither	O
	Little Artificial	O
G_5	Very Soft	O
	Neither	×
	Little Soft	×
G_6	Very Comfortable	O
	Neither	×
	Little Comfortable	O

8 CONCLUSION

This study investigated the relation between the degree of human feeling expressed by the 6 words and the 32 electronic sounds via questionnaire. The physical parameters to design electronic sound and the human feeling evaluation words were obtained.

We applied the design of experiments to reduce the number of evaluation tests. The quantitative relation between the physical sound parameters in Table 1 and the human feeling evaluation words in Table 4 were obtained by the questionnaire tests.

The experimental data obtained by evaluation tests about 32 electronic sounds were stored into artificial network. The network can automatically evaluate electronic sounds based on human feelings.

The results obtained were as follows;

- The words to express the human feeling were selected.
- The way to decrease the load of a subject was investigated.
- The data about the physical sound parameters and Kansei were obtained.
- The electronic sound evaluation system based on Kansei was obtained. The percentage of correct answer by the evaluation experiment was 67%.

These results are available for support to design electronic sounds based on Kansei.

REFERENCES

Arai Y. and K.Miura, ``Evaluation on the Signal Function of Frequency Modulation Sounds," 321/322, 1990 Acoustic Society of Japan Lecture Meeting, Shibuyaku, Tokyo, 1990.

Taguchi G., et al., Design of Experiments, Maruzen Publishing, 1976.

Nishizato S., Scaling of Categorical Data, Asakura Press, 1982.

Arai Y. and K.Miura, ``Sound Quality Evaluation of Cordless telephones by the three-way scaling method, and the correlation analysis between psychological measure and physical measure," 43.88.Si, THE JOURNAL OF THE ACOUSTICAL SOCIETY OF JAPAN, Shibuyaku, Tokyo, 1994.

Sakai H. and T.Nakayama, Auditory Sense and Sound Psychology, Corona Publishing Co., Ltd., 1978.

JUSE, Sensory Evaluation Handbook, JUSE Publishing, 1973.

Okuno T., et al., Multivariate Analysis, JUSE Publishing, 1971.

A STUDY OF COMPUTER AGENTS TO HELP YOUNGER CHILDREN

Yoshie Soutome Sagata* Katsunori Shimohara Yoshinobu Tonomura***

* *NTT Human Interface Laboratories, 1-1 Hikarinooka Yokosuka-Shi Kanagawa 239-0847 Japan*
** *NTT Communication Science Laboratories, 2-4 Hikaridai Seika-ChoSohrakugun Kyoto 619-0237 Japan*

Abstract: Image services that enable people to meet in cyberspace are attracting considerable attention as important services in the multimedia era. We focus on how computers and the Internet can be used by younger children to stimulate their imagination and creativity. Our aim is to create a new computerized editing system that enables children to make stories similar to a picture-story show. This paper describes on-going experiments with younger children in which they produce moving pictures on computers and points out some considerations for constructing systems geared towards younger children. *Copyright ©1998 IFAC*

Keywords: User Interfaces, Agents, Education, Human factors,MAN/machine interaction

1. INTRODUCTION

Children aged four-to six-years old who go to nursery school or kindergarten have a chance to play and learn in a stimulating environment. Playing in the school yard, talking with friends and teachers, dancing and singing, etc. help develop social skills. Imaginative play and creative activities are also central to a child's world. Blocks, crayons, colored paper and musical instruments are still as useful as they always were for these purposes, however, the computer is now included as a tool for creative play. At least it is in theory. Creative use of a computer is usually too difficult for kids because such use requires specialized training even for adults. In Japan, there are few nursery school teachers who handle a computer creatively and the number of computers in nursery schools and kindergartens is quite limited.

Under such constraints, we should consider what kids can do with computers in comparison with traditional tools as well as identify most important things children can do with a computer.

This paper describes on-going experiments with younger children, in which they produce moving pictures on computers, and points out some considerations for constructing and supporting such systems for kids. We worked for one and a half year with thirty-eight local nursery school children (ages from four to five). Our goal was to create a new computerized editing system that enabled children to make stories similar to a picture-story show. In the second section, we outline previous research on creating new tools for kids.The third section describes the experiments we conducted with four-year olds in the nursery school. The fourth section gives our suggestions for supporting the use of computers in a nursery school and making new tools for kids. The fifth section is a conclusion and outlines of future work on this subject.

2. RELATED WORK

A lot of research on creative computer tools for kids has already done. Philippe P. Piernot et al. created the Penpal which is a portable communications device for children aged four to six(Philippe P. Piernot, 1995). The Penpal is used as a communication tool, and a learning tool. The system employs easy-to-use 'messagecards' that can store six fixed images. A combination of messagecards and 'Penpal Dock' allows a child communicate with distant friends. Penpal function is limited to communication with friends.

Shannon L. Halgren et al. developed a new tool that makes movie-making simple for kids (Shannon L.Halgren, 1995). In making their software, they considered what design was best suited for easy use. Their words "Don't push an adult interface on kids" and "Don't push adult metaphors on kids" were very interesting to us.

Ron Oosterholt et al. focused on interaction design and human factors that supported their product(Ron Oosterholt, 1996). Through interaction with researchers and designers, they considered metaphors for kids. However they said that it is difficult to define the child-as-user requirements clearly.

Allison Druin et al. developed "Kidpad" which produces zoom images and drawn images and allows children to discuss their work on computer (Allison Druin, 1997). Children were able to use this tool in interesting ways at the elementary school level. KidPad was developed by taking into account children's suggestions on what they thought the system should include.

3. PRACTICE IN A NURSERY SCHOOL

Experiments are currently being conducted by our group at a nursery school in Yokosuka. About 200 children, from one to five years of age, attend the school, and there are 30 teachers and helpers, making it the third largest nursery school in Kanagawa Prefecture. The school operates from 7:30 AM to 6:30 PM, including lunch time and teatime. The main activities are playing instruments, drawing and singing, which are done between 10:00 AM and 11:30 AM. For the experiments reported here, we used two Performa 5270's and one "SoftBoard" [1]. One of the authors (Sagata) and another woman (Ms. Yuka Sawahata: designer) provided computer support. We have been conducting experiments with 37 four-year olds one day a week since November 1996. The principal of the nursery school suggested that we choose older children (four-year olds) to do computer experiments, because they are able to express their opinions in words to teachers and therefore would be the best source of constructive feedbacks. The experiments had three steps. The first step (November 1996 - June 1997) involved determining how children used the mouse and how best interest children in the computer experiments. The second step (May 1997 - September 1997) involved the children drawing their favorite things by computer and us animating the pictures (including audio) based on the children's suggestions.The third step (November 1997 - March 1998) was to make picture-story shows for each child using the drawings from second step. We have already some of our findings in (Yoshie Soutome Sagata, 1997).

3.1 *Analysis os a child's ability and first experience with a computer*

On November 20th 1996, we met 37 four-year olds at the nursery school. We showed then how to use the computers and what they could can do with them. Despite initial surprise, after a few minutes they were enjoying themselves with some software and mouses. The children laughed and talked about what was occuring on the computer screens.

In January 1997, we had the children make a picture story so we could examine how they drew pictures and created stories. Our main goals at this stage in the experiment were to examine how children constructed original stories and to show the children that their stories could be animated. We then would have them draw characters for an 'electric picture story'.

The general guidelines were as follows: Initially, pictures were drawn with color crayons on white paper. The children then described their finished drawings to the teachers and the support staff, who wrote down their descriptions. Children were encouraged to continue drawing until they had finished their story. Questions were answered by the teachers and staff. Within two hours, 25 children had finished drawing picture stories between three and five pages long.

The staff digitized each show as a simple movie on a computer, We showed the content of the the picture stories as text data. Each story was played back to the children. If they had a suggestion about their movie, we incorporated their suggestion into the movie in the front of them. The children enjoyed watching their favorite movie. An example drawn by Keisuke is shown Fig. 1. This story was so interesting for all the children that they said it was the best.

[1] ©Copyright 1993-1995 by Microfield Graphics, Inc.

1. There was a cake house. Keisuke and his mother went to the house.
2. Keisuke ate candies only, and Mother ate chocolate only.
3. Mother was caught by man who lived in the house.
4. Keisuke went up stairs to help mother.
5. Keisuke remembered he brought a cord. He tied the man with the cord.
6. The man became small.
7. Keisuke went back his home with his mother.

Fig. 1. Keisuke's story

3.2 *Character drawing with computer*

Our step was to have the children draw with computer and add motion and sound to their drawings. The aim of the second step was to let the children create an original work through our experiment. Unfortunately, the number of computers and the time available to use them were too limited. The children didn't have the time to master mouse operation well enough to express themselves in an original drawing. So we decided to introduce a new instrument called a "SoftBoard", which can be used to input line data and trace line data. It was easier for the children to draw with SoftBoard than with a mouse because it was similar to using a pen. The children drew contours of their characters or objects on SoftBoard. They then did a line drawing that was colored in with the computer. SoftBoard's width and height was almost too big for the children. At first they couldn't express their ideas on SoftBoard. However, after a month, they became skilled with it. The new instrument allowed one child to draw a human face for the first time in his life. The principal of the school and teachers were surprised at his drawing.

Usually, it took 5 - 7 minites for each child to finish the drawing process. Then, they recorded the sound and described the type of action they wanted to be animated. The activity was done in groups of two to four children because they could relax during the do computer experiment and weren't overworked. The support staff constructed a simple movie based on the children's instructions. These were recorded as objects in the show and were used as characters in the 'electric picture-story show'. The children's expectations grew when the staff explained this action to the children. All drawings done by the children are Fig. 2.

Fig. 2. The cast of characters drawn by the children

3.3 *Electric picture-story show*

As a final experiment, the children made a picture-story show with three sheets of colored paper. The purpose of third step was for each child to make a picture-story show using their own and a friend's character. At a nursery school, children can collaborate with friends. We wanted to know if this worked well with computer.

We told the children how to make a picture-story show: Characters were limited to the 37 characters drawn in the second step (see Fig. 2). Backgrounds could be added, if desired. After finishing, the children had to explain their drawings to teachers or support staff, who wrote down the content. Children were encouraged to keep drawing until the end of their story. Questions were answered by teachers and staff.

After two hours, over 30 children had finished drawing their picture story. Two children wanted

Kouji's show

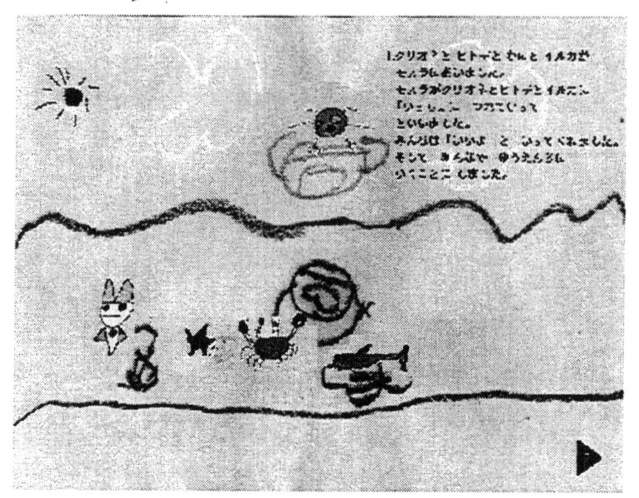

Part 1
A Kurione (a very small shellfish without a shell), a starfish, a crab and dolphin met a mussel.
The mussel said, `Please go out with me.". They said `OK!".
They decided to go to an amusement park.

Part 2
In the park, the dolphin and the starfish rode on the coaster.
The crab and the Kurione rode on the Ferris wheel.
After the mussel rode on the merry-go-round, he went for a swim in the Slider Pool.

Part 3
Everybody ate lunch in the park and had a good time.

Fig. 3. An example of a picture story show

Fig. 4. Title image of software

to work on their stories at home and they returned their shows to us the following week. The staff digitized the shows as a simple movie on the computers, and then showed the movie to the children. The children's verbal story lines were recorded as sound data. Next, the motion they wanted was recorded as motion data. We show an example in Fig. 3.3.

3.4 *Presentation to teachers and parents*

These stories were arranged as one software package that allowed children to show all electric picture stories with only the click of a mouse. The title image is shown Fig. reff3.

The principal suggested that we showed our package to the parents. We agreed and arranged dates and times that were convenient for the parents. Thirty minutes were provided for each family.

Only 30 percent of all families had a personal computer at home, so this presentation was first experiment to see computer software package geared for children. Two or three families joined one presentation. They enjoyed watching the stories the children chose. The software used mouse only, and parents as well as the children were able to review the stories.The children taught their parents how to use it smoothly. The parents were surprised by the children's computer abilities. Many parents wanted the experiment to continue the following year.

At the same time, we showed the stories to the teachers. Most were amazed that children's drawings turned into animation, and furthermore, they thought it was a wonderful world that the children had created.

After the presentation,we placed two computers at the entrance of the school to allow the younger children to see what was going on. They enjoyed our presentation as well.

4. SUGGESTIONS OF SUPPORT'S PART

Our findings can be summarized as follows. We learned that children can make an electoronic picture-story show with the help of computer support staff. Children were able to express their story, animate their characters, provide character voices and design a show stage. The support staff digitized the picture-story based on the children's input. We then showed the result to the children. If someone didn't like their own show, we modified that show until the creator was satisfied.

Stimulating the children's imagination and creativity by using computers was important; it made them want to finish the work by themselves. We found that it was possible when we used a very limited editing system whose functions included two types of drawing, erasing, cutting, pasting, and recording an object as characters. Too many functions tend to cause distraction in younger children so they can't keep their minds on their work.

We emphasize the following three points. First is that careful thought on how one introduces a computer to children is necessary because children remember for a long time what they used that first time. Second is that children are good at cooperative work (we emphasized work pairs or groups of three). Children working alone were not able to relax, whereas a few children working together generated a lot of excitement. The support staff was there to answer the children's questions and help with the computer work. Third is that cooperation between teachers and computer support staff is important because the teachers know children's nursery life. They were especially helpful in selecting the children's group and always gave useful advice for us. Holding a meeting about once a week was needed to keep the experiments on track.

5. CONCLUSION AND FUTURE WORK

This article describes our experiments with 38 four-five year old children at nursery school for term of one and half years. We created a simple system for animating using children's drawings that required minimal creative input from the trained staff and could be operated by the children themselves in a simplified form. Three ideas were crucial to the success of this endeavor: the first impression of the children, collaboration among the children and cooperation between teachers and computer support staff. In the future, we will continue our collaboration with the school involved in this research and continue to investigate ways of producing easy-to-use creative software for children

For optional analysis, We would like to do more experiments with younger children. An idea is to let them send E-mail with drawings over the Internet. Perhaps collaboration with children in other countries will be possible.

ACKNOWLEDGMENTS

We thank Miss Yuka Sawahata for her support and for her design of the computerized memorial album for the nursery school. We also Ms. Yukiko Sekiguchi who is the principal of the nursery school and all teachers for giving us the oppotunity to do experiments with such young children. Finally, we thank our colleagues for their discussions.

6. REFERENCES

Allison Druin, Jason Stewart, et.al (1997). Kidpad:
A design collaboration between children, technologists, and educators. *Proceedings of CHI'97.* pp. 463–470.

Philippe P. Piernot, et.al. (1995). Designing the penpal: Blending hardware and software in a user-interface for children. *Proceedings of CHI'95.* pp. 511–518.

Ron Oosterholt, Mieko Kusano, Govert de Vries (1996). Interaction design and human factors support in the development of a personal communicator for children. *Proceedings of CHI'96.* pp. 450–457.

Shannon L.Halgren, Tony Fernandes, Deanna Thomas (1995). Amazing animation: Movie making for kids design briefing. *Proceedings of CHI'95.* pp. 519–524.

Yoshie Soutome Sagata, Katsunori Shimohara, Yoshinobu Tonomura (1997). A study of computer agents to help yonuger children. *Proceedings of the Thirteenth Symposium on Human Interface.* pp. 603–608.

DESIGNING BETTER INTERFACES: CAN WE LEARN FROM OTHER SECTORS ?

James L. Alty, LUTCHI Research Centre, Department of Computer Science,
Loughborough University, Loughborough, LE1, 3TU, UK,
j.l.alty@lboro.ac.uk

Abstract:. The importance of having a machine-operator interface at the right level of abstraction is stressed and the roles of multiple media in this process are highlighted. The relevance of work in other areas of activity such as the cinema, music and the arts is highlighted and some examples of possible influences are discussed. *Copyright © 1998 IFAC*

Keywords: Interfaces Design, Music, Cinema, Creativity, Multimodal, Multimedia

1. INTERFACE DESIGN IN PROCESS PLANTS

In recent years, the key and pivotal role of the human-machine interface in the effective control of large process plants, has been established beyond doubt. From the hard-wired mimic boards of the 80's, we have moved to the programmed work-stations of the 90's. This new software approach has opened up more effective ways of assisting operators, both through the presentation of information in a more understandable form, in the appropriate context, and through better support based upon more underlying intelligence (Alty *et al*, 1994, (Alty and Bergan, 1995). These two aspects, of course, are both essential and complementary since beautifully presented interfaces, not backed by an intelligent environment, could be misleading and dangerous, and poor interfaces for excellently designed systems can also lead to difficulties and inefficiencies.

It is the responsibility of interface designers to explore improved methods for presenting information since systems continue to become more and more complex, with fewer operators responsible for much more. The addition of expert advisory modules and sophisticated multiple displays have not always resulted in easier-to-use interfaces.

2. INTERFACE DESIGN AND COMPLEXITY

But what is an easy-to-use interface? Buxton, 1997, has pointed out that the goals of an interface designer should be more concerned with efficiency than with ease-of-use. Some tasks are intrinsically hard and cannot be made easy by any interface design (Buxton quotes learning to play the violin as an example). What designers should be striving to achieve is the minimum acceptable complexity demanded by the task. There should be no additional complexity added by the computer implementation.

Achieving the minimum acceptable complexity for a task requires an appropriate abstraction for the task in hand. For example, consider a Chemist who is required to examine the three-dimensional shape of a complex organic molecule. The Chemist could be provided with a detailed listing of the co-ordinates of all the molecules, together with a complex description of the electron density at any point. Such information would require many pages of print-out. Its analysis would be difficult and time consuming. Consider an alternative presentation of the same data - a true three-dimensional model of the molecule using colour coding to plot out the electron densities. The Chemist would be able to appreciate in seconds what would take hours with the paper print-out, and the risk of error would be greatly reduced. The level of abstraction is optimal for the task required. Of course, if the Chemist required very accurate electron density data, then the model might not be appropriate.

The process of abstraction is concerned with the coalescing of domain states into larger chunks, thereby reducing the complexity of the search problem, Williams *et al*, 1996. Some media readily afford this process, others do not. The representation of button icons for activating features in WORD cannot be readily abstracted, whereas points displayed in a graph can be abstracted to higher levels such as clustering or trends. Williams et al coined the term " expressiveness" to represent this higher level abstracting ability. In this view a medium which supports many levels of abstraction is "more expressive", than one which does not. However, the appropriate level of abstraction needed to support user problem solving will depend upon the experience and expertise of the user. Thus the trick in allocating media effectively is to match the expressive requirement of the user with the maximum expressiveness of the medium used. We say "maximum expressiveness" because too expressive a medium could inhibit the problem solving process. People find graphs useful because they are at the right level of expressiveness for many problems. A more expressive medium would be natural language, but the description of a graph in natural language usually is less useful than the graphical form, which is less expressive.

Spreading information across multiple media channels is another way of reducing complexity. Let us say an operator has two print outs of a manual procedure. The operator knows that there are minor differences (which are important for present purposes) but does not know where they are. The two documents are placed side by side, and the operator performs a line by line visual comparison. This operation would be slow and very error prone. However, the problem could be also be solved another way using an additional operator: the first operator could audibly read out the first document whilst the second operator visually followed the other. This procedure would be quick and far less error prone. Here, the use of two contrasting media has transformed a difficult problem into a routine task.

These examples show that good interface design, in the first case using an appropriate visual transformation, and in the second case, good use of multiple media, can make a hard problem easier.

Other examples include using a map as opposed to a directory, or a graph instead of a table for certain tasks. Note that we are not saying that a graph is always better than a table - it depends upon the problem, and one can think of cases where the reverse is true.

3. THE INPUT MEDIA PROBLEM

Part of choosing the right abstraction is ensuring that the data can be assimilated or created by a human being without more effort than is required by the problem. As far as output media are concerned, we are not doing too badly. The text on a video screen is not the same as a book but is reasonably similar. Coloured diagrams can now have the same quality as printed diagrams and we can animate them, providing a medium which is more versatile that the medium being replaced. Moving video is now close to the resolution required to provide an equivalence to the real world. Much audio output can be made very realistic (including speech), though the auditory medium is not extensively used in control rooms. There are limitations however. Current video conferencing is not very successful at the moment because of a lack of visual feedback of body language. Some recent experiments contrasting video conferencing with a telephone link using audio alone, showed that the video conference took 30% more time to transmit the same information. We have learned to pick up clues in a telephone conversation from the voice of the caller, but this is not possible visually with the limited video screens which are currently provided for video conferencing.

There are two weak output medium areas –use of haptic media (touch) and kinesthetic (movement) feedback. But, apart from these, the output of a computer can be quite similar to that of a human being.

When we consider the input media currently available, a very different picture emerges. At present we provide extremely clumsy interfaces using keyboards, mice, and function keys. This is far removed from how human beings communicate - using speech, gesture and touch in combination. On current computer systems, voice input is exceedingly error prone, and gesture and touch are almost non-existent.

This problem of input media has two immediate consequences. Firstly, all operators have to undergo training to use these devices, and secondly it is impossible to communicate emotion to a computer using current input techniques. The future must lie in voice and gesture input but progress has been slow. It may be that we are not approaching the problem from the right perspective. We have tended to develop either voice recognition systems OR gesture recognition systems, whereas if we learned from human beings we would only develop combined systems. The human approach utilizes two or more media simultaneously, and exploits the redundancy of information between them. If the voice is indistinct, the gesture may carry us through, and vice-versa. Even if both are indistinct we have knowledge of the task to rely on. We are currently developing a simple voice recognition and gesture system which exploits redundancy between the channels and has extensive task knowledge as well (Mills and Alty, 1998).

4. THE IMPORTANCE OF SOUND

Sound has always been difficult to use in control rooms because of its intrusive nature. Indeed, because of this, it has generally been used for important alarms only. Sound however, is an essential part of human activity. It often combines with the visual to produce a "wholeness" of experience (Marmollin (1992), has argued that multimedia involve the whole mind). Much groupware, for example is impoverished because of the lack of a realistic auditory approach. Audio is not thought trivial by those in the traditional media. The audio activity in modern films, for example, is at a similar level to the visual activity. At a recent ICAD conference, the audio designer from Steven Spielberg's "Starwars" film gave an example of the first eight minutes of the film, before and after, audio dubbing. The effect was dramatic. It made clear that many of the effects that one had assumed were carried by the visual channel are actually achieved through the sound channel.

For interface design, we are not necessarily talking about prominent foreground sounds, they could be rather subtle background sounds, say, of plant operation (many of which the operators of 20 years ago actually heard in their adjacent control rooms).

Early on, Gaver, (1986) suggested the use of Auditory Icons in interfaces. These are well-known, often natural, sounds which have common associations (such as the siren on a police car). They therefore can be used in a similar way to visual icons. However, such elements have a limited information carrying capability.

Music is the most sophisticated of the output sound media and has some similarities with language. It has therefore has been suggested that it could be used to communicate output information from a computer application. Blattner (Blattner et al, 1989) has demonstrated the use of what she calls structured Earcons which are short musical phrases (based on simple musical motifs) to communicate error information. Such motifs (or jingles) are often used in public address systems to precede messages and alert listeners. They differ from auditory icons in that there are usually no natural associations.

Because of the MIDI system (Moog, 1986), most computers can now support the full range of musical output, since all MIDI commands can be created directly within programming languages (for example C++, Basic and Pascal). This enables complex interactions using music to be developed. One possible approach would be to create sounds from a running program in an equivalent way to that of visualisation, in a process which has been termed "audiolisation".

Successful demonstrations of mappings between computer algorithms and music have been reported (Alty, 1995). The internal workings of the algorithm are mapped to musical structures. So, in an audiolisation of the Bubble Sort Algorithm, movement up the list, swapping elements, and the current state of the list, were all mapped into different instruments and rhythmical structures. Disambiguation was further assisted through stereo output. Experiments have already demonstrated that people can understand algorithms from the musical output alone, but more work is needed to determine which types of musical mappings are most appropriate.

Musical mappings have also been suggested to aid computer program debugging. Recently, Vickers and Alty (1996), have constructed a variant of the PASCAL compiler which adds musical elements to the common structures in the program (i.e. FOR LOOPS, IF statements etc..) without altering their functions. The program therefore runs, as normal, but creates a continuous musical output whilst running, which can be used for debugging purposes. It is too early to say how useful such approaches will be, but initial results are encouraging. Although such ideas might appear far fetched, it must be remembered that in the early days of computing,, sound output was often used in debugging both hardware and software. Many of the ICL 1900 series computers had computer consoles with audio output, and operators did use them effectively, without training.

5. THE CINEMA

As we move towards using more media in our interfaces, we need to draw inspiration from others who solve different but related problems. For example, the "good" practices of film directors may have important lessons for computer interface designers. May and Barnard (1995), have examined the domain of cinematography and explored the relevance of their practices. A film director has to make a film easy to follow, yet has to cut from one shot to another in a similar way to which interface designers move from screen to screen. Thematic continuity and structural and semantic interpretations have to be maintained across film edits, some of which are quite radical in nature. The cuts have to be radical because there is not time for the camera to pan from side to side or move from place to place as the action shifts. There are rarely gradual shifts or fadeouts. Frequently views are changed dramatically.

May and Barnard carried out an interesting analysis of the implications for interface design of cinematographic techniques using a model of cognition called ICS (Interacting Cognitive

Subsystems (Barnard, 1985). ICS consists of a set of mental representations in which information is progressively transformed as it flows through. Sensory information received via the visual and auditory channels produces Object and Morphological representations. These are then turned into Propositional representations which describe the relationships between elements. Once in this form, information can be fed back to the subsystems, providing top-down information to be blended with the bottom-up information arriving at the senses. There are several cyclical flows of information as bottom-up and top-down information flows affect each other (i.e. what a person sees and hears is a combination of sensory input and higher level expectancy). The point is that the subsystems cannot disagree for long. If inputs disagree, one must be ignored or blended in with the other to provide a single consistent cognitive representation. Many perceptual illusions result from this blending.

Whilst watching a film, the viewer has to blend the visual and auditory input with the continuing narrative of the film. Usually this is not a problem because the sound, visual and current narrative are all highly blended anyway by the director. But potential problems can occur. For example, within-scene changes often involve a camera position change. In between-scene changes, the visual aspect can change abruptly. The prime constraint for a director is to avoid destroying the viewer's illusion that they are "there", and an abrupt visual scene change might do that. As May and Barnard point out "The skill of the film maker lies in constructing cuts so that they are not noticed".

Directors therefore use certain factors to maintain continuity. The trick is to maintain some aspects of the propositional representation over the scene change. So, the subject of a new shot will be placed in the same place on the screen as the subject of the previous shot. Meanwhile the camera can move in, zoom back, or even change viewing position. Furthermore, the director will cause the viewer's attention to be directed to the common subject just

Figure 1 Preparing for the scene change

before the change, by making an actor on the screen look at the place or point to it. The viewer's attention shifts to the common area, the scene is changed, but the new common area contains expected information so the cut is seamless.

In Figure 1, in the first sequence, the gun leads expectancy to the left of the screen, but the man shot appears in the next clip on the right, so the viewer's eyes are in the wrong place. In the second sequence he appears shot where he would be expected to be.

Another well-known sequence is used to represent a person walking along. If a shot shows someone going off the screen left, they will always be shown in the next scene coming on right. If seen going away from the audience, the next shot will show the subject coming towards them. Another example is the problem of showing two people talking in the back seat of a moving car. The problem is that when the shot changes to show the face of the person talking, the scenery in the background reverses its movement (see Figure 2). Taking the shot from the front of the car, with the rear window as the backdrop usually solves the problem. Then there are no scene reversals.

There could be useful guidelines for interface designers from the work in cinematography. Perhaps we should direct the viewers attention to a particular point on the screen before changing the screen content.

Figure 2 The Back-seat Problem

6. LITERATURE AND ART

Designers have tended to neglect the possible contribution from Art and Literature. For example, it is only recently that designers have realized that metaphor is important in interface design. The power in metaphor, as every poet knows, comes from what does not match, as well as what matches, and this distinguishes it from the model or analogue. One thinks of those lines of Auden made famous in the film "Three Weddings and a Funeral":

> "The stars are not wanted now: put out every one;
> Pack up the moon and dismantle the sun"

The power of such metaphor is quite remarkable. Clearly metaphor is important for learning but it may also be useful in guiding creative problem solving as well.

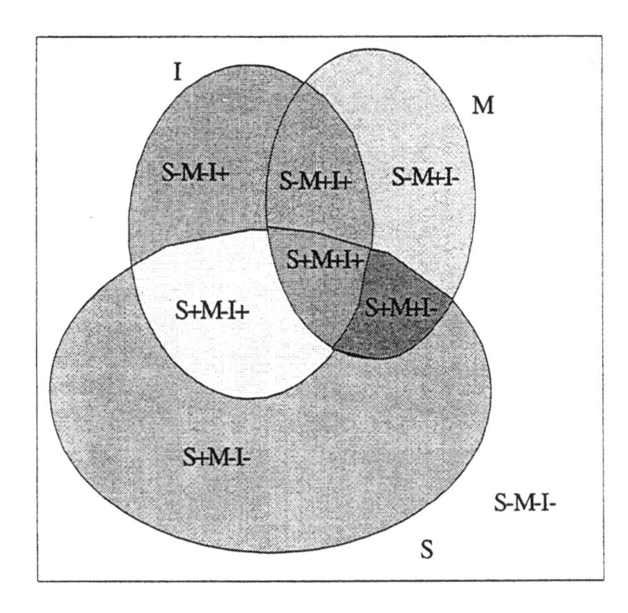

Figure 3 A Model of Metaphor at the Interface

Two of the most ubiquitous metaphors used in Human Computer Interaction are the "Desktop Metaphor", where many housekeeping functions are mapped to the control of papers on a desktop and the "Windows" metaphor whereby users have window-like views onto different application. These metaphors have been successful in enabling users to control a number of applications simultaneously, switching control and even data between windows.

We have defined a Metaphor Model for examining metaphor in interface design (Alty and Knott, 1997). For a particular computer implementation, let a metaphor M have an implementation I and be installed in a system with functionality S. The area I represents the ways in which the designer actually chose to implement the metaphor. There is a complex relationship between S, M and I, diagrammatically shown in Figure 3. The number of distinct areas is eight, namely, S+M+I+, S+M+I-, S+M-I+, S+M-I-, S-M+I+, S-M+I-, S-M-I+, and S-M-I.

The model allows a designer to "exercise" any metaphor implementation. Each operation should be examined and classified into the various eight areas listed above. Areas which should be paid particular attention are S+M-I+ (Inconsistencies), S+M+I- (Implementation Exclusions) and S-M+I- (Conceptual Baggage, additional concepts brought in by the metaphor). The other areas are less problematic for the following reasons:

> S+M+I+ these are "correct" implementations
> S-M-I- These are not relevant
> S+M-I- These are functions not connected with the metaphor
> S-M+I+ These are surface functions only
> S-M-I+ Tailoring operations only

S-M+I- conversion to S+M+I+ is a powerful design tool. In a similar way, implementation exclusions S+M+I-, are missed opportunities. Their adoption strengthens the metaphor, make the interface easier to learn and remember, and provide a more unified approach. Such situations are usually the result of extensions to a system, or lack of sufficient thought about implementation.

Many will argue that metaphor has no place in process control interfaces. Traditional approaches have concentrated upon the analogy and the model, which have closer one-to-one mappings than metaphor. To turn our backs on metaphor is to miss an opportunity. Metaphor enables users to make dramatic leaps of understanding, way beyond what similes, analogies and models can do. It is the lack of a complete match that makes metaphor so powerful. It clearly has a use in training, and in ancillary services for the operator, but may help in difficult problem solving situations as well. The open-endedness of metaphor and is impreciseness could provide new and interesting solutions in difficult problems

7. SUPPORT FOR CREATIVITY

Traditionally, the creative arts have not been well supported by computer technology. It is an area where many of the players (with some notable exceptions) are either computer-phobic or perceive the computer as a threat to artistic creativity. But it is also true that many technologists do not understand the nature of the support required for the artistic domain. During the last few years however, this situation has changed with the arrival of a number of important tools to support artistic activity. Such tools, in themselves, are not very creative, but they are able to transform what is possible for the creative mind. Tools have become available to the artist, the poet and the essayist. In music, probably more so than in the other arts, computer technology has made a large impact in a relatively short time.

The importance of these tools is that they are *designed in the domain characteristics of the user*, and that is what has made them so useful. It is hard to convey the sense of freedom which such tools engender. The tools are not, of course, creative in themselves, but they remove much of the drudgery of creation and allow artists and designers to appreciate the immediate effects of their creative efforts. In music, they remove one of the biggest stumbling blocks to any unknown composer - that of obtaining feedback (at least to a reasonable approximation) of the sounds created. Score creation is also very important. It is a time consuming process, but one of the main advantages is in removing mistakes from the score. Composers can hear what they have written - generated directly from the score.

Most computer advisory programs either offer passive support, providing information and carrying out calculations at the request of the operator, or perform sophisticated reasoning (as in an expert system). What we really need is an advisory program which lies between these extremes, which supports the creativity of the operator, resonating with the operators line of reasoning and suggesting modified lines of enquiry. We can learn much here from artistic activity where there are a number of such support programs (for example in musical composition). Such programs provide sophisticated information selection and manipulation facilities

It is often an observed fact that when technology is used to improve practices in a particular application area, the results of that application can often be reflected back into the technology itself. Thus, display technology has provided artistic designers with much better tools for creating and manipulating images, and as a result, creative designers have contributed to the aesthetic design of computer interfaces. This symbiosis has also happened in music. The development of the MIDI system (Moog, 1986), has opened up fascinating new possibilities for computer interface design.

Thus, as is so often in human endeavour, technology that was originally developed to assist creative musicians is now depending upon creative musicians to further improve the use of the technology. What has already become apparent is that the computer interface designers using music will have to rely heavily on advice and guidance from composers. The symbiosis moves into the next cycle.

8. CONCLUSIONS

In order to continue our progress in developing better process control interfaces, we need to reach out and exploit the experiences of others in related or distinct disciplines. The final paper will give detailed examples.

9. REFERENCES

Alty, J.L., Bergan, J., and Schepens, A., The design of the PROMISE multimedia system and its use in a chemical plant, in *Multimedia systems and applications* (R. A. Earnshaw, 1994), Academic Press, London, 53-75.

Alty, J.L., Can we use Music in Human-Computer Interaction? , in *People and Computers X* (Proc. HCI'95), Kirby, M.A.R., Dix, A.J., and Finlay, J.E., (eds.), Cambridge Academic Press, (1995), 409 - 424.

Alty, J.L., and Bergan, M., Multi-media interfaces for process control: Matching media to tasks. *Control Engineering Practice 3*, 2 (1995), 241-248.

Alty, J.L., and Knott, R.P., Proc. of Annual Conf. on Manual and Automatic Control, (Johannsen, G., ed.), Kassel, 1997, to be published.

Alty, J.L., and Vickers.P., The CAITLIN Auralisation systm: Hierarchical Leitmotif Design as a clue to Program Comprehension, in *Proc. ICAD'97*, (1997), Santa Fe Institute.

Auden, W.H., *Collected Poems*, Twelve Songs, IX, Faber and Faber Ltd.

Barnard, P.J., (1985), Interacting Cognitive Subsystems: A psycholinguistic approach to short term memory. In A Ellis (ed.), Progress in the Psychology of Language, Vol. 2, Lawrence Erlbaum: London, pp 197 – 258.

Blattner, M., Sumikawa, D.A., and Greenberg, R.M., (1989), "Earcons and Icons: Their Structure and Common Design Principles", Human Computer Interaction, Vol. 4, No. 1, pp 11 - 44.

Buxton, W., (1996), Proc. of OzCHI,

Gaver, W., (1989), "Auditory icons: Using Sound in Computer Interfaces", Human Computer interaction, Vol4, No. 1, pp 67 - 94.

Holland, S., Interface Design Empowerment: A Case Study from Music, in *Multimedia Interface Design in Education*, Edwards, A.D.N., and Holland, S., eds., *NATO ASI series F*, Springer Verlag, Berlin, (1992), 177 - 194.

Marmollin H., Multimedia from the Perspective of Psychology, in Kjelldahl, L., (ed.*), Multimedia: Systems Interactions and Applications*, Springer-Verlag, Berlin, (1992), 39 - 52.

May, J., and Barnard P., in Human-Computer Interaction, Proc. INTERACT95, (Nordby, K., Helmersen, P.H., Gilmore, D.J., and Arnesen, S.A., (eds.), Chapman-Hall, London, pp 26 – 31.,

Mills, K., and Alty, J.L. *Lecture Notes in Computer Science*, Springer-Verlag., 1998.

Moog, R., (1986), "MIDI: Musical Instrument Digital Interface", J. Audio Engineering Soc., Vol. 34, pp 394 - 404.

Rigas, D., and Alty, J.L., The Use of Music in a Graphical Interface for the Visually Impaired, Proc. of INTERACT'97 (Howard., S., Hammond, J., and Lindegaard, G., (eds.)), Sydney, (1997), 228-235.

Williams, D., Duncumb, I & Alty J.L., (1996), Matching Media to Goals: An Approach based on Expressiveness, *Proc HCI'96, People and Computers XI*, (Kirby, M.A.R, Dix, A.J., & Finlay, J.E., eds.), ,Cambridge University Press, 333 – 347

INTRODUCTION OF ACTION INTERFACE AND ITS EXPERIMENTAL USES FOR DEVICE CONTROL, PERFORMING ART, AND MENU SELECTION

Yu SHIBUYA, Kazuma TAKAHASHI, and Hiroshi TAMURA

Kyoto Institute of Technology, Matsugasaki, Sakyo-ku, 606-8585, Kyoto, JAPAN
Email: shibuya@dj.kit.ac.jp

Abstract: Action Interface is a new type of man-machine interface which uses a user's action as an input method. Users are able to control machines or to perform art with their body movement, i.e. their action. The system of Action Interface introduced in this paper basically consists of video camera, PC, and display. Some ideas are proposed to improve the response time and accuracy of Action Interface. Action Interface is usable to control device, e.g. to change camera view. To make an interactive games with action interface is also interesting. Furthermore, Action Interface is usable for menu selection with proper menu design.

Keywords: cameras, computer system, human-centered design, human-machine interface, image processing, image sensors, input equipment, interfaces, multimedia

1. INTRODUCTION

Action Interface is a new type of interfaces between human and machine. Users are able to control machines, to perform art, or to interact with machines by means of their body actions or gestures. For example, a user action takes to control camera view, to play a game or a virtual musical instrument, or to select a menu item.

There are several commercial or experimental uses of human action for performing arts. The MANDALA© VR system (Vivid) is a video gesture-based VR system. In that system, a player's body image is captured, then combined with computer graphics, and finally displayed on the monitor in front of the player. The player can make a tune with virtual musical instruments displayed on the monitor. Trans Plant (Christa, 1995) is an artistic use of combined users' live body image and computer graphics. Users' images are also captured, combined with computer graphics, and displayed to them. The system monitors players' spatial position and draws artificial plants dynamically depending on their movement. However, these systems are too expensive and complex for moderate people.

Recently, information technology has been improved extraordinarily. Many Personal Computers (PCs) have multimedia input-output facilities and the processing speed of such PC is fast enough to treat multimedia data. Nowadays, it is possible to configure an Action Interface with PCs. Reality Fusion provides an interface that enables users to interact with a computer simply by using hand or body motion (Reality Fusion, 1997).

In this paper, Action Interface is introduced and several applications of it are explained. They are a video camera control, a virtual musical instrument, and some menu selections. Kubodera and Takeda developed a hand position recognition interface (Kubodera, 1997) that was similar with our Action Interface. However, they used only Red level of RGB color expression to detect user action and therefore its application field is restricted. Artificial retina chip is a processing chip to detect moving object with using monochrome image (Miyake, 1996). It is suitable for pattern and motion recognition. On the other hand, in our Action Interface, most of process is done by software, so that it is flexible and adaptable for a lot of kinds of application.

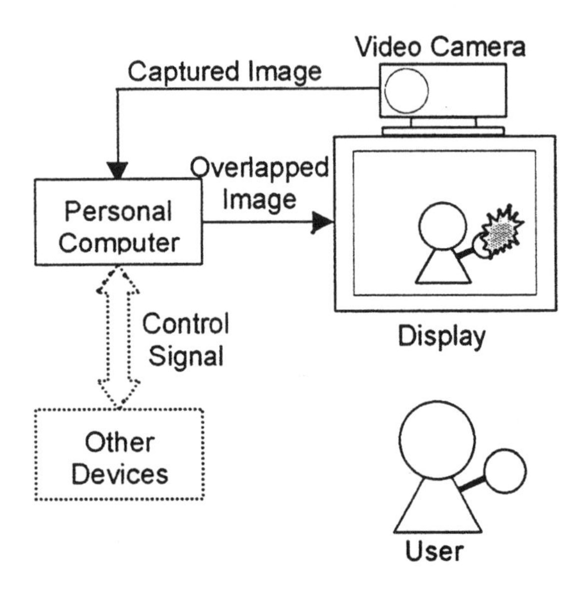

Fig. 1. System configuration of Action Interface.

2. SYSTEM CONFIGURATION

In Fig. 1, the system configuration of Action Interface is described. The system basically consists of a video camera, a PC, and a display. A user's live body image is captured by the video camera and processed by PC in real time. Depending on the usage, computer graphics are overlapped on the user's body image and displayed to the user.

The system detects a user action by monitoring a color change at the pre-located position. Of course, there are several methods to detect users' action using their images. Among of them, detecting a change of color might be one of the easiest ways. It is straightforward idea to check a value of RGB level of color but it doesn't work well. In order to detect user's action faster and more accurately, some ideas are introduced. They are described as followings.

- Using HSL format for expressing color instead of RGB format.
- Monitoring only a desired region not whole part of captured image.
- Using a mosaic image to shorten the response time.

When an image is captured with a video camera, color of image is expressed in Red, Green, and Blue level respectively. However, RGB color level changes sensitively depending on the environment. For example, even at the same condition except the brightness, the RGB color level is dynamically changed depending on the brightness. Such changing of brightness is often occurred in general environment. In order to avoid such unexpected sensitivity, HSL format is used to express the color. In the HSL format, color is separated into three elements; they are Hue, Saturation, and Lightness. If the brightness changes slightly, both saturation and lightness levels change widely but the hue level does not change dramatically. Therefore, in order to monitor the changing of color, it is better to use HSL format to express color.

Quick response is important to avoid the user's stress during the interaction with machines. In order to make the image processing faster, monitoring region is restricted to a part of images. Furthermore, before checking the color changing, captured image resolution was reduced. For example, in Action Interface, size of a whole captured image is 320-pixels width and 240-pixels height (320x240 pixels). Inside of this image, several monitoring regions are set and the image of each region is transferred to mosaic image. Using mosaic is equivalent to reducing the number of pixels of the monitoring regions. As a result, it is enough fast to respond to the user's action.

When the user makes an action, the system should make some feedback to that user. As such feedback, in one case of Action Interface, if the user touches the button drawn on the display then its shape or color is changed and the sound is played simultaneously.

3. EVALUATION OF RESPONSE TIME

Using a mosaic image was expected to shorten the response time of Action Interface. In order to verify it, that time was measured with varying the size of mosaic. For this measurement, a square formed sensor point was located on the display. When the user touched this point the color of the square was changed. With monitoring the recorded video image of the user display, the duration between the touch and the color change was measured. Results are shown in Fig. 2. Using large sized mosaic shorten the duration of checking sensor point while it takes a long time to make mosaic images. As shown in Fig. 2, 5X5 pixels might be practical size of mosaic image in this experiment.

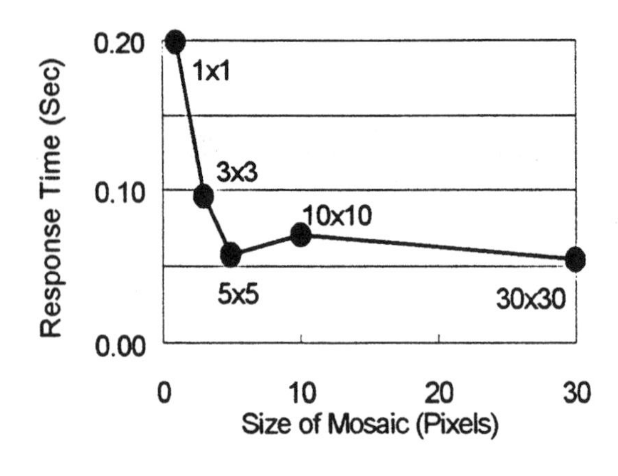

Fig. 2. Response time. Size of original monitoring region is 30x30 pixels. Beside each of plots, the size of mosaic is drawn in pixels.

Fig. 3. Camera control with Action Interface.

4. APPLICATIONS OF ACTION INTERFACE

Several applications of Action Interface have been realized. Some of them will be explained in followings.

4.1 Device Control with Action Interface

A device control is introduced as the first application of Action Interface. Some devices, e.g. video camera, videocassette recorder, video disc drive, etc., are able to be included in the system and controlled. A video camera control system is explained here. In this system, the user can zoom in and out, pan, and tilt a video camera using Action Interface. As shown in Fig. 3, a user can control the video camera by touching the control button on the display. They need no particular equipment, e.g. a wireless remote controller. They just locate in front of the capturing video camera and they can control the remote site video camera. Furthermore, the controlling button on the display is able to relocate by the users themselves, so each user can adapt that location, as they like. This kind of camera control is usable for TV conference system. If the participants use Action Interface, they can control the video camera with their action.

4.2 Entertainment Use of Action Interface

Device control with Action Interface is just substitution of Action Interface for the traditional controller. Action Interface is also expected to use for entertainment or performing art. Because of that Action Interface is using human action as an input method so that it might be fun for the users to make interaction with computer or with other users by moving themselves. Two kinds of entertainment use of Action Interface will be introduced. One is a bouncing ball game and another is playing virtual instrument.

In Fig. 4 shows a snap shot of playing a bouncing ball game. The purpose of this game is to keep the bouncing balls inside of the display without falling down to the bottom of that. When a player touches a ball then the ball bounce up and user get a point with each touching. Similar with our real world, the bounced up or thrown up ball has initially vertical velocity, then decrease its velocity during going up, and finally it turns to fall down. Each player competes against other players by getting higher points.

Another entertainment use of Action Interface is shown in Fig. 5. In this figure, a player makes a tune with a virtual musical instrument. Each button has similar function with a keyboard of a piano. A player makes contact with that button to play sound. However, it is not so easy to play this kind of virtual musical instrument well. Similar with a real musical instrument, many experiments or practices are needed to play it well.

In the both entertainment uses of Action Interface described above, most of players play them with fun while they could not play virtual music instruments well. They delighted in sounds or images, which is simultaneously changing with their action. Unfortunately, there is not enough feedback, e.g. there was no tactile feedback, players sometimes complained about that point.

Fig. 4. Bouncing ball game.

Fig. 5. A virtual musical instrument.

(a) Menu button

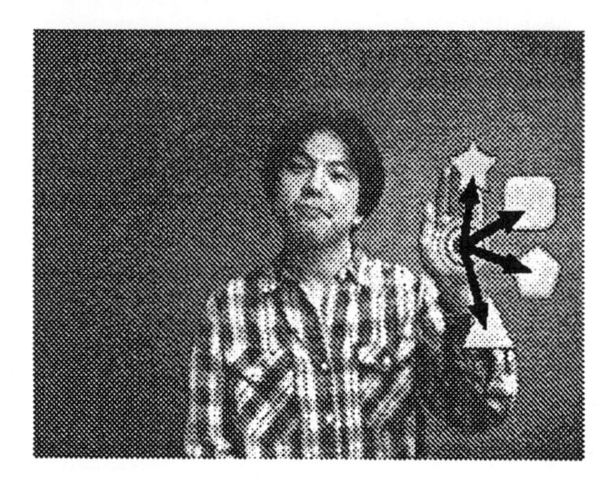

(b) Quasi-pie menu

(c) Pull-in selection

Fig. 6. Procedure of pull-in selection. At first, (a) a menu button is displayed, then (b) menu items appear surrounding the menu button and the menu button changes to the decision button, finally (c) the user can select the desired menu item by pulling it into the decision button.

5. MENU SELECTION WITH ACTION INTERFACE

In the previous chapters, simple interactions using Action Interface were introduced. If a user touches a button on the display then a particular event happen, e.g., a camera begins to pan or tilt, or the sound comes out. However, in order to make practical interfaces, it is important to include a menu selecting function in Action Interface. Because of the restriction of the display space, most of menus used in computer desktop system are difficult to use in Action Interface. A pie menu was expected to be possible to use in Action Interface. When the user touches a menu button on the screen, menu items appear surrounding that button. In Action Interface, it is not so good to overlap the menu item on the user body so that it is not complete pie menu but is quasi-pie menu.

In usual menu operation, at least two procedures are needed. They are selecting a menu item and deciding that selection. In Action Interface, it is easy to select the menu item on the display by touching one but it is not so easy to decide the selection. Because, for the simplicity of the system, the system only monitors the color changing, it is difficult to distinguish the deciding action from the selecting action. In order to distinguish between them, a pull-in selection procedure is introduced. This procedure is shown in Fig. 6. When a user touches the menu button, menu items appear surrounding the menu button and the menu button is changed to the decision button. Then the user selects the desired menu item and that menu item is highlighted. If the user wants to decide the selection, it is simply done by pulling the menu item into the decision button that located at the center of the quasi-pie menu.

By using pull-in selection, it is usually enough to make a menu system with Action Interface. However, hierarchical menu is needed for more advanced use. In order to get over such problem, Raijin Menu was introduced.

5.1 Raijin Menu

Fig. 7 explains how Raijin Menu works. In Raijin Menu, if the user select the menu item which includes further menu items, selected menu item shifts to upper left side and further menu items are appeared on the same position where previous menu items were displayed. Then, if the user selects menu items including further menu items, there happen same procedure with previous case. Until the final selection, such procedure is continued. With this manner of menu selection, Action Interface is usable for hierarchical menu. "Raijin" is the God of thunder in Japan and behind of him there were drums surrounding him. The arrangement of those drums is similar with the allocation of the shifted menu item so that this menu was named Raijin Menu.

Fig. 7. Raijin Menu.

(a) Menu icon

(b) Menu items

Fig. 8. Karugamo Menu. (a) Menu icon always follows the user's movement. (b) If the user touches the icon then menu items appear.

5.2 Karugamo Menu

With using Raijin Menu. Action Interface is usable for hierarchical menu operation but user must stay at the pre-fixed location of the display. In order to avoid such restriction, a following menu, named Karugamo Menu, was introduced. "Karugamo" is Japanese name of one kind of duck and that babies are always running after their parents. Karugamo Menu is the menu that follows the user's movement as "Karugamo" does. This is the origin of this name.

Fig. 8 shows how Karugamo Menu works. In Karugamo Menu, until users come into the display there is no menu button or icon on the display. When a user comes into the display, there suddenly appears the icon of Karugamo Menu. And this icon follows user movement during the user stay in the display. If the user touches the icon then the pop-up menu is displayed. Of course, Raijin Menu is able to combine with this dynamic menu allocation.

6. CONCLUSION

In this paper, a new type of interface called Action Interface was introduced and some experimental applications were explained. Action Interface is usable not only for entertainment but also interaction between human and machine. For the practical use of Action Interface in menu system, Raijin Menu and Karugamo Menu were proposed. A character-input method with Action Interface has been also developed. By combining this input method with above menu selection or dynamic menu allocation, it should be possible to develop a suitable dialogue system using Action Interface.

Any way, Action Interface has advantages over other interfaces. It is easy to configure the system, non-touch interface, and human-centered system. However, it has also some disadvantages. One of the them is lack of tactile feedback.

Up to this day, Action Interface is supposed to be used by one user. On the next step, it is expected to use this interface for multi-user interface and make interaction between them. Furthermore, to make the system to be simple, a single video camera is used for Action Interface, so that it is difficult to detect the 3 dimensional user action. Therefore, it is also planned to use two or more video camera simultaneously and make a 3 dimensional Action Interface in near future. Talking about the application of Action Interface, until now, it has been expected to use for entertainment or multimedia presentation. However, in future, it might be also usable for monitoring or controlling industrial processes.

ACKNOWLEDGEMENT

This study is partly supported by Grand-in-Aid for Scientific Research from the Ministry of Education, Science, Sports, and Culture of Japan under Grand No. 09838020.

REFERENCES

Christa, (1995), http://www.mic.atr.co.jp/~christa/link6.html

Kubodera, A., T. Takeda (1997), Creation of Games Software using Action Interface, *Human Interface N&R*, **12**, pp217-222.

Miyake, Y., W. T. Freeman, J. Ohta, K. Tanaka and K. Kyuma (1996), A gesture controlled human interface using an artificial retina chip, *IEEE-LEOS*, **1**, pp292-293.

Reality Fusion, (1997), http://http://www.realityfusion.com/

Vivid, http://www.vividgroup.com/vivid/

DESIGNING AN INTERFACE FOR CHOREOGRAPHERS

Tom Calvert[1], Sang Mah[2] and Zeenat Jetha[3]

Computing Science and Engineering Science
Simon Fraser University
Burnaby, BC, V5A 1S6, Canada.

Abstract: This paper describes the design, implementation and evaluation of computer-based tools to assist choreographers in the composition of a dance. The compositional task is similar to that in many design situations and the evaluation methodology is similar to that used to study operator behavior in supervisory control of complex industrial systems. By studying the strategies adopted by the user in carrying out the complex compositional or design tasks it is possible to design better interfaces to support these tasks. *Copyright © 1998 IFAC.*

Keywords: Design, Computer graphics, Computer interfaces,
Human machine interface.

1. BACKGROUND

This paper discusses the design, implementation and evaluation of computer-based tools to assist choreographers in the composition of a dance and relates this experience to parallel work designing interfaces for complex industrial control systems. In contrast to the other performing arts, dance has made relatively little use of technology. Typically, choreographers work with live dancers in a studio to work out the composition of a new piece over an extended period (sessions spread over weeks or months). In the mid-eighties, after working with computer-based systems for dance notation for a number of years (Calvert et al, 1982) our group at Simon Fraser University began the design of a new system to support the choreographic process. The goal was to provide a computer-based interactive tool with which the choreographer could work out patterns of movement for multiple dancers. After prototyping the movement on the computer, the choreographer could then refine it working with live dancers in the studio. Typically there would be a number of iterations between work in the studio and work on the computer, but the creativity of the choreographer is enhanced since it is possible to explore many more possibilities than in a situation where every move must be worked out in the studio.

The resulting system, *Life Forms*™, has met most of our goals and is widely used by choreographers around the world (Calvert et al, 1991; Calvert et al, 1993a; Calvert et al, 1993b; Calvert and Mah, 1996). This paper describes the process used to evolve the user interface, discusses a study which examines how a system like this is used for creative composition and suggests how the interfaces for systems such as this can be improved.

[1] Now with Technical University of BC, Surrey, BC, Canada. {calvert@tu.bc.ca}
[2] Now with Credo Interactive Inc., Burnaby, BC, Canada. {sangm@credo-interactive.com}
[3] Now with Philips Research Labs Ltd, UK.

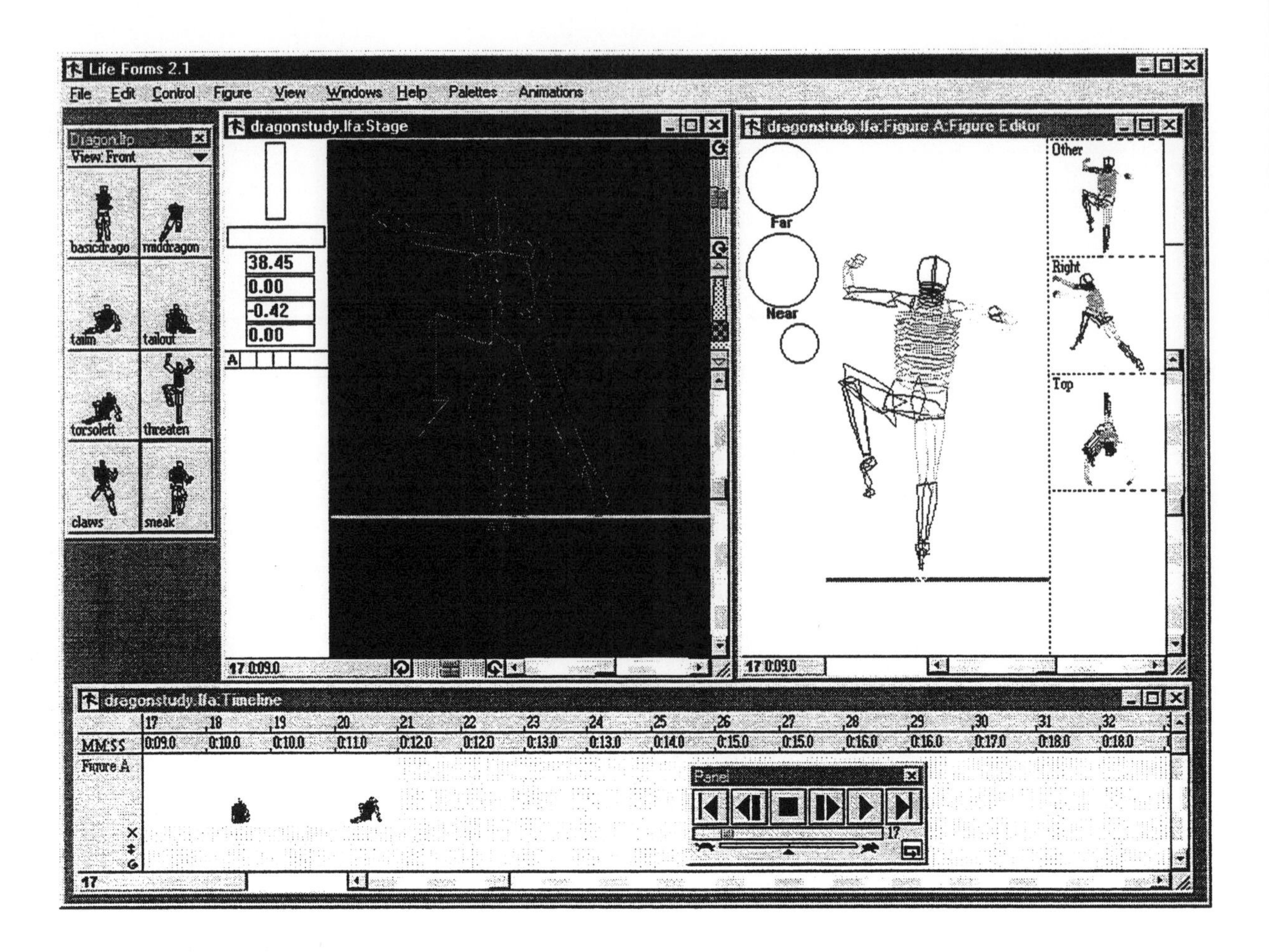

Figure 1. The three Life Forms windows: Stage, for composition of different figures, Figure Editor for creating stances and Timeline for editing the timing.

2. USER INTERACTION WITH A CREATIVE TASK

As with many other design tasks, creative composition is highly interactive and any graphical user interface must meet the accepted criteria for such interactive tasks (Caroll and Rosson, 1985; Shneiderman, 1992). Beyond this, creative design and composition tasks share specific features (Koestler, 1964; Simon, 1969). These include:

1. Typically, there is no one "best solution" and different artists will evolve different end results since the criteria for an optimal design are esthetic and thus somewhat subjective.
2. The process is hierarchical, starting with a concept which is first implemented at a high level and then filled in at a detailed level.
3. Although the process is hierarchical, there is need to be able to work with detailed movements at the same time as modifying the overall concept for the piece. This leads to the need to examine "detail in context".
4. New insights can be gained by examining the composition from different points of view (e.g. space vs. time).

3. COMPOSING A DANCE

While there is no unique approach to dance composition there are a number of common elements which any system must support. Any composition starts with a concept which develops into a plan - possibly in the form of a crude storyboard. Then the components of the composition must be developed, through development of the movement sequences for each dancer. The individual sequences then have to be integrated with each other and placed in an environment which provides appropriate background scenery, etc. Finally, there is a live performance or in the case of a computer-based composition, the models of objects and movement must be rendered as a series of frames which can be displayed in real time.

4. EVOLVING THE LIFEFORMS INTERFACE

Life Forms was initially developed on Silicon Graphics Iris workstations and SGI remains the platform for our research. However, to make the system more accessible we have developed versions for the Apple Macintosh and for Windows95/NT. The system is available commercially from Credo Interactive Inc (http://www.credo-interactive.com).

As shown in Figure 1, the system provides the user with three alternative environments: the body editor, the stage, and the timeline. In the body editor the choreographer develops the body stances which are the elements of the movement sequences; individual stances are saved in menu's which are displayed graphically and sequences are also stored in a library. The stage provides the environment in which movement sequences are assembled for the individual dancers and coordinated in space. The timeline view provides for coordination of the timing between dancers. A more detailed description can be found in other publications (Calvert et al, 1993a; Calvert and Mah, 1996).

The evolution of the interface has been highly iterative - beginning from a very simple concept, many, many alternatives were suggested and evaluated before selecting the few that have been implemented. Many of the observations of the effectiveness of the interface evolved from users of the system: choreographers and animators, as well as members of the design team. New York choreographer Merce Cunningham worked with the system for over three years and used it to create several new works. He noted when he began working with *Life Forms*, "The thing that interested me most, from the very start was not the memory -- it wasn't simply notation -- but the fact that I could *make* new things" (LA Times, May 15 1991).

Lessons learned from this user centred evolution of the interface were reported in our InterCHI93 paper (Calvert et al, 1993b). The key lessons were:

1. An interdisciplinary development team is necessary
2. The role of users - it is crucial that users be part of the development team but there is also a need for external users; master users play a special role.
3. Time from conception to implementation can be a problem - unfortunately some new features that evolve from the user centred process take a long time to implement - this disrupts the iterative process which should involve turn-around times of only a few hours or at most a day.
4. Discontinuities in development are sometimes necessary - although much of the development is iterative and incremental it is sometimes necessary to adopt a new, discontinuous approach
5. Impact of *Life Forms* on the way choreographers work - many of the choreographers who used the system during its development reported that it changed the way they worked; this compromises their role as users.
6. Users prefer direct interaction - the users clearly preferred to be able to interact directly in the GUI with the elements of the composition; thus real-time interaction is essential.
7. Access to explicit knowledge - choreographers make a lot of use of stances and sequences which they and others have used before; it is important that the interface support this.
8. Procedural generation of movement sequences - the compositional process can be greatly improved if procedural generators are available for repetitive movements such as walking and running (Bruderlin and Calvert, 1989 and 1996).
9. Simple abstract models may be as useful as more realistic models - for this tool which is primarily to plan a physical dance photorealistic body models are unnecessary and indeed they can impede the creative process.

5. STUDYING THE CREATIVE PROCESS

5.1. General

The lessons learned from the evolution of the interface continue to provide a useful basis for design of the system but many questions remain. Specifically, although *Life Forms* had proved to be a very useful tool for working choreographers and students, there were only anecdotal reports on how well it supported the creative process. Thus, in 1993 one of us (Zeenat Jetha) designed and carried out a study to investigate how well the *Life Forms* system and its interface supported creative composition of a dance (Jetha, 1993).

The design for this study was based on our experience in evaluating human interaction in supervisory control of complex systems such as power networks or telephone networks. Since the situations which can arise in such complex interactive systems are hard to predict, a methodology has been developed in which the user's actions are captured on video tape and in computer logs for a variety of real problem solving situations. This data is then analysed using a technique called protocol analysis (Ericson and Simon, 1980) - this involves creating an annotated script describing the user actions and from this deducing the user strategies. The interface can then be evaluated in terms of how well it supports these strategies.

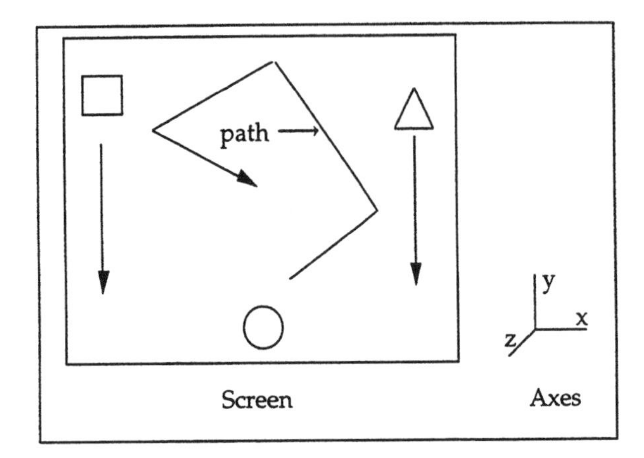

Figure 2: The Shapes Animation

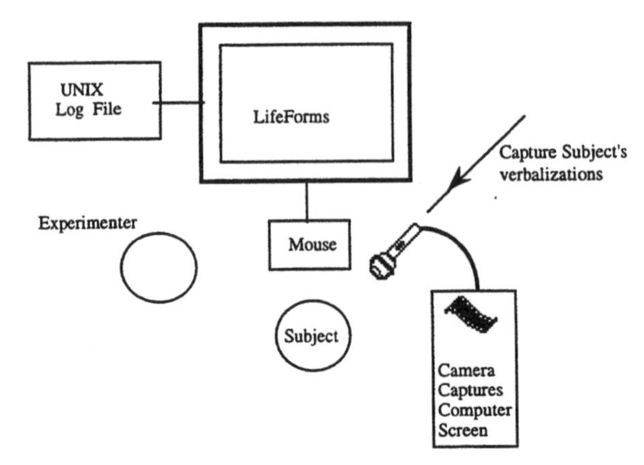

Figure 3: The Shapes Experiment. While creating a new dance, the choreographer and the screen are captured on video and keystrokes are logged.

5.2. Defining the task.

The first step in this study was to choose a composition task. For a first pilot experiment, several dancers were asked to recreate a short piece in *Life Forms* based on a video sequence. The video sequence, for this design composition assignment showed three live human subjects performing three different but simple movement patterns: a walking sequence, a star jump, and a sitting sequence. The task was designed in such a way as to encourage the subjects to utilize all of the main tools provided by *Life Forms*. As such, it was helpful in evaluating the ease of use and the adequacy of the various *Life Forms* tools, but it did not provide any evaluation of how the system supported creativity in composition since the dancers were merely replicating a given video sequence.

It is difficult to study how users interact with a system which is designed to support the creative process since, by definition, it is expected that each user will create something different. In order to provide some basis for comparison we decided to define the task with a short, relatively abstract video animation. Then each subject was asked to create a short *Life Forms* dance based on whatever this animation meant to them.

For the task definition video three shapes were animated to give subjects a movement idea: a purple cube, a blue pyramid and a turquoise sphere. Starting in a triangular configuration (Figure 2) with the cube in the upper left, the pyramid in the upper right, and the sphere at the bottom centre of the screen, the three shapes become alive.

First, the cube and pyramid spin on the spot around the y axis followed by a single flip around the z axis. Then they translate to the bottom of the screen, while the sphere zig-zags around the screen on the x-y plane. The animation ends with the three shapes in another triangular configuration, only this time, the cube and pyramid are at the bottom of the screen and the sphere is in the middle.

5.3. Methodology of the study with 8 dancers.

Second and third year students enrolled in the Dance Program at Simon Fraser University were asked to participate as subjects for this experiment. All subjects were given seven 60 - 90 minute training sessions on how to use the various tools in *Life Forms*. Subjects not familiar with computers were given extra time to develop their skill. For the last tutorial, after all subjects were familiar with the various tools in *Life Forms*, subjects were asked to create a composition on *Life Forms* as practice for their final test. Since the test would be limited to two hours, this last tutorial was also limited to two hours. On the day of the test, subjects saw the design task, or the shapes animation, for the first time and were asked to use the shapes animation as a basis for their composition.

Based on the animation of the simple shapes, each of the eight subjects composed a piece using *Life Forms*. The subjects were encouraged to verbalize their intentions and actions. The visual experimental result (the computer screen that subjects were working with) was video-taped, while the human interactions with the computer were captured in a UNIX log file (Figure 3).

Subsequently, to develop a protocol analysis (Ericson and Simon, 1980), all the video tapes were annotated using the *Timelines* application (Harrison and Baecker, 1992; Owen, 1993). The experimenter used *Timelines* to capture the VCR counter (hours:minutes:seconds) for particular events or intervals displayed on the TV screen when the corresponding event or interval button is clicked in the *Timelines* window on the Macintosh screen. The basic setup is illustrated in Figure 4.

5.4. Results

Using the *Timelines* system, the videotapes and computer logs were used to create an annotation of

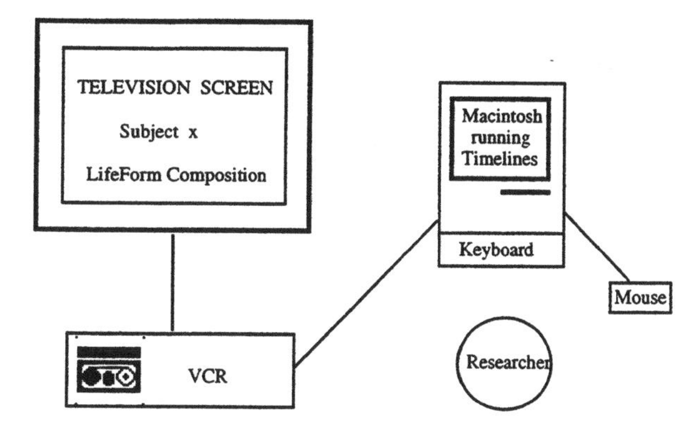

Figure 4: Analysis of video and log data using Timelines.

each experiment. The first annotation pass through all the video tapes was fairly coarse, simply identifying how long each subject spent in the three main views of *Life Forms*: the body editor, the stage view, and the timeline. Figure 5, below, presents the results for all eight subjects. These graphs show the views of *Life Forms* (y axis) with respect to time (x axis). Each bar signifies the amount of time spent in each view and the order in which the intervals and events occur. For example, in Figure 5, subject 6 initially started in the stage view then enters the body editor. After spending some time there, he returned to the stage view, then moved into the timeline, and so on.

All subjects started in the stage view because *Life Forms* always opens in the stage view. Further analysis on all graphs reveals that all subjects also end in the stage view, where sequences can be combined together, positioned and played. Moreover, transitions from the timeline to the body view, or vice versa, though possible, rarely occur.

The graphs show that subjects adopted two general approaches in carrying out the design task - they either moved through the views linearly (subject 2, 4, 5, 6, and 7), or they frequently switched among the views (subject 1, 3 ,8). Furthermore, those who worked sequentially either created one entire sequence in the body editor and utilized predefined sequences from the *Life Forms* library, or they created everything in the body editor and did not use any library sequences (but used library stances). To gain more insight into the details of the choreographic process there was detailed of how two typical subjects from each group approached the composition task.

6. CONCLUSIONS

It is clear from even the top level analysis of the composition task that different choreographers take very different approaches to creative composition.

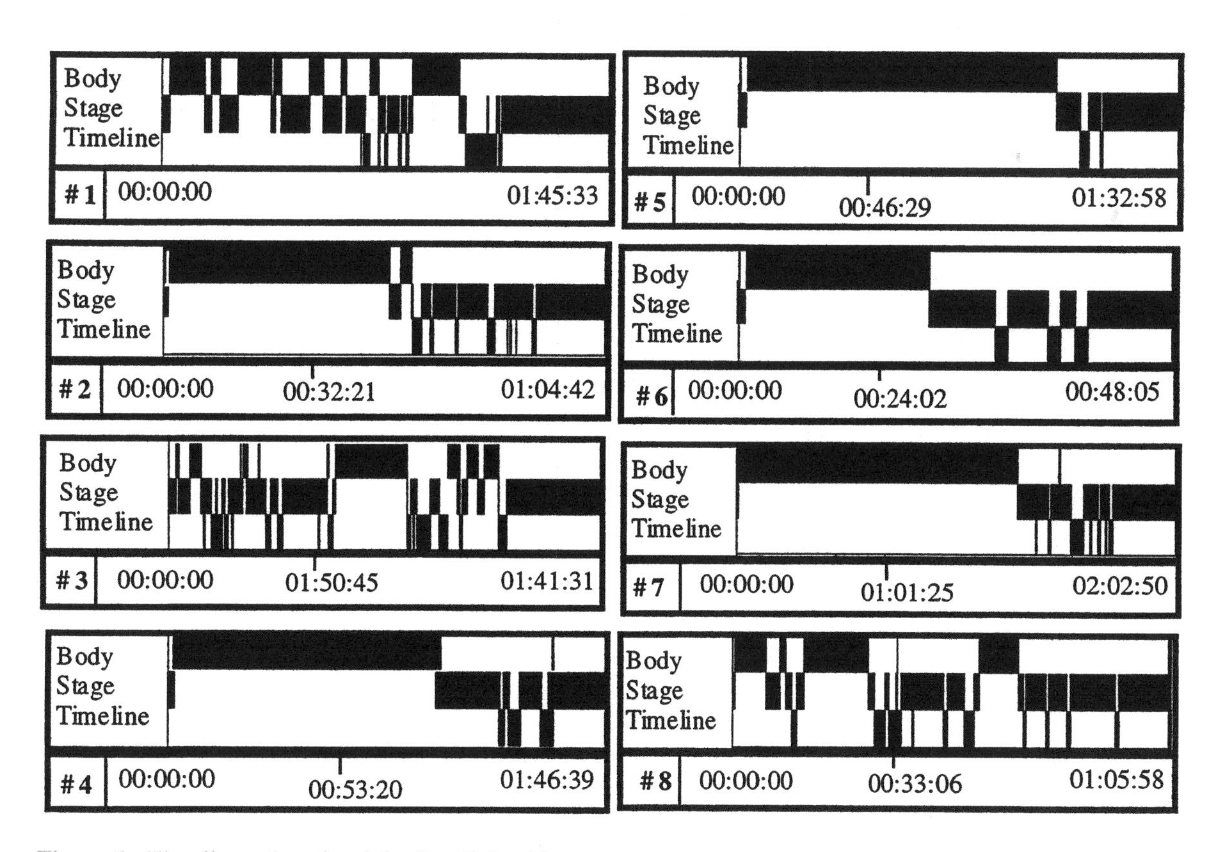

Figure 5: Timelines plot of activity for all 8 subjects.

We believe that the very different strategies can impose quite different memory loads (Green, 1990), on the choreographer and that the interface should be modified to better support this memory load. More detailed analysis of the annotated scores allows many other aspects of the compositional process to be identified.

It is interesting to consider what the experience gained in these studies can tell us about user interface design for other problem areas such as supervision and control of large systems. These are also tasks where operators have significant memory load and where it is necessary to examine "detail in context".

ACKNOWLEDGEMENTS

The authors gratefully acknowledge research grant support from the Social Sciences and Humanities Research Council of Canada, the Natural Sciences and Engineering Research Council of Canada and Project HMI-1 of the IRIS Network of Centres of Excellence.

REFERENCES

Bruderlin, A. and T.W. Calvert, (1989). "Goal-Directed, Dynamic Animation of Human Walking". *Computer Graphics (SIGGRAPH 89)*, vol. 23, pp. 233-242.

Bruderlin, A. and T. Calvert, (1996). "Knowledge Driven, Interactive Animation of Human Running", *Proc. Graphics Interface 96*, pp: 213-221.

Calvert, T.W., J. Chapman, and A. Patla. "Aspects of the Kinematic Simulation of Human Movement". (1982). *IEEE Computer Graphics and Applications*, vol. 2, pp. 41- 50.

Calvert, T.W., C. Welman, S Gaudet, T. Schiphorst, and C. Lee. (1991). "Composition of Multiple Figure Sequences for Dance and Animation". *The Visual Computer*, vol. 7, pp. 114-121.

Calvert, T.W., A. Bruderlin, J. Dill, T. Schiphorst, and C. Welman, (1993a). "Desktop animation of multiple human figures", *IEEE Computer Graphics and Applications*, pp: 18-24.

Calvert, T.W., A. Bruderlin, S. Mah, T. Schiphorst and C. Welman, (1993b). "The Evolution of an Interface for Choreographers", *Proc. InterCHI Conference*, pp:115-122.

Calvert, T.W. and S.Y. Mah, (1996). "Choreographers as Animators: Systems to support composition of dance", in *Interactive Computer Animation*, N. Magnenat-Thalmann and D. Thalmann (eds), pp: 100-126, Englewood Cliffs: Prentice-Hall.

Carroll, J.M. and M.B. Rosson, (1985). Usability specification as a tool in interative development. In H.R. Hartson (Ed.), *Advances in Human-Computer Interaction*. Norwood: Ablex.

Ericson K.A. and H.A. Simon. (1980). "Verbal Reports as Data." *Psychological Review*, 67, pp. 215-251.

Green, T.R.G., (1990). Limited Theories as a framework for human-computer interaction. In *Mental Models and Human-Computer Interaction I*, D. Ackermann and M.J. Tauber (editors), pp. 3-39, New York: Elsevier.

Harrison, B.L. and R. Baecker, (1992). "Designing Video Annotation and Analysis Systems", *Proc. Graphics Interface'92*, Vancouver, pp.157-166.

Jetha, Zeenat, (1993). *On The Edge of the Creative Process: An Analysis of Human Figure Animation as a Complex Synthesis Task*, BASc Thesis, School of Engineering Science, Simon Fraser University, Burnaby, BC, Canada.

Koestler, A., (1964). *The Act of Creation*.

Owen, R., (1993). "A System for Analyzing Time Based Data", University of Toronto. Report for Ontario Telepresence Project and the Institute for Robotics and Intelligent Systems.

Shneiderman, B. (1992), *Designing the user interface: Strategies for effective human-machine interaction*. 2nd Edition. Baltimore, MD: John Hopkins University Press.

Simon, H.A. (1969), *The Sciences of the Artificial*, MIT Press, Cambridge, MA.

REAL-TIME INTERACTIVE PERFORMANCE WITH COMPUTER GRAPHICS AND COMPUTER MUSIC

Yoichi Nagashima

Laboratories of Image Information Science and Technology
1-1-8, ShinsenriNishi-machi, Toyonaka, Osaka, 565-0083 JAPAN

Abstract: This is a report of some applications of human-machine systems and human-computer interaction about experimental multimedia performing arts. The human performer and the computer systems perform computer graphics and computer music interactively in real-time. As the technical point of view, this paper is intended as an investigation of some special approaches: (1) real-time processing and communication system for performing arts, (2) original sensors and pattern detecting techniques, (3) distributed system using many computers for convenience to compose, arrange and perform. *Copyright © 1998 IFAC*

Keywords: Computers, Human-machine interface, Interactive, Multimedia, Pattern recognition, Performance, Real-time, Sensors, Systems design

1. INTRODUCTION

The research called PEGASUS project (Performing Environment of Granulation, Automata, Succession, and Unified-Synchronism) had produced many experimental systems of real-time performance with many original sensors, and have composed and performed many experimental works at concerts and festivals. The new step of this project is aimed "multimedia interactive art" by the collaboration with CG artists, dancers and poets.

2. THE CONCEPT OF MULTIMEDIA

The best account for the concept of multimedia art can be found in Fig.1 that is the conceptual system block diagram of this project. In this system there are some types of agents in a computational environment like UNIX X-Windows system. Each agent is produced as a client process. Input images and input sounds are sampled in real-time via cameras and microphones. The graphic outputs are connected via projectors or display monitors. The output sound consists of that from direct DSP computed by computer and from MIDI-controlled synthesizers. The 'control' agent exists in the center of the system. This agent manages 'control messages' and sends them to the sound agency and the graphics agency in time layer, spatial layer and structural layer. The messages input to these agencies may be divided into four types: (1) traditional 'scenario' of artists: time scheduling, spatial mapping, characteristics of motion, etc, (2) sensor information of the performance: sensor fusion of event triggering and continuous parameters, (3) real-time sampled sound: as a material for granular synthesis and granular sampling, (4) real-time recorded images: as a material to generate CG -- pixel, texture, motion, etc. The 'sound agency' section organizes the 'world model' of sound. It contains many agents, for example, a database about musical theory and music psychology, sound synthesis level generator, note level generator,

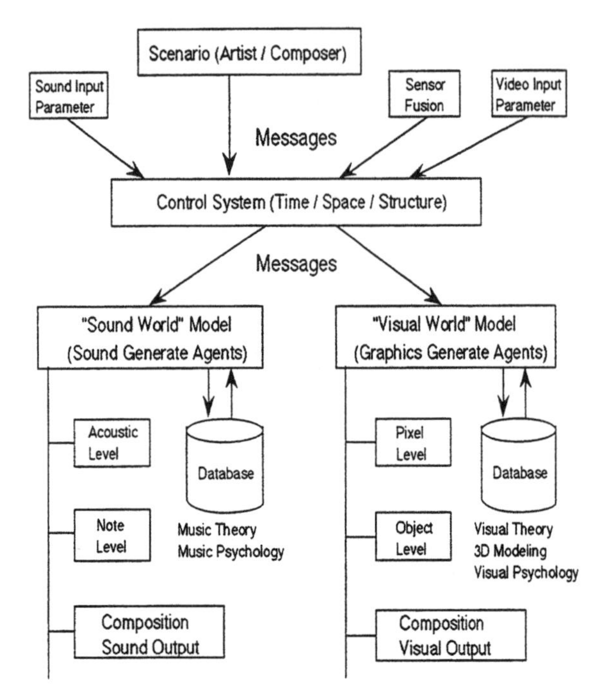

Fig.1. Conceptual system block diagram of the PEGASUS project, multimedia interactive art.

phrase level generator, and sound distributing generators. These agents receive the control messages and send or receive information to and from each other as interaction. There are also many types of agents in the 'graphics agency' section: including pixel level, object level, modeling level, and so on (Nagashima, 1995a).

3. EXPERIMENTAL WORKS AND SYSTEMS

In this part there are some reports and discussions about performances as the application of multimedia interactive arts which were realized at concerts and events in Osaka, Kobe, Kyoto, Tokyo and Seoul(Korea) during 1996 to 1997. These performances were the "live visual arts" for the graphic artist and were the "live musical arts" for the music composer, but these could not be produced by each artist only. At the presentation, some demonstrations of videos will be included.

3.1 "Asian Edge"

The multimedia interactive art called "Asian Edge" was composed in 1996 and performed at the JACOM (Japan Computer Music Association) concert in Kobe in July, 1996. This work was inspired with Asian feelings, Asian culture and the integration of dance, graphics and performance. At this performance, there requested these equipments: SGI Indy<1> (for computer sound with MIDI) running original software produced by the composer to generate many sounds in UNIX environment, SGI Indy<2> (for computer graphics with MIDI) running original software to generate real-time 3D graphics controlled by sensors, Macintosh<1> running MAX patch to manage sensor fusion information, Macintosh<2> running MAX patch to control synthesizers, sampler S2800i, synthesizer K4r<1> and K4r<2>, effector SE50, microphone for performer, special Lute (original sensor) produced by the composer, MIBURI and SNAKEMAN (original special sensors produced by composer), original MIDI merger, original MIDI filters, original MIDI video switcher, four video players for BGV controlled via MIDI, four video cameras controlled via MIDI, three video projectors to display graphics and 4 channels stereo PA system (see Fig.2).

All materials of computer sound part are arranged from recorded Asian instruments, and signal processed with original softwares written by C in SGI IRIX(Unix), and multiple player software is also produced. Each sound file is not played by fixed sequence, but triggered by MIDI real-time control by sensors. Indy workstation for sound part and graphics part are controlled with MIDI. In addition, back ground videos and live videos of performance with four video cameras are real-time switched by MIDI-video switcher and three video projectors with the performance. Performer also speaks and sings

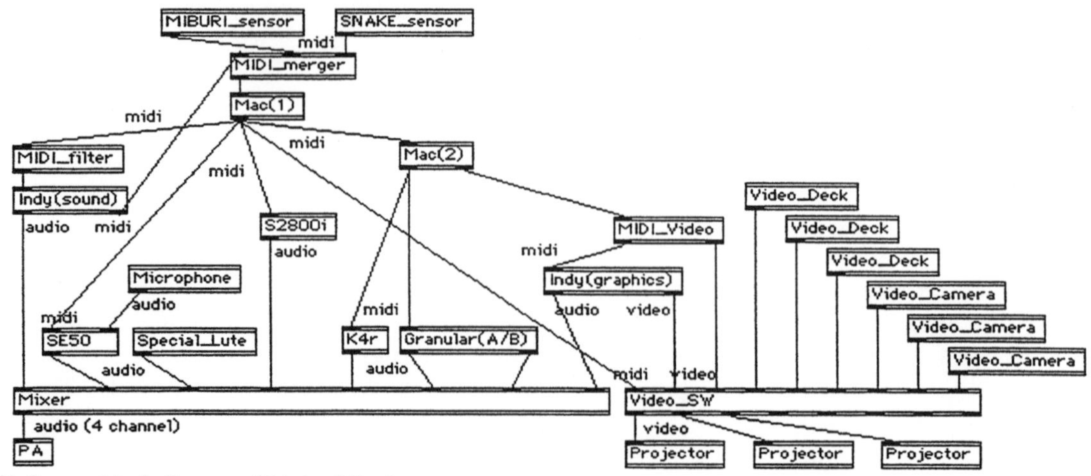

Fig.2 System block diagram of "Asian Edge".

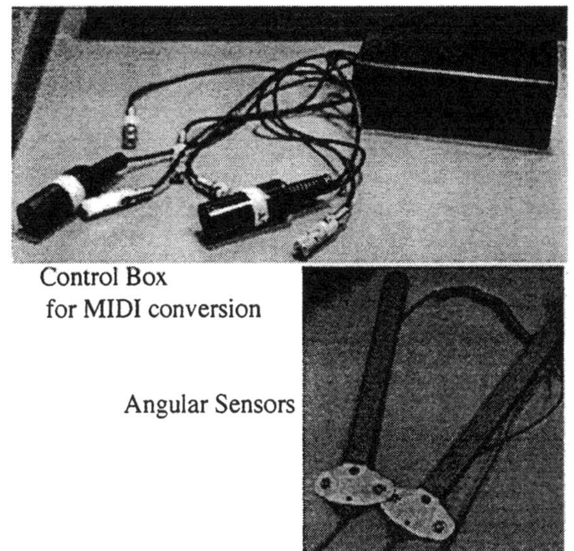

Fig.3 "MIBURI" sensor.

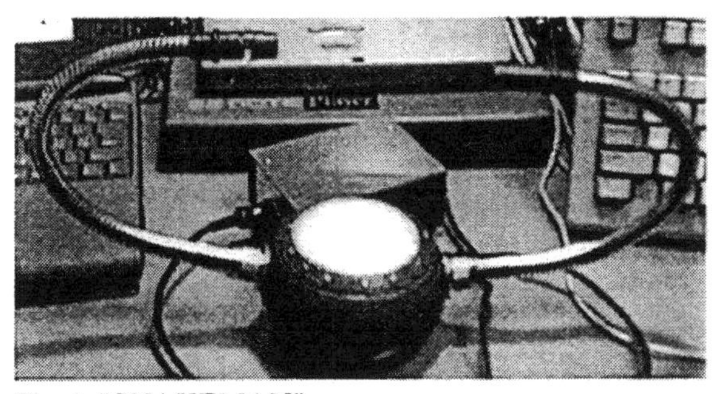

Fig.4 "SNAKEMAN" sensor.

some phrases, this live voice and pre-sampled and processed sounds are modified by effector or generated by sampler. MIBURI sensor detects the motion and the body-language of performer, SNAKEMAN sensor generates trigger actions, and special Lute is performed as traditional musical instrument.

The MIBURI sensor has 6 angular sensors for human joints. Normally these sensors detect both wrists, both elbows and both shoulders. The angular sensor module outputs voltages corresponding with its current angle. The control box supplies power for sensors and converts these inputs to MIDI messages

Fig.5 Performance of "Asian Edge".

with A/D converter and microprocessing unit. The MIDI message which means "motion language of the performer" is received by MAX software of Macintosh computer, and pattern detection and recognition algorithm work to generate live computer music and live computer graphics (Fig.3).

The SNAKEMAN sensor detects the cutting speed of infrared beam. This sensor may detect from 500msec to 30sec range with logarithmic compression for 7bits MIDI data representation. The MIDI message which means "scene change" is received by MAX software of Macintosh computer, and the control algorithm works to change the scene to generate live computer music and live computer graphics (Fig.4, Fig.5).

3.2 "Johnny"

The multimedia interactive performance called "Johnny" was composed in 1996 and performed at the Japan - Germany Media Art Festival in Kyoto in October, 1996. This work was inspired with "Battle and Session" of music, graphics and dance. On the

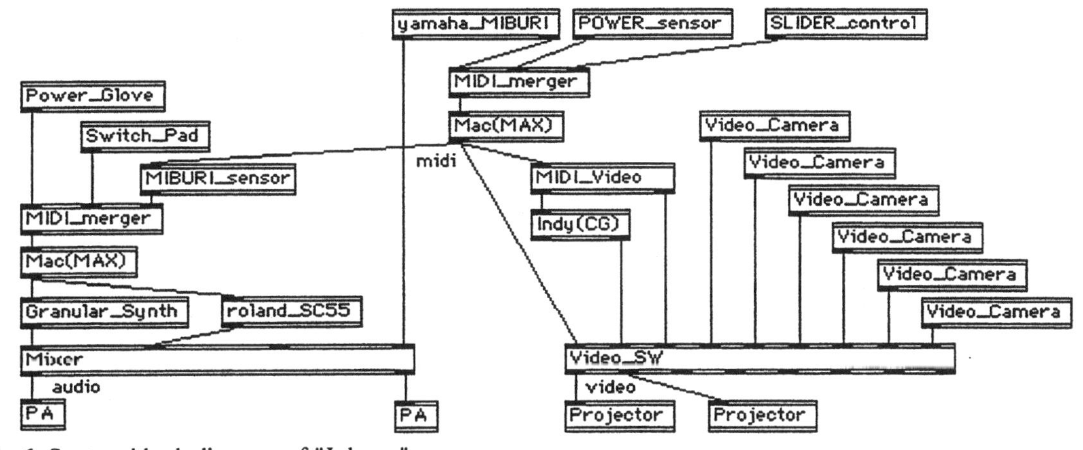

Fig.6 System block diagram of "Johnny".

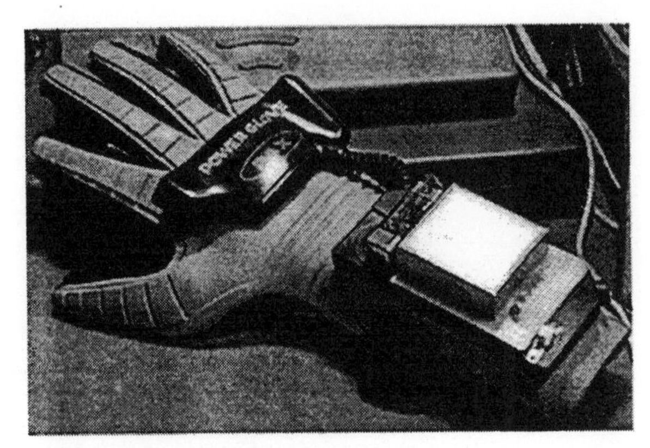

Fig.7 "MIDIGLOVE" sensor.

stage there were 3 performers: (1) dancer who wears MIBURI sensor to detect body language for conducting and solo performance, (2) CG artist who controls the video mixer to generate live video and (3) musical performer who has MIDIGLOVE sensor and plays MIDI control pads to arrange the background music in real-time. Fig.6 is the system block diagram of "Johnny".

The MIDIGLOVE sensor is developed by arrangements with PowerGlove which is a controller of Nintendo Entertainment System by the composer. There is a small microprocessing unit on the glove to detect the status of fingers, and the 4-bits decoded information is transmitted via 250MHz wireless. The MIDI message from the receiver module is received by MAX software of Macintosh computer, and pattern detection and recognition algorithm work to generate live computer music and live computer graphics (see Fig.7).

3.3 "Brikish Heart Rock"

The interactive computer music called "Brikish Heart Rock" was composed in 1997 and performed at the concert in Kobe in October, 1997. This is live computer music for two performers: a flute player and a sensors' player. The flute part is independent of the computer system electrically, and the player can play any improvisational phrases with the score. The sensor player performs two original sensors: TOUCH sensors pad and the "MiniBioMuse" sensor. In this piece, the MIDI outputs of "MiniBioMuse" was not used, the audio output was only used.

The main concept of this work is the "session" mind of rock and jazz musicians. Performers may play any phrases, sounds and rhythms with the real-time generated BGM part from MAX patches. The BGM band starts as simple 8-beats rock patterns, and grows with 16-beats or euro-beat and randomly insert some faking rhythms. Performers must pass these fake rhythms with playing no gestures, so this rule is one kind of a game. The "MiniBioMuse" player may move both arms and hands as an improvisational performance like a dance. The analog output of the sensor is the noise signal of muscles, so the sounds are real-time processed with the effector. The duration of this piece is not fixed because the two performers and the operator of the computers may continue any scenes, and may repeat any BREAK patterns with their improvisation.

The TOUCH sensor is developed by the composer. This sensor has 5 electrostatic sensing pads for human electricity. This sensor is also an installation of human interface, so it has 80 LEDs to demonstrate 16 * 5 display messages of touch events. The microprocessing unit on the board detects the status of touch performance, and the sensing information is transmitted via MIDI. The MIDI message from the sensor is received by MAX software of Macintosh computer, and pattern detection to generate live computer music (see Fig.8).

The "MiniBioMuse" sensor is also developed by the composer as a part of composition. This sensor has 3 electrode bands: two are sensing pads and one is common ground to reduce noises. This sensor has a microprocessing unit in it, and detects electric pulses of muscles, and generates not only direct analog signals but also converted MIDI information (see Fig.9).

Fig.8 Performance of "TOUCH" sensor.

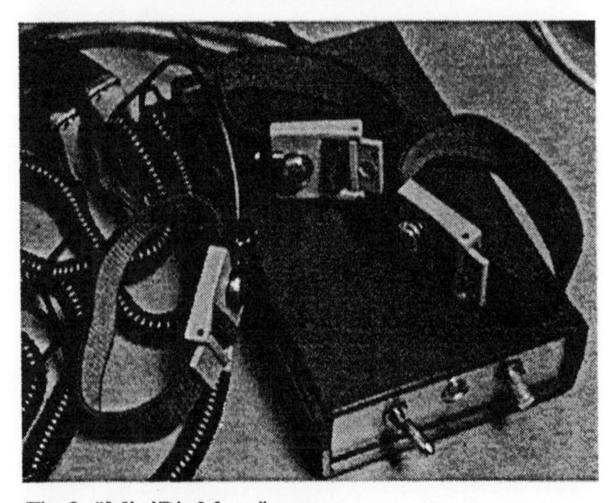

Fig.9 "MiniBioMuse" sensor.

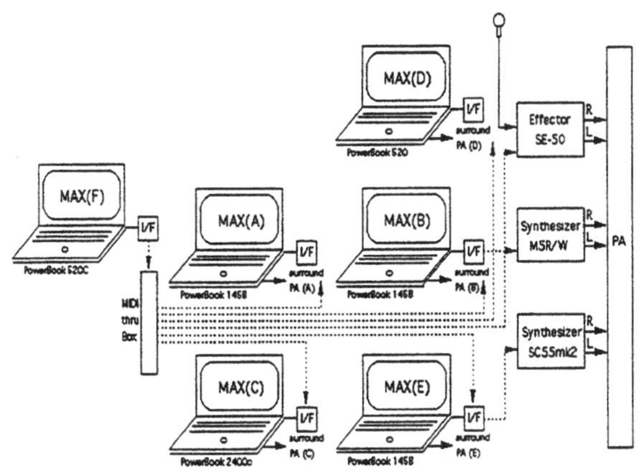

Fig.10 System block diagram of "The Day is Done".

3.4 "The Day is Done"

The interactive computer music called "The Day is Done" was composed in 1997 and performed at the concert in Kobe in October, 1997. This is live computer music which focuses the "voice" as a material of music. This piece contains many types of "voices": human voices, signal processed natural voices, and computer generated voices of Macintosh. The lyric "The Day is Done" written by W.Longfellow is only used for its fantastic sounds and images. The other concept of this piece is the keyword "environment". Many natural sounds are used for the environmental atmosphere: stream, waterfall, rain, seashore and waves. These images are inspired by the Joumon-Sugi (7000 years old Japanese cedar) and the Yakushima island in Japan.

This piece contains two musical elements. The back-ground part is pre-processed and fixed to DAT or CD. The signal processing compositions are: SGI Indy workstation standard "soundeditor", original tools for signal processing, and original multi-playback tool and UNIX shell scripts. The real time part requires 6 Macintosh computers (running MAX with "speak" object produced by Mr. Ichi in Japan), and two performers: vocal (mezzo soprano) and "computer percussionist" who plays 6 Macintosh click-buttons in real-time. The two performers use stop-watches to adjust their timings, but they can perform with free feeling of improvisation.

3.5 "Atom Hard Mothers"

The multimedia interactive art called "Atom Hard Mothers" was composed in 1997 and performed at the concert in Kobe in October, 1997. This is live computer music with live computer graphics and live video images which focuses the concept of "multimedia game". The two performers play each special sensor as instruments, and the output information of these sensors controls musical and graphical objects in real-time.

The music part of this piece contains three types. The background sound part is pre-processed with SGI Indy and fixed to DAT or CD, and this part is generated only from voices of a 'bell-ring' cricket. The live-background music part is real-time composed and played with MAX algorithms, and this part may be played differently at each performance. The performers' parts are generated with their sensors: special HARP sensor, SNAKE sensor and MIBURI sensor (see Fig.11).

The graphic part of this piece contains three types, and these sources are real-time switched with performances and projected to the screen: pre-processed background image video (Hi-8), real-time generated computer graphics and live images of performers on the stage. The computer sounds and computer graphics are brought up by the two performers as "mothers", and grows up with improvisations. This piece is an interactive art version of the TAMAGOTTI game in a sense.

The HARP sensor is developed by the composer as a part of composition. This sensor has 3 horizontal beams and 13 vertical beams of LED-photoTR pairs with optical fiber sensor modules. The control system of it detects these 16 light beams crossing events, and generates MIDI messages of the status. The MIDI message from the sensor is received by

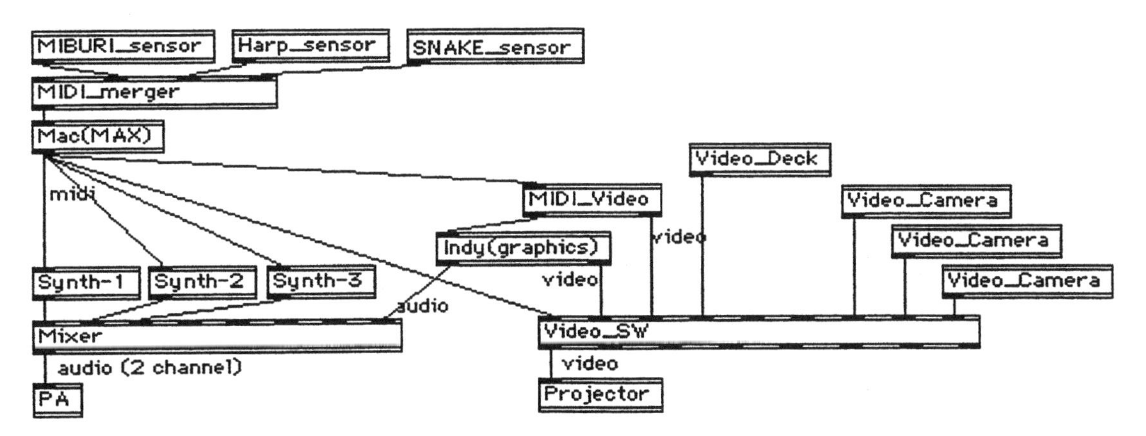

Fig.11 System block diagram of "Atom Hard Mothers".

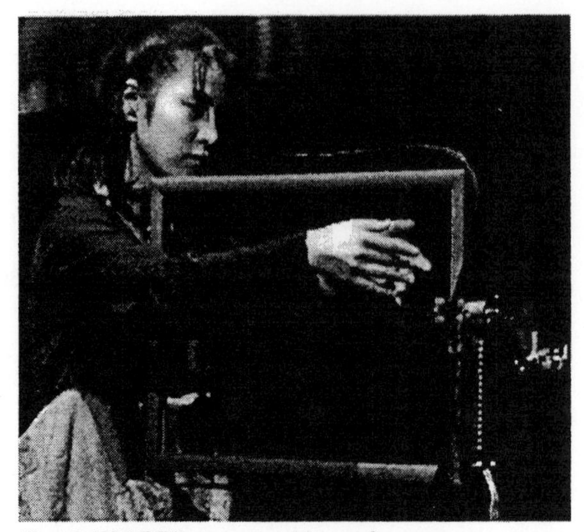

Fig.12 "HARP" sensor and its performance.

MAX software of Macintosh computer, and generates live computer music and live computer graphics (see Fig.12).

4. DISCUSSION

For live performance like these multimedia arts, it is most important to manage "events" with "timing" and "actions" in algorithmic environment. In the algorithmic composition, there are many structural components to perform: for example, a group of sequential events, triggered events with sensors, masking condition for other events, waiting sequences for triggers, and so on. It is difficult to handle this complex status with traditional environments for composition and performance like commercial sequencers or "timeline" object of MAX. This drives us to the question how to deal with many kinds of components and how to compose and perform easily. This project is developing the compositional environment with multi-media, and reports a new idea to solve this problem.

There have been developed some types of modules to construct interactive performance of algorithmic composition. These modules contain many inputs and outputs to connect with other modules like the patches of MAX. Composers can use the modules as universal parts of the composition in this environment, and it is easy to construct a big structure with the concept of hierarchy. Some of the input nodes are: "start trigger", "sensor input", "sub sequence finish", "sensor masking time", "stop trigger", "stop masking time", "scene id", "initialize", and so on. The output nodes are: "trigger to next", "start sub-sequence", "stop sub-sequence", "status ID", "scene ID", and so on. These modules can be used hierarchically and recurrently, and it is easy to see the construction of the relation of compositional information (see Fig.13). One of the advantages of this idea is an ability of debugs for the composition. Each scene described with the proposed module has much independence and this enables the "partial rehearsal" in composition or performance (Nagashima, et al., 1995b).

5.CONCLUSION

It is important for research of human-machine systems and human-computer interaction to experiment as applications. The live performance like multimedia performing art is suitable for its real-time condition. Some new projects for new works are running.

REFERENCES

Nagashima, Y. (1995a). Multimedia Interactive Art: System Design and Artistic Concept of Real Time Performance with Computer Graphics and Computer Music. *Proceedings of Sixth International Conference on Human Computer Interaction,* **volume 1**, pp.89-94. Elsevier, Yokohama Japan.

Nagashima, Y., Katayose, H., Inokuchi, S. (1995b). A Compositional Environment with Interaction and Intersection between Musical Model and Graphical Model --- "Listen to the Graphics, Watch the Music" ---. *Proceedings of 1995 International Computer Music Conference*, pp.369-370. International Computer Music Association, Banff Canada.

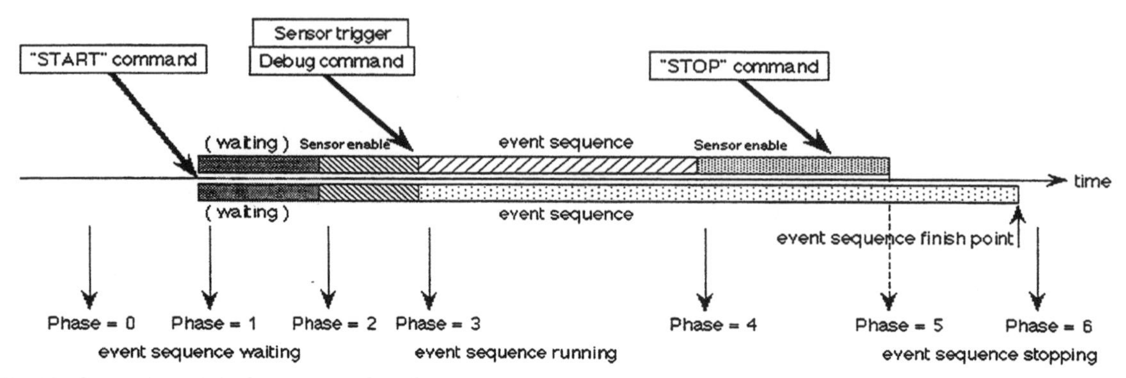

Fig.13. Control module for enevts of performance.

ASSESSING HUMAN-CENTRED DESIGN PROCESSES IN PRODUCT DEVELOPMENT BY USING THE INUSE MATURITY MODEL

Kari Kuutti[1], Timo Jokela[2], Marko Nieminen[3], Pirkko Jokela[4]

[1] University of Oulu, Department of Information Processing Science
[2]Nokia Mobile Phones, Oulu
[3]Helsinki University of Technology, TAI Research Centre
[4]TeamWARE, Helsinki
FINLAND

Abstract: The paper discusses about the evaluation of the quality of human-centred design processes in product development. It introduces the concept of usability maturity model to be used as an instrument in evaluating human-centred processes, and presents experiences from two assessments, where a novel usability maturity model developed by an EU research project INUSE was used. *Copyright © 1998 IFAC*

Keywords: Evaluation, Human-centered design, Process models, Products industry, Software engineering

1. INTRODUCTION

Besides the functionality and technical quality, the success of information technology products is also dependent on their usability. This aspect is becoming more and more important, when the pressure of increasing competition is narrowing the technical differences between products. The increasing importance of usability aspects means increasing importance of the human-centred activities within the software design process, and thus also a need to assess their quality and effectiveness. So far, however, the methods and tools for such assessment have been rare and their adequacy for the purpose not much tested. This paper discusses about the experiences of using one such tool — the INUSE Usability Maturity Model — in practical assessments.

1.1 Reference Process Definitions - what is a maturity model?

There are two perspectives to study quality: one related to the end product itself and another related to the process, where the end product is produced. The first one is relatively straightforward to apply: does a system or product possess quality properties? There exist international standards such as ISO/IEC 9126 that define dimensions of product quality. With respect to one such dimension, usability, the standard ISO/IEC 9241 defines what desired usability properties are, and testing the usability of systems and products against either standards or otherwise set norms has become a widely used practice.

The product-oriented perspective to quality has one defency, however: it does not tell how a good result can be achieved, and thus it is not very useful in planning how a production work should be done or improved. Therefore, a complementing process-oriented perspective is necessary. A process-oriented perspective assumes that the ability of a process to produce quality is related to certain persistent characteristics of that process. A process-oriented perspective is thus useful e.g. in attempts to improve a production process, or in evaluating the capability of a contractor to fulfill the quality demands. The process-oriented perspective to quality has gained much popularity for example in the form of ISO 9000 standards.

Examples of evaluation-oriented process models that can be applied in software development organisations are CMM (Paulk et al 1995), Bootstrap (Kuvaja et al. 1994), existing ISO/IEC standard 12207 (Software Life Cycle Processes), and the SPICE model on Software Process Assessment (El Emam et al. 1998), that will in a near future be accepted as a new standard ISO/IEC 15504. Further, under preparation is a draft for ISO DIS 13407 -- Human-centred Design Processes for Interactive Systems (ISO 1996b). These can be characterised as reference process definitions (Nyström 1997). ISO (1996, p.1) characterises a reference model as follows: "A reference model defines, at a high level, the fundamental objectives that are essential to good software engineering. The high-level objectives describe what is to be achieved, not how to achieve them.". Reference processes can be used for various purposes. Curtis & al. (1992) present that goals and objectives for software process modeling are: (1) facilitate human understanding and communication ("represent process in form understandable bu humans"), (2) support process improvement ("identify all the necessary components of a high-yield software development process"), (3) support process management ("develop a process-specific software process to accomodate the attributes of a particular project such as its product, or organisational environment"), (4) automated guidance in performing process ("define an effective software development environment"), (5) automated execution support ("automate portions of the process").

The reference model that was used in trial assessments was a Usability Maturity Model (UMM) developed by the INUSE project (EC IE 2016) (INUSE 1998), a model of the processes which make a system human-centred. The UMM is an ISO 13407-based, ISO 15504-compliant process model.

Other, partially overlapping reasons with the ones mentioned above for applying reference processes in companies can be

- getting a maturity level ranking
- benchmarking own activities against competitors or other "state-of-the-art" organisations
- improving existing practices in certain important activity areas
- goal setting for key business activities in the future
- characterising and structuring certain new approaches in development environment.

Obviously, many of the reasons mentioned above may exist simultaneously when a company decides to apply a reference process model. The reasons for applying the Usability Maturity Model in the case organisations of this study were in getting a maturity level ranking, goal setting for key business activities in the future, and improving existing practices in certain activity areas. As the Usability Maturity Model itself was in a developing stage, its testing was of central importance in these first evaluations.

Reference processes include lists of important activities that should be applied in general in an organisation that aims at operating in a qualitative way (see figure 1 "The structure of the INUSE model used in the first assessment"). In addition to list of important activities only, reference processes may include a concept of maturity, or *capability*. According to ISO (1996, p.27) "the scale represents increasing capability of the performed process from performance that is not capable of fulfilling its goals through to performance that is capable of meeting its goals and sustaining continuous process improvement". Capability attempts to approach the level of quality at which the organisation acts.

1.2 Assessment of Company-specific Practices

In order to transform the "high level objectives" (ISO 1996) of the reference processes to practical activities, practitioners must establish lower level practices. This stresses for more detailed information about the organisation under evaluation. One needs to be able to map the existing practices within the organisation with the high-level objectives of the model. This may require pre-studies about the current state in the organisation. During the pre-studies, e.g., company-specific concepts will be translated into model-specific concepts, or the quality documentation may be consulted. At the same time, persons that need to be interviewed are selected. All this information can be then used while planning and executing the assessment sessions.

The capability level (or "maturity level") of an organisation is assessed in *assessment sessions* (also "*appraisal*"; Masters & Bothwell 1995). Basically, assessments are sessions in which the staff of the target organisation are interviewed. An appraisal team with a team leader will conduct these sessions asking specific questions about specific processes and base practices within the organisation. The results of the interviews are then evaluated and ranked by the appraisal team.

During the staff interviews, lots of practical information about the current work practices and operation structures will become visible. The results of the interviews are then used to evaluate the missing parts when compared to the reference process definition. Thus, in addition to capability level assessment only, the information gathered in interviews can provide good starting points for later organisational development. Thus, remarks that have been presented in the interviews and appraisal team discussions should be documented for later utilisation.

1.3 Background of the assessment experiments

The Department of Information Processing Science at the University of Oulu has been involved in the development of process maturity models already long time. The department has been one of the central participants in the development of BOOTSTRAP, the major European software process maturity model, and it has also been actively involved in the SPICE project work and the preparation of the SPICE/ISO 15504 standard. When it was found in 1996 that several Finnish companies were interested in assessing the maturity of their human-centered product development processes, a joint project (named KAYPRO) was established between the department and companies in the beginning of 1997. The goal of the KAYPRO project was first to develope, experiment and refine a assessment model that can be used to assess the maturity of human-centred design processes in organizations, and then continue to develop another model how organizations can use the results of assessments and stepwise improve further these processes.

Introducing usability and human-centered design into organizations in industry has proved to be a challenge. While smooth and disciplined change is desired, introduction of human-centred design often - on the contrary - means conflicts and minsunderstandings in an organisation. An assessment tool which provides objective data can be seen as one means of communicating the status of the organisation in human-centred design. Another - and perhaps even a more important - viewpoint is that an assessment model could be used as a tool for planning a step wise improvement plan, to goal being transforming an organisation towards human-centeredness through controlled and smooth changes.

During the first months of the project work it was found that there already existed an European project (IE 2016 "INUSE", funded by European Union and coordinated by Lloyd's Register in London) which had partially similar goals with KAYPRO. A contact was established between the projects, and it was concluded that a cooperation would benefit both sides. A decision was made to use the usability maturity model developed in the INUSE project in the assessments planned for KAYPRO.

The INUSE usability maturity model is noteworthy because it is the first such model developed in international cooperation (large companies like IBM and Philips have their own in-house developed models) and it is also the first model aiming for compatibility with two drafts of relevant standards, ISO 15504 ("Software Process Assessment") and ISO 13407 ("Human centred design processes for interactive systems"). Although the INUSE project itself has also done experimental evaluations using the model, this far the KAYPRO assessments are the most complete ones done with the final versions of the model.

Both of the conducted assessment were experimental in the sense that the aim of the assessment was not only to find out the level of maturity of the assessed part of the organization, but also to test the adequacy of the used model. In both cases the focus of evaluation was at the level of development projects, and the corresponding units or organizations were not evaluated as wholes. It is also worth of noticing that both of the companies involved have already several years history in promoting usability and human-centred design.

2. THE FIRST ASSESSMENT

The first assessment took place at Nokia Mobile Phones R & D unit in Oulu, Finland, where one of the development projects was assessed Nokia Mobile Phones is an international producer of cellular phones which are heavily software-dependent electronics products in which user interface has an important role.

The model used in the assessment was INUSE working draft model dated 30/09/96. This was based on the SPICE/ISO 15504 process model in the following way. The SPICE model has two dimensions: the process model and the capability scale.

The SPICE process model is constructed hierarchically: it divides the development process first in five Process Categories. Each Category consists of several Processes, and thus the whole model contains altogether 30 processes. Each of the processes contains several Base Practices, which are finally practical descriptions what organization members do in their work. In assessments the way how the work is done in an organization is compared with the base practices of each of the processes to find out how well the scope of the process model is covered. The capability scale refines the process evaluation. It consists of five levels that measure how well existing processes are planned and managed.

In this draft version of the INUSE model the human-centredness was integrated into the SPICE structure by adding specific, human-centred base practices under the SPICE processes The purpose of the Figure 1 is to illustrate of the resulting structure of the model. From the whole SPICE model only the processes belonging to the Engineering Process Category (one of the 5 process categories) are listed (ENG.1 through ENG.6). The process ENG.1 (Develop System Requirements and Design) is further refined into base practices. There are two sets of base practices: first the base practices belonging to the original SPICE model; second the base practices for user-centered design, as additions brought by the INUSE model. In this draft version of the INUSE model, the different base practices for usability and human-centred design were distributed under the rest of 29 processes in a similar manner.

SPICE Categories
Customer Supplier Process Category
Engineering Process Category
ENG.1 Develop system requirements and design
SPICE Base Practices:
- Identify system requirements
- Analyze system requirements
- Describe system architecture
- Allocate requirements
- Determine release strategy
- Communicate system requirements
Additional INUSE Usability Base Practices:
- Describe the context of use
- Task analysis
- Allocate functions
- Generate the user and organizational requirements
- Define usability parameters
ENG.2 Develop software requirements
ENG.3 Develop software design
ENG.4 Implement software design
ENG.5 Integrate and test software
ENG.6 Integrate and test system
ENG.7 Maintain system and software
Support Process Category
Management Process Category
Organization Process Category

Figure 1. The structure of the INUSE model used in the first assessment

Only the usability/human-centred base practices of this model were used in the evaluation. An ongoing user interface development project was as the target of the assessment. The assessment took a full week and it was conducted in a manner normal to SPICE evaluations. On Monday there was a preparation meeting for the assessment team, which due to the experimental nature of the assessment was quite large, consisting of 6-8 persons, consisting of both process improvement and usability experts. On Monday there was also a briefing meeting for those to be interviewed during the week. From Tuesday to Thursday altogether 8 persons working in the project were interviewed according to the model, each session lasting 2-3 hours. Each evening the team had a meeting to collect the experiences. On Friday morning the results of the assessment were collected into a graphical form, and in the Friday afternoon the results were reported back to the project.

The experiences from the first assessment were mixed. Although from the beginning of interviews it was apparent that the model was working and producing meaningful results, those members of the assessment team without a SPICE background had difficulties in maintaining an overall view. With the help of the lead assessor who had a long experience from SPICE assessments the team was anyhow able to construct a coherent view how the human-centred issues were taken into account in the evaluated project. A major problem, however, was found to be the communicability of these results back to the project team. Although the structure of the model was explained in the briefing session and again in the beginning of the feedback session, the members of the evaluated project remained confused and largely unable to find out how the results were related to their work. This was judged to be due to the complexity of the INUSE model structure, and distribution of the assessed INUSE practices over it. As a result, a restructing of the model was proposed to the INUSE project, to make the model more understandable.

3. THE SECOND ASSESSMENT

The second assessment took place at TeamWARE Tampere unit, where two development projects were evaluated. TeamWARE is an international developer of office productivity software. The way how assessment was conducted was in principle similar than earlier, and altogether 11 persons were interviewed during the week.

The model used in the second assessment was INUSE working draft model dated 24/10/97. The 'human centred design' content of this model does not differ much from the earlier model, but the model is organized in a different way. Instead of distributing the human centred base practices into the SPICE processes, a new process category called 'Human-Centred Design', consisting of seven human centred processes (HCD.1 to HCD.7), was added to the SPICE model. The new structure of the model is illustrated in Figure 2, where base practices belonging to one of the processes (HCD.4: Understand and Specify the Context of Use) are listed.

SPICE Categories

Customer Supplier Process Category
Engineering Process Category
Support Process Category
Management Process Category
Organization Process Category
Human-Centred Process Category (INUSE addition)

HCD.1 Ensure HCD content in system strategy
HCD.2 Plan the human-centred design process
HCD.3 Specify the user and organizational requirements
HCD.4 Understand and specify the context of use

- *clarify system goals*
- *identify user's tasks*
- *identify user's attributes*
- *identify organizational environment*
- *identify technical environment*
- *identify physical environment*

HCD.5 Produce design solutions
HCD. 6 Evaluate design against requirements
HCD.7 Facilitate the human-system implementation

Figure 2. The structure of the INUSE model used in the second assessment

The central reference for this structure has been to use the process structure defined in the draft of the standard ISO 13407 as the core organizing principle (the processes HCD.2-6 are taken directly from ISO 13407).

The results of the second evaluation were more clearly structured than those in the first evaluation. The model was found to be working and producing meaningful and useful results, and the new structure of the model clearly helped the team in managing the assessment process and collecting the results coherently. The greatest difference, however, was during the communication: the restructuring of the model was a lot easier to communicate to the members of the projects that were assessed. The time used to introduce the model in connection with the second assessment was about the same than it was in the first experiment (one hour briefing before the assessment, another in the beginning of the feedback experience session). On the contrary compared with the earlier experience, now this time it was enough so that the members of the evaluated projects were able to understand and to link the results to their daily work practices. It was to start a meaningful and productive discussion about the results and potential improvments the processes in the feedback situation, and this discussion relatively rapidly converged into suggestions for practical improvements.

4. CONCLUSIONS

There are several aspects in the current INUSE model for assessing the maturity of usability design processes that still need further development and refinement, but our experience is that already now it is a working and useful tool for identification of possibilities for organizational improvement.

While the INUSE model seems to work rather well when assessing the capability of human centered processes, it is evident that assessing the capability of processes is not enough. At least in case of organisations which are less matured - where usability is only recently introduced or has only a limited impact - there are apparently additional dimensions to be considered for assessing. Human-centeredness is not only a process issue. It is also an awareness, attitude, skills, empowerement and resources issue. For example: an organisation may have some processes of high capability but the usability driven results may be diminished in decision making - due to the inadequate awareness of human-centeredness of the decision makers.

Actually, the INUSE proposes an additional model - an attitude scale - to recover these other dimensions (INUSE 1998b). In the KAYPRO experiments some limited experiments were done using this scale as well. However, the finding was that the scale is not well balanced: it combines too wide a set of different viewpoints: processes, attitude, training, etc. Anyway, it seems clear that it is a most relevant idea to develop assessment of other dimensions as well. The scale developed in the INUSE project offers a good starting point for such development. At the moment, the KAYPRO project is working with an idea to have three dimensions of human-centeredness: organisational awareness, human-centered infra-structure, and processes. The maturity should be assessed from all these viewpoints. And on the other hand, when introducing human-centered design, all these aspects should be developed in an balanced way.

REFERENCES

Curtis, B., Kellner, M. & Over, J. (1992): Process modeling. In *Communications of the ACM*, September 1992. Vol. 35, No. 9. Pp. 75-90.

El Emam, K., Drouin. J-N. & Melo, W. (eds.) (1998) *SPICE : the theory and practice of software process improvement and capability determination.* Los Alamitos (Calif.) : IEEE Computer Society

INUSE (1998) *Usability Maturity Model: Processes*, Lloyd's Register project IE2016 INUSE Deliverable D5.1.4p.

INUSE (1998b) *Usability Maturity Model: Human-Centredness Scale*, Lloyd's Register project IE2016 INUSE Deliverable D5.1.4s.

ISO (1996): *Software Process Assessment. Part 2: A Reference Model for Processes and Process Capability.* 26-July-1996. ISO/IEC/JTC/SC7/WG10/N102. Working draft for information purposes only.

ISO (1996b) DIS 13407, *Human-centred design processes for interactive systems.* ISO/TCI59/SC4/N376. Working draft for information purposes only.

Kuvaja, P. (et al.) (1994) *Software Process Assessment and Improvement. The Bootstrap Approach.* Oxford: Blackwell.

Masters, S. & Bothwell, C. (1995): *CMM Appraisal Framework*, version 1.0. Technical Report CMU/SEI-95-TR-001. ESC-TR-95-001. February 1995. Software Engineering Institute, Carnegie Mellon University. Pittsburg, Pennsylvania, USA.

Nyström, T. (1997): *Comparison of Software Reference Processes Definitions.* Master's thesis, Helsinki University of Technology, Department of Information Processing Science.

Paulk, M.C. [et al.] (1995) *The capability maturity model : guidelines for improving the software process* / Carnegie Mellon University, Software Engineering Institute. Reading (MA) : Addison-Wesley.

SHOPFLOOR SYSTEMS BASED ON HUMAN SKILL AND EXPERIENCE

E. A. Hartmann, A. Westerwick

Department of Computer Science in Mechanical Engineering and Centre for Research in Higher Education (HDZ/IMA), Aachen University of Technology

Abstract: New organisational concepts in manufacturing - e.g. groupwork - imply changing roles and tasks for shopfloor employees. The importance of Human Oriented Design increases as human experience and skill becomes the core focus of group-oriented work organisation. The concept of socio-technical systems can give a theoretical framework for the derivation of design guidelines within the context of groupwork-oriented technology design. Three examples of technological systems for shopfloor applications - a CNC controller, a shopfloor information system and a work planning system - are described which were designed to meet these guidelines. Some methodological remarks on co-operative software design are added. *Copyright© 1998 IFAC*

Keywords: Organisational factors, Human Oriented Design, CNC, Work Planning Systems

1 INTRODUCTION

In the metal processing industries – as well as in other industrial sectors – a new paradigm of organisational design and human resources management has emerged in recent years. Quite different from the vision of „unmanned factories" predominant in the 1980s, the new approaches encompass semi-autonomous work groups, project-oriented co-operation across different departments and hierarchy levels within the company and a strong emphasis on learning and development with respect to humans as well as organisations (Hartmann, 1997).

Given this situation, the concept of socio-technical systems design appears to be of an even greater practical relevance. Although this theory has quite a long history (e.g. Emery, 1959), *practical* examples of *deliberate* design of *technology* according to this perspective are still rather rare. In this paper, tree such systems – a CNC controller, a manufacturing information system, and a work planning system – are described with respect to their relevance for the theory of socio-technical systems as well as for practical needs in industry.

2 DESIGNING SOCIO-TECHNICAL SYSTEMS

A socio-technical system may be conceived of as consisting of three non-trivially interlinked (sub)-systems: individual persons, organisations, and technology (Strina & Hartmann, 1992; Fig. 1). Whenever one of the subsystems is changed, the others will be affected as well in a usually complex fashion. Thus, design has to be a deliberate joint design of the socio-technical system as a whole.

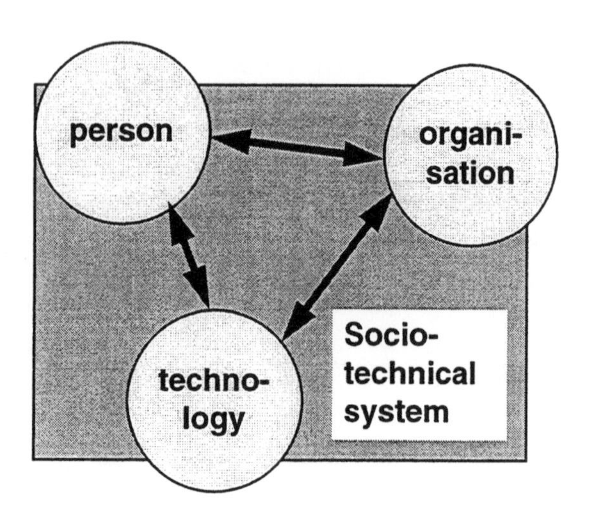

Fig. 1: Socio-technical system

From the theory of socio-technical systems and the vision of learning organisations we may derive some general design criteria. These criteria relate to the relations between the subsystems (Hartmann et al., 1994).

Concerning the relation between individual persons and the organisation, task design is the crucial point. Work tasks should be configured and designed in a way to enable the working person to make use of the full range of his/her skills and at the same time to require continuous development of these skills and experiences. On the other hand, if individual learning processes are to have any impact on organisational learning, the continuous redesign of work processes must be a part of the workers´ tasks.

The main design rule for the relation between organisation and technology is that technology should fit the organisational structures of the company – e.g. decentralised work processes in semi-autonomous groups – and should never by its own internal structures impose certain organisational structures on the company.

In the relation between individual persons and technology, the design of technical systems should on the one hand fit the skills, experiences and ways of thinking and acting of the workers, and on the other hand allow and foster learning by offering flexible modes of action for a step-by-step increase in competence.

In the following, three technical systems are discussed which were designed to fit these criteria.

3 THE CNCplus-CONTROLLER

In a project funded by the German Federal Minister of Science and Technology, a new controller for CNC lathes was developed to fit the needs of skilled workers. This controller, put on the market by the German company Keller in co-operation with the French Company NUM, combines high-tech performance with unique features to be a flexible tool for the worker (Keller & Reuter, 1993). It has a CNC programming system which allows shopfloor programming in a standard ISO programming code, but at the same time offers quite different ways of manufacturing (Fig. 2). The program may also be generated in a graphic-interactive style by using geometry elements to define a workpiece and then describe a workplan using the same words and concepts the workers are used to from their working experience with no need of a single line of program code to be put in; the NC-code will be generated automatically according to the workpiece geometry and the workplan. Even more, the CNCplus-machine allows conventional turning and a mixture of conventional and CNC machining using record-playback features or turning with an electronic hand-wheel „virtually“ on the screen in the computer-generated simulation. All these modes of operation may be mixed within the programming of one work-piece so that the worker may choose the optimum combination of his own skills and automatic support.

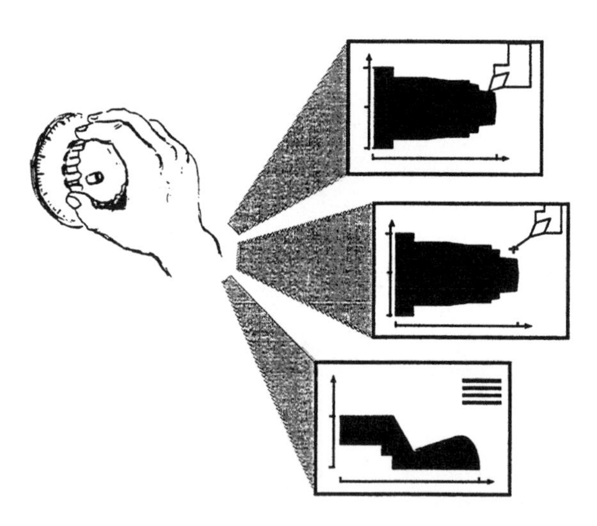

Fig. 2: Modes of operation of the CNCplus controller

It should be obvious how this machine fits the criteria of socio-technical design mentioned above. In practice, this technology meets especially the needs of companies producing customised products in small lot sizes. It fits in heterogeneous work groups, where machinery is shared by people with different skills and experiences. It is also of great value in mixed production/learning environments like production-integrated „learning islands“. Meanwhile, other companies have provided similar products which reflects a new attitude towards shopfloor technology design.

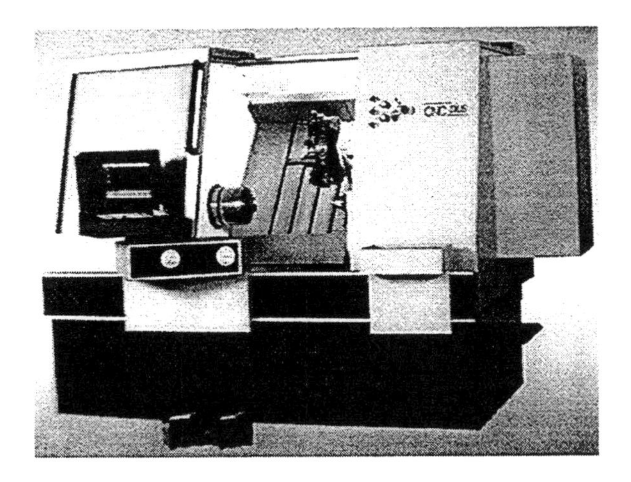

Fig. 3: CNCplus lathe by REALMECA

In a second project funded by the same Federal Minister, this concept was transferred from turning to milling. Also, several new skill-oriented features were added like force-feedback handwheels and joysticks to give the worker a „feel of the cut" (Daude et al. 1997; for similar developments in Japan cf. Nakazawa & Nemoto, 1994). But unlike the first project, which focused on the human-machine interface for programming, the second project opened a wider perspective on computer support for mechanical manufacturing in group-oriented organisations. This led to an information system and a work planning system to be described below.

4 InfotiF: AN INFORMATION SYSTEM FOR THE SHOPFLOOR

InfotiF is a client-server system to provide information and communication within the shopfloor and between the shopfloor and other departments (planning, programming, design etc.). It was developed at the Laboratory for Machine Tools and Production Engineering (WZL), Aachen University of Technology, and may be integrated in a PC-based CNC controller or installed in some other Personal Computer. Various kinds of electronic material – like workpiece drawings or CAD-Data, workplans, order lists, CNC programs, machine set-up information etc. – can be accessed by the worker directly at his workplace. Perhaps more important, shopfloor workers can communicate efficiently on these materials with other departments (Daude et al., 1997). This is very important for learning organisations, as may be shown by two examples:

In a modern organisation, there should be ways to allocate programming tasks between (central) programming offices and the shopfloor in a flexible way. To work efficiently in practice, this needs shared information for all people concerned, like: Which program is actually used one the machine right now? Which is the latest program version for a given workpiece? Why have these specific changes been made to this program? Why did this program not run effectively? With InfotiF, users can add remarks and comments to the programs and automatically make them available for all communication partners. Experiences with specific manufacturing situations can be stored for common analysis and transfer to new colleagues.

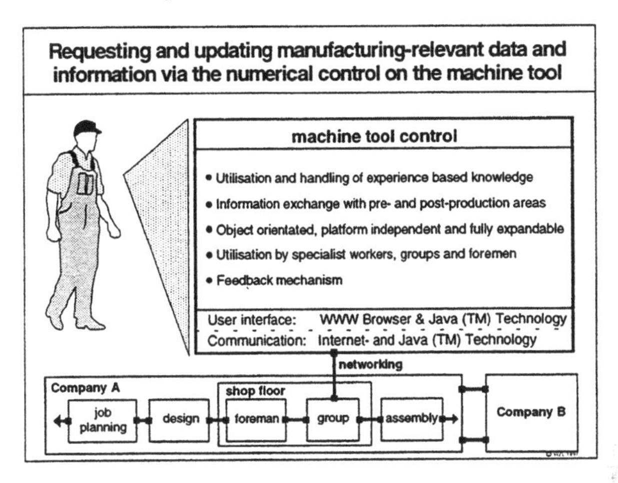

Fig. 4: Outline of InfotiF-System (Westerwick et al., 1998)

Furthermore, specifically in customised, small lot size production, the design of workpieces often is not optimal in terms of design for manufacture and design for assembly. In many cases a skilled shopfloor worker can figure out a (slightly) changed design which does not impair function but allows more efficient manufacturing. With InfotiF, he/she can give direct feedback to the designer, so that the both together can agree on a better solution. These suggestions and experiences can also be stored for further learning.

5 FIPS: A SKILL-BASED WORK PLANNING SYSTEM

Perhaps the most critical point in the transition from a Taylorist to a group-oriented organisation is the design of work planning procedures. Many companies – especially small and medium sized enterprises – experienced that conventional planning systems did not fit their need for decentralised, experience-based planning. FIPS – developed at the Centre for Research in Higher Education and Department for Computer Science in Mechanical Engineering (HDZ/IMA), Aachen University of Technology – gives shopfloor workers a simple and efficient means to plan orders and allocate them to machines. FIPS is an acronym for „Innovative

Planning System for Skilled Workers" (In German: **F**acharbeiterorientiertes, **i**nnovatives **P**lanungs**s**ystem)

Following a similar philosophy as the CNCplus controller, the user may choose between automatic and manual planning. The system can be used as a standalone system or as part of a network with de-central workplace systems for group planning and a base system for work planning between groups (Fig. 5).

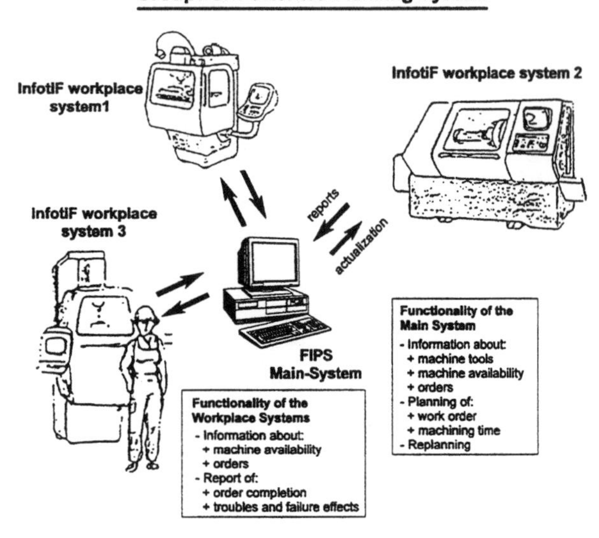

Fig. 5: FIPS work planning system

In the order list of FIPS, basic data of the orders - rank order and duration of the manufacturing steps - are entered (Fig. 6).

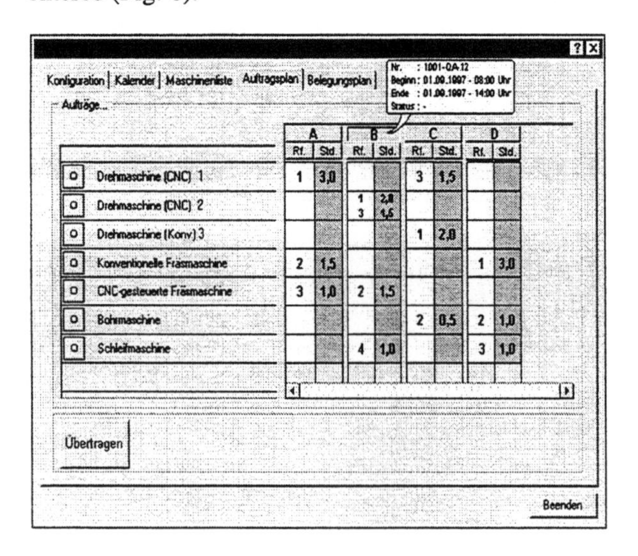

Fig. 6: Order list of FIPS system

On the basis of the order list information, the allocation plan is generated (Fig. 7). The allocation plan can also be used for manual planning and re-planning in case of changes in the planning parameters (e. g. priority of order, availability of machinery or personnel etc.).

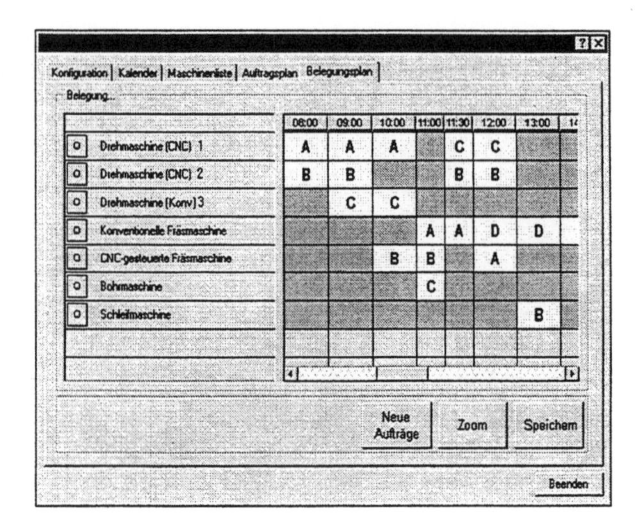

Fig. 7: Allocation plan of FIPS system

It is planned to integrate a multimedia computer based learning (CBT) tool to be used in courses on group-oriented work planning.

6 A WORD ON METHODOLOGY

Technology design for individual and organisational development does not only imply new *products*, but also new *processes*, i. e. technology design methods. All the systems described here have been designed involving representatives of all the people being concerned with this technology, including specifically the actual users. So co-operative or participatory design is the crucial paradigm. Within this paper, it is impossible to cover methodological detail, which has been published elsewhere (e.g. Daude et al. 1997; Hartmann 1994a,b, 1995; Hartmann et al., 1994). The following is merely a rough outline of design principles employed in designing the systems discussed above.

Carroll et al. (1991) proposed an inductive methodology for human-machine-systems, called the „task-artifact cycle" (Fig. 8).

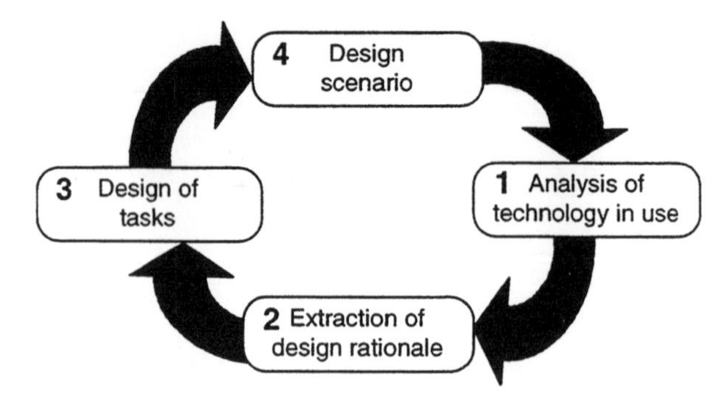

Fig. 8: The Task-Artifact-Cycle (Carroll et al., 1991)

This cycle consists of four steps: (1) Analyse work with existing technology (tasks and system analysis) (2) Find out the underlying „philosophy" (design rationale) of the technology-in-use (3) Re-design the task structure of the users according to ergonomic and organisational principles (4) Design a scenario of a new technology-in-use supporting this task structure.

This approach gives a basic orientation for evolutionary design, which systematically uses the knowledge embedded in the existing technology. It also supports co-operative design, as it takes the working experience of the users as a starting point for the design process. Design scenarios are also best set up and evaluated in a co-operative process.

To be practically applicable, however, this strategy has to be supplemented by more specific tools. Elaborating on work by Johnson-Laird (1983), Hartmann (1994a,b; 1995) suggested a taxonomy of mental models as a means to provide for cognitive compatibility of man-machine interfaces. Mental models are elementary modes of human perception, memory, and thought. Physical models correspond more closely to our perception of the „real" world, whereas conceptual models represent linguistic knowledge. Physical models may be distinguished from each other by how they depict temporal and spatial aspects of „real world" objects (Fig. 9). The three modes of operation of the CNCplus controller (Fig. 2) correspond to three different mental models (from top to bottom in Fig. 2: kinetic, sequential, and spatial).

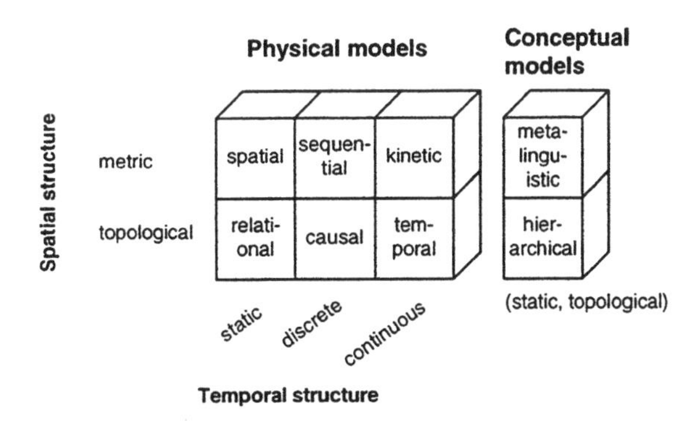

Fig. 9: A taxonomy of mental models (Hartmann, 1994a)

In human-machine system design, this taxonomy helps to answer the following questions: (1) Which kind of mental model of their working environment (e. g. workpiece, tools, machinery) do the users employ for which kind of task? (2) How should the human-machine interface of the technology-to-be-designed look like to be compatible with these mental models?

As has been shown in the CNCplus-project, this taxonomy fits nicely into the task-artifact cycle approach. The taxonomy may be used in analysing existing technology as well as giving a framework for design scenarios.

Finally, a tool is needed to „translate" these general design considerations into terms specifically appropriate for software design. Object-oriented analysis and design (OOA/OOD) techniques provide excellent means for this purpose and additionally support structured and modular software design (Tschiersch et al., 1996). In the design process of the FIPS system, a complete object-oriented approach was employed from analysis to design and implementation. The class diagram of the basic structure of FIPS is depicted in Fig. 10.

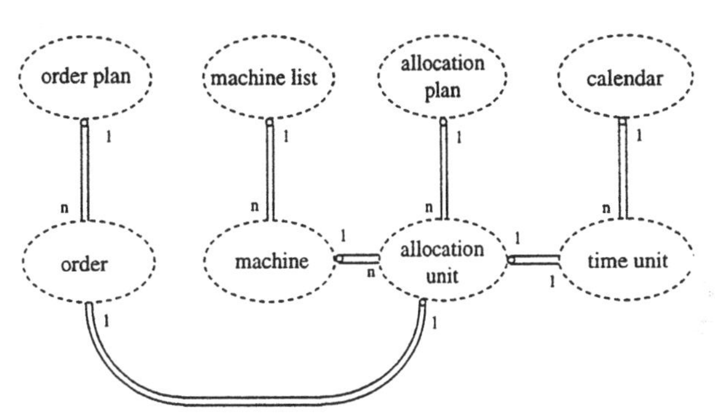

Fig. 10: Object-oriented structure of FIPS system

REFERENCES

Carroll, J. M., W. A. Kellogg & M. B. Rosson (1991): The Task-Artifact Cycle, in: J. M. Carroll (Ed.): *Designing Interaction.* Cambridge University Press, Cambridge.

Daude, R., C. Wenk, A. Westerwick, K. Henning & M. Weck (1997): Supporting skilled workers at shop floor machine tools, *Proc. 6th IFAC Symposium on Automated Systems based on Human Skill*, Aachen/Ljubljana.

Emery, F. E. (1959): *Characteristics of Socio-technical Systems*, Tavistock Institute Document No. 527, London.

Hartmann, E: A. (1994a): Designing human-machine interfaces to match the user's mental models – an issue of interdisciplinary research and education, *Proc. Int. Conf. Multimedia Vaasa*, Vaasa.

Hartmann, E. A. (1994b): Cooperative design of software as a method of computer aided ergonomic analysis – the case of Computerized Numerical

Control systems, *Proc. 12th Triennial Congress of the International Ergonomics Association.* Toronto.

Hartmann, E. A. (1995): *Eine Methodik zur Gestaltung kognitiv kompatibler Mensch-Maschine-Schnittstellen, angewandt am Beispiel der Steuerung einer CNC-Drehmaschine.* Augustinus, Aachen.

Hartmann, E. A. (1997): Shaping Organisation and Technology in Germany, *Proc. 6th IFAC Symposium on Automated Systems based on Human Skill.* Aachen/Ljubljana.

Hartmann, E. A., P. Fuchs-Frohnhofen, R. Sell. & K. Henning (1994): Lernen und Fertigen: Ein integriertes Konzept zur Personal-, Organisations- und Technikgestaltung, in: K. Henning, V. Volkholz, W. Risch & W. Hacker (Eds.): *Moderne Lern-Zeiten.* Springer, Berlin/Heidelberg/New York/...

Johnson-Laird, P. N. (1983): *Mental Models.* Cambridge University Press, Cambridge.

Keller, S. & W. Reuter (1993): CNCplus: High-Tech für den Facharbeiter, *Technische Rundschau,* **Sonderheft „CNC-Steuerungen"**, pp. 39-42.

Nakazawa, H. & H. Nemoto (1994): Human-oriented manufacturing system, *Transactions of the Japan Society of Mechanical Engineers,* 93-1485.

Strina, G. & E. A. Hartmann (1992): Komplexitätsdimensionierung bei der Gestaltung soziotechnischer Systeme, in: K. Henning & B. Harendt (Eds.): *Methodik und Praxis der Komplexitätsbewältigung.* Duncker & Humblot, Berlin.

Tschiersch, I., A. Westerwick, P. Fuchs-Frohnhofen & E. A. Hartmann (1996): Möglichkeiten und Grenzen objektorientierter Programmierung – am Beispiel CNC-Steuerungs- und Programmiersoftware, in: H. Rose (Ed.): *Objektorientierte Produktionsarbeit.* Campus, Frankfurt / New York.

Westerwick, A., M. Louha, C. Wenk (1998): Shopfloor Information Support: Developing User-Oriented Shopfloor Software, in: C.T. Leondes (Ed.): *Computer Aided and Integrated Manufacturing Systems, Techniques and Applications.* Gordon and Breach International Series in Engineering, Technology and Applied Systems; Newark, N.J., USA

SUPPORT OF COMPONENT ASSIGNMENT TO FUNCTIONS USING PAST CASES IN A FUNCTION-ORIENTED CAD SYSTEM

Akio GOFUKU*, Yuji SEKI*, and Yutaka TANAKA*

* *Department of Systems Engineering, Okayama University,*
3-1-1 Tsushima-naka, Okayama, 700-8530 Japan
E-mail: fukuchan@apollo2.mech.okayama-u.ac.jp

Abstract: This paper describes a support sub-system of the CAD system the authors have been developing for the component assignment to functions by effectively using past design cases, where two case retrieval techniques are implemented. The one called 'function flow simplification matching' simplifies the function flow of a case and compares it with the function flow represented by the designer. This matching technique realizes flexible case retrieval and can promote the idea of designers. The other is the utilization of typical component hierarchy data abstracted from past similar plant designs. The data navigates designers to determine the structure step-by-step. The applicability of the component assignment support sub-system is demonstrated by some design trials of a fluidized refuse incineration plant. *Copyright © 1998 IFAC*

Keywords: Computer aided design, Functional modeling technique, Component assignment, Case retrieval, Conceptual design

1. INTRODUCTION

Functional models are useful in design, diagnosis, and maintenance activities because they can organize and provide access to why a component is in a system(AAAI, 1993). In the design activities, Chandrasekaran, et al. discuss the applicability of functional modeling techniques to the various phases of design activities(Chandrasekaran, *et al.*, 1993). Yoshikawa, et al. propose a concept called intelligent CAD (III CAD)(Yoshikawa, *et al.*, 1989), where a future CAD should be Integrated, Intelligent, and Interactive.

Recently, it is pointed out that a design considering not only constructing plants but also plant operation and maintenance is necessary(JSPS, 1993). In the study, a design is defined to ensure easy and safe operation as well as to produce products in the minimum production cost.

The authors have been developing a CAD system to represent in a well-organized manner and to use efficiently the information about goals, functions, and structure of a design to plant design and operation(Gofuku, *et al.*, 1996a; Gofuku, 1996; Gofuku, *et al.*, 1996b; Gofuku, *et al.*, 1997). The previous studies discussed techniques to represent efficiently and systematically the information of designers' intention coupled with the structural information of engineering systems(Gofuku, *et al.*, 1996a), to derive the information related with behavior from a functional model(Gofuku, 1996; Gofuku, *et al.*, 1996b), and to use design purpose in the management of plant anomalies(Gofuku, *et al.*, 1997).

In a top-down design approach, the determination of components which realize functions is an important design step after the analyses of goals and functions of a design target. A designer is requested to have wide and deep understanding of various components in this step because there are usually several optional components which can realize a set of functions. The past similar designs can offer much information about the relations between functions and structure because they are considered to be the results of many investigations from various points of view. This study proposes a technique to support the determination of components or parts to realize functions using effectively past cases.

2. REPRESENTATION OF DESIGN INFORMATION

The information of a design is expressed in a four layers design model(Gofuku, *et al.*, 1996a) by extending a functional modeling framework of Multilevel Flow Modelling (MFM)(Lind, 1990; Lind, 1994). The model is composed of goal, function-goal, function-structure, and structure layers. The design information is represented graphically in these layers.

The design goals, sub-goals, and their hierarchical relations is represented in the goal layer. Quantitative specifications of each goal is also described in a frame format. The function-goal layer expresses relations between goals and functions. The relations among functions to achieve a goal are also expressed as a function flow. This layer is almost the same as the MFM excluding the relations between functions and components. The function-structure layer represents the relations between functions and components to realize them. The structure layer expresses an abstract configuration of components as a system diagram.

3. COMPONENT DETERMINATION SUPPORT USING PAST CASES

3.1 Support system of component determination

The determination of components which realize a set of functions requires much knowledge about both components and systems and many experiences of design. The authors consider that the results of past designs are good examples and can offer much information related with the component determination suitable to realize functions.

Based on the idea mentioned above, the authors have developed a technique to support the component determination using past similar cases and implemented it as an important sub-system of the CAD system(Gofuku, *et al.*, 1996a). It supports designers by retrieving and indicating some past similar cases from a case base when a designer tries to determine the structure of a part of a design after intentional aspects are described in the goal and function-goal layers.

The case base is composed of two types of information about past designs. The one is design information of each past design represented in the four layers MFM model. The other is a component hierarchy data which is made by analyzing ordinal component configuration of past same type of systems. Figure 1 shows a part of an example of the component hierarchy data. In the figure, the 'AND' symbol means that the lower components in the hierarchy need to compose of the higher component. On the other hand, the 'OR' symbol means that one of the lower components should be selected.

A case in the case base stores a four layers design model of past designs or some parts of them. The information in the function-structure and structure layers of a case is output when the information in the goal and function-goal layers match between the case and the specification by a designer. In this retrieval of past similar cases, the authors develop a case matching technique called function flow simplification matching to flexibly find the candidates and to give a matching distance to each retrieved case. The technique will be described in Sub-section 3.2.

The design steps using the CAD system with the developed case retrieval technique is shown in Fig. 2. Firstly, a designer expresses his/her design intention in the goal and function-goal layers (Steps 1, 2 and 3). Then, the CAD system retrieves similar design cases by comparing the information expressed in the goal and function-goal layers with those of cases in the case base (Step 4). The designer selects one of the indicated candidates (Step 5). The design information is automatically revised

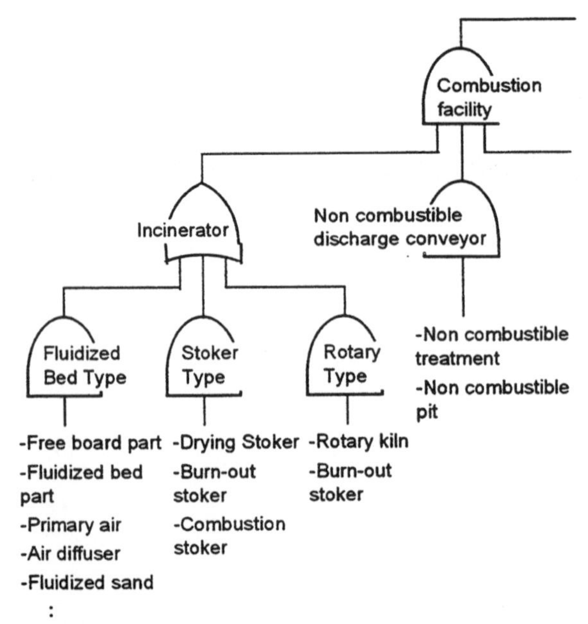

Fig. 1 Example of component hierarchy data.

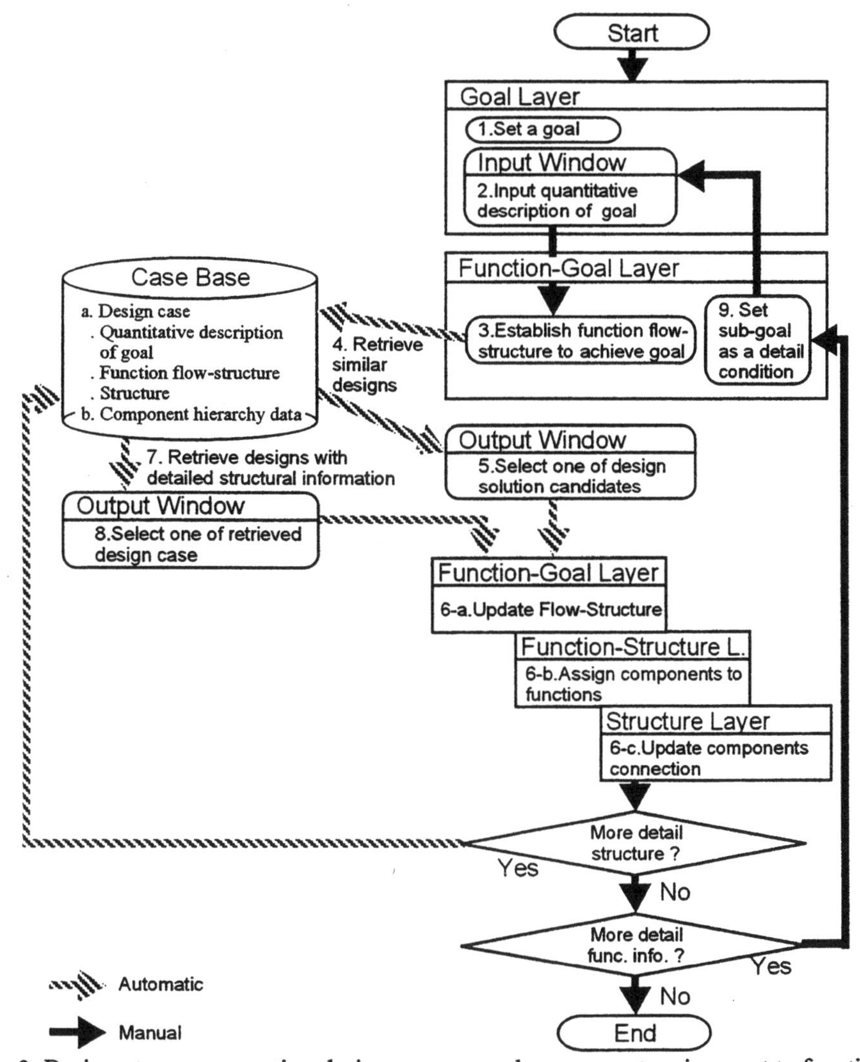

Fig. 2 Design steps representing design purpose and component assignment to functions.

according to the selected case (Step 6). If a designer wants to go into the detail of a component, he/she can ask the CAD system to retrieve some past design cases including the sub-components of the component and can select one of the retrieved cases (Steps 7 and 8). The designer can also express detailed design intention by adding a sub-goal as a condition of a function (Step 9) and can repeat the steps 2 to 6.

3.2 Case retrieval by function flow simplification matching

In order to realize a flexible case retrieval, the authors develops a technique to match the function flow structure expressed by a designer to the simplified function flow structure of a case by the simplification called function flow simplification. The case retrieval by function flow simplification matching is reasonable because a tank with inlet and outlet pipes which is modeled as a series of transport, storage, and transport functions can be regarded to realize a transport function like a pipe from a coast modeling point of view, for example.

Figure 3 shows the patterns of the function flow

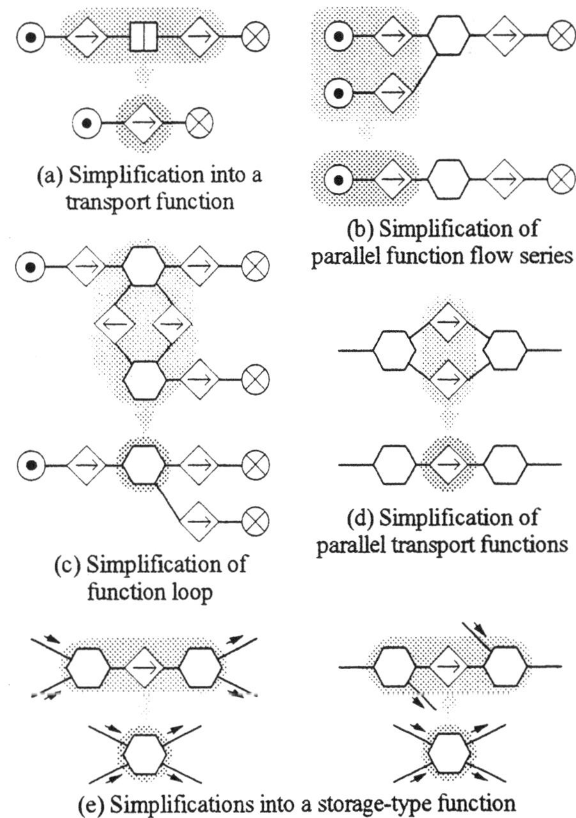

(a) Simplification into a transport function
(b) Simplification of parallel function flow series
(c) Simplification of function loop
(d) Simplification of parallel transport functions
(e) Simplifications into a storage-type function

Fig. 3 Example of function flow simplification.

simplification. For example, the function flow composed of two transports and one of storage, balance, or conversion as shown in Fig. 3 (a) can be simplified as a transport function as a whole. An example of function flow simplification is shown in Fig. 4. The alphabet in the parenthesis is indicated the applied simplification pattern. Suppose the top function flow is the function flow of a case. It is simplified to the second one by the simplification pattern (c) of regarding the function loop as a storage function. The simplified function flow is compared to the function flow represented by a designer and it is retrieved if these function flows are identical. By applying the function flow simplification, almost all function flows are simplified to the simple flow composed of source, transport, and sink functions like the bottom function flow in Fig. 4.

The priority of function flow simplification is determined as follows by considering that a designer will express his/her intention by a rather simple function flow and he/she can easily detect and represent multiple sources and sinks in a function flow. The priority of each simplification pattern is

Pattern (a) > Pattern (c) > Pattern (d) > Patterns (e) and (f) > Pattern (b). (1)

The matching distance of each case is given by the number of the function flow simplification and the

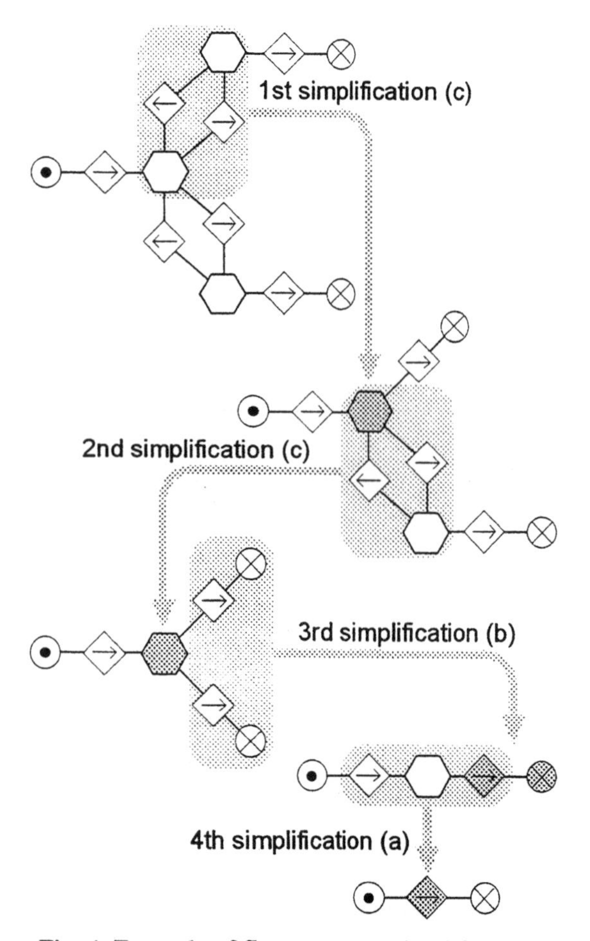

Fig. 4 Example of flow structure simplification.

retrieved cases are shown in the order of the smaller matching distance.

By the retrieval technique, cases which have complicated structure information can be retrieved from a simple specification of intentional aspects of a design.

3.3 Usage of component hierarchy data

In the determination of a combination of sub-components, component hierarchy data summarized from past design cases are very useful because they can offer typical configuration of sub-components. The developed CAD system uses the component hierarchy data to navigate a designer from a macroscopic viewpoint to a microscopic one. That is, it retrieves the cases which structure include the sub-components of the component designated by him/her.

4. EXAMPLE DESIGN PROBLEM

The applicability of the case retrieval is demonstrated by structure determination through a conceptual design of a refuse incinerator plant. The authors implemented the case retrieval technique on an engineering workstation by a Smalltalk-80(ParcPlace-Digitalk, 1995). Fifty three cases extracted from several types of refuse incineration plants are constructed and stored in the case base. The design problem is to design a refuse incineration plant which processes garbage of 200 t a day.

Firstly, the 'Goal 0' shown in Fig. 5 to burn refuse is specified as a main goal of the plant. Then, the energy flow structure to achieve the goal is established. The flow structure represents the transportation of the chemical energy in refuse, conversion into thermal energy, and discharge of the converted energy.

Form the goal setting and representation of energy flow functions, the CAD system retrieves two cases of 'Case 1' and 'Case 5' by the function flow simplification matching as shown in Fig. 6. The figure shows the design overlapping information of function-goal and function-structure layers due to the page space. The 'case 1' matches to the specified function flow after first simplification of function flow of the case (matching distance: 1), while the case 5 after second simplification (matching distance: 2). The 'case 1' is indicated to the designer with higher priority. The structure of 'Case 1' is a common one to refuse incineration plants. However, the structure of 'Case 5' is a fluidized incineration plant.

In this design trial, the designer selects the 'Case 1'

considering both the priority of the retrieval (matching distance) and the quantitative specification of the goals of the cases. The quantitative specifications of the goals in 'Case 1' and 'Case 5' are to burn refuse and to keep combustion temperature of refuse at 600 - 700 degree, respectively. The CAD system automatically revises the representation of design information referring 'Case 1' as shown in Fig. 7.

The design procedure after the selection of 'Case 1' is outlined here. The designer adds the sub-goal of supplying refuse of 200 t a day as the condition of the energy conversion function and represents a mass flow structure to achieve the goal. The CAD system retrieves five cases. The designer selects 'Case 29' among the cases. Then, he/she wants to represent a detailed structural information of the component 'Incinerator' which realizes the energy conversion function and the three associated transport functions. The CAD system retrieves 'Case 6' and 'Case 9'. The designer selects 'Case 6' and obtains the final representation of design information as shown in Fig. 8.

The authors also tries to continue the design with the same design intention as the design mentioned above after selecting 'Case 9'. The final structure of the plant is the same as the one shown in Fig. 8. However, the design information shows more detailed goal-function hierarchy than that of selecting 'case 6'.

By the design trials, the CAD system is proved to effectively support an important design step to bridge between design intention and structure.

5. CONCLUSIONS

This paper describes a support sub-system of component determination to realize functions by effectively using past design cases, where two case retrieval techniques are proposed. The one called 'function flow simplification matching' simplifies the function flow of a case and compares it with the function flow specified by the designer. This matching technique realizes flexible case retrieval. The other is the utilization of component hierarchy data abstracted from past similar plant designs. The data helps to navigate a designer to determine the details of design structure step-by-step.

The component determination support technique is implemented in the CAD system the authors are now developing. The technique is investigated by some design trials of a fluidized refuse incineration plant using the CAD system.

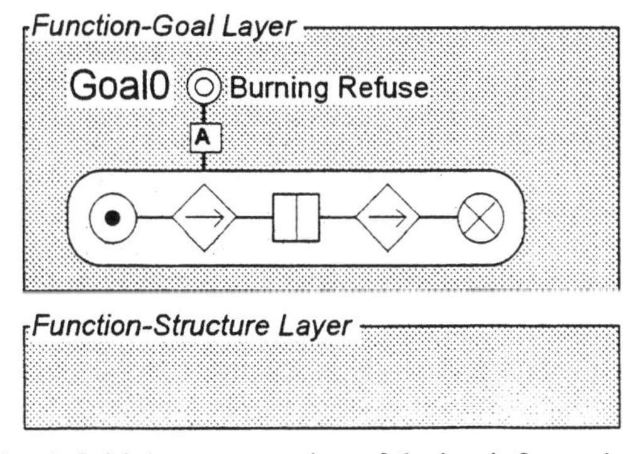

Fig. 5 Initial representation of design information.

REFERENCES

AAAI (1993). AAAI-93 Workshops Summary

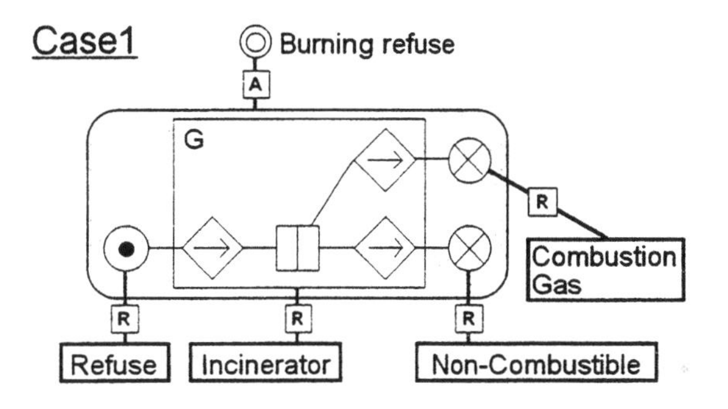

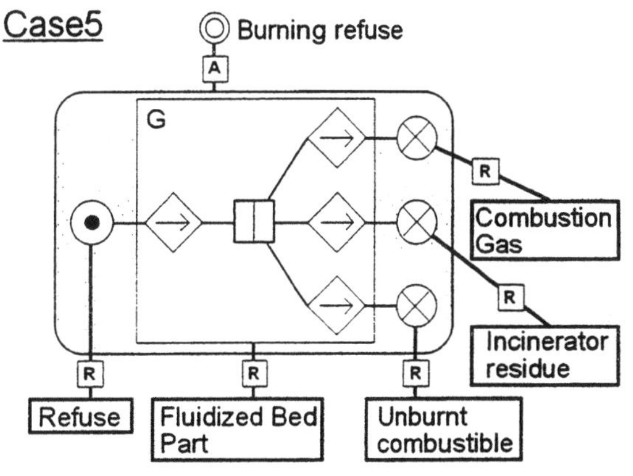

Fig. 6 Retrieved cases.

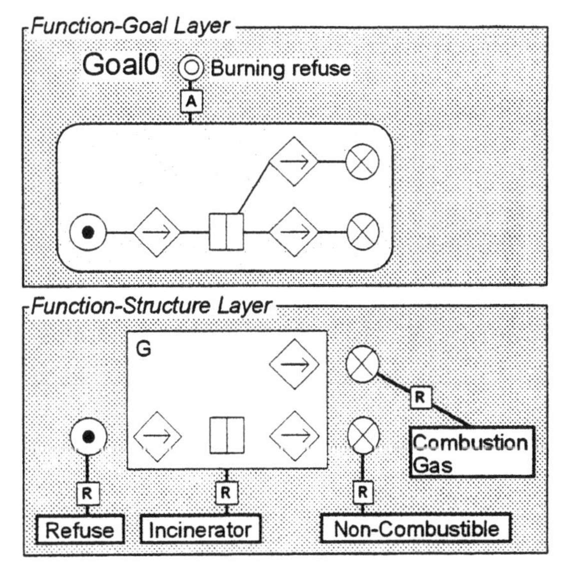

Fig. 7 Design information after applying 'Case 1'.

Reports, *AI MAGAZINE*, **15**, (1), pp. 63-66.
Chandrasekaran, B., Goel, A. K. and Iwasaki, Y. (1993). Functional Representation as Design Rationale, *COMPUTER*, **26**, (1), pp. 48-56.
Gofuku, A., Seki, Y., and Tanaka, Y. (1996a). Representation of Goals-Functions-Structure Information for Efficient Design of Engineering Systems, *Proc. Int. Symp. on Cognitive System Engineering for Process Control, CSEPC'96*, November 1996, Kyoto, Japan, pp. 238-245.

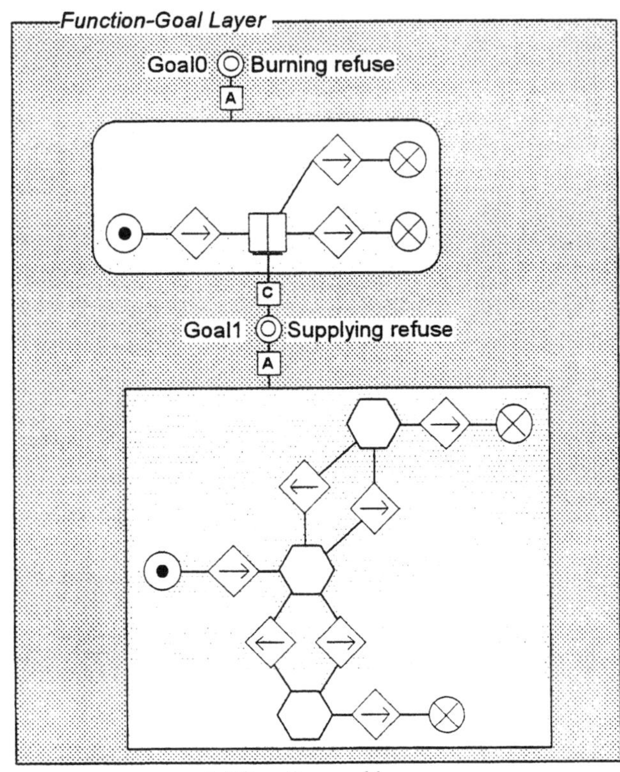

(a) Function-goal layer

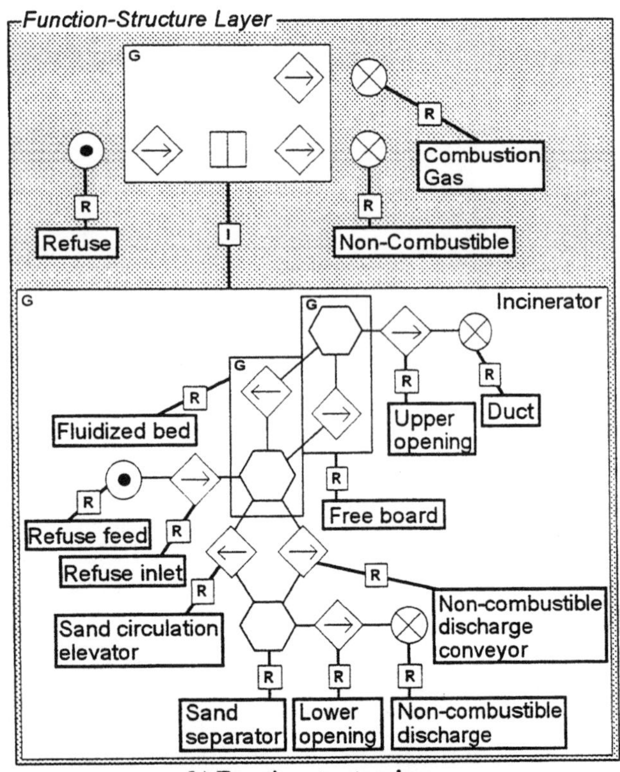

(b) Function-structure layer

Fig. 8 Final design information1.

Gofuku, A. (1996). Deriving Behaviour of an Engineering System from a Functional Model, *J. Japanese Society for Artificial Intelligence*, **11**, (1), pp. 112-120. (in Japanese).
Gofuku, A., Seki, Y., and Tanaka, Y. (1996b). Generating Multi-Level Dynamic Calculation Program from a Functional Model, *CD-ROM Proc. IASTED Int. Conf. on Modelling, Simulation and Optimization*, May 1996, Gold Coast, Australia, 242-175.PDF.
Gofuku, A. and Tanaka, Y. (1997). Development of an Operator Advisory System - Finding Possible Counter Actions in Anomalous Situations, presented in IFMAA 5th Int. Workshop on Advances in Functional Modeling of Complex Technical Systems, July, 1997, Paris, France.
JSPS (1993). Design Guidelines Based on the Operation, *Technical Report of the Committee (No. 143) of Process System Engineering, Japan Society for the Promotion of Science*. (in Japanese).
Lind, M. (1990). *Representing Goals and Functions of Complex Systems - An Introduction to Multilevel Flow Modelling*, Institute of Automatic Control Systems, Technical University of Denmark, Report No. 90-D-381.
Lind, M., (1994). Modeling Goals and Functions of Complex Industrial Plants, *Applied Artificial Intelligence*, **8**, pp. 259-283.
ParcPlace-Digitalk Inc. (1995). *VisualWorks Revision 2.0 (Software Release 2.5) User's Guide*.
Yoshikawa, H. and Tomiyama, T. (1989). *Intelligent CAD (I) - Concept and Paradigm*, Asakura-shoten. (in Japanese).

A HUMAN-COMPUTER INTERACTIVE SCHEDULING SYSTEM IN PARALLEL MACHINE SYSTEMS

Wen-Hung Kuo* **and Sheue-Ling Hwang****

*Department of Industrial Engineering and Management, Chien-Hsin College of Technology and Commerce, Chung-Li, Taiwan, R.O.C.
**Department of Industrial Engineering, Tsing Hua University, Kuang Fu Road, Hsin-Chu, Taiwan, R.O.C.

Abstract: A human-computer interactive scheduling system is developed with an automatic scheduling function. This function is designed based on the schedulers' experiences. By conducting an experiment, the total setup time and scheduling time of the subjects with/out the automatic scheduling function are compared. The results revealed that the performances of scheduling were better with the automatic scheduling function. Moreover, a discussion of total setup time between four common dispatching rules and experimental data is presented. For the further study, we can develop a system which can properly balance the work of the computer and the human by providing the assisting functions. *Copyright © 1998 IFAC*

Keywords: interactive, parallel, human, computer, automatic, and function

1. INTRODUCTION

In the CIM systems, the scheduling task is very important during the production process. The steady state of the production system depends on the effective and real-time scheduling system. In the past, the goal of researches is to accomplish the scheduling task as soon as possible. Therefore, most studies focus on the whole computerized systems, for example, simulation-based methods, mathematical modeling systems and heuristic methods. These researches indeed solve some real-time scheduling problems, but they can not completely deal with the complicated problems with rush orders, orders change and machine breakdowns during the production process. Especially, the computerized systems appear more deficient in the decision making and problem recognition in more altering states of the production systems.

According to the preceding statements, we need to reconsider the position of CIM systems. That is to say, does a CIM system need to be a whole computerized system or a human-computer interactive system? Corbett (1996) indicated that the manufacturing systems could gain considerable competitive advantage through the full utilization of the skills, flexibility, creativity, and knowledge of the human. Based on this viewpoint, we propose a human-computer interactive scheduling system in a parallel machines system. We let computers deal with the parts which they can definitely do and leave those computers cannot definitely deal with to human schedulers. In addition, we offer an automatic scheduling function through the collection of human schedulers' experiences.

2. STRUCTURE OF HUMAN-COMPUTER INTERACTIVE SCHEDULING SYSTEM

To take advantage of human-computer interaction, we develop a model of human-computer interactive scheduling system based on the process of scheduling and the experiences of human schedulers, as shown in Fig. 1. Kuo and Hwang (1998) elicited some useful experiences of human schedulers from this scheduling process. The general steps of scheduling process were described as follows:

(1) Arranging the priority of each order according to EDD rule.

(2) Scheduling the order roughly according to the priority.

(3) Selecting machines that produce the same subtype of product first to serve the order. If there are two or more choices, one should select machines whose available time is close to the processing time of this order. If the available time of machines selected is not enough for the processing time of the order, one should select machines that produce the same type of products and do not produce the same one as those of the following orders. Otherwise, the quantities which are not scheduled should be left to the next level scheduling.

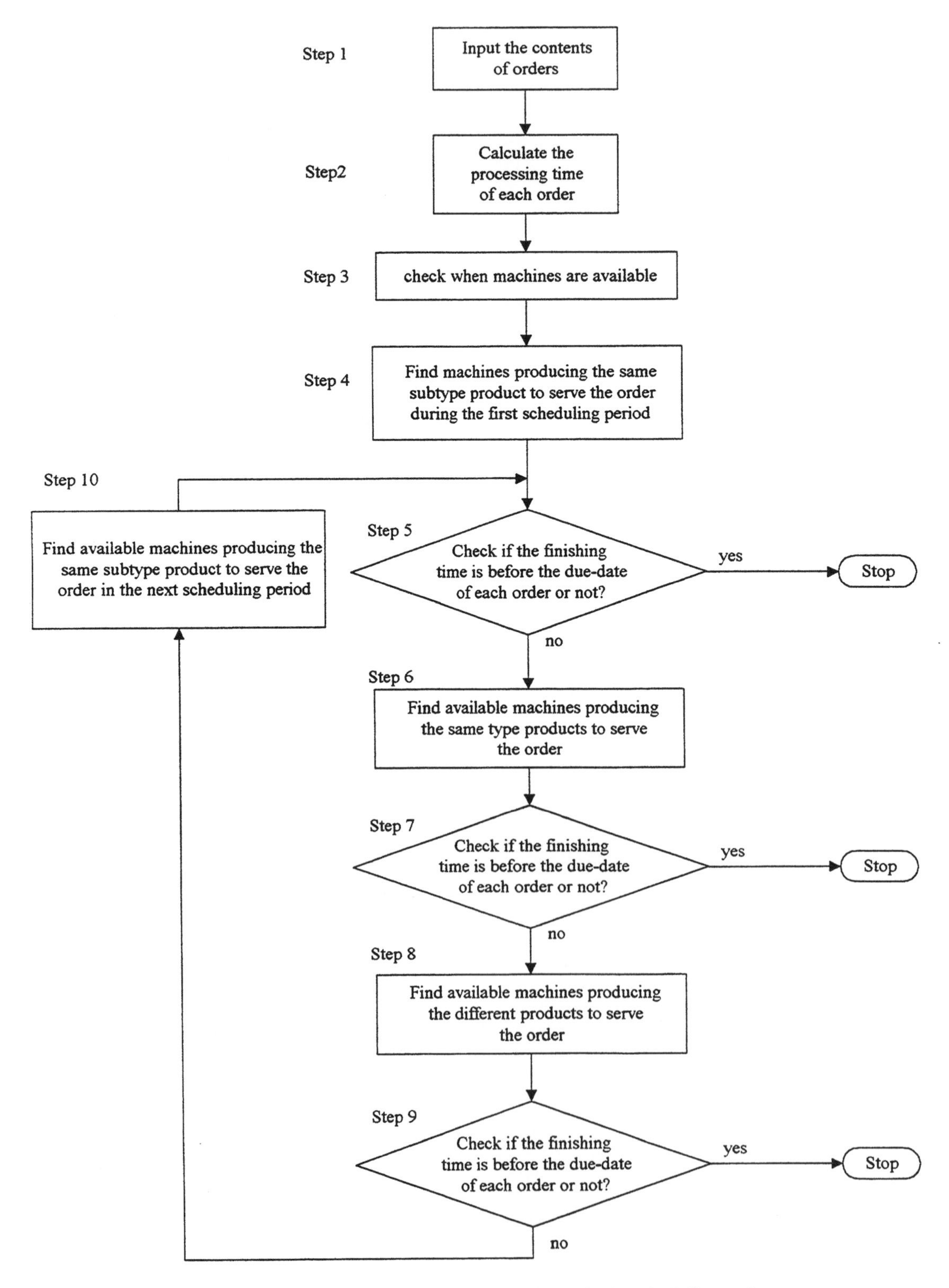

Fig. 1. A model of human behavior in scheduling tasks

The designing principle of the human-computer interactive scheduling system is that computer deals with the definite part and human scheduler deals with the part that computer could not definitely do. Based on this principle, a model of the human-computer interactive system is developed, as shown in Fig. 2. The system also provides an automatic scheduling function which will be described in the next section.

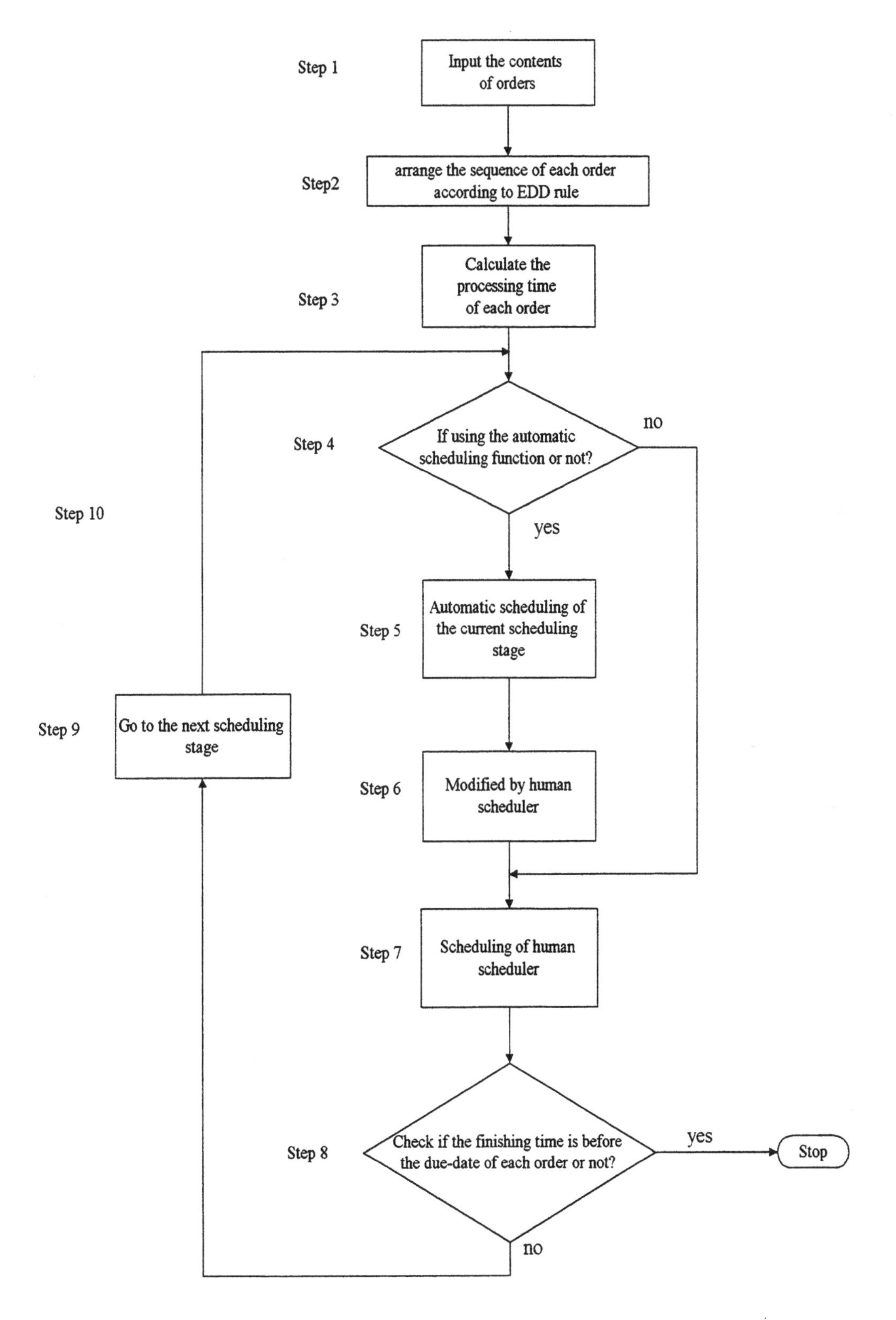

Fig. 2. The structure of the human-computer interactive scheduling system

3. THE AUTOMATIC SCHEDULING FUNCTION

The steps of automatic scheduling function are described as follows. In the first scheduling stage, the function selects the first order in the EDD sequence and then arranges it to the machines that are producing the same subtype product. If the quantity of this order is too large to be arranged to the available machines, the function will arrange the quantity of the order as many as possibly and leave the excess quantity of the order to human scheduler to deal with it later. Then, it will continue to deal with the next order according to the same rule until it finishes tackling the final order. After finishing automatic scheduling task in the first scheduling stage, the human scheduler can modify the result of automatic scheduling and arrange the left orders to the suitable machines. If the scheduling tasks are not finished in the first scheduling stage, all left orders will be transferred to the second scheduling stage. The automatic scheduling function of this stage is a little different from that of the first scheduling stage. It is that the function will arrange the orders to the machines which are producing the same subtype product first, and then arrange the left orders to the machines which are producing the same type products. After these processes, the human scheduler will modify the result and arrange the left orders to the proper machines if necessary.

4. THE PRODUCTION SYSTEM AND EXPERIMENT

4.1 A human-computer interactive scheduling system in a production system

The production system which we consider is simplified from one department of a semiconductor company. For the convenience of study and without losing reality, the system is considered as follows. There are three product types: a, b, and c in the system. The a-type product has three subtypes: a1, a2, a3, and either b-type or c-type products have two subtypes: b1, b2, c1, c2. There are ten parallel machines in the production system.

Based on the structure of the human-computer interactive scheduling system, we design an interface of the system. In this interface, we focus on providing useful information to reduce the workload of schedulers.

4.2 Experiment

An experiment was conducted to evaluate the effects of the human-computer interactive scheduling system using automatic scheduling function and without using this function on scheduling performance.

Subject. Sixteen subjects were selected from the graduate students at the Institute of Industrial Engineering at National Tsing Hua University. All these subjects were required to have taken some courses about manufacturing process or production management.

Task. Five sets of orders corresponding to five different circumstances of the production systems were used in the experiment. Subjects had to make a schedule of each order according to the corresponding circumstance of the production system. The first priority in the scheduling problem is to make finishing time before the due-date of each order. Also, one should not spend too much time in the machine setup. Each subject was asked to make a schedule of the same order twice, and subject could use the automatic function in one time but could not use it in the other time.

Design. The experimental design was two factors with repeated measures on two factors. The dependent variables were (1) Setup time — the total time subjects spent in the machine setup in the final schedule, and (2) Scheduling time — the time subjects spent in making the schedule. The independent variables were Method (M) and Circumstances(C), as followed.

- M1: Scheduling system with automatic scheduling function;
- M2: Scheduling system without automatic scheduling function;
- C1: Five orders in the set and having more combinations of suitable order-machine pairs;
- C2: Five orders in the set and having fewer combinations of suitable order-machine pairs;
- C3: Ten orders in the set and having fewer combinations of suitable order-machine pairs;
- C4: Ten orders in the set and having more combinations of suitable order-machine pairs;

Procedure. After reading the explanatory document of scheduling task, all subjects were asked to make a schedule twice according to the following procedure. In the first trial, they could use the automatic scheduling function to accomplish their tasks, but in the second trial, they could not use this function. In the beginning, a preliminary test was given to each subject. All subjects were given the same order set corresponding to the circumstance of production system for the preliminary test. Subjects could make a schedule for this order set without time limit. They had to fully realize the scheduling processes. After this test, each subject had to take the other four order sets corresponding to the circumstances as tests. Subjects were asked to finish their scheduling tasks as soon as possible. They also had to make the finishing time before the due-date of each order as well as they could. In addition, they had to minimize machine setup time. The final schedule and scheduling time of each order were recorded.

5. RESULTS

The experimental results and related discussion are presented in two parts.

5.1 Comparison of simulation and human-computer interaction method

Table 1 presents the data about the average setup time of experiments and total setup time which are obtained from the simulation just four common dispatching rules for the same circumstances as those testing the subjects.

Table 1 Total setup time for common dispatching rules

Circumstance	1	2	3	4
EDD	1.8	2.6	2.2	3.6
SPT	1.8	1.6	2.2	2.1*
FCFS	4.3*	1.8	1.5	2.4*
LCFS	1.3	2.6	3.0	2.7
without ASF**	1.59	1.79	2.86	2.73
using ASF	1.28	1.63	1.78	2.04

Note: * means that the finishing time after the due-date of some order in the final schedule, i.e. the schedule is not good using this dispatching rule.
** ASF means automatic scheduling function.

From the Table 1, the means of setup time of human-computer interactive system using auxiliary function are better than the setup time of simulation method using four common dispatching rules in most circumstances. It is found that the simulation method using dispatching rule, FCFS, can not take all orders into account in the final schedule for circumstance 1 and those using dispatching rules, SPT and FCFS, for circumstance 4 cannot, either.

5.2 ANOVA

The first analysis dealt with the effects of human computer interactive scheduling system using automatic scheduling function and without using this function on total setup time. The interaction of method and circumstance was significant ($p < 0.01$). Therefore, the simple main effects were further analyzed. The results showed that the scheduling methods were significant ($p < 0.01$) in circumstance 3 and 4 and the circumstances were significant ($p <0.01$) in both scheduling methods.

The second analysis concerned the effects on scheduling time. The analysis of variance for scheduling time revealed that the interaction of method and circumstance were significant ($p < 0.01$). Therefore, the simple main effects were also analyzed. The results showed that the scheduling methods were significant ($p < 0.01$) in circumstance 3 and 4 and the circumstances were significant ($p <0.01$) in both scheduling methods.

6. DISCUSSION

6.1 Effects of scheduling methods

In the statistical analysis, the simple main effects of the scheduling methods are significant in circumstance 3 and 4 and those of the circumstances are significant in both scheduling methods. These results indicate that the human-computer interactive scheduling system with the automatic scheduling function can improve the performances of the scheduling tasks. In addition, the phenomenon is more evident in more complicated circumstances (more orders). This means that the aiding function is more efficiently in more complicated circumstances. In fact, the humans must deal with many orders in a real production system. Therefore, the previous results support us to develop the human-computer interactive scheduling system from the direction of analyzing the behaviors of the humans in their scheduling tasks. Moreover, the discussions of statistical results reveal that the prototype of the scheduling system, which was proposed by Kuo and Hwang (1998), is indeed improved.

6.2 The interactive scheduling system and simulation method

From the comparison of simulation and human-computer interactive system, it is found that most means of setup time using aiding function in the system are less than those of simulation methods using the four common dispatching rules. These results reveal that the human-computer interactive system can be applied to scheduling problems effectively. In addition, the average setup time using aiding function is the least for all methods that are considered in circumstance 1 and 4. These results mean that the human-computer interactive scheduling system is more efficient in more combinations of proper order-machine pairs in interim of the scheduling process. In other words, the scheduling aid helps when the tasks get complex. It seems the automatic scheduling function may reduce the mental workload in complex tasks from the primary task performance (e.g. scheduling time).

Because the system elicits some useful experiences from human schedulers to form an automatic scheduling function and leaves the uncertain part to human schedulers to make decisions during the scheduling process, it can complete the scheduling task in a single run or a few runs. Therefore, the schedulers may not spend too much time in the scheduling tasks if they are familiar with the system. Of course, we can spend less time in the scheduling task if we let the computer deal with the part which needs complicated calculation. This may drive us to take the concepts of AI method to form more assisting functions in the human-computer interactive system. These discussions provide a way to solve the problem of simulation-based method that may consume too much time for one run in some conditions. Furthermore, the human-computer interactive system just gets the states from the shop floor and uses a computer program to make a schedule in each decision point. This method is quite different from the simulation-based method. It does not need to store the status of the previous simulation run in the memory of computer. Therefore, the problem of simulation-based method that the simulation status of each decision point must be saved effectively for the usage in the next decision point does not exist anymore. From the above discussions, A much better scheduling system can be proposed if we properly combine the advantages of simulation-based method and those of human-computer interactive method.

7. CONCLUSION

In this study, the first contribution is that a human-computer interactive scheduling system is developed based on the experiences of human schedulers in their scheduling tasks. Second, we not only propose the structure of the human-computer interactive scheduling system but also take this system into a real application. Third, we show an alternative way of human-computer cooperation in scheduling tasks. It is that the task of the human scheduler is not just monitoring or controlling but really incorporating the scheduling tasks. Finally, we propose the principle of forming an automatic scheduling function. That is, human schedulers must be able to realize the whole automatic scheduling process of this function and they can give minor modification of the result during the scheduling process.

To improve the current scheduling system, some directions for future researches may be focused on:

(1) Combining the advantages of AI methods and simulation-based methods into human-computer interactive systems.

(2) Instead of developing the entirely automatic scheduling system, properly balancing the work of the computer and the human and providing the auxiliary information to assist human schedulers in proper situations.

REFERENCES

Corbett, J. M. (1996). The development of user-centered advanced manufacturing technology: new design practice or new marketing rhetoric? *The International Journal of Human Factors in Manufacturing*, 6 (2), 79-87.

Kuo, W. H. and S., L. Hwang (1998). A prototype of a real-time support system in the scheduling of production systems. *International Journal of Industrial Ergonomics*, 21, 133-143.

HUMAN-ORIENTED CNC MACHINE TOOLS TO MATCH HUMAN SATISFACTION

H. MATSUNAGA and H. NAKAZAWA

Department of Mechanical Engineering, School of Science and Engineering, Waseda University, Tokyo, Japan

Abstract: A methodology of synthesis based on human-oriented idea should be urgently constructed. One of the important aspects of human-oriented design is how to make an artifact with which humans are satisfied. In this paper, element technologies of a human-oriented CNC lathe which have been developed based on the concept of human-oriented manufacturing system (HOMS) were described, and one of them was evaluated using satisfaction measurement system based on measuring electroencephalogram (EEG).

Keywords: Human-centered design, Human-machine interface, Human factors, Systems design, Systems methodology, CNC, User interfaces, Manufacturing systems

1. INTRODUCTION

In such fields as design, manufacturing, environmental maintenance, recycling and safety, humans and technologies closely relate each other. Therefore, a methodology of synthesis based on human-oriented and human-respected ideas should be urgently constructed. It is necessary to systematize human based synthesis with emphasis on human factors.

Design and development proceed chiefly by emulation of prior art (Carroll, et al., 1991). That is, a new technology has been developed by analyzing existing, in-use technology, modifying or maintaining it, and operating it. Most technological developments were repetitions of inductive design and artifacts have been evolved based on the technological background in the era. Because the technology is the one for humans, the artifacts made from the technology should be human-oriented. The aim of the human-oriented design is how to make an artifact with which humans are satisfied.

In manufacturing system today for example, however, automation does have to constrain human ability and undermine human dignity. It is important to understand the human role in manufacturing system is neither to be an incorporated component of the system nor an excluded one. The alternative human role is to interact with the automated manufacturing systems to obtain high flexibility and productivity and to improve product quality.

In this paper, element technologies of a human-oriented CNC lathe which have been developed based on the human-oriented idea were described, and one of them was evaluated using satisfaction measurement system based on measuring electroencephalogram (EEG) (Matsunaga and Nakazawa, 1997a, b).

2. HUMAN-ORIENTED MANUFACTURING SYSTEM

Since the Industrial Revolution, the manufacturing industry has been aiming at the realization of unmanned automation of manufacturing systems and human history has enjoyed the wealth of economy. Today, however, various contradictions have become apparent in sectors which are considered to have been successful in adopting unmanned automation. Thus, with systems becoming increasingly complex, the investment in plant and equipment has grown enormously, causing the break-even point to rise; systems have grown highly inflexible, even while the increase in productivity has failed to match initial expectations; because there is no generation-to-generation transfer of technical know-how, difficulties arise when the time arrives to design the next-generation manufacturing system; without the feeling of joy that accompanies the act of creation, workers find it difficult to seek motivation in their work; the resulting products are lacking in terms of added value; and so on.

To solve these problems, a Human-Oriented Manufacturing System (HOMS) (Nakazawa, 1993) has been proposed as the next-generation production system. HOMS has the following three functional requirements:

(1) A manufacturing system in which humans have pride and pleasure in making;
(2)A manufacturing system which is tender to humans; and
(3)A manufacturing system which makes a profit.

This is not a negation of automation. HOMS aims to combine automation with human abilities such as know-how, implicit knowledge, heuristics, and intuition that are uniquely to humans. With HOMS, the worker will feel pride and joy in the creation of products, and feel satisfaction in the act of making. When superior human qualities not replaceable by machines are introduced into manufacturing, products will have higher values added onto them, and it will be much easier to accommodate small-lot production of diverse products. Furthermore, by incorporating human factors, the system should be able to evolve and increase its productivity.

3. A DESIGN METODOLOGY OF HUMAN-ORIENTED MANUFACTURING SYSTEM

3.1 A manufacturing system in which humans have pride and pleasure in making

To provide the design basis for HOMS, a survey of the satisfaction of workers in a foundry was conducted in 1992 (Nakazawa and Nemoto, 1995). One of the surprising results is that amenity or tender conditions are not necessarily sufficient for the workers' satisfaction. It is found that there are seven satisfaction factors for the pleasure of creating (manufacturing): diversity, creativity, subjectivity, feedback effect, technical skill, interest, and social contribution. Thus, in order to satisfy the first functional requirement of HOMS, a system must be designed to satisfy each of these factors.

As a basic function, a means by which the operator can intervene with the automated system must be provided. In other words, the automated system must possess a function for manual operation or means of easily changing work parameters, through which the operator can fully express such human capabilities as innovation or self-direction. This does not mean that the operator is kept busy at the production line at all times. When the manual control is finished, the automatic system continues production under those conditions.

Whether operating the manual function or changing parameters, the operator must be able to see the process being carried out by the system. Moreover, the results of operator intervention must be fed back to the operator. This is why a black box is undesirable. In short, the operator must be able to see the process and the black box made transparent. The feedback function is necessary also in that the operator can self-check his/her improvement in skills.

3.2 A manufacturing system which is tender to humans

This does not simply mean that the manufacturing environment should be set up in a clean and orderly manner, or that it should be made comfortable for the operator. Rather, what is more important is that the manufacturing system conforms to the human operator instead of vice versa.

Of course, in order to remove dirty, hard and dangerous images of the manufacturing industries, amenity of the manufacturing systems must be realized. This means not only air conditioning colouring and light design, but also that non-ergonomic labour, mental stress and physiological pain must be avoided, and labour must not infringe individual home life. Also the manufacturing systems guarantee the human dignity.

3.3 A manufacturing system which makes a profit

Economic profit can not be neglected even in HOMS. Economic profit among other things is one of the most important objectives. But HOMS is superior to the conventional systems with respect to producing high value products as well as productivity because of the incorporation of human abilities such as skill, experience, judgment, heuristic solving problem and so on. HOMS also has the possibility to make up for the lack of flexibility.

Economy of scale should be changed to economy of high value from now on. The necessities of life, on the other hand, must be mass produced at low cost, which can be produced at low level automation. But things other than this should be produced by manufacturing systems like HOMS.

3.4 Cognitive compatibility

Cognitive compatibility is important in the design of human-oriented systems. The various information on the tool controlled by an operator and its object shall be presented in a form that can be perceived, recognized and memorized by the human operator at a highly intuitive level. All human capabilities can never be replaced by computers, and these are naturally based on information available to humans. Thus, for the human operator to be able to express those capabilities to the fullest, it is important that various information concerning the tool being manipulated and its object be perceived and recognized at a highly intuitive level, and that it be memorized in a state that is most conductive to human performance.

4. HUMAN-ORIENTED CNC MACHINE TOOL

CNC machine tools are representatives of automated systems. Hence, CNC machine tool is selected and examined how it could be made into a human-oriented system. Since the human-machine interface is the first thing that comes under consideration when the human operator intervenes with an automatic system, human-oriented interfaces for a CNC lathe were developed as follows:

(1) Drawing recognizable automatic manufacturing system (DREAMS)
(2) Machining phenomenon visualizing system
(3) Cutting resistance torque feedback handle
(4) Operator intervention error correction device

These systems are briefly described below together with the design concept.

4.1 Drawing recognizable automatic manufacturing system (DREAMS)

Figure 1 shows schematic of the system. This is an integrated manufacturing system that interprets a handwritten drawing on the knot scale, that has been input through an image scanner or a touch panel, then automatically writes an NC program, which machines the product (Nakazawa and Kanayama, 1987). This is a human-oriented system in that a machine interprets human expressions and uses it to write tedious NC programs in place of the human. Another human-oriented feature is the editing function that can incorporate the operator's ideas and experience into the automatically produced program.

By using the machining parameter setting function, automatic cost estimation function, and machining simulator, the operator can write an improved NC program based on experience and ideas. Because human abilities are taken full advantage of to produce better products with higher added values, the operator will feel satisfied that skills are being put to good use.

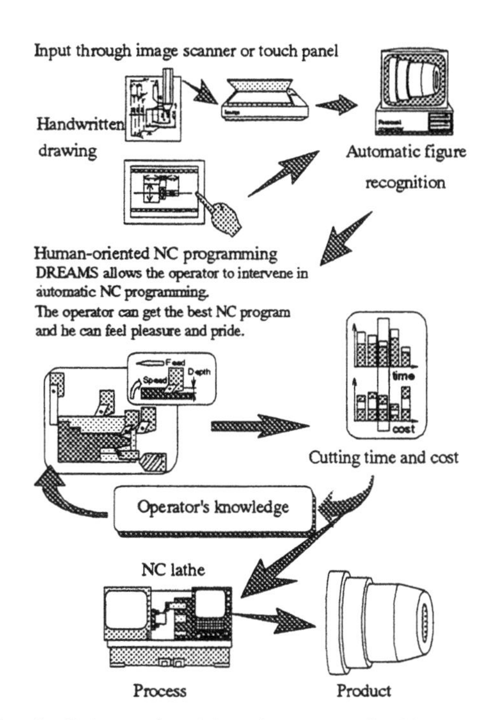

Fig. 1. Schematic of drawing recognizable automatic manufacturing system (DREAMS)

4.2 Machining phenomenon visualizing system

Figure 2 shows system configuration. By taking in the visual image of the cutting point using a CCD camera and reflecting mirror, and synchronizing with the rotating speed of the workpiece, a static image of the spinning workpiece can be captured and displayed on the monitor (Nakazawa and Sugaya, 1996). Static images of the entire circumference can be displayed by shifting the synchronization phase.

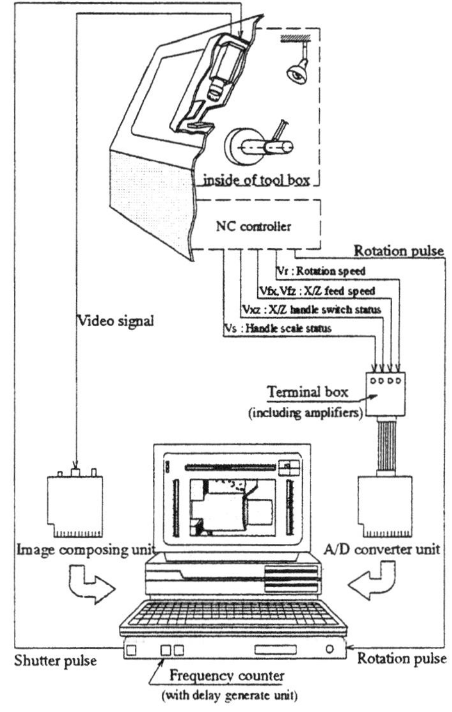

Fig. 2. Configuration of machining phenomenon visualizing system

This system transmits the objects as an intuitive visual image that requires no translation, while maintaining a point of contact between the speed of the machining phenomenon and the cognitive processing speed of humans. It enables the operator to clearly grasp various machining phenomena, including the surface conditions of the workpiece or the shape of chips, which had previously been difficult to capture with the naked eye. The operator can use the override function to achieve the optimum machining condition based on such information. It is possible using this system to see the cutting path or machining progress. This makes it suitable for on-site training as well.

4.3 Cutting resistance torque feedback handle

By taking the tool a part of the body, humans are able to psychologically immerse themselves into the object via the tool. If in CNC lathe work, where it is not possible to touch the object being machined, the information obtained by the tool part directly in contact with the object is fed back to the operator in the form of physical sensory information, it becomes possible to narrow the psychological distance between object and operator. Hence, a jog handle for manual control, in which the cutting torque is fed back, was developed (Nakazawa and Izumi, 1996). Figure 3 shows the system configuration. This is achieved by extracting the current signals of the main spindle motor (which are approximately proportional to the cutting torque) and inputting them to a magnet powder brake to impart to the handle a resistance torque proportional to the cutting torque.

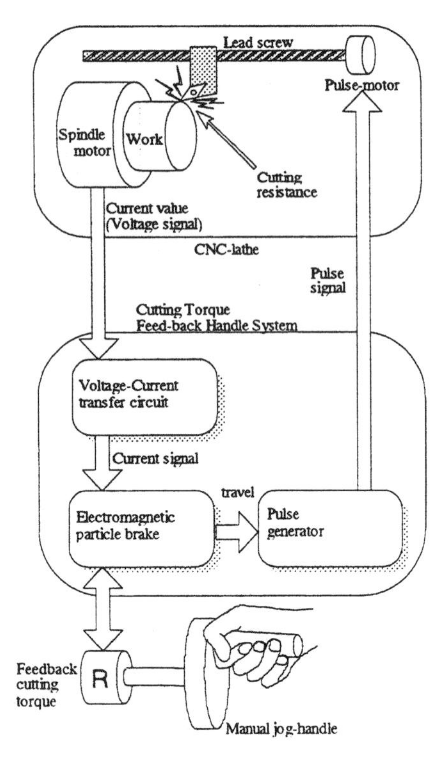

Fig. 3. System configuration of cutting resistance torque feedback handle

When this system was installed in a CNC lathe and tested by inexperienced and skilled operators of CNC lathes, the typical response was that "the reaction force helped to understand the actual machining conditions and served as a guide in setting the machining conditions".

4.4 Operator intervention error correction device

Feedback control techniques have generally been used when constructing systems for correcting machining errors. However, for various reasons, direct feedback from the cutting point has not been realized in practice. Thus, an error correction device incorporating feedback control, with which humans can intervene easily, was developed (Nakazawa and Kanegae, 1996). Figure 4 shows the system. After machining, errors existing at a workpiece are measured by a laser displacement gauge, then input to a personal computer. Instead of simply reversing the errors for the finishing process, the operator develops a correction curve that has been input by a mouse based on his/her ideas and experience.

It has been shown that this considerably improves the machining precision compared to correction by a simple reversal of errors. Moreover, while conventional feedback control theory allows for correction only within the same processing stage, this system has the advantage that the machining error measured prior to the finishing stage can be used for correction at the final finishing process.

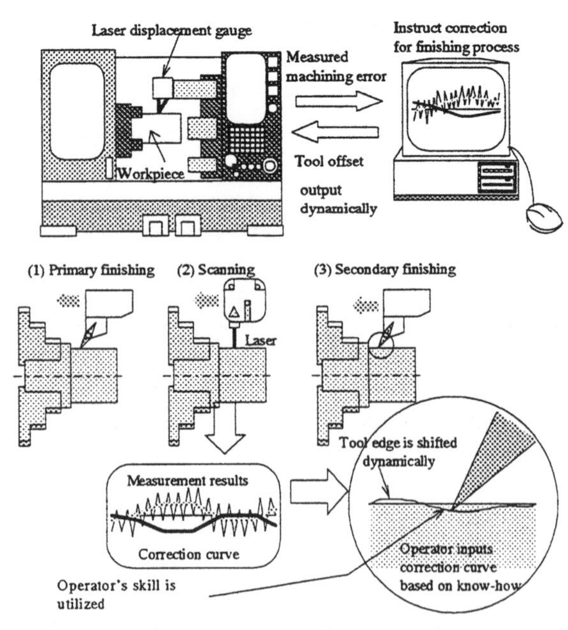

Fig. 4. Schematic diagram of operator intervention error correction device

5. SATISFACTION MEASUREMENT SYSTEM

Since the state of mind is supported by the brain, it

appears on EEG as well as in subjective experience. To measure satisfaction, it is necessary to obtain the relationship between the EEG and the subjective experience. Then, an indicator for measuring satisfaction was obtained, which combined psychological and physiological information (Matsunaga and Nakazawa, 1997a, 1998b). First, a language-based model of satisfaction was constructed to evaluate subjective satisfaction properly. Second, frontal EEG was measured from bipolar derivation when subjects were doing various tasks. The results showed that relative power of α band has a positive correlation with subjective satisfaction while that of β band has a negative correlation.

This reflects asymmetries in frontal EEG activity, and there have been many researches on the relationship between frontal EEG and emotion. Wheeler et al. (1993) reported that the left frontal region is associated with positive emotion while the right associated with negative emotion. Now the relation between our result and other study results is being examined, and positive results are being obtained.

A satisfaction measurement system has been developed that employed neural network (Matsunaga and Nakazawa, 1997a, b). Although an established model does not exist in the relation between psychological and physiological states, it is possible using neural network to acquire experienced knowledge and reflect individual differences. The input signals of learning were selected to be normalized power spectrum, which reflects not only the appearance rate but also other unknown information like the shape of power spectra. The training data were (1.0, 0.0) for satisfaction, (0.0, 1.0) for dissatisfaction respectively. Satisfaction was given by how many primary unit was represented in the sum of the two outputs.

Figure 5 shows the system configuration. The system consists of detective part, signal processing part which derives the power spectrum using FFT, and the neural network which determines the satisfaction level. The system is equipped with a human-oriented feedback device using micro vibration, with which the worker can work passing his/her condition unnoticed by other people. Analysis time for satisfaction data can be chosen from either 8.53s or 4.27s.

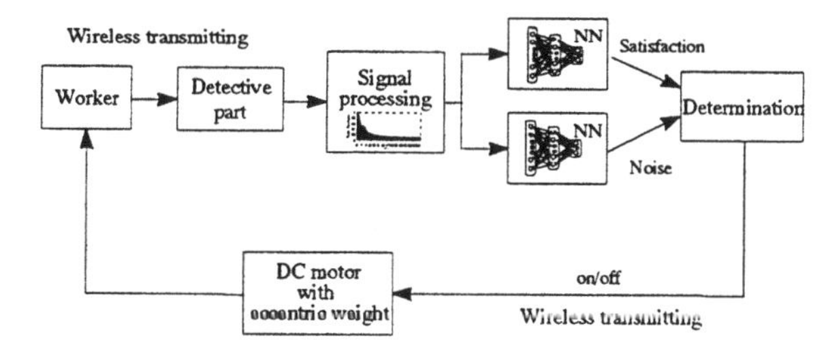

Fig. 5. Configuration of the satisfaction measurement system

To confirm the validity of the system, measurement results are needed for various situations. Measurements were thus taken for various operations, showing the system was able to successfully evaluate satisfaction (Matsunaga and Nakazawa, 1998a). The system will be brought to market in cooperation with the company Techna electronics (Tokyo, Japan).

6. EVALUATION OF HOMS USING SATISFACTION MEASUREMENT SYSTEM

Subjective evaluation showed the effectiveness of a human-oriented CNC lathe. On the other hand, evaluation using physiological information is more objective and offers the changes in real-time. Hence, from among the element technologies, satisfaction measurements were taken when inexperienced and skilled male operators of machine tools were using operator intervention error correction device.

Table 1 shows mean values of change rate of satisfaction between operation and two-minute resting prior to the work. Analysis time was 4.27s. Subject J, K, and L attended the experiment twice one week apart. Parentheses in the table indicates rank order among operations. It indicates that satisfaction is higher when most of the subjects are using their skills to make the correction ("Correction input" in the table) than NC machining that operator only sees the process is taking place ("NC machining"), but that satisfaction is not always higher when the error, resulting from the subject's intervention, is displayed ("Error display") than NC machining. As for subject L, he is skilled in engine lathe and dislikes the computer system, which matches the measurement result.

The error is not always a good one for the operator to have expected, which could probably affect satisfaction degree. Figure 6 shows the relation between satisfaction and error. In the figure, all errors existing at a workpiece are averaged. It demonstrates the negative relation between satisfaction degree and error size. Figure 7 shows an example of error results of subject K, showing that satisfaction is higher when the error result is good.

Table 1 Mean values of change rate of satisfaction between operation and rest

Subject	Age	Experience of machine tools	Operation: Correction input	Operation: NC machining	Operation: Error display
A	24	inexperienced	1.442 (1)	0.992 (2)	0.940 (3)
B	22	inexperienced	0.742 (1)	0.524 (3)	0.538 (2)
C	25	inexperienced	0.922 (3)	0.968 (2)	1.001 (1)
D	22	inexperienced	1.483 (1)	1.226 (3)	1.523 (2)
E	22	inexperienced	1.243 (1)	1.027 (2)	0.775 (3)
F	22	inexperienced	1.552 (2)	1.318 (3)	1.562 (1)
G	22	inexperienced	0.926 (1)	0.826 (2)	0.605 (3)
H	22	inexperienced	0.669 (2)	0.584 (3)	0.973 (1)
I	23	inexperienced	1.021 (2)	0.997 (3)	1.031 (1)
J	23	4 years skilled	0.894 (1)	0.732 (2)	0.612 (3)
			0.488 (1)	0.327 (3)	0.427 (2)
K	24	5 years skilled	0.622 (2)	0.474 (3)	0.685 (1)
			0.541 (1)	0.380 (2)	0.305 (3)
L	51	33 years skilled	0.724 (3)	0.981 (1)	0.737 (2)
			0.448 (3)	0.621 (1)	0.577 (2)

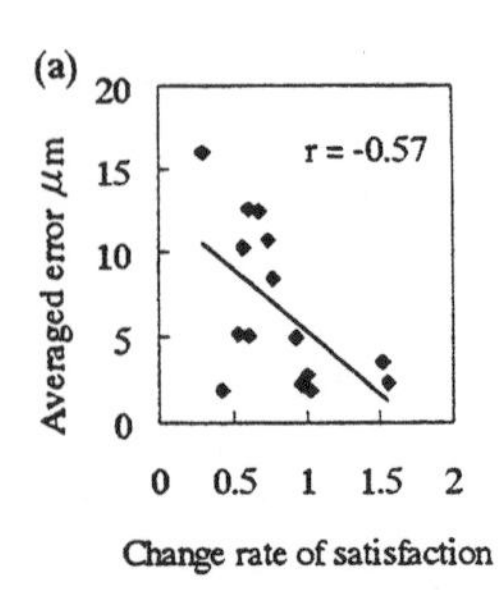

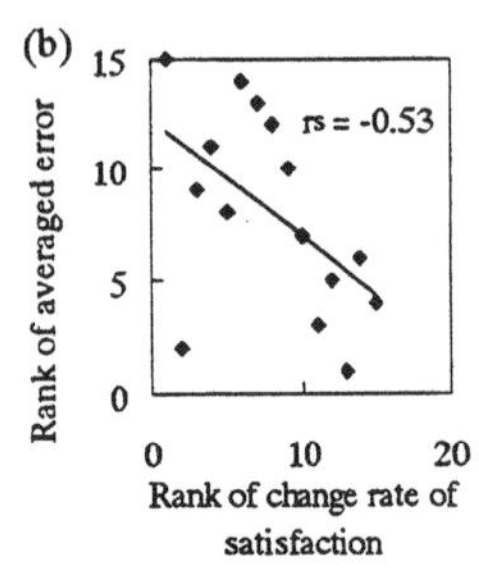

Fig. 6. Relation between satisfaction and machining error. (a): change rate of satisfaction vs averaged error; (b): rank of change rate of satisfaction vs rank of averaged error

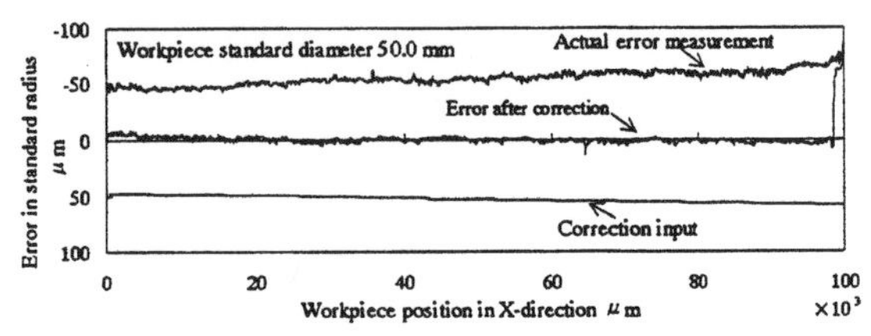

(a) Error display when satisfaction was high (rank 1)

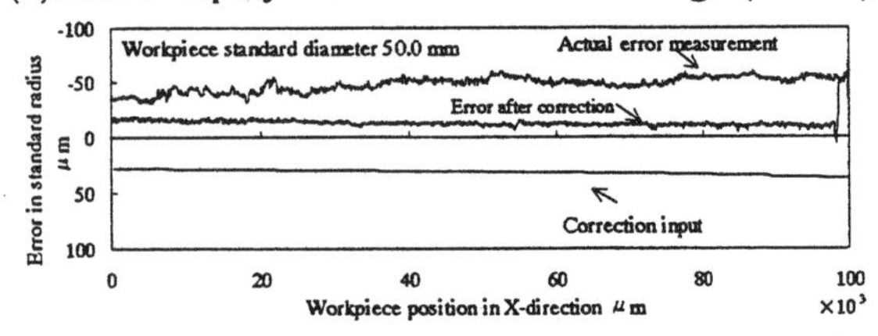

(b) Error display when satisfaction was low (rank 3)

Fig. 7. Comparison of error display when satisfaction was high and low (subject K)

The measurement results showed that satisfaction increases when the subject is creativity intervening and also when seeing the good result. This points to the importance of the seven factors of workers' satisfaction mentioned above. Conversely, workers will feel no satisfaction in monotonous operations that do not meet the satisfaction factors. Thus, to allow the operator to feel satisfaction, it becomes necessary to design a system with an interface which operation meets those satisfaction factors.

7. CONCLUSION

The element technologies of a human-oriented CNC lathe which have been developed based on the concept of HOMS were described, and one of them was evaluated by using satisfaction measurement system. The result proved that a human-oriented interface which allows the operator to intervene in a creative manner increased their satisfaction. Now a human-oriented CNC lathe, with which the developed element technologies will be installed, is being totally developed in connection with the Japanese national project "Science of Human Based Synthesis" which is supported by the Research for the Future Program of the Japan Society for the Promotion of Science.

REFERENCES

Carroll, J. M., W. A. Kellogg and M. B. Rosson. (1991). The Task - Artifact Cycle. In: *Designing Interaction* (J. M. Carroll (Ed .)). Cambridge University Press, Cambridge.

Matsunaga, H. and Nakazawa, H.(1997a). A Study on Human-Oriented Manufacturing System (HOMS) -Measurement of Workers' Satisfaction by Psychological and Physiological Information, *Proc. 13th Triennial Congress of the International Ergonomics Association*, Vol.1, 507-509.

Matsunaga, H. and Nakazawa, H.(1997b). A Study on Human-Oriented Manufacturing System (HOMS) -Development of Satisfaction Measurement System (SMS) and Evaluation of Element Technologies of HOMS using SMS, *Proc. Int. Conf. on Manufacturing Milestones toward the 21st Century*, Tokyo, 217-222.

Matsunaga, H. and Nakazawa, H.(1998a). Design method considering human satisfaction. In: *Automated Systems Based on Human Skill* (J. Cernetic and D. Brandt (Ed.)). Elsevier Science. (in press)

Matsunaga, H. and Nakazawa, H. (1998b). Experimental study for satisfaction measurement - Relation between subjective satisfaction and frontal EEG measured from bipolar derivation, *Jpn. J. Ergonomics*. [Japanese] (will be published)

Nakazawa, H. and Kanayama, K.(1987), Development of Manufacturing System by Automatic Input of Drawings of Turning parts, *J. Jpn. Soc.Pre.Eng.*, **54**,No.1,151-156[Japanese]

Nakazawa, H.(1993). Alternative Human Role in Manufacturing, *AI & Soc.*, Springer-Verlag London, London, 7, 151-156.

Nakazawa, H. and Nemoto, H.(1995). Human-Oriented Manufacturing System -Workers' satisfaction, *Jpn. J. of Advanced Automation Technology*, **7**, No.1, 1-6.

Nakazawa, H. and Sugaya, I.(1996). Study of Human-Oriented Interface for CNC machine Tools, *JSME International J.*, Series C, **39**, No.2, 397-403.

Nakazawa, H. and Kanegae, T.(1996). Study of a Human-Oriented Manufacturing System - Human-Interactive Control for CNC Machine Tools, *Trans. Jpn. Soc. Mech. Eng.*, **62**, No.593, C, 410-415.[Japanese]

Nakazawa, H. and Izumi, T.(1996). A Study on Human-Oriented Manufacturing System - Study on Cutting Torque Feedback Handle, *Trans. Jpn. Soc. Mech. Eng.*, **62**, No.601, C, 3719-3724.[Japanese]

Wheeler, R. E., Davidson, R. J. and Tomarken, A.(1993). Frontal brain asymmetry and emotional reactivity -A biological substrate of affective style, *Psychophysiology*, **30**, 82-89.

MODEL-SHOP FACTORY DESIGN - HUMAN INTEGRATED FACTORY PLANNING AND PROCESS OPTIMIZATION

Dr. Matthias Hartmann, Jörg Bergbauer

Fraunhofer-Institut für Fabrikbetrieb und -automatisierung
(Fraunhofer-Institute for Factory Operation and Automation)
Sandtorstr. 22, 39106 Magdeburg, Germany

Abstract: Aim to discuss of this paper is the method, the functionality and the experiences of industrial use of Virtual Reality to support the participative approach in the factory design. It describes the system developed at the Fraunhofer IFF and its use as basis for an innovative experimentation field to support the factory planning process. In this connection the speeding up and the optimization of the planning process by supplying all involved people a common environment for their decision making process will be pointed out. *Copyright © 1998 IFAC*

Keywords: Factory Planning, Virtual Reality, Participation, Experimentation Field, Model-based Cognition Process

1. INTRODUCTION

1.1 Current situation

The current industrial situation could be specified by catchwords like »turbulent environment«, »flexibility and acceleration of processes« or »short product cycles« and »continuous optimization«. In view of the increased dynamic of the enterprise surroundings, concepts for the continuous adaptation of the enterprise structures are required (Hartmann, 1996). In this case, proactive acting and the ability to organizational learning is particularly important due to the fact that in the case of reactive adaptation to changes the possibilities are strongly restricted. The previous procedure to learn and optimize by real experiences according to the trial-and-error-principle can be realized no more because it represents often a painful, slow and high-cost way for enterprises.

The planning and organization of modern production systems under this turbulent circumstances can no more be understood as isolated ranges of tasks of individual disciplines. Rather, an jointly operation of all in the planning process involved persons is required. This necessity derives itself from the fact, that factories in their nowadays form are complex, cross-linked, intransparent and dynamic systems (Daenzer and Huber, 1992). Because of the multitude of different elements, e.g. machines as well as information and organizational rules, which are acting or reacting totally different in the case of environmental impacts, a complex system structure is derived which could be hardly understood by a single discipline. This state is complicated by the fact that, due of the relationships of these elements, a network is resulting which is of great importance for the functioning of the factory (Daenzer and Huber, 1992). Effects on the entire system as well as on individual subsystems which are resulting by manipulating and changing of a single element or the connection of the elements to each other are, due to the interplay of a multitude of different factors, partially or completely intransparent for the actor (Dörner, 1993). However, system changes are resulting automatically because factories are no inflexible but high-dynamical systems, which have to adapt themselves permanent at the changing environmental impacts. I.e. they show a certain intrinsic dynamic. The success of planning regards mainly from the consideration of these factors. However, in most cases the individual planning participants have no complete knowledge and/or a wrong idea of the essential system attributes. Only by integration of the knowledge and the ideas over all disciplines a solid realization of the planning task can be guaranteed (Hoffmann, 1996).

1.2 The participative approach

The complexity of the planning task and the pressure on the development speed are leading to the necessity for a better utilization of unused potentials and to a systematic production of ideas, impulses and concepts (Dörner, 1993). New innovative solutions which innovative factories are demanding form the planners only can be created by intensive creative efforts. In this connection the creativity is a product of knowledge and the ability of imagination (Higgins and Wiese, 1996). From this results the fact that a productive creativity is impossible without knowledge; the extensive the knowledge the bigger is the probability to think up al lot of patterns, variants of combination and new ideas. For the realization of the alternatives and solutions worked out together there is a lack of imagination by all individual disciplines; the common knowledge remains mostly unused and sterile. The efficiency of the creative productivity depends to what extent the planning team is able to unite knowledge and imagination ability.

However, in order to be able to integrate all the different disciplines into the organization process, a certain basic knowledge of systematical mastering of the planning task must be provided. Basis for this kind of interdisciplinary working is a general dialog platform which helps to develop different solutions and to reach clear and expressive decisions. Only this allows to make sure that the common knowledge will be processed for practical and active using. In this case, a vivid and understandable representation of the planning intention as well as of the actuating variables is required. An essential principle to illustrate these complex connections as well as the »real« system, exists in the use of models in the course of so-called »factory experimentation fields« (Bergbauer, 1998). With the aid of computer generic models the process of learning could be activated by playful thinking over and testing of consequences of specific strategies and measures as well as by confrontation with other strategies and results.

2. Factory Experimentation Field

2.1 Virtual Reality as an user interface

To support these participative procedure the Fraunhofer-IFF pursues with its »Model-Shop Factory Design« the approach, to make available an factory experimentation field for the planning participants that serves as platform for mutual dialog and the common generation of different solution alternatives. Therefore, the Fraunhofer-IFF sets Virtual Reality for the model formation and visualization onto the effort of an understandable representation of all elements and performing connections of a factory. In this case Virtual Reality is an alternative interface between human and machine which allows a new way of working with the computer. It realize a real-time interaction with three dimensional computer data and produce for the user with the aid of special immersive technologies a subjective sensation of an apparent real environment (Bauer, 1996). Has the planner only been up to now a simple observer of his models, now he will become to an active element of an artificial environment in which he can move around without restrictions and with which he can come in interrelation.

Fig. 1: Participant in the factory experimentation field

Especially for the planning and design of factories the immersion, i.e. the dive in into the virtual world, discloses new possibilities to increase the efficiency of the development process. The clear and impressive visualization of the problem task and the different possible solutions can be used excellently for the development of alternatives. Abstract data and information which have been not comprehensible for the planner up to now and which are very important for the planning process can be visualized by insertion of icons and graphics as well as by using colors with different intensities. By it the degree of the visual perception of the user could be increased highly. This is in so far of importance as the human sense does respond more creatively to pictures than to words or other sensory perceptions; 70 percent of the impressions which the human brain is picking up are things what the man sees, 25 percents are things what he hears and the last 5 percent of impressions are distributed to the other senses like smelling, touching and feeling (Hoffmann, 1996). By using Virtual Reality this fact can be made utilizable by visualizing every thing, i.e. objects, information etc., which should be noted by the brain.

2.2 The model-based cognition process

The fundamental idea of the factory experimentation field of the Fraunhofer-IFF is based on the model-based cognition process (Wiendahl, 1996). This process starts with the modeling of the real system. To understand how the factory should be organized (organization units), how it should be run (operations, control) and how it should look (layout) an integrated modeling of the factory (plants and buildings) and the processes is necessary for a successful realization. For that purpose an universally valid and general model of the future factory will be created and visualized in Virtual Reality by extracting all necessary information from all the different models used by the involved planning disciplines. A homogeneous model results as a platform for the interdisciplinary dialog and the discussion of all participants. These VR-model serves as a common planning basis over all planning stages and can be used continuously for all involved disciplines (Mezger, et al., 1996).

Because of the three dimensional visualization of these model by using Virtual Reality as an user interface all participants are able to enter the new building and facilities already long time before the realization and to have experience with the new factory even though it is not really existing. Due to this fact the observer has the option to give up his position as passive onlooker and to cause active changes. The observer can interact with and within the factory model. By seeing, hearing and feeling, the actor can discover and interpret the model and make decisions out of experiences. This allows the development of an information potential by verification through interaction of all participants. As a result, many knowledge carriers are linked into the extraction and the processing from information on decision-making, which leads for an efficient workmanship of distributed knowledge. Mistakes in the model are uncovered by common experiments and simulations. This allows to reach unambiguous statements and the derivation of unequivocal decisions. Interpreting these results, the model could be validated or the consequences transferred into the real system to improve it.

Looking exactly on this procedure it could be notified that the working method within the experimentation field is divided into tow major parts. One is the »knowledge acquisition« part, in which experiences can be made and information defects are shown in an early stage. Through discovery and experience knowledge is generated within the virtual model to build up a knowledge database. The »knowledge utilization« part is to transfer the benefits, if needed, into the real system.

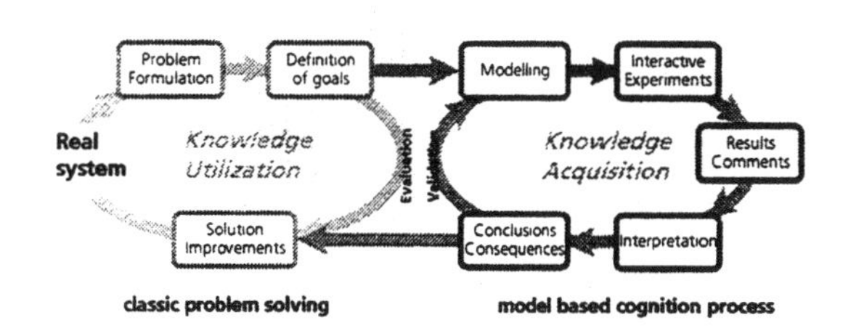

Fig. 2: The model-based cognition process

3. MOD!FACT

Basis of the factory experimentation filed of the Model-Shop Factory Design is the innovative software MOD!FACT, developed by Fraunhofer-IFF. The subrack panels suffice for projecting different layout variants up to the integrated organization of complex factories and plants, inclusive for the evolution of alternative development concepts within the framework of general development from the first one.

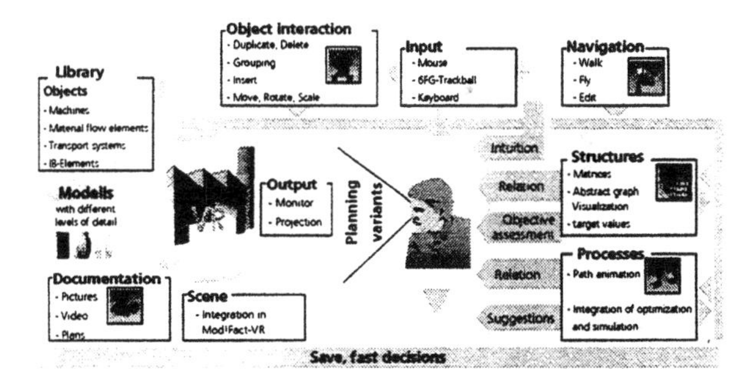

Fig. 3: Functionality of MOD!FACT

Essential sign of MOD!FACT is - in addition to the effort of Virtual Reality - an integrated concept overall, that the usual procedure for layout design and improvement units with innovative methods for the formation of spatial-organizational structures to a continuous tool. In this case, improvement in the information and communication relationships is in the foreground in addition to the optimization of the flow of materials and an afterwards arrangement and structuring of the resources. Effects after this, according to the design of the building, one can vary and check it in a conditional manner. The results won in the respective stages are available for all planning stages.

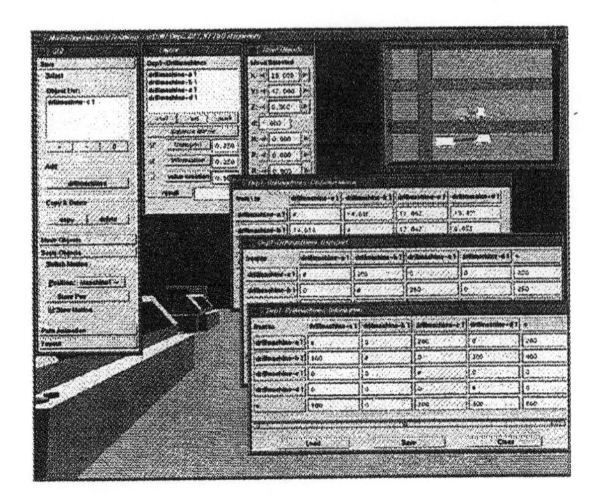

Fig. 4: Structuring with MOD!FACT

Substantial part of MOD!FACT are specific intelligent model libraries, e.g. for resources, buildings, conveyer, in which all objects required by planning are filed. In addition to pure graphical data, further information as weight, load, hygienical stage or output as well as important details for planning like attractor fields and their strength are saved there. Within the framework of the separate planning states the corresponding objects can be called up by the user and brought by intuitive interaction into the planning environment. Essential relationships between the elements which result only on account of their arrangement can be determined and specified directly in the experimentation field. In this case, the selection and grouping of the objects is actively checked by MOD!FACT. So the system will not allow to the user to place e.g. a machine into a production field with adequate load or in which already are placed other types of machines with opposite attractivity respectively.

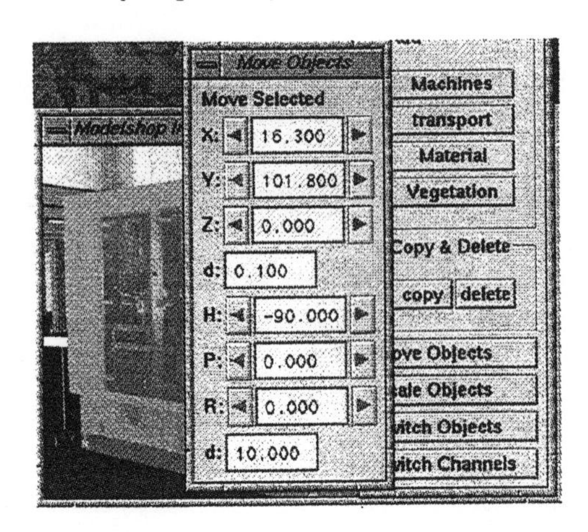

Fig. 5: Library and object insertion with MOD!FACT

After insertion of all objects structure formations begins. The orientation on that occasion is lying in the process tasks, which are defined by different relationships (material, information and communication flow) as well as on restrictions (formation of spatial-organizational structures) defined by the user. With the aid of the usual procedures for layout planning and optimization, which are integrated in MOD!FACT, as well as with intuitive methods for structuring, the optimization of the material flow with arrangement of the resources on the one hand and the improvement of communication and information connections on the other hand are in the foreground. In addition, MOD!FACT has an integrated structure simulator which accepts the data from model formation and values them dynamically. The link-up of simulation and animation directly with Virtual Reality allows a clear and significant representation of complex processes. In the scenery, progressing processes not only can be judged by qualitative criteria but also be evaluated to quantitative factors. In this case, the use of Virtual Reality as an interactive user interface allows manipulating acting processes directly in the model. Modifications immediately become visible and can be analyzed and evaluated without further expenditure.

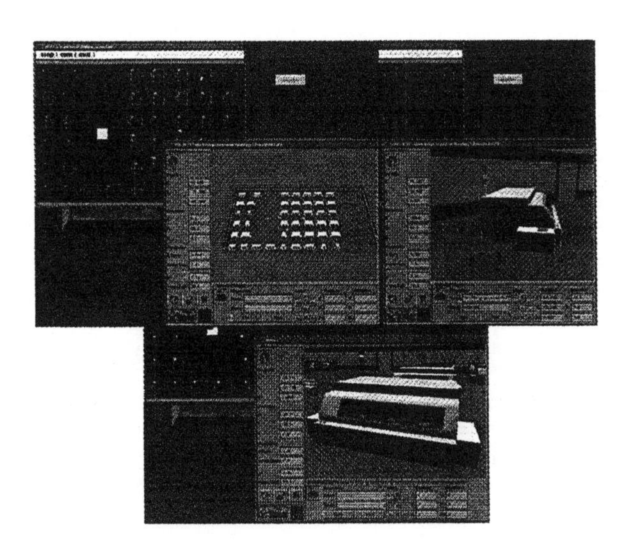

Fig. 6: Structure simulation with MODIFACT

After structure generation the users has the possibility to interact with the objects directly in the experimentation field. Therefore he could change the layout as well as move or delete parts of the building, like walls or windows, to make better conditions for an optimal configuration. Manner and scope of interaction depends of implemented possibilities of interaction in Virtual Reality. In the same way he can create different alternatives. A specific effect in this case is, that all the time the structure simulator is running in the background. If the user modifies the situation or the characteristics and qualities of some objects direct in the experimentation field and/or he inserts new ones or deletes some, the effects are considered instantly in the process simulation. The user can observe immediately the effects of his action and receives for qualitative valuation the respective keys from the system.

To sum it up it can be said that MOD!FACT provides an interactive factory experimentation field which allows by the aid of computer generic models in the technology Virtual Reality to activate learning processes by thinking over and investigating of consequences of specific strategies and measures as well as of confrontation with other strategies and results. Due to the intensive collection of information about the prevailing state of the system as well as the extensive acquisition of knowledge about the systems structure, the persons involved in the factory planning process are able to predict the future state of the system on the one hand side and to estimate the effects of practicable interventions on the other.

4. EXAMPLES

4.1 Honey factory – Planning a new production site

Objectives - Aim of this project was to support the design and realisation process for a new factory for the treatment of honey by the Model-Shop Factory Design. The models to be elaborated should represent the different solution alternatives worked out by all people involve to the planning process (architect, utility planer etc.) in an understandable way. They could be used like a platform for a dialogue between the different disciplines of planning in which new ideas could be discussed and worked out Moreover they should offer the possibility, to allow diverse experiments by modifications of partial aspects (moving machines, changing the material flow, etc.), in order to develop and correct the current models.

Realization - Modelling the new factory formed the basis for the common design of the factory by investigating every knowledge carrier as every architect, utility planer, plant supplier, decision maker but also the own employee. Therefore, the integrated illustration of the new factory was in the foreground in order to be able to represent the future situation vividly and to allow an evaluation with regard to building equipment, room use, operations and systems engineering.

Starting point for the model preparation was the reunion and preparation of the plans and of the information elaborated in the different planning groups. On the basis of this material, the generation of a first coarse model in Virtual Reality occurred. Therefore the planned building substance in connection with the already available buildings was to be visualised on the one hand as soon as to represent the required machines and plants in accordance with the planning state. In the following process of the project the information from the current factory organisation process was treated and integrated into the model correspondingly. The achieved planning state was valid to check it and to verify it together with all planning participants and to modify if necessary.

Fig. 7: VR-model of a honey production site

Results - In short term succeeding workshops and in an iterative procedure (check up - modification - check up - etc.) a precise and verified planning state was achieved, which was borne by all participants. A knowledge increase that led to a sufficient state of cognition at shortest time was implemented by the interpretation of the results elaborated into the workshops, to derive consequences for the organisation of the new factory and to take secured decisions within the framework of the current organisation process.

4.2 Pharmaceutical Enterprise – Redesign of a production unit

Objectives - The goal of this project was in the redesign of an available production field of a pharmaceutical enterprise. A part of the production should be shifted into another already existing manufacturing hall. To increase the acceptance of a future solution as well as of its consistent transformation the present solution should be judged and improved by including the affected employees. The result should be a layout, elaborated together. In this case, the development of the motivation factor by integration of all employees was considered to be essential.

Realization - For the carrying out of the project the relevant information (physical dimension of machines, operating cycles, necessary equipment etc.) for modelling was gathered and analysed, starting from the real system, by integration of the employees as soon as the goal were determined. The information and results fixed together with the employees represented the basis for modelling the future production field with the aid of the Model-Shop Factory Design and MOD!FACT. The level of detail necessary for model formation and the communication degree were defined on this basis. In addition it was essential for an exact layout planning to generate authentic illustration with precise dimensions. The VR-model was generated in such a way that it could be measured by means of communication of the shared team members on the formulated aims. Different solution scenarios were tested by communication of all shared collaboration with the model, whose results represented statements about the model. The interpretation of this results lead to consequences concerning the modelling of the future system.

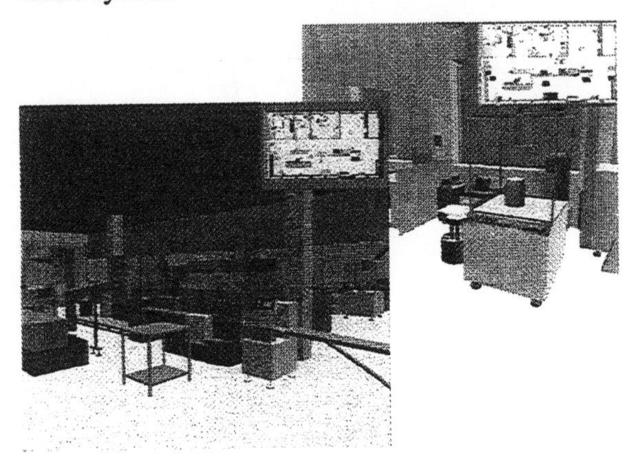

Fig. 8: VR-model of a pharmaceutical production unit

Results - Due to the use of Virtual Reality as tool for factory planning a multitude of different layout variants could be generated with the help of the affected employees. Therefore it was important to represent the future concept in an easy understandable way and to give the participants the possibility to modificate the VR-model on their own. The overcome of employees inhibition could increase the motivation for co-operation significant. Because of the willingness to realise the experiment together the experiences and knowledge of every individual employee was integrated in the planning process. In contrast to the common planning process the use of Virtual Reality as tool for factory planning achieved more realisation and also more security in planning process.

REFERENCES

Bauer, C. (1996). *Nutzenorientierter Einsatz von Virtual Reality im Unternehmen.* Computerwoche Verlag GmbH, München.

Bergbauer, J. (1998). Virtual Reality als Experimentierfeld zur *Fabrikplanung. In: Wirtschaftsfaktor VR.* Arcitec, Graz.

Daenzer, W.F. and Huber, F. (1992). *Systems Engineering - Methoden und Praxis.* Verlag Industrielle Organisation, Zürich.

Dörner, D. (1993). *Die Logik des Mißlingens – Strategisches Denken in komplexen Situationen.* Rowohlt Verlag, Hamburg.

Hartmann, M. (1996). *Erfolgreich produzieren in turbulenten Märkten.* Logis Verlag, Stuttgart.

Higgins, J.M. and Wiese, G.G. (1996). *Innovationsmanagement – Kreativitätstechniken für den unternehmerischen Erfolg.* Springer Verlag, Berlin.

Hoffman, H. (1996). *Kreativität – Die Herausforderung an Geist und Kompetenz.* printul Verlagsgesellschaft mbH, München.

Mezger, M., Bergbauer, J. and Ballerstein, H. (1996). Fabrikplanungsprozesse mit neuen Methoden. In: *Dezentrale Fabrikplanung – Neue Methoden und Mitarbeiterbeteiligung.* VDI-Verlag GmbH, Düsseldorf.

Wiendahl, H.P. (1996) Grundlagen der Fabrikplanung. In: *Betriebshütte – Produktion und Management.* Springer Verlag, Berling.

MAN-MACHINE COOPERATION IN NETWORK ERA FOR INFORMATION RETRIEVAL, FILTERING AND ACCESS

Hiroshi Tsuji and Hiroshi Yajima

Systems Development Laboratory, Hitachi, Ltd.
8-3-45 Nankohigashi, Suminoe, Osaka, 559-8515 Japan.
e-mail : tsuji@sdl.hitachi.co.jp

Abstract: This paper presents an integrated proxy server that supports information sharing among intranet users. The presented proxy server includes the following functions to cooperate with users: indexing for visited sites, URL classification into relevant categories, search term presentation by thesaurus, page caching for first access, and data transformation. These functions are designed to improve the accessibility to the Web. Experimentation from implementation is also discussed. Copyright © 1998 IFAC.

keywords : Information Retrieval, Information Flows, Group Work, Indexes
Cache, Internet Access, Proxy Server

1. INTRODUCTION

To search for URLs, keyword-based search engines (Cheng, 1996)) are powerful for many users. However, due to the increased number of WWW (World Wide Web) sites, users are uneasy with the large volumes of search results. It is difficult for them to get a relevant URL from the thousands of possibilities found by a search engine. On the other hand, "bookmarks" for storing favorite URLs are useful for revisiting sites that are frequently updated. However, it is not easy to share bookmarks among users. The search engines are implemented in a Web server while bookmarks is implemented in a Web client.

It should be remembered that there are proxy servers (Luotonen et al. 1994) that transfer messages between Web servers and Web clients. This provides chance to realize a groupware function in a proxy server because clients in the same organization are generally connected to the same proxy server, as shown in Figure 1 (Nishikawa, et al. 1998a).

From the viewpoint of Web information sharing, this paper presents a proxy server with the following functions: indexing for visited sites, URL classification into categories, thesaurus for keyword selection, caching for the first access, and data transformation. These functions are designed to improve the accessibility to the WWW. This paper will describe the man-machine cooperation in these

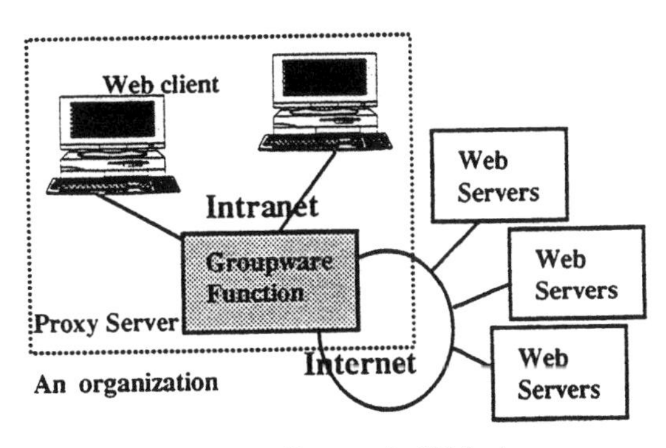

Figure 1 Proxy Server in Web Access.

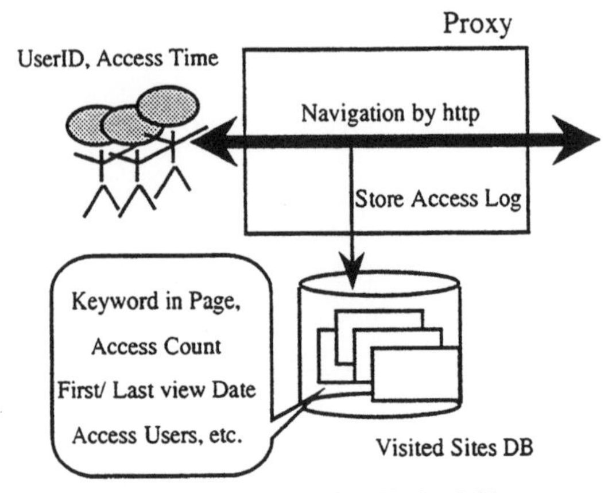

Figure 2 Indexing for Visited Sites.

functions.

2. INDEXING FOR VISITED SITES

When a user accesses a WWW site, our proxy server makes an index for the site. Such an index allows a user group to later search the site by keywords. The system keeps records of who accesses the site, when a user visits, and the dates of the first view and most recent visits. The system also updates the count of visits for each site. The conceput is illustrated in Figure 2.

The information stored in the visited sites' database in the proxy server is useful for developing a variety of query functions. For example, a user can learn the popular sites within an intranet group. Such a function can be regarded as collaborative filtering because group users generally have the same interests. Example queries might include:

(1) Sites I visited last month that include the terms "Hitachi" and "Internet",
(2) Sites that either Dave or Simon visited last week,
(3) Sites that more than six persons visited this month,
(4) Sort the sites by visit frequency,
(5) Sort the sites by most recent view dates of Taro's access.

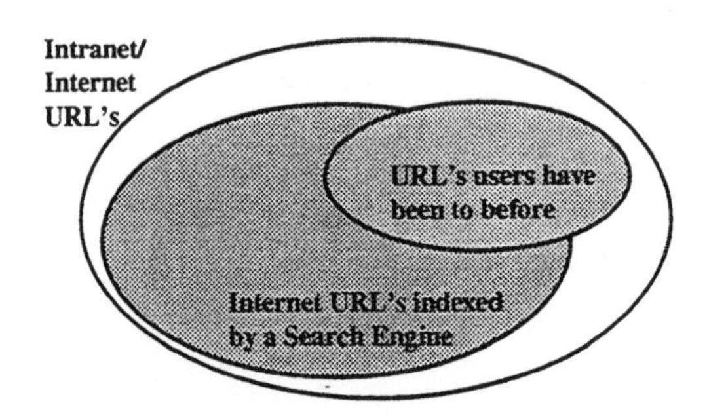

Figure 3 URLs in WWW.

On the other hand, some URLs will be deleted by evaluating the visited count and most recent view date (for example, if the access count is less than two a month), which keeps visited sites database to a manageable size.

It should be noted that the presented system collects URLs that at least one user has visited, while most search engines (Cheng, 1996) try to automatically collect as many URLs as possible from throughout the world. The relationship between them is shown in Figure 3. Then there is man-machine cooperation in URL collection.

3. URL CLASSIFICATION

As the system continues to collect URLs into its visited sites database, there may be thousands of URLs eventually gathered. From the experience, an active user collects more than fifty URLs in a week. A research group of five researchers collects about one thousand URLs in a month.

In general, these URLs should be classified into categories. The presented visited sites database has a framework for classification. While a user can classify URLs into categories, it is a time consuming task. Sometimes, he may misclassify or fail to classify an URL. Furthermore, each user in a group may have different criteria for classification.

Therefore, the presented system provides a function for automatic classification by sites and keywords. This function, called FLUTE, uses a knowledge base generated from training examples. The basic idea. discussed elsewhere (Mase et al. 1996b), is shown in Figure 4. The conceptual classification rule generated from training examples is as follows:

If the site page includes term xxx,
the site can be regarded as category yyy
by credit z(xxx, yyy)

where the value "credit z(xxx, yyy)" represents the degree of connectivity between the term and category. The value is assigned between zero and one. If there are two terms x1 and x2, the confidence for the category is calculated as follows:

$$Confidence(y) = credit(x1,y) + credit(x2,y) - credit(x1,y)*credit(x2,y)$$

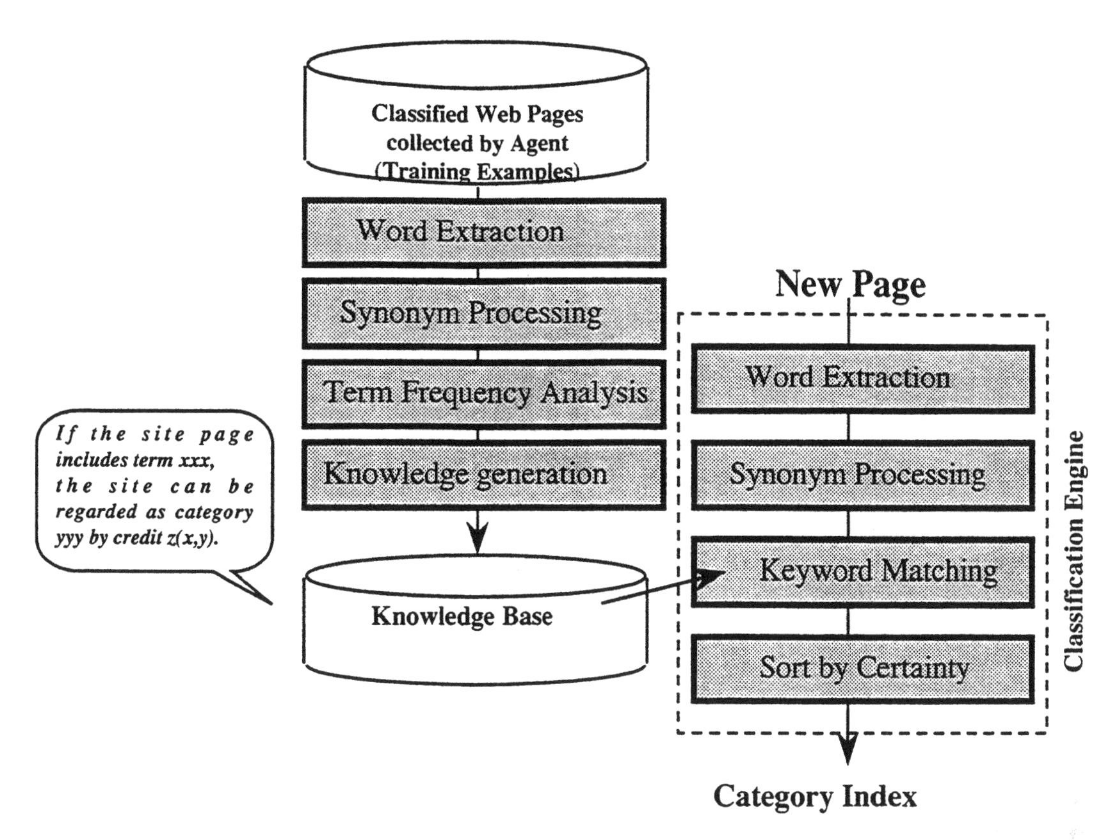

Figure 4 Components for Page Classification FLUTE.

To assign the credit of the relationship between category and keywords, not only term frequency but also the HTML tag is an important clue. For example, bold type font or large size font indicates the important keywords in a web page. The linked pages in the same site are also evaluated to derive keywords.

Table 1 summarizes the experiment conducted for page classification by FLUTE. A detailed evaluation and experimental simulation for classification will be discussed elsewhere.

Table1 FLUTE Evaluation Example

Categories	16 (borrowed from http://hole-in-one.com/
Training Data per Category	500 pages
Sample Data per Category	200 pages
Stop words	3287 words
Classification Rules	80,000 rules
Recall for Training data	78.83%
Recall for Sample data	37.85%

While the classification engine shown in Figure 4 sometimes provides a unique category, it may provide alternatives (more than two candidates) or give up to present a relevant category (should be classified into a new one). Because the current technique in FLUTE is based mainly on keyword frequency, it may fail to provide correct classification. There may be noise or misses. If the site is classified into an irrelevant category, it is noise. If the site is not classified into a relevant category, it is a miss. Even in such cases, the user can modify classification made by the system later. Thus, there is also man-machine cooperation in URL collection.

4. THESAURUS FOR KEYWORD SELECTION

For keyword-based search, it is a difficult task for users to select relevant terms. In general, it is well known that most users (more than ninety percent) specify only one term for a specific query. A one-keyword query retrieves too many noise sites or misses the target sites.

Accordingly, a thesaurus is useful to remind terms and to refine or expand search results. Thesaurus (Aitchison et al. 1987) is a dictionary that includes the relationships between terms, such as conceptual hierarchy (e.g., animal and dog). The broader terms contribute toward expanding the user's search while

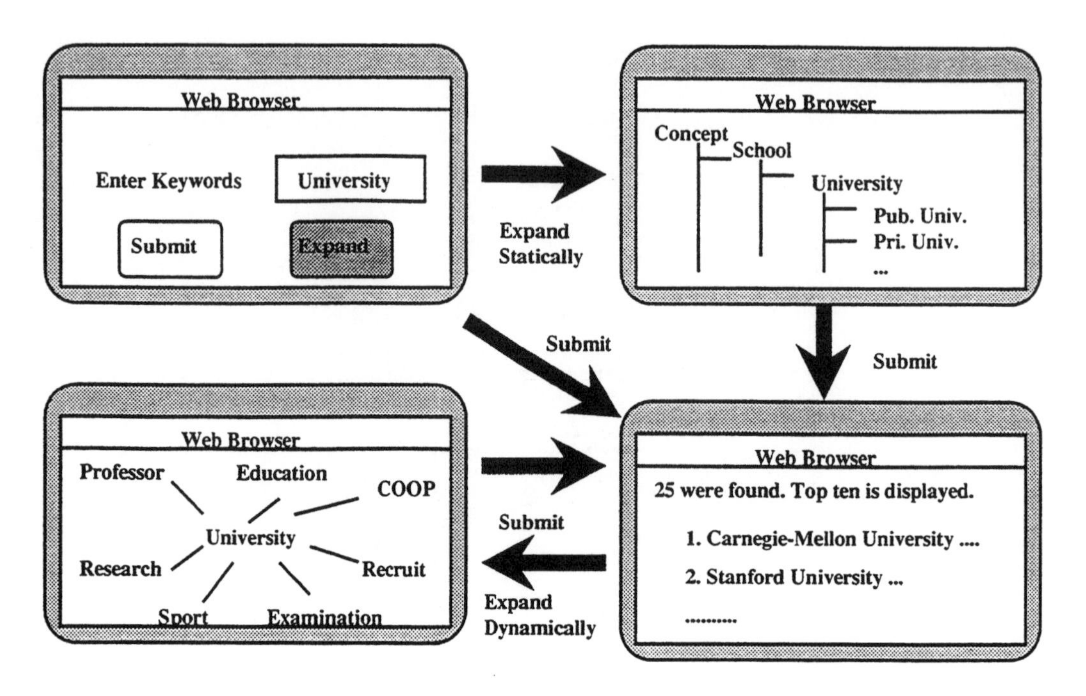

Figure 5 Thsaurus for Term Selection Aid by THEATER.

the narrower terms refine the search.

The thesaurus manager called THEATER presents not only the conventional broader and narrower terms but also co-occurent terms (Tokuda, et al. 1998). Co-occurent terms are pairs of words that appear in the same page frequently (for example, the term "university" has such co-occurent terms as "professor", "examination", and "research"). This concept is shown in Figure 5. THEATER is implemented in a proxy server and independent of any specific search engine.

Table2 THEATER Evaluation

	all counts Without THEATER	all count with THEATER	
Is-a relation	132	179	35.6%
Has-a Relation	8	11	37.5%
Synonym	19	21	10.5%
Compound word	119	211	77.3%

increased rate of words counts Avg 48%
increased rate of compound words count Avg. 77%

Consequently, concurrent pairs are generated from the HTML pages automatically collected by user access. On the other hand, the broader and narrower terms should be edited by a human expert. For intranet users, THEATER also provides a customizing function and a group maintenance function. Therefore, this is also man-machine cooperation.

One concern is how co-occurent terms contribute to improving search performance. Table 2 illustrates the experiment done on thesaurus usefulness where there is detailed discussion (Tokuda et al. 1998). The experiment counts the number of related terms generated for a given term for reminding human subjects of concept with and without THEATER.

Table 2 shows the kinds of terms THEATER reminds the user of. The reminded term is classified into Is-a relation, Has-a relation, synonym, and compound word related to the given term. Because the increase rate in compound terms is especially larger that the others, THEATER should store the compound words.

5. CACHING FOR FIRST ACCESS

Because many users in a group access the same URL repeatedly, it is reasonable to store the pages not in the client computer but in the proxy server. Furthermore if there is a clear requirement to get access, it is also reasonable to collect pages in the proxy server before users access the site. Thus caching that stores copies of original pages in the proxy server is an important function for improved Web access performance.

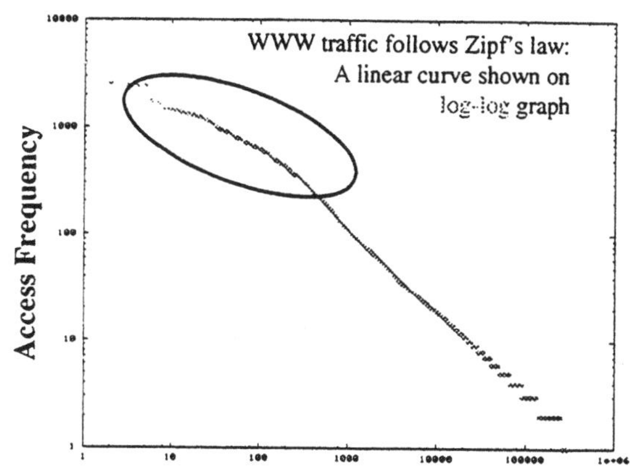

Figure 6 Order of Access Frequency

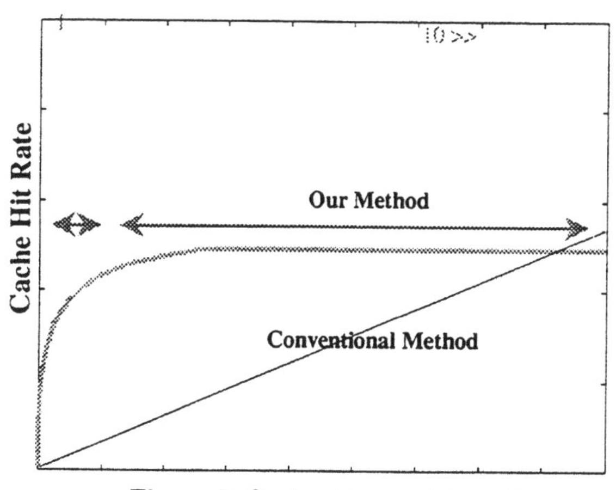

Figure 7 Cache Storage Capacity

While a conventional cache is a temporary file to allow users to revisit the least recently visited pages fast, the presented proxy server provides a couple of related functions. The system caches the pages that are visited many times for a specific period (Nishikawa et al. 1998b). This frequency-based approach reduces memory capacity for caching by using Zipf's law (Santlon, 1988). Figure 6 implies that sites with higher order of access frequency would be cached while most sites would not be cached. Figure 7 illustrates that the approach is robust if the cache storage capacity is limited.

There are several approached to caching:

(1) As popular sites in a group can be identified from an access log, the presented system pre-fetches and updates these pages for caching during idle times if the pages are modified.

(2) Search agents can collect pages for caches beforehand if the user specifies the initial pages, the number of links to follow, access time interval, and so on.

(3) Cache storage facility can also have an archiving function that continues to keep previously visited pages even if the original pages are modified. For example, this function is useful for checking news sites.

6. DATA TRANSFORMATION

Sometimes users would like to have a different format from the original HTML pages. On the other hand, some pages are prohibited from access because their contents. These requirements come not only from readability and access performance but also from the area of ethics.

For example, a user may want a machine to translate and summarize the pages (Mase et al. 1996a). He or she may also want to access the web pages by reducing large-sized objects such as images and videos (Shimada, 1997). Such functions are implemented in the presented proxy server.

Such functions serve important purposes:

(1) Original pages should be viewed if needed.

(2) The intellectual property rights of original pages should be respected.

7. CONCLUSION

The presented functions are integrated into a proxy server. Figure 8 shows the proxy server, which exploits man-machine cooperation for information retrieval, filtering and access. The integrated proxy server can be regarded as a groupware system for intranets.

Currently, this approach is aimed toward users organized in a company or similar organization. In the future, more diverse and informal user groups will be the target of the groupware system.

ACKNOWLEDGEMENT

This paper concerns the survey of the authors' research project. The research results owe to all colleagues of Kansai Systems Laboratory of Hitachi, Ltd. Special thanks are also due to Mr. D. Toth, Mr. S. Chen and Mr. H. Yagi of Hitachi Computer Products, America who presented proxy platform and indexing for visited sites.

REFERENCES

Aitchison, J. and A. Gilchrist (1987). *Thesaurus Construction*, Aslib.

Cheng, F.(1996). *Internet Agents*, New Riders.

Luotonen, A. and K. Altis(1994). World-Wide Web Proxies; *http://www.w3.org:80/History/1994/ WWW/Proxies/ Overview.html*.

Mase, H. et al(1996a). Computer-aided News Article Summarization, *Proceedings of 9th International Conference on Industrial & Engineering Applications of Artificial Intelligence & Expert Systems*, pp.627-632.

Mase, H., H. Tsuji, et al (1996b). Experimental Simulation for Automatic Patent Categorization, *Proceedings of Advances in Production Management Systems*, 377-382.

Nishikawa, N., et al (1998a). Proxy Server for the Control of Internet Information Flow, Systems, Control and Information **42**, 3, pp163-168.

Nishikawa, N. K. Yoshida, et al (1998b), Memory based Architecture for Distributed WWW Caching Proxy, *Proceedings of 7th World Wide Web Conference,*.

Shimada, T., et al (1997). Interactive scaling control mechanism for World Wide Web systems, *Computer Networks and ISDN systems* **29**, 1467-1477.

Salton. G. (1988) *Automatic Text Processing*, Addison Wesley.

Tokuda, T. et al (1998): Experimental Evaluation for Associative Keyword Reminder by Thesaurus, *Proceedings of Analysis, Design and Evaluation of Man-Machine Systems*, 16-19 September , Kyoto, JAPAN.

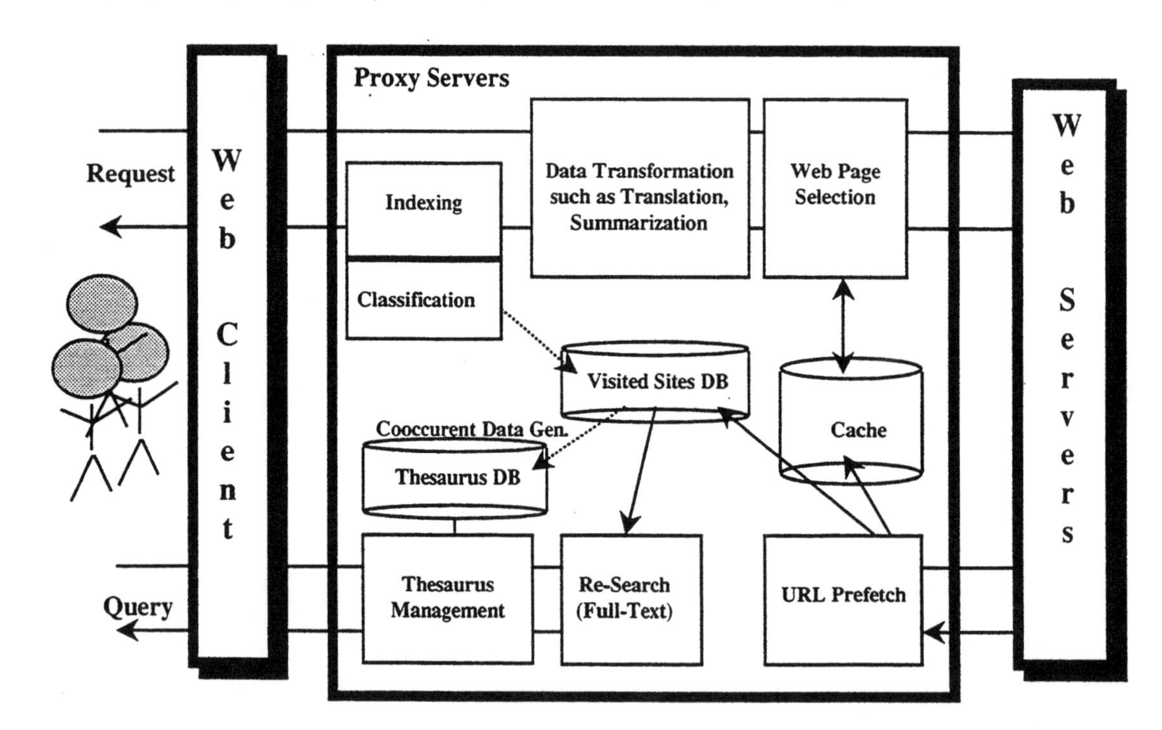

Figure 8. Target System Configuration Image

COOPERATIVE DIAGNOSIS SUPPORT TOOL INCLUDING A MULTI-POINT OF VIEW MODELLING

David Jouglet, Pascal Renaut, Sylvain Piechowiak, Frédéric Vanderhaegen.

LAMIH - URA 1775 CNRS
Le Mont Houy BP 311
59304 Valenciennes Cedex FRANCE
E-mail : {Surname.Name}@univ-valenciennes.fr

In collaboration and with the agreement of the CNET, French acronym of National Centre of Telecommunication Study and France Telecom corporation.

Abstract: System modelling is the first step of diagnosis research. Up to now, the automatic diagnosis domain was focused on technical applications, without taking into account human capacity reasoning. On the one hand, the difficulty to get an exhaustive expertise, and on the other hand, the high calculating time are the weak points of automated methods. This point argues to evolve a new vision of man-machine diagnosis.
Human diagnosis studies underline reasoning on multi-representation. Topological, functional, probabilistic and case based points of view are the complementary human mental representation of system misbehavior integrated in the proposed approach. The validation of a tool including this multi-point of view modelling gives first results about the approach. *Copyright © 1998 IFAC*

Keywords : Multilevel modelling, Mental model, Man / machine interaction, Computer-aided diagnosis, Decision making.

1. INTRODUCTION

This paper presents the interest and the bases of a multi-point of view modelling adapted to diagnosis task. The basic principle of this modelling is a multi-representation of information : a topological, a functional, a probabilistic, a case based and a tests based points of view. In fact, strategies handled by operators are various and depend on the situation. Diagnosis support has to be designed related to mental models commonly used by operators, in order to inherit of human diagnosis qualities and to free human operators from any constraints when using this support tool.

A case where human is highly implicated in system's behaviour was chosen to apply our approach : telephonic troubleshooting repairing service of France Telecom.
Preferential human-machine co-operation modes are identified, an analysis of a set of experiments will aim at designing new support tools.

2. HUMAN-MACHINE CO-OPERATION

Human-machine co-operation concept implies both human and machine decision-makers. Therefore, related to vertical and horizontal co-operation modes, different classes of support tool can be designed: the preventive tool class, the curative tool class and the regulative tool class (Vanderhaegen, et al., 1996), figure 1.

Both preventive and curative classes are related to a vertical co-operation mode because only one decision-maker, i.e. the human operator or the machine, acts alone on the process, but can take into account possible assistance from the other decision-maker. On the other hand, the regulative class is an application of the horizontal co-operation mode.

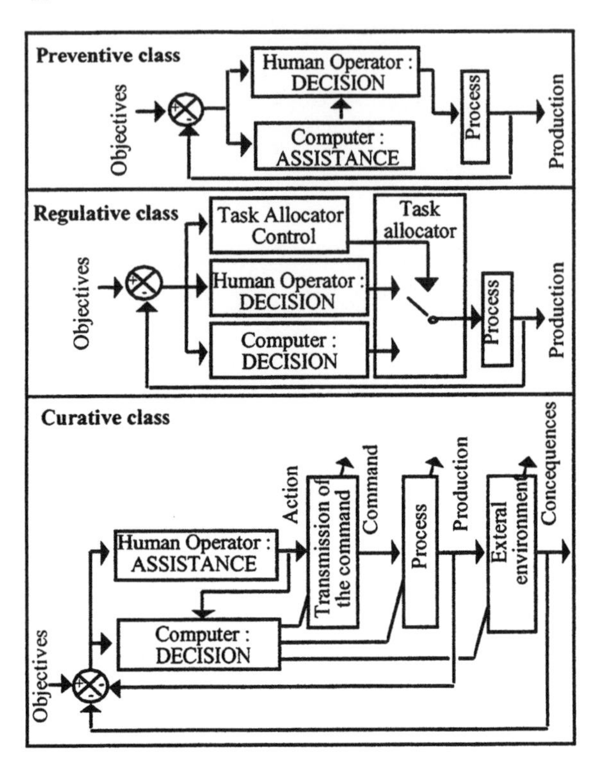

Fig. 1 : Example of computer based co-operation.

The preventive tool class concerns pre-action based tools. This kind of tools supports decision provides human operators who act alone on the process with advice. For instance, this aims at avoiding human error and can be useful to support alarms filtering or diagnosis.

The regulative tool class concerns tools operating during action. This class requires an organisation in which assistance tool and human operator are on the same decisional level. Control and supervisory tasks can then be distributed between human and computer in order to relieve human operator of overloaded situations. The corresponding task sharing policy is either subjective in a manual allocation, or objective in an automated allocation, related to the task characteristics or to the human behaviour, in order to obtain an optimal regulation of system performance. For instance, this aims both at reducing the risk of human error and at regulating workload.

The curative tool class concerns post-action based tools. This class aims at recovering human error after action. This allows human operators to have feedback about any erroneous actions, and adds advice about intervention and fault management. The computer based tool can act directly on the transmission of the command, on the process or on the external environment. For instance, those interventions are over-ruling commands or emergency stops. This approach to fault management is thus an intelligent watchdog which filters human action according to evaluation criteria such as a list of possible erroneous actions, or cognitive modelling.
In the case of diagnosis support tool, a preventive structure in vertical co-operation characterises the situation.

3. INTEREST OF THE MULTI-POINT OF VIEW

An industrial context related diagnosis can be achieved with regard to a compromise between two opposing solutions. On the one hand, diagnosis can be achieved automatically : human can be considered as an information sensor. On the other hand, diagnosis can be achieved by human operator without any decision support tool. Both cases illustrate two classes of research on diagnosis : artificial intelligence and human factors community.

3.1 The multi-point of view and diagnosis

The processing of computer failure diagnosis can be divided into three consecutive steps (Falzon, 1989) (Figure 2) :

- Problem presentation and understanding.
- Hypothesis production about the kind of failure due to observations.
- Hypothesis validation, i.e. explanation and failure repairing.

Failure is a modification of the system behaviour which implies a misfunctioning of the system. The misfunctioning of the system produces observable events called symptoms. Then the symptoms can explain failures.

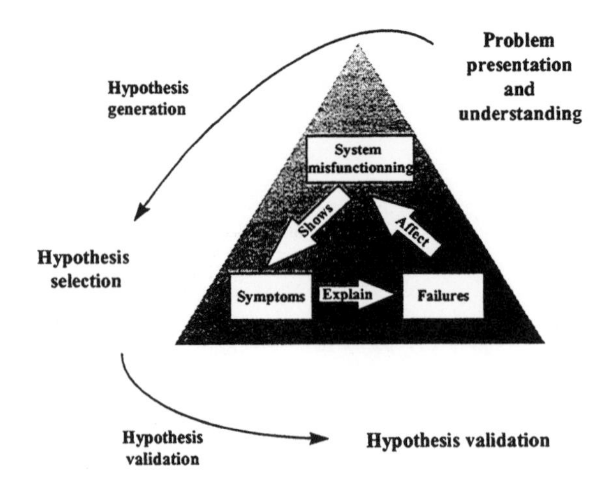

Fig. 2 : Diagnosis steps and hypothesis handling

System misbehavior based approaches, called « *experiential approaches* » (Reiter, 1987) are based on symptom and failure list integrating the links between them. For instance, the case based approaches aims at analysing the past according to relations between failures and observed symptoms during repairing.

On the other hand, normal system function based approaches, called « *model-based diagnosis* » focus on describing in details the system and its behaviours (De Kleer, 1987) (Reiter, 1987). If the observed behaviour is different from the expected one, the diagnosis task is made up of identifying which component can explain this observation. This activity is called the « discrepancy detection problem » (Davis 83). Qualitative models can also be used in such an approach. They are made of parameters representing physical characteristic values and the relations between those parameters.

3.2 The diagnosis and telephonic context

In the case of telephonic troubleshooting repairing service at France Telecom, the construction of a normal system function model seems to be difficult. As a matter of fact, in order to solve this modelling problem and to reduce computing time, a second alternative can be developed : the multi-point of view modelling.

The real system on which the diagnosis task is focused is composed of :

- Many technical sub-systems, e.g. France Telecom's distribution facilities and customers equipment.
- An user (i.e. the customer) integrated in the possible diagnosis, e.g. he/she can produce a failure on the system or be the failure himself/herself.

Each part of the system integrates its own failures. The diagnosis purpose is to identify the origin of the disturbance, for example trouble due to customer fault or technical network failures.
In the real situation, operators are able to solve diagnosis problems with high adaptability. This human quality may complement the automatic approaches. Therefore, it is important to investigate on operator's abilities to solve problems. The approach propose a reasoning among several mental models.

4. BASES OF THE MULTI-POINT OF VIEW

Different mental models can be used by human operators related to the working situations : functional, structural and case based models (Vicente, et al., 1995) (Rasmussen, 1986) (Reason, 1993).

4.1 Principle of multi-point of view

Moreover, in order to refine the final diagnosis, a set of exhaustive tests can also be applied. Therefore, the multi-point of view approach will be based on five mental representations(Figure 3) :

- The topological view integrates the structure characteristics.
- The functional view synthesises the functions of the system.
- The case based view itemises the typical situations that might occur.
- The test based view is a list of exhaustive tests on each possible failure.
- The probabilistic view classes the occurrence of each failure.

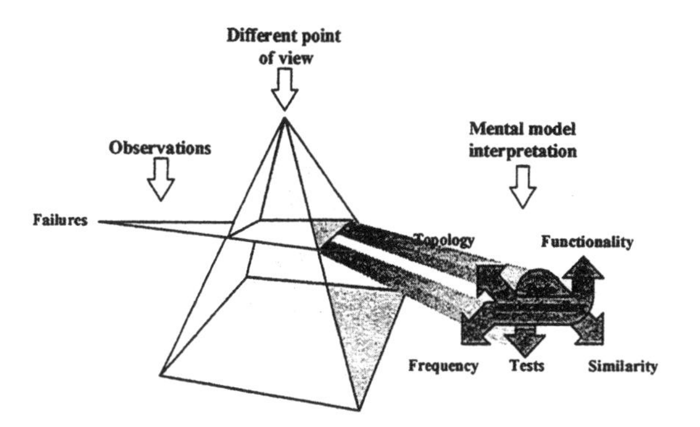

Fig. 3 : Multi-point of view failure interpretation.

The five points of view are the link between three important characteristics of man-machine diagnosis :

- The mental models of human operator as different ways to understand information in the situation.
- The observations produced by system malfunction that can appear as symptoms of failures.
- The information's representation of the situation integrated into a support tool.

The views are different ways to understand observations and to make correspondence with mental models.

4.2 Support tool including approach

With regard to the support tool integrating this architecture, views are kept coherent by using a common failure set. Each view is modelled as a hierarchy, from symptoms on the top to failures on the bottom. Links between nodes express consistency : each manifestation corresponds with a sub-set of failures that can explain it. Operator can use the views to lay down hypothesis about failures or observations about the suspected situation. Sub-sets of failures can be eliminated by human operator by excluding observations (because they are not relevant) or selected by focusing on them (presence of observations).

The set of all possible failures used in approach is unique and linked to all views. Each failure is covered by at least one view. Most of the failures are covered by several views, there is generally not only one interpretation of an observation.

By focusing and excluding hypothesis on the views, the operator acts directly on the failure set to select the most relevant ones only (Figure 4). For a given point of view, the corresponding set of suspected failures P={p1, p2, p3, p4, p5, p6}, can be modified by excluding or focusing a symptom, the set is then modified. A mask of failure related to actions is used to maintain coherence between views. A treatment on action's history and their failure's mask permits to undo any action at any time without order constraint.

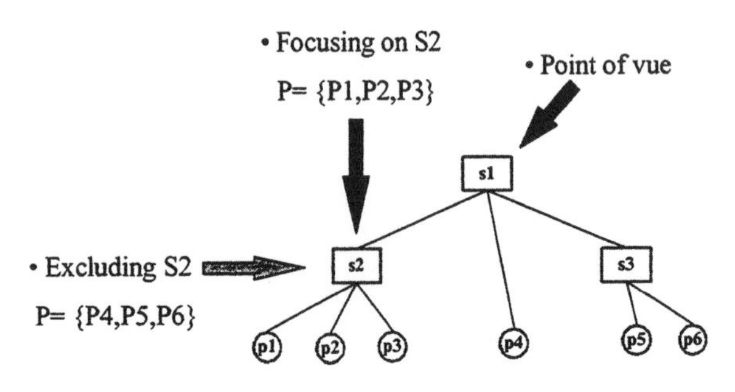

Fig. 4 : Excluding and focusing in a point of view on a sub-set of failures

Human diagnosis qualities such as gathering, filtering, understanding and choosing information are exploited. The tool is able to assist human in maintaining information coherent and in giving different presentation of the problem with regard to different points of views.

4.3 *The control panel of the support tool*

Above this treatment of information for diagnosis, a control panel allows to the operator all the sub described actions on views.

The solution chosen to handle the multi-point of view was a visual control panel (described in (Jouglet, et al., 1998b)). A particular attention was focused around usability of the tool.

The main characteristics are :

- No order constraint for the operator in use of the different views.
- Available controls to select information in each point of view easily .
- Integration of a feedback on the relevant failures list and selectable observations at any time.
- A special area and controls to resume operator actions and backtrack on any part of reasoning (any action can be undo without re-selecting all consecutive ones).
- All the views on the same screen.
-

4.4 *The expertise integrated in the tool*

An identification of failures to be taken into account on each model was realised (Jouglet, et al., 1997). A list of 49 failures related to phone troubleshooting service. It concerns external process functioning problem, i.e. it includes customer related problems and phone network related problem (Figure 5):

- Customer equipment problems such as problems for using a given service, problems with the local phone configuration, problems with the physical telephone, problems with the plugs or the sockets.
- Phone line problems such as problems with concentrators, with cables or with switching system.
- Correspondent related problems, such as problems for customer because of the phone configuration made by the correspondent, problems with phone installations.

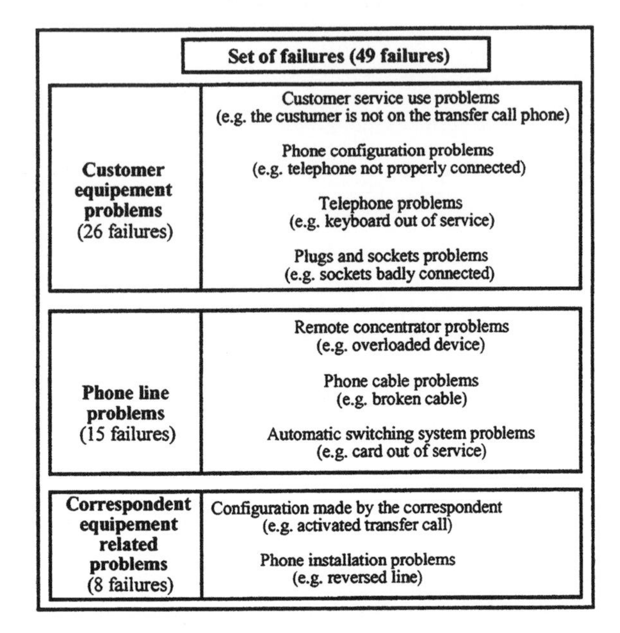

Set of failures (49 failures)	
Customer equipement problems (26 failures)	Customer service use problems (e.g. the custumer is not on the transfer call phone) Phone configuration problems (e.g. telephone not properly connected) Telephone problems (e.g. keyboard out of service) Plugs and sockets problems (e.g. sockets badly connected)
Phone line problems (15 failures)	Remote concentrator problems (e.g. overloaded device) Phone cable problems (e.g. broken cable) Automatic switching system problems (e.g. card out of service)
Correspondent equipement related problems (8 failures)	Configuration made by the correspondent (e.g. activated transfer call) Phone installation problems (e.g. reversed line)

Fig. 5. Gathering of phone network troubleshooting.

A convenient framework for human hypothesis management is provided by the multi-point of view based tool and aims at preventing human errors (Jouglet, et al., 1998a). Experimentation underlines advantages and disadvantages of this kind of tool.

5. THE FIRST RESULTS

The tool was tested with 10 experts of telephonic network troubleshooting experimenting 15 different scripts. The experimental protocol integrates a training stage (during half a day) at first. Then tests phases (about 2 minutes) and debriefing phases (5 to 10 minutes) have been alternated to collect subjective data with each operator (Figure 6).

The situations and the failures are field based ones but the customer was simulated (he was playing a script associated to failures). Important data were recorded, as actions of operators on the control panel and verbal exchange between operators and customers. Subjective information about the experience was collected by interviewing the operators after each experiment.

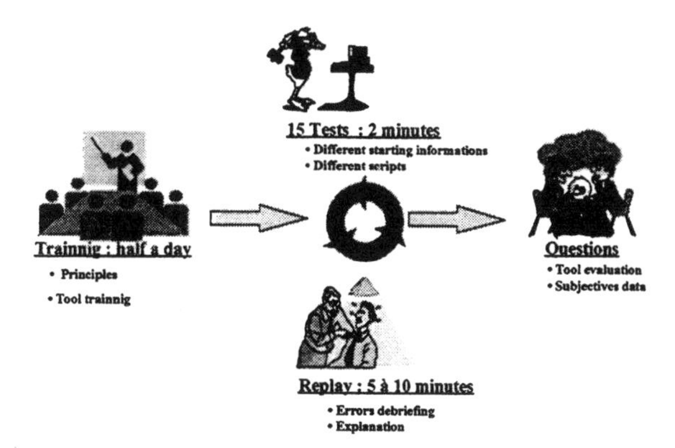

Fig. 6 : The experimental protocol

The analysis of data produced first conclusions about the proposed support tool. The tool facilitates diagnosis activity (about 70% of exact diagnosis) and gives a frame to errors prevention and recovery.

The results of diagnosis are distributed among four categories of experiments processing (Figure 7) :

- The first concerns exact diagnosis without any correction or recovery on human reasoning (51,3% of all tested scripts) : in this case the couple operator-tool was totally efficient.
- The second and the third include cases in which an error was detected, for 18% of all diagnosis operators are able to resolve the problem, for 22% of them it was impossible for him to find the exact one (for multiple causes).
- The last category concerns non detected errors or situations in which operators were not able to recover their errors (8% of all tested scripts).

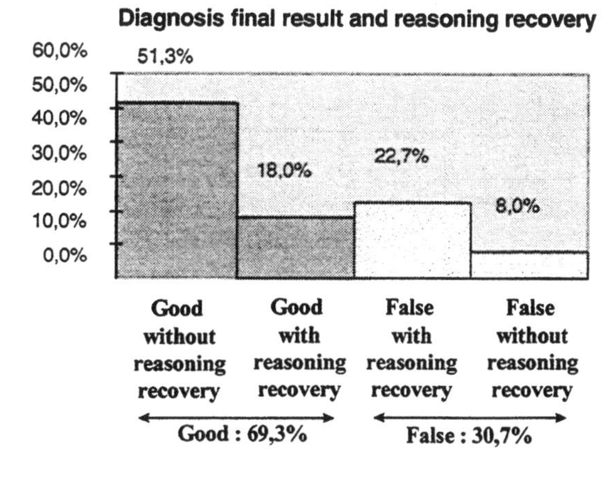

Fig. 7 : Different diagnosis results

The 30% of false diagnosis show that in some case a preventive support is not sufficient : recovery and detection has to be enhanced. In this situations, operators stand by new information or evaluate their reasoning to backtrack. Then most of errors occurs when the risk taken when laying down an hypothesis is too high (e.g. because operator has no ideas on possible failures).

According to the different selected observations and laid down hypothesis, 4 of the 5 views are well used (Figure 8). The functional and test views are the most used ones (40% and 36%). Topological and similarity views are used in particular cases when a specific context of information appears. Frequency view is quite never used, in fact probabilistic information is taken into account implicitly by operator. Operators do not need to select explicitly this kind of information on the control panel. The four most used views have been described as very important by operators. Those views allow them to display several aspects of the problem related to a given situation.

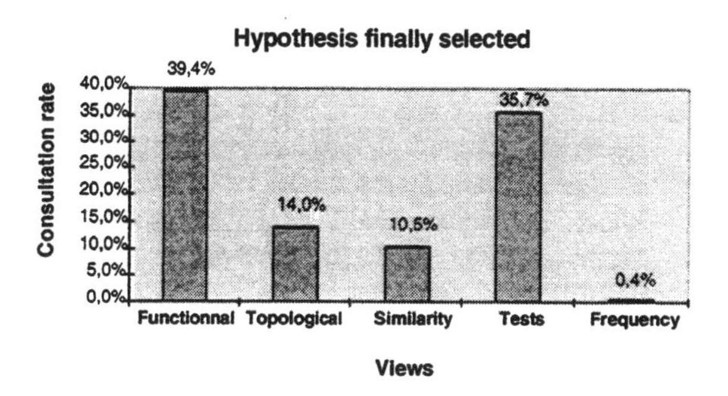

Fig. 8 : Consultation of the different views integrated in the tool.

The control panel used to manage hypothesis contributes to obtain positive results during experimenting the multi-point of view based tool. A lot of the integrated elements are useful but can be refined. The main results of control panel's evaluation, is the very good acceptance of operators and a good transfer of knowledge between human and machine.

6. CONCLUSION

This paper has detailed the multi-point of view approach based on five most important mental representations of the situation. This approach was applied to phone network troubleshooting and a diagnosis support tool based on multi-point of view has been tested. The purpose of this tool is to allow human operators, using different problem solving strategies to manage the diagnosis coherence. The first results show that this tool facilitates human diagnosis activity and validates a multi-point of view approach. Four of the five views integrated were well used and participate in the diagnosis process.

The efficiency of the couple operator-tool underlines the interest of man-machine co-operation in diagnosis. Some important parameters of human operators and situations have to be noticed and their influence on diagnosis reasoning was analysed. The test of the tool gave a lot of information about human errors causes and some ways to increase robustness. Beyond the proposed multi-point of view approach, thc study of data collected on man-machine interaction in diagnostic task will precise the guideline to specify efficient support tool to diagnose a system.

The first conclusions are incentive but not sufficient to specify a new class of support tool able to reduce human errors. Over human errors avoidable with this kind of approach, some situations still cause problems. The preventive context of multi-point of view support tool has to be completed by anticipating human errors. An active automated reasoning module is actually developed to support operator in critical situations.

7. REFERENCES

Carroll J.M. (1984). Mental models and software human factors : an overview. In :*Research report RC 10616*. Yorktown.

Davis R. (1983). Reasoning from first principles in electronic troubleshooting, *International Journal of Man-Machine Studies,* **n°19**, p. 402-423.

De Kleer J., Williams B.C. (1987). Diagnosing Multiple Faults. *Artificial Intelligence,* **n°32**, p. 97-130.

Falzon P.(1989). Cognitive Ergonomics of Dialogue Presses Universitaires de Grenoble, Grenoble.

Gentner D. et Stevens A.L. (1983). *Mental models.* Hillsdale, NJ : Lawrence Erlbaum Associates.

Jouglet, D., Piechowiak, S., Vanderhaegen, F. Renaut P. (1997), Decision support system which is cooperative with its user, Contractual report CNET France Telecom/University of Valenciennes, Contract number 94 1B.Valenciennes.

Jouglet, D., Piechowiak, S., Vanderhaegen, F (1998a). Toward a multi-model based co-operation to support human-machine diagnosis : The case of phone network system. In : Computational Engineering in Systems Applications CESA'98 IMACS Multiconference (P. Borne, M. Ksouri, A. El Kamel), p 582, Hammamet.

Jouglet D., Piechowiak S., Vanderhaegen F., Renaut P. (1998b). Decision support system which is cooperative with its user, Contractual report CNET France Telecom/University of Valenciennes, Contract number 94 1B.Valenciennes.

Rasmussen J. (1986). *Information processing and human-machine interaction.* Amsterdam : North-Holland

Reason J. (1990). *Human error.* Cambridge University Press.

Reiter R. (1987). A theory of diagnosis from first principles, *Artificial Intelligence* **n°32**,p 57-95.

Rouse W.B. et Morris N.M. (1986). On looking into the black box : prospects and limits in the search for mental models. *Psychological Bulletin*, 100, 349- 363.

Samurcay R. (1995). Conceptual models for training. In cognition and human computer coopération. (Hillsdale), Lawrence Erlbaum Associates.

Vanderhaegen F., Telle B. and Moray N.(1996), Error based design to improve human-machine system reliability, In Computational Engineering in Systems Applications CESA'96 IMACS Multiconference, Lille, France, July 9-12, pp. 165-170.

Vicente K.J. , Christoffersen K., Pereklita A. (1995). Supporting operator problem solving through ecological interface design. *IEEE Transactions on Systems*, Man and Cybernetics, **n°25**, p 529-545.

GLOBAL DEMANDS OF NON-EUROPEAN MARKETS FOR THE DESIGN OF USER-INTERFACES

M. Romberg, K. Röse, D. Zühlke

University Kaiserslautern, Institute for Production Automation

Abstract: Nowadays the interaction between different cultures has become a common occurrence. The Institute for Production Automation of the University Kaiserslautern investigated the emerging markets in the Far East to evaluate the differences of eastern cultures in machine operation. This evaluation identifies the momentary requirements and needs. Most of the developing countries, have prior needs than a well designed human-machine-interface. Nevertheless the study pointed out, that user-interface-design has to take into account the social and cultural differences of these markets.

Keywords: cultural aspects of automation, human-machine-interface, international surveys, standards, behavior, tests, machine recognition

1 INTRODUCTION

One of the byproducts of increasing globalization is the intensified communication and economic exchange between countries. This is facilitated through modern forms of transportation and telecommunication which make it possible to bridge large distances in relatively short periods of time. Thus, interaction between very different cultures has become a common occurrence. Yet, the different rituals, practices and rules of communication, nonetheless still seem foreign and strange for those newly confronted by them. For Europeans, the Far East is particularly challenging in this regard.

On the other hand, economic access, low incomes, the desire to catch-up to western levels of consumption, the advantageous conditions for establishing firms, and the large population levels, make the far eastern regions especially attractive production sites and sales markets. Whether to sell products or produce components inexpensively, many investors from industrial nations are interested in the Far East. However, in order to be successful, detailed knowledge of the needs, similarities, differences, and technical possibilities of these areas is necessary. It is essential to possess a well-developed customer orientation for success on these markets. The goal of the project INTOPS[1] is to contribute to the knowledge of potential Asian markets by investigating which requirements „new markets" potentially pose for the design of machine controls in comparison with the USA.

2 HUMAN MACHINE COMMUNICATION

In essence, machine operator interfaces are a form of communication, and one which has become an important topic of research due to the increasingly complex and rich functional spectrum found in today's machines (see Zühlke, 1996). In Germany, under the rubric Human Machine Communication, it is intensively studied in colleges and universities. The focus of the research is developing user-oriented operating systems, that is operating systems adapted to the abilities and tasks of the users, in order to facilitate and optimize operation. To achieve this it is necessary to define a generally understandable coding

[1] Global demands of non-European markets for the design of user interfaces. Project funded by the German minister for Research and Technology.

scheme for information so that the information transmitted by the machine lets the operator know what the current status of the process is. This coding scheme is compiled in so-called style-guides in which the terminology, design attributes of the coding scheme, the way the information is grouped, etc. are all laid down. These style-guides were first created for Germany and then with the introduction of CE guidelines were extended for European use. In defining the terminology it was found that, in part as a result of the differing languages, foreign cultures place different requirements on the information coding scheme. Also, as we know, communication does not take place entirely at the level of speech.

As made clear by many news reports recently, problems in understanding often emerge particularly in communications with far eastern discussion partners due to differing intercultural forms of communication. Communicating non-verbally through body language and gestures is determined by the cultural and social framework, which in turn, influences how information will be interpreted. For this reason it is important to examine how various cultural influences on communication effect the human-machine interaction and to see to what extent machine operating systems can be designed that take the idiosyncrasies of different cultures into account (see **Figure 1**).

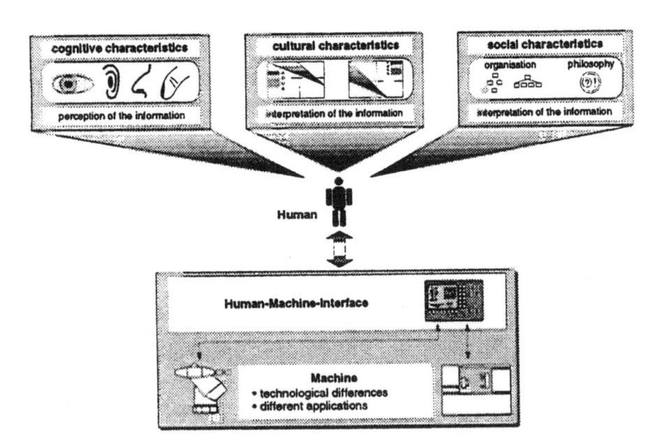

Figure 1: Characteristic elements influencing the interpretation of perceived information by operating machines (Romberg, 1997)

3 INVESTIGATED NATIONS

This project investigated this issue in 5 countries rated as important export markets for German machines and process lines[2]. The countries in the study were found to have the following characteristics:

- The USA are currently the strongest sales market for German machines outside Europe and has displayed consistently high sales figures over a period of several years.
- China is especially important as a future market, given its enormous potential in terms of population, resources, etc. and its development in the last 4 years (since 93/94). The sales volume in China for machine tools has reached that of the US. and an end to growth is not yet in sight. Difficulties lie, however, in China's very different culture compared to that of Germany.
- South Korea has developed quickly in the past several years up to the point of being almost comparable to highly advanced industrial nations. The strongly expanding economy with a growing product spectrum demands equipment which, as the statistics show, to a growing extent are bought in Germany.
- India is an important potential market with high levels of resources; large increases in sales volumes are also evident. The new more capitalistically oriented government is encouraging import by decreasing tariffs and simplifying regulations. Particularly relevant for the study are the obvious cultural differences to Germany.
- Indonesia has ambitious goals for its future and has the reputation for being very positively inclined to Germany. It is characterized by a large growth potential and high levels of training thereby representing an important market for German exports despite current drops in orders.

Four to five industrial sites were chosen in the selected countries from the following sectors: automobile producer, automobile supplier, plastics or synthetic manufacturer or company with a product reflecting national priorities. The differing company types broadened the scope of machines that could be observed, but limiting the diversity of company types allowed comparisons across countries. Within the companies a range of interviewees - management, machine operators, maintenance people - were questioned to acquire a range of interaction with the machine operating system. The interviews were conducted using a standardized set of questions on qualification structures, work organization, thought processes, training procedures, machine purchase practices, experience and problems with technology to allow comparability, but with open-ended answers to enable flexibility according to company situation and interview partner. A variety of survey methods for the identification and evaluation of the meanings of color and symbols were also carried out.

[2] Statistical data on the import and export of machine building products in these countries were gathered from publications of the German Machine Tool and German Machine and Process Line Associations (see VDMA 1995).

4 RESULTS

4.1 *Major results*

The results of the study (see Romberg, 1997) confirmed that an orientation to the needs of the customer is extremely important in order to have success on these markets. One particularly significant finding was that each individual market, perhaps even each individual customer, makes differing demands on the product. However, there is broad-based agreement on certain, especially elementary, requirements for the machine. Therefore one outcome of the study was the development of a „needs pyramid" (see **Figure 2**). The needs of the lower levels of the pyramid have to be fulfilled to reach the next level.

- The first level, and thus the primary need, is to achieve **process ability** and meet elementary company prerequisites (climate, energy source).
- At the second level the need for continuing **process reliability** has to be satisfied (service, parts).
- The third level involves the need for completely dependable basic functions - in other words, **basic functionality**.
- The concerns of the fourth level is to satisfy all of the needs involving **special functions**.
- The fifth level is the need for free **configurability** on the part of the customer.

4.2 *Process stability*

The significance of human-machine communication begins from the third level onwards and increases quickly in importance at the fourth and fifth levels. The levels one and two involve mainly technically oriented points.

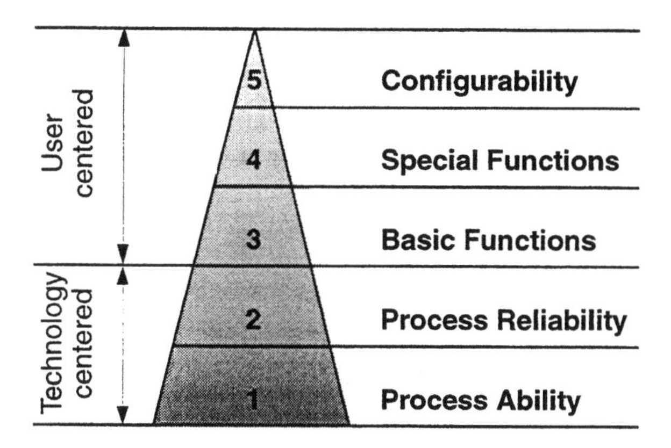

Figure 2: Requirement oriented development model „needs pyramid" (Romberg, 1997)

In three of the five countries visited (China, India and Indonesia), the technically oriented needs of levels one and two have not been satisfactorily fulfilled. Electronic control components of machine tools are not reliable either under the existing conditions of energy fluctuations or climatic conditions (Level one). Defective components cannot be acquired locally and can only be exchanged by highly qualified service technicians who often have to come from abroad. Thus the maintenance of process reliability (Level two) is unsatisfactory. This shows that user-friendly operation of the machines is not the main problem in these countries at this time.

4.3 *Life cycle oriented optimization of machine operation*

German machines were criticized due to their complexity and difficulty to repair. Complexity is defined as a large number and interdependence of operating functions, which the semi-skilled machine operator does not utilize. Difficulty in repair means that the machine can only be serviced by specialized technicians. In one of the companies that was visited, in which German products are manufactured under license, the products in some areas were also criticized for being difficult to assemble. These are weaknesses which damage the good reputation of German products and erode customer satisfaction. To achieve and keep customer satisfaction, it is necessary to satisfy all of those coming into contact with the product throughout its life cycle, from the assembler, the maintenance department, the operator, etc.

4.4 *Modularization*

To improve the design of the machine controls, proposals can be made in the areas of modularization of functionality, information coding and the design of operating sequences. Given their broad functional range, German machines have the reputation of having a high performance potential and therefore a wide utilization spectrum. In many countries, however, the machine operator do not have the types of training, nor is work organized in the plants in such a way, as to allow the use of these diverse functions. Therefore, for many machine operators, the performance capability of the machines turns into simply intransparent complexity. In order to come to terms with this complexity, without limiting the performance capacity of the machine, a modularization of the machine and especially the operating system, is recommended. In this way, the machine producer can coordinate the control possibilities and the functionality of the machine to the qualification level of the operators and the product spectrum of the company. This modularization also allows the machine producer to respond flexibly to customer wishes and only to sell the components and functions that are really needed. The potential for configurability, however, makes it possible to adapt the performance ability of the machine to the increasing experience of the machine operators and the growing needs of the company. If the modular functions are made up of operable units, then the needs of the customer will be optimally met.

4.5 Color and icon coded information

The findings in the area of information coding should be viewed with some caution given the small sample size for such a broad range of cultures. Therefore the results should be understood as representing tendencies derived from the statements of the surveyed respondents. However, the assumption that there are different requirements in the various markets in terms of information coding was confirmed.

With regard to the use of colors for information coding, red and green appear to be understood worldwide for production purposes (see Norm DIN EN 60073) the same way they are in Europe. The study could not proof, that foreign cultures relate a different meaning to a color. Hence, it shows that the meaning of a color depends very significantly of its context. This can also be stated for European cultures. For instance the color *red* can mean love, power, threat or alarm. One explanation for this is that under all conditions the same emotional excitement is generated. Based on these results it can be assumed that the state of excitement for technical important conditions is the same for Asians as for Europeans. For positive, normal conditions like e.g. „machine is operating according to plan" the colors green or blue where assigned. In Europe the color yellow is used for coding the condition „warning" as an interstate between normal operation and alarm. In India, Indonesia and the USA everybody interviewed agreed to that. However China and Korea do not differentiate that precisely between the operating status. All people interviewed used for statement e.g. „the machine is operating then the condition is *normal* or the machine is stopped then the condition is *alarm*" the according colors green and red (see **Figure 3**).

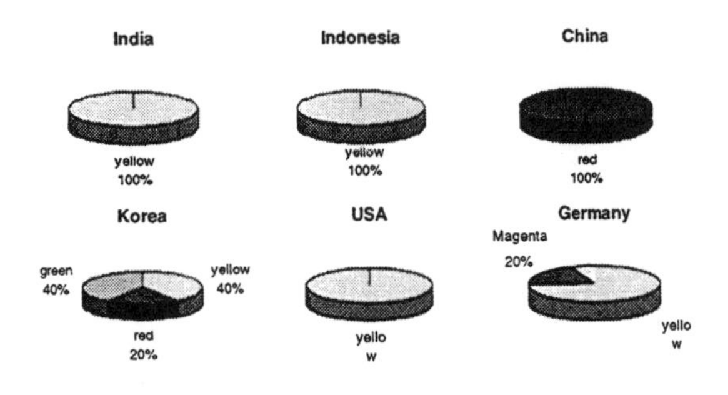

Figure 3: Colors assigned to the operation condition „Caution" for each country

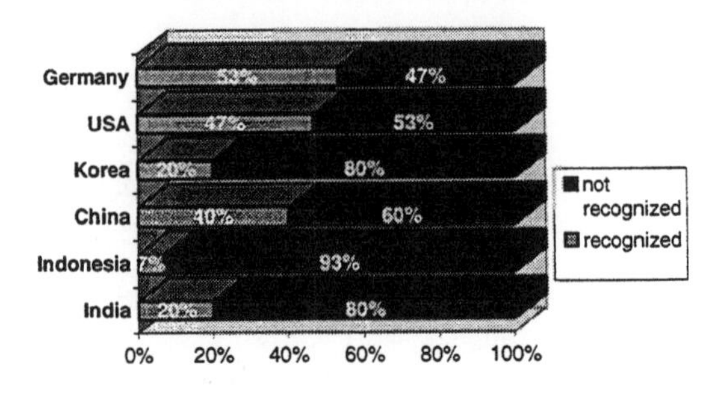

Figure 4: Recognizability of abstract icons

Furthermore, groupings of information, control elements and compatibilities of place, direction, and grouping also seem to be transferable from Europe into Asia.

Symbols are, however, in contrast to widespread opinion, not independent of cultural context. It was repeatedly demonstrated that symbols could not communicate information independent of language. The tests showed that standardized symbols (see e.g. Norm DIN 30600) with widespread acceptance in Germany are not understood or are misunderstood in other parts of the world. Abstract symbols and abbreviations on little used control elements caused the most confusion; pictorial symbols were recognized more often but tend to be linked to particular cultural contexts (see **Figure 4** and **Figure 5**). Thus, it is important that symbols used in technology do not contradict a specific cultural meaning, as for example in the coding of hand signs. In order to assure clarity, it is recommended that a symbol be accompanied by an explanation in the appropriate language. Currently, it appears that using symbols to replace translating labels in the language of the country is not possible. In most cases, the interview partners expressed the preference that manuals and control system labels be offered in their national language.

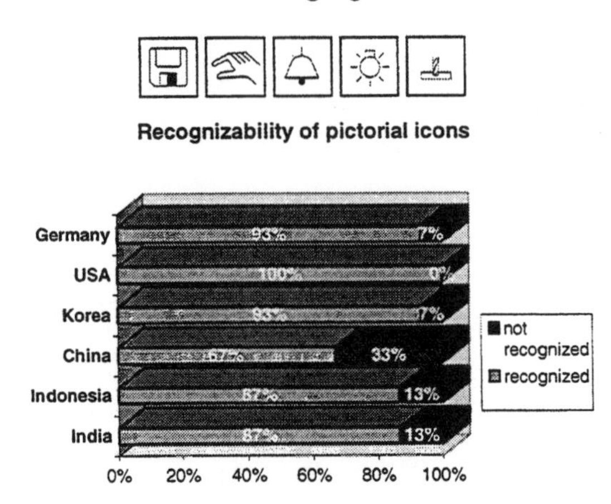

Figure 5: Recognizability of pictorial icons

4.6 Support

Apart from the design of the control panel, it was found that the support and design of diagnostics and error correction has to be conceived differently for the foreign customer; the machine operators and maintenance personnel need more detailed and better organized information.

4.7 Training and work organization

Another aspect of the project was an investigation of how existing training practices and work organization influence contact with and use of modern technologies. In addition to the practice of using memorization as a pedagogical method, another characteristic of training in other parts of the world is the lack of skilled workers in the German sense. Without such workers, companies use semi-skilled or unskilled workers to operate production lines. The resulting deficits in technical knowledge are compensated through specialized further training programs in the company or in schools. Often developing basic craft skills are simply not included in such programs. Many aspects of machine operation or maintenance tasks are closed off to machine operators with this kind of training.

5 CONCLUSION

It is difficult for European developers to take the specific requirements of such users into account when developing new products and functions - mainly because they don't know much about such users or their needs. Without this knowledge, they design machines for highly skilled workers or engineers. Japanese developers are better acquainted with the training backgrounds of these customers because their educational system, as that of many other countries, is based on a model of the American system. This is one reason why the machine interfaces that they develop receive such a positive resonance from many users.

This study led to the conclusion that, in order to meet the needs of future machine users worldwide, producers of machine tools have to design interfaces that are easier in terms of user orientation, but not in the reduction of functions or operating options. These findings are particularly relevant given possible new definitions of the qualification requirements of CNC machine users in the future. The following strategies are proposed as ways to better understand the future needs of users:

- Think Global - Act Local; Information that German developers lack regarding foreign cultures can be obtained through close cooperations between customers and producers in the form of subsidiaries or joint ventures that operate locally.

- Use of networks; German machine producers should build up and maintain personal networks with Asian customers. Asians are used to comprehensive service and close cooperation between partners; tightly knit groups and relationships play a large role in the private sphere as well as in business.

The solutions offered in this project in response to the conditions found in the countries and companies visited should contribute to European development potential and open new market opportunities for machine producers in these very important export markets.

REFERENCES

KRAUß, L.: Charakterisierung von Designelelmenten zur Gestaltung von Bedienoberflächen unter Berücksichtigung des fernödstlichen Kulturkreises. Kaiserslautern, Universität, Maschinenbau und Verfahrenstechnik, Diplomarbeit, 1997

Norm DIN EN 60073 01.1994: Codierung von Anzeigegeräten und Bedienteilen durch Farben und ergänzende Mittel *(Coding of indicating devices and actuators by colors and supplementary means)* : Deutsche Fassung der IEC 73

Norm DIN 30600 Teil 2 07.76: Bildzeichen : Übersicht *(Pictograms : Overview)*

Romberg, M. u.a.: Anforderungen außereuropäischer Märkte an die Gestaltung der Maschinenbedienung *(Global demands of non-European markets for the design of user interfaces)* : INTOPS. Kaiserslautern, Universität, Maschinenbau und Verfahrenstechnik, Projektabschlußbericht, 1997

VDMA (Hrsg.): Statistisches Jahrbuch für den Maschinenbau *(Statistical yearbook for mechanical engineering)*. Frankfurt : Maschinenbau Verlag GmbH, 1995

ZÜHLKE, D. (Hrsg.) ; VDI/VDE-Gesellschaft Mess- und Automatisierungstechnik (GMA) (Veranst.): Menschengerechte Bedienung technischer Geräte *(Operation of technical devices suitable for human)* : VDI-Berichte : 1303 (1. Fachtagung „Menschengerechte Bedienung technischer Geräte“ Kaiserslautern 1996 - 09 - 17/18). Düsseldorf : VDI, 1996

DESIGN GUIDELINES OF ICON PICTURES BASED UPON RELATIONSHIP BETWEEN COMPREHENSIBILITY OF ICON FUNCTION AND DESIGN FACTORS

Kazunari Morimoto, Takao Kurokawa and Ryo Takaoka

Graduate School, Kyoto Institute of Technology
Matsugasaki, Sakyo-ku, Kyoto 606-8585, JAPAN

Abstract: For characterizing iconic pictures, four design factors were proposed: concreteness, minuteness, dimensionality and functionality. The relationship between comprehensibility of icon function and these factors was analyzed based on the results of experiments for obtaining subjective evaluation of comprehensibility and subjective factor values of various icons. The results showed that the subjective factor values are useful in evaluating quality of iconic pictures. On the basis of the results guidelines for designing icons were proposed.

Keywords: Picture elements, Evaluation, Ergonomics, Man/Machine interfaces, User interfaces

1. INTRODUCTION

Graphical representation of system states and functions can improve interaction between users and systems (Carr, 1998). Icons used in graphical user interfaces are the most familiar elements expressing system information. The objective of icon pictures is twofold; 1) to convey what is drawn and 2) to inform what it means (icon function). Many of the icons in use, however, do not give enough information about what is drawn, and hence users cannot understand their meanings in human-computer interaction (Morimoto et al. 1994). Although there are a lot of guidelines for designing interfaces, they fail to show clear and concrete methods of drawing icon pictures.

The appearance of an icon is composed of elements that are both visual and semantic. If appropriately designed, they lead to understanding of the picture and meaning of the icon (Byrne, 1993). But it is hard to select elements to be drawn and unify them into an iconic picture in order to let it attain the objective as stated above. In designing iconic pictures designers have to take so many factors into consideration from the cognitive and semantic point of view. It is this that makes designing icons very difficult. The situation should be simplified by separating visual expression and semantic one. Though they are not independent each other and cannot completely be separated, the visual expression is roughly said to consist of visual elements such as color and shade, the number of lines used, the omission of details, while the semantic expression includes the selection of objects to be drawn, relative size of the objects and the attachment of labels. In this article we will focus our interest on visual expression of icons, putting away connotational semantics on the designing side for future studies.

The aim of this article is to define four fundamental design factors that specify the appearance of icons, to explore the relation between comprehensibility of icon function and psychological values of the factors and to propose clear guidelines for drawing iconic pictures. In Section 2, the design factors are defined and Experiment I is described where comprehensibility of the function of icons drawn in accordance with ranking of the factors was subjectively evaluated. Section 3 describes Experiment II that was carried out in order to derive design guidelines of icons with respect to two of the factors and the proposed guidelines are given in Section 4.

2. EXPERIMENT I

2.1 Four design factors and designing icons

There are many factors specifying the design details of iconic pictures. They include the size of icons, the number of colors used and the skill in drawing. But these are superficial and secondary factors in designing icons. After due consideration of properties intrinsic in the appearance of icons, we extracted four factors as the most fundamental that can control it and give useful guidelines for designing of icons; they are concreteness, minuteness, dimensionality and functionality.

Concreteness: This factor defines the degree of concreteness of iconic expression. In terms of this factor drawings vary from extremely abstract to highly concrete.
Minuteness: The minuteness factor controls the degree of details of iconic expression. Drawings poor in this factor have just contours of the drawn objects.
Dimensionality: Dimensionality is defined as the extent of depth expression of the object under the restriction of two dimensions. Objects rich in dimensionality are expressed with shadows, shades and perspective.
Functionality: The object has its own function, and it can be represented in the drawing. For example the drawing of a pencil alone has little functionality, but a pencil drawing with traces like curves and letters can show its function to some extent.

These factors can have qualitative and quantitative ranks in their own range. We defined five ranks in each factor from lower one (Rank 1) to higher (Rank 5). They serve as the rules in drawing a given object. The factors are not completely independent one another as easily understood.
Concreteness: The Rank 1 picture in this factor is drawn with just outlines of the object. In Rank 2 some lines or curves are added to the picture of Rank 1 in order to increase its concreteness, and the Rank 3 has simple depth expression or lines in addition. In Rank 4 some attachments or further lines are represented on the picture of Rank 3. By adding depth clues like shades concreteness is augmented in Rank 5.
Minuteness: As the rank goes higher, the detail of the picture is increased by changing the number of lines and the dot density in shades. The object of Rank 1 is drawn with just outlines as in Rank 1 of concreteness. In Rank 5 the picture contains many lines and shades.

Dimensionality: In order to represent the dimensionality perspective, shades and shadows are effectively used. The Rank 1 and 2 pictures are a 2D front or side view of the object. 3D clues are introduced in Rank 2 and the higher ones. In Rank 2, 4 and 5 shadows are expressed, perspective is begun to be adopted in Rank 3, and shades are drawn in Rank 5.
Functionality: The Rank 1 picture is roughly the same as the picture of Rank 3 in dimensionality. In Rank 2 some lines or attachments related to the object's function is added. The object has some traces of having worked in Rank 3 and is just working in Rank 4. In Rank 5 a hand manipulating the object is attached.

2.2 Methods

Ten objects were chosen for designing icons among stationery and things used in offices: "pencil", "magnifying glass", "printer", "document", "trash can", "folder", "paintbrush", "eraser", "book" and "compass". Each of them was then represented in twenty black and white drawings (5 ranks × 4 factors) on a white background as exemplified in Fig. 1. Therefore, we had a total of 200 icons. They were put to use as icons in Experiment I. Their size was 11 mm × 11 mm.

Experiment I consisted of two sessions. In the first session, subjects subjectively evaluated the comprehensibility of the naturally assigned function of each icon on a five-point scale using the working screen shown in Fig. 2. Then they rated each icon's quality concerning a given factor using the rating method of ten points. Figure 3 shows a typical screen layout for rating, where the subjects dragged each icon to the position on the rating scale. We used the ratings to give the icon psychological value in each factor. Ten subjects who had experienced to use icons with some systems participated in Experiment I.

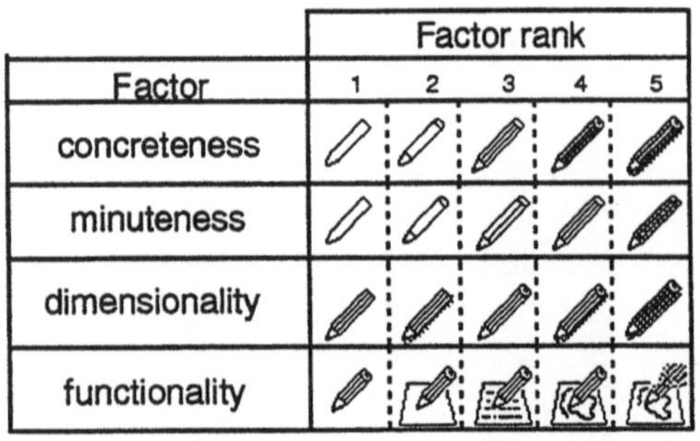

	Factor rank				
Factor	1	2	3	4	5
concreteness					
minuteness					
dimensionality					
functionality					

Fig. 1 Icon pictures of "pencil" drawn according to the definitions of the four design factors.

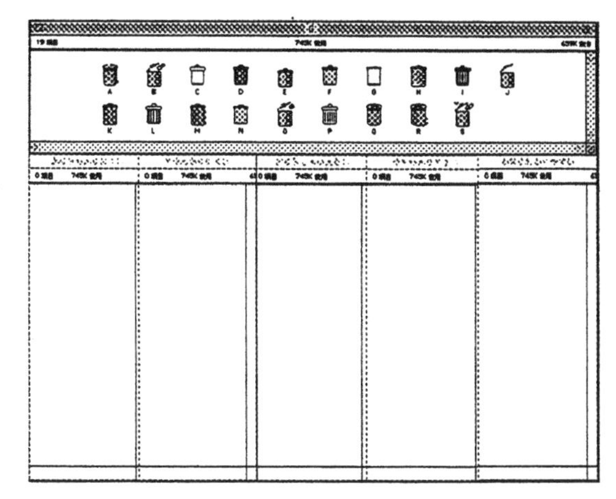

Fig. 2 Working screen presented to the subjects for subjective evaluation of comprehensibility of the icon function.

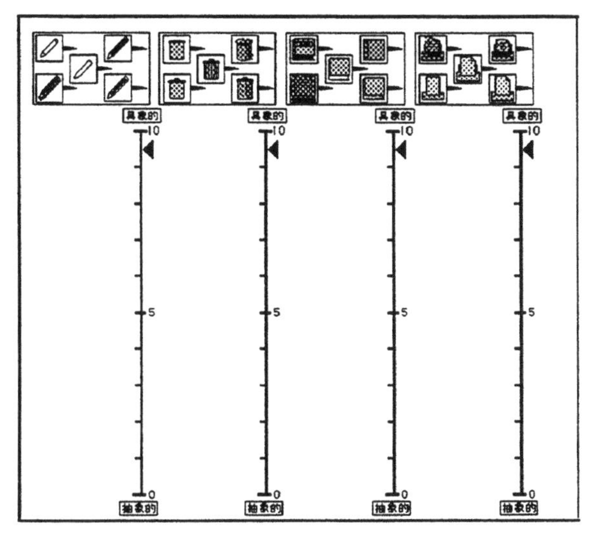

Fig. 3 Typical screen presented to the subjects for rating.

2.3 Results

Figure 4 shows the relationship between comprehensibility and the psychological factor value of the "book" icons. We could confirm Psychological intervals among pictures of icons represented according to the design rules were not equivalent. This finding assists that psychological distance between pattern drawn by a rule was different with an icon.

Although the comprehensibility of the "book" icons showed a tendency to increase with the rank in concreteness, the shades added in the picture of Rank 5 were not effective in understanding the icon function. The same was found in the other icons in the factor. In the case of minuteness, the comprehensibility was highest in Rank 3 or 4. The higher concreteness of icons was, the higher their minuteness was, because concreteness and minuteness are highly dependent. This fact caused that the results of subjective evaluation were similar between both factors.

3D representation of the icons did not always have positive effect on recognition of the icon functions. Only in case of 3D objects such as "book", "eraser" and "printer", 3D cues were effective in understanding the icon functions. Flat objects should not be drawn with shades and shadows.

As for functionality, the comprehensibility tended to increase with the rank in case of "book". However this was not found in the "pencil", "trash-can" and "eraser" icons. Though Authors had thought that the comprehensibility of the icons with a manipulating hand drawn (Rank 5) would be higher than the others, we had values lower than the other ranks. Concerning this result we supposed that it was difficult to recognize the hand drawn because of the small icon size and to understand the relation between the icon function and the hand drawn.

3. EXPERIMENT II

3.1 Icons designed and two design factors

Four design factors used in Experiment I did not always depend on each other. For instance the picture of the best icon for the factor of concreteness was not evaluated on the factor of dimensionality.

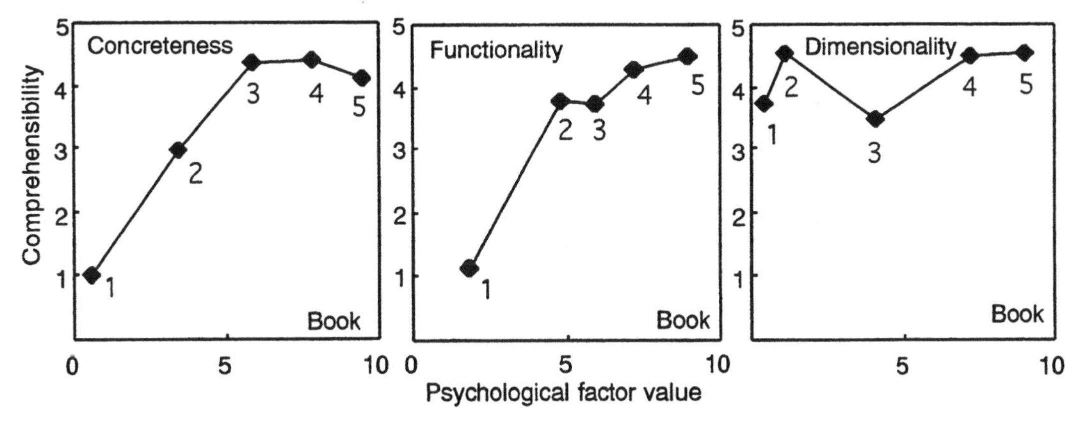

Fig.4 Comprehensibility as a function of psychological valuein case of "book" icons. The figures neighbouring data points indicate the rank in each factor.

Unfortunately the pictures of the most comprehensive icon couldn't determine. In order to propose design guidelines of icon only one icon picture must be derive from design factors adopted.

The results of Experiment I showed three dimensional representations of icons were less effective on recognition of the icons' meaning. However the pictures of dimensionality were drawn independent to the factor of concreteness. Therefore high correlation in comprehensibility wasn't found between concreteness and dimensionality. On the other hand the relation between concreteness and complicatedness was very high. Therefore concreteness and dimensionality were adopted as design factors of icons tested in Experiment II.

Nine kinds of objects used in the experiment were "pencil", "printer", "document", "trash can", " folder", "paintbrush", "eraser", "book" and "compass". Each icon was designed to be assigned to one of five values of concreteness factor and to one of four values of dimensionality factor according to the design rules developed by the authors. Icons drawn with combination of the two design factors were represented in Fig. 5.

Design rules of concreteness are as follows. The picture of Rank c1 is drawn with outlines of an object, and Rank c2 is added some lines to that of Rank c1. Rank c3 is painted gray color on an object to that of Rank c2. Rank c4 is added some lines to the picture of Rank c3. Rank c5 is a photograph of real objects recorded by digital camera. Design rules of dimensionality are as follows. The picture of Rank d1 is drawn without perspective representation. Rank d2 is drawn with perspective representation. Rank d3 is added shadow to that of Rank d1. Rank d4 is added shade to that of Rank d3. All pictures were drawn at the area of a regular square as well as Experiment I.

3.2 Methods and apparatuses

Two experiments are carried out. First the comprehensibility of functions of icons was tested by five rating scales as well as Experiment I. Second psychological factor values of icons were evaluated

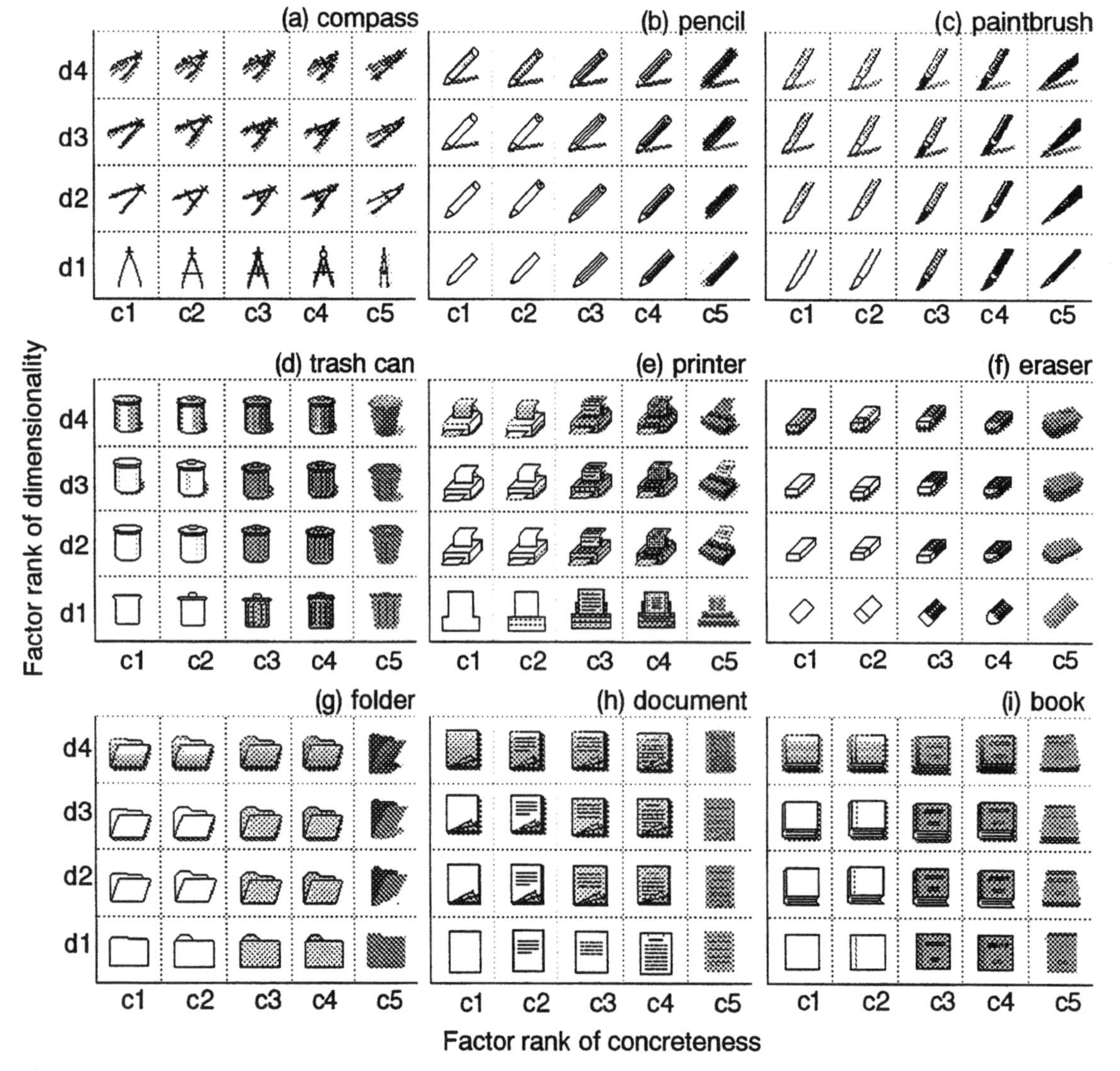

Fig.5 Icons used in the experiment II .

using the rating of ten points. Experimental apparatuses were Macintosh (Apple) and a color display of high resolution.

3.3 Subjects

Forty subjects participated in the experiments to evaluate the comprehensibility of functions and the psychological intervals among pictures of icons designed above. Twenty subjects were novice users who have not been experienced to operate a computer. On the other hand the other subjects have been usually used a computer of UNIX, Macintosh or Windows.

3.4 Results and discussions

Figure 6 shows the relationship between concreteness and dimensionality of subjective factor values of trash can icons. The psychological factor value of Rank c5 in concreteness was shorter than that of Rank c3 and c4. This suggests the photographic pictures are not viewed as a real object. On the other hand the psychological factor value of Rank c5 in dimensionality was shorter than that of the other icons.

Figure 7 showed the concreteness of icon pictures increased in the factor values of concreteness except Rank c5 which icon was drawn based on the photo images of the real object. It is difficult to draw precisely a picture of photo images of an object in small area. Therefore some retouching to the photo images was carried out to make similarly to real objects. Nevertheless the concreteness of the icons of Rank c5 was evaluated lower than the icons of Rank c3 and c4. The other finding was that the comprehensibility of icons drawn with outline only was low.

Subjective factor values of icons in dimensionality increased in rank values. This result showed that the pictures represented according to the design rules stood in rows of psychological intervals. Comprehensibility of the icons except the Rank c5 increased with degree of concreteness and dimensionality. In many cases the comprehensibility of the Rank d2 was higher than that of the Rank d1. This suggests it is important to draw icons with a bird's-eye view method. Furthermore the comprehensibility of the Rank d3 was higher than that of the Rank d1 and d2 in "trash can", "printer" and "eraser". This suggests the appropriate usage of shadow and shade contribute to obtain higher comprehensibility. On the other hand there is significant caution in use of shadow or shade to "pencil", "folder" and so on.

Figure 8 showed the correlation between novice and expert users about the comprehensibility of trash can and folder icons. The correlation coefficients of icons except folder icon were above 0.89. On the other hand the correlation coefficient of folder icons was 0.64. The following tendency was obtained in folder icons. Comprehensive values of expert users were higher than that of novice users. The reason of

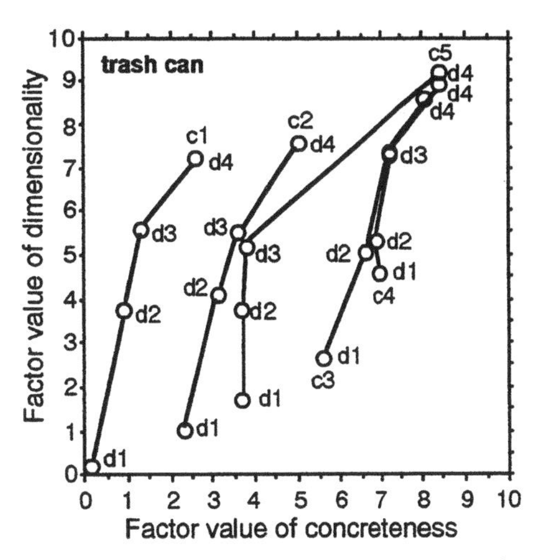

Fig. 6 Relationship between concreteness and dimensionality in subjective factor values.

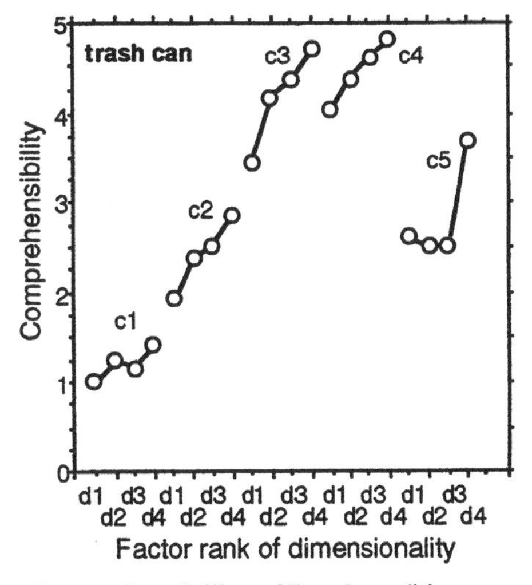

Fig.7 Comprehensibility of "trash can" icons.

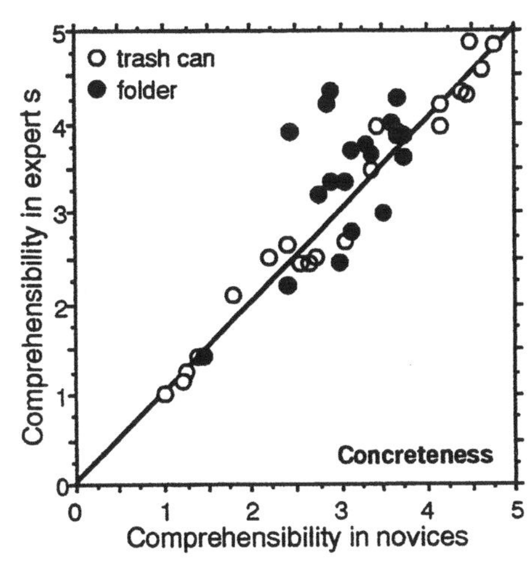

Fig. 8 Relationship between experts and novices in comprehensibility of "trash can" and "folder" icons.

this results obtained would be estimate that novice users did not know the folder or not use it. Therefore they felt lost in evaluation. However this result suggests the degrees of comprehensibility except the folder icons dose not depend deeply on experience of computer operation.

In order to clarify the relation between characteristics of icon pictures and comprehensibility, nine icons used were divided into three groups; thin, three dimensional and flat objects. "Compass", "pencil" and "brush" were included into the thin objects' group. "Trash can", "printer" and "eraser" were in the three dimensional objects' group. "Folder", "document" and "book" were in the flat objects' group. Figure 9 showed the mean values of comprehensibility of the three icon groups. As for dimensionality, comprehensibility of icons tends to increase with growing up the factor value as well as concreteness, and the highest comprehensibility was obtained at the Rank d4 of each three group. Comprehensibility of icons with shade became low. The influence caused by difference of factor values of dimensionality was very small in thin objects of Rank c3 and c4. For three-dimensional objects, comprehensibility of icons depends on the factor values of dimensionality. Icons of Rank d4 in three-dimensional objects were easy to become understand. These findings were not gotten in Experiment I.

4. PROPOSED GUIDELINES

Horton (1994) insists to draw an icon concretely and clearly as possible, on the other hand Meyhew (1992) points out to draw icons abstractly. Not only the relation between those two design factors used in Experiment I was not independent. Therefore it is difficult to present the most significant design factor for designing recognizable pictures of icons. However relationship between design factor values

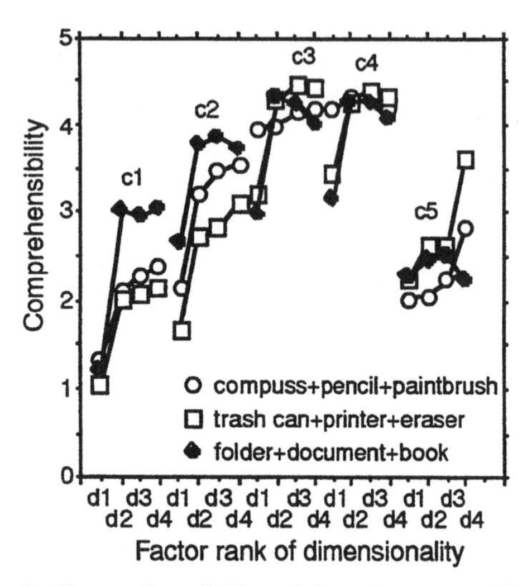

Fig. 9 Comprehensibility of three icon sets; thin, three-dimensional and flat objects.

and design elements of pictures of each design factor analyzed in Experiment II. Following guidelines to draw pictures of icons was derived from the results of the analysis and the design rules.

(1) Not to use excessive representation of reality.
(2) Not to draw an object with outline only.
(3) Not to draw with shadows on flat or thin objects.
(4) Not to use the shades preferably.
(5) To use 3D representations to an object with many sides.

5. CONCLUSIONS

The experimental results showed that the resultant psychological distance measure was significant in evaluating function of icons in the factors of concreteness, minuteness, dimensionality and functionality. According to the results, guidelines for expressing iconic pictures were proposed.

Comprehensive icons make possible users lower cognitive stresses interacting with computer. In order to accomplish good iconic communication we must investigate thoroughly about the design factors according to the cognitive and psychological engineering. For further extension of this research examination about relation of concreteness and functionality is necessary.

ACKNOWLEGMENTS

Authors wish to thank Mr. T. Yamamoto and Mr. Y. Morimoto who were under graduate students of Kyoto Institute of Technology for their cooperation in this work. This research was supported in part by a Grant-in-Aid for Scientific Research from Ministry of Education (No. 09680396).

REFERENCES

Byrne, D.M. (1993). Using icons to find documents: Simplicity is critical, INTERCHI'93, 446-453.

Carr, D. (1998). Interaction object graphs: An executable graphical notation for specifying user interfaces. In: *Formal Methods in Human-Computer Interaction* (Philippe, P. and Fabio, P. (Ed)), 141-155, Springer

Horton,W.K. (1994). The icon book, John Wiley & Sons.

Mayhew, D.H. (1992). Principles and guidelines in software user interface design, Prentice Hall.

Morimoto, K., Kurokawa, T., Nishimura, T. and Torii, T. (1994). Analysis of cognitive factors of icons based on subjective evaluation, Progress in Human Interface, 3, 2, 65-72. (In Japanese)

CONSTRUCTION OF A WWW-BASED CAI SYSTEM BASED ON COGNITIVE DESIGN GUIDELINE

F. OBAYASHI, H. SHIMODA, H. YOSHIKAWA*

*Kyoto University, Graduate School of Energy Science, Kyoto, Japan

Abstract. Firstly, "Cognitive Design Guideline" is proposed for designing WWW-based CAI system by applying effective concepts and models in education, which have been proposed in cognitive psychology. A new CAI system was developed as an augmented material of lecturing "Human Interface Science" by Internet. This was designed based on the proposed cognitive guideline and the field experiment was conducted to compare this new CAI system with the conventional CAI system. It was confirmed that the new CAI system be better to promote learning process than the conventional one. *Copyright ©1998 IFAC*

Key Words. Computer aided instruction, Cognitive science, Education, User interface, Communication environment, Intelligent systems

1. INTRODUCTION

Recently with explosive expansion of Internet around the world, there have been a lot of WWW-based CAI systems developed not only for classroom teaching but also for remote teaching outside of schools.
But usually, those conventional CAI systems only show the contents of learning very simply, and they are not so well organized as to be an effective teaching system.
In this paper, the authors would like to propose a new concept of designing WWW-based CAI system based on the "Cognitive Design Guideline". With regards to the derivation of the "Cognitive Design Guideline", the authors conducted on literature review in the field of cognitive psychology, to find useful models and concepts especially effective for computerized education.
Based on the "Cognitive Design Guideline", the authors developed a WWW-based CAI system to teach basic knowledge on "human interface science", for master course students of their belonging graduate school, and then they have conducted on a field study to evaluate the effectiveness of the design methodology by the intercomparison of the new WWW-based CAI system and that of conventional type (only hyper text configuration).

2. Problems in Conventional WWW-based CAI System

WWW-based CAI system has many merits, namely it offers educational environment without restriction of time and space and so on. But we must consider its specific characters such as hyper link structure, one way presentation on CRT display, application of multimedia, and asynchronous distributed educational environment.
The points of issue of conventional WWW-based CAI system lie in the following questions: A) Hard to comprehend, B) Lack of motivating to learn, C) Lack of promotion of knowledge acquisition, D) Unsuitable for all kinds of user's preference, E) Hard to read, F) Unusable to operate, G) short of interaction among students or students with teacher.
What we wish to show in this paper is overcoming these problems and constructing more effective educational system.

3. Application of concepts of cognitive science to CAI System

With regards to the derivation of the "Cognitive Design Guideline", the authors conducted on literature review in the field of cognitive psychology, to find useful models and concepts especially effective for computerized education.
We explain how we should apply knowledge on cognitive psychology for reducing specific guideline by one example of applying "Schema" for Design Guideline. Schema is a general concept that represents the structure of data in one's memory. Regarding education, the schema theory says that a student works on the object of study with the suitable schema which he has and he composes a meaningful interpretation, and then he is led to learning. When the students who have various backgrounds begin to study a new field, they often feel difficult to understand it at first. The schema theory explains its cause as the following way: he does not apply the suitable schema or does not have it. So if the teacher leads the student so that

the student can organize suitable schema successfully, then the student can understand the subject more easily.
Therefore, the CAI system should prepare for the introduction part and indicate the context, the viewpoint, the subject, or the analogy, so that the student can apply their original schema suitably or make it naturally.
For natural acquisition of knowledge, it should apply schematheory, method of presenting information, and theory of ATI[1]. And for the purpose of enhancing users' motivation for learning, it should apply meta-cognition[2], feeling of control, and so as to acquire knowledge more positively, it should adopt theory of learning by teaching. Furthermore, it should put affordance[3] for intuitive operability of CAI interface. By this way, we should derive the "Cognitive Design Guideline".
The reduced Guideline is composed of four parts: Firstly, (i) Flowchart of making Introduction Part of CAI system (Fig. 1), and then, three descriptive guidelines of (ii) Guideline of instruction (see Table 1), (iii) Guideline of Supporting Learning (Table 2), and Guideline of Page Design (Table 3). The specific points of applying the Guideline to cope with the problems in conventional CAI as pointed out in section 2 are summerized as shown in Table 4.

4. Cognitive Design Guideline for Active Participation in Learning

Here we will explain three examples of applying ideas of "Virtual Student", "Virtual Class", and "Experiential Learning" for Active Participation in Learning.
In the usual WWW-based CAI, the students just see or read the object passively, so it is hard to master the subject.
So the new CAI system should implement "active role" function that the student can enhance his understanding on the subject by his teaching what he has learned to other students. The CAI system provides the section of teaching to "virtual student" by CGI (Fig.2). In this section, the student teaches her what he has studied and by this way, he can deepen his understanding; the student answers the question she asks or corrects her false understanding. Every student has his own virtual student. By this interaction with virtual student, the student can monitor himself through the computer as the reflector.
And furthermore, after he has finished his teaching to virtual student, he will visit the "Virtual Class" where all virtual students attend. In this class, the virtual teacher asks questions and each virtual student answers them. Each virtual student presents what she has been taught by each real student. The student inspects this class and he can monitor other students' thoughts. By this way he can promote a greater diversified understanding. And moreover, he will be motivated to study more through competition to other student. To take other example of special feature of this system, it provides Experiential Learning environment. This part allows user to experience cognitive psychological experimentation and enhance his essential comprehension of subjects. It is produced by Java applet (Fig.3).

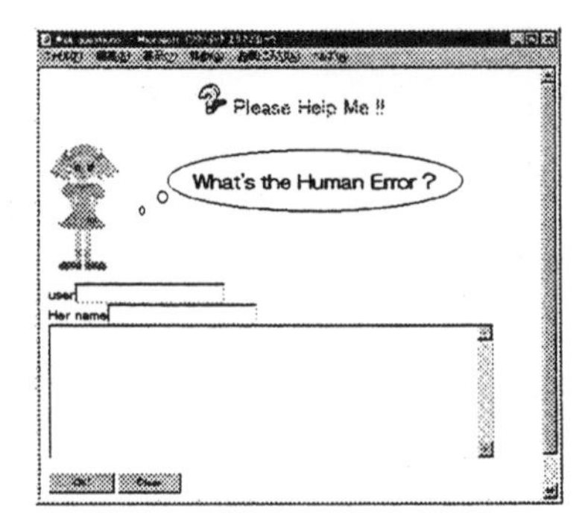

Fig. 2. Teaching to Virtual Student

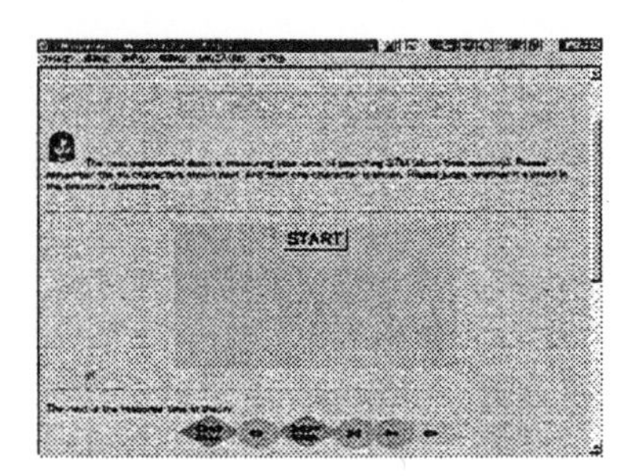

Fig. 3. Experiential Learning part

5. Construction of WWW-based CAI System

Based on the "Cognitive Design Guideline", the authors developed a WWW-based CAI system to teach basic knowledge on "human interface science", for master course students of their belonging graduate school. This CAI system consists of five parts : Introduction Part, Learning Part, Advanced Learning Part, Exercise Part, and Function Part. Exercise Part prepares practice problems section and Virtual Student section. This CAI system analyzes the result of exercises and offers his best suitable review course. This system has 175 pages in all (Fig.4).

[1] Aptitude Treatment Interaction

[2] cognition of one's cognitive process, its outcome, etc

[3] fundamental character of design of object

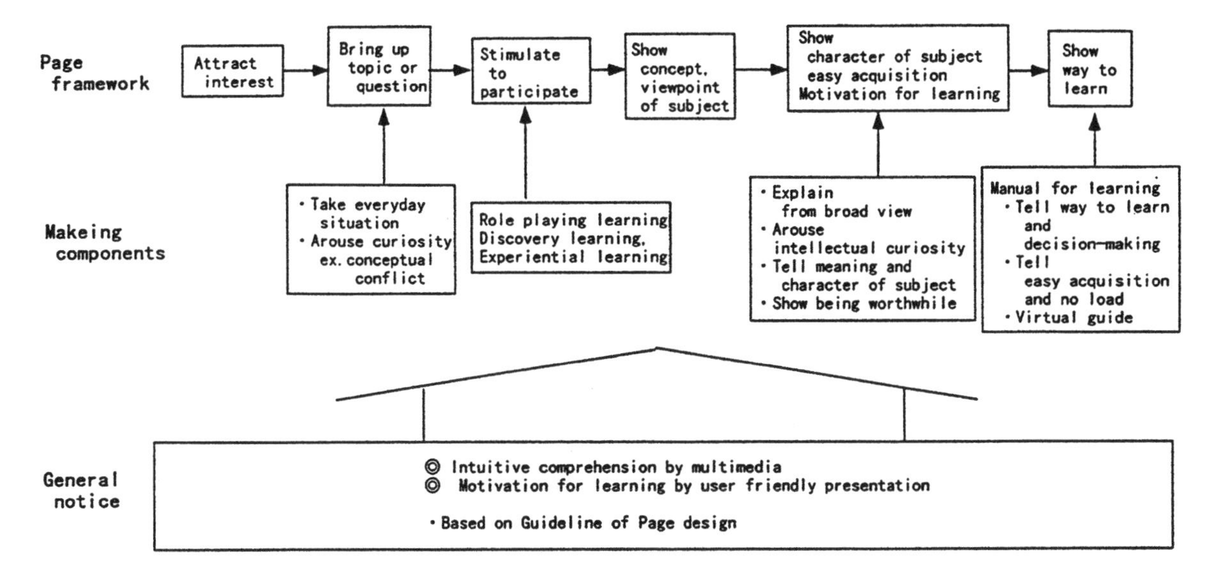

Fig.1. Guideline of Introduction

Table 2. Guideline of Supporting learning

Guideline	Details
1.For acquire knowledge	1. Virtual Student and Classroom 2. Conference room on WWW: For free talk and for discussion some themes 3. Open questions and answers corner, FAQ corner 4.Above corners1-3 are available by pen name. But use only members 5. Discovery learning and Experiential learning environment 6. Exercise corner ・analyze result and present suitable learning course ・consider to keep user's will to learn 7. Above corners opens another window and enable user to reference to Learning Part
2.Enrich interaction	1. Enable teacher to grasp user's history of learning 2. Between teacher and users : open questions and answers corner, open opinion corner, giving assignment and receiving corner 3. Among users : mailing list, conference room on WWW, Virtual classroom 4. Between user and computer : load Virtual guide
3.Enrich function of supporting learning	1. Function of searching keywords 2. List and link of keywords 3. Present exterior related information and link 4. Note page on WWW for each user 5. NAVI – MAP: Clickable map of whole learning pages
4.Cope with gap of user's surroundings of network	1. Download service of CAI system by ftp 2. Distribute CAI system by removable media (CD-R, etc)

Table 3. Guideline of Page design

Guideline	detail
1. Improve operability	1. Apply affordance to icons, etc 2. Apply bookmetaphor, etc 3. Indicate location at head of page 4. Show links in every page to NAVI-MAP, before and next page, and top page 5. Show links of standard course, and review course at end of page
2. Offer information friendly	1. Fully worked-out design familiar characters or topics, etc 2. Friendly wording 3. Virtual guide : mediator between user and computer. Voice guidance and receptionist of question
3. Present character base information	1. Organize information to module 2. Cut down characters 3. One module to one screen 4. Fully worked-out design to easy to see 5. Make full use of non verbal message 6. Practical use of icons, etc

Table.1. Guideline of instruction

Procedure	Details
A. Constructing General teaching structure	1. Make chapter 2. Organize main framework based on schema i. Introduce：Call or make schema ii. Promotion of learning： Re-organizing schema Adjusting schema iii. Give schema of expert iv. Use same schema repeatedly 3. Construct components · Enable user comprehend and organize contents easily · Arouse intellectual curiosity :ex. conceptual conflict · Prepare discovery learning, experiential learning environment 4. Organize contents to compact module 5. Make standard learning course and adapted courses for every state of learning
B. Making Components	1. Organize hyper-link Taking notice not to lose way in hyper-link 2. Prepare introduction part for every new subject · Summons or making schema by means of showing context, Call or making schema By indicating context, viewpoint, subject, analogy · Bring up examples from everyday situation · Arouse curiosity by multimedia 3. Make full use of image learning · Practical use of non verbal messages · For abstract concept： Apply multimedia or experiential learning system · Promote making mental image： Make good use of design, image figure 4. Bring up instances from everyday situation
C.Making pages	1. Enable user to acquire more easily in beginning of learning 2. Indicate outline or meaning for Contents links 3. Summary page for each chapter · Help user to organize contents and to make mental image · Show placement taking a broad view. give schema of expert 4. Load Virtual guide： Tell purpose and outline of subject by voice 5. Give one point advice at end of page Show meaning from broad view, schema of expert 6. Help user to organize contents by visual design · Characterize page design in every chapter or theme · Make good use of theme color for keywords 7. Base on Guideline of Page design

Table 4: The Outline of Reducing Cognitive Design Guideline

Problems in conventional CAI	Assumed reason	Useful notions applicable from cognitive science	Application of specific guidelines
A:Hard to understand			
A-1:Hard to follow new fields	Students who have various background learn new specialized subjects	Schema theory	Guideline of Introduction, Instruction: A-2
A-2:Poor method of teaching	It does not match with original mental models of students	Schema theory, Methods of offering information, Non verbal message, Discovery learning	Guideline of Introduction, Instruction
A-3:Hard to Comprehend abstract concept	Limitation by verbal explanation	Non verbal message, Experiential learning	Guideline of Instruction:B-3
A-4:Must reconstruct Contents after grasping whole	Student does not have expert schema	Schema theory	Guideline of Instruction:A-2, C-2,3,5
B1:Not be willing to study	Uninteresting, Not motivated	Meta-cognition, Knowledge of motivation, Feeling of control	Guideline of Introduction, Instruction:C-1, Page design:2
	Feels difficult	Feeling of control	
B2:Hard to continue to study	Lost motivation	Meta-cognition, Knowledge of motivation, Influence of others' existence in study	Guideline of Instruction, Supporting learning:1
	Result fell short of his expectations	Theory of meta-cognition, Influence of others' existence in study	Guideline of Supporting learning: 1-6
C:Difficult to acquire	Only watching	Learning by teaching, Interaction with others, and so on	Guideline of Supporting learning:1
D:Not suitable for learning styles of students	Not suitable for all tastes	ATI(Aptitude Treatment Interaction)	Guideline of Instruction:A-3 Supporting learning:2
E:Hard to read	Not designed to suit to characters of CRTdisplay	Non verbal message, Methods of offering information	Guideline of Page design:3
F:Unusable to operate	Not designed to suit to characters of computer and WWW	Affordance	Guideline of Supporting learning:3,4 Page design:1
G:Short of interaction	Weak point of asynchronous distributed environment	–	Guideline of Supporting learning:2
H:Awkward to join discussion and to ask question	Feels small with his unclear understanding	Influence of others' existence in study	Guideline of Supporting learning:1

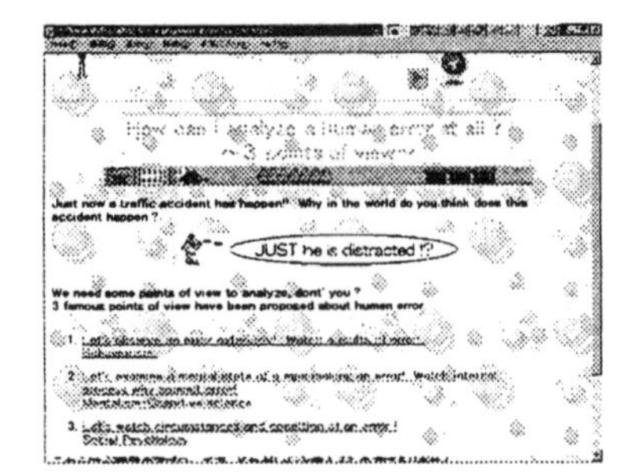

Fig. 4. one page of this CAI system : suitable learning course

6. Field experiment and discussion

6.1. *Field experiment*

We conducted on a field experiment to compare this new CAI system based on the the "Cognitive Design Guideline" with the conventional CAI system based on hyper text. This field experiment was conducted on subjects of as many as twenty students in the graduate school with versatile fields of majoring. The method of field experiment was that each subject experienced two different types of CAIs in turn, and afterwards the subject was asked to full in the questionnaires sheets which was prepared for hearing their subjective impression on their self-learning course. Further they evaluated the two systems by scoring five grade, with respect to the following six questions. (1) " Could you follow the lecture smoothly ? " (2) "Did you feel like to learn this in the beginning ? " (3) "Could you comprehend the contents easily ? " (4) "Could you hold your will to learn this ? " (5) "Didn't this study impose a burden on you ? " and (6) "How do you evaluate this CAI system totally as a learning environment ? ".
For the purpose of test of significance, we examined the results by sign test as a kind of statistical testing and it was confirmed that the new CAI system got higher valuation than conventional one. Table 5 shows the average marks of the results of these questions by scoring five grade. From Table 5, it is clear that the new system gets higher scores than conventioal one for each question item.
Therefore, the result of our experiment clearly shows that the new CAI system based on the "Cognitive Design Guideline" be better to promote learning process than the conventional process.

Table 5 The average marks of the results of these questions by scoring five grade

	(1)	(2)	(3)	(4)	(5)	(6)
Type						
Conventional	2.1	2.2	2.0	1.9	1.7	2.3
New	3.9	4.0	4.1	4.1	3.3	4.1

6.2. *Some problems pointed out to be considered*

From the result of the experiment, a few problems still to be improved in the present system are pointed out. They are :
1) Need to investigate more effective support of interaction
With regard to interaction, some problems are brought to light such as lack of immediate reply to the asked question to teacher, reluctance to ask trivial question, and problem by anonymity. And discussion in virtual conference room does not work well without interference.

2) Need to investigate how to organize hyper link
It is necessary to make new guideline for organizing hyper link and the way of its navigation more effectively.

7. Conclusion

In this paper, the "Cognitive Design Guideline" for designing WWW-based CAI system was proposed, and the new CAI system based on the "Cognitive Design Guideline" was developed for supporting the lecture on the "Cognitive Science on Human Interface" of their graduate school.
The field experiment conducted to compare this new CAI system with the conventional CAI system based on hyper-text, resulted in that the new CAI system be better to promote learning process than the conventional one. Therefore, it was confirmed that the proposed Design Guideline was fundamentally effective for organizing WWW-based CAI system for Internet education.
In the future, we will improve the "Cognitive Design Guideline" and go on to new concept of CAI system which will be an adaptive system to user's mental state, his personal character in studying, and so on.

8. REFERENCES

Saeki, H. (1982). *Reasoning and Understanding, Cogntive Psychology Lecture 3.* University of Tokyo Press

Hatano, G. (1982). *Learning and Development, Cogntive Psychology Lecture 4.* University of Tokyo Press

Imaei, K. (1992). *Computerization of Education and Cognitive Science.* Hukumura Press.

Sugai, K. (1989). *Invitation to CAI.* Doubun Press.

Norman, D. (1990). *The Psychology of Everyday Things.* Basic Books Inc.

EXPERIMENTAL EVALUATION ON ASSOCIATIVE KEYWORD REMINDER BY A THESAURUS

Tamayo Tokuda, Hisao Mase, Hiroshi Tsuji, and Yoshiki Niwa*

Systems Development Laboratory, Hitachi, Ltd.
Kansai Bldg. 8-3-45 Nankou-Higashi, Suminoe, Osaka, 559-8515 Japan.
**Advanced Research Laboratory, Hitachi, Ltd.*
2250 Akanuma, Hatoyama, Hiki-Gun, 350-0395 Japan
e-mail : ta-toku@sdl.hitachi.co.jp

Abstract: There is a need for search support as the amount of information on the network grows remarkably. Thesauri can be expected to suggest useful keywords for users search, however, their efficiency as keyword reminders is not well-known. This paper describes an experimental evaluation for clarifying the efficiency of thesauri as a term suggestion source, and gives consideration to the design guideline of the user interface when related terms are recommended by thesauri. *Copyright©1998 IFAC*

keywords : Information retrieval, Searches, Terms, Documents, Efficient evaluation

1. INTRODUCTION

In the current network era, the amount of the information on the network has been growing remarkably (Cheong, 1996). Since information retrieval (IR) systems have become popular, general users specify keywords to retrieve related information from the World Wide Web. As the search scale grows, however, obtaining search results that include many unnecessary records is a critical problem for users. The users must refine the search results again and again by adding other search terms or using logical operators (and/or/not).

However, it is often not easy for users to imagine new suitable search terms from their vocabulary. Imagining suitable search terms for their work depends on their vocabulary and how much they know about the search target. To alleviate this problem, support for users' search is necessary (Mase et al., 1997). Term suggestions are believed to be useful in IR, and Thesauri can provide the users keyword reminders.

There are many IR systems with a keyword reminder function. A well-known search engine on the Web is called Altavista (http://altavista.digital.com/). It has a keyword suggestion function that refines the many needless records. Some digital library projects developed the information infrastructure for the keyword recommendation feature, such as SONIA of Stanford University (http://www-diglib.stanford.edu/) , (Koller and Sahami., 1997), and DLI of Illinois University (http://dli.grainger.uiuc.edu/default.htm), (Schatz et al., 1995).

However, an evaluation of the thesauri efficiency is not well-known. In order to support a search with thesauri, it should be clarified whether thesauri are useful or not for the purpose. Furthermore, if thesauri are useful, then the designer of an IR system with an association support function must learn what kinds of data are helpful for users, and what presentation style is effective. The purpose of the experiments is to provide design guidelines for how such terms are presented to users. This paper describes the evaluation of obtaining the above knowledge.

2. TERM SUGGESTION FOR INFORMATION RETRIEVAL

Whether or not term suggestion contributes to IR remains obscure. Since terms suggested by the IR system are expected to contribute to a search, the support mechanism should recommend useful terms to users.

Traditional thesauri remind users of relevant terms by providing "broader terms" and "narrower terms" (Aitchison and Gilchrist, 1987). "Animal" is a broader term of "dog", and "apple" is a narrower term of "fruit". Broader and narrower terms contribute to expansion or refinement of the user's search. However, broader and narrower terms do not perfectly cover all domains. These terms are sometimes too general or too specific depending on the situation. General thesauri have general terms that are not suitable for application to a specific domain, and generating specified thesaurus data for the domain is not realistic.

To solve this problem, some researchers have proposed methods for creating relevant term pairs from a text corpus using co-occurrence of terms (Niwa et al., 1997a, 1997b; Schatz et al., 1995). Co-occurrent terms are a pair of related terms appearing frequently in the same component, that is, in a sentence, a paragraph, or the whole text. These pairs of terms depend on the target domains. Co-occurrent terms can supplement domain-depending vocabulary that broader and narrower terms can not cover. Since a pair of terms and the relation between them has structure, co-occurrent terms can be regarded as a kind of thesaurus data (Tokuda, et al. 1996). Managing the co-occurrent terms with broader and narrower terms may reinforce the thesaurus vocabulary as a keyword reminder.

3. EXPERIMENTAL EVALUATION

Two kinds of experiments are performed: experiment A and B for twenty subjects. Subjects do similar tasks under different conditions in these two experiments.

Tasks: Subjects are asked to write down relevant terms associated with a given term, and do so with as many terms as possible within three minutes. They use the worksheet shown in Figure 1, where the given term is written in the center of the sheet.

3.1 Used Tool

The thesaurus management system called "Theater" (Tokuda, et al.,1996) is used as a keyword reminder.

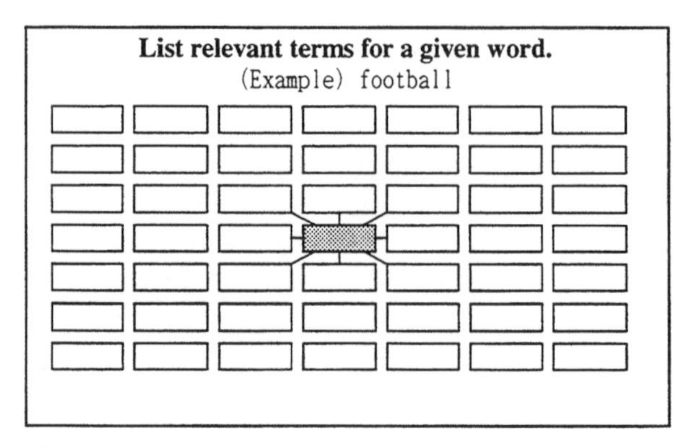

Figure 1. Worksheet sample.

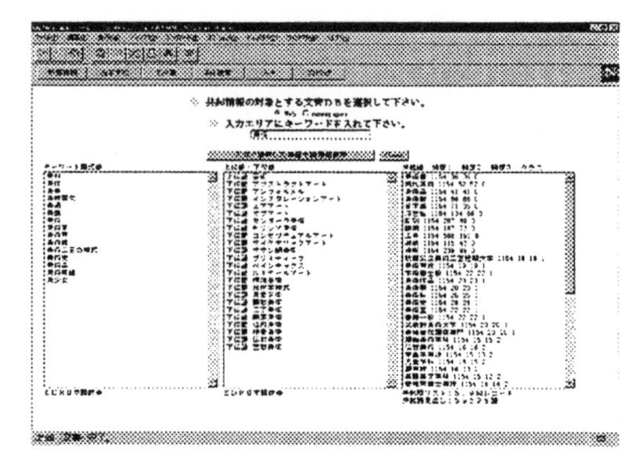

Figure 2. Example window of Theater.

As shown in Figure 2, Theater provides three kinds of thesaurus data for one given term: broader and narrower terms, alphabetically neighboring terms, and co-occurrent terms. Theater uses the EDR thesaurus dictionary (Japan Electronic Dictionary Researches Institute, Ltd), which holds about 400,000 terms and 800,000 broader/narrower relations (http://www.iijnet.or.jp/edr/). Theater also includes 60,000 pairs of co-occurrent terms extracted from 45,000 Web pages. The frequency of co-occurrence is also counted for two terms when both of them appear in a Web page.

If the user clicks any one term displayed, then he/she receives the thesaurus data on the term. Therefore, the user can navigate through Theater to obtain term relations. Theater helps users associate terms on a term by term basis.

Alphabetically neighboring term list: The left column in Figure 2 provides terms sorted in alphabetical order. For a given term "intensive", for example, Theater displays "intensity", "intent", etc. In this paper, since Theater has Japanese terms, it shows neighboring terms in Japanese.

Broader and narrower term list: The center column in Figure 2 provides broader and narrower terms. For a given term "fruit", for example, Theater displays narrower terms such as "apple" and "orange", and broader terms such as “food”.

Co-occurrent term list: The right column in Figure 2 provides co-occurrent terms. For a given term “university”, for example, Theater displays “professor”, “examination”, “recruit”, etc. A pair of co-occurrent terms have the frequency of the term occurrence and the frequency of term co-occurrence. Terms in this column are sorted by these frequencies, that is, users can look at the terms in order of validity.

3.2 Conditions

While some subjects are interested in specific topics, others are not. Such differences in interest may influence the ability of term selection. Therefore, users' interests in the topics are interviewed beforehand and the experiment is done with the following conditions.

Topics: Subjects are given a task term relevant to one of the fourteen topics, which come from categories in the Web search engine called Hole-in-One (http://hole-in-one.com/).

User Interests: Subjects' level of interest in each of the fourteen topics is asked before the experiment. According to the interview, the interest level for each topic is classified into three levels: high, low, or intermediate.

3.3 Experiment Process

Table 1 is the summary of the two experiments. Each experiment requests three tasks (one for each of the three levels of user interest). Subjects are observed during the operation and interviewed after each task. The following two experiments are performed.

Experiment A: The process is the following.
(1) First, subjects are asked to write down as many associated terms as possible on the worksheet WITHOUT Theater within three minutes.
(2) Just after that, they are asked to add associated terms WITH Theater to the same worksheet within another three minutes.

Table 1. Experiment methods outline

Experiment Parameters	A	B1	B2
Term Assigned	3 Terms each (high/medium/low interest)		
First Task	Without Thesaurus		
Second Task	With	With	Without
Time interval between 1st and 2nd	Soon	10 days	
Human Subjects	20	10	10

(3) An interview about the task execution is done to confirm their behavior after the experiment.
(4) The subjects' behavior is observed and recorded during the experiment.

Experiment B: The process is the following.
(1) First, subjects are asked to write down as many associated terms as possible on the worksheet WITHOUT Theater within three minutes.
(2) Then, twenty subjects are divided into two groups: an “experimental group” and a “controlled group”. Ten subjects are in each respective group. Subjects perform a similar experiment under different conditions according to the group they belong to. The controlled group performs an experiment for a standard comparison, and the experimental group performs an experiment to test the factors.
(3) The second experiment is executed after ten days interval. Experiment B1 is carried out for the experimental group, and experiment B2 is for the controlled group. The experimental group is given the same topic as the first, and asked to write down the associated terms on a new worksheet WITH Theater within six minutes. The controlled group is given a task identical to the first experiment.
(4) An interview about the task execution is done to confirm their behavior after the experiment.
(5) The subjects' behavior is recorded during the experiment.

4. RESULTS AND CONSIDERATION

Each subject's response differed in quality and quantity. Differences may have resulted from factors such as task hurdle level, the way of thinking, and interest in the task topic.

To evaluate these factors, the following are measured:
(1) The average number of terms each human subject filled in,
(2) The average number of terms according to subjects' interests,
(3) The semantic relation between a given term and the terms filled in,
(4) The recall ratio of terms (used in experiment B1 and B2 only).

4.1 Average Number Of Terms Filled In

Experiment A: Figure 3 shows the number of written terms without Theater and those with Theater. According to this experiment, the average number of associated terms increased by 48% (14.5->21.1) when the thesaurus data was shown to the subjects

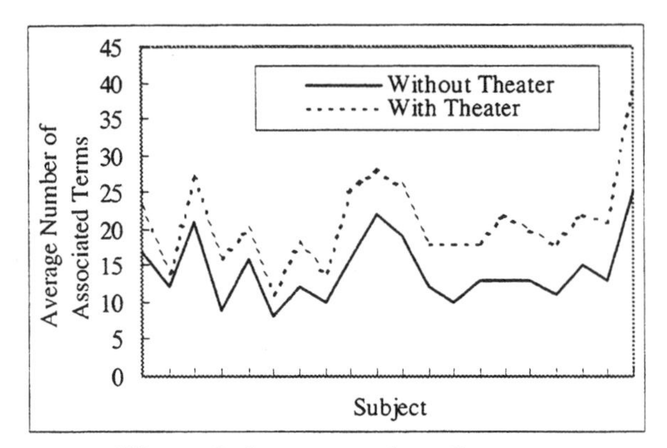

Figure 3. Average number of terms. (Experiment A)

Table 2. Average value of associated terms

###	value
Average number of terms in 1st	14.5
Average number of added terms in 2nd	6.6
Average increased rate	48%

(Table 2). This implies that the thesaurus data allows end-users to obtain new associative terms.

Experiment B: Figure 4 and Figure 5 show the average number of written terms for experiment B1 and B2, respectively. According to Figure 5, each subject wrote down almost the same number of the terms as that in the first experiment. Therefore, a ten day interval is reasonable for removing the influence of the first experiment. Figure 5 includes one piece of data, which can be ignored as an exception. In experiment B1, the number of associated terms with Theater increased by 23% (15.6->19.2), as shown in Table 3. Among ten subjects, two subjects did not associate more terms than in the first experiment. This implies that term suggestion in experiment B1 does not work well. There seems to be room for changing the presentation style considering the differences in results of experiment A and B. This will be touched upon in 4.4 Comparison Between Experiments.

4.2 Average Number Of Terms According To Users' Interests

Figure 6 shows the average number of associated terms in experiment A, based on the subjects' interest level. There is no difference between the results. The result in experiment B is similar to that in experiment A.

The given task topics could be too abstract or the task itself could be too obscure for the subjects. If a given task topic was more concrete, the results could be different. Another experiment based on the topic's term quality is necessary to clarify the difference between interest levels.

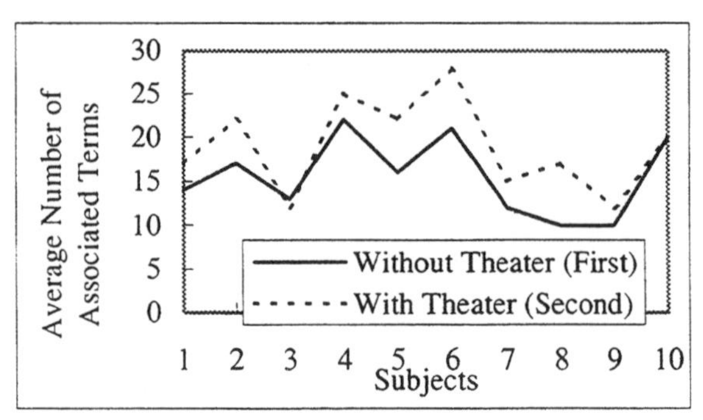

Figure 4. Average number of terms. (Experiment B1: experimental group)

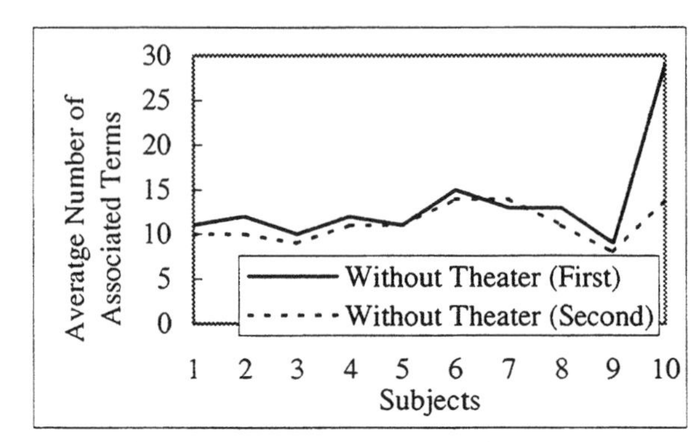

Figure 5. Average number of terms. (Experiment B2: controlled group)

Table 3. Average value of associated terms

###	value
Average number of terms in 1st	15.6
Average number of terms in 2nd	19.2
Average increased rate	23%

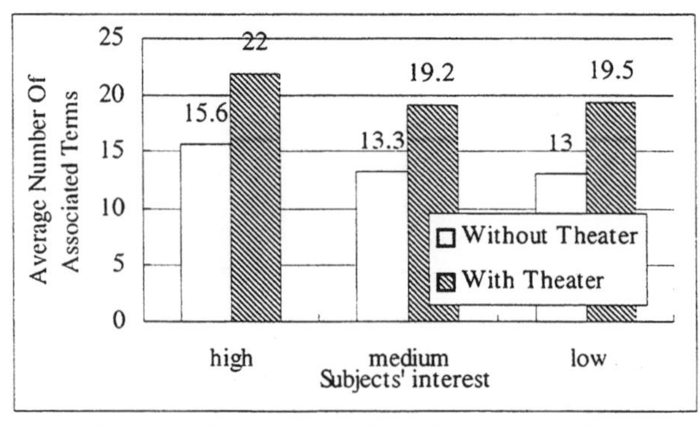

Figure 6. Average number of terms based on subject interest. (Experiment A)

4.3 Semantic Relation Between A Given Term And The Terms Filled In

The terms written by the subjects in experiment A are classified into four kinds of relations: Is-a relation, Has-a relation, Synonym, and Compound term. Of course, there are many terms which are not classified into these four relations. Table 4 shows a summary of into these four relations. Table 4 shows a summary of the results. The rate of increase in the written number of compound terms is especially higher than the others. If there is no typical change in the term qualities, the increased rate should be around 48%, which is the increased rate of associated terms. This shows that Theater reminds users of compound terms because the increased rate is 77%. The terms which have the given task term as a sub-string seem to be very useful, such as "life insurance" or "unemployment insurance" for a given term "insurance".

4.4 Comparison Between Experiments

In experiment B, the increase in the number of associated terms with Theater is only 23%, which is less than the value of experiment A (48%). Furthermore, from observing user behavior, subjects in experiment B seemed to not use Theater at all, while those in experiment A seemed to use Theater enthusiastically.

From these factors, the related term should be suggested after the user has finished the association. If relevant terms are shown initially, the user is disturbed and effective support for association is not accomplished.

4.5 Recall Ratio Of Terms (Used In Experiment B1 And B2 Only)

Table 5 shows recall ratios, which are defined as the rate of common terms written in the first and the second experiments. If subjects write the terms twice, there should be semantic differences between associated terms.

Table 4. Relation between given term and terms filled in Experiment A

###	Terms filled in (first task)	Terms added (second task)	Percentage of Increase
Is-a Relation	132 terms	47 terms	35.6%
Has-a Relation	8 terms	3 terms	37.5%
Synonym	19 terms	2 terms	10.5%
Compound Word	119 terms	92 terms	77.3%

Table 5. Recall ratio of terms: analysis of differential terms (Experiment B)

###	Experiment set	Control set
Average Recall	40.6%	43.0%
Max. Value of Recall	71.4%	75.0%
Min. Value of Recall	12.5%	12.5%

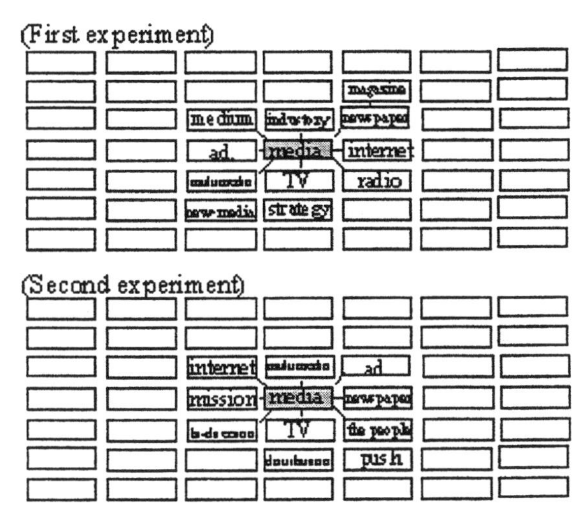

Figure 7. Sample result of subject's work

For example, Figure 7 shows the result of a subject's work. The subject is from the controlled group and did experiment B2. The number of associated terms are almost the same in the first and the second experiment(11/10), and five terms (newspaper, internet, TV, ad., multi-media) are recalled. His recall ratio is about 45% (5/11), which is the average value.

The recall is much lower than expected. The value of the recall ratio is almost the same in both experiments, and the subjects did not seem to use Theater at all in experiment B1. Therefore, the result value possibly depends on their own recognition of the given term. The value may be related to human knowledge of the specified term.

5. EXPERIMENTAL OBSERVATIONS

From our observations of human behavior and the interviews, the following was obtained:

When the suggestion should be provided: If related terms are provided initially, human association may be interrupted. In the analysis of the subjects' comments, they recognized the usefulness of Theater in experiment A, however, there were no such comments in experiment B. Therefore, subjects can use Theater effectively after they complete associating terms. Theater should be activated on user demand to promote human association.

How the suggestion should be done: Most subjects refer to the co-occurrent terms. However, co-occurrent terms include useful data as well as noisy data. Subjects pointed out the problems:

(1) The load of selection from a lot of suggested terms.
(2) The co-occurrent term list is difficult to use, because it is not so organized.
(3) Suggested terms do not make a frame and a direction of the current association.

Here, the third comment concerns on experiment B.

When Theater suggests terms, it should group the co-occurrent terms and the contrivance into a cluster to make it easier to use.

Tool usefulness: Subjects provided the comments:
(1) Theater is useful when comparatively unfamiliar terms are given.
(2) Subjects feel a duty to use Theater when it is presented.
The first comment is interesting considering the term suggestion interface. Subjects regard Theater as useful for unfamiliar terms, although the experimental result presents there is no typical difference based on the interest level. Although given task terms are roughly classified into three interest levels in the experiment, if subjects' interests in the task term are measured in more detail, the keyword suggestion may reflect the result. The second comment concerns conclusion (1), which is the way the presentation should be improved.

Cluster of the suggested terms: Theater generally presents more than fifty co-occurrent pieces of data. At present, there is no clustering algorithm in Theater and this causes difficulty in selecting terms from the columns (Figure 2). It might be useful to arrange compound terms into one cluster so that many compound terms in Theater include the given task term as a sub-string, and grouping may make it easier to see the term list.

Synonyms and antonyms: Synonyms are important and useful for selecting relevant terms. Although synonyms and antonyms are not at all associated with the experiment results. It is easy to generate co-occurrent term pairs from existing data. However, automatic synonyms, antonyms, and derivation are not studied enough and remain for future work.

6. CONCLUSION

This paper has presented an experimental evaluation for term suggestion by a thesaurus. The thesaurus system, called Theater, consisted of traditional broader and narrower terms as well as co-occurrent terms which were extracted from existing documents. Theater is expected to work as a keyword reminder.

The thesaurus tool can be recognized as a useful keyword reminder because of results in experiment A, which show 48% increased association. Some attention is necessary for using it. Considering the experiment results and the comments from subjects, the following observations can be made:

Co-occurrent terms can be useful while there is a possibility to spread users thinking out. An algorithm that clusters the data is required.

The experiment shows that terms suggesting timing influence the efficiency of the tool. People should first associate terms by themselves, and after that, terms suggested by the system are shown. It is better for Theater to display terms only at the request of a user.

Compound terms, which include the given term as sub-string are newly recognized as relevant terms by using Theater from an analysis of the relation between a given term and the associated terms filled in on the worksheet. Such compound terms can be a key for making a cluster from a variety of co-occurrent terms. It is useful in this case that the users' retrieval purpose is comparatively obscure.

The observations described above can be also regarded as a guideline for user interface design.

ACKNOWLEDGEMENT

The authors would like to express sincere thanks to Dr. H. Kinukawa and Dr. H. Yajima. Special thanks is also due to members of the Kansai Systems Laboratory, who were volunteers for subjects.

REFERENCES

B. Schatz et al. (1995), Interactive Term Suggestion for Users of Digital Libraries: Using Subject Thesauri and Co-occurrence Lists for Information Retrieval, *ACM International Conference on Digital Library Conference*

F. Cheong,.(1996), *Internet Agents*, New Riders

H. Mase et al., (1997), Knowledge Spiral Interaction for Information Search, In: *Proceedings* of *Human Computer Interaction '97*

J. Aitchison and A. Gilchrist (1987), Thesaurus Construction (2), (Aslib, London)

Koller, D. and Sahami, M.(1997) Hierarchically Classifying Documents Using Very Few Words. In: *Proceedings of the Fourteenth International Conference on Machine Learning (ICML-97)*

T. Tokuda et al. (1996), Architecture of thesaurus management system for network era (In Japanese), *In: Proceedings of the Fifty third Annual Conventions (IPSJ)*, **3**, 163-164

Y. Niwa et al. (1997a), Topic Graph Generation for Query Navigation: Use of Frequency Classes for Topic Extraction, In: *Proceedings of NLPRS'97, Natural Language Processing Pacific Rim Symposium*, 95-100

Y. Niwa et al. (1997b), Interactive Support of Query Refinement by Dynamic Word Co-occurrence. In: *Proceedings of the 17th International Conference on Computer Processing of Oriental Languages (ICCPOL'97)*, 383-386

MODELLING SYSTEMS USING SYSTEM THEORY AND NEURAL NETWORKS - APPLIED TO HUMAN WALKING

P.H. Wewerinke * F. Brandt ** M. Millonzi ***

* *Faculty of Mathematical Sciences, University of Twente, P.O. Box 217, 7500 AE Enschede, The Netherlands*
** *Goudappel en Coffeng, Deventer, The Netherlands*
*** *Signaal, Hengelo, The Netherlands*

Abstract: A new modelling approach is presented in this paper to describe partly known systems. Basically, the approach combines system theory models (describing the known part) and neural networks (describing the unknown part, which input-output relationship has to be learned based on data).
This modelling approach is applied to the process of human walking. The known characteristics of this complex process are modelled by two simplified models of nominal walking and of balancing the body. In addition a neural network is included to provide a complete and realistic description of human walking.

Keywords: System modelling, system theory, neural networks, human walking

1. INTRODUCTION

This paper consists of two parts. First, a new modelling approach to describe partly known systems will be discussed. Basically, the approach combines system theory models and neural networks to describe partly known systems. The known part is accounted for by the system theory model and the unknown part by the neural network (which needs data to learn this unknown part of the system). This approach is discussed in the next section.

This modelling approach is applied to the process of human walking. Human walking is a complex process. The known characteristics are modelled by two (simplified) models of nominal (open loop) walking and of balancing the body in the presence of disturbances. To the extent these models are not correct and not complete, a neural network is included to provide a complete and realistic description of human walking. This is discussed more specifically in section 3.

2. MODELLING PARTLY KNOWN SYSTEMS

Traditionally, system theory, especially linear system theory, has been successful in modelling a variety of systems (processes). However, serious problems arise if the system is not precisely, or not completely, known. This can be related to the structure of the system, the order and the parameters involved. The result can be unacceptably large model errors.

In case nothing (or little) is known about the system itself, it is possible to use measurements of system behavior to describe this behavior in terms of the relationship between input and output data of the system. A classic example of such a time series analysis is an ARMA model as an input-output model of the system.
More recently, neural networks have become popular to describe systems based on input-output data. In these cases, in principle nothing is (assumed to be) known about the system structure and the model only reproduces a more or less precise relationship between the input and the

output of the system.

In case the system is partly known (e.g. based on physical laws, economic principles, etc.) both system theory fails to model the system correctly and the input-output models, such as a Neural Network (NN), can not utilize the advantage of knowing partly the relationships involved in the system (structure). So, in this case, it seems plausible to combine both approaches and describe a system in terms of system theoretic models and (e.g.) NN. This idea is successfully investigated in (Brandt, 1996). Figure 1 summarizes the situation. The part of the system that is known is described by the system theory model. In case nothing is known, the system can be approximated by (e.g.) a second-order linear system.

The study described in (Brandt, 1996) considers both linear and nonlinear systems, deterministic and stochastic systems, completely known and completely unknown systems and partly known systems (the order, a given parameter and part of the dynamics). All the different configurations were modelled by means of a NN only and by means of a system model (an assumed second-order approximation or a model of the known part of the system) in combination with a NN. The results of these two approaches were compared in terms of the model errors (performance) and the computational effort (iterations), as the costs involved.

The results show that for the two types of NN investigated (a feedforward NN and a recurrent (with feedback) NN (Elman network)), for all the conditions considered the Elman NN yielded superior performance. This is not surprising given the dynamic systems involved as a recurrent NN explicitly describes the temporal relationship.

The results show that in general (independent of all the factors considered) the best modelling procedure is to use a system model for the known characteristics of the system in combination with a NN (describing the unknown relationships, based on the data).

The more is known about the system (order, parameter(s), subsystems) the better modelling performance is obtained, both for linear and nonlinear systems, and both for deterministic and stochastic systems. This is not surprising; however, the nontrivial result is that the combination of the (approximation of the) system model and the NN yields the best (possible) results for all conditions investigated.

In case the system is completely known it is evident to use the model of the system. In case the system is stochastic the standard approach to estimate the state of the system is by means of an extended Kalman filter (EKF). Preliminary results of the study suggests that the EKF performance can be improved by combining the EKF with a NN. This makes sense because the EKF is based on an approximated (linearized) model of the system (higher order terms in the Taylor series approximation are neglected). It seems that for smaller system noise and/or smaller measurement noise the improvement of the NN becomes smaller. This is plausible as for smaller noise the effect of the EKF modelling errors becomes smaller and (only) the modelling errors can, potentially, be compensated for by the NN.

3. MODELLING HUMAN WALKING

Human walking is a complex process from a theoretical mechanical point of view. It is still not (precisely) known how man is walking. Yet several preliminary modelling attempts to mimic human walking, especially aimed at biped robot walking, have been reported ((Shih and Gruver, 1992),(Shih *et al.*, 1993),(Goddard *et al.*, 1992),(Kajita *et al.*, 1992),(Taga, 1994), (Golubev and Degtyareva, 1993)). However, no realistic model of human walking is available yet. This would be useful to study problems of disabled people, balancing problems (with a model including the dynamics of the semi-circular canals and otholits), and applications in several sports and robotics.

In this part of the paper a model of human walking is discussed following the afore-mentioned modelling approach. The model structure is summarized in Figure 2. The model consists of three components.
The first component describes nominal (open loop) human walking. In addition, because of the inherent unstable cinfiguration of a standing human body (like an inverted pendulum) and because of disturbances, it is essential to balance the body in order to maintain its (unstable) vertical orientation. This function is performed by a balancing model. These two models represent the known characteristics of human walking, although the models are relatively simple and based on many assumptions. However, a useful model of human walking is obtained, especially if this model is supplemented with a neural network to enhance the realism of the model and compensate for the model deficiencies. Of course this can only be done if data are available based on which the neural network can be trained.

The human body will be considered as a mechanical and controlled system consisting of 13 components (links). This is shown in Figure 3. The links consist of feet, lower legs, upper legs,

lower arms, upper arms, neck, body and head. The geometry of these components are simplified to rectangular plates, cylinders, an elliptical cylinder and a sphere, respectively.
In the following, the three submodels will be discussed briefly.

3.1 *Nominal Walking*

The periodic motion of walking consists of two different phases: the single-support phase (only one foot on the ground) and the double-support phase. This distinction is important because of the different mechanics (DOF, forces, etc.) involved. The dynamics are described in (Millonzi, 1997) by the two Newtonian equations for the rotational and translational (forward) motions. The rotation of the center of gravity (c.g.) of the body with respect to the position of the forward foot is given by

$$M_t = \frac{d}{dt}[I_t \dot{\alpha}_{cg}] \tag{1}$$

where $I_t = I_t(\alpha_i)$, $i = 1, \cdots, 13$ and M_t is the total moment due to the gravity force and, in the double-support phase due to the push off force. Various assumptions can be made concerning the various DOF (α_i). This is discussed in (Millonzi, 1997).

Also the assumption is made that the forward motion is constant. This motion is a result of the horizontal push off force and can be related to (expressed in) α_i. So basically the model provides all the necessary (13) relationships between the 13 DOF (α_i). As an input the angle α_1 of the upper pushing off legg is chosen (which results from the forces of the toe, ankle and upper leg.) The model outputs are the remaining α_i.

In (Millonzi, 1997) model simulation results are discussed for several configurations (bodies of different length and weight) in terms of the nine geometric angles as a function of time during one complete step (consisting of a single-support phase and a double support phase) of 1.5 seconds. Also (potential and kinetic) energy considerations are included.

3.2 *Balancing*

In order to stabilize and to maintain the vertical body orientation it is necessary to control the body. It is assumed that the vertical orientation of the body (defined as angle φ) is controlled by ankle moments (u) of the standing leg(s).
A first onder approximation of this problem is the control of an inverted pendulum (mass m and length ℓ); this is summarized in Figure 4. The linearized dynamics of the inverted pendulum (IP) is given by

$$\ddot{\varphi} = a\varphi + bu \tag{2}$$

where $a = \frac{3g}{2\ell}$ (positive!) and $b = \frac{3}{m\ell^2}$.

The controller consists of two parts: a control command (in the brains) u_c which is proportional to the tilt angle φ and the angular rate $\dot{\varphi}$

$$u_c(t) = K\varphi(t) + L\dot{\varphi}(t) \tag{3}$$

with K and L optimal feedback gains (explained in the following), and a delayed ankle moment response u according to

$$u(t) = u_c(t - \tau) \tag{4}$$

This delay equation can be approximated by a Padé-approximation.

Assuming this Padé-approximation, i.e. $e^{-\tau s} = (2/\tau - s)/(2/\tau + s)$ and combining eqs. (2) to (4) yields

$$\dot{x} = Ax, \quad x(0) = x_0 \tag{5}$$

with $x = \text{col}\,(\dot{\varphi}, \varphi, u)$ and $A = \begin{bmatrix} 0 & a & b \\ 1 & 0 & 0 \\ c_1 & c_2 & c_3 \end{bmatrix}$

where

$$\begin{aligned} c_1 &= K - 2L/\tau \\ c_2 &= La - 2K/\tau \\ c_3 &= Lb - 2/\tau \end{aligned}$$

Now, K and L are selected such that u stabilizes the system (i.e. eigenvalues of A should have negative real parts) and yields maximum damping. This can be solved analytically.

The eigenvalues will have negative real parts if the Routh-Hurwitz criterion is met. This implies the following inequalities

$$L < \frac{2}{\tau b} \tag{6}$$

$$K > \frac{a}{b} \tag{7}$$

$$K < \frac{2L}{\tau} - \frac{a}{b} \tag{8}$$

$$K < \frac{2\frac{L^2 b}{\tau} - \frac{4L}{\tau^2} - La}{Lb - 4/\tau} \tag{9}$$

This is shown in Fig. 5. The middle of the admissible (shaded) region represents the optimal values (yielding maximum system damping) for K

and L, indicated with K_0 and L_0. The analytical expressions are

$$L_0 = \frac{4}{\tau b} - \frac{1}{b}\sqrt{\frac{8}{\tau^2} - 2a} \qquad (10)$$

$$K_0 = \frac{bL_0^2 - \frac{2}{\tau}L_0 - \frac{2}{b}a}{\tau b L_0 - 4} \qquad (11)$$

Equations (10) and (11) shows that K_0 and L_0 are a function of the system parameters a, b and the human operator parameter τ. Also from equations (6) to (9) it follows that τ should be smaller than $\sqrt{2/a}$.

Using these results the closed loop response can be computed.
A typical model result is shown in Figure 5 for an adult human body stabilizing an initial disturbance (in which case the human body with moment of inertia I replaces the IP).

3.3 *Neural Networks*

The third submodel concerns a neural network. A neural network is trained (using data of human walking) to learn the relationship between input variables (angle α_1 of the upper pushing off leg and the ankle moment u of the standing leg) and the output variables (remaining α_1, determining the body shape, and the tilt angle φ). Essentially, the neural network is aimed at describing the walking behavior that is not accounted for by the afore-mentioned submodels of nominal walking and balancing.

Assume that the desired walking behavior (e.g. corresponding with the real walking behavior which possibly can be obtained from measurements) is given by α_i, $i = 1, \cdots, 9$. The total model output α_{m_i} is the sum of the nominal walking model input (α_{n_i}), the balancing model output (α_{b_i}) and the NN output (α_{N_i}). Now, the NN will have to be adjusted such that the difference(α_{e_i}) between α_i and α_{m_i} will be minimized.

This adjustment procedure can be based on various learning schemes. For dynamic systems an efficient NN is a recurrent (e.g. Elman) network, which structure involves also feedback connections between the neurons. A standard concept to adjust the weightings of the NN is adapted, i.e. the adaptive 'back propagation' method. For details the reader is referred to the literature, e.g. (Brandt, 1996).

4. CONCLUDING REMARKS

A new modelling approach is presented in this paper to describe partly known systems. The known part of the system is described by means of a system theoretical model; the remaining, unknown part of the system is accounted for by a neural network. The latter input-output model requires measurements to train the network. It is shown in (Brandt, 1996) that this combination of model structures yields the best description of partly known systems.

The foregoing modelling approach is applied in the paper to the process of human walking. This complex process is described by a combination of a biomechanical model of nominal human walking, a model to balance the human body in the presence of the gravity force and disturbances and a neural network to complete the realistic description of human walking.
Although the separate models have been developed and demonstrated in simulations to provide meaningful models of human walking and balancing, the complete modelling structure has still to be tested.

5. REFERENCES

Brandt, F. (1996). Modelling systems using system theory and neural networks. Master's thesis. Dept. of Applied Mathematics, University of Twente. (in Dutch).

Goddard, R.E., Yuan F. Zheng and H. Hemami (1992). Control of the heel-off to toe-off motion of a dynamic biped gait. *IEEE Trans. on Systems, Man and Cybernetics.*

Golubev, Yu.F. and Ye.V. Degtyareva (1993). Modelling of the dynamics of a walking robot by the small-parameter method. *Journal of Computer and System Sciences International* **31**(6), 138–148.

Kajita, Shuuji, Tomio Yamaura and Akira Kobayashi (1992). Dynamic walking control of a biped robot along a potential energy conserving orbit. *IEEE Trans. on Robotics and Automation.*

Millonzi, M. (1997). Dynamics of human walking. Master's thesis. Dept. of Applied Mathematics, University of Twente. (in Dutch).

Shih, Ching-Long and W.A. Gruver (1992). Control of a biped robot in the double support phase. *IEEE Trans. on Systems, Man and Cybernetics.*

Shih, Ching-Long, W.A. Gruver and Tsu-Tian Lee (1993). Inverse kinematics and inverse dynamics for control of a biped walking machine. *Journal of Robotic Systems.*

Taga, Gentaro (1994). Emergence of bipedal locomotion through entrainment among the neuro-musculo-skeletal system and the environment. *Physica D* **2**(4), 190–208.

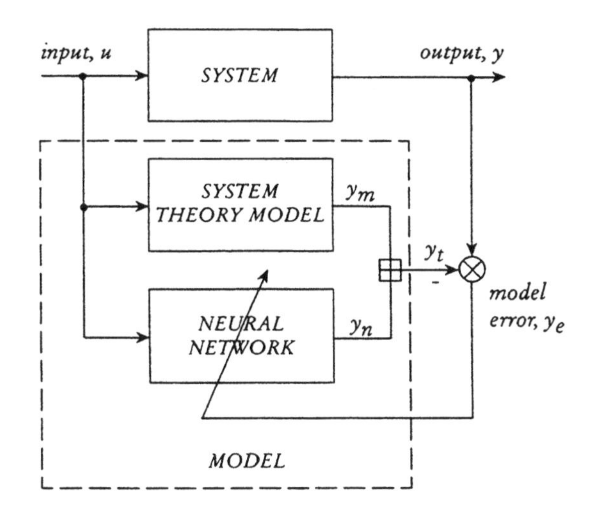

Fig. 1 Block diagram of a system theory model and a neural network

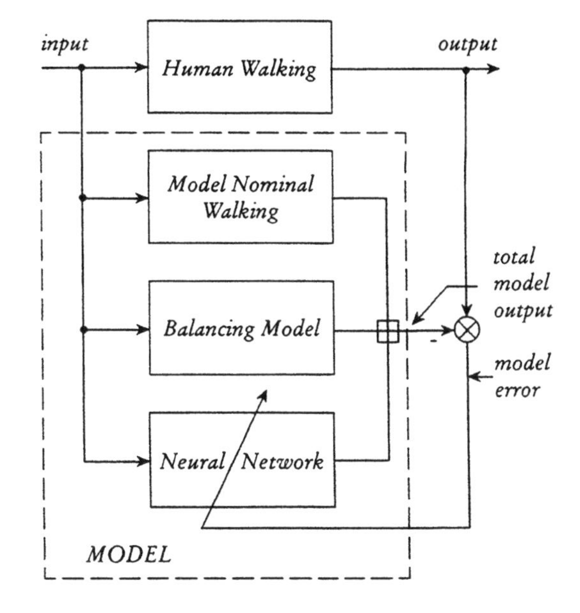

Fig. 2 Modelling human walking

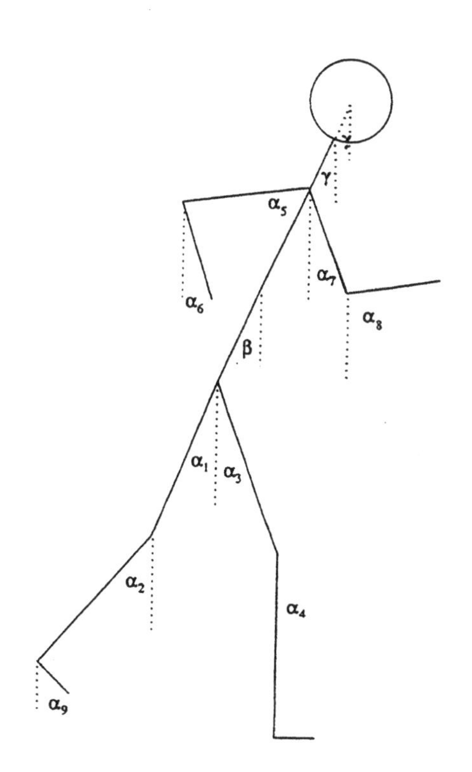

Figure 3: Human body geometry

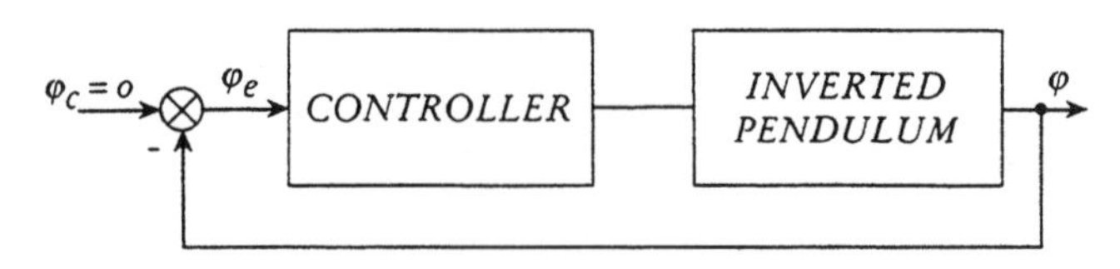

a) Block diagram control task

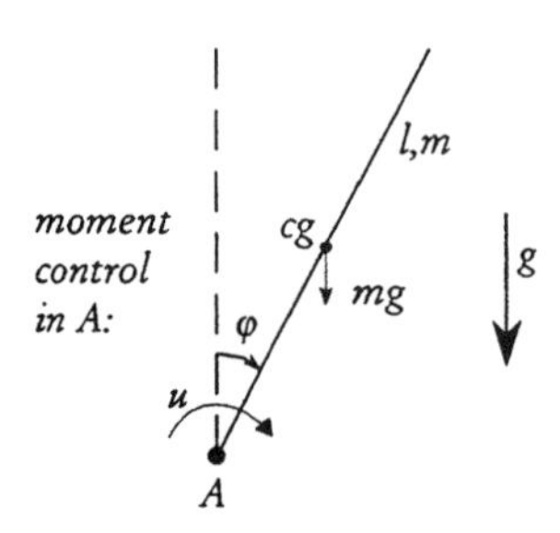

b) Control situation

Fig. 4 Control vertical orientation

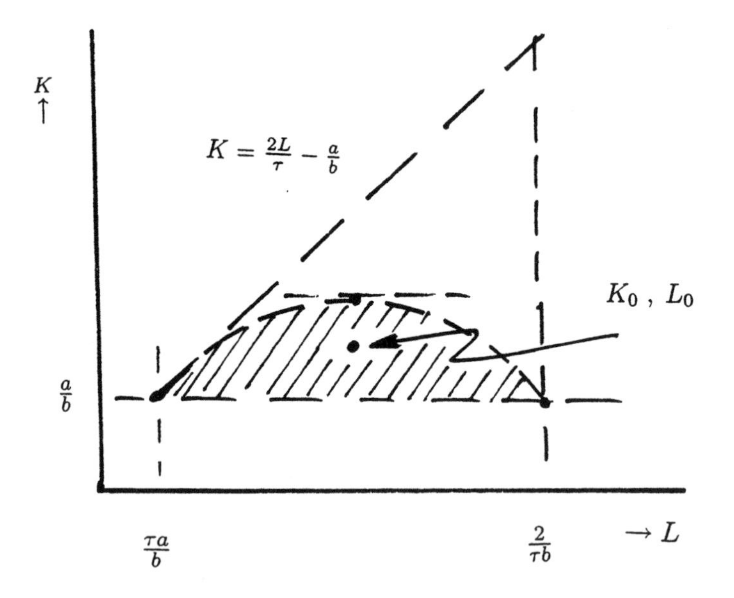

Figure 5: Optimal values for the feedback gains K and L

MODEL-BASED VISUALISATION OF TECHNICAL SYSTEMS BY MEANS OF FUZZY TECHNOLOGIES

Salaheddin Ali

University of Kassel, Laboratory for Human-Machine Systems, D-34109 Kassel, Germany

Abstract: A novel design technique for the construction of human-machine interfaces is presented. This technique strives after increased user orientation adapting human-machine interfaces to the cognitive structures of human users. Different models of the human knowledge that are available in analogous or conceptual form have to be considered, both in the design of a graphical user display and in the design of the underlying information management system. Fuzzy logic is used for translating natural language procedures acquired from operators into objects in knowledge bases. Thus, technical systems can be considered and controlled from different viewpoints.

Keywords: process control; human-machine interface; knowledge-based systems; object-oriented software design, fuzzy logic.

1. INTRODUCTION

In the following, an integrated approach with new design methods for human-machine interfaces in process control will be presented. All methods and techniques developed during this research work supported partly by the German Research Foundation (DFG) were evaluated and tested with a simulation of a chemical distillation process. This simulation was developed at the Institute of System Dynamics and Control Engineering of the University Stuttgart. It represents a typical, highly complex system and is thus very suitable for demonstrating the methodology.

Technical systems are mostly visualised by conventional user displays based on topological representations and flow diagrams (VDI/VDE 3695, 1986). For large and complex installations, views in these kinds of user displays are split up according to system/sub-system hierarchies. Such views become intricate if complex plants have to be visualised. The design method and technology of this study is based on different models of human operators and the technical process. These models are acquired from experienced operators through task and process analysis must be used in integrated way to achieve a powerful work environment for human operators.

Presentations of the different models containing analogous and conceptual knowledge are considered, both in the graphical design of the user display as well as in the construction of software structures realised in the from of object-oriented modules (Johannsen, 1993). With this type of visualisation, adapted to the cognitive structures of human operators, process control and supervision can be supported on different cognitive behaviour levels (see Rasmussen, 1986; 1987). A high transparency is accomplished and thus, it is possible to visualise malfunctions and process violations in an intuitive manner.

2. MODELS OF KNOWLEDGE PRESENTATION AS FUNDAMENTAL DESIGN MATERIALS

In the human-machine interface, different models of the technical process and of the operators, describing their supervisory- and control behaviour, are used to represent the process on different levels of abstraction and from different viewpoints. The different models are applied to both the design of the graphical user display and the development of the information management system of the human-machine interface. Both sub-systems are adjusted to the cognitive structures of human operators. Experiences with human-machine interfaces have shown that the visualisation of process states through the graphical user display can be a substantial aid to operators (Ali, 1997). It is essential that the human operator can observe the state of a technical process without wasting a lot of time in observation and identification of process variables. The knowledge that is required for this development is acquired through task- and process analyses and contains the following models:

- A model of the human information processes involving human decision making, adopted from (Rasmussen, 1988). This model is not explicitly used in the design of the user interfaces, but it is applied in the design philosophy.

- A state-task model containing a hierarchical order of elements obtained by task and process analysis (see Fig. 1). A technical system is first divided into *critical subsystems*. Relevant process states are determined for these subsystems. The state-task model defines the mapping between these process states, and the tasks that have to be carried out by the operator. A similar description is represented in Matern (1984). The manner in which such model can aid a human operator in his task is discussed at the hand of the information- and decision-*ladder* model by Rasmussen (1988), which describes human cognitive processes as a sequence of information processing stages A human-machine interface using the state-task models should relieve and support the operators by their activities of information processing, such as *observe information and data*, *identity present state of the system* and, last but not least, *Define task; select change of system condition*. The human-machine interface mediates to the operators not only the final result of the knowledge states system state and task computed by the information processing system but also the necessity for doing tasks. As shown in Fig. 1, the state-side includes elements about the state of the technical systems. On the task-side, levels containing elements about tasks, activities and actions are modelled. These task elements are categorised in task classes such as process goals, strategies and security classes.

- A qualitative causal model of the technical system (Funke, 1992). This model consists of the important process variables that are sufficient to describe qualitative process behaviour (the qualitative part of the model) and the causal relations between these variables (the causal part). It will be shown that causal models are partly used in the graphical design and as a basis to define the different knowledge elements in the fuzzy knowledge base.

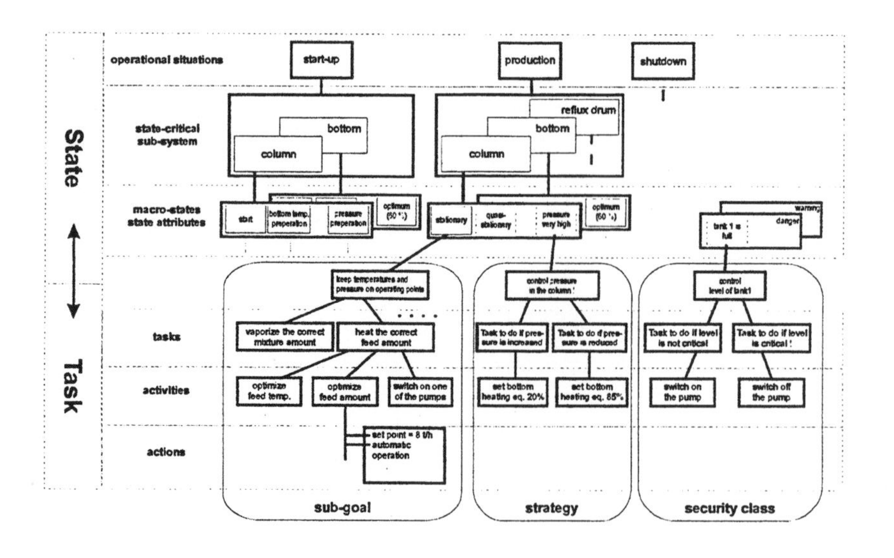

Fig. 1. State-task model of the distillation column

3. APPROXIMATE KNOWLEDGE-BASED PROCESS VISUALISATION WITH FUZZY LOGIC

The different models mentioned above are used to design knowledge objects located and managed by the information processing system as well as the corresponding graphical modules that will be presented on the user display. These models contain both descriptive or declarative knowledge, and operational or procedural knowledge. This knowledge is mostly available in the form of natural language. For the construction of qualitative models, this knowledge must be recorded and converted into a computer representation. Fuzzy logic is especially suited for this second step (Zimmermann, 1991).

Followings describe the reasons why fuzzy logic is suitable for converting and processing the knowledge used in approximate knowledge-based human-machine interfaces. For more detail information see (Ali, 1998):

- The knowledge acquired from human operators is available in a colloquial form. The application of a means, i.e. fuzzy logic, for the direct conversion of this knowledge into a computer representation leads on the one hand to shorten the development time. On the other hand it will be possible to create new kinds of display elements which correspond to the ways of human thinking (Ali, 1998).

- On higher levels of process control and supervision, users think not only in specific values, but also in linguistic terms (Kahlert, 1995).

- Undisputedly, display elements with symbolic information contents represent a central constituent of advanced user displays (Johannsen, Ali, Heuer, 1995; Johannsen, Ali, Van Paassen, 1997). To mention graphic items for the representation of states, alarms, fulfilment-grades of process goals and strengths of the urgency for doing tasks and actions (Ali, 1997).

- A further important problem with the use of graphical elements for the visualisation of process states or fulfilment-grades is the characterisation of value ranges of process variables used to describe these process information. Display elements, which are based on such values, strongly depend on the way of interpretation of the appropriate intervals.

The fuzzy rules used within the objects of the knowledge base take linguistic variables as their input. To prepare these input variables, values from the controlled process are converted into fuzzy sets. The values of the output variables of the knowledge base are fulfilment-grade values for process goals as well as strength values for states, necessity for determining which tasks are relevant.

Different inference mechanisms for calculating approximate values about the whole human-machine system are realised within the information management system. In the following, one of these mechanisms will be discussed. This method converts the fuzzy result set as it is calculated by the inference machine into a concrete value. Such values are then used to animate graphical objects on the user display, in this example for the visualisation of fulfilment - grades of process goals. Fig. 2 illustrates the principle of the inference mechanism at the hand of the example *process goal of the feed*.

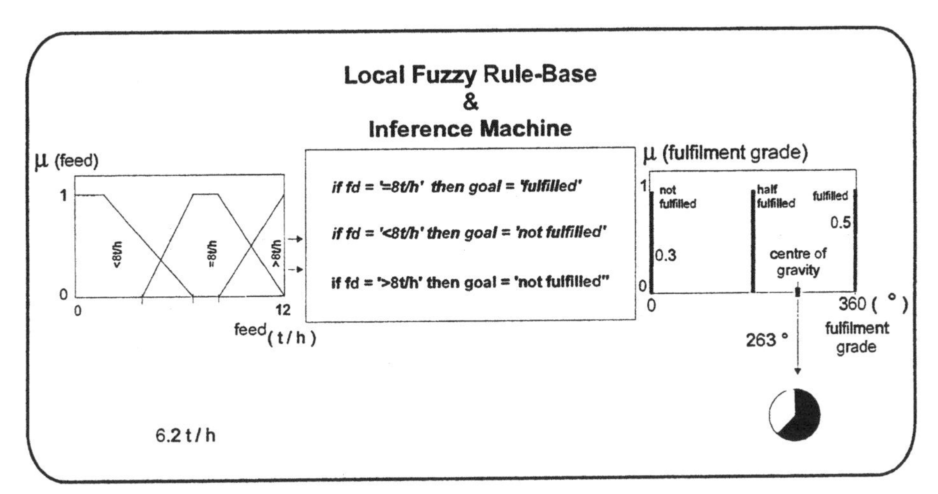

Fig. 2. Inference process for output variables of fuzzy knowledge-base for visualisation of a fulfilment - grade.

For the defuzzification of fuzzy result sets, the centre-of-gravity method is used. This results in concrete values called fulfilment - grades. The fulfilment-grades of the goals on the user display are approximate. This means that a goal can get values with different strengths, e.g. *60 % fulfilled.* The goal value is represented on the display by a small pie-chart, displayed in the (circular) goal icon.

4. APPROXIMATE KNOWLEDGE-BASED PROCESS VISUALISATION

During this research work, a complete human-machine interface containing both a software management system and a graphical user display has been developed. This tool allows, technical systems to be considered and controlled through different representations such as topology, causal coherence, process goals, states, necessity for doing tasks and security classes. In this section, some visualisation aspects of the approximate knowledge-based user display, the *PRISMA* user display, will be presented. Fig. 3 shows the *PRISMA* user display during the visualisation of approximate states and goal fulfilment - grades of the chemical process *distillation column*. In the upper portion of the screen, various information about the process such, as operational situations and security classes, as well as states and process goals of the critical subsystems of the technical system are presented, by means of an overview. This overview window cannot be covered by other displays and, thus, the operators are always informed about the current process state. The goal icons visualised by pie-chart symbols mediate whether the goals of the influence-function units in the different critical subsystems are fulfilled or not. As illustrated above, the calculation of every pie-chart value is based on fuzzy rules and is executed by a local fuzzy inference machine.

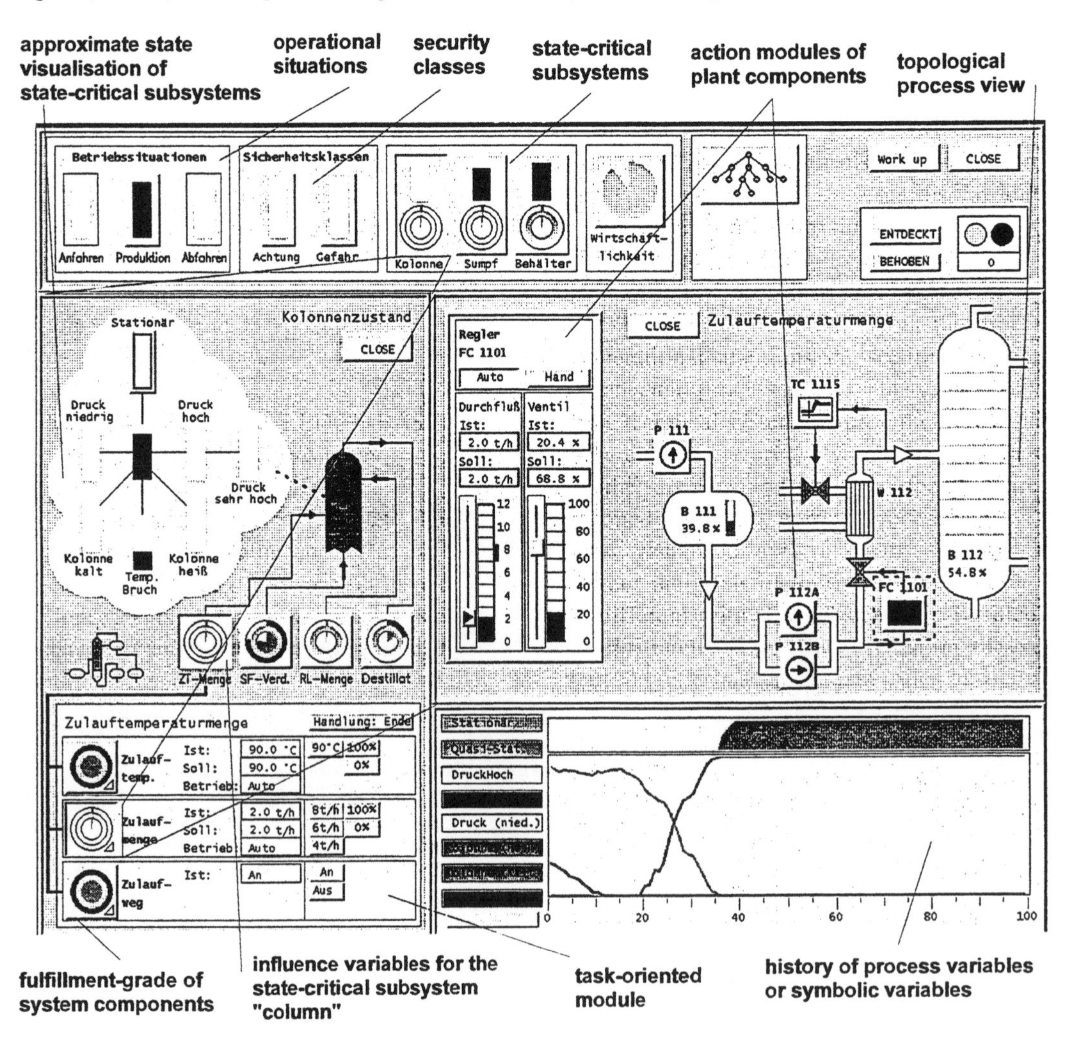

Fig. 3. The knowledge-based user display *PRISMA* during visualisation of approximate states and goal fulfilment-grades of the chemical process *distillation column*

Details of a critical subsystem can be enlarged to get more information about the macro-states (sub-states) and necessities for doing tasks. Moreover, detailed information about the fulfilment-grades of component functionalities within a function unit can also be visualised. These icons can be used to activate the corresponding topological views of the process. The operator can select any given macro state represented by a gauge icon to gather information about the rules underlying this specific macro-state. These rules are presented in the window shown on the lower right side of Fig. 3.

Another possible representation form supported by the *PRISMA* user display is the task-dependent state visualisation managed by a rule-based module called *security classes*. Within this object-oriented module, process violations describing illegal or undesirable combinations of process values are modelled. These combinations of process values are not related to one specific state. This means that these process violations can occurred during all operational situations. Fig. 4 represents the task-dependent state visualisation supported by the *PRISMA* user interface by means of the example *security class: warning*. As shown in Fig. 4, two process violations with different strengths are visualised by a graphical rule-based process view located on the left-middle side of the desktop. Approximate values for terms of security classes represented in the overview of the technical process are calculated by means of the MAX operator (Zimmermann, 1991). The MAX operator combines all results of fuzzy rules belonging to a class to a single value.

Furthermore, locations of process violations can be achieved by sub-premises of rules represented by icons. These sub-premises containing process variables and values margins can be selected to activate topological views, and so to go to the location of the components that must be controlled and handled. As mentioned above, this state visualisation is task-dependent. This means that after completion of correct actions, a process violation will disappear although the state variables are still have undesirable values.

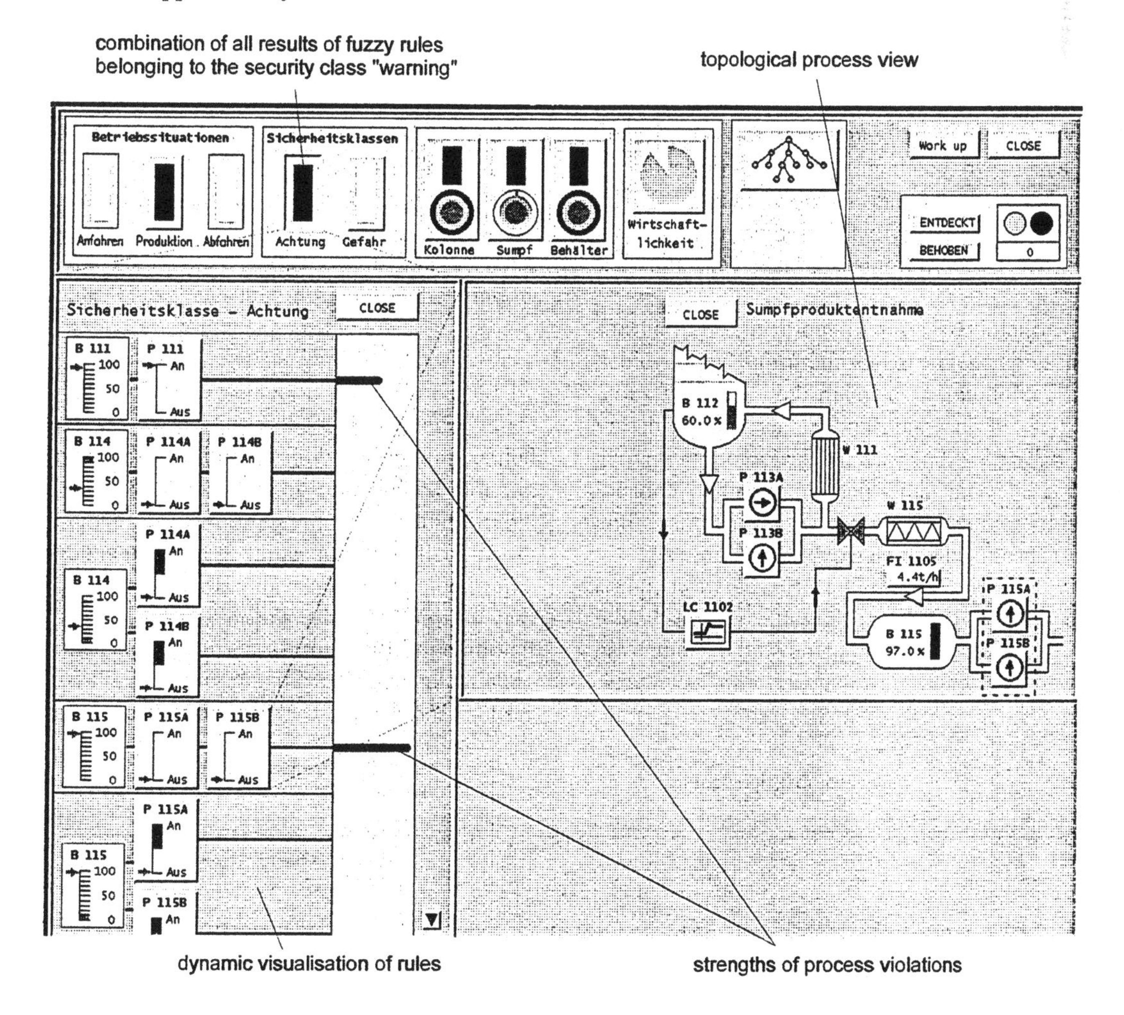

Fig. 4. The knowledge-based user display *PRISMA* during task-dependent state visualisation

5. EVALUATION

The usability of this human-machine interface was evaluated in a comparative investigation (Kopecny, 1996). Two other user displays using different concepts were used for the evaluation after software and cognition-ergonomic aspects, which reflect the requirements at man-machine interfaces for dynamic processes. The first user display is oriented on the guideline VDI/VDE 3695 (1986) and structures the information topological and hierarchical such as outlines, sub-systems, areas, groups and measurement-points. The other user display visualises functional coherence and is based on the multilevel flow modelling theory (MFM) by Lind (1988, 1993). These user displays were implemented to be acquainted with the status of man-machine interfaces in process control and for evaluation purposes. The results of the laboratory experiments show that approximate knowledge-based process visualisation is very suitable for the support of human operators during their control and monitoring activities (see Ali, 1998; Kopecny, 1996).

6. CONCLUSION

This contribution presents a new kind method for the design of human-machine interfaces in process control. By means of this method a graphical user display called *PRISMA* was developed and compared with other two displays. Through these displays that based on different design philosophies, the same technical process *distillation column* can be controlled and monitored. The purposes for developing these user displays were to analyse supervision problems in process control as well as to use them as reference systems in the evaluation mentioned above. The visualisation technique achieved through this method is called *approximate knowledge-based process visualisation*. The main goal of this visualisation is the optimum integration of the operator in the whole control loop of the human-machine system. This will lead to better work satisfaction of the human as well as more security and economic efficiency.

REFERENCES

Ali, S. (1998). Approximative wissensbasierte Prozeßvisualisierung auf Basis der Fuzzy-Logik. Dr.-Ing. Dissertation. Universität-Gh Kassel.

Ali, S. (1997). Wissensbasierte Prozeßvisualisierung mit Fuzzy-Logik. In: K.-P. Gärtner (Hrsg.), Menschliche Zuverlässigkeit, Beanspruchung und benutzerorientierte Automatisierung, 183-196. 39. Fachausschußsitzung Anthropotechnik der (DGLR) am 21. und 22. Oktober 1997 in Karlsruhe. Bonn: DGLR, ISBN 3-922010-98-9.

Funke, J. (1992). Wissen über dynamische Systeme: Erwerb, Repräsentation und Anwendungen. Berlin: Springer.

Johannsen, G. (1993). Mensch-Maschine-Systeme. Berlin: Springer.

Johannsen, G., Ali, S., Heuer, J. (1995). Human-Machine Interface Design Based on User Participation and Advanced Display Concepts. Post HCI 95 Conference, Seminar on Human-Machine Interface in Process Control, Kyoto, July 1995.

Johannsen, G., Ali, S., van Paassen, R. (1997). Intelligent Human-Machine Systems. In: S. G. Tzafestas (Ed.), Methods and Applications of Intelligent Control, 329-356. Dordrecht, The Netherlands, Kluwer Academic Publishers.

Kahlert, J. (1995). Fuzzy Control für Ingenieure: Analyse, Synthese und Optimierung von Fuzzy - Regelungssystemen. Braunschweig: Vieweg.

Kopecny, A. (1996). Untersuchung und Bewertung einer kognitiv-kompatiblen Mensch-Maschine-Schnittstelle für einen chemischen Prozeß. Unveröffentlichte Diplomarbeit am Labor für Mensch-Maschine-Systeme, Universität Gh Kassel.

Lind, M. (1981). The use of flow models for automated plant diagnosis. In Rasmussen, J. und Rouse W. B. (eds.) Human Detection and Diagnosis of System Failures. New York: Plenum Press.

Lind, M. (1988). System concepts and the design of man-machine interfaces for supervisory control. In: L.P. Goodstein, H.B. Andersen, S.E. Olsen (Eds.), *Tasks, Errors and Mental Models*. Taylor and Francis, London, pp. 269-277.

Lind, M. (1993). Multilevel flow modeling. *AAAI'93 Workshop on Functional Reasoning*. Washington, July 9-15 1993.

Matern, B. (1984) Psychologische Arbeitsanalyse. Springer-Verlag, Berlin.

Rasmussen, J. (1986) Information Processing and Human-Machine Interaction, North Holland, New York.

Rasmussen, J. (1988) A cognitive engineering approach to the modeling of decision making and its organization, in W.B. Rouse (ed.), Advances in Man-Machine Systems Research, Vol. 4, pp. 165-243.

VDI/VDE (1986) Preformatted displays on video display units for the control of process plants, Volume VDI/VDE Richtlinie (Guideline) 3695, VDI-Verlag, Düsseldorf.

Zimmermann, H.-J. (1991). Fuzzy Set Theory and its Applications. Boston, MA.: Kluwer Academic Publishers.

TOWARD A METHOD TO ANALYZE CONSEQUENCES OF HUMAN UNRELIABILITY - APPLICATION TO RAILWAY SYSTEM

F. Vanderhaegen[1], B. Telle[1,2], J. Gautiez[1]

[1] Laboratoire d'Automatique et de Mécanique Industrielles et Humaines (LAMIH)
University of Valenciennes - URA 1775 CNRS
Le Mont Houy BP 311 - 59304 Valenciennes Cedex FRANCE
Tel: 0033 (0)327 14 12 34 - Fax: 0033 (0)327 14 11 83
Email: {vanderhaegen, telle, jgautiez}@univ-valenciennes.fr

[2] Institut National de REcherche sur les Transports et leur Sécurité - Evaluation des Systèmes de Transport Automatisés et leur Sécurité (INRETS-ESTAS)
20 rue Elisée Reclus - 59650 Villeneuve d'Ascq FRANCE.

Abstract: This paper presents a descriptive method to analyse the consequences on system safety due to human unreliability. System functions have to be identified and human implication to achieve some of them is defined in terms of procedures, i.e. lists of tasks to be performed related to work contexts in normal and abnormal system functioning. A failed task can be caused by three different behavioural factors: an acquisition related failure, a problem solving related failure and/or an action related failure. Consequence analysis consists of identifying consequence level of human unreliability. This method is then applied to the rail traffic. *Copyright © 1998 IFAC*

Keywords: Error analysis, Evaluation, Human factors, Human reliability, Rail traffic.

1. INTRODUCTION

Human reliability is defined as the probability for a human operator (1) to perform correctly required tasks in required conditions and (2) not to perform tasks which may degrade system performance (Swain and Guttmann, 1983). A human reliability analysis aims at assessing this probability. An human error analysis is the opposite, i.e. it consists of calculating the probability that an error will occur when performing a task (Miller and Swain, 1987). An error is a deviation related to a reference. Usually, human error analysis is oriented toward safety analysis with regard to the controlled process. Different methods can then be used.

As a matter of fact, different methods to assess human error can be included into the design process. Firstly, machine centred methods can be adapted (Villemeur, 1988), such as FMECA (Failure Mode, Effects and Criticality Analysis) or the Fault Tree method. Even though those machine centred methods can be adapted for human behaviour assessment, they are difficult to apply because they require data on human behaviour which cannot be tested in the same way as technical components. Other specific methods were built to analyse human reliability or human error, such as:

- TESEO method (Tecnica Empirica Stima Errori Operatori) (Bello and Colombari, 1980).
- THERP method (Technique for Human Error Rate Prediction) (Swain and Guttmann, 1983).
- SHERPA method (Systematic Human Error Reduction and Prediction Approach) (Embrey, 1986).
- CREAM method (Cognitive Reliability and Error Analysis Method) (Hollnagel, 1991).

Results which are obtained by different human centred methods often differ one from another (Reason, 1990). Moreover, some of them require a strong operational data base on human error in order to assess or estimate the probability of human error occurrence. Most industrial applications cannot use those methods because this data base does not exist or is incomplete. Furthermore, rather than use those human centred safety analyses to specify on-line human support tools, they are used to improve off-line error prevention supports such as work environment ergonomics or training. Indeed, despite a good ergonomic design and an excellent training repeat human errors is not taken into account. Therefore, a new approach is proposed in order to consider both off-line and on-line prevention support specification into the global system development during the safety analysis phase. This method is called APRECIH, a French acronym for Preliminary Analysis of Consequences of Human Unreliability. It is divided into four main steps: a functional analysis step, a procedural and contextual analysis step, a human task feature identification step, a consequence analysis step.

With regard to an operational point of view, even though future railway systems will allocate an important human role for train safety, they have to reinforce this safety role with appropriate on-line support tools. The method to analyse consequences of human unreliability is then applied for rail traffic control.

The paper presents the principles of the APRECIH method and proposes a feasibility study to apply this method on a part of the rail traffic control process.

2. THE APRECIH METHOD PRINCIPLES

Before analysing consequences, different steps are required. Moreover, the global consequence analysis is based on unreliability caused by human behavioural factors.

2.1. The main method steps

In order to analyse the consequences of human unreliability on system safety, this method is divided into four steps: a functional analysis step (step 1), a procedural and contextual analysis step (step 2) , a task feature identification step (step 3) and a consequence analysis step (step 4), Figure 1.

The functional analysis of the system includes different analyses: process analysis, human activity analysis and feedback experience analysis. Those analyses aim at identifying system functions and human tasks. Functions allocated to human operators define the human tasks required by the controlled system. Therefore, the main objective of this step is to identify the matrix which indicates the interactions between human tasks and the system functions. As example, related to the system function F1, F2, F3 and F4, the human operator has to realise the F2 function performing three main human procedures: P1, P2 and P3.

The procedural and contextual analysis consists of identifying the work contexts and the corresponding human tasks, analysing safety rules and determining human safety procedure related to those contexts. As example, two contexts related to the procedure P1 have been identified: C1 and C2. For each one, a procedure, i.e. a list of tasks, is required. C1 requires three tasks T1, T2 and T3 to be performed. On the other hand, C2 requires one modification: the list of T1, T4 and T3 has to be performed to achieve the function F2.

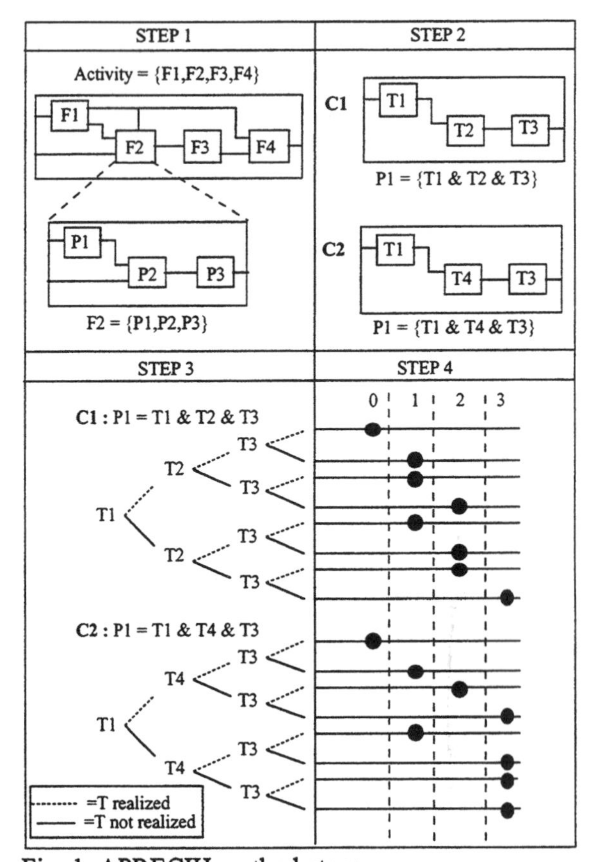

Fig. 1. APRECIH method steps.

The task feature identification aims at evaluating the possible task state (e.g. correct task, erroneous task) and at analysing causes of unreliability. Those causes of erroneous task are analysed according to possible scenarios of human unreliability which is presented as combining between three behavioural factors (Telle *et al.*, 1997):

- acquisition related failure (i.e. data sensor failure),
- problem solving related failure (i.e. erroneous processing of information or/and processing of erroneous information),
- and action related failure (i.e. correct action or erroneous action).

Finally, the consequence analysis consists of combining human unreliability factors, i.e. acquisition related failure, problem solving related failure and action related failure and their consequences on the system. The result is a matrix with the causes of unreliability and the consequences on the system. The more the consequence level is, the more catastrophic the consequence is. The identified unreliable scenarios are due to one of unreliable human factor or are combinations between different unreliable human factors.

2.2. Unreliable behavioural factors.

The task feature identification step tries to determine propagation mechanisms of human error with regard to three-component behavioural factors: acquisition related failure, problem solving failure and action related failure, Figure 2:

- Acquisition related failure. This is often the first cause of reduced reliability, because it generates an erroneous perception of the controlled environment. It is failure such as attentional failure and has the same consequence, i.e. a lack of information.
- Problem solving related failure. This is a failure of information processing related to the elaboration of an action plan and concerns both erroneous processing of information and processing of erroneous information. For example, because of lack of attention, the perceived information might not correspond to the real state of the process. In such a way, it is possible to generate an unsafe situation by performing an unrequired action plan related to the real state of the process.
- Action related failure. One must distinguish the action, its consequences for the process, and the output information which is generated. With regard to the agent who performs an action, this action is correct when it corresponds to a required action according to the input information whatever the quality of this information (i.e. wrong or correct with regard to the system state), otherwise the action is erroneous. Nevertheless, considering the consequence of an action, a criterion is needed to describe an action as correct or erroneous. Indeed, related to process, an action can be unsafe while the human operator who performs it believes that it is correct.

Each failure can generate an error related to the process objectives. Before acting on the process, human operators may perform several cycles of acquisition related failure, problem solving related failure and action related failure and interact with other human operators and technical tools. Therefore, some relations between components of unreliability can be identified.

The three components of the model can be linked. For instance, the consequence of an acquisition related failure can involve problem solving related failure or an erroneous action; an unrequired action can be caused by problem solving related failure; and an erroneous action can generate an acquisition related failure. The relations between components of unreliability can be internal or external. Internal relations are inter-relations and are related to physical, psychological and cognitive state of a human operator. External relations concern the links with other human operators.

Several intra-relations can then be distinguished, Figure 2:

- A failure can generate a failure of the same kind.
- A failure can generate an omission.
- A failure can generate another kind of failure.
- A failure can generate a cascade of failures.

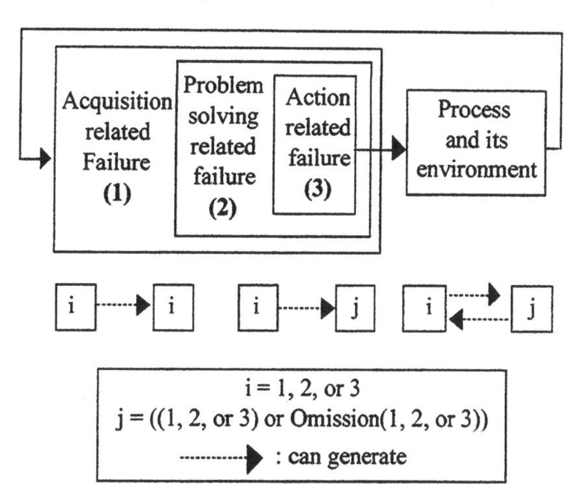

Fig. 2. Relations between unreliable human factors.

Such relations between failures can be serial cycles and parallel cycles. They can also be applied between different entities of a human-machine system. Therefore, other relations between unreliable components of humans and machines can be defined. For instance, there are inter-relations between human agents. They can be made through verbal communication (e.g. directly or using a communication tool such as telephone or radio), or through non-verbal communications (e.g. manual transfer). A model of the unreliability of one agent can be connected to the model of unreliability of another agent through dialogue interfaces. Non-verbal communication can also be realised through a screen or a printer via network, radio or radar. Therefore, human-machine interactions and possible acquisition related failures of technical tools have to be taken into account. Moreover, even if the definitions of error seem to mean that only humans can make errors (Resaon, 1990), the proposed model of unreliability may consider either a human or a machine. Therefore, if one considers the case of intelligent automated systems, notions of intention and error may be introduced. A technical tool can then make acquisition related failures, problem solving related failures and/or action related failures depending on its reasoning capacity and its connection with other system components.

With regard to those different inter-relations and intra-relations between the three unreliable human factors, a consequence scale has then to be defined. Moreover, a calculation mode has to be determined in order to assess the consequence level for each identified unreliable scenarios.

A feasibility study to apply the APRECIH method on railway system is made at the University of Valenciennes by the LAMIH and the INRETS-ESTAS team. Firstly, the study consists of applying the method on the whole rail traffic control process and of focusing on the shunting function, taking into account one human operator. An example of consequence assessment is then proposed to identified unreliable scenarios related to unreliable factors of this operator.

3. APPLICATION TO THE RAIL TRAFFIC

The functional analysis step of the APRECIH method is applied to the entire rail traffic control function. The other steps related to procedural and contextual analysis, to the task feature identification and to the consequence analysis are focusing on the shunting function.

3.1. Functional analysis.

Structured Analysis Design Technique (IGL, 1989) was used to identify the system functions. In such representation, the main railway system function is to realise a transport mission by means of railway system and competent staff, according to the specification of the demand to transport passengers or goods by train, Figure 3. The inputs of this function are the real needs required by customers and the feedback when missions are realised in order to update the planning of the rail transport services. A mission consists of driving passengers or goods from a departure point to an arrival one. Different sub-functions can then be identified taking into account the inputs and the outputs for each one, the data flow between them, and the human integration to realise them. Figure 3 lists some of those sub-functions of the global railway system.

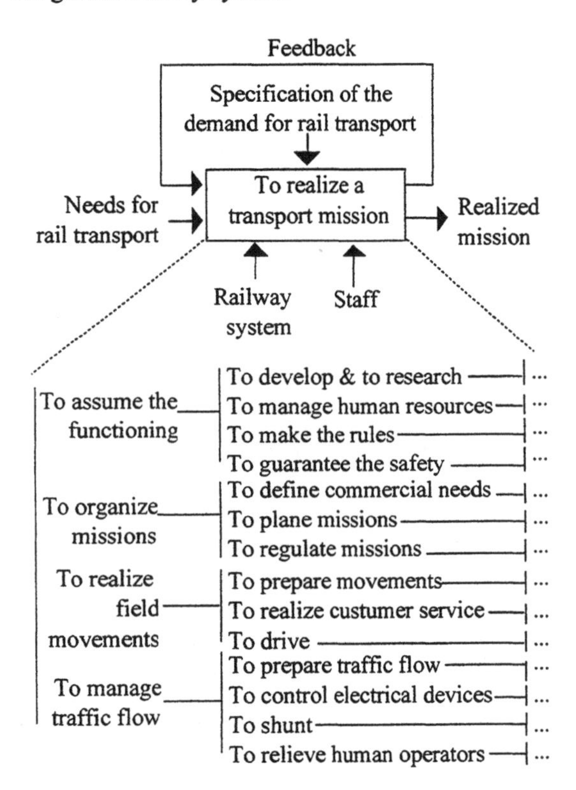

Fig. 3. Functions of railway system

For each sub-function, the implication of humans can be determined. For instance, with regard to the shunting function which is divided into four sub-functions: to realise required routes (F1), to cancel reserved routes (F2), to realise required authorisations (F3), to cancel performed authorisation (F4). When performing those functions, standard procedures are required to human operators who have to control the train crossing (P1), the train operation (P2), to inform the train arrival to the adjacent stations (P3), to control the train schedule (P4), to supervise the trains during their moving (P5). Interaction between those standard procedures and the system function are listed on Table 1.

Table 1 Interaction between human activities and system functions (0: no interaction, 1: interaction)

Procedure\Fonction	F1	F2	F3	F4
P1	1	1	0	0
P2	1	1	1	1
P3	1	1	0	0
P4	1	1	1	1
P5	1	1	1	1

3.2. Procedural and contextual analysis.

With regard to contexts, the procedures to be followed to control railway system can differ from the procedures of normal situations. The contexts concern incidents, breakdowns or particular railway configurations which orient the specific operations to be performed.

The incidents can be obstacles on rails, danger or presumption of danger, shock or abnormal movements, wanderings of cattle, problem on level crossing, problem on electrical devices, problems on train during moving. A breakdown can be a problem when realising a route or a problem which makes the establishment of a route or of an authorisation impossible. Different configurations of railway equipment (e.g. of trains, of rails, of shunting equipment) require different tasks to be performed.

For example, related to the shunting function, four contexts can be identified: Normal functioning (C1), abnormal busy tracks (C2), lateness of train (C3), or dangerous moving of trains (C4). The procedures P1, P2, P3 and P4 can be not required depending on the context, Table 2.

Table 2 Relations between procedures and contexts (1: procedure to achieve, 0: no procedure to achieve)

Procedure\Context	C1	C2	C3	C4
P1	1	1	0	0
P2	1	0	0	0
P3	1	0	1	1
P4	1	0	1	0
P5	1	0	0	1

For each context, the list of tasks to be performed for a given procedure can be listed. Table 3 gives an example for the procedures P1, P3 and P4 on contexts C1, C2 and C3

Table 3 Relations between tasks, procedures and contexts

	List of tasks
t1	To perceive a train arrival
t2	To control the route state on screen
t3	To realise the required route
t4	To control on screen the train moving
t5	To free the reserved route
t6	To wait the track freeing
t7	To cancel constraints on route realisation
t8	To take into account data related to a train
t9	To call the arrival station
t10	To transmit data
t11	To evaluate the lateness
t12	To write data
t13	To evaluate the causes of a lateness
P1 on C1 = t1 + t2 + t3 + t4 + t5	
P1 on C2 = t1 + t2 + t6 + t7 + t3 + t4 + t5	
P3 on C1 = t8 + t9 + t10	
P3 on C3 = t8 + t11 + t9 + t10	
P4 on C1 = t8 + t12	
P4 on C3 = t8 + t11 + t13 + t12	

3.3. Task feature identification.

With regard to the characteristics of human behaviour which were composed of an acquisition phase, a problem solving phase and an action phase, the main required behavioural factor to perform each task can be determined. Table 4 indicates this main factor to achieved the tasks of the procedures P1, P3 and P4 according to the contexts C1, C2, C3.

Table 4 Identification of behavioural factors implicated on the task performing

Code of the task	Main human behaviour
t1	Acquisition
t2	Problem solving
t3	Action
t4	Acquisition
t5	Action
t6	Acquisition
t7	Action
t8	Acquisition
t9	Action
t10	Action
t11	Problem solving
t12	Action
t13	Problem solving

The procedures to achieve the global rail traffic function are composed of sequential and high dependant tasks to be performed. The first step of the consequence analysis is to defined possible state of each unreliable factor (i.e. acquisition related failure, problem solving related failure, action related failure) regarding each human task.

3.4. Consequence analysis.

Three minimum states can be considered: correct, incorrect or omitted. Indeed, an acquisition, a problem solving and an action can be realised correctly, realised but incorrectly, or not realised at all, i.e. omitted. Therefore, considering those three possible states and the consequences related to the process safety, matrix such as Table 5 can be obtained.

Three levels of consequences have been defined. The level 0 means without catastrophic consequence, the level 1 means problematic consequence and the level 2 means catastrophic consequence. It is important to note that when t8 is incorrect, i.e. data were perceived not correctly, the state of t10 depends on the deviation between those perceived data and the data that are transmitted. Moreover, impossible scenarios may be identified. For instance, as t9 and t10 depend on t8, the omission of t8 implies only the omission of both t9 and t10. Moreover, as t10 depends on t9, the omission of t9 implies only the omission of t10. The omission of t10 is also impossible if t8 and t9 are realised correctly because the human operators of the adjacent stations are also implicated when exchanging data by communicating. A problem on t8 may cause catastrophic events when for example the human operators of the adjacent stations have to control a manual level crossing or when they have to manage a single-track line. Those impossible scenarios are indicated on Table 5 with the character "-".

Table 5 Example of a matrix combining consequences on system safety and human unreliability (procedure P3 on context C1)

t8	t9	t10	Consequence
correct	correct	correct	0
		incorrect	1
		omitted	-
	incorrect	correct	1
		incorrect	2
		omitted	2
	omitted	correct	-
		incorrect	-
		omitted	2
incorrect	correct	correct	2
...	...	...	...

Because of the combinatory problem, a computer based tool is developing in order to try to simplify the elaboration of this consequence analysis, taking into account relations between tasks in order to identify scenarios that may be impossible and to identify catastrophic combinations of unreliable human factors. This tool will be able to review the consequence analysis when, for example, assistance tools are specified in order to support human acquisition, human problem solving and/or human action. Therefore, those assistance tools might aim at limiting consequences of human unreliability.

4. CONCLUSION

The paper has proposed a new approach to consider human reliability. This approach, called APRECIH, is based on the system safety related consequences of human activity when controlling this system. The first part of the paper has developed the principles of the method, i.e. (1) the four main steps to achieve consequence analysis of human unreliability and (2) the unreliable human factors and their relations which can cause catastrophic consequences. The second part has presented an application of this method to the rail traffic control process, and more precisely to the shunting function.

A definition of an automated tool is developing at the University of Valenciennes in order to (1) facilitate and refine the elaboration of the consequence analysis, and (2) limit catastrophic consequences proposing a specification guideline to prevent human error.

This guideline consists of proposing different possible prevention supports:

- Off-line error prevention supports: training programme, awareness programme, ergonomic improvement, rules modification, specification reports.
- On-line error prevention supports: assistance tools, automated tools, human-machine interfaces.

Moreover, the method has also to take into account feedback of operation level in order to identify usual unreliable scenarios related to human activity. Therefore, it will be useful to determine the system modification priorities depending on the occurrence of acquisition related failures, problem solving related failure and erroneous actions. Moreover, those priorities may be guided by other criteria, e.g. consequence occurrence frequency based criteria, economical based criteria, political based criteria.

REFERENCES

Bello, G. C. and V. Colombari (1980). The human factors in risk analyses of process plants: the control room operator model, TESEO. *Reliability engineering*, **1**, 3-14.

Embrey, D.E. (1986). *SHERPA: A Systematic Approach for Assessing and Reducing Human Error in Process Plants*. Human Reliability Associated Ltd.

Hollnagel, E. (1991). The phenotype of erroneous actions: implications for HCI design. In: *Human-Computer Interaction and Complex Systems* (G. R. S. Weir and J.L. Alty, Eds.), pp.73-121. London Academic Press.

IGL Technology (1989). *SADT: a language to communicate*. Eyrolles, Paris.

Miller, D. P. and A. D. Swain (1987). Human error and human reliability. In: *Handbook of Human Factors* (G. Salvendy, Ed.), pp.219-250. John Wiley & Sons.

Reason, J. (1990). *Human Error*. University of Manchester, Dept. of Psychology.

Swain, A.D. and H.E. Guttmann (1983). *Handbook of Reliability Analysis with Emphasis on Nuclear Plant Applications*. Technical Report NUREG/CR-1278, NUclear REGulatory Commission, Washington D.C.

Telle, B., F. Vanderhaegen, N. Moray (1997). *Human centered dependability for railway system design*. World Congress on Railway Research, Florence, 16-19 November, pp. 859-869.

Villemeur, A. (1988). *Dependability of industrial systems*. Paris: Eyrolles.

FAILURE DIAGNOSIS SUPPORT USING BOND GRAPH WITH MEANS-ENDS HIERARCHY STRUCTURE

Takehisa Kohda, Takayoshi Tanaka, and Koichi Inoue

Dept. of Aeronautics and Astronautics, Kyoto University
Kyoto 606-8501, JAPAN

Abstract: Failure diagnosis is generally composed of three activities: (1) failure detection, (2) cause identification, and (3) protective action. This paper proposes the use of bond graphs with a means-ends hierarchy structure for the support of the entire failure diagnosis process. Bond graphs represent the system behavior from the viewpoint of energy flow, which makes it easy to model not only physical phenomena or components, but also abnormal events. Though the relations among state variables represented by the system bond graph can support the detection of an abnormal event and the identification of its cause, they cannot support the evaluation of its effect on the system function explicitly. Using means-ends hierarchy structure, the diagnosis result in the state variable space is transformed into the system functional space, which can help the system analysts select an preventive action and predict its effect. An illustrative example of a water flow control system shows the details and merits of the proposed method. *Copyright©1998 IFAC*

Keywords: Bond graphs, Diagnosis, Failure detection, Hierarchical structure, Systems engineering

1. INTRODUCTION

Since an accident in a large-scale system such as nuclear power plants and airplanes causes a serious damage in both economics and human life, failure diagnosis systems are installed to maintain its safety. A failure diagnosis is generally divided into three activities: (1) failure detection, (2) cause identification, and (3) protective action. For the failure detection, the reference state under normal conditions must be specified in advance. If some abnormal event occurs, a deviation indicates its occurrence, Then, the cause of the deviation must be identified based on the expert knowledge or the cause-effect relations among process states. Since the stored knowledge or experience cannot be applied to a new system, the system behavior model should be applied, which can be easily modified based on the design assumptions. The causal relation among state variables should be obtained from the system behavior model. To perform an appropriate action on the plant based on the diagnosis result, its effect on the system state or function must be estimated. In a complex system where a component plays several roles, so many interactions exist among components. To estimate the effect of a protective action on system functions, the relation between process variables and system functions must be clarified.

This paper considers the use of bond graphs (BGs) (Karnop et al. 1990) for the support of the entire failure diagnosis process. Using BGs, not only the normal system behavior, but also abnormal events can be easily modeled similarly by considering the correspondence between BGs with physical phenomena (Kohda et al. 1993). Further, since BGs model the system behavior from the viewpoint of energy flow and integrate various systems such as electrical and mechanical ones in a unified way, they are suitable for representing complex systems. For failure detection and location of failure causes, BGs were applied to process plants (Tsach et al. 1982) and nuclear plants (Lorre et al. 1994). BGs were also applied to failure analysis (Kohda et al, 1993, Kohda and Inoue 1996, Kohda et al. 1998). However, while BGs can describe the system behavior with causal relations among state variables, they cannot consider the relation of system behaviors with sysstem functions explicitly, which plays an important role in the evaluation of failure effect or the selection of an appropriate protective action. A system has several functions and their relations are usually represented as a means-ends hierarchy such as in multilevel flow models (MFM) (Lind 1982). MFM can describe various relations among goals, functions, and components, explicitly and consider conservation of mass, flow and energy, but the cause-effect relations among state variables are not described clearly. From the viewpoint of failure diagnosis, the relation between monitored or examined values of process states and system functions should be described clearly. An additional information is necessary to identify the cause of failure or failed component.

This paper proposes the use of BGs with a means-ends hierarchy structure to support the entire failure diagnosis process. Using an example of a simple water flow control system, the details and merits of the proposed method is described.

2. SUPPORT OF FAILURE DETECTION

This section describes how to develop a system behavior model using BGs and how to obtain the reference state under normal conditions and how to evaluate failure effect.

2.1 System Bond Graph

Consider a system bond graph (SBG) for a water flow control system shown in Fig. 1. The system is composed of three subsystems: (1) an electrical system driving a motor, (2) the motor connected to a pump, (3) the pump filling the water into the tank. The water finally flows from the tank into the outside through the pipe. Table 1 shows the correspondence between BG elements and physical components or phenomena to be considered in the analysis. Combining these elements with bonds representing energy flow, the SBG is obtained as shown in Fig. 2. Thus, an SBG can be constructed easily by combining component BG elements representing physical phenomena according to the energy flow.

2.2 System State Equations

For simplicity, we make the following assumptions on the SBG.

1. The SBG is composed of basic BG elements: Source of Effort (SE), Source of Flow (SF), Resistance (R), Capacitance (C), Inertance (I), Transformer (TF), Gyrator (GY), 0-junction (0), and 1-junction (1). The characteristic function for each element is shown in Table 2.
2. Deviation is defined as a difference between the current and the steady state values.
3. A failure can be represented as either a deviation of input variable (disturbance) or a deviation of component function (component failure).

For the SBG under assumption 1, state variables are displacement (q, integral of flow with respect to time) of a bond connecting to a C-element and momentum (p, integral of effort with respect to time) of a bond connecting to an I-element, and input variables are effort (e) of a bond connecting to an SE and flow (f) of a bond connecting to an SF (Karnop et al. 1990). Based on the causality of the SBG, all e's and f's can be expressed in functional forms of state variables and input variables (Kohda et al. 1988). Using these functional expression, the reference state under normal conditions can be determined from the state variables and input variables.

For the SBG in Fig. 2, the input variables are e's of bonds 1 and 20, and the state variables are p's of bonds 3, 7 and 11. and q's of bonds 9 and 17. Represent them as U_1, U_2, X_1, X_2, X_4, X_3 and X_5, respectively, and all e's and f's in the SBG can be expressed in their functional forms as shown in Table 3, where $R_i\{\}$, $C_i\{\}$, $L_i\{\}$, $GY_i\{\}$ and $TF_i()$ denote characteristic functions of R_i, C_i, I_i, G_i and TF_i, respectively.

From the definitions of q and p, the system state equations are obtained as:

$$\frac{dX_i}{dt} = \begin{cases} e_{b(i)}, & \text{if } X_i \text{ represents momentum,} \\ f_{b(i)}, & \text{if } X_i \text{ represents displacement.} \end{cases} \quad (1)$$

where b(i) denotes the bond number corresponding to state variable X_i. Using the system state equations, the dynamic behavior of system state variables can be evaluated.

For the flow control system, the system state equations are:

$$\frac{dX_1}{dt} = U_1 - R_1(L_1(X_1)) - GY_1(L_2(X_2)) \quad (2)$$

$$\frac{dX_2}{dt} = GY_1(L_1(X_1)) - R_2(L_2(X_2)) - C_1(X_3) \quad (3)$$

$$\frac{dX_3}{dt} = L_2(X_2) - L_3(X_4) \quad (4)$$

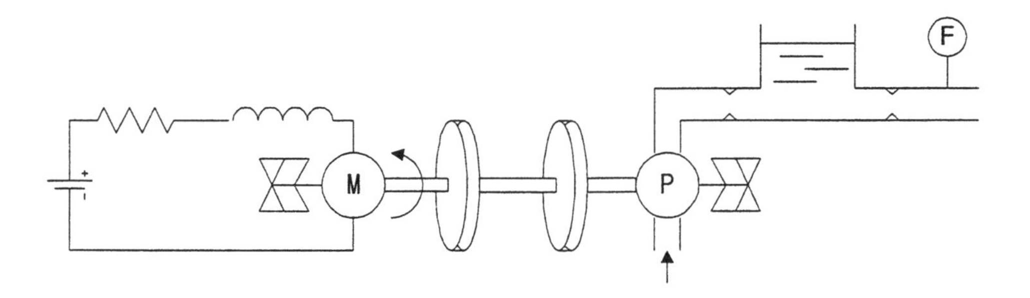

Fig. 1 Flow control system

Table 1 Correspondence between BG elements and physical components

BG element	Physical component	BG element	Physical component
SE_1	Electric voltage source	R_3	Pump shaft fiction
R_1	Resistance	TF_1	Pump function
I_1	Inductor	R_4	Flow resistance
GY_1	Electric motor function	C_2	Water tank
R_2	Motor shaft friction	R_5	Flow resistance
I_2	Motor inertia	SE_2	Outside pressure
C_1	Shaft compliance	1-Junction	Serial circuit, Joint
I_3	Pump inertia	0-Junction	Parallel circuit, Joint

R_1 I_1 R_2 I_2 I_3 R_3 R_4 C_2 R_5

2 3 6 7 11 12 15 17 19

SE_1 —1→ 1 —4→ GY_1 —5→ 1 —8→ 0 —10→ 1 —13→ TF_1 —14→ 1 —16→ 0 —18→ 1 —20→ SE_2

9

C_1

Fig. 2 System bond graph for flow control system

$$\frac{dX_4}{dt} = C_1(X_3) - R_3(L_3(X_4)) - TF_1(R_4(TF_1(L_3(X_4))) + C_2(X_5)) \qquad (5)$$

$$\frac{dX_5}{dt} = TF_1(L_3(X_4)) - R_5(C_2(X_5) - U_2) \qquad (6)$$

2.3 Reference System State

Since system failure effect is usually defined as the deviation from the steady state under a normal condition as shown in assumption 2, the steady state must be obtained as a reference system state firstly. Since $dX_i/dt=0$ for all state variables at the steady state, the steady state conditions can be obtained as:

Table 2 Characteristic function for bond graph elements

BGE	Symbol	Function	BGE	Symbol	Function
SE	$SE \xrightarrow{i}$	$e_i = E(t)$	TF	$\xrightarrow{i} TF \xrightarrow{j}$	$e_j = e_i/n$ $f_i = f_j/n$
SF	$SF \xrightarrow{i}$	$f_i = F(t)$		$\xrightarrow{i} TF \xrightarrow{j}$	$e_i = n \cdot e_j$ $f_j = n \cdot f_i$
R	$R \xleftarrow{i}$	$e_i = R(f_i)$	GY	$\xrightarrow{i} GY \xrightarrow{j}$	$f_j = e_i/m$ $f_i = e_j/m$
	$R \xleftarrow{i}$	$f_i = R(e_i)$		$\xrightarrow{i} GY \xrightarrow{j}$	$e_i = m \cdot f_j$ $e_j = m \cdot f_i$
C	$C \xleftarrow{i}$	$e_i = C(\int f_i dt)$	0	$\xrightarrow{i1} 0 \xrightarrow{in}$, $i2$	$e_{ij} = e_{ik}$ for any ij, ik $\Sigma f_{ij} = 0$
I	$I \xleftarrow{i}$	$f_i = L(\int e_i dt)$	1	$\xrightarrow{i1} 1 \xrightarrow{im}$, $i2$	$f_{ij} = f_{ik}$ for any ij, ik $\Sigma e_{ij} = 0$

Table 3 Functional expressions of efforts and flows

i	Effort i	Flow i	i	Effort i	Flow i
1	U_1	$L_1(X_1)$	11	$C_1(X_3)-R_3\{L_3(X_4))$ $-TF_1(R_4(TF_1(L_3(X_4))+C_2(X_5)))$	$L_3(X_4)$
2	$R_1\{L_1(X_1))$	$L_1(X_1)$	12	$R_3\{L_3(X_4))$	$L_3(X_4)$
3	$U_1-R_1\{L_1(X_1))-GY_1(L_2(X_2))$	$L_1(X_1)$	13	$TF_1(R_4(TF_1(L_3(X_4))+C_2(X_5)))$	$L_3(X_4)$
4	$GY_1(L_2(X_2))$	$L_1(X_1)$	14	$R_4(TF_1(L_3(X_4))+C_2(X_5)$	$TF_1(L_3(X_4))$
5	$GY_1(L_1(X_1))$	$L_2(X_2)$	15	$R_4(TF_1(L_3(X_4))$	$TF_1(L_3(X_4))$
6	$R_2(L_2(X_2))$	$L_2(X_2)$	16	$C_2(X_5)$	$TF_1(L_3(X_4))$
7	$GY_1(L_1(X_1))-R_2(L_2(X_2))-C_1(X_3)$	$L_2(X_2)$	17	$C_2(X_5)$	$TF_1(L_3(X_4))-R_5(C_2(X_5)-U_2)$
8	$C_1(X_3)$	$L_2(X_2)$	18	$C_2(X_5)$	$R_5(C_2(X_5)-U_2)$
9	$C_1(X_3)$	$L_2(X_2)-L_3(X_4)$	19	$C_2(X_5)-U_2$	$R_5(C_2(X_5)-U_2)$
10	$C_1(X_3)$	$L_3(X_4)$	20	U_2	$R_5(C_2(X_5)-U_2)$

$$\begin{cases} e_{b(i)} = 0, & \text{if } X_i \text{ represents momentum,} \\ f_{b(i)} = 0, & \text{if } X_i \text{ represents displacement.} \end{cases} \quad (7)$$

The steady state conditions are also represented in functional forms of state variables and input variables. If characteristic functions for components under their normal conditions and input variables are specified, the corresponding state variables, the reference state, can be determined from eq. (7).

3. SUPPORT OF CAUSE IDENTIFICATION

For the identification of failure cause, the relation between failures and their effect on the state variables plays an important role. Two types of support are proposed: one is failure effect analysis at the design stage to obtain this kind of information beforehand, and the other is estimation of state conditions based on the monitored values.

3.1 Failure Effect Analysis

Even though characteristic functions or input variables may change due to a component failure, the steady state conditions, eq. (7), must be satisfied by modified state variables at a new steady state. The failure effect can be analyzed by comparing new and old steady states or state variables. At the initial stage of a system design, the qualitative information such as correlations between component failures and state variables are more effective in considering the preventive action. A qualitative method using a tree graph is given to analyze component failure effect.

Qualitative Expression From assumption 2, the effect of a component failure on a state variables can be expressed as decrease (-), no change (0) or increase (+). Characteristic functions can be also expressed qualitatively as increasing(+), decreasing(-) or unrelated (0) around the steady state. The deviation of a characteristic function from its normal one is represented in a similar way.

Tree Graph To obtain qualitative deviations in the state variables satisfying steady state conditions (eq. (7)), the constraint relations among state variables, input variables, and characteristic functions are represented as a tree graph (TG). Fig. 3 shows a TG for the SBG in Fig. 2. A node denotes a characteristic function of a BGE, an operator, a state variable or an input variable. A branch expresses a relation between two nodes: its sink node is a kind of input to its source node. An operator node * ("+" (addition) or "-" (reduction)) has two nodes: left and right one, and its output is calculated as (left node value) * (right node value). A function node has an input node, whose output is calculated as (qualitative expression of characteristic function) * (input node value).

By propagating qualitative fixed values along the TG, all the node values must be solved in such a way that a constraint condition on a node representing a junction must be satisfied by its related nodes. The same qualitative value can propagate from a failed node to its output node, while the opposite value can propagate to its input node. In the water flow control system, the effect of an increase in the motor shaft friction, which is represented as + in R_2, can be obtained as: X_1=+, X_2=-, X_3=-, X_4=-, and X_5=-. Thus the output flow, which is represented as $f_{20}=R_5(C_2(X_5)-U_2)$, decreases due to this failure.

3.2 State Condition Estimation

In the failure diagnosis, the cause identification is equivalent to the estimation problem of system state condition that satisfies the monitored value of the system state variables. Using a TG, state condition estimation can be performed in the same way as failure effect analysis. The cause identification problem can be transformed into a problem where unknown state variables and characteristic function must realize the monitored value of state variables in the TG. In this way, the SBG can be used for not only failure effect analysis, but also the identification of failure causes.

4. SUPPORT OF PROTECTIVE ACTION

In planning an appropriate protective action, its effect must be considered based on the system function or its goal. Using means-ends hierarchy structure, the system behavior represented by BGs is linked to the system function.

4.1 Means-Ends Hierarchy Structure

Especially in a large scale system where several functions interact one another, their relations must be clarified first. Means-ends hierarchy structure is one viewpoint to classify system functions. If function A needs function B, A is considered as a means to achieve B. In this way, various system function is classified into a kind of hierarchy. From the viewpoint of the failure diagnosis, this hierarchy corresponds to the priority among system functions. The effect of a failure on the hierarchy of system functions determines its severity, in other words, what kind of preventive action or protective action to take.

4.2 Goal State

Whether a system function is achieved or not can be usually judged based on a specific system state. For example, in the water flow control system, the main function is to maintain constant water flow. Achievement of the system function is checked by its output flow. In other words, the abnormality in the output flow means the failure of water flow control system. By defining its goal state for each system function, the system functional space can be related to the system state space described by BGs. Thus, the effect of a protective action or preventive action on the system state can be transformed into its effect on the system function.

5. CONCLUSIONS

This paper proposes the use of BGs with means-ends hierarchy structure to the support of failure diagnosis process. Using BGs modeling errors can be reduced, Based on functional expressions of process variables obtained from the SBG, failure effect analysis can be performed qualitatively as well as quantitatively. TG expression of the system state equations is used for not only failure effect analysis, but also state condition estimation. Using means-ends hierarchy structure, the current system state can be evaluated from the viewpoint of system functions.

REFERENCES

Karnopp D.C., et al. (1990) *System dynamics: A unified approach (2nd ed.).* John Wiley & Sons, Inc., New York.

Kohda T., et al. (1988) Simulation of bond graphs with nonlinear elements by symbolic manipulation. *Bulletin of Mechanical Engineering Laboratory, Japan*, **49**.

Kohda T., et al. (1993) Identification of system failure causes using bond graphs, *Proc of 1993 International Conferenee on Systems, Man and Cybernetics.* Vol. 5, pp. 269-274

Kohda T., and K. Inoue (1996) System failure analysis based on energy flow concept model - a bong graph approach -, *Probabilistic Safety Assessment and Management, ESREL'96 - PSAM III*, Vol.3 , pp. 1878-1883

Kohda T., et al. (1998) Disturbance propagation analysis based on system behavior model. to appear in *Proc. of ESREL'98* .

Lind M. (1982) Multilevel flow modeling of process plant for diagnosis and control, *Riso-M-2352.*

Lorre J.P., et al. (1994) SEXTANT: An interpretation system for continuous processes. *Proc. of SAFEPROCESS '94*, pp. 655-660

Tsach U., et al. (1982) A new method for failure detection and location in complex dynamic systems. *Proc. of the 1982 American Control Conference* Vol. 1, pp. 277-282

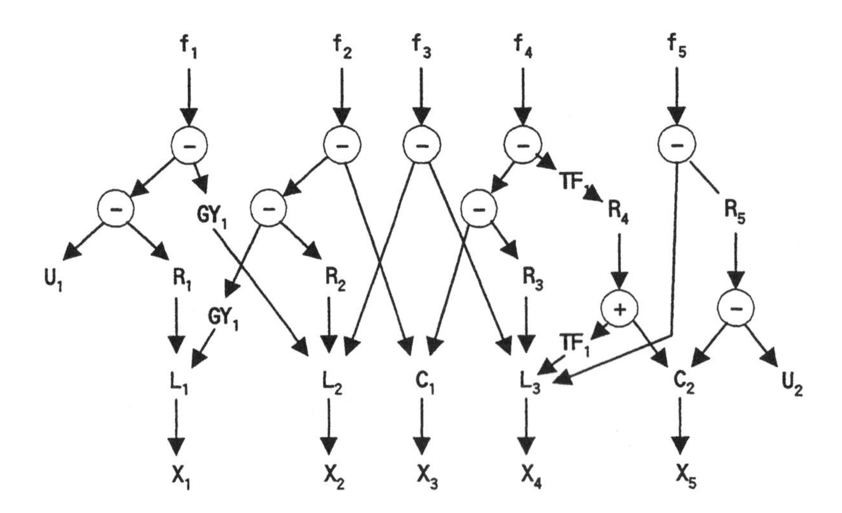

Fig. 3 Tree graph representing state space equations

INFORMATION AIDING OF REAL-TIME SYSTEMS FOR HUMAN FAULT DIAGNOSIS TASKS

Chao-Hua Wen and Sheue-Ling Hwang

Department of Industrial Engineering
National Tsing Hua University
Hsinchu, Taiwan, R.O.C
Tel: 886-3-5742694
Fax: 886-3-5722658

Abstract: This research aims to develop and evaluate a real-time diagnosis assistant system that based on operator behavior recognition, hierarchical neural networks and expert explanation systems. The independent variables of were aiding levels and scenario types. The results showed that either partial or entire aiding approach could improve the performance significantly, and there was no significant difference between the partial aid and the entire aid. However, there was a significant effect of the level of aiding on the transitions of diagnostic strategy. These results revealed that with a direct and short information aiding, one could save diagnostic time and test steps. *Copyright © 1998 IFAC*

Keywords: Fault diagnosis, man-machine interaction, neural networks, and expert systems.

1. INTRODUCTION

Fault diagnosis is a prerequisite in automation systems from the perspectives of factory safety and manufacturing cost. Flexible production including many different products demands higher operator skills and the operator will play a more important role in the future (Wennersten, *et al.*, 1996). Diagnosticians perform a high level of mental activity because they are the final authorities in taking appropriate actions to avoid or to recover from system failures. Moreover, the operator experiences a tremendous amount of temporal stress. Under above circumstances, the likelihood for making decision errors is high (Sheridan and Hennesy, 1984).

Up to now, despite artificial intelligence related efforts to develop an automated intelligent diagnosis system (Karsai, *et al.*, 1996; Becraft and Lee, 1993), full automation of fault diagnosis remains rare in actual plants. It is generally assumed that the computer is better at performing numerous precise and rapid calculations, while the human operator is the best at making associative and inferential judgements (Sheridan, 1981). Therefore, a support system should (1) assist diagnosticians to quickly and accurately enter the process of fault diagnosis in uncertain and unforeseen circumstances, and (2) offer operators correct information about diagnosis tasks to release their mental workload from temporal constraints and to avoid the failure of automatic systems.

Parasuraman (1997) suggested that understanding the factors associated with each of these aspects of human use of automation can lead to improved system design, effective training methods, and judicious policies and procedures involving automation use. However, introducing an assistant approach cannot accurately predict the operator performance during fault diagnosis. Yoon and Hammer (1988) recommended that the aiding information should be compatible with a human's information process. More thoroughly understanding the strategies used by humans and the difficulties they encounter may facilitate the development of aiding methods for human operators in fault diagnosis tasks.

Up to the present, although many researches focus on development of real-time diagnosis systems, there are still insufficient discussions on human operator's role and responsibilities in terms of real-time system. In this research, FDAS integrated both of neural networks and expert systems, which involves fault diagnosis, explanation and human behavior analysis. Experiments were conducted to evaluate the effectiveness of the FDAS. A flow chart model for fault diagnosis dissects operator's behavior to discuss the different action between unaided diagnosticians and aided diagnosticians. This study assumed that human operators might require the different aiding information and adopt the varied diagnosis strategies during facing the dynamic real-time aiding systems.

2. THE ARCHITECTURE OF FDAS

This section presents the development of the FDAS. First of all, The simulation of heat transfer salt system was retrieved from a previous study (Wen and Hwang, 1997). Fig. 1 summarizes the architecture of FDAS that is based on a combination of hierarchical neural networks and expert systems. The experimental apparatus included two Pentium 586-120 personal computers, which were connected by way of the RS232 communication. One simulated the heat transfer salt system of the melamine plant, another was FDAS that provided aiding information for diagnostician. FDAS monitors the simulated system and analyzes human operator's behavior to offers him/her three kinds of visual information aiding and one auditory alarm. FDAS contains three major modules in terms of physical architecture, namely, hierarchical neural networks, rule-based explanation expert system, and diagnostic behavior recognition and inference. The methodology employed to do so is discussed below.

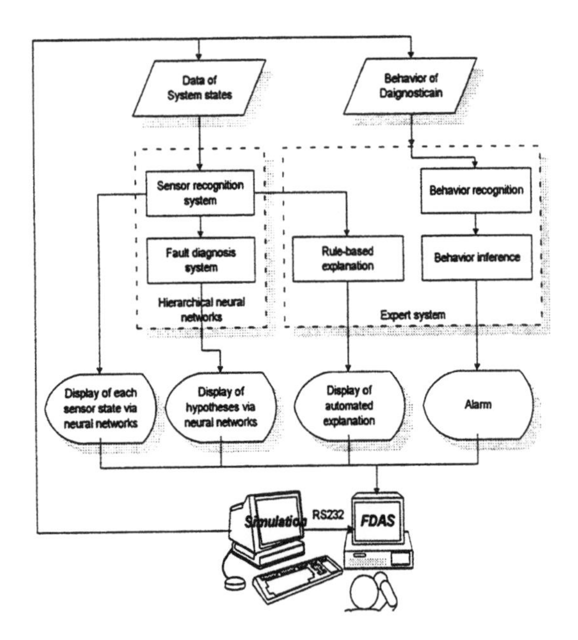

Fig. 1. The architecture of the fault diagnosis assistant system

2.1 *Hierarchical neural networks*

This research applies the back-propagation algorithm that is in widespread use for fault diagnosis in process controls. Hierarchical neural networks (HNN) were trained to model the normal and abnormal patterns of the heat transfer salt system. The first level is the sensor recognition system (SRS) which identifies the status of each sensor from the behavior of ten past value. The trend of sensor can be generally classified into normal, high, low, and unstable. Display of those results generated by SRS is different with the original alarm of simulation. SRS supplies the prediction capability to operator for taking action early. And SRS outputs also are the source data of rule-based expert system. The second level is the fault diagnosis system (FDS) which identifies the fault types to generate the fault representation and provide the final message to the diagnostician.

2.2 *Rule-based explanation expert system*

An operator often has insufficient time to read the operating manual or to refer the off-line/on-line help in real-time systems. From previous section, the final output message of FDS is displayed on the VDT in simple and short form only. Intuitively, the operator needs explanations concerning the diagnostic results of HNN. In other words, we should overcome the shortcoming of hierarchical neural networks. This approach focuses on integrating expert systems into the hierarchical neural networks. The rule-based explanation expert system used to explain the reasoning methodology of hierarchical neural networks.

2.3 *Diagnostic behavior recognition and inference*

We analyzed the 144 diagnostic behavior data of 12 participants from previous study (Wen and Hwang, 1996), and found that two diagnostic behavior characteristics of participants have high occurrence probability in lower performance group. First, they did not concentrate their attention, that is, their cursor position changed frequently for 3 seconds. Second, they stuck to their attention, and their visual field focuses on a single and narrow block for 9 seconds. They even did not take any action. Therefore, these behavior characteristics were considered and designed in FDAS. When the first characteristic presents, a slow tone was given to the diagnostician. If the second characteristic occurs, a quick tone will reminds the diagnostician to do something.

3. RESEARCH METHODOLOGY

3.1 *Experiment*

Participants: All subjects who were undergraduate students at Tsing-Hua University voluntarily participated in this experiment, and those have taken at least one fundamental chemical course. Thirty participants were divided into none aided, partial aided and entire aided groups equally.

Experimental tasks: A couple of days before the experiment, a reference manual about the operating and diagnostic rules of the system was given to the subjects and they were asked to study it thoroughly. The fourteen simulated scenarios derived from historical data in a real plant contained single fault problems and multiple fault problems. The first scenario was set as a practice problem to make subjects familiar with operating this simulated system. The second scenario was used to test the participants' familiarity of simulated system. The scenario 12, 13 and 14 were multiple fault problems; the rests were single fault problems. All the answers to the scenarios could be found in the reference manual, and the participant could consult the manual during the experiment. The participants started to diagnose whenever the simulated system was shutdown, and the participants were instructed to find the fault in the minimum diagnostic time with the minimum number of tests. As soon as the participant was confident of his/her hypothesis, he/she verified the hypothesis by using mouse to point and click that component. The overall procedures of verification were recorded by computer system. The difference among three kinds of experimental conditions was the level of aiding information. In none aided group, participants diagnose all scenarios without any aiding information. In partial aided group, participants could refer the hypotheses that were generated by the HNN and hear the alarm which came from operator behavior recognition. In entire aided group, FDAS displayed sensors' states obtained from sensor recognition system and showed automatic explanation system to support diagnostician other than the partial aided group.

Experimental design: The independent variables were three groups (none aided, partial aided and entire aided) and scenario type (single fault and multiple faults). Two dependent variables, the diagnostic time and test steps were recorded in each scenario. In this study, an unbalanced factorial experiment with repeated measures on scenario type was applied. In order to analyze and compare the transition of diagnostic strategies, participant recalled his/her diagnostic procedures after finished each scenario. Finally, all participants were asked to answer a simple subjective questionnaire about usability of FDAS.

4. RESULTS

4.1 Diagnosis Performance

We recorded each subject's diagnostic time and test steps and analyzed those data by unbalanced two-factor factorial analysis of variances with repeated measures. The results of the ANOVA are summarized in Table 1 and Table 2.

Table 1. Results of ANOVA of the diagnostic time

Source	Type III SS	*df*	MSE	F	p-value
Group	130476.27	2	65238.14	13.59	0.0001
Subject (Group)	129586.01	27	4799.48	—	—
Scenario	53905.67	1	53905.67	43.32	0.0001
Group × Scenario	18604.42	2	9302.21	7.47	0.0026
Subject × Scenario (Group)	33601.56	27	1244.50	—	—

Table 2. Results of ANOVA of test steps

Source	Type III SS	*df*	MSE	F	p-value
Group	184.18	2	92.09	16.24	0.0001
Subject (Group)	153.15	27	5.67	—	—
Scenario	122.68	1	122.68	46.10	0.0001
Group × Scenario	55.30	2	27.65	10.39	0.0005
Subject × Scenario (Group)	71.85	27	2.66	—	—

On diagnostic time, there was a significant interaction effect between the group and the scenario type ($F_{(2,27)} = 7.47$, p-value < 0.005). And on test steps, there also was a significant interaction effect between the group and the scenario type ($F_{(2,27)} = 10.39$, p-value < 0.005). Fig. 2 and Fig. 3 reveals that the aiding effect is much more significant for multiple faults scenarios.

The main effects of group were significant on diagnostic time ($F_{(2,27)} = 13.59$, $p < 0.001$) and test steps ($F_{(2,27)} = 16.24$, $p < 0.001$). The results revealed that both aided groups spent less diagnostic time ($M_{partial} = 37.473$; $M_{entire} = 35.917$; $M_{none} = 74.726$) and fewer test steps than none aided group did ($M_{partial} = 1.775$; $M_{entire} = 1.700$; $M_{none} = 3.025$).

The main effects of scenario type were significant on diagnostic time ($F_{(1, 27)} = 43.32$, $p < 0.001$) and test steps ($F_{(1,27)} = 46.10$, $p < 0.001$). Subjects diagnosed faster in single fault scenarios ($M_{single} = 42.307$) than in the multiple faults scenarios ($M_{multiple} = 70.567$), and subjects used fewer test step in the single fault scenarios ($M_{single} = 1.830$) than in the multiple faults scenarios ($M_{multiple} = 3.178$).

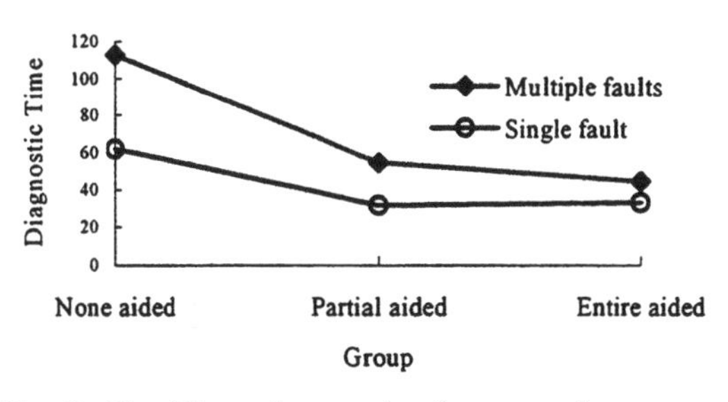

Fig. 2. Significant interaction between the group and the scenario type on diagnostic time

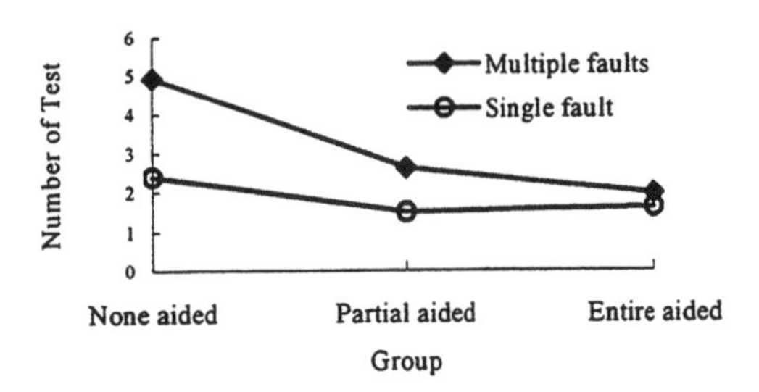

Fig. 3. Significant interaction between group and scenario type on test steps

4.2 Transitions of diagnosis strategy

This paper used a fault diagnosis model, showed in Fig. 4, to analyze the diagnostic process of participants. For example, one would go through the thinking route (p1, p2, p3, d1, d2, d3, p5, d5, d6, d7, d8, and t1) when he/she was familiar with the fault scenario and smoothly solved the problem. It is assumed that subject's thinking routes could be classified into the thinking route table (Table 3), and subject's solution for each scenario might be combination of some different thinking routes. One change of thinking route was called a transition. The data of transitions of diagnosis strategy were analyzed by unbalanced two-factor factorial analysis of variances with repeated measures. The results of the ANOVA are demonstrated in Table 4. The main effects of group were significant on transition ($F_{(2, 27)} = 4.78$, $p < 0.05$). The results indicated that the partial and entire aided group had fewer transitions ($M_{none} = 0.800$; $M_{partial} = 0.600$; $M_{entire} = 0.433$). The main effects of scenario type were also significant on transition ($F_{(2, 27)} = 13.95$, $p < 0.01$). The results proved that the single fault scenario let participants do fewer transitions ($M_{single} = 0.496$; $M_{multiple} = 0.956$).

Table 4. Results of ANOVA of the transitions

Source	Type III SS	*df*	MSE	F	p-value
Group	9.8074	2	4.9037	4.78	0.0167
Subject (Group)	27.6889	27	1.0255	—	—
Scenario	14.2370	1	14.2370	13.95	0.0009
Group × Scenario	1.9852	2	0.9926	0.97	0.3910
Subject × Scenario (Group)	27.5556	27	1.0206	—	—

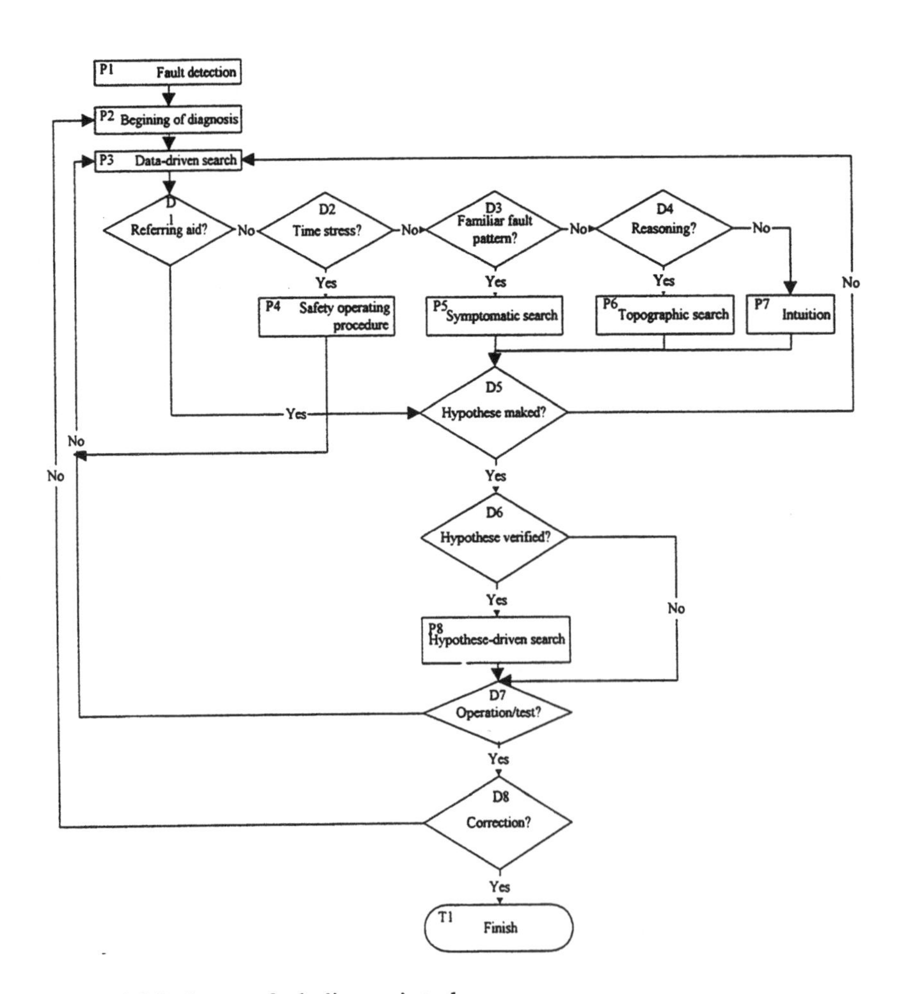

Fig. 4. A flow chart model for human fault diagnosis tasks

Table 3 Thinking routes table

Route	Procedure	Classifying point
1	p1-p2-p3-d1-d2-d3-[p5]-d5-d6-d7-d8	Symptomatic search (hypothesis made)
2	p1-p2-p3-d1-d2-d3-[p5]-d5-d6-[p8]-d7-d8	Symptomatic search (hypothesis made) and hypothesis-driven search for verifying
3	p1-p2-p3-d1-d2-d3-d4-[p6]-d5-d6-d7-d8	Topographic search (hypothesis made)
4	p1-p2-p3-d1-d2-d3-d4-[p6]-d5-d6-[p8]-d7-d8	Topographic search (hypothesis made) and hypothesis-driven search for verifying
5	p1-p2-p3-d1-d2-d3-d4-[p7]-d5-d6-d7-d8	Intuition (hypothesis made)
6	p1-p2-p3-d1-d2-d3-d4-[p7]-d5-d6-[p8]-d7-d8	Intuition (hypothesis made) and hypothesis-driven search for verifying
7	p1-p2-p3-d1-d2-[p4]	Automatic safety operating procedure
8	p1-p2-p3-d1-d2-d3-[p5-d5-p3]	Symptomatic search (no hypothesis made)
9	p1-p2-p3-d1-d2-d3-d4-[p6-d5-p3]	Topographic search (no hypothesis made)
10	p1-p2-p3-d1-d2-d3-d4-[p7-d5-p3]	Intuition (no hypothesis made)
11	p1-p2-p3-[d1-d5]-d6-p7-d8	Referring information aiding
12	p1-p2-p3-[d1-d5]-d6-[p8]-d7-d8	Referring information aiding and hypothesis-driven search for verifying
13	p1-p2-p3-[d1-d5-p3]	Referring information aiding (no hypothesis made)

[ ] Denotes the key process

5. DISCUSSION AND CONCLUSIONS

FDAS presented here is efficient to improve the diagnostician performance in real time systems. The evaluation shows that the operators who used FDAS spent less diagnostic time and fewer test steps. In terms of the level of information aiding, direct and short information for real time diagnostic tasks is enough to save the diagnostic time and test steps. Although the alarm of behavior recognition system had negative subjective evaluation, the sounds could urge the diagnosticians to find the fault.

Stankovic and Ramanritham (1992) proposed that a real-time system must meet four requirements, that is, speed, predictability, reliability and adaptability. These requirements used to discuss the feasibility of FDAS as follows,

1. Speed: FDAS has the advantage of high parallel computing of neural networks to "compute" the cause. Because the inferring engine of rule-based explanation system came from the results of neural networks, FDAS save the excessive computation times and ambiguous diagnosis in model-based expert system.

2. Predictability: Before the system alarms, the FDAS can provide the correct message to diagnostician. When there is a hidden fault in the system, such as in an unstable situation, the aiding system can judge the status of the system by the SNN to recognize the trend of sensor before the sensor showing over upper or lower limitation.

3. Reliability: Human supervisory controllers have the capability, and often the freedom, to strategically manage their interaction with automation in an effort to keep both workload and system performance at acceptable levels (Rouse, 1977; Sheridan, 1987; Tulga and Sheridan, 1980). Although the accuracy of aiding system is not 100% in this study, due to the capability of human, operators can strategically manage their interaction with aiding systems in an effort to keep appropriate workload and system performance.

4. Adaptability: FDAS is flexible to be adapted for new system faults. If the FDAS is necessary to be revised, only the DNN needs to be trained again without modifying the SNN. In this study, training DNN is much easier than SNN. Therefore, the FDAS can be easily generalized for all process control systems.

However, there is some comparison of end user interface of three real time artificial intelligence toolkits for real time power system alarm processing (Hasan et al., 1994). There is still rarely design advice for real-time assistant systems in related fields. Therefore, more study is required about how to design the information aiding in real-time systems.Future work on FDAS will redesign the display form of rule-based explanation.

ACKNOWLEDGEMENTS

The authors would like to thank the National Science Council of the Republic of China for financially supporting this work under Contract No. 86AFA04C0120004.

REFERENCES

Becraft, W. R. and P. L. Lee (1993). An integrated neural network/expert system approach for fault diagnosis. *Computers & Chemical Engineering*, 17(10), 1001-1014.

Hasan, K., B. Ramsay and I. Moyes (1994). Object oriented expert systems for real-time power system alarm processing. Part I. Seletion of a toolkit. *Electric Power Systems Research*, 30, 69-75.

Karsai, G., S. Padalkar, H. Franke, J. Sztipanovits and F. Decaria (1996). Practical method for creating plant diagnosis applications. *Integrated Computer-Aided Engineering*, 3(4), 291-304.

Parasuraman, R. (1997). Human and automation: Use, misuse, disuse, abuse. *Human Factors*, 39(2), 230-253.

Sheridan, T. B. (1987). Supervisory control. In: *Handbook of Human Factors* (G. Salvendy. (Ed)), 1243-1268, New York, Wiley.

Sheridan, T. B. and R. T. Hennesy (1984). Research and modeling of supervisory control behavior. Report of a workshop, Technical Report, Committee on Human Factors, National Research Council, Washington, DC 20418, USA.

Sheridan, T. B. (1981). Understanding human errors and aiding human diagnosis behavior in nuclear power plant. In: *Human Detection and Diagnosis of System Failure* (J. Rasmussen and W. B. Rouse. (Ed)), 19-35, N. Y., Plenum.

Stankovic, J. A. and K. Ramamritham (1992). *Advances in Real-time Systems*, IEEE Computer Socisty Press, Los Alamitos.

Wen, C. H. and S. L. Hwang (1997). Development of an online aiding system in fault diagnosis tasks. *International Journal of Cognitive Ergonomics*, 1(3), 211-229.

Wen, C. H. and S. L. Hwang (1996). An automatic modeling and analysis system for human fault diagnosis tasks. In: *Proceedings of the 4th Pan Pacific Conference on Occupational Ergonomics*, 29-32, November 11-13, Taipei, Taiwan.

Wennersten, R., R. Narfeldt, R. Granfors, A. Sjokvist and S. Sjokvist (1996). Process modeling in fault diagnosis. *Computers & Chemical Engineering*, 20, S665-S670, Supplement, European Symposium on Computer Aided Process Engineering-6, Part A.

Yoon, W. C. and J. M. Hammer (1988). Deep-reasoning fault diagnosis: An aid and a model. *IEEE Transactions on Systems, Man, and Cybernetics*, 18(4), 659-675.

A CASE-BASED REASONING APPROACH TO FAULT DIAGNOSIS IN CONTROL ROOMS

C. Skourup* and J. L. Alty**

**Department of Engineering Cybernetics, Norwegian University of Science and Technology, N-7034 Trondheim, Norway (Charlotte.Skourup@itk.ntnu.no)*
***LUTCHI Research Centre, Department of Computer Science, Loughborough University, Loughborough, LE11 3TU, UK*

Abstract: This paper discusses the problem of transferring information and knowledge between members of the operating staff in process plants. In particular, the paper concentrates upon what we call unwanted situations. These are problems which are time-consuming and troublesome to solve but not necessarily critical. The experience gained from solving such problems is often an unused resource. A case-based reasoning approach is suggested to store operator experience and knowledge concerning these situations. When an unwanted problem occurs, it is then related to the stored experience. In the initial phase, the approach focuses on acquiring knowledge from operators to describe the situation and later on, to include the human operator in the matching procedure. *Copyright © 1998 IFAC*

Keywords: Process control, knowledge-based systems, man/machine interaction, cognitive systems, user interfaces.

1. INTRODUCTION

Existing support systems for process control generally incorporate explicit knowledge about the underlying process. Most of these systems use different types of models of the process to predict problems and to suggest solutions. However, domain knowledge in process control is often incomplete, uncertain and inconsistent. In addition, some problem situations can be classified as complex problems, that is, where the connection between the problem and its solution is difficult to explain and predict. Hence, there are problems which traditional support systems exclude from their solutions space. Such problems often involve implicit knowledge in the minds of individuals which is not usually accessible to the rest of the operating staff. The existence of implicit knowledge makes the creation of complete models of the process an unrealistic goal.

This approach introduces a common knowledge repository for process plants, that stores information and knowledge, with a focus on implicit knowledge held by operators about unwanted situations. The main purpose of such a repository is to share knowledge between operating staff. The system consists of a knowledge base and an operator interface. The knowledge base contains both domain knowledge, and information and knowledge acquired from operators reflecting their individual experiences. The case-based reasoning paradigm is a technique which can handle incomplete domain knowledge and complex problems. The system uses case-based reasoning to structure and reuse information and knowledge. The operator interface performs communication between the operator and the knowledge base, guides the operator on what symptoms might be appropriate, and provides a matching process to assist the operator in obtaining an appropriate match to the stored cases.

The rest of this paper gives a detailed description of how the system acquires information and knowledge, and describes the matching procedure. An interesting feature of the matching process is its use of visualisation techniques. Section 2 defines the concept of an "unwanted" situation. Section 3 provides an introduction to the case-based reasoning paradigm. In section 4, a case-based reasoning approach for process control is described. The section outlines the structure of the knowledge base, the functionality of the approach and introduces the importance of the user interface. Section 5 concludes the paper.

2. PROBLEMS IN PROCESS PLANTS

In recent years it has become common practice to provide decision support systems for operators in process plant control rooms. These advisory systems help and support the operators in many different ways. Much of the research carried out to-date in developing such systems, however, has been targeted at providing support for the operator during critical situations where the consequences could be damage to the process equipment and/or putting human life at risk (Alty and Bergan, 1995; Alty and Schepens, 1994; White, 1993). The system described in this paper, however, focuses on providing advice for the operating staff in solving problems called "unwanted" situations and unwanted situation is a situation where at least one observation of the system is unexpected or outside normal behaviour, but the situation is not critical to the process or safety. Such unwanted situations are troublesome and are frequently not solved. The symptoms are usually transitory and often return after a given period of time. This may happen a number of times before the problem is solved. Occasionally, the problem remains unsolved even after a long period of time. Thus, the time it takes to solve the problem can vary from a few minutes to several weeks. In some cases, only one operator solves the problem whilst in other cases many staff members from different shifts are involved in the problem solving process. Even though an unwanted situation is not critical to the process, it often results in added costs as a result of non-optimised operation of the plant and in the extra time the operators spend trying to solve the problem.

Operators acquire knowledge about faults and their solution during the problem solving process, and such knowledge is often individual to each operator. It is well known that this type of knowledge can exist in two different forms; as implicit knowledge and as explicit knowledge. The former is knowledge which is powerful and useful but is difficult to explain (for example, how to ride a bicycle). Explicit knowledge, on the other hand, is knowledge whose reasoning process is understood and can easily be communicated to others during the decision making process. Although implicit knowledge held by operators is used frequently by themselves in isolation, it is an unused resource from an organisational viewpoint. The conversion process between implicit and explicit knowledge, where externalisation converts implicit knowledge into explicit knowledge has been studied by Nonaka (1994). There are few accepted approaches to assist this externalisation process. The use of metaphors and predefined questions are two techniques which have been used.

Implicit knowledge has special characteristics. Often a set of symptoms is linked directly to a solution with no intermediate reasoning. Explicit knowledge, on the other hand, is based upon solution strategies which are completely, or at least partially, understood. Both types of knowledge can be understood in terms of Rasmussen's decision ladder model (Rasmussen, 1983). The rest of this paper is mainly concerned with the use of implicit knowledge in problem solving, though this is not meant to imply the absence of explicit knowledge in the techniques used by operators.

3. THE CASE-BASED REASONING PARADIGM

The case-based reasoning paradigm is a method for problem solving. A reasoner, that is a human being or an intelligent system, retrieves previous experiences similar to a new problem and reuses them to solve the new problem (Aamodt and Plaza, 1994; Kolodner, 1993; Riesbeck and Schank, 1989). A case represents previous experience that has taught a lesson to the reasoner and may help in solving future problems. A case is composed of different components - a description of a situation or a problem, the solution, and an evaluation of the solution process (Kolodner, 1993). Hence, cases can form both problem situations and optimal situations that are of interest for future performance. The case-based reasoning cycle includes four main steps. These are retrieval, reuse, revise and retain (Aamodt and Plaza, 1994.). The first step retrieves previous experienced cases that match the new case. The retrieved cases are similar, or at least partially match, the new case. The matching procedure determines which dimensions of a case are important to focus on in judging the similarity. In the second step, the reasoner reuses information and knowledge from previous cases to solve the new problem. The third step evaluates the proposed solution. Finally, the case-based reasoning cycle adds the experience of the new case to the case base so that it can be used in future problem solving. This last step includes indexing of cases. The memory organises cases so that the features of future input problems can be matched against these indexes. In addition to information and knowledge related to cases, the reasoner often incorporates some sort of domain knowledge. Often, a case-based reasoning

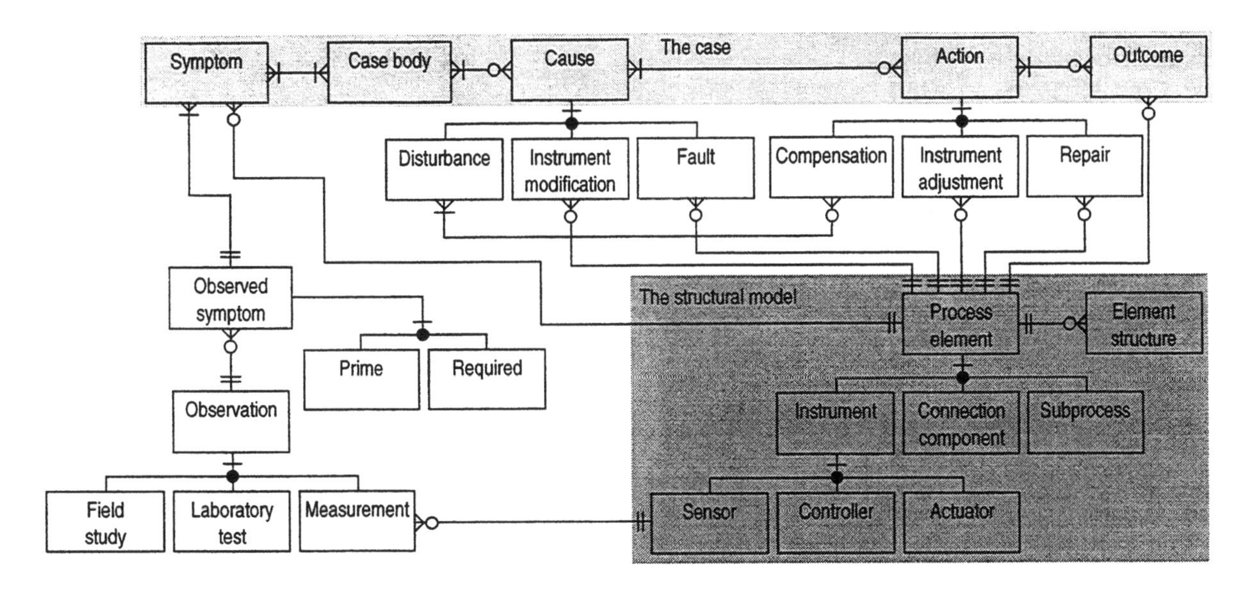

Fig. 1. A data model of the case, the structural model and their relationships.

approach can be characterised as being either "knowledge-intensive" or "knowledge-poor" (Aamodt and Plaza, 1994). In a knowledge-intensive system, knowledge about the domain is available and is integrated as a main part of the approach. Hence, there is less need for cases. A knowledge-poor approach is characterised by incomplete, uncertain or inconsistent domain knowledge. However, cases substitute domain knowledge in that they reflect different types of situations taken place in the domain. The cases approximately cover the entire domain. Therefore, the case-based reasoning paradigm is, unlike model-based and rule-based reasoning, able to operate in "weak-theory" domains (Porter, *et al.*, 1990).

In process control, operators already use the case-based reasoning paradigm (often without realising it). They retrieve previous similar experiences and adapt the solution to the new situation. Normally, the individual experiences of particular operators are unavailable to the rest of the operating staff members. However, if a common knowledge repository of previous experiences could be created, then knowledge concerning unwanted situations would become available to the entire operating staff.

4. A CASE-BASED REASONING APPROACH FOR PROCESS CONTROL

In process control, the solution of an unwanted problem situation is usually not predictable because the causes of the problem often include factors that cannot be incorporated in a general model of the process plant. The case-based reasoning approach circumvents this difficulty by retrieving previous similar cases from the case base. The matching procedure is based on the similarity between two cases with somewhat similar symptoms. Thus, the connection between a problem and its solution becomes less important. Furthermore, knowledge of most process plants can be characterised as belonging to the "weak-theory" domain (Porter, *et al.*, 1990) and therefore, it is impossible to make a general model of the plant that is able to reason about problem causation in all situations. This problem is suitable for a knowledge-poor case-based reasoning approach that focuses on cases reflecting situations that cover different aspects of the domain instead of domain knowledge (Aamodt and Plaza, 1994).

The heart of many case-based reasoning systems is the capability of adapting the stored solutions to the needs of the current problem to be solved, using adaptation rules, since cases will not always result in an exact match. This case-based reasoning approach interacts with the user during the adaptation process as it is natural for humans to adapt (Ferguson, *et al.*, 1992). Hence, the operator transfers individual knowledge concerning the situation to the knowledge base.

4.1 THE CASE-BASED REASONING KNOWLEDGE BASE

The purpose of the case-based reasoning knowledge base is to organise previous cases describing unwanted situations and store with them their related solutions, together with domain knowledge. Cases and domain knowledge together serve as a common memory for all the operators when solving problems. The knowledge base has three parts: a case base, a structural model of the plant and some simple qualitative rules about function. Figure 1 illustrates the data model of the case and its components, the structural model and the relationships between these.

Operating staff members make observations of the

IF $T_{in, eva}(NH3) \uparrow$
THEN
$T_{out, eva}(FW) \uparrow$
OR $(P_{out, eva}(NH3) \uparrow$ AND $L_{within, eva}(NH3) \downarrow)$
OR $L_{within, eva}(NH3) \uparrow$

Fig. 2. An example of a simple qualitative rule.

process. In general, there are different types of observations; measurements, laboratory tests and field studies. A subset of observations forms the features of a case (in this work, these are called symptoms). A symptom is an observation made by a staff member that contributes to a complete description of the state of the problem situation. A case is described by one or more symptoms together with a description of the unwanted situation where at least one of the symptoms is unexpected. In addition, the case representation contains the cause, the action(s) and the outcome of the actions. This representation includes the basic components of a case (Kolodner, 1993).

The domain knowledge includes a structural model of the process plant. The process consists of process elements, that are subprocesses, connection components and instruments. A subprocess represents a unit of the process that contributes to the production of the final product. Examples are an evaporator and a boiler. Connection components connect the subprocesses together. Instruments includes sensors, controllers and actuators. An instrument relates to a subprocess or a connection component, for example, how to measure the temperature in a mixer or to control the flow in a pipe.

To assist the matching process, domain knowledge is augmented with sets of simple qualitative rules for all the subprocesses. The rules describe how a change in a variable, for example a temperature, influences other variables connected with the subprocess in a qualitative way. The rules are used in two ways:

- to extend the number and the quality of symptoms related to a case. Often the operator will have a small number of initial symptoms which might generate too many matches in the knowledge base. The qualitative rules suggest other symptoms which ought to be looked for.
- to refine the matching process. If too many matches are found the rules can be used to "tighten" the match conditions.

It is important to note that the rules are not meant to predict or diagnose the cause of the unwanted situation, since it is almost impossible to model the processes completely. They simply assist the operator in exploring the problem space. Figure 2 illustrates an example of a qualitative rule describing how an increasing temperature within an evaporator influences other related variables.

4.2 SYSTEM OPERATION

The operator triggers the knowledge-based system by supplying the system with one or more symptoms. The accuracy and flexibility of the matching procedure depends upon the number of supplied symptoms and on the number of matching cases. A situation described only by a few symptoms will often match with too many cases in the case base. Alternatively, too many supplied symptoms may result in no match.

To retrieve fewer previous cases but more similar cases, the system uses the qualitative rules in the domain knowledge to prompt the operator for important related symptoms. This usually results in an increase in the number of identifying symptoms. The rules represent qualitative relationships between changes in variables, and provide indications of expected changes in other variables in the process, which the operator should look for. This prompting therefore refines the pattern of symptoms related to each case.

The matching process is carried out using the refined symptom set. At this stage, the user interface presents to the operator a "solution space" in which the operator can move through the space and see the results from alterations to the matching conditions (for example, how many more cases might be included if a certain condition is relaxed or tightened). The small number of resulting matches then provide information on how the problem might be solved.

Of course, all problems will not result in solutions, and some unwanted situations will be solved by other means (for example, when stripping down components or trying to solve other problems). In these cases, new information is added to the knowledge base as a solved case, and will immediately be available for future matching.

4.3 THE IMPORTANCE OF THE USER INTERFACE

The user interface performs a key role in supporting the interaction between the operators and the knowledge base (Alty and Bergan, 1995; Vicente, 1997). The first task of the interface in this approach allows the operator to extend and encode relevant observed symptoms. The other key role of the user interface is to make the matching process more visible to the operators, exploiting their superior pattern matching capabilities.

The matching process returns an unordered set of previous cases somewhat similar to the new problem. The match, so far, is based on exact information

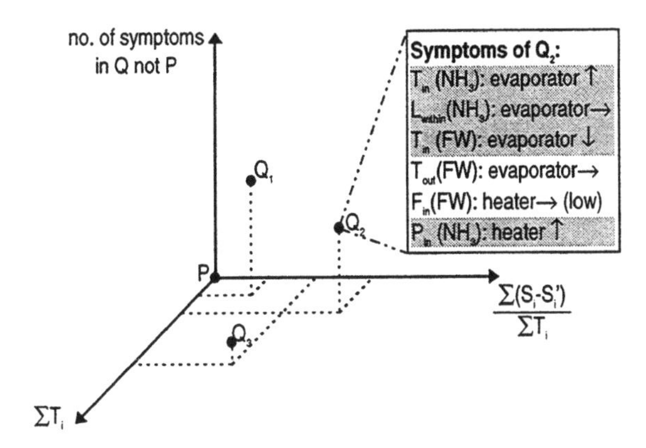

Fig. 3. The solution space where the new problem situation is located in origin and the retrieved cases in the space indicating the similarity in three dimensions.

given by the user. However, importances of symptoms and tolerances of numerical values such as temperatures, may vary dependent on the situation. Members of the operating staff are the only one who are able to judge these importances and tolerances in the situation context.

The solution space is a 3D visualisation of the new problem situation, P, and its similarities to the retrieved cases, Q_i. Equivalent to visualisation of information retrieval systems, the user interface should present relationships among cases, among symptoms, and between cases and symptoms (Krohn, 1996). The three dimensions of the solution space represent:

- the number of non-overlapping symptoms between P and Q_i
- the difference of similar symptoms related to the tolerance
- the sum of the tolerances

The number of non-overlapping symptoms, that is symptoms existing in Q and not in P, is a measurement of the similarity between the two cases. The difference between two symptoms proportional to the tolerance of the symptom in the new problem reflects the relationship among symptoms, whereas the third dimension, to some extensions, expresses the relationship between a case and its symptoms. As smaller the tolerance is, the more similar are two symptoms related to the cases. Figure 3 illustrates the solution space visualising three previous cases, Q_1, Q_2 and Q_3. The symptoms of Q_2 are shown and those symptoms outlined in gray also exist in the new problem, P.

The user interface allows the operator continuously to add symptoms to the new case and to alter the tolerances of each symptom. As the solution space reflects the similarities between the new case and the retrieved cases, an alteration may result in another picture of the similarities and hence, another previous case may be the one nearest to the new problem. At any time, the operator is responsible to make a decision about which previous case, if any at all, has a solution that would solve the new problem situation by adapting it. The solution space makes the matching procedure visible to the user and furthermore, it induces the operator to share important knowledge concerning the specific situation.

5. CONCLUSION

The approach discussed in this paper describes the first step in developing techniques to acquire individual knowledge held by operators and to make such knowledge available to the rest of the staff. The approach has adapted the case-based reasoning paradigm to match the characteristics of the environments in process plants. Especially, the design of the user interface for making the matching procedure between a new problem situation and previous cases visible to the operators applies special emphasis. The problems presented in the paper have been derived from interviews with operators and other personnel at three different process plants.

ACKNOWLEDGEMENTS

This research is supported by the Norwegian Research Council and is a part of the research program, ORD (Operatørkommunikasjon, Regulering og Distribuerte datasystemer), taken place at Department of Engineering Cybernetics, NTNU. Both authors thank the Training and Mobility Programme of the European Commission for the financial support which enabled them to spend time at Institute for Process and Production Control Technology of the University of Clausthal under the COPES project. In particular, they are grateful to members of the Institute for technical support and helpful discussions and ideas.

REFERENCES

Aamodt, A. and E. Plaza (1994). Case-based reasoning: Foundational issues, methodological variations, and system approaches. *Artificial Intelligence Communications*, **7**, 39-59.

Alty, J. L. and M. Bergan (1995). Multi-media interfaces for process control: Matching media to tasks. *Control Engineering Practice*, **3**, 241-248.

Alty, J. L., J. Bergan and A. Schepens (1994). The design of the PROMISE multimedia system and its use in a chemical plant. In: *Multimedia systems and applications* (R. A. Earnshaw, (Ed)), 53-75. Academic Press, London.

Ferguson, W., R. Bareiss, L. Birnbaum, and R. Osgood (1992). ASK systems: An approach to the realization of story-based teachers. *The Journal of the Learning Sciences*, **2**, 95-134.

Kolodner, J. L. (1993). *Case-based reasoning*. Morgan Kaufmann, San Mateo, California.

Kolodner, J. L. and L. M. Wills (1996). Powers of observation in creative design. *Design Studies*, **17**, 385-416.

Krohn, U. (1996). *Visualization for retrieval of scientific and technical information*. Ph.D. dissertation. Papierflieger, Clausthal-Zellerfeld, Germany.

Nonaka, I. (1994). A dynamic theory of organizational knowledge creation. *Organization Science*, **5**, 14-37.

Porter, B. W., R. Bareiss and R. C. Holte (1990). Concept learning and heuristic classification in weak-theory domains. *Artificial Intelligence*, **45**, 710-746.

Rasmussen, J. (1983). Skills, Rules and Knowledge: Signals, Signs, and Symbols, and Other Distinctions in Human Performance Models. *IEEE Transaction on Systems, Man, and Cybernetics*, **13**, 257-266.

Riesbeck, C. K. and R. C. Schank (1989). *Inside Case-Based Reasoning*. Lawrence Erlbaum Associates, Hillsdale, New Jersey.

Vicente, K. J. (1997). Operator adaptation in process control: A three-year research program. *Control Engineering Practice*, **5**, 407-416.

White, G. R. (1993). The operator is the process manager. *Control Engineering*, **Jan.**, 57-70.

HELP DESK SYSTEM WITH DIAGRAMMATIC INTERFACE FOR INDUSTRIAL PLANTS

Yoshio Nakatani and Misayo Kitamura

Industrial Electronics & Systems Lab., Mitsubishi Electric Corp.
1-1, Tsukaguchi-Honmachi 8, Amagasaki, Hyogo 661, Japan
TEL: +81-6-497-7656 FAX: +81-6-497-7727
misayo@sys.crl.melco.co.jp nakatani@soc.sdl.melco.co.jp

Abstract: A new method is proposed for case retrieval in a case-based help desk system for large-scale industrial plants by using hierarchical 2D/3D diagrams as a keyword-input method for customer support engineers to search for trouble cases with less effort than conventional keyword retrieval methods. The users can select devices and their functions in the diagrams, and the selected device/function names are automatically entered into the corresponding columns of the keyword template. This template is used to find trouble cases having similar indices to the keywords. *Copyright © 1998 IFAC*

Keywords: Man/machine systems, man/machine interaction, man/machine interfaces, artificial intelligence, cognitive systems

1. INTRODUCTION

A new method of case retrieval in case-based help desk systems is proposed by using 2D/3D diagrams as a keyword-input method, in order to help users search for trouble cases with less effort than conventional keywords retrieval methods. We have developed case-based customer service support systems for "Level 1" analysts who directly answer customer calls about troubles of the products. The analysts are, however, generally reluctant to accept these kinds of systems; this is partly because they are not familiar with computer systems (especially in Japan) and it is stressful for them to input Japanese characters as the keywords by keyboard typing when they are required to provide real-time interactive responses to high volumes of calls. Although alternative methods which produce keyword lists have also been proposed (e.g., Shimazu and Takashima, 1997), they require the system developers to deeply analyze and predict a wide variety of potential keywords in advance. For the users, it is not easy to find appropriate keywords within a large list, even if the list has some structure.

To solve this difficulty, a diagrammatic case retrieval method is proposed by using 2D/3D diagrams (e.g., system configuration diagrams, design plans of devices, 3D computer graphics images, photos, video images) in place of keyboard typing. By selecting a target device in the diagrams with a mouse, pen, or track ball, the users can get a list of previous trouble cases with the target device very easily. Malfunctions of the devices can be selected from the function list of the devices, which can

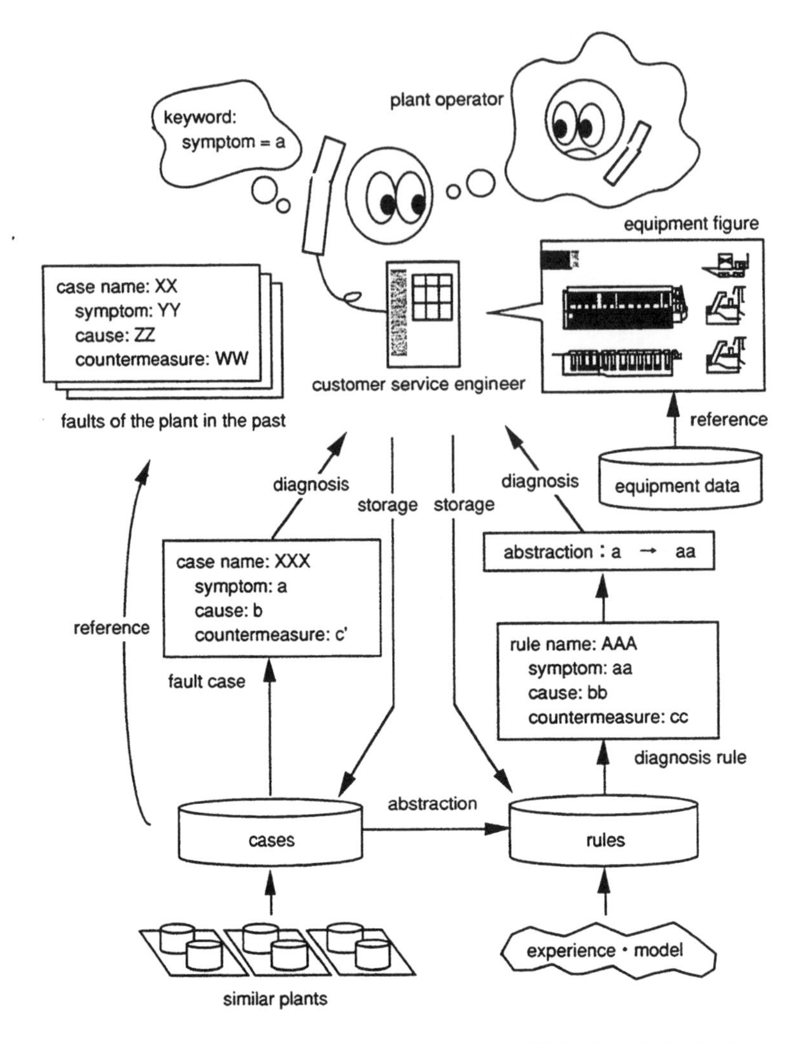

Fig.1. Use of the systempresented here within the HelpCare help desk system

also be obtained from the device function design tables. Moreover, by integrating this system with an equipment information system, the users can know the current conditions and previous trouble cases of the devices in the diagrams.

The proposed method has the following features:

(1) It is easier for the users to find target devices in the system configuration diagrams.

(2) A free keyword method may result in a variety of names for the same keyword, while the proposed method can unify keywords.

(3) Using the diagrammatic keyword-input method, various kinds of design plans can be reused, which were produced when the plant was designed. This can reduce production costs for each plant.

(4) The users can easily find relevant devices around the target device at a glance. This enables the users to understand the current trouble well.

On the other hand, the proposed method has the following drawbacks to be solved:

(1) In exchange for ease of input, flexibility of case retrieval (e.g., similarity search by using a wild card) is partly sacrificed.

(2) In the case of large plants, a correspondingly large set of (more than one thousand of) diagrams is required. It may take time for users to find the desired diagram within the set.

2. DESIGN OF CASE BASE

The proposed method is implemented in a case-based help desk system, HelpCare (Nakatani, et al., 1996), written in Java. HelpCare uses both rules and cases (Figure 1). The rules are collected from trouble-shooting manuals, and the cases are collected from trouble-shooting reports. In order to collect a sufficient number and variety of case, the cases are collected from plants that have similar composition. When the user inputs symp-

Attribute	Example
case ID	FB92####
plant name	Plant △
order name	BV45××
date	920806
engineer	Mr. XX
symptoms	The sealing water of the No.2 washing sludge pump ...
causes	Insufficient sealing water was supplied to the ...
countermeasures	The sleeve was replaced, and the packing ...
memo	The Teflon-immersed packing is NIPPON XXX ...

(a) Example of natural language mode case

Attribute	Example
rule name	#152: Leakage of sludge through the gland
symptoms	The sealing water of the washing sludge pump was ...
causes	Insufficient sealing water was supplied to ...
countermeasures	The sleeve is to be replaced, and ...
memo	......
case list	FB92###1, FB92##13, FB94###7

(b) Example of natural language mode rule

Fig. 2. Example of case and rule

toms as keywords, the system searches for the cases and rules, in parallel, whose symptoms are most similar to the keywords specified.

Two modes of representation of the cases and rules are adopted: natural language mode and index mode. There are two reasons. First, a simple representation is required because the diagnosis knowledge must be easily understood for quick solution of the claims, while a detailed representation is required because users often want to understand cases written by the other users. Natural language mode is for detailed understanding, and index mode is for quick understanding. Second, the continual addition of new cases to the case base causes delay of case retrieval and difficulty of case maintenance. The system must keep all cases because it is required to maintain trouble-shooting reports. By forgetting the verbose index mode cases, retrieval speed can be maintained.

The cases and rules have similar representation as shown in Figure 2. The case consists of nine attributes: case ID, plant name, order name, fault occurrence date, engineer name, symptoms, causes, countermeasures, and memo. The symptoms, causes, countermeasures, and memo can be written in natural language.

Cases and rules each have three kinds of indices, corresponding to the symptoms, causes and countermeasures. For example, each trouble generally includes more than one symptom. Each Symptom is written in one symptom template. The example, "The sealing water of the No. 2 washing sludge pump was clogged and caused the sludge to leak through the gland" is written in two symptom indices: "The sealing water of the No. 2 washing sludge pump was clogged" and, "The sludge leaked through the gland." The symptoms are classified into three categories: (abnormal) action, condition, and location. The causes are classified into seven categories: inferior part, loss, wrong setting, wrong operation, foreign matter, unknown, and other. The countermeasures are classified into nine categories: divert, exchange, modify, add, buy, remove, clean, report, and other. The attributes of the indices vary according to their categories. For example, when the symptom class is "abnormal action," the index consists of six attributes: class, equipment name, part name, malfunction of the part, cause number, and case ID. The malfunctions include emergency stop, unable to operate, and so on. The cause number is used to distinguish among multiple causes. When many causes are found as a result of diagnosis, the users assign a distinct integer to the cause index, and the corresponding symptom and countermeasure indices. Figure 3 gives an example of indices of a trouble that has two causes.

Users can search for cases and rules by using the symptoms, causes, and countermeasures as keywords. This reflects the user's way of remembering the previous cases from various viewpoints. When the user input symptom keywords which represent a current trouble into all or part of the attributes of the keyword templates, the system searches for similar cases and rules whose indices partially match the keywords, and lists them according to their similarity. The users can refer to the detailed trouble shooting reports of the retrieved cases and their relevant diagrams/photos/faxes.

The proposed diagrammatic retrieval method focuses on the use of diagrams, and automatically produces keywords from the user's selection of devices in the diagrams. HelpCare has the function to search for similar

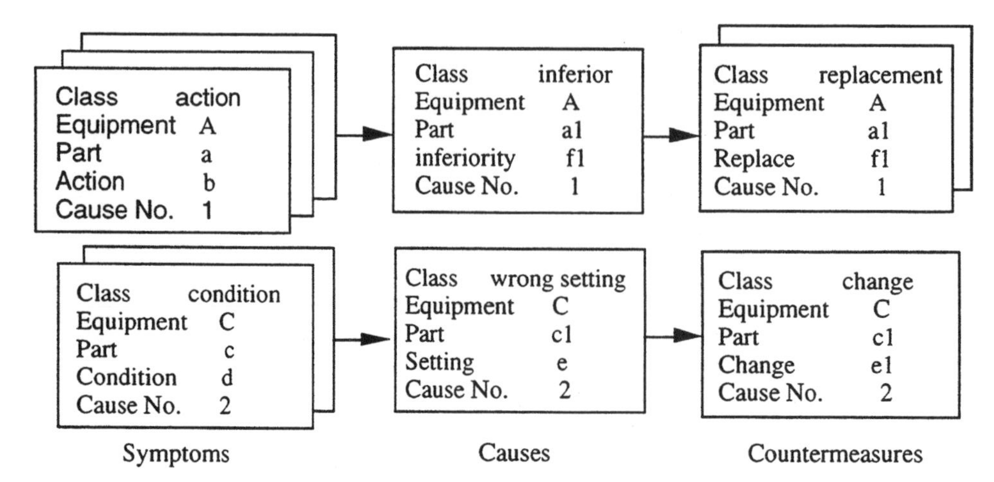

Fig. 3. Example of index mode case

cases/rules by automatically replacing the values of the keywords with similar concepts based on Hyperscript (a kind of structured list of concepts as shown in Figure 4). This functionality is maintained in the proposed diagrammatic retrieval method. Figure 5 denotes a screen image of HelpCare, showing a retrieved case and its related device diagram.

3. DIAGRAMMATIC CASE RETRIEVAL

Diagrams, such as system configuration diagrams and design plans, are used as the interface to specify the keywords for case/rule search. Most diagrams are made in the earlier stage of plant design. The diagrams are classified hierarchically so that the user can easily locate the desired diagram. The classes are linked together by the following mechanism:

(1) Structure --- The diagrams are classified from the viewpoint of their relations, such as sub-diagram, cross-section, device-function relation, and so on. Figure 6 shows an example of the diagram structure.

(2) Marks --- Some devices and parts have marks, such as rectangles and lines, each of which denotes the corresponding lower (more detailed) diagrams and cross-sections. When a mark is selected, the corresponding sub-diagram is displayed.

(3) Buttons --- The names of the device and parts are shown on the buttons. When a button is selected, the corresponding menus, such as the keyword-selection menu and the (mal) function list of the device, are displayed.

In Figure 6, the user can display the lower diagrams by selecting marks in the upper diagram. Parenthesized parts mean that they can be entered into the corresponding slots in the keyword template from the diagram, when the user selects them from the keyword input menu of the buttons.

Figure 7 shows an example of a screen image of the sys-

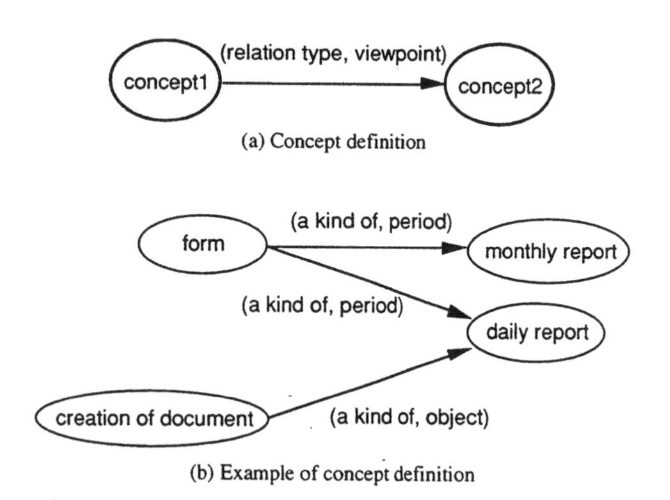

Fig. 4. Representation of Hyperscript

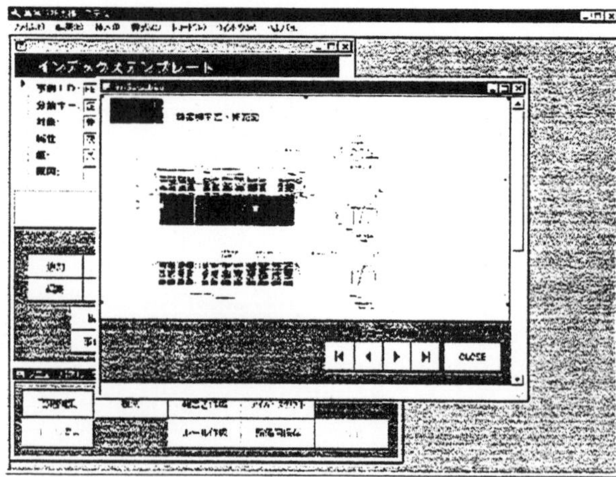

Fig. 5 Example of screen image of HelpCare

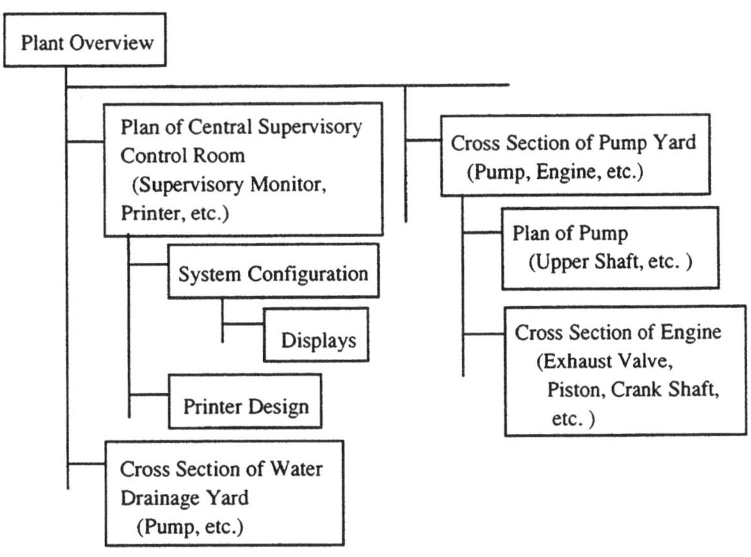

Fig. 6 Example of diagram structure

tem. In this figure, the sub-diagram list popped up when the user selected a rectangle around the device. The user can input the device name into the keyword template by selecting a button on which device names are written.

The proposed method provides the jump function, by which skilled users can directly jump to the target diagram from the top (plant overview) diagram. A structured list of the diagram configuration (consisting of only the diagram names of Figure 6) is provided so that users can select a target diagram by selecting the diagram name. This method, however, shares the disadvantage of the keyword list method- it is not easy for novice users to find a desired diagram.

Figure 8 shows the process of specifying keywords by using both the diagrammatic method and the usual keyword method. When the user decided to do incremental diagram selection, and when he/she has selected a certain button in a certain diagram, the corresponding names of the target equipment and part are entered into the keyword template. The user can input a malfunction name manually or select a malfunction name from the function list. After doing a search, if the result is not satisfactory, the user can change the mode to template modification mode, modify the template, and retry the search. When the template involves inappropriate keyword, the user can modify it by selecting an appropriate button again.

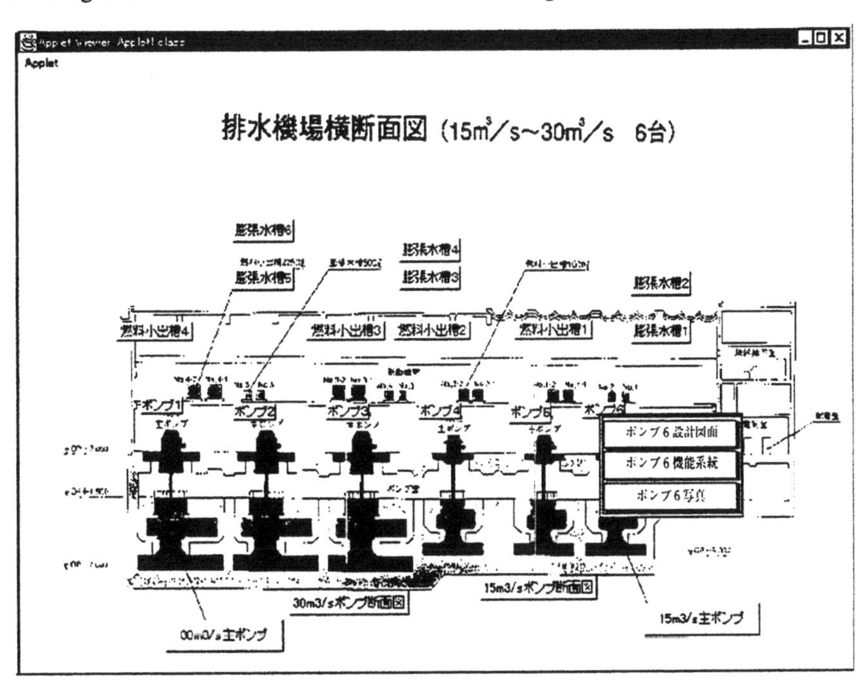

Fig. 7 Example of display

4. EVALUATION

To evaluate the usefulness and effectiveness of the diagrammatic method, a prototype system is implemented. The target domain is a supervisory control system for a water drainage plant. Some diagrams are collected from paper plans, which were read in by a scanner, and others from electrical CAD diagrams. The function lists are manually constructed because any appropriate lists could not be obtained. The proposed method is implemented in HelpCare using the Object-Oriented GUI construction support tool, GhostHouse (Kitamura and Sugimoto, 1995), which substantially reduced implementation costs by providing the graphical editing environment of marks, buttons, and other graphical elements. For example, it is very easy to create marks on the read-in paper diagrams and to create links between related diagrams using GhostHouse.

Kohno et al. (Kohno, et al, 1994) and Schaaf et al. (Schaaf, et al., 1995) proposed diagrammatic case retrieval methods of diagrammatic cases in the domains of analogue LSI configuration and CAD plans of architects, respectively, while the proposed method realizes diagrammatic case retrieval method of textual cases. This is the most important unique feature of our method.

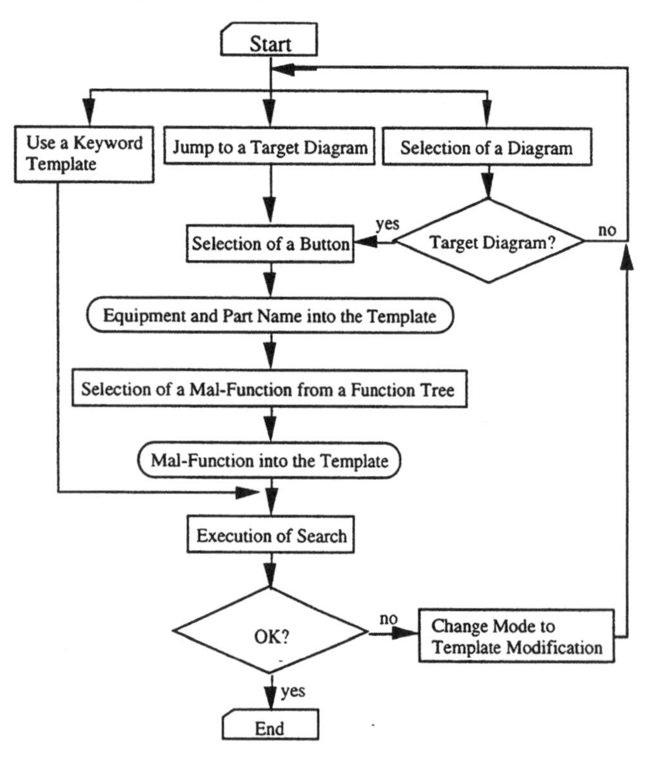

Fig. 8. Algorithm to specify the keywords

We are currently evaluating the effectiveness and usefulness of our method with various levels of users. Thus far, we have received favorable evaluations, except that the CAD diagrams are easier to read than the read-in plans, and the simple configuration diagrams are easier to read than the design plans with many variables and additional lines.

5. FUTURE WORK

The proposed method can be applied to various kinds of systems. For example, when this method is applied to central supervisory control systems, they can provide trouble cases from their supervisory control display. Another target is on-site equipment maintenance support systems which use 2D/3D equipment diagrams or actual equipment video images as case retrieval keys. We are currently designing AR (Augmented Reality)-based on-site equipment maintenance support systems which can retrieve trouble cases through actual video images (captured by a mobile computer) of the target device.

REFERENCES

Kitamura, M. and A. Sugimoto (1995). GhostHouse: A Class Library for Generating Customizable Graphical User Interfaces, Trans. of Information Processing Society of Japan, **Vol. 36**, **No. 4**, pp. 944-958, in Japanese.

Kohno T., S. Hamada, D. Araki, S. Kojima and T. Tanaka (1994). Error Repair and Knowledge Acquisition via Case Based Reasoning, J. of Japanese Society for AI, **Vol.9**, **No.3**, pp. 408-416, in Japanese.

Nakatani, Y., M. Tsukiyama and T. Wake (1996). Plant Fault Diagnosis by Integrating Fault Cases and Rules, 9th Intl. Conf. on Industrial & Engineering Applications of Artificial Intelligence & Expert Systems (IEA/AIE-96), pp. 103-108.

Schaaf, J.W., M. Nowak and A. Voß (1995). Using Gestalten to Retrieve Cases. In: Advances in Case-Based Reasoning: EWCBR-94 Selected Papers, Lecture Notes in Artificial Intelligence 984 (Haton, J-P. et al. (Eds.)), Springer Verlag, Berlin, pp. 136-150.

Shimazu, H. and Y. Takashima (1997). Help Desk Builder: a commercial CBR tool, its design decision and its function, IJCAI'97 Workshop on Practical Use of Case-Based Reasoning, pp. 53-62.

VISUALISATIONS IN PROCESS CONTROL

Chr. Rud Pedersen[1] and Michael May[2]

*[1] Seven Technologies A/S, Købmagergade 26, Post Box 81,
DK-1003 Copenhagen K, Denmark. e-mail: crp@sevent.dk*

*[2] Danish Maritime Institute, Hjortekærsvej 99,
DK-2800 Lyngby, Denmark. e-mail: mim@danmar.dk*

Abstract: The problem of analytical assessment of visualisation is separated into a problem of information mapping and a problem of cognitive and perceptual support. These problems are analysed and discussed in this paper. It is argued that presentational modalities (i.e. semantic invariants) must be identified in order to be able to construct a taxonomy of graphical displays and their potential to visualise certain types of information. The problem of assessing displays analytically is illustrated by two new displays developed for a water treatment plant. In these displays, one graphical type is embedded into another. Common graphical modalities are identified from displays used in the process industry in general. *Copyright © 1998 IFAC*

Keywords: man/machine interfaces, displays, process control, supervisory control, cognitive science, structured analysis

1. INTRODUCTION

Two important but unresolved problems in the analysis and design of Man-Machine Systems are (1) the cognitive support for multi-modal interactions and representations, and (2) the semantic and cognitive principles for selecting presentations for given representations in the design of man-machine interfaces. Representations refer to what is shown on the display, i.e. the display content and presentations refer to how the selected representation is visualised, i.e. the form of the display. The first problem deals with choosing a representation of the domain, which enhance the user's understanding of the possibilities available and support problem solving within the working domain. That is, the selected display content must match the task of the user. Skill, rule and knowledge based operator behaviour (see Vicente and Rasmussen, 1992) are concepts which relates to this problem.

The focus of this paper is on the latter problem. Information mapping, which considers the semantic part of problem (2), has become actualised by recent attempts to construct Intelligent Multimedia User Interfaces (IMUIs) with the ability to chose between different relevant presentations of the same information (or at least to suggest the best ones to the user). The relevance of these attempts are not limited to the development of IMUIs, but is relevant in general for design of interfaces to man-machine systems.

The cognitive principles of problem (2) deal with how the user perceives the developed interface and what the user deduces from this perception. These cognitive principles are mainly concerned with how items appear on the screens and how they are interrelated. The cognitive principles involved in this can be described by gestalt principles and semiotics. Fig. 1 on the next page, illustrates the relations between these main problems. The user's control actions are shown for completeness, but not dealt with in this paper. Information mapping in Ecological Interface Design is where differential equations are mapped into a geometric figure (see Vicente and Rasmussen, 1990).

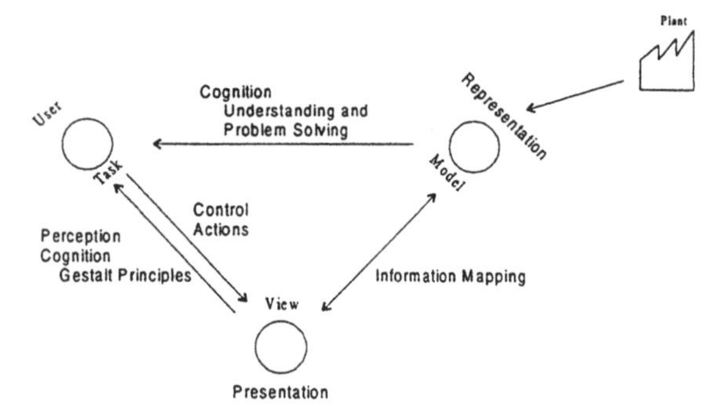

Fig. 1. Main problems in the relation between user, representation and presentation of a working domain (in this case in process control).

In order to address that question a description of available representations and the different types of presentations are needed. Being more specific the types of information existing in a given domain and the available visualisations must be identified. In short a taxonomy of information types and presentation objects is required. The taxonomy must besides mapping representations to presentations also consider the perceptual and cognitive dimensions of the presentations. Such a taxonomy will help the display designer in selecting the right presentation objects for a certain display content, where the display content is derived from an analysis of the user's task.

The display designer must be aware of the perceptive and cognitive aspects of a presentation object to be able to assess advantages and disadvantages of the object for a given representation. These perceptive and cognitive aspects must be stated clearly so the display designer can consider them explicitly and not as today in an ad hoc manner.

The explication of modalities together with perceptual and cognitive effects of visual means will not only work as a guideline for the display designer but it might also reveal paths for development of new presentation objects which deals with the unintended effects or limitations of existing visualisations.

An example is used to illustrate the problems of analytical display assessment before some preliminary work on the problem of information mapping in the domain of process control are presented with a limitation to the visual media, i.e. to visualisation of information. Later principles for structuring the cognitive and perceptual part of the problem are outlined before the conclusion.

2. AN EXAMPLE OF INFORMATION MAPPING IN PROCESS CONTROL

To illustrate the problem of analytical assessment of displays an example from a water treatment plant is used. The focus is on the problem of information mapping but also the aspects of cognition and especially perception will be commented.

Simplified the main problem in water treatment plants is to remove nitrate (NO_3) and ammonium (NH_4) from the water coming into to the plant. The chemistry of the removal process in an alternating plant (see Thornberg, et. al., 1993) is that during a denitrification phase (no oxygen) nitrate is transformed to nitrite (N_2), which disappears in the air. During a nitrification phase, where oxygen is added, ammonium is transformed to nitrate. Hence denitrification decreases the nitrate concentration and the ammonium concentration increases due to the inflow of ammonium. During nitrification the nitrate concentration increases and the ammonium concentration decreases.

The process is automated meaning that a control program manages the shift between the denitrification and nitrification phase. A time delay exists (20 min.) for the on-line measurements of nitrate and ammonium. The plant is supervised by humans 8 hours per day during normal working hours. One of the tasks of the production manager is to monitor the process and optimise it by adjusting parameters to the control system. From an interview with the production manager he explains that they look at yesterdays plants performance when they come in the morning. Asking how he judge the plant performance, the answer is: by looking at the trend curves for NO_3, NH_4 and O_2, they should looked the ones shown in Fig. 2, if not we check the PH-value in the inlet, return sludge flow etc. Asking to more details about the trend curves it appears that the operator uses the form of the curves in his judgement rather than reading the specific value for a given parameter. When deviations from normal are observed they are often caused be equipment failure, so it is seldom that the production manager adjusts the parameters to the control system.

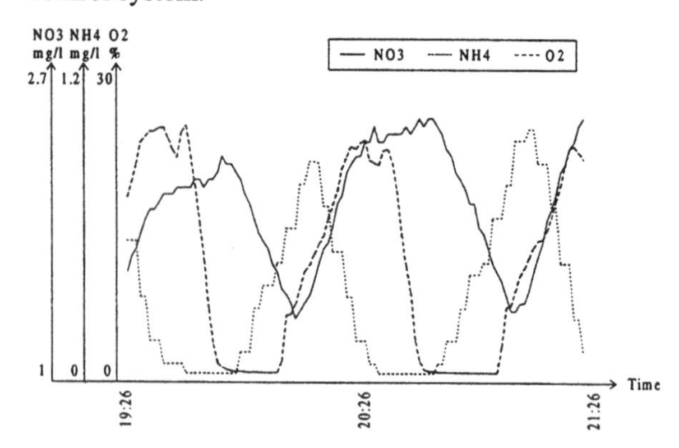

Fig. 2. Main display for a water treatment plant. From these trend curves the production manager judge the plant performance. (On the supervisory screen colours are used to distinguish the curves, not different line styles as shown here.)

According to an expert on water treatment plants, who analyses data from several different plants, there are cases where the operators could improve performance. Therefore the aim was to develop displays which show the plant performance in a more understandable manner.

The normal or expected value could be shown in the trend curves making it easier to observe deviations from the normal operation point. One problem with this approach is the load of the plant depends on the weather (high load in the start of a rain fall) and on the industry in the surroundings of the plant (e.g. high load once a week when tanks are cleaned, low load during week-ends and vacations). A high load will be observed as higher concentrations of nitrate and ammonium.

Another problem is that placing other curves representing the normal values for each trend curve in Fig. 2. will make it even more difficult to perceive, because it is already crowed by the three curves.

An individual y-scale is chosen for each variable, because their operation ranges are different. Table 1. shows the admissible and typical range of the main variables (the typical range is calculated as 3 standard deviations from the mean value).

Table 1. Main variables and their ranges.

	Admissible Range		Typical Range		
Variable	Min.	Max.	Min.	Max.	Unit
NO3	0	6.4	1	2.3	mg/l
NH4	0	3.3	0	0.8	mg/l
O2	0	36	0.3	30	%
rotor	0	4			-

The rotor variable is the one which can be indirectly manipulated through the parameters to the control system and the result can be seen in the concentrations of nitrate (NO_3), ammonium (NH_4) and oxygen (O_2).

The development of these variables over time are the ones which must be visualised. This information can be mapped into many different forms. Two are shown and discussed in the following.

In Fig. 3 the oxygen concentration is the primary variable plotted as a function of time. The nitrate and ammonium concentrations are the secondary variables shown by the markers for the oxygen concentration. Visually the form of the trend curve is perceived together with the form of the markers

	NH4 mg/l	NO3 mg/l
	0.01	0.01
|	0.8	0.01
—	0.01	2.3
○	0.8	2.3

	Rotor Speed
●	0-0.5
●	0.5-3.5
●	3.5-4

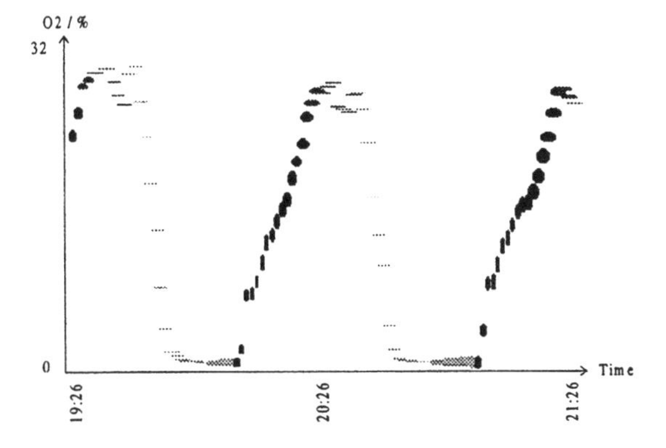

Fig. 3. Oxygen as a function of time. Ellipses are used as markers where the ammonium and nitrate concentration is indicated by the height and width accordingly. The status of the rotor is mapped into colours (here grey scales).

(vertical markers when oxygen concentration increases, and horizontal when it decreases). Making the reverse mapping from form to content this means: when the nitrate concentration (width) exceeds a given limit, the nitrification phase is started, that is the rotors are running (black colour) to add oxygen. The nitrification phase should stop as soon as the ammonium concentration (height) is low enough. When the ammonium concentration is low the markers are flat. Some fluctuation can be seen on the tops in Fig. 3, indicating that even though the ammonium concentration is low the rotors are running - wasting energy. (This is corrected at the plant without use of this display.)

With this display the production manager needs to learn two things: (1) flat markers means that the rotor should stop (light grey colour) (beginning of denitrification phase) and (2) high markers means that the rotor should start (black colour) (beginning of nitrification phase).

This graphical construction is a rather dense and complex graph, that makes some information directly perceivable, but does so at the dispense of other types of information (the numerical values of the NO_3 and the NH_4 concentrations). Also it has the disadvantage of focusing the attention on the level of oxygen (the primary graph), which does not inform the production manager about the load of the plant.

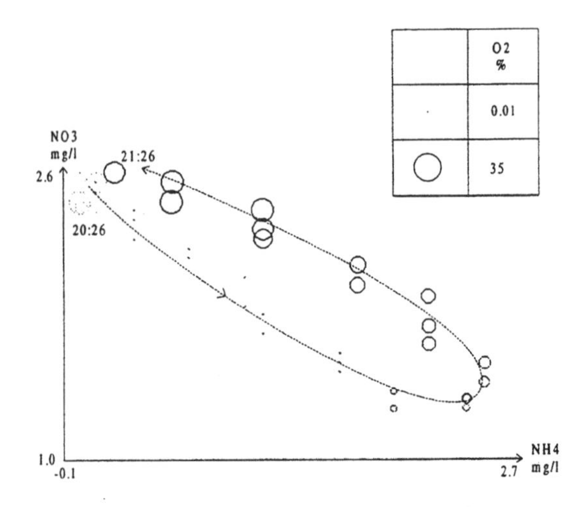

Fig. 4. The main variables nitrate (NO_3) and ammonium (NH_4) are plotted in a xy-plot over time. Colours (here grey scale) are used to visualise the time. The size of the circular markers visualises the amount of oxygen. The dotted line is here added to make the development of time plain.

In Fig. 4 another mapping of the nitrate, ammonium and oxygen variables over time is shown.

Notice that the status of the rotor is not visualised, meaning that the control point is not visible to the production manager. Instead the focus is on the information which indicates the load of the plant. A higher load means higher nitrate and ammonium concentrations, which is visualised by the track moving toward the upper right corner of the graph. This is the situation shown in Fig. 4. The phase shift can be identified as "the turning point" for the track or by the size of the circular markers, where the size indicates the oxygen concentration. With this display it will be possible to indicate good operating areas and areas in which different actions must be taken.

Rings are used instead of filled circles, making it easier to see the markers when several measurements are nearly identical. The time is mapped into a colour-gradient fading from black (latest measurement) to white (oldest measurement) in Fig. 4. When measurements are similar it is difficult to distinguish new measurements from old, because of the colour coding. Therefore only 50 measurements (one denitrification and one nitrification phase) are shown in this graph.

In conclusion the embedding of secondary graphical types into primary graphical types (graphs in this present case), makes it possible to visualise several variables which will be percieved as one form. By chosing the right mapping between the variables (the information) and the visual dimensions, it might be possible to improve process displays. The question is how the right mappings are found. In the next section we will discuss some general principles involved in the mapping of information in visualisation.

3. INFORMATION MAPPING AND GRAPHICAL TAXONOMY

Scientific visualisation or visualisation in general is concerned with two levels of problems. At one level visualisation is a problem of modelling a working domain and choosing representations as part of a model of that domain. At another level, the level of information mapping, the model and its different data types has to be presented using combinations of modalities (cf. Fig. 1). Any model has several possible presentations, i.e. a model can always be presented using different combinations of graphical objects, relations and events.

The distinctions of modality and media has been suggested by several authors in an attempt to understand the invariant semantic types (modalities) that can be use for expressing information across different channels of communications (media) (Stenning & Inder 1995, May 1993, Bernsen 1994). Natural language (a modality) can be expressed in the accoustic media (speech) as well as in the graphical (text) or the haptic (braille text for the blind) media. Similarly, a flow chart (a modality inherited from conceptual diagrams) can be expressed in the graphical media and the haptic media (though in the haptic version, only simple charts can be expressed because of the imposed linear reading of the chart), but acoustic flow charts are not possible.

With a delimitation to the graphical media (excluding true multimedia presentations) for simplicity, a taxonomy of graphical modalities available for presentation of information is still needed. Graphical modalities cannot simply be classified as the objects known in different working domains, e.g. process control, because the used displays are complex combinations of modalities, when seen from a semiotic point of view (Bertin, 1983; May, 1993; May, 1998). In order to be able to analyse the different combinations of modalities and infer their properties with regard to information-mapping (their "potential" for presenting different types of information), it is necessary to consider each modality in isolation. It should be noted that most modalities are not usefull individually, but must be combined with others.

A part of a taxonomy in the graphical media contains the following modalities: *images, maps, symbols, structural diagrams, graphs, conceptual diagrams and texts.*

Images as well as texts can be usefull and informative individually, whereas maps, graphs and diagrams generally will have to be combined with other modalities to be useful.

It is beyond the scope of this article to present the taxonomy as a whole, but the basic principles is that each modality comes with different potentials for information mapping, and these properties are inherited when simple types are combined into multimodal presentations.

Before we return to the displays for the water treatment plant it is illustrated how modalities can be "reused" and combined in different ways in different working domains.

A flow chart is basically a conceptual diagram based on an iconic presentation of potential movement, which is visualised as a path between points in the plane (May, 1998). Flow charts are as such a generic type, that can be used to present many different types of information in different working domains (flow of information, flow of fluids etc.). However, in order for a flow chart to be interpreted in any domain, it will at least have to be annotated with text, i.e. combined with another modality. In process control the flow chart is used together with other combinations of modalities than text. The nodes have been substituted by pictorial symbols ("icons") representing valves, tanks etc.; that is the mimic diagram, which represents the topological layout of the plant. All though the mimic diagram usually is considered as one integral object, it is useful to consider it as a combination of modalities, resulting from substitution of one modality (pictorial symbols) into the parts of another (flow chart), in order to develop at taxonomy for information mapping. The process of substituting modalities is shown in Fig. 5.

Returning to the displays for the water treatment plant the two parameters (NO_3 and NH_4 concentrations) in Fig. 3 are expressed as horizontal extension and vertical extension respectively yielding an elliptic shape. In Fig. 4 one parameter gives a circle.

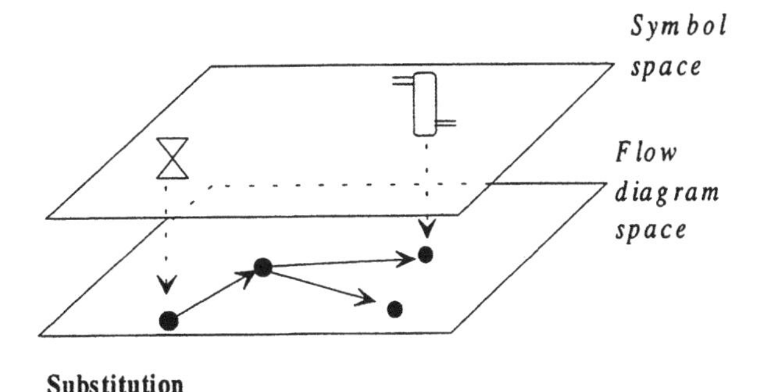

Fig. 5. "Iconic" flow charts, known as mimic diagrams in the domain of process control, as resulting from substitution instances of one modality into parts of another.

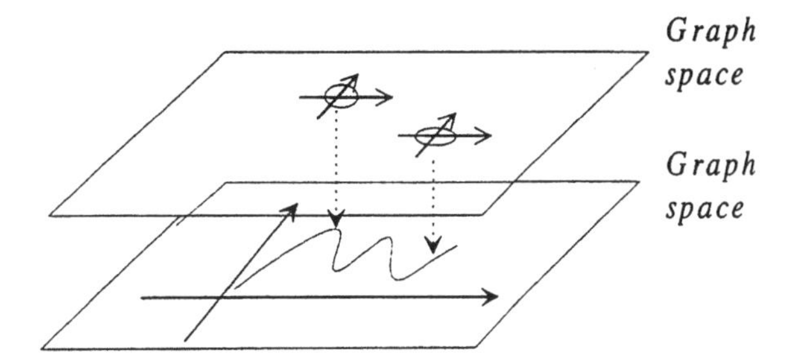

Fig. 6. Graph with embedded graphs as points of the graph (cf. the displays in Fig. 3 and Fig. 4).

These natural mappings of quantity to spatial extension are the basis of graphs based on co-ordinate systems in general, but in the display shown, the qualitative distribution of parameter values has been presented visually by interpreting the extensions as the axes of an elliptic object. In theory any number of axes could be used resulting in a rounded polygon shape or a star. The basic idea is that several parameters can be evaluated at the same time based on perception of shapes (Lindsay, 1990).

The polygon graphs are embedded in the parts of a presentation, that is itself a graph, see Fig. 6.

The two graphs differs in the way they visualised their information, which in both cases are continuous ranges. In the primary graph the position of the marker is used to visualise the information, where as the shape is used in the secondary embedded graph.

Knowledge about human cognition and perception is needed to assess how many axes there can be placed in the embedded graph. Moreover the form must be considered: should be visualised as a star or as a rounded polygon ? (with two parameters as in the water treatment plant as a cross or as an ellipsis ?).

3.1. Cognitive and Perceptual Support for Visualisations

The mapping of information types into different presentations for visualisation purposes should follow a set of principles. The most fundamental of these is that any mapping should have "cognitive support". Cognitive support of information mapping can consist in one of several possible forms of support:

(1) a natural association (e.g. high temperature → red colour)

(2) a metonymic association (whole→ part, e.g. start process operation → push button)

(3) a metaphor (whole→ whole, e.g. desktop metaphor: office desk → workspace of screen)

(4) an "image schematic association "(Johnson 1997) (e.g. vertical schema: up is more, down is less in mapping of quantities)

(5) a symbolic convention (e.g. pictographic symbols in flow charts).

Perceptual support of visualisation should be based on this kind of primary cognitive support for the selected presentations. Perceptual support is found in the perceptual dimensions used to present a selected combination of modalities in a give media. In the graphical media important dimensions are shape, position, size, orientation, colour (hue), brightness, and texture (Bertin, 1981). Which one of these dimensions there can be used for perceptual support of the presentation is however determined by the semantic type, i.e. the modality. In a graphical image for instance, the exact position and shape of objects are important dimensions for the communication of information whereas in a conceptual diagram - such as the flow chart – the position of objects are not informative. In the flow chart the connectivity (the topology) of objects carries the information and the positions of objects can be changed arbitrarily.

4. CONCLUSION

The separation of visualisation into to a problem of information mapping and into considerations of the cognitive and perceptual support seems to make it possible to develop taxonomies of graphical objects, which might be used in analytical assessment of visualisation. Though several problems and aspects needs to be solved and analysed before such a taxonomy can be proposed.

By identifying the graphical modalities from existing displays it has been possible to state the capabilities of the modality to present specific types of information.

It has been possible to decompose the newly developed displays for a water treatment plant into the common graphical modalities listed in the paper, though it should be noted that other graphical modalities than the ones mentioned might exist.

ACKNOWLEDGEMENT

Marinus Nielsen, Krüger A/S provided useful insight on the processes and working procedures involved in water treatment plants.

REFERENCES

Bernsen, N. O. (1994). Foundations of multimodal representations. A taxonomy of representational modalities. *Interacting with Computers*, Vol. **6**, No. 4.

Bertin, J. (1981). Graphics and Graphic Information Processing. Walter de Gruyter, Berlin. ISBN 3-11-006901-6.

Bertin, J. (1983). Semiology of Graphics. University of Wisconsin Press, Madison.

Johnson, M. (1987). The Body in the Mind. The Bodily Basis of Meaning, Imagination and Reason. University of Chicago Press, Chicago.

Lindsay, R. W. (1990). A Display to Support Knowledge Based Behavior. In Proceedings of the Topical Meeting in Advances in Human Factors on Man/Computer Interactions: Nuclear and Beyond, pp. 266 – 270. Nashville, Tennessee, 1990.

May, M. (1993). Representations and homomorphisms, a taxonomy of representations for HCI. Esprit Basic Research Project Action P6296 (GRACE) Deliverable 2.1.

May, M. (1998). Diagrammatic reasoning and levels of schematization. In Martin Skov (Ed): Iconicity. Copenhagen 1998, NSU Press.

Stenning, K. and R. Inder (1995). Applying Semantic Concepts to Analyzing Media and Modalities. In: J. Glasgow, N. Hari Narayanan and B. Chandrasekaran (Eds.) (1995): Diagrammatic Reasoning. Cognitive and Computational Perspectives. AAAI & MIT Press, Menlo Park CA

Thornberg, D. E., M. K. Nielsen and K. L. Andersen (1993). Nutrient Removal. On-line Measurements and Control Strategies. 6th IAWPRC Conference on Instrumentation, Control and Automation of Water Treatment and Transport Systems. Hamilton, Ontario Canada, June 21-25, 1993.

Vicente, K. J. and Rasmussen, J. (1990). The ecology of Human-Machine Systems II: Mediating "Direct Perception" in Complex Work Domains. Ecological psychology, **2(3)**, pp. 207-249.

Vicente K. J. and Rasmussen J. (1992). Ecological Interface Design: Theoretical Foundations. IEEE Transactions on Systems, Man and Cybernetics, Vol. **22**, No. 4 July/August 1992, pp. 589-606

ADVANCED AUTOMATION FOR POWER-GENERATION PLANTS -PAST, PRESENT AND FUTURE -

K. Kawai, Y. Takizawa and S. Watanabe

Toshiba Corporation, Tokyo, Japan

Abstract: Toshiba is designing next-generation control room which promises safer and more stable operation based on use of "safeware" and an integrated design concept. This paper discusses advanced methods and technologies for automating power-generation plants. First of all, the historical development of plant-automation technologies are summarized , then, Toshiba $GSXP_{TM}$ (Global System for neXt Power Plant) as a typical present-day system is introduced. Finally, as an example of a next-generation system applying advanced system technologies, three on-going studies and future control-room design are introduced. *Copyright ©1998 IFAC*

Key Words: man-machine interface, open/distributed architecture, multi-agent, advanced control algorithms

1.INTRODUCTION

The computerized power plant automation in Japan has started around 1968, when the first DDC (Direct Digital Control) system such as turbine run-up control was designed and put successfully into operation. Since then, the scope and the depth of power plant automation has been expanded step-by step. The current utmost level of power plant automation is well explained in a paper describing the Shiriuchi Power Station Unit #2 of Hokkaido Electric Power Company (Konishi, et al., 1997), for example.

The computerized automation system is usually introduced into new thermal power plants in Japan. This automation system is also introduced to existing thermal power plants, when the replacement of control/computer systems are planned.

In the early days of plant automation there are a lot of discussion who, operator or the automation system, shall be the final decision-maker? What sort of tasks shall be retained by the operator, considering the operator's backup capability in case of automation system failure? This problem is called "out-of-the-loop unfamiliarity" (Wickens, 1992). Introduction of automation system could deteriorate the human operator's skills, especially those in case of system accidents (Perrow, 1984). These problems have been solved based on the firm automation design concept and evolutionary systems approach, which further will lead to tomorrow's advanced automation system.

Today's popularity and success of power plant automation are analyzed as follows :

(1) User needs for the automation system are identified/defined as ;
   (a) to reduce the number of operating staffs,
   (b) to save the life-expenditure of the plant equipment,
   (c) to make the start-up or shutdown time constant or as less as possible.
(2) Operation experience of highly reliable automation system and its highly available field test results.
(3) Interface problems among users, plant equipment suppliers and automation system manufactures are minimized since the automation system is provided by major turbine or boiler supplier.
(4) Functional allocation is properly made between plant automation system and the main controllers such as APC (Automatic Power

Plant Control) system, ABS (Automatic Burner Management) system , EHC (Electro-Hydraulic Control) system and SEQ (Sequence Control) systems and so on.

(5) A group of background functions (engineering tools) are utilized during factory testing and throughout field commissioning .

These items described above are regarded as the basic factors of the popularity of today's successful plant automation. The most important factor, however, is the evolutionary approach of the plant automation development. Its scope has been evolved as follows;

(1) Turbine run-up control,
(2) Boiler start-up control,
(3) Auxiliaries start/stop control,
(4) Unit shutdown control,
(5) Unit normal mode control,
(6) Emergency mode control.

2.AUTOMATION PHILOSOPHY AND SYSTEM DESIGN PRINCIPLE

The design goal of power plant automation in the era of DSS (Daily Start-up and Shutdown) operation can be summarized as follows;

(1) Safe, smooth and reliable start-up and shutdown operation,
(2) Reduction in plant equipment life-expenditure,
(3) Reduction in time required for start-up and shutdown operation,
(4) Reduction of operator's workload during star-up and shutdown operation.

Toshiba developed event-oriented expert system named COPOS (Computerized Optimum Plant Operation System) to overcome the limitations of automation-software production based on flow-chart method. A piece of automation knowledge is expressed in one of "plant tables" in COPOS world(Tanaka, et al., 1975).

In order to design the plant automation system properly, the "automation philosophy" shall be established as "system design principle" and well documented. It is also necessary to share that information among users, plant suppliers and automation system manufacturer. This design document shall cover such items as :

(1) Automation objective, scope and its depth,
(2) The specification of plant equipment to be automated,
(3) Operator's role and man-machine communication interface,
(4) Hardware/software "fail safe" consideration,
(5) Target of automation availability and method of failure localization.

3.HISTORICAL TRENDS OF AUTOMATION TECHNOLOGIES

3.1 Man-Machine System Design

Historical Development Of Man-Machine Interface; The typical automation system in the early days of TOSBAC computer series are shown in Table-1. The relevant man-machine interface devices are also shown in chronological order.
The main topics in man-machine interface during 1970s and early 1980s can be summarized in the following five stages :

Stage 1 ASC (Automatic Start-up and Shutdown Control) console and TW (digital printer) were provided (Hachinohe Unit 3).
Stage 2 Monochromatic CRTs were introduced (Atsumi Unit 1).
Stage 3 Color CRTs with character display were introduced (Shin-ainoura Unit 2).
Stage 4 Color CRTs with graphic display and VAS (Voice Announcement System) were introduced (Hirono Unit 1).
Stage 5 Operator friendly software functions were introduced (Shinkokura Unit 5).

The scope of plant automation system is also shown in Table-1. The step-by-step expansion of its scope can also be understood. In the early 1980s, the scope of plant automation was expanded to a one-man operation level. Along with the scale expansion and an increase of system complexity, the most important factor in system design became the transparency of the automation logic to operating staffs.

To achieve the design target of transparency to the operators, new functions, such as "logic chart" displays, and "automation progress" displays were introduced (Kawai, et al., 1984) . These new features became a part of standard software for successive plants, assisting the operator to decrease his mental workload and determining the entire plant status quickly.

Man-Machine Equipment Hardware; All the man-machine interface devices such as ASC console, graphic color CRTs, computer-driven alarm windows, sequential control master-selection panels and so on are gathered together in automation board. Once this automation board is introduced into a new power generating unit in 1980, all the users adopted this concept in the succeeding plants. Since mid 1980s engineering workstations and, full-graphic CRT with multi-windows capability are introduced. In late 1980s, adopting large projection screen (70"-100") started in central control room as an information sharing device among power plant operators.

Table 1 Historical Development of Man-Machine Interface for Plant Automation

No.	COM. OPERT.	PLANT NAME UNIT No. CAPACITY	POWER COMPANY	COMPUTER MEMORY PI/O	SYSTEM SOFT	SCOPE OF AUTOMATION	MAN-MACHINE INTERFACE					REMARKS
							TW	CNS		CRT		
								ASC	OP	ALM	INF	
1	1968	HACHINOHE UNIT-1 250 MW	TOHOKU	T-7000/40 16KW(MAIN) 48KW(BULK) AI 295 DI 200 DO 160	SPAC-I	TURBINE	1	-	1	-	-	
2	1971	ATSUMI UNIT-1 500 MW	CHUBU	T-7000/20 32KW(MAIN) 163KW(BULK) AI 579 DI 276 DO 196	SPAC-II	TURBINE BOILER	1	1	1	-	-	MONOCHR OMATIC CRT
3	1976	SHIN-AINOURA UNIT-2 500 MW	KYUSHU	T-7000/25 36KW(MAIN) 512KW(BULK) AI 850 DI 1288 DO 512	COPOS-25	TURBINE BOILER AUXILIARIES SHUTDOWN	1	1	1	1	1	COLOR CRT (CHARACT ER DISPLAY)
4	1980	HIRONO UNIT-1 600 MW	TOKYO	T-7000/25 64KW(MAIN) 768KWx2 (BULK) AI 1016 DI 1886 DO 1024	AD-COPOS	TURBINE BOILER AUXILIARIES SHUTDOWN NORMAL MODE EMERGENCY MODE	1	1	3	1	4	GRAPHIC CRT VAS (ANALOG)
5	1983	SHIN-KOKURA UNIT-5 600 MW	KYUSHU	T-7/70 1MBx2 (MAIN) 12MB+30MB (BULK) AI 976 DI 1600 DO 656	COPOS-7	TURBINE BOILER AUXILIARIES SHUTDOWN NORMAL MODE EMERGENCY MODE DSS	2	2	2	1	5	GRAPHIC CRT VAS (DIGITAL)

AI: Analog Input

DI: Digital Input

DO: Digital Output

CNS: Console

ASC: Automatic Start and Stop Console

OP.: Operator console

TW: Type writer or digital printer

ALM: Alarm use

INF: Information use

VAS: Voice Announcement System

T-7000: TOSBAC 7000 (24bit)

T-7/70: TOSBAC 7/70 (32bit)

3.2 *Enhanced Plant Monitoring*

The TMI (The Three Mile Island) accident in 1979 triggered various checks on the system safety considerations and man-machine system design guidelines throughout control communities in the world. Plant monitoring functions, which had been regarded well established, were also evaluated again. Safety parameter display system was introduced into many nuclear power plants.

In the thermal power generating units, the enhanced plant monitoring functions which suppress excessive alarm messages were developed. The purpose of this software is to assist the operator with reliable and effective power plant monitoring in emergencies as well as during normal operating stages. During the normal operating stages, the conventional alarm system is useful enough for the operator. In emergencies, however, too much information is usually produced to the operator and there is sometimes a case in which suitable action is not taken due to the floods of messages.

In order to improve the situation, such concepts as "alarm zone No.", "alarm message activity" and "SP pointer (suppression pointer)" are introduced (Kawai, et al., 1982). The concept of "alarm zone No." is defined to be the degree of penetration into the alarm region. Its incremental part is called "delta", i.e. "significant change value". If the process input changes its value by "delta", a worse or better message (WRS/BTR) is produced. "SP pointer" is controlled by the operator's request or by event-initiated digital calculations, which show plant emergencies such as boiler trip or turbine trip. "Alarm message activity" is defined as the measure of the number of alarm messages on the alarm CRT.

If an alarm message exists below the SP pointer, the message is suppressed. If it exceeds the SP pointer, some outstanding format change (such as color change) will occur to enable the operator to easily grasp the process status at that time.

Although some alarm messages are suppressed, all the message are outputted to the printer via the historical log buffer. This is indispensable to analyze printer information for post trip analysis of plant malfunction.

4. PRESENT INFORMATION AND CONTROL SYSTEM

GSXP$_{TM}$, Toshiba's open/distributed C&I system, has been applied to several commercial power plants. To date, it has produced cost-effective results and has proved to be a highly reliable system (Kawai, 1997a,b).

4.1 System Configuration

Typical GSXP$_{TM}$ system is divided into three main components;

(1) A monitoring and automation subsystem,
(2) A Digital control subsystem,
(3) A Field LAN subsystem.

The $GSXP_{TM}$'s monitoring and automation subsystem provides ease-of-operation functions using touch-sensitive screens on operators' terminals and control panels. The system's plant-automation functions assure safe and efficient operation requiring only a minimum crew of operators. Alarm-handling functions provide operators with real-time data to quickly identify plant anomalies before they cause component or system failures. Alarm messages are provided to operators in prioritized order and are supplemented by lists of recommended countermeasures to correct conditions that caused the alarms.

Toshiba developed the $GSXP_{TM}$ system using de-facto standard technologies such as open-distributed, UNIX-based computers and an FDDI (fiber-distributed data interface) network system. Several of the system's operator stations are dedicated to monitoring plant operations, while dual-function servers are used for handling system alarms and logging plant data. A high-speed, duplicated LAN links all system computers. Thus, the failure of a single computer or the primary LAN will not affect overall system reliability.

The $GSXP_{TM}$'s high-performance, digital-control subsystem assures sophisticated data management and advanced control capabilities, including various conventional PID controls.

A field LAN subsystem, which is backed up with duplicate system configuration, transmits field data to digital control systems and function severs via analog- and digital-input modules. This system significantly reduces metal-cable requirements and provides improved system maintainability.

4.2 Man-Machine Interface

The Operator-Station/Server system supports and enhance many C&I functions, such as monitoring plant status, plant automatic control, performance calculation and various CRT-based operations. Plant operators can quickly obtain required information about all plant functions, then perform control tasks using the same CRT terminals. Basically, operators need only press their terminal's touch-sensitive screen buttons for plant operation without inputs via a keyboard. Operator-Station/Server system is different from usual client/server system, in which a client cannot act without a server. The availability of the operator-station, thus, has much more improved.

Since there are no alarm windows on backup panels, operators acknowledge abnormal-status alarm message sent directly to operator's terminal. This system provides a high-performance alarm function that speeds operator comprehension by classifying alarm messages according to rank, such as a serious trouble or a slight trouble, and/or mechanical system group, such as turbine system, boiler system, etc..

5.NEXT GENERATION MAN-MACHINE SYSTEM

Regarding future trends for improving plant information and control systems, it is vital to discuss the following requirements (Kawai, et al., 1998):

(1) How to apply and optimize dynamic plant-configuration technologies such as improved sensors and actuators that incorporate artificial-intelligence (AI),
(2) How to improve training for plant operators to better incorporate these new technologies, and how the roles of operators will change as a result of increased application of computers and other technologies,
(3) How to apply ever-improving computer technologies.

To improve a plant's human-machine interface, it is important to distinguish between their interaction levels, such as "direction," "suggestion" and "information exchange". As shown in Table 2, based on this perspective, five areas as key requirements for future information and control systems are listed.

As shown in Figure 1, Toshiba proposes a systems approach that uses advanced technologies to satisfy the three requirements listed above. Integrated systems technologies are the most promising approach to such a large scale, nonlinear dynamic processes like the power-generation plants. Leveson wrote in her book .... "This book attempts to convince the computer science and engineering communities that a different approach is possible and should be tried. The title of the book, *Safeware*, expresses the impossibility of separating the various aspects of the system in dealing with safety issues....(Leveson, 1995).

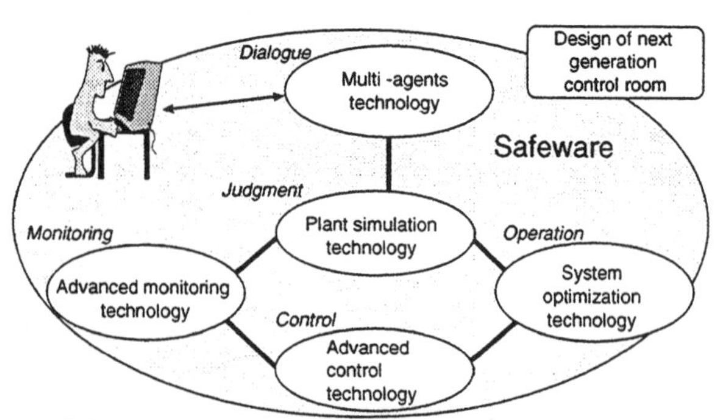

Fig.1 Advanced technologies for responding to the needs of information and control systems

Table 2 Requirements for future information and control systems

Item	Goal
Control	To improve system operability and environmental maintainability by applying advanced control algorithms
Monitoring	To reduce the time required to accurately assess dynamic operating conditions
Operation	To achieve highly reliable, efficient automation of plant operation that requires a minimum number of operators
Judgment	To improve system support for plant operators by supplying them with real-time plant information and suggested responses to all possible anomalies
Dialogue	To achieve highly sophisticated interaction between humans and machines

5.1 Advanced Control Technology For Improving System Operability And Environment Maintainability

Toshiba is developing an advanced system to control the process for decomposing a combined-cycle plant's nitrogen-oxide (NOx) by using two methods: generalized predictive control (GPC) and linear quadratic regular (LQR) (Nakamoto, et. al., 1995). NOx contained in a gas turbine's exhaust flow is decomposed by use of anmonia (NH_3), whose flow rate is adjusted to keep the NOx rate at an operator-designated set point. The control system has a cascade scheme that includes NOx flow control designed by a GPC method and NH_3 control designed by an LQR method. Experimental results on a commercial power plant prove the system's practicability as well as its improved control performance. The fluctuation of NOx flow rate became less than one fifth compared with conventional control system, for example.

5.2 Reward Strategies For Adaptive Start-up Scheduling

A power plant's start-up schedule is designed to minimize start-up time while limiting maximum turbine-rotor stresses. Improving a plant's start-up scheduling can be seen as a combinatorial optimization problem with constraints. To achieve an efficient and robust search model, Toshiba proposes using an enforcement operator to focus the search along the boundary, as well as other local-search strategies such as a reuse function and a tabu search used in combination with genetic algorithms (GA) (Kamiya, et al., 1997). To further increase search efficiency and satisfy on-line search performance, the integrating GA with reinforcement learning algorithms has been developed. The following points resulted from our simulation analysis:

(1) Shorter start-up time than design basis (about 12% reduction in warm start-up mode)

(2) Positive potential for on-line application (average calculation time is 1-8 sec using SPARC Station 20)

5.3 Plant Information Navigator Using Multi-agent Technology

To improve the speed and efficiency of dialogue between a plant's humans and the machines they operate, multi-agent technology is seen as a promising methodology (Kawai, 1996). As show in Table 2, several kinds of information must be presented to operators concurrently. Use of multi-agent will help integrate overall plant control in the following way: a block of information processing will be assigned to a single agent while other agents are used to control supervisory information. Figure 2 illustrates sample dialogue content that would occur between an operator and the Process Information Navigator after a plant-status anomaly has been detected.

5.4 Next Generation Control Room

To achieve an optimal operational environment, the following factors should be considered from the earliest stage of planning a plant's control system.

(1) To maintain the benefits of conventional BTG(Boiler Turbine Generator) panel, the large display panel should be divided into two parts: variable and fixed display areas. The variable display area provides specific plant information according to the plant's current operating conditions. The fixed display area provides plant overview information and identical system information continuously, regardless of a plant's current conditions.

(2) Improving pattern-recognition capabilities among operators to speed reaction times to alert and alarm warnings.

(3) To avoid cognitive overload among a plant's operators, the amount and format of the information they receive should be well within an average operator's information-processing limitations.

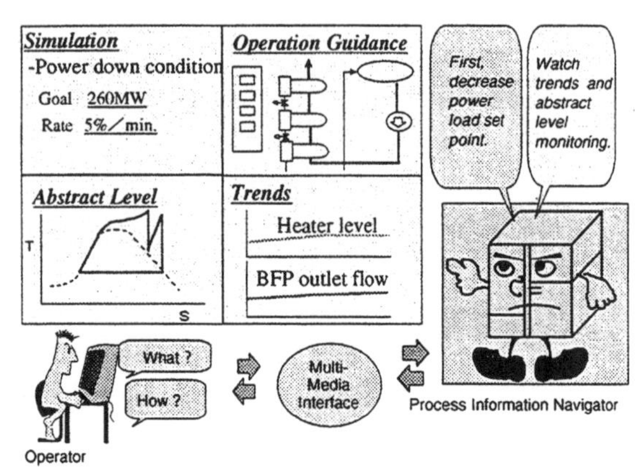

Fig. 2 Example of displayed messages and dialogue between a plant operator and the Process Information Navigator

Figure 3 shows an image illustration of a next-generation control room. The control system's support engineers are stationed behind the central control room's large display panels, maximizing human-human communication between the plant's operators and its engineers.

Fig.3 An image illustration of next-generation plant control room

6.CONCLUSION

Toshiba's $GSXP_{TM}$ system applies state-of-the-art technologies to information and control systems used at power plants. $GSXP_{TM}$ features an open/distributed architecture that improves overall plant efficiency and speed. Based on probable future trends in the power-plant industry, Toshiba proposes five requirements for future information and control systems. As examples of the advanced system technologies, NOx decomposition process control, optimization of start-up scheduling problem and improving dialogue between human and machine are introduced. Finally a concept of next generation control room making use of advanced system technologies and philosophy of "safeware" was presented. The results of these studies satisfy most development targets and required countermeasures for designing future automation technologies used in power-generation plants.

REFERENCES

Kamiya, A., S.Kobayashi, K.Kawai (1997). Reward strategies for adaptive start-up scheduling of power plant, Proceedings of IEEE International Conference on Systems, Man, and Cybernetics, pp.3417-3424

Kawai,K(1982). A Multi-level alarm information processing system applied to thermal power plant, IFAC MMS Symposium , Baden-Baden.

Kawai,K., S.Takishima, M.Tsuchiya (1984). Operator friendly man-machine system for computerized power plant automation. IFAC world Congress, Budapest.

Kawai, K., (1996) Knowledge Engineering in Power-Plant Control and Operation, *Control Eng. Practice*, Vol.**4** No.9, pp.1199-1208

Kawai, K (1997a). Technological trends in advanced information and control systems for thermal power plants (in Japanese), *Toshiba Review*,Vol.**52**, No.7, pp4-6.

Kawai,K (1997b). An intelligent multimedia human interface for highly automated combined-cycle plants. *Control Eng. Practice*, Vol.**5** No.3, pp.401-406.

Kawai,K, Y.Takizawa, S.watanabe (1998). Advanced control and monitoring system-development of information and control technologies toward the 21st century(in Japanese), *Toshiba Review*, Vol.**53**, No 2, pp52-54.

Konishi, T., I.Sugimori and T.Watanabe (1997). Application of Advanced Digital Control System to Thermal Power Plants, 1997 Joint Power Generation Conference, Vol.**2**, pp.213-220, ASME.

Nakamoto, M., K.Shimizu, K.Nagata and T.Kubota (1995). Generalized Predictive Control for a NOx Decomposition Process of a Combined Cycle Power Plant, IFAC Control of Power Plants and Power Systems SIPOWER'95.

Leveson, N.G. (1995). *Safeware-System safety and computers*, Addison-Wesley

Perrow, C.(1984). *Normal accidents*, Basic Books.

Tanaka,S, .Ohta, .Minoura,Y.Kogure(1975). New concept software system for power generation plant computer control -COPOS, PICA conference.

Wickens, C.D., (1992) *Engineering psychology and human performance*, Harper Collins.

INFORMATING PROCESS CONTROL SYSTEMS - KNOWLEDGE-BASED OPERATION SUPPORT

Hannu Paunonen
Jaakko Oksanen

Valmet Automation Inc., P.O. Box 237, FIN-33101 Tampere, Finland
Hannu.Paunonen@valmet.com, Jaakko.Oksanen@valmet.com

Abstract: The increased level of automation and changes in organizations are changing the nature of process control. These developments are setting new requirements on process control systems. Technologies available today make it possible to support organization related functions like communication, co-operation, co-ordination and knowledge management. The systems of the future will contain tools that support functions inside the organization as well. This paper presents an organization related framework for the development of process control systems. Example tools and a case study are also presented. *Copyright © 1998 IFAC*

Keywords: Organizational factors, process control, knowledge transfer, human-machine interface.

1. INTRODUCTION

There is a lack of consistent theory today which combines the requirements of practical decision making and the possibilities offered by modern control systems. This paper introduces a framework for developing and engineering generic process control systems. It is based on the decision making needs of an organization in practical process control work. In this respect it guides activities which are performed daily in thousands of plants and in thousands of control system engineering projects. It also adds organization related aspects to the traditional man-process view.

The rise in the level of automation has changed the work of the production organization. One person can now control very large process entities. Operators have been detached from the process (Bainbridge 1983). The work has become more theoretical (Zuboff 1989) and organizations themselves are changing. Maintenance, development and production activities are merging (e.g. Croon 1994). Control systems are required to support activities on a high abstraction level. However, hasty on-line process control work with its special requirements still has to be done. A disturbance, for example, may call for very quick decision making and action.

Generic control systems, such as distributed control systems (DCS), are an efficient means of distributing tools for operational support. These systems are developed as products and consequently they implicitly gather knowledge about requirements and best practices in process control work.

As for knowledge support, the expert systems were expected to make their way to practical process control work and practical engineering projects. However, that never happened on a large scale. The reason was the economic inefficiency caused by hard-to-get expert knowledge and the tailoring work needed at the engineering phase.

Intelligent functions have emerged embedded in generic control systems as automating controls. These include fuzzy controls, neural networks, and self-organizing maps etc. However, they don't replace the need for normative knowledge in process control situations, which have become increasingly complicated and need expertise from several fields and perhaps from several plants. A lot of this kind of knowledge is available in the form of documentation. This can originate at the design phase or it can be generated through experience of operating the plant. A process control system can provide this knowledge for operational situations while still keeping the amount of engineering work low. A system accumulates and mediates this expertise. It offers knowledge support for the operator when other experts are not available, e.g. during night shifts. The knowledge support in process control systems is based on simple tools, web browsers for instance, and the contents and the structure of the knowledge is simple. The knowledge system itself does not infer; instead, it supports the operators. Control systems are entering the organization. They are part of the organization's memory (Auramäki&al. 1998).

This paper is based on a number of studies and development projects performed over several years by Valmet Automation Inc. and research institutes. The background theories behind this research are cognitive science, activity theory, information system science and automation. The work has included interviews, observations and system development.

In this paper first the content of practical process control work is introduced in chapter 2. Chapter 3 explains the concept of the generic process control system. The framework for informating process control work is presented in chapter 4. Chapter 5 introduces example tools of process control systems and chapter 6 presents a knowledge management system in a pulp mill.

2. PROCESS CONTROL WORK

A night shift operator in a chemical plant notices that the temperature in Reactor 1 is too high. He tries to identify the cause to compensate for the disturbance because soon the product will no longer match the specifications . He memorizes the variables that the temperature depends on. He also tries to deduce from the variables if there is a block in the pipes or if there has been a change in the quality of raw materials. The operator uses the system information to form a complete picture about the disturbance. He also discusses with his colleagues in adjacent departments to find if there have been changes made and to warn them about the coming disturbance. The night shift cannot solve the problem but they note all the information about the phenomenon for discussions with chemical experts and maintenance people. Also the next shifts must be informed. An expert used the system's statistical tools to analyse the problem. When the group of people finally finds the cause some production is already lost. On the basis of this case the production personnel decide to use the reactors in a different way to avoid this problem in the future. A memorandum was written and distributed to all personnel to inform them about this. Also a new control function was implemented in the system.

The case above reveals the nature of practical process control work and thus gives hints for the system support. It shows that in addition to process data and operator actions, the work also involves knowledge handling, communication, and development of performance, for example. The whole organization is involved.

The nature of process control work is quite different from office work. It sometimes calls for quick action but in such situations there may also be a need for deep knowledge. To be able to solve problems plant staff need to move on different levels of abstraction (Rasmussen 1986). Even though the level of abstraction has risen in the operator's work the operator frequently has to return to the lowest level of abstraction. Since one person can control wide process areas the amount of detailed knowledge and information may be enormous. The operator cannot memorize it all. Hence the system should make it available for him in a form and structure that matches the requirements of hasty decision making situations.

Systems should provide production personnel with *"semi-analytical"* tools which contain deep knowledge in efficient and shallow structures linked right to the context. A system is also an important base for developing the operation further. It reveals needs, supports analysis, distributes actions and stores results.

3. GENERIC PROCESS CONTROL SYSTEMS

Process control systems are used for control and for production information management in immediate process control tasks. One example is the distributed control system (DCS). They contain functions like automatic controls, process graphic displays, alarming, analysis tools, reporting, process documentation, etc. A generic process control system (which a typical DCS also represents) contains general purpose functions customized for a process control context. They are designed to facilitate efficient engineering of the generic system to an individual process environment.

It is important to understand the role of generic process control systems as a basis for developing the organization's work support. Generic systems contain features that support the process control work and that are easy to implement in control system design and implementation projects.

For a design engineer the system with its functions forms a high level "language" to support the work and decision making of the production personnel. It contains concepts such as pictures, alarms, PID-controllers, guides linked to an alarm. Efficiency of engineering is an important requirement for the language (Korhonen 1991).

From the operator's point of view the system contains features that are common for all process environments. They contain efficient tools to control a plant handling tens of thousands of measurements and thousands of pictures. Tabbing a special key, for example, will produce a process graphic picture where the new alarm resides.

The systems have been developed as products in cyclic developmental generations. Hence through feedback they have accumulated embedded knowledge about real work requirements which are common to several production environments. When the product (system) is spread throughout thousands of projects the embedded knowledge is spread along with the system both in implicit and in explicit form. For this reason user interfaces have also been developed to match the work. Figure 1 depicts the development information flows during a delivery project and the system life cycle.

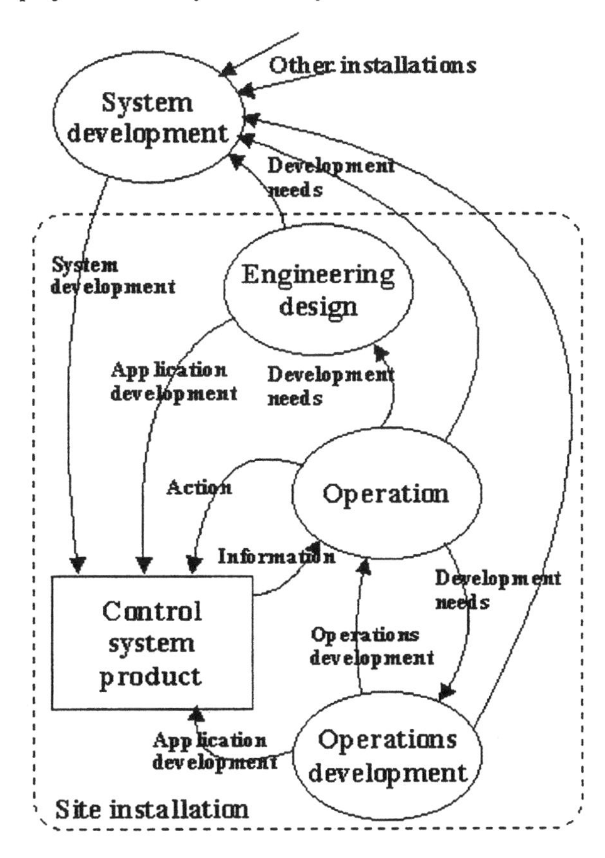

Fig. 1. Development and engineering of a process control system

4. A CONTROL SYSTEM INFORMATING AN ORGANIZATION

If the aspects of practical process control work and its decision making are considered, the roles of the systems can be distinguished. A control system supports the organization in (Paunonen 1997):

- monitoring
- predefined tasks
- disturbance control
- information exchange
- knowledge management
- continuous development
- learning.

In the list above the multidimensional field of process control work is squeezed into one dimension. New organization related aspects have been included in addition to the traditional views prevailing in present system development practice.

Monitoring is the organization's means of staying aware of the process and detecting deviations from the goals as early as possible. It is a prominent mode of work when controlling a highly automated process (Rijnsdorp 1991). In the role of the process monitoring tool the system provides the organization with integrated patterns about the total situation and different views of the process (Sheridan 1988). Monitoring can also be automated with alarms.

Running the production process requires actions by the organization which are formulated as predefined tasks. When supporting these tasks the process control system facilitates operations, automates the tasks and assists in performing them. The system can also offer guidance for correct performance of tasks. Guidance is needed, for example, if the task is complicated, performed seldom or if the operator is not familiar with the task.

The processes may enter an unwanted state, a disturbance. Automation is not good at handling disturbances and therefore their control is the organization's responsibility. Disturbances are difficult situations; normally they cannot be anticipated and, in addition, may produce situations which are totally new for the organization. Disturbance control is another significant mode in process control work. When assisting in disturbance control the system has to alert the operators on duty about the need for action. Furthermore, the system must lead operators to the cause of the disturbance and assist when they select the required action. Stored knowledge is a valuable base for supporting disturbance control.

Almost any process control work is team work. However, distances in process plants are long and people are not working at the same time. The system offers the means to transfer information, e.g. what the future plans of production management are, what operators are doing in other departments and what changes have been made in previous shifts.

Running the process requires expertise and knowledge. The system acts as a storage and mediator of the knowledge needed for process control decisions. The origin of the stored knowledge may be in the design phase (design knowledge) or it can be stored during operation based on the organization's own experience (experience knowledge) (Kaarela&al. 1993). Examples of such system-based knowledge are process descriptions, predesigned guides for tasks or descriptions of past disturbances. For example, guides for specific tasks can originate from a disturbance which is first reported and then generalized as a guide to control operation in the future. A system works as part of the organization's memory.

During operation the organization continuously develops the process, the systems and the organizational functions. As a development tool, the system is used to reveal needs for improvement. It also acts as a store for the results of development as new functions or as explicit knowledge to support new ways of operation. The system itself, especially the displays, must be adaptable to the work. In this way it implicitly accumulates the organization's experience.

The personnel circulates and there is a continuous need to train new operators for their jobs. Some tasks may occur quite seldom and the skills cannot be kept up without regular training. Reports and guides also act as on-the-job learning material. Other learning tools include simulators and training programs.

5. TOOLS

This chapter introduces example tools implemented in Damatic XD*i* (trademark of Valmet Automation Inc.) control system. They comply with the framework introduced above.

Monitoring display

The display in Figure 2 represents a tool for monitoring the overall situation of a pulp mill recovery boiler.

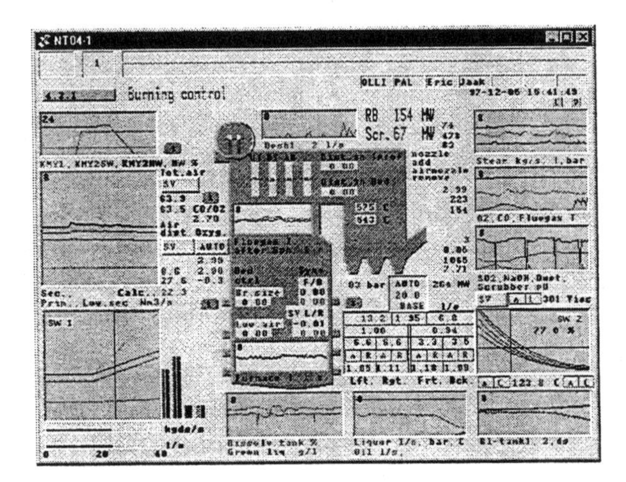

Fig. 2. Monitoring display: Overall situation monitoring

Efficient patterns like trend curves and profiles have been utilised in this display for early detection of an abnormal state. Its content is based on a few representative variables which efficiently tell the operator the status of the process

Process replay

Process replay is a tool for analysing disturbance situations to find the cause. The tool stores and freezes situation information during the disturbance. Operators and engineers can replay it back and forth on a process graphic display to see the changes in the process as they really happened. Such a replay tool has enabled the organization of a chemical plant to identify and remove several causes of disturbances related to the burner. Consequently it worked as a source of development during operation. The stored situations have successfully been used for training as well.

Message pages

A control system contains a tool for leaving and sending messages to other users of the system: other shifts, maintenance and production management.

Experience database

The future challenge will be how to gather and mediate the experiences of the organization for further use in on-line situations. Experience database is a tool to gather the experience knowledge of an organization and to facilitate its utilization as supporting knowledge in process control situations. The experience database includes a diary, disturbance descriptions and disturbance guides. In on-line situations the user can search for supporting knowledge using different methods depending on the nature of the problem at hand. The purpose of such knowledge is to give hints for solving a problem.

Mobile control

Until today process control has been restricted to fixed locations. Wireless communication with portable and wearable terminals has made it possible for an operator to move freely and still have all the information and controls with him. In disturbance and maintenance situations this kind of system also provides the user with all the manuals needed right at the location.

Remote expertise

Continuous improvement and difficult disturbance situations call for a combination of deep theoretical knowledge and practical experience. To speed up the recovery the best possible expertise must be combined together. The expertise may not be found inside one mill or even inside one company.

Combining the expertise of research institutes and process supplier experts with the mill's engineers and operators on site may provide the best solution. Tools for this are remote access to on-line data (process displays) and video conferencing.

Process help

Process help is a tool for informing and mediating the knowledge in the mill. Process helps are based on the Internet and WWW technology integrated in the control system. The descriptions can be stored for example in an HTML format. Figure 3 presents an example of the tool. Chapter 6 contains a case description of the implementation project of the tool as well as experiences of its use. In the case the helps are used as a loop description.

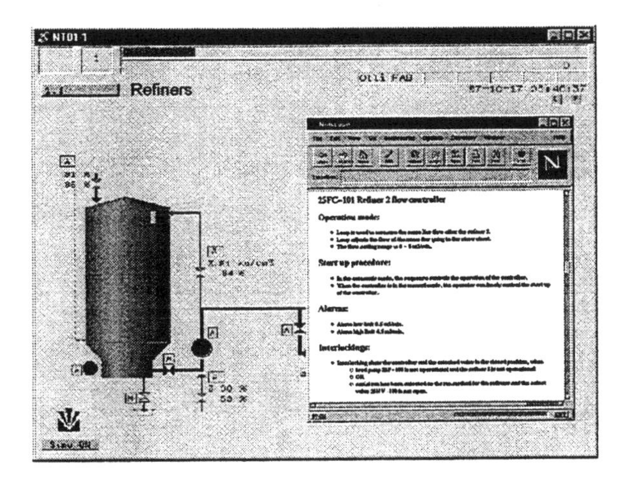

Fig. 3. Process help linked to process graphics

Multimedia training tools

In addition to on-the-job learning the production personnel need additional training to improve their skills and understanding. Multimedia training tools provide new opportunities for continuous learning.

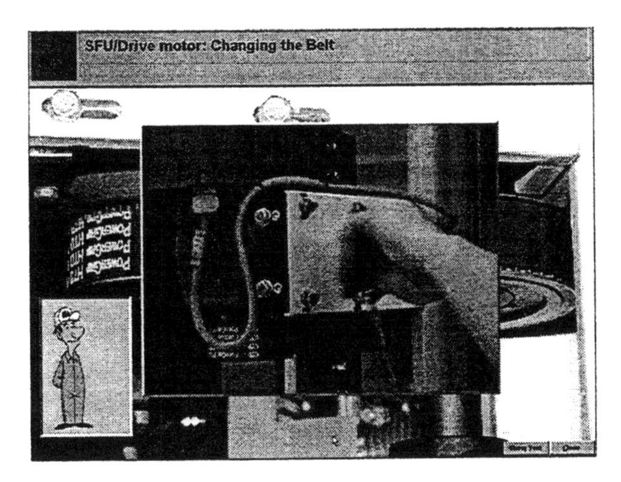

Fig. 4. Multimedia training tool.

Figure 4 shows a display of a training tool for paper quality measurement and a control system. This multimedia training application helps the user to gain a better understanding of the hardware maintenance principles and makes it easier to carry out maintenance tasks on-site.

6. CASE: KNOWLEDGE TRANSFER WITH LOOP DESCRIPTIONS

In a pulp mill the engineering project produced a textual description about every loop, i.e. control loop, motor control, sequence control etc. (Oksanen&al. 1998, Kerkelä&al. 1997).

The descriptions were written during process design and automation design. Both the designers and the operators considered the descriptions to be very useful. There were two crucial aspects which made this solution successful:

- the descriptions served both automation design and operation
- the descriptions were linked with the context in operation tools allowing instant access.

To be able to support mill operations, descriptions were made accessible for the operators in the control room. To find the desired piece of information easily from among all 6000 descriptions the display element and the corresponding description were linked together. By clicking the display element, for example a pump symbol, the operator gets the description directly from that pump to the monitor in a separate window. He can keep the process display and the description simultaneously on view.

The guides functioned as a mediating language between the designers and the operators. The superintendent said: *"Without descriptions we would have never had such good comments from the operators."* And a designer said: *"This was an exceptional project because we didn't need to explain the functions of automation to the operators in the control room".*

During and immediately after a start-up a central part of the operators' work is to keep the plant going. All kinds of jams, faults and fluctuations cause interruptions in the "pipe-like" process. Some situations are sudden. One operator said: *It is important to get the needed guide straight from to the item. If we had to browse or search we would not use the descriptions".*

Figure 5 presents the statistics on how often these documents were retrieved at the control room terminals (the figures don't include paging of documents after opening). The start-up of the plant took place in several steps, department by department, mostly between October and December, causing abundant requests for knowledge support. In May and October the increased need for knowledge was caused by the shutdown and process disturbances. The rise in August is due to people returning from summer holidays.

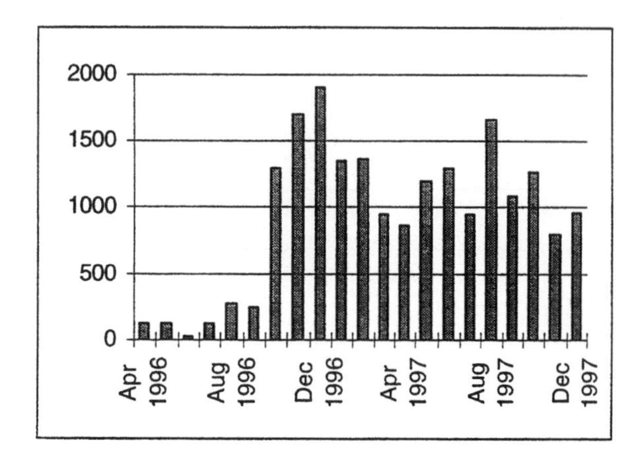

Fig.5. Monthly statistics on the use of loop descriptions.

The descriptions were mainly used for short-term problem solving. Continuous development of the process and automation generated a lot of changes to the process and automation and thus to the descriptions as well. The documentation gathered and mediated this longer term knowledge all over the organization.

The pulp mill started in record time by world standards. The target quality was achieved after one week of production and the nominal production was achieved in a month. Based on the good experience from the example mill the control system's process help tool was developed further. From the user's point of view the usability of the system has been improved: now the system can support several precise helps at the same time. The process helps have now become a normal part of the automation delivery. Also the process supplier offers process descriptions in textual format.

7. FUTURE AND CONCLUSIONS

Today persons running the production have become knowledge workers. The control systems have to support practical work which is more knowledge intensive than before. The systems are used more consciously as learning and knowledge management tools of the organizations. A challenge for the future is to make the systems gather experience knowledge from the production organization and to employ it in operational situations.

8. ACKNOWLEDGEMENTS

The authors wish to thank their colleagues for co-operation and discussions in this development work during several years. Their comments and ideas have been invaluable. Special thanks are due to UPM-Kymmene Inc. in Kaukas and Kaipola, Enso Inc. in Oulu, Technical Research Centre of Finland in Espoo and Oulu, University of Jyväskylä, University of Oulu, Tampere University of Technology. Thanks also to Valmet Inc. and Valmet Automation Inc. for offering the opportunity to do this challenging work with inspiring colleagues.

9. REFERENCES

Auramäki E., Kovalainen M. (1998). In search of organizational memory in process control. To be published in: *Co-operative Process Management*, (Waern Y. (Ed.)) Taylor & Francis.

Bainbridge, L. (1983). Ironies of Automation. *Automatica*, **Vol. 19**, No. 6., pp.767-773.

Croon, I. (1994). Future Challanges for pulp and paper industry. Summary of the conference. Control Systems 94. In: *Conference on control systems in the pulp and paper industry.* STFI/SPCI. May 31 - June 2. Stockholm, pp. 11-12.

Kaarela, K., Huuskonen, P., Leiviskä, K. (1993). The role of design knowledge in industrial plant projects. In: *4th International conference on cognitive and computer sciences for organizations,* Montreal, May 4 to 7.

Kerkelä, K., Oksanen, J., Paunonen, H., Skyttä, T., Huhtanen, M. (1997). WWW/Intranet pohjainen sellutehtaan käytön tukijärjestelmä. *Proceedings of Automation Days -97.* September 23.-25. 1997, Finnish Automation Society, Helsinki.

Korhonen, R. (1991). *Framework for improving quality and efficiency in automaiton design.* Tampere University of Technology, Publication 85, Tampere.

Oksanen J., Paunonen H. (1998). WWW/Intranet Based Operation Support System. To be published in: *Proceedings of the Control System '98 Conference*, 1-3 September 1998, Porvoo, Finland.

Paunonen, H. (1997). *Roles of informating process control systems.* Tampere University of Technology, Publication 225, Tampere.

Rasmussen, J. (1986). *Information processing and human-machine interaction, an approach to cognitive engineering.* North-Holland, New York.

Rijnsdorp, J.E. (1991). *Integrated process controls.* Elsevier Science Publishers B.V. Amsterdam.

Sheridan, T.B. (1987). Supervisory Control. In: *Handbook of human factors* (G. Salvendy (Ed.)). John Wiley&Sons, New York.

Zuboff, S. (1988). *In the age of the smart machine.* Basic books, New York.

THE ROLES OF THE MULTIMEDIA COMMUNICATIONS ON THE PLANT

Katsumi TAKADA, Hiroshi TAMURA and Yu SHIBUYA

Faculty of Engineering and Design, Kyoto Institute of Technology
Matsugasaki, Sakyo- ku, Kyoto, 606 JAPAN
E-mail: takada@hisol.dj.kit.ac.jp

Abstract: Modality choice of participants in media conferencing is analyzed in this paper. Firstly, the role of multimedia in plant operations is mentioned. Secondly, the problem of communication by telephone among the plant operators is described. Thirdly, model conferencing is done in nine combinations of media, which include text, sound and image. The following conclusions were obtained from the results, 1) the task efficiency depends on the combination of media, 2) sound should be used in simple task, and 3) image should be used when solving a complicated problem. *Copyright ©1998 IFAC*

Keywords: multimedia, multi-modal, human error, model conferencing, plant operation

1. INTRODUCTION

Multimedia systems are increasingly being installed for supervising and controlling plants of various sizes. From the plant operational point of view, multimedia should not be a simple aggregation of various media. They should constitute an integrated entity of plant information as a comprehensive whole, including not only logical but also emotional effects of operations. Also multimedia is expensive and time consuming in development. Too much and complex information display makes it difficult for operators to find really urgent data. Thus the real merits and demerits of multimedia need to be studied carefully.

Actually many attempts have been made at evaluating the merits of introducing multimedia from the human point of view. However, from the point of view of work performance, or accuracy and response speed, positive effects of introducing multimedia have not been confirmed. Naturally, in certain fields of applications, there are many cases in which certain media have special important effects. For example, in the presentation of new model cars and modes of fashion, high quality color display with the capability of moving image presentation is essential. On the contrary, there are many cases where expensive multimedia systems which are installed for reasons of special needs of efficiency in certain cases, are not actually used at all.

Multimedia might have different functions for different people working together in the work place. The people might have different knowledge and training backgrounds. Although their tasks are sometimes separated, they have to share a common understanding of the situations. The team structure often changes by shift.

From a point of view described above, the following three points seem to be important fact to evaluate multimedia systems.

(1) Multiplicity of expression.
(2) Diversities of modality
(3) Task efficiency

Here multiplicity of expression will help to decrease misconception and misunderstanding. Diversities of modality might provide users the best choices of modality. Say the task efficiency; it is never the most efficient for all to choose a modality. The modality which most users like to select as the first choice is often not the most task efficiency. For example, a user might select speech as first modality, but its efficiency is, in many cases, less than the user's expectations.

The purpose of this paper is to examine the task efficiency in relation to multiple choice of modality. Firstly, a communication analysis of plant operators at trouble is introduced. Secondly, the cognitive classification of the human errors (Reason, 1990) and the roles of the multimedia in plant operations are discussed. Thirdly, the model conferencing are done in nine combinations of media (including text, sound, and image). Finally, the efficiencies of the multimedia systems are related to multiple choice of modalities.

2. TROUBLE CALLS IN A WATER PURIFICATION PLANT

The water purification plants in a small-scale city in Kyoto prefecture are located in three places, and supply 36,000m^3 water per day in totality. Twenty people, who are classified into three ranks, run the plants; three experts, eight technicians and nine assistants. The experts and technicians do routine work like design of the plant or repair of the equipment. The assistants attend to night and holiday work, and operate the system in a limited ranged.

The control systems of each plant consist of a personal computer and a programmable logic controller for feedback control.
Overview information of a plant is displayed at the built-in graphic panel, and detailed information of the plant is displayed on the personal computer screen. It automatically gives warning in case of machine trouble. When an assistant finds machine trouble, usually depending on the warning signal, he may call to a superior in order to get instructions to meet the warning. And if the telephone is not sufficient media for communication, the superior will go to the site directly.

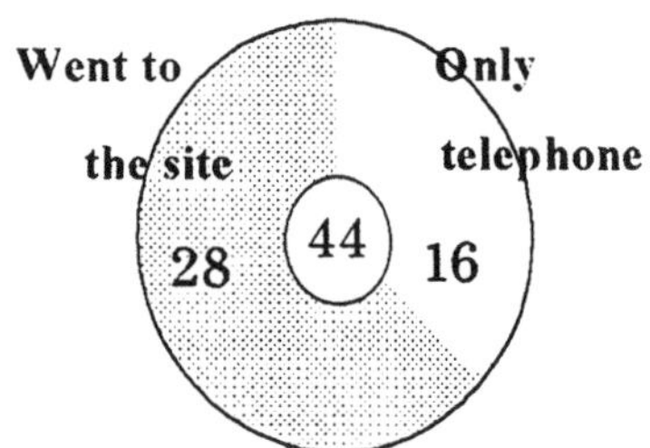

Fig.1. The number of total troubles in 1997

Figure 1 shows the number of total troubles registered in 1997. In major cases, the superior went to the actual site. The reason is that they were not able to catch the meaning of the warning or to explain the restoration by telephone communication.

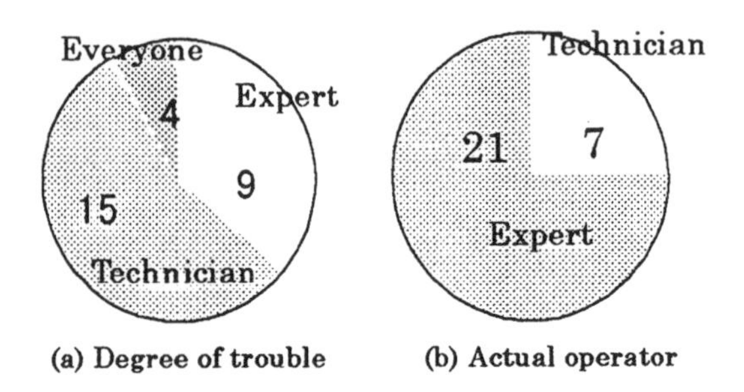

Fig.2. The details of the number of troubles

Figure 2 shows the details of the number of troubles in 1997. The numbers of calls by an assistant for experts were more than the number of calls for technicians. Experts settle most of troubles. The assistants' actions are sensible because they believe experts are reliable in solving troubles. This situation makes reliable experts busier and makes less use of technicians. This is a waste of the manpower and a kind of HE by the assistant. Also assistants were only referring to the value on the panel, and they seldom refer to the value to other information. If assistants are able of grasp the trouble situation better themselves and the communication of the assistant and the expert is done smoothly, the experts might be saved the trouble of going to the site for restoration.

After all, the following factors are causes in these problems. (1) The current graphic panel is not sufficient to provide detailed information to the assistants, and (2) telephone is not sufficient to communicate in complicated situations. Thus we have to study new methods in order to correct these defects.

3. THE ROLES OF MULTIMEDIA IN PLANT OPERATION

In the field of man machine interface design for plant operation, the results of examine work achieved by (Rasmussen, 1986) are worthy of special mention. His studies help to know the actual condition of HE in the plant. They are referred to for this paper. In this chapter, the relation between multimedia systems and communications at a plant will be mentioned briefly.

3.1 The multimedia system in plant operation

The recent studies of HE in the plant include: the improvement of operator's skills, the training of the team, the interface design and so on. But a final concrete measure has not been determined.

In the multimedia systems in the plant, the following functions seems to be needed. Firstly, it does not make a mental load for the operators. Secondly, the necessary information is acquired at once. Finally, it makes an optimal operation environment.

Therefore the multimedia systems have to integrate all kinds of information (such as perceived information based on skill, logical information based on rule, conceptual information based on knowledge.) into a comprehensive whole.

A simple example of the three kinds of information is shown in Figure 3. It shows the simple purification system.

This system consists of pre-treatment equipment, eight valves, one pump and three filters. Raw water constantly flows into the pre-treatment equipment, the pre-treatment water is pumped up into two filters, and the third is a spare.

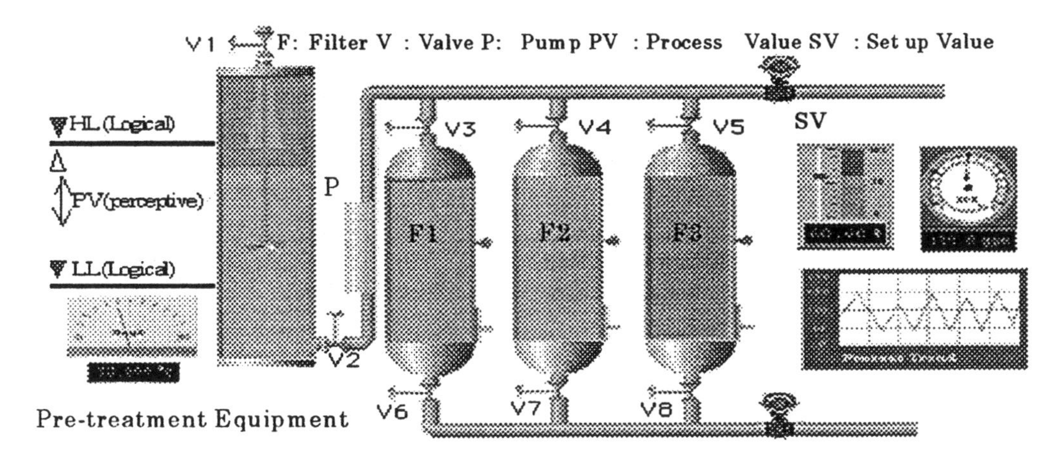

Fig. 3. Simple purification system

As for the display method with the water level of the pre-treatment facilities, the following three types of information exist with three kinds of operations.

(1) *Perceptual information:*
When the quantity of inflow water and outflow water is in constant fluctuation. In order to maintain the correct balance of water levels, the operator must continuously adjust valves. This is a skill-based operation

(2) *Logical information:*
When the water level reaches high level (HL) or low level (LL), which are critical levels for the facilities, the operators have to adjust valves and adjust water level to normal. This is a rule-based operation.

(3) Conceptual information:
When the water level reaches above HL or below LL, and can not be returned to the normal level by the valve operations, the operators have to operate the equipment with an understanding of the possible situations. This is knowledge-based operation.

It is necessary to find the relative information for each kind of operation (Figure 4).

For example, it is effective to combine the multimedia systems with key word retrieval, also to visualize numerical value information.

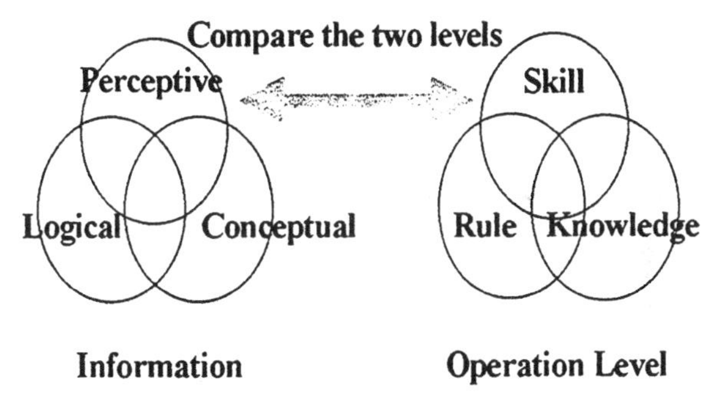

Fig.4. Comparison indication with operation level

Furthermore the graphical user interface using the computer graphics becomes effective as the output of the total set of the systems. However, integrating multiple modals on a screen at the same time make the display area crowded, and sometime the sound is noisy. This might cause stress for the operators. Therefore, it is necessary to study how to use multimedia systems in the plant.

In this paper, the model conferencing are done in nine combinations of media (including text, sound, and image) in order to examine the task efficiency in media conferencing. And the results of some measurement on experiment are discussed.

4. EXPERIMENT

On the basis of the above considerations, several sets of experiments using the conferencing model were done.

4.1 The purpose

The purpose of these experiments is to examine the task efficiency in relation to multiple choice of modality

4.2 The scenario

The conferencing model using a game was made. The scenarios of the game are as follows.

(1) Two examinees can communicate with one another using multimedia system.

(2) One examinee has a document describing fifteen Japanese phrases (Figure 5(a)), and the other examinee has their antonyms (Figure 5 (b)).

(3) The task of the examinee is to find the spelling of the antonym on the other, and to write it on the one's own document (Figure 5 (c)).

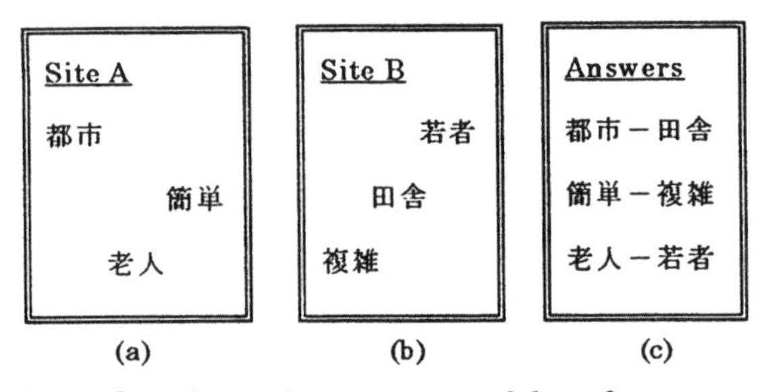

Fig.5. Question and answer on model conference

(4) The Japanese phrase (it is called the phrase after this) consists of three elements ("reading", "content" and "spelling"), and every examinee has a different ability to understand the phrase. Therefore, the result of pre-experiments and reference books fix difficulty degrees of the phrase, and they are assigned to every document with equal degree.

4.3 The concept

The concept of the conferencing model is to relate a difficulty of the phrase to the information level (the perceptual, the logical or the conceptual). Thus the difficulty of the phrase is the degree of examinees understanding the phrase. The contents of this are as follows.

[Perceptual level]:

Operation: When an examinee understands the antonym of the phrase, his operations are to confirm the antonym of his partner.

Target: Which media is the best to quickly perceive?

[Logical level]

Operation: When an examinee grasps the meaning of the phrase but does not understand its antonym, his operations are to find the antonym of his partner systematically from the meaning of the phrase.

Target: Which media is the best to find an answer systematically?

[Conceptual level]

Operation: When an examinee could not find the meaning of the phrase, his operations are to find an antonym in drawing out the synthetic knowledge of his partner.

Target: Which media is the best to draw out the knowledge?

4.4 The examinee

The examinees were 18 male students separated into 9 groups of two. They played the games on the nine kinds of media combinations.

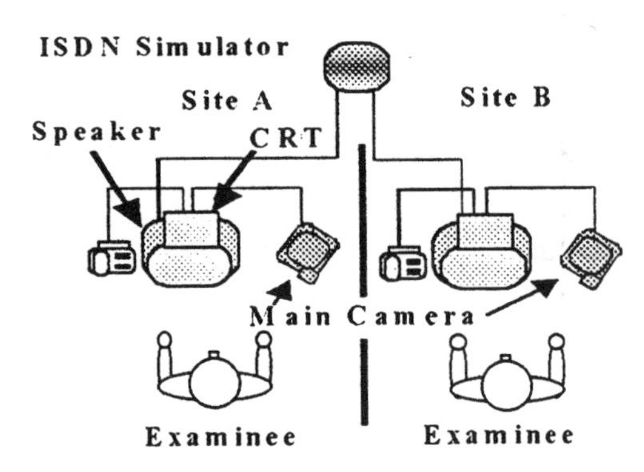

Fig.6. The system configuration

Firstly, an examinee starts to communicate with his partner, but their choice of communication media might change according to the progression of the dialog. The important factors for examinees are as follows.

① Can the examinee grasp the dialogical situations of himself and his partner?

② Can the examinee understand easily the information form of his partner?

③ Can the examinee correctly convey the messages to his partner by using his own media?

4.5 Media environments

Examinees played the game nine times with the nine combinations of transmitting media, which were shown in Table 1. Thus, examinees could use all media as the receiver in every combination of media. Here the sound means all acoustics which contains a speech.

Table 1 The combinations of media

Case	ONE EXAMINEE	THE OTHER
1	Sound	Sound
2	Text	Text
3	Sound, Text	Sound, Text
4	Sound, Text, Image	Sound, Text, Image
5	Sound	Text
6	Sound, Image	Sound
7	Text, Image	Text
8	Sound, Image	Text
9	Text, Image	Sound

The examinees exchange the information freely with one anther. And measure the following five factors.

1. The task achievement time
2. The changing point of main modality (R. Zhang, H. Tamura (1997))
3. The Show-mind and Watch-mind
   Show mind: The feelings of trying to tell the information to others, it is realized by showing action.
   Watch mind: The feelings of trying to get the information of the others, it is realized by watching action. (H. Tamura, R. Zhang (1997))
4. How to coordinate each examinee
   Coordination: The mutual understanding and cooperation of actions between the human examinees who must compensate for mutual contradictions.
5. Post agreement utterance (PAU)
   The authors have noticed that a lot of discussion occurred after an agreement through media experiments using conferencing models. (K. Takada, H. Tamura, Y. Shibuya (1995))

5. RESULT

(The task achievement time)

Figure 7 shows the task achievement time from case1 to case 4, in which both examinees use the same media.

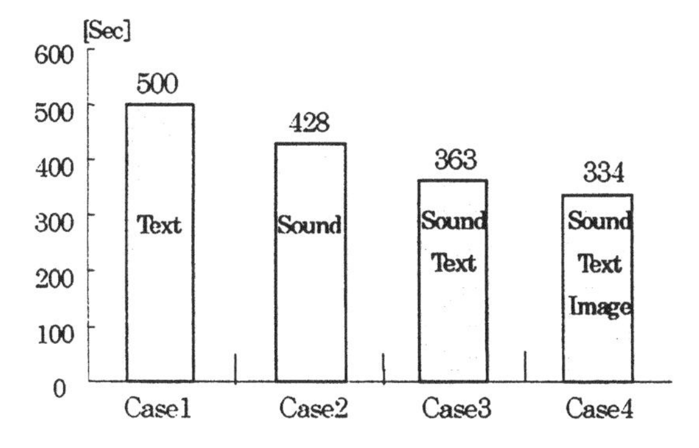

Fig.7. The task achievement time (a)

The significant difference of the task achievement time is shown between text (Case1) and sound (Case2, Case3) ($p<0.05$), also between a single media (Case1, Case2) and multimedia (Case3, Case4) ($p<0.05$).

Therefore, the multimedia makes task achievement time decrease. This result seems to be the multiplier effect of multimedia in the media communication. We assume that rich media could transmit a more profound expression than poor could.

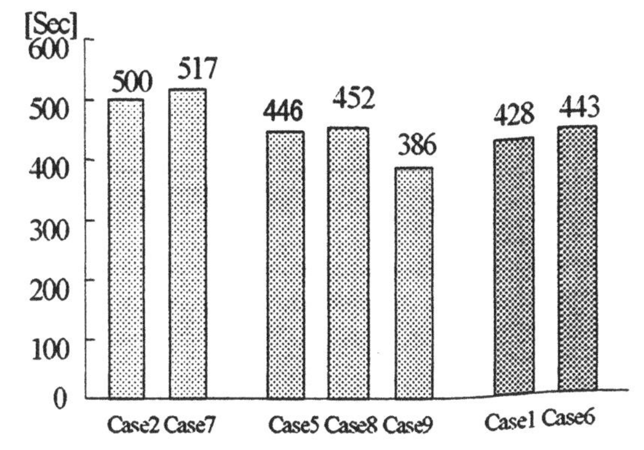

Fig.8. The task achievement time (b)

Figure 8 shows the influence of the image in task achievement time. One examinee in case 7 has image as same circumstances in case 2.

An examinee who can use sound to transmit information at case 8 has image as same circumstances at case 5. An examinee who can use text to transmit information at case 9 can use image as same circumstances at case 5. One examinee at case 6 can use image as same circumstances at case1.

From Figure 8, even if only one person can use an image, the task achievement time doesn't decrease. When comparing with first two cases (Case 2, Case 7) and other five cases, the sound was thought to be more important than the image in the decrease of the task achievement time. ($p<0.05$)

Also we confirm the following things about the sound usage in media conferencing by the video analysis.

① The sound user, who can use sound to transmit information, could recognize an antonym faster than the others did.
② The sound user tried to use sound first.
③ If one examinee noticed that the other examinee couldn't use sound, he tried to use the text. But after making a role share with one another, he used sound again.
④ An examinee issues indication in using sound or image, and gets data using the text.
⑤ When sound was the main modality, post agreement utterance appeared frequently.

(The choice of modality)

Figure 9 shows the choice of main modality of every degree of task. In Figure 9 rank1 is perceptual level task, rank2 is logical level task and rank3 is conceptual level task. Each classification was due to the video analysis based on the definition in chapter 3. But the rate of rank3 became little contrary to the intention before experiment. Communication may have enabled each examinee to find an antonym easier. It may also be that since examinees were able to use a process by elimination and every examinee has a different ability to understand the phrase.

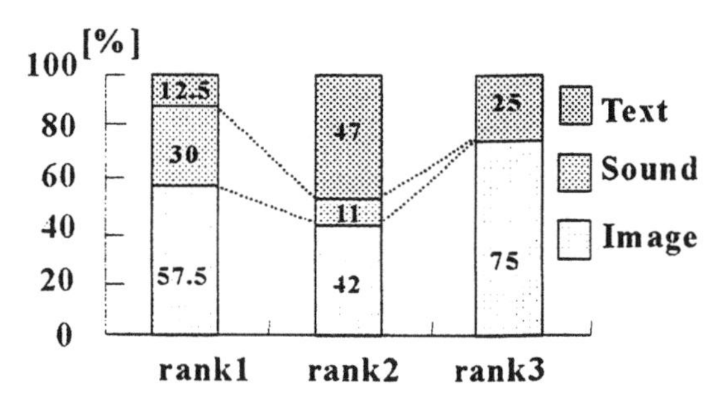

Fig.9. The choice of main modality

And image is being most abundantly used as the main modality. Though much of sound becomes the main modality in rank 1, the tendency becomes weak as rank rises. ($p<0.1$) Also we confirm the following things about the choice of modality in media conferencing by the video analyzing.

① Usually the main modality chosen by the examinees was sound, but this choice was also influenced by the kind of task requested.

② If the information became complicated, many examinees used the image and text.

This result will be related to the study of (J. L. Alty, 1993) and (Zhang, 1997). Alty suggested that multimedia interfaces were particularly useful for aspects of the interface which were difficult to understand. And Zhang referred that multimedia user could automatically integrate their action modalities to select a suitable way to communicate effectively during the system.

(The others)

We confirm the following things by means of the video analyzing.

① In the case of both examinees not being able to use sound, they send mutual data to one another.

② Show action appears well by the image user (an examinee who can use image to transmit information).

③ Watch action appears frequently when user gets information by using image. .

6. CONCLUSION

In this paper, firstly, we mention an example in regard to HE and the role of multimedia in plant operations. Secondly, the problem of communication using a telephone among the plant operators is described. Thirdly, model conferencing is done in nine combinations of media. Finally the result of media conferencing led to the following conclusions.

(1) As for the multimedia communication, the effect depends on the combination of media.
(2) Media should be chosen by the kind of task. Even if it is only just added, it usually can't make the task achievement time decrease.
(3) The main modality is changed by the kind of task requested. Sound should be used in simple task.
(4) The image should be used in case of solving a complicated problem.
(5) The text is used additionally to sound or image.

And we assume that multimedia systems are the most effective candidate to decrease HE in the plant from the result of the experiment. In this paper the use of the multimedia system at the communication levels has been referred to, but next the use of it in the actual plant will be discussed.

Acknowledgement
This study is partly supported by Grand-in-Aid for Scientific Research from the Ministry of Education, Science, Sports, and Culture of Japan under Grand No. 09838020.

REFERENCE

J. Rasmussen: "Information processing and human-machine interaction: An approach to cognitive engineering", North-Holland Series in system Science and Engineering, New York, USA, 1986

Reason: "Human Error", Cambridge, UK: Cambridge University Press, 1990

K. Takada, H. Tamura, Y. Shibuya: "The model of Media Conference", advance in human factors/ ergonomics, vol. 20A, 1995, pp.305-310

R. Zhang, H. Tamura: "Show action in TV Conferencing", advance in human factors/ ergonomics, vol. 21A, 1997, pp.11-14

H. Tamura, R. Zhang: "Remote and Local Works in Media Communication", advance in human factors/ ergonomics, vol. 21B, 1997, pp.419-422

K. Takada, H. Tamura, Y. Shibuya: "Post Agreement Utterance: Communication to Media Conference", advance in human factors/ ergonomics, vol. 21B, 1997, pp.41-44

J.L. Alty, M. Bergan, P. Craufurd, C. Dolphin: "Experiments using Multimedia Interfaces in Process Control: Some Initial Results", Computers and Graphics, vol. 17, No 3, 1992, pp.205 - 218.

MODELLING RECOGNITION-PRIMED DECISION MAKING UNDER TIME PRESSURE AND MULTIPLE SITUATIONAL CONTEXTS

Tetsuo Sawaragi * and **Osamu Katai** **

* *Dept. of Precision Engineering, Graduate School of Engineering, Kyoto University Yoshida Honmachi, Sakyo, Kyoto 606-8501, Japan.*
** *Dept. of Systems Science, Graduate School of Imformatics, Kyoto University Yoshida Honmachi, Sakyo, Kyoto 606-8501, Japan.*

Abstract. In this paper, instead of designing the automation system that is intended to replace the human operator completely with it and to exclude him/her out of the loop, we introduce an idea of *interface agent* as a sophisticated *associate* for a human operator. Since its decision making style must be close to the human's proficient *naturalistic* decision making, we formalize its resource-bounded reasoning under critical time pressure by joining a machine learning method for concept induction and a classical decision theory. *Copyright © 1998 IFAC*

Keywords. Man-machine systems, decision support theory, machine learning, interfaces, human-centered design.

1. INTRODUCTION

Recent popular concept of *human-centered automation design* has stressed the importance of the design philosophy of "people are in charge" or "human-in-the-loop" (Rouse, 1988). For this *idealistic* and very broad concept, however, Sheridan presented ten alternative meanings with a restrictive qualification saying that the real potential is somewhat questionable (Sheridan, 1997). As Sheridan depicted, we would like to stress that a human-centered design principle has to be able to answer to the following critical issues;

- how to make the human-autonomy and the mechanical autonomy (i.e., automation systems) coexist letting them keep a friendly and sharable partnership
- how to avoid the human from the flood of data from the plant and of computer-based advice
- how to allow the human operator's response variability for enhancing and maintaining his/her proficient skill levels and learning

In the following of this paper, we at first introduce an idea of interface agent as an alternative to the conventional stand-alone, isolated automation systems. Then, we will characterize that such an interface agent's reasoning tasks and its decision making styles are quite different from the ones of conventional expert systems and from conventional decision support systems in that they are severely bounded to the *realtime contexts*. This is close to the human's *naturalistic* decision making style called *recognition-primed decision making* (Klein, 1993). We present a methodology for designing such an interface agent and formalize its resource-bounded reasoning under critical time pressure by joining a machine learning method and a classical decision theory.

2. INTERFACE AGENT AS A HUMAN ASSOCIATE

An interface agent is a semi-intelligent computer programs that can learn by continuously "looking over the shoulder" of the user as he/she is performing actions against some complex artifacts and is expected to be capable of providing the users with adaptive aiding as well as of alternating the activities instead of a human (Maes, 1993; Maes, 1994). In this sense, an agent has to coexist with a human user so that it can evolve by itself as a human user's proficient level improves. It also has to be able to stimulate a human user's creativeness coordinately by changing its role dynamically as a human's associate, rather than to replace the human user with itself.

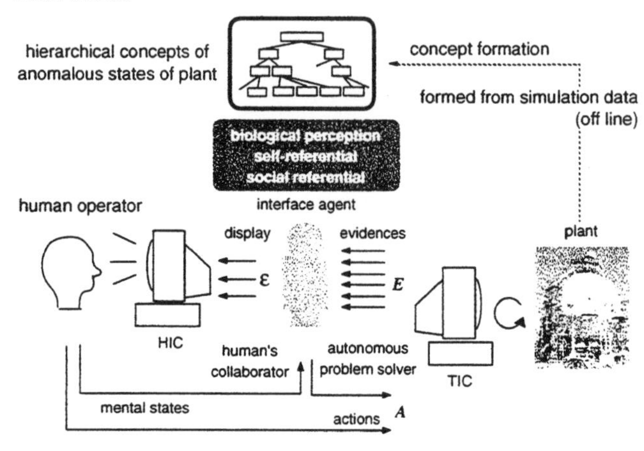

Fig. 1. An interface agent situated within the human-system interactions.

As for the design of the interface agent, we need to introduce some novel principles different from the conventional orientation. Currently there occurs a rapidly developing paradigmatic shift from the conventional *rationalistic* orientation toward an *ecological* view that the proper units of knowledge or sources of intelligence are primarily *concrete, embodied, lived*, concentrating on how to deal with the domain of *immediate present* where the concrete actually lives (Winograd and Flores, 1986; Lave, 1988; Suchman, 1977). A similar paradigmatic shift is also ongoing in the field of decision research (Klein et al., 1993); a shift from a *classical normative decision making paradigm* toward a *naturalistic decision making paradigm*. The latter has concentrated increasingly on the proficient experts' situation assessment ability and their ways of looking at a situation and quickly interpreting it using their highly organized base of relevant knowledge. That is, these are abilities to recognize and appropriately classify a situation. Hereafter, we call this style of decision making as a *recognition-primed decision (RPD)* model after Klein (Klein, 1993).

The distinguishing feature of the RPD model is to attempt to describe how people bring their experience to bear on a decision and how they are able to continually prepared to initiate workable, timely, and cost effective actions. Especially we are interested in modeling their capability to act proficiently under severe time pressure (i.e., under emergency); to identify the situation quickly and accurately and to act promptly with less time and effort to act.

In the following of this paper, we investigate into a recognition-primed decision model of an interface agent, that is embedded inside the human-artifact interactions and has to work as an intelligent associate for a human user/operator in a time-critical situation like at an emergency. This is schematically illustrated in Fig.1. Herein, different from the conventional supervisory agent's task of seeking for optimizing the isolated control task, such an agent has to be able to maintain its identities as an organism living within the multiple contexts looking inwards to consider the the nature of memory and perception, and looking outwards to consider the nature of social action with a human operator.

3. DYNAMIC CATEGORIZATION OF PLANT ANOMALIES

Important aspect of our interface agent is how to organize an "appearances" of the world (i.e., a complex plant), which is different from a mere collection of objective features but is specific to a particular human operator depending upon current time-criticality he/she is forced to work. To be a human-friendly associate for a human operator, an interface agent has to be able to present the current status of the plant so that he/she can help his/her situation awareness and can remind him/her of the recovering operation to be adopted. In time-critical, high-stakes situations, the time required by people to review information, and confusion arising in attempts to process large amounts of data quickly, can lead to costly delays and errors. Numerous psychological studies have provided evidence that human's decision quality may degrade with increases in the quantity and complexity of data being reviewed, and with diminishment in the time available for a response, while the rate of his/her performing tasks can be increased by filtering or suppressing irrelevant information.

Fig.2 shows the situation under which an interface agent is supposed to work. This illustrates a particular type of anomaly "shutdown of ANG-lines" that is generated by a simulator for a gas production plant and is used in our previous publication (Sawaragi et al., 1996a). This shows that the anomaly has occurred and the observed trend data begin to change from the normal states. An interface agent encountering such an anomalous situation has to be able to do the following as well as to assist the human operator to do the following; to *identify* the occurring anomaly type, to *predict* the consequences, and to *determine* and *execute* the appropriate recovery operation. Moreover, these multiple tasks must be finished until the deadline comes, or in more severe cases, they must be done as soon as possible, wherein value of the action effectivity may be degraded as the time elapses.

Under such a time-critical situation, an interface agent must be able to flexibly organize the appropriate appearance of the plant status discriminating among what is *now* relevant and what is not for assisting an operator's *situation awareness*. In our system, a taxonomy of all possible plant anomalies is organized in a hierarchical fashion using a machine learning technique called *concept formation* (Fisher, 1986) as shown in Fig.3 (Sawaragi et al., 1996a). Wherein, the root node represents a class of concepts covering all possible anomaly types, and the leaf nodes represent the individual anomalous cases (i.e., a hierarchy via *is-a* and *subset-of* relations). In terms of this hierarchy, determination of the appearance of the plant concerns with how to determine the appropriate categorization of the plant anomalies out of the hierarchical taxonomy (Fig.3); to find a set of exclusive subtrees whose extensions are regarded as equivalent. The category of the hypothesis on the plausible anomaly types can be defined with a variety of abstraction levels of the hierarchical taxonomy. We call each possible categorization within the taxonomy a *conceptual cover*, that is a categorization of all possible anomalies into mutually exclusive and exhaustive classes.

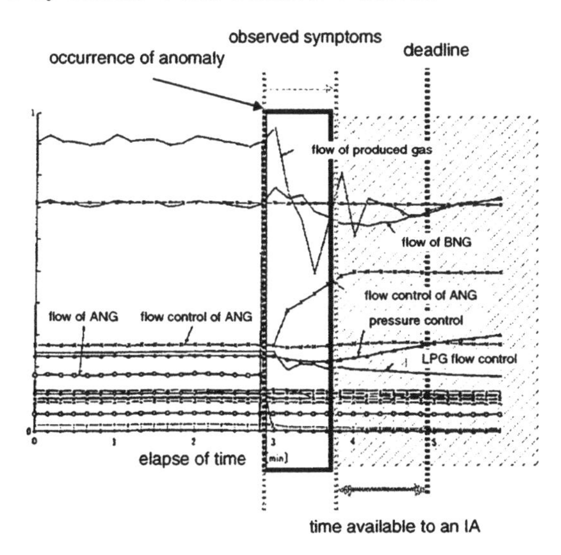

Fig. 2. An example of plant behaviors for an anomaly "shutdown of ANG-lines".

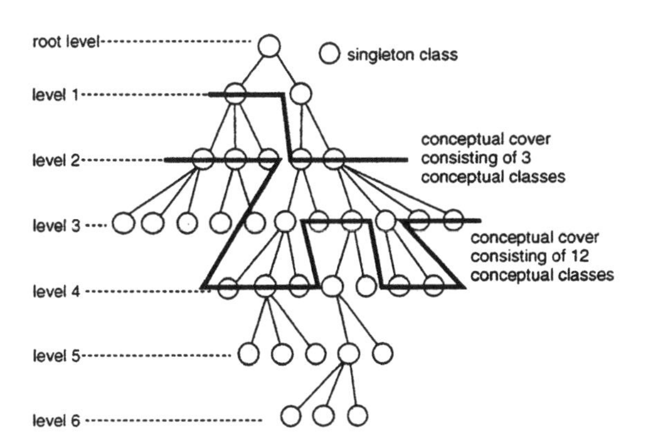

Fig. 3. Conceptual cover of anomalous state concepts: a categorization adopted by an interface agent for its decision framing.

4. RECOGNITION-PRIMED DECISION MAKING MODEL

In order to rationalize the agent's selecting the appropriate categorization to be presented to the human operator and also to be used for its own problem solving (i.e., diagnosis), we at first consider about a general probabilistic reasoning model (i.e., an *influence diagram* (Howard and Matheson, 1983)) as an agent's decision model that derives an appropriate action inferring the most plausible plant anomaly type by getting partially observable symptoms from evidential observations obtained so far. This is illustrated in Fig.4. Wherein, a

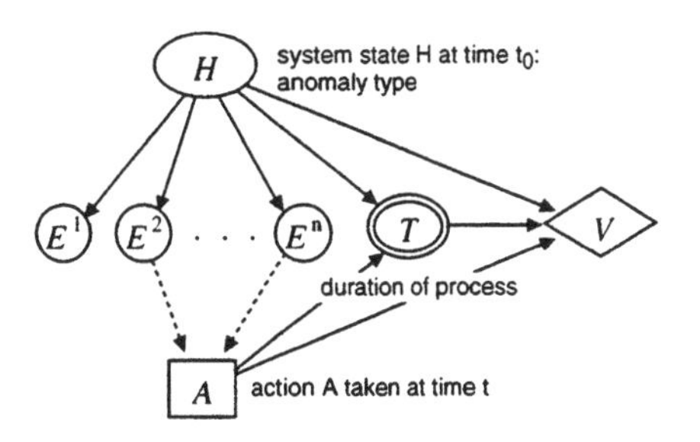

Fig. 4. Interface agent's resource-bounded reasoning represented by an influence diagram.

chance node H denotes a hypothesis node representing a state of the plant taking the value of anomaly types. Nodes E^i's represent a set of symptoms available to an interface agent, each of which corresponds to the features observable in individual instrumentations of the plant. As knowledge of plant anomalies, an agent has acquired dependency among the classes of anomaly types and their associating evidences as well as the prior probabilities of occurrence of those hypothetical classes through the concept formation process, all of which are stored in a taxonomy hierarchy (Sawaragi et al., 1994). Based on this apriori knowledge, getting new observations from the plant, an agent updates its prior belief on probable anomaly types based on a Bayes theorem. Then, these posterior beliefs are used for calculating the expected utilities of the available options defined in the decision node A based on the knowledge that are defined in the diagram defining the interrelationships among nodes of H, A, T and V. The option having the maximum expected utility would be determined as a recommendation for an interface agent to adopt. A diamond node is a special type of an oval node and is called a value node representing an agent's comprehensive utility. This shows that an agent's utility is determined by the types of anomaly and the adopted action as well as by the delay. This reflects the fact that some anomaly types require an immediate recovering operations as soon as possible and if it is delayed its utility would drastically decrease (i.e., a *time-dependent utility* (Horvitz, 1991)).

To illustrate an idea of time-dependent utility, here we assume only two anomalies for simplicity, H_1 and H_2, and two actions, A_1 and A_2. We can define an utility table as sown in Table 1. For instance, $u(A_1, H_1, t_0)$ de-

utility at time $t = t_0$	H_1	H_2
A_1	$u(A_1, H_1, t_0)$ time variant	$u(A_1, H_2, t_0)$ time invariant
A_2	$u(A_2, H_1, t_0)$ time invariant	$u(A_2, H_2, t_0)$ time invariant

Table 1. Utility table at time $t = t_0$.

notes an utility obtained when an action A_1 is adopted immediately after the occurrence of the anomaly H_1. Actually the validity of some actions may vary while the time elapses. Here we assume that the utility of an action A_1 for an anomaly H_1 monotonically decreases according to the elapse of time as illustrated in Fig.5(a) (i.e., $u(A_1, H_1, t) < u(A_1, H_1, t_0)$ for $t_0 < t$), and the utilities of $u(A_1, H_2, t), u(A_2, H_1, t)$ and $u(A_2, H_2, t)$ do not vary, but are constant. Fig.5(b) and (c) illustrate how such a time-dependent utility effects on the recommendation of actions to be adopted. Here a horizontal axis represents a posterior belief on H_1's occurrence when evidences $\boldsymbol{E}$ are observed. The lines denote how the expected utilities for actions A_1 and A_2 vary according to the beliefs of H_1. p^* denotes a belief with which actions A_1 and A_2 are indifferent, and when H_1 is confirmed more strongly than p^*, an action A_1 is preferred to A_2. Therefore, for the evidences $\boldsymbol{E}$ obtained at time t_1 just after t_0, the recommended action is A_1. Fig.5(c) shows the similar relations after some time has passed. At this time $t = t_2$, $u(A_1, H_1, t_2)$ decreases, so the indifferent belief p^* shifts towards the right. Consequently, the recommended action for the same evidence $\boldsymbol{E}$ as the ones at time t_1 turns out to be A_2. In our system, we categorize the utilities defined for all the combinations of hypothesis and actions into one of the three classes of time-dependent utilities as shown in Fig.5(a), each of which corresponds to the utility for the hard-deadline, the utility for the soft-deadline and the time-invariant utility, respectively.

In terms of this decision model, a definition of a domain of a hypothesis node corresponds to a conceptual cover showing an appearance of the plant to an interface agent. This definition reflects the granularities of an agent's recognizing anomaly classifications, so we call this style of decision making as a recognition-primed decision making. The qualities of each possible conceptual cover must be evaluated in terms of the effectivity brought about by the adopted actions (i.e., *expected value of categorization* (Poh et al., 1994)). This can be formulated as follows.

Given a particular conceptual cover z, we denote the recommendation obtained by solving a model constructed using this by A_z^*, and define an expected value of categorization w.r.t. a conceptual cover z as an expected utility calculated for the recommended option A_z^* and denote this by $EVC(z)$. Denoting a set of symptoms as a vector $\boldsymbol{E}$ and neglecting the changes of utilities according the elapse of time, A_z^* and $EVC(z)$ can be represented as follows;

$$A_z^* = \arg\max_A \sum_j u(A_i, H_j^z) p(H_j^z \mid \boldsymbol{E}, \xi) \qquad (1)$$

and

$$EVC(z) = \sum_j u(A_z^*, H_j^z) p(H_j^z \mid \boldsymbol{E}, \xi), \qquad (2)$$

where H_j^z represents a set of hypothesis classes of anomaly

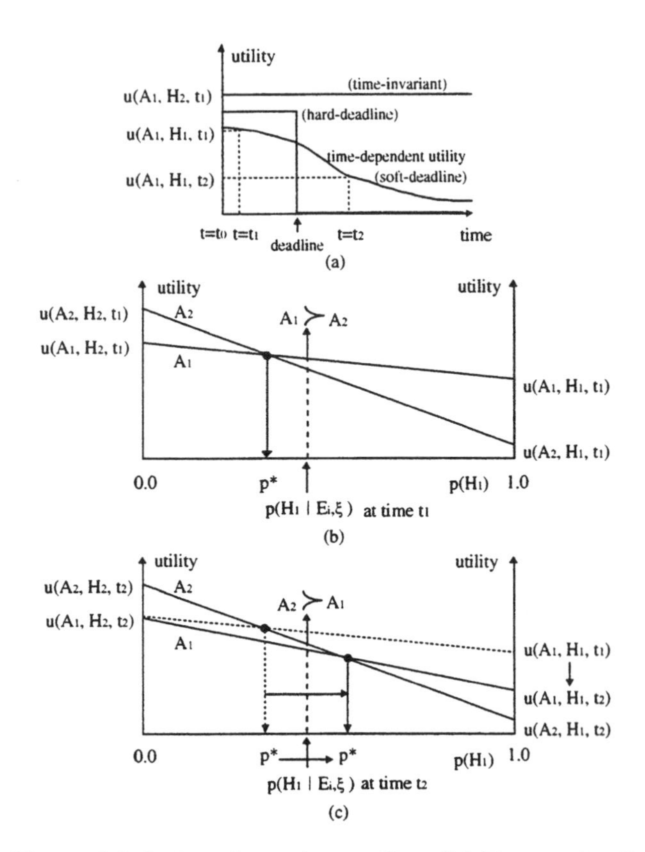

Fig. 5. (a) A time-dependent utility. (b) Expected utilities for two actions at time $t = t_1$. (c) Expected utilities for two actions at time $t = t_2$ $(> t_1 > t_0)$.

types making up the conceptual cover z, and $p(H_j^z \mid \boldsymbol{E}, \xi)$ is the probability over them, given observations $\boldsymbol{E}$ and background state of information ξ. Note that for each conceptual covers z, a corresponding decision model may be constructed by retrieving the quantified conditional dependencies that are learned in the concept formation process and by joining them into a model.

5. AGENT'S MANAGING COMPLEXITY UNDER TIME-CRITICALITY

As an intelligent associate, an agent aims to support an operator to focus his/her attention into the actual candidates of the anomalies rather than enforcing him/her execute a recommended action. That is, the agent presents the appropriate conceptual cover highlighting a set of possible anomalies while hiding its competitive other classes of anomalies behind (i.e., distinction between "figures" and "grounds" in psychological terms). At the same time, this conceptual cover makes up the agent's decision models (i.e., a hypothesis node H) that calculate the beliefs on the occurrence of possible anomalies based on the currently available evidences.

Here, we set the following criteria concerning with the determination of the appropriate conceptual cover:

(1) An agent should take into account of the expected delay of the operator's action execution in response to the presented information.
(2) An agent should aggregate anomalies into more abstract classes if the expected utilities of all possible actions for those anomalies are negligible.
(3) An agent should decompose a class of anomalies into more precise classifications if their probabilities of occurrence are high and the recommended action to be adopted varies depending on the anomalies within that class.
(4) An agent should avoid the risk of hiding the correct anomaly into the "grounds" and of highlighting the wrong one in the "figures".

The selection of the appropriate conceptual cover is important not only to present an affordance-rich information to a human operator but also to consider about an interface agent's own problem solving activity of constructing its decision making model and solving that model. At this time, an agent has to deal with tractability of decision making inference at the expense of decision quality. That is, an agent has to determine the appropriately manageable model by finding the definition of the hypothesis node as an agent's categorization of the environment (i.e., a plant status). Wherein, a resource-constrained agent has to be able to enhance its decisions by expending some resource to consider the tradeoff between the expected benefit of using more detailed models to increase the expected utility of action and the resource costs entailed by computing decisions with a more detailed model. Note that this activity of finding a better conceptual cover for model modification must be done by the interface agent and this is also a time-consuming activity. Therefore, an agent has to take account of the side-effects produced due to the change of conceptual cover as well as the cost performing the model refinement operation.

6. FORMULATING AN AGENT'S RESOURCE-BOUNDED REASONING UNDER MULTIPLE CONTEXTS

Based on the above considerations, we formulate a procedure for determining the appropriate conceptual cover by explicitly introducing the time cost incurred to the delay of the operator's executing actions and to the agent's own decision making. Let us define a delay cost $C_d(z)$ that is incurred to the time needed for the operator to respond to the presented conceptual cover z [1]. We further introduce a computational time cost $C_c(z)$ that is incurred to the agent to solve its decision model (i.e., calculation of posterior beliefs by probabilistic reasoning) that is constructed using the conceptual cover z. Then, we extend the definition of $EVC(z)$ towards a *net expected value of categorization* w.r.t. a conceptual cover z and denote this by $NEVC(z)$, which is defined using the time-dependent utility as follows [2];

$$NEVC(z) = \sum_j u(A_z^*, H_j^z, C_T(z)) p(H_j^z \mid \boldsymbol{E}, \xi), \quad (4)$$

where $C_T(z)$ denotes the total cost of $C_d(z)$ and $C_c(z)$.

In principle, an agent should consider all possible conceptual covers and pick up the one with the maximum net expected utility of categorization. In practice, however, such an approach would be highly expensive in terms of computational cost. We take an approach to this problem by iteratively changing the level of abstraction within the conceptual cover used for model construction starting with a particular conceptual cover z_0. Then, to this initial conceptual cover, the following two operations are considered to either increase or decrease the level of specificity of the concepts concerned as illustrated in Fig.6. Wherein, a *specialization* operator is to modify a concept in the conceptual cover by replacing it with the set of its most general subclasses located in the schema hierarchy, while a *generalization* operator is vise versa.

Given a conceptual cover z_0, we will denote a model modification operation by s_i and the resulting conceptual cover by $z_i = s_i(z_0)$.

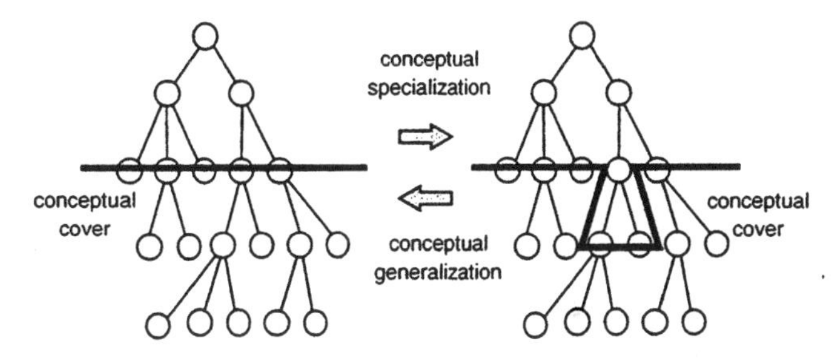

Fig. 6. Conceptual specialization and generalization.

Here, we can define a expected value of model modification by operation s_i on a conceptual cover z_0, $EVMI(s_i)$ as the improvement of utility due to the modification;

$$EVMI(s_i) = EVC(s_i(z_0)) - EVC(z_0) \quad (5)$$

With the account of the change of delay cost and computational cost due to the change of conceptual cover, a *net expected value of model improvement*, $NEVMI$, can be defined analogously to the extension of EVC towards $NEVC$ as follows;

$$NEVMI(s_i) = EVMI(s_i) - \Delta C_T(s_i) - C_g(s_i), \quad (6)$$

where $\Delta C_T(s_i) = C_T(z_i) - C_T(z_0)$ represents the change in the costs of delay and computation due to the change

[1] This cost may vary according to the proficient levels of the operator. The above modeling can be simply extended to include such a variation.

[2] For the simple implementation, it is usually assumed that the time dependent utility function u is of the additive form and EVC is in units of cost, i.e.,

$$NEVC(z) = EVC(z) - (C_d(z) + C_c(z)) \quad (3)$$

of conceptual cover to be adopted, and $C_g(s_i)$ is the cost performing the model modification operation. The determination of the optimal sequence of the operations can be approximated by taking a greedy, single-step model improvement operation with the greatest $NEVMI(s_i)$ until all the operations have $NEVMI(s_i) \leq 0$. In this way, an interface agent can identify the best conceptual cover.

The above formulation is to reflect the agent's deliberation tasks under multiple contexts. In time-critical, high-stakes situations, the time required by people to review information, and confusion arising in attempts to process large amounts of data quickly, can lead to costly delays and errors. Therefore, an agent should take into account of those issues, and this is essential to establishing a human-centered design. In addition to the considerations on the collaborative partner, an agent has to be also able to reason about its own multiple activities needed to perform that basic task. In other words, the situated interface agent has to be able to control itself and to reason about the self being aware of the multiple contexts or backgrounds surrounding an agent's current task at hand.

7. CONCLUSIONS

In this paper, we presented a formulation of an interface agent's resource-bounded reasoning for displaying information to the human operator under severe time pressure. The design principle for such an intelligent artifact that is supposed to collaborate with a human must be fundamentally different from the conventional supervisory control systems. In the latter, the supervisory strategy has been dealt with as an optimization problem in some way; how to allocate the tasks to the human and the automation systems. Wherein, objects to be optimized are isolated from the optimizing agent, and it has been assumed the infinite computational resources are available to the agent. But this is not true in the design of human-machine symbiotic systems under time pressure. The agent should be embedded within the interactions with both internal and external processes, and must seek for bounded-optimality with a capability of referring to its own activities (i.e., a *self-referencing* capability).

REFERENCES

Fisher, D. (1987). Knowledge Acquisition via Incremental Conceptual Clustering, *Machine Learning*, **2**, pp.139-172.

Horvitz, E. and Barry, M. (1995). Display of Information for Time-Critical Decision Making, *Proc. of the Eleventh Conference on Uncertainty in Artificial Intelligence*, Montreal.

Gibson, J.J. (1979). *The Ecological Approach to Visual Perception*, Houghton Mifflin Company, Boston, MA.

Horvits, E. (1991). Time-Dependent Utility and Action under Uncertainty, *Proc. of the Seventh Conference on Uncertainty in Artificial Intelligence*, Los Angels, pp.151-158.

Howard, R.A. and Matheson, J.E. (1983). Influence Diagrams, in Howard, R.A. and Matheson, J.E. (Eds.), *The Principles and Applications of Decision Analysis*, Strategic Decision Group, Menlo Park, CA.

Klein, G.A. et al. (1993a). *Decision Making in Action: Models and Methods*, Ablex Pub. Corp., Norwood, NJ.

Klein, G.A. (1993b). A Recognition-Primed Decision Model of Rapid Decision Making, in Klein, G.A. et al. (eds.), *Decision Making in Action: Models and Methods*, pp.138-147, Ablex Pub. Corp., Norwood, NJ.

Lave, J. (1988). *Cognition in Practice: Mind, Mathematics and Culture in Everyday Life*, Cambridge University Press, N.Y..

Maes, P. and Kozierok, R. (1993). Learning interface agents, *Proceedings of the Eleventh National Conference on Artificial Intelligence*, pp.459-465.

Maes, P. (1994). Agents that Reduce Work and Information Overload, *Communications of the ACM*, 37-7, pp.30-40.

Poh, K.L., Fehling, M.R. and Horvitz, E.J. (1994). Dynamic Construction and Refinement of Utility-Based Categorization Model, *IEEE Trans. of System, Man, and Cybernetics*, Vol.**24**-11, pp.1653-1663.

Rouse, W.B. (1988). The Human Role in Advanced Manufacturing Systems, in Compton, D. (Ed.), *Design and Analysis of Integrated Manufacturing Systems*, National Academy Press, Washington, D.C..

Sawaragi, T., Iwai, S., Katai, O. and Fehling, M.R. (1994). Dynamic Decision-Model Construction by Conceptual Clustering, *Proc. of the Second World Congress on Expert Systems*, pp.376-384, Lisbon, Portugal.

Sawaragi, T., Takada, Y., Katai, O. and Iwai, S. (1996). Realtime Decision Support System for Plant Operators Using Concept Formation Method, *Preprints of International Federation of Automatic Control (IFAC) 13th World Congress*, Vol.L, pp.373-378, San Francisco.

Sawaragi, T., Tani, N. and Katai, O. (1996). Evolutional Concept Formation for Personal Learning Apprentice by Observing Human-Machine Interactions, *Proc. of CSEPC (Cognitive System Engineering for Process Control)*, pp.122-129.

Sheridan, T.B. (1997). Human-Centered Automation: Oxymoron or Common Sense?, *Proc, of IEEE Int. Conf. on System, Man and Cybernetics*, Vancouver, Canada.

Suchman, L.A. (1977). *Plans and Situated Actions: The Problem of Human-Machine Communication*, Cambridge University Press, N.Y..

Winograd, T. and Flores, F. (1986). *Understanding Computers and Cognition: A New Foundation for Design*, Ablex Pub. Corp., Norwood, N.J..

SA*TEST*: THE 1997 ROTTERDAM EXPERIMENT ON VTS OPERATOR PERFORMANCE

Erik Wiersma [1], Nathasja Mastenbroek [2] and Johan H. Wulder [2]

[1] *Delft University of Technology, Safety Science Group, Kanaalweg 2 B, NL-2628 EB Delft, the Netherlands*
[2] *MarineSafety International Rotterdam, P.O. Box 51290, NL-3007 GG Rotterdam, the Netherlands*

Abstract: Traffic management and control systems are deployed in all modes of tranportation to ensure a safe and efficient traffic flow. In maritime transportation, traffic control is performed by Vessel Traffic Services (VTS). In search for optimal support for VTS operators, tools are developed constantly. A performance based measurement of situation awareness of VTS operators is used to evaluate changes in the support system. This paper describes a laboratory experiment in a full scale interactive VTS simulator which aims at establishing a base-line measurement of situation awareness for VTS operator performance. A scoring system for VTS operator performance accounts for the specific characteristics of VTS operator work, such as accuracy and relevance of information. *Copyright © 1998 IFAC*

KEYWORDS: Vessel Traffic Management, Situation Awareness, Laboratory Experiment

1. INTRODUCTION

During recent years the quantity and complexity of traffic has grown in every mode of transportation. This has led to a large increase in the work load of all Traffic Operators. For maritime transportation, Vessel Traffic Service Operators (VTS operators) are responsible for safe and efficient handling of traffic in and out harbours, rivers and approach areas. They monitor the vessel traffic, provide information, advise ships and coordinate between ships if necessary. The maritime VTS operators are well trained nautical experts, with an extensive nautical experience. The VTS operator training is nationally harmonised, formed according to the IMO VTS Guidelines (1985). The Dutch legislation concerning the VTS operators (Scheepvaartverkeerswet, artikel 9 lid 2) obliges VTS operators to take a so-called "refreshment" simulator training once every 3 years.

Due to ever increasing quantity and complexity of vessel traffic, efforts have been made to decrease the VTS operator work load by a number of newly developed electronic tools. However, not all tasks of a VTS operator can be supported electronically or replaced. In order to improve the resulting man/machine interface, the European Commission initiated a research project.

In this project several European institutes are involved, amongst which MarineSafety International Rotterdam and the Delft University of Technology. The project is primarily aimed at providing a theoretical framework which can serve as an evaluation tool for modifications in the VTS (support) system. The evaluation results can also provide ideas for the design of improved VTS support systems. The experiment described in this paper is a part of this project.

2. MEASURING VTS OPERATOR PERFORMANCE

A new methodology for measuring VTS operator performance needed to be developed, because at the start of the project there was no methodology available to evaluate changes in a VTS support system. New tools were tested by using them in practise. In order to evaluate the (cost) effectiveness of these changes prior to their implementation, Wiersma and Heijer (1996) have proposed a test bank which takes VTS operator performance as a base-line for measurements. Situation awareness has been suggested as one of the measures of VTS operator performance (Wiersma and Heijer, 1996). A method SA*TEST* for measuring VTS operator situation awareness has been developed and tested (Wiersma *et al.*, 1997).

Situation awareness can be measured in many ways. Several taxonomies of situation awareness methods have been proposed, see for instance (Endsley 1995a) and (Pew, 1996). Pew (1996) breaks down the situation awareness methods in four categories:

- Direct system performance methods
- Direct experimental methods
- Verbal Protocol analysis
- Subjective measures

SA*TEST* is a direct experimental method, based upon the Situation Awareness Global Assessment Technique (SAGAT) developed in aviation research (Endsley, 1995b). Scenarios of maritime traffic are run on a simulator. At specified moments, the simulator is stopped and questions pertaining to the traffic situation are asked. The objective of SA*TEST* is to get a complete recollection of the traffic situation, which can consist of up to thirty ships, with a certain number of interactions and potential conflicts. Individual scores are evaluated on the basis of the actual traffic situation, taking into account the specific characteristics of VTS operator work, such as accuracy and relevance of information (Wiersma and Mastenbroek, 1997)

3. EXPERIMENT SET-UP

SA*TEST* has been used in several experiments. The first of these experiments took place in Rotterdam. The objective of this experiment was to get a base-line measurement of VTS operator situation awareness in a traffic situation that was of moderate difficulty, with a certain number of conflicts and complex traffic situations. The traffic situation in this scenario is regarded to reflect actual VTS work conditions in the Rotterdam area well. Hypotheses were proposed on differences between more or less experienced operators regarding their ability to reproduce information, taking into account relevance of information and anticipation of conflicts.

In the experiment 21 VTS operators with different levels of experience participated. The operators were divided into three groups: non-experienced VTS operators (0-3 years experience), experienced VTS operators (more than 3 years) and VTS operators who are also qualified as VTS examinors (average 20 years of experience).

For this experiment a simulation run from the VTS operator refreshment training courses was used. Because these refreshment runs are regularly evaluated by experts, both relevance and complexity are ensured. The run chosen is placed in an artificial area, with a representative number of ships as well as the number and complexity of (potential) conflicts. The conflicts that develop have to be solved by the operator in order to avoid collisions. The conflicts differ in complexity and number of ships involved. The scenario run takes about 45 minutes.

During the experiment the simulation was "frozen" four times. Three times the simulation was stopped prior to a potential conflict situation, several minutes before the point of no return, at which a collision is unavoidable. Figure 1 shows one of these situations. Although for an inexperienced observer the circled ships may be far apart, an experienced VTS operator will recognise that these ships are in a potential conflicting situation, some ten minutes ahead. Operators usually do not wait for conflicts to exist long and try to solve them before they get critical.

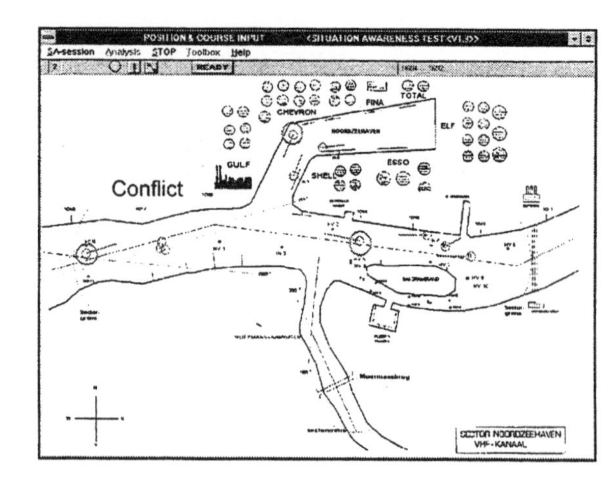

Figure 1. Potential conflict situation 8 minutes ahead of point of no return

The simulation was stopped in this pre-conflict situation. For different operators the simulation was stopped at different times (but all severla minutes before the point of no return). Thus an understanding could be established at what particular moments VTS operators started considering a situation a potential conflict. The fourth stopping moment was placed regardless of conflicts, was the same for all operators and functioned as a base-line.

4. DATA ANALYSIS

The data of the experiments have been analysed with two methods:

- Qualitative analysis by ranking the filled in traffic displays;
- Quantitative analysis of completeness and accuracy of ship positioning in traffic displays.

4.1 Ranking of traffic displays

The analysis was performed by a qualified examiner of VTS trainees, and trainer in VTS refreshment courses (from which the scenario was taken). This was to ensure that the analysis was performed by someone very familiar with the scenario and the traffic development therein.

Ranking of the traffic displays was performed by comparing the filled in score forms with each other and with the correct answers (taken directly from the scenarios). During the ranking the person performing the ranking was asked to think loud about the considerations which led to the results obtained. Thus an insight could be obtained of the criteria that played a role in the ranking process.

The ranking was repeated after two months to validate the consistency of the scoring.

4.2 Completeness and accuracy of traffic displays

The most simple quantitative analysis performed was comparing the number of vessels and their location and course vector, indicated in the traffic displays, with the actual number of vessels and their actual location and course vector.

The remaining data filled in by the subjects, regarding the information about the ship, was only used for identification of the ships. This information was not separately scored for completeness and accuracy.

Although this analysis is fairly straightforward when performed by hand, automation of this analysis is not as simple. Automatic identification of the vessels is complicated by inaccurate locating vessels, and by the fact that operators were asked to give their personal label of the ship, which does not necessarily correspond with the full name.

More complex and elaborate methods of quantitative data analysis are currently being developed. These include combining ships in larger groups, or "chunking", and using path analysis to evaluate deviations in longitudinal and lateral position.

5. RESULTS

As expected, differences can be found between performance of VTS operators with different levels of experience. As experience grows, operators anticipate conflicts earlier, and solve them as soon as they encounter them. This result is more pronounced in the qualitative analysis than in the quatitative analysis.

The processing of operators' answers and the allocation of scores is complicated by the fact that the VTS operators do not consider situations as conflicts anymore, if they have acted upon them already (although ships may still be on a conflicting course). At first it may look as if they have missed a potential conflict situation, but closer analysis shows a difference in interpretation by different operators of the same perceived situation.

Operators differ in what they consider relevant information for a given situation. The relevance attributed by operators does not always correspond with the relevance attributed by the designers of the scenario.

Lack of situation awareness of particular information shows up in the results as missing (forgetting) elements in the situation or inaccurate recollection of information. This is especially the casewith information which, according to the operator is less relevant for his task as traffic controller, like ships leaving the traffic area, which have passed the conflict area. With a simple quantitative scoring system, without regard of the relevance for a (potential) conflict of a specific ship, a situation may receive a lower score than justified.

From the verbal comments during the qualitative analysis, it can be concluded that reproduction of a complete situation is not considered the most important aspect when evaluating traffic scenes. This can be illustrated with the figures in table 1.

This table shows the (qualitative) ranking of the scenarios, for a number of stops, with the actual number of ships reproduced in the different trials. As can be concluded from table 1, the mere recollection of ships alone, does not gain one a higher ranking.

Table 1. Ranking scores compared to number of ships scored

Stop time	Rank	No. of ships	Stop time	Rank	No. of ships	Stop time	Rank	No. of ships
5	1	10	14	1	12	35	1	17
5	2	10	14	2	11	35	2	17
5	3	9	14	3	12	35	3	18
5	4	13	14	4	11	35	4	14
5	5	8	14	5	12	35	5	18
8	1	11	17	1	14	37	1	18
8	2	10	17	2	13	37	2	16
8	3	11	17	3	14	37	3	15
8	4	10	17	4	12	37	4	15
8	5	8	17	5	12	37	5	17
10	1	8	20	1	13	40	1	18
10	2	11	20	2	14	40	2	15
10	3	10	20	3	13	40	3	17
10	4	11	20	4	13	40	4	18
10	5	9	20	5	12	40	5	11

From the verbal comments during scoring, the most important aspects in the qualitative evaluation of the traffic scenes can be elicitated. They have been ranked according to importance for evaluation of a traffic scene:

1. **Recognition of conflicts** (indication of all ships invloved, speed and course indicated);
2. **Accuracy** with which ships in the conflict are reproduced;
3. **Overview of the complete traffic area.** Of ships that are not involved in a conflict situation, an exact location was not considered necessary, for these ships a global indication was sufficient;
4. **Accuracy** with which ships, not invloved in conflicts, are placed;
5. **Indication of focus of attention.**

A clear difference can be noticed between VTS operators in their strategy of remembering data and positioning the ships. While some operators remember the position of the ships as such, others reconstruct the position of a ship based on information from communication or the last position of the ship that was observed.

This last result opens new fields of understanding VTS operators' strategies in performing their work and raises questions for further research. New experiments will focus upon these issues.

The qualitative analysis (ranking) was repeated after two months, with a correlation of .7934 ($p<.001$).

6. QUALITATIVE VERSUS QUANTITATIVE ANALYSIS

When quantitative analysis is limited to straightforward analysis, such as the mere completeness of traffic recollection without interpretation of the traffic scene, there is not much correlation between the qualitative scoring and the quantitative scoring. This is caused by the fact that an expert interpretation of a traffic scene is much more than an account of the location of individual vessels. The verbal reporting of the scoring process made clear, that examiner was focusing on at least two things that can not be captured in a straightforward analysis.

6.1 Chunking ships together

The traffic scene is observed in "chunks" larger than single vessels. This corresponds with findings from the field of chess, like the famous research by de Groot (1965). Chess masters are much better in recalling a given situation on a chess board than novice chess players. The experienced players do not recall single pieces, but remember the position of pieces in relation to other pieces (Chase and Simon, 1973). The relation between the work in the field of chess and situation awareness has been noted before, for instance by Durso *et al.*(1996).

The same combining of individual elements to larger chunks can be observed when VTS operators reproduce a traffic scene: Ships which are interacting are placed in exact position compared to each other, while the total "chunk" of ships may be located somewhat off. Expert judgement of the responses accounts for this effect.

For instance figure 2 depicts a scene from SA*TEST*. The three ships (circled) in the left part of the picure are anchored for work throughout the scenario. Although many VTS operators place the ships more towards sea or towards the crossing, they are always remembered together.

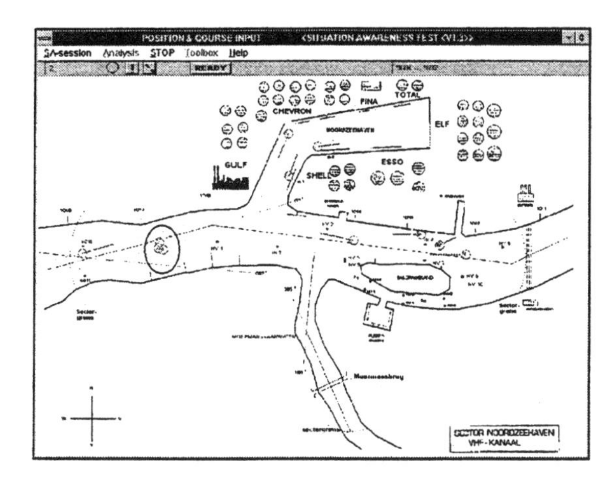

Figure 2. Ships are recalled in "chunks"

6.2 Location of ships along path

Often operators locate ships not exactly at the point where they should be placed. The inaccuracy however, is not arbitrary. Inaccuracy at right angles of the path the ship is supposed to take, is much smaller than inaccuracy along this path. This follows from the fact that, in the mind of the VTS operator, the ship is not a mere point at a certain place in time. A collection of information items is stored with the ship, including name, trajectory, communication et ceteras. Although a VTS operator may not remember the exact location of the ship, he does have in mind its path, and places it somewhere along this line. Therefore, a quantitative analysis of inaccuracy must include path information. Currently we are working on this analysis.

7. CONCLUSION

The evaluation of VTS operator performance is largely a *terra incognita.* The experiment described in this paper is a first attempt to study the matter thoroughly. The experiment seems to be successful. As expected, not all questions can be answered in one such experiment and more work is needed. Situation awareness has proven to be a useful concept for this type of research and the method developed seems to be solid enough to use.

Qualitative data analysis gives a good understanding of the process and fairly consistent results. This understanding of relevant aspects from this analysis will be used in oncoming projects.

Quantitavie analysis should be more elaborate than used in this experiment. Simple quantitative analysis like the ones carried out in this experiment do not provide a solid base to evaluate acomplex situation like a traffic scene with multiple interactions and complex potential conflicts. More elaborate methods of analysis will be needed and are being developed currently.

8. ACKNOWLEDGEMENT

This research is part of the COMFORTABLE-project which is performed within the 4th Framework Programme of the European Commission, DG VII: Transports. The experiments could only be carried out with the voluntary participation of the VTS operators of the Rotterdam Port Authorities.

REFERENCES

Chase, W.G. and Simon, H.A. The mind's eye in chess. In: Chase, W.G. Ed. (1973) *Visual Information Processing.* Academic Press, New York, pp. 215-378.

Durso, F.T., Truitt, T.R., Hackworth, C.A., Crutchfield, J.M., Nikolic, D., Moertl, P.M., Ohrt, D. and Mannaing, C.A. (1996) Expertise and chess: A pilot study comparing situation awareness Methodologies. In: Garland, D.J. and Endsley, M.R eds. Experimental analysis and measurement of situation awareness. Embry-Riddle Aeronautical University Press, Daytona Beach Fl. USA., pp. 295-304.

Endsley, M.R. (1995a) Measurement of situation awareness in dynamic systems. *Human Factors* **37 (1)** 65-84.

Endsley, M.R. (1995b) Measurement of situation awareness in dynamic systems. Human factors, 37(1), pp. 65-84.

Groot, A.D. de (1965) *Thought and choice in chess.* Second ed. 1978 Mouton, the Hague.

Pew, R.W. (1996) The state of situation awareness measurement: Circa 1995. In: Garland, D.J. and Endsley, M.R eds. *Experimental analysis and measurement of situation awareness.* Embry-Riddle Aeronautical University Press, Daytona Beach Fl. USA., pp. 7-16.

Wiersma, E. and Heijer, T. (1996) *Human factors models and methodologies for vessel traffic service operator performance.* Report to: E.C./D.G. 7, Delft University of Technology, Delft.

Wiersma, E., Heijer, T. and Hooijer, J. (1997) SATEST, a method for measuring situation awareness in Vessel Traffic Service operators.

In: Soares, C.G. ed., Advances in safety and reliability. Vol. 2. pp. 901-906.

Wiersma, E. and Mastenbroek, N. (1997) Measurement of VTS operator performance. In: Brandt, D. ed., Automated Systems *based on human skill: Pre-prints*. pp. 59-62.

RISK RECOGNITION AND COLLISION AVOIDANCE BY VTS OPERATORS

Michael Baldauf [1] and Erik Wiersma [2]

[1] *Wismar University, Department of Maritime Studies Warnemünde, Richard -Wagner - Str. 31, D-18119, Rostock, Germany*

[2] *Safety Science Group, Delft University of Technology, Kanaalweg 2 B, NL-2628 EB Delft, the Netherlands*

Abstract: To ensure the safety and efficiency of vessel traffic VTS operators collect and analyse traffic and environment data in a shore based VTS centre. If a traffic situation deviates from safe state they have to send messages to certain vessels in order to reach back this state. Although clear criteria and limit values for collision avoidance are still missing VTS operators work safe and reliable. Within this paper some results gained during experimental trails and regarding the operators' intuitive situation assessment and the limit values they use to interact on situations with danger of collision will be presented. *Copyright © 1998 IFAC*

KEYWORDS: Vessel Traffic Management, Collision Avoidance, Situation Assessment, Field Experiment, Laboratory Experiment

INTRODUCTION

The term Vessel Traffic Service (VTS) describes several shore based services for seagoing vessels in a monitored area. According to the Guidelines of IMO and IALA a VTS is defined as "*a service implemented by a competent authority, designed to improve the safety and efficiency of vessel traffic and to protect the environment. The service should have the capability to interact with the traffic and to respond to traffic situations developing in the VTS area*". Safety of traffic can be seen in general terms as that state of traffic where vessels reach their destinations without damage. Serious damage may be caused by collisions or grounding. Thus one important task of a VTS is to contribute to collision avoidance in the area monitored. For these reasons operators in a VTS must detect dangerous encounter situations reliably and in time. After recognising the risk, they have to interact by giving information or assistance to the vessels involved in the dangerous situation. For this purpose operators may choose between providing information, warning, advice or instruction. The actual message type depends on the service the VTS provides to the vessels, which is different in different areas but moreover it depends on the risk level of a concrete situation. Any interaction of a VTS operator should clearly express the level of obligingness to the vessel concerned. But clear criteria and uniform limit values at which time or distant and by means of a which kind of message type a VTS operator should interact on a dangerous encounter situation to contribute to collision avoidance are still missing.

In order to assess the safety of traffic the operators collect and analyse relevant data. For this purpose they have several technical sensors at their disposal such as radar, visibility indicators, tide graphs, ship data bases et cetera. The main interface operators use for shore based situation assessment is a radar screen, to which automatic collision warning systems are added. Beside these warning devices the operators use for situation assessment their experience and their own intuition mainly.

RISK MODEL

Investigations were carried out to compare the mental risk assessment on board the vessels and by VTS operators in shore-based VTS (Baldauf 1997a). First a risk model based on the International Rules for Preventing Collisions at Sea (COLREGS), was created and parameters and criteria were defined by Hilgert and Baldauf (1996). In this model a distinction is made between four levels of risk (see table 1).

Table 1 Model for assessment of collision risk in any encounter situation.

Risk level	Criteria
Level 1 risk of collision is developing	$CPA < C_A$ and $RNG > R_A$
Level 2 risk of collision exists	$CPA < C_A$ and $R_M < RNG \leq R_A$
Level 3 danger of collision is developing	$CPA < C_A$ and $R_C \leq RNG \leq R_M$
Level 4 danger of collision exists	$CPA < C_H$ and $RNG < R_A$

(abbreviations explained at the end of the paper)

Limit values were derived by means of a questionnaire and compared with those possible from a theoretical point of view. The model was found useful for description and for application to on-board risk assessment. The model takes into account the COLREGS, the physical background (e.g. hydrodynamic interactions between passing ships, the ship's manoeuvrability) as well as the behaviour and the habits of navigating officers.

APPLICATION OF THE RISK MODEL FOR VTS

Investigations were performed to analyse the impact of the potential application of the risk model for triggering enhanced collision warnings by the VTS system (Baldauf 1997b).

The warnings for shore-based risk recognition currently implemented in Germany are based on conventional algorithms and thresholds, which use very large safety margins (CPA = 900 m and TCPA = 10 min). The warnings do not correspond with the risk assessment of the VTS operators nor with the risk assessment on board the vessels. For example, if a VTS operator would act on basis of a triggered collision warning, the shore-based service would not be accepted at the vessels. Furthermore in areas with high traffic density the number of warnings increases to such an extend that it becomes practically unusable.

Traffic analysis and investigation of occurred warnings, that were performed in the selected area of the German Bight, confirm the statements mentioned above.

For the improvement of observation results a questionnaire was designed. Its special focus was the use of the automatic collision warnings for initiating a VTS interaction in concrete traffic situations. The idea was to get more reliable insight into what VTS operators do, if a conventional collision warning is triggered by the system automatically, and how their assessment of a situation compares to the automatic warnings. One example of the results obtained is given in figure 1.

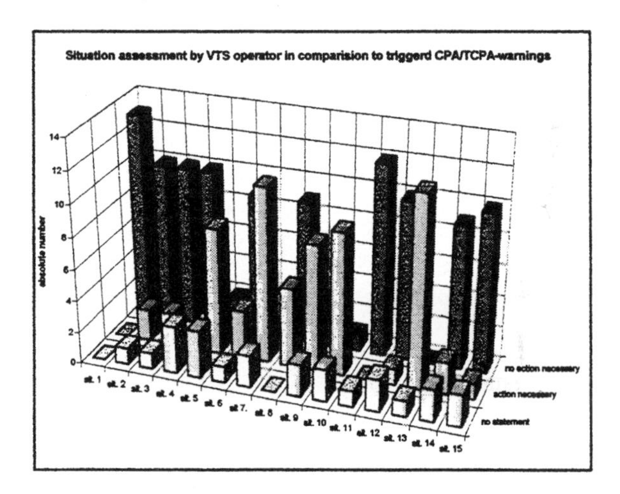

Fig. 1: Results of VTS operator questionnaire according to the usefulness of the automatic collision warnings presently implemented in the system

From figure 1 it can be concluded that in most cases the assessment of the VTS operators was different to the risk recognition by the triggered warnings. A more detailed classification of situations presents the results in a better way and leads to a surprising but nevertheless logical result. The situations to be assessed by the operators were divided into overtaking, meeting and crossing situations as defined formerly for the risk model and consequently according to the COLREGS. This is the same way of situation assessment as navigating officers on board do. This confirmed the assumption that an application of the risk model for triggering collision warnings could be more helpful than conventional warnings.

EXPERIMENTAL ANALYSIS OF RISK MODEL

Experimental trials were performed using PC-simulation runs of real traffic scenarios in the German Bight area, to compare the use of conventional warnings and the impact of new and enhanced collision warnings. First the developed risk model was applied for different real traffic scenarios and a significant reduction of the number of warnings was achieved in comparison to the original state of the art. A reduction of 47% was reached using risk level 2 with TCPA-limit = 8 min for triggering a warning automatically. That means a significant reduction of the operator's workload seems to be possible.

EXPERIMENT INVOLVING VTS OPERATORS

In order to complete the work an experiment was planned to test the risk model with VTS operators. This experiment was partly motivated by Kolrep (1996). He performed experiments to investigate the mental processes of pilots in air traffic control and to model such processes according to Anderson (1993).

The aim of the VTS-operator experiment performed here was to investigate the potential impact of the new warnings on the operator's daily work under different conditions. For that reason two different conditions were tested: "no alarms available" and "enhanced alarms available" respectively. The risk model was also tested using VTS operator performance as a measure, as suggested by Wiersma and Heijer (1996). In the experiment SA*TEST* was used, a method for measuring VTS operator performance, based on measurement of situation awareness, developed by Delft University of Technology and MarineSafety International Rotterdam (Wiersma *et al.*, 1997).

In SA*TEST* VTS operators are placed in a traffic simulator. At specified moments the simulation is stopped and operators are requested to reproduce the traffic situation. VTS operator scores are derived taking into account accuracy and relevance of information reproduced (Wiersma *et al.*, 1997). In addition to this, operators are also requested to indicate the future traffic situation ten minutes ahead.

The scenario used for operator experiments was a real traffic scenario of high density, slightly changed and an encounter situation was added additionally which was similar to a real situation that had led to an accident in the area. (One of the reasons of the real accident was probably that the situation was overlooked by the operator, due to a high number of warnings at that particular moment.) The initial situation of the scenario is shown in figure 2.

As can be seen from figure 2, traffic is dense and a lot of different ships as well as a large number of potential conflicts must be monitored at a time. Conditions of good visibility, no current and tidal influences and insignificant winds of force 1 Beaufort's scale were assumed for the scenario. The scenario was divided into two sessions: one scenario with high and one with low dense of traffic. The second scenario was produced by eliminating some targets from the scenario.

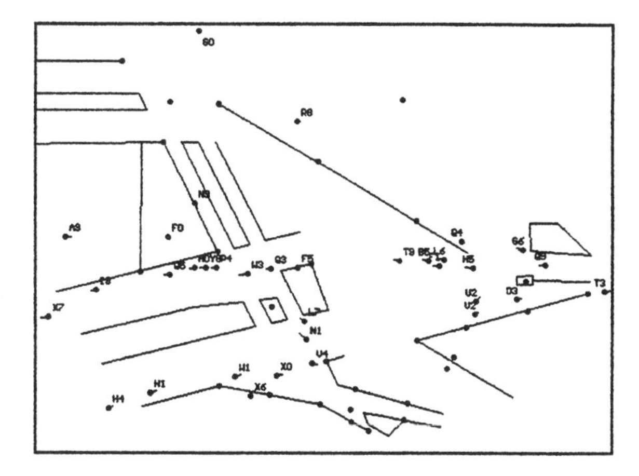

Fig. 2: Initial situation of the high density traffic scenario used for VTS operator experimental trails

Twenty VTS operators participated in the experiment. They can be categorized in operators familiar and not familiar with the sea area of the German Bight which was used for the scenario. Each the operators performed two sessions: one time the enhanced collision-warnings were implemented and the second session was performed using the other scenario and without any supporting warning.

For the purpose of this experiment the main functions of the VTS operator working place necessary for shore based collision avoidance were simulated.

RESULTS

The experiments delivered a wide range of answers for VTS operator actions on same situations. The basic thinking observed during the experiment of every operator was that they suppose that an existing or developing conflict situation will be solved after the first information given to a vessel involved in a certain situation.

Concerning the traffic monitoring in the German Bight VTS area the following statements can be derived from the experiments:

- The focus of the operator's observation is the crossing area of both the TSS "Terschelling-German Bight" and "Jade Approach" clearly. Every operator starts his data collection with vessels staying in or going into this area.
- A connection of the vessel's positions relating to the data collection results in a growing spiral polygon starting north-westerly of the buoy "SZ-N" and further covering the separation zone with the established racon "Jade-Weser" further continuing to the deep water roads
- Although manoeuvring capabilities of the vessels are not directly available in the VTS centre, the

operators will take it into account for situation assessment. They estimated these capabilities by the ship's dimensions roughly.

- A prioritisation in case of more dangerous situations simultaneously will be performed by the operators basing on the ship's dimensions and the cargo information (dangerous goods acc. IMDG-Code) available in the centre.
- One can observe at every experiment, that operators act on a traffic situation relative early - in comparison to the risk model developed at risk level 2. The operator's interferes were performed in defined steps. Always the give-way vessel was contacted firstly. The second contact were performed at different times acc. to the test persons. The last interfering was in most cases a message addressed to the stand-on vessel. Then VTS operator give such vessels information about the give-way vessel and that she will probably not manoeuvre added by an advice to navigate with care and according to the "International Rules for Preventing Collisions at Sea " (COLREGS) and prepare a potential own manoeuvre to avoid a collision.
- The times of interaction in case of risk escalation , e.g. risk level 3 or 4 of the created risk model is not homogeneous. A wide spectrum of times exists. From this point of view the implementation of warnings seems to be necessary on principle.
- To identify the level of the urgency of a certain interfere the IMO/IALA-message types were used:
    - information
    - warning, advice and
    - instruction.
- It was observed, that the higher level of urgency was expressed by a given same information repeatedly but louder than before if necessary partly.
- Operators detect and act on dangerous encounter situations near the ELBE approach (easterly part of the scenario) later than on similar situations in the crossing area. This is valid for both the classes of operators. The automatic warning release an operator action in several cases.
- For detection of dangerous situations operators use the monitoring of CPA data or the bearing ruler.
- The experiment proved that the 10-min-vector for collision risk monitoring was commonly used during the experiment. This technical aid was apparently used for the first selection and marking of targets, which should be monitored especially.
- The situation assessment by operators is influenced by former traffic courses. From the ship's names it would be recognized, that there is an approach of two scientific vessels and an occurred alarm was confirmed by a contact to the vessels.
- A significant influence of the operator's situation assessment by automatic warnings can be stated for these experiments in that manner, that the operators not familiar with the German Bight traffic depends more on this information than operators form this area.
- The reproduction of the last traffic situation (final situation of simulation run) has shown repeatedly: very often the operator starts to indicate the situation's **course** instead of the last position of targets. Vessels not relevant for safety during the course of the scenario were not reproduced by the operators.
- The prioritisation of indicated emphasis of observation for the coming 10 min is based on a estimation of the danger's potential and the expected extension of potential damages (dimensions of vessels, dangerous goods aboard a.o.). It depends on time remaining but also on the behaviour of the vessels in the past course (e.g. vessel that crosses a separation zone without any necessity and has impeded vessel 'XY' will be observed with more or special attention with respect to ... (another vessel, comply with routeing guides or so)).

With respect to shore based collision avoidance and the use of relevant message types (information, warning, advice and instruction) defined by IMO and IALA an analysis of the times and kind of interaction were performed. Figure 3 shows exemplary the times and kind of interactions performed by the VTS operators who are familiar with traffic and sea area. The depicted results were gained during the sessions with high traffic density and without collision warning.

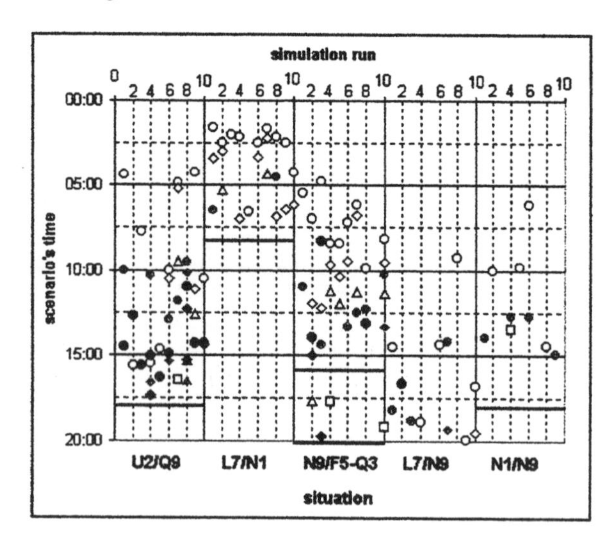

Fig. 3: Spectrum of times and kind of interactions performed by VTS operators familiar with the German Bight traffic during the high traffic density scenario (without implemented collision warnings)

The graph shows all the interactions performed by each of the operators, to avoid a collision during the several trails. The empty symbols indicate information, symbols filled grey indicate advice or warnings and black filled symbols indicate instructions given by an operator at an certain time. Each the first information, warning/advice or instruction is depicted as circle, repetitions are depicted as rhombus (1. repetition), triangle (2.) and square (3.).

The situation U2/Q9 lead to a collision, the other situations were solved by manoeuvres of one of the ships involved, so that the CPA-value increased from a certain time on. The black lines included in the graph mark the times of closest approach or collision and increasing CPA respectively.

As can be seen from figure 3 the message type mostly used for interaction was information. The other message types warning/advice and instruction were hardly used. This complies with the assumption of the operator's feeling, that he is an assistant for the officer on watch, while the captain onboard is responsible for any action to be taken to avoid a dangerous situation.

If the operator observed that a vessel did not react on the information supplied, most of the times he repeated this information before he started to use the next, more urgent, message type. The operators were able handle more than one ship in complex situations simultaneously (see e.g. time slot form 10:00 to 15:00 min scenario time). But it seems that complications may arise if these situations are more far away from each other, as e.g. one situation in the centre of the area and another one at the area's edge.

In case of situations with a give-way and one stand-on vessel the operators first act on the vessel which was obliged to give-way. In case the operator could not observe a sufficient action by the give-way vessel after having provided such a vessel with an advice then he informed also the stand-on vessel. That means that VTS operators take into account the rules for collisions avoidance which are relevant for each of the individual vessels.

With respect to the level of the danger of collision within a certain situation the message types will not be applied homogeneously. As e.g. the interactions belonging to the collision situation U2 - Q9 (This situation - added to the real scenario and create as much as possible similar to a real casualty with same constellation of courses at this position - ends in a collision at scenario time 18:30 min.) were performed at an relative high level of danger. Taking into account the operator's actions for data acquisition in comparison to the other encounter situations, which happened near the crossing area of the Traffic Separation Schemes this situation was detected later. The warnings and instructions, given to the vessel which was obliged to give way (U2), were given sometimes too late with respect to the remaining time to check the situation on board the ship and start an evasive manoeuvre.

Figure 4 shows the actions performed by the VTS operators during the second session (scenario with low traffic density and implemented collision warnings).

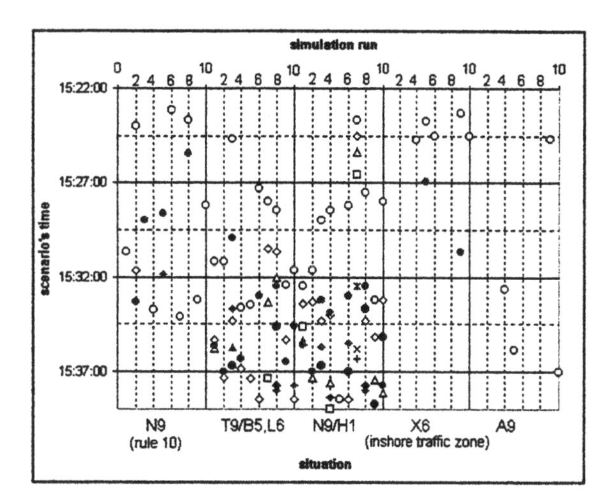

Figure 4: Spectrum of times and kind of interactions performed by VTS operators familiar with the German Bight traffic during the low traffic density scenario (collision warnings are implemented)

In comparison to the first session the scenario used for the second session contained only two situations with danger of collision. However, the number of information and advice items is increased. With respect to the different types of danger it can be seen furthermore, that VTS operators now also act on situation with lower risk levels. In the second scenario VTS operators interferes besides situations with danger of collision also on situations were vessels do not comply with the Mariner's Routeing guide as e.g. navigating inside of the Inshore traffic zone (target X6), crossing the roadstead (A9) or violates the separation zones of the established TSS.

During the experiments with operators not familiar with German Bight in several cases the implemented new warnings initiated operator's action ("Oh yes, now I would call the vessel "M5" and ask for his intended actions with respect to the vessel coming from west."). The German Bight operators mostly confirmed the warnings given or they had taken action already.

During the experiments with implemented warnings none of the non-familiar operators has overseen the dangerous encounter situation which was similar to the situation that had led to the accident in real life.

CONCLUSION

The model created for the assessment of encounter situations is useful for navigators on-board and VTS operators ashore as well. From the experiments it can be concluded that a supporting collision warning seems to be necessary in order to keep the correct message type corresponding to the risk levels. In this way the use of the created risk model for triggering automatic collision warnings improve the man/machine system installed in shore-based VTS. However, the implementation of warnings in the man/machine-interface for traffic monitoring may not lead to an information overflow. Warnings triggered by the system will be accepted well, if they are transparent and comprehensible. The improvement of warnings or alerts may have significant impact on the effectiveness of the whole system and in case of the VTS, on the safety and efficiency of the traffic in a certain area as well.

Because of the great difference in time and kind of the performed interactions it is still necessary to determine clear criteria and limit values for defined VTS interactions into dangerous encounter situations. This will affect the process of world-wide harmonization of VTS systems especially in the fields of training and education but also in the design of man-machine systems.

ACKNOWLEDGEMENT

This research is part of the COMFORTABLE-project which is performed within the 4th Framework Programme of the European Commission, DG VII: Transports. The experiments could only be carried out with the voluntary participation of the VTS operators of the German VTS centers and the kind support of the responsible authorities.

REFERENCES

Anderson, J. R., 1993, *Rules of the mind.* Erlbaum, Hillsdale, NJ.

Baldauf, M., 1997a, Traffic situation assessment and shore based risk recognition. In: Soares, C.G. ed., *Advances in Safety and Reliability* Vol. 2 pp. 893-900.

Baldauf, M., 1997b, *A method for shore based detection of situations with danger of collision in VTS areas* Sub-workpackage Report to E.C. Task 27 WP 02 (Draft), Wismar University, Dept. of Maritime Studies Warnemünde.

Hilgert, H. & Baldauf, M., 1996, A common risk model for the assessment of encounter situations on board ships. in Zhao, J. et al (ed.): *Maritime Collision and Prevention.* Chiavari Publishing, Surrey, England 1996, Proceeding on the International Conference on Preventing Collision at Sea, in Dalian, China 22-25 September 1996

Kolrep, H., 1996, Klassifikation von Unterstützungs-Systemen aus kognitiv-psychologischer Sicht - Untersuchungen mit Experten und Anfängern in der Flugsicherung. In Rothe, H.-J. & Kolrep, H. ed: *Psychologische Erkenntnisse und Methoden als Grundlage für die Gestaltung von Mensch-Maschine-Systemen.* ZMMS-Forschungsbericht 96-3, Technische Universität Berlin, Zentrum Mensch-Maschine-Systeme, Berlin

Wiersma, E. and Heijer, T., 1996, *Human factors models and methodologies for vessel traffic service operator performance.* Report to: E.C./D.G. 7, Delft University of Technology, Delft.

Wiersma, E., Heijer, T. and Hooijer, J., 1997, SATEST, a method for measuring situation awareness in Vessel Traffic Service operators. In: Soares, C.G. ed., *Advances in safety and reliability.* Vol. 2. pp. 901-906.

LIST OF ABBREVIATIONS

CPA	distance at closest point of approach
C_A	safe passing distance for situation assessment
C_H	hydrodynamic safe passing distance
IMO	International Maritime Organisation
IALA	International Association of Lighthouse Authorities
R_A	upper range boundary for "risk first begins to apply"
R_C	critical range (utmost lowest action limit for evasive manoeuvre)
R_M	manoeuvring range (distance at which an action requested by the rules have to be completed at the latest)
RNG	actual distance between two approaching vessels
TCPA	time to closest point of approach

INTERACTION OF AUTOMATED AND NON-AUTOMATED VEHICLES

Marita Irmscher, Thomas Jürgensohn, Hans-Peter Willumeit

Institute of Road and Rail Transport, Berlin University of Technology
Gustav-Meyer-Allee 25, Sekr. TIB 13
13355 Berlin, GERMANY
Phone: (49 30) 314 72990
Fax : (49 30) 314 72505
E-mail : irmscher@zmms.tu-berlin.de

Abstract: Novel car technologies like driver assistance systems may have unforeseen and unintended consequences on overall performance of the vehicle-driver-environment system. As an example the question about the interaction of vehicles with and without automatic support systems is discussed, for further investigations computer simulation of typical and possibly critical traffic situations, where both human conductors and support systems are involved, is recommended. For this purpose a model that accounts for the cognitive capabilities of human drivers is developed.

Keywords: Artificial intelligence, Agents, Models, Cognitive science, Decision making, Computer simulation

1 INTRODUCTION

Car driving exhibits various aspects of human interaction between man and machine that are under research in the field of man-machine systems. The introduction of driver assistance systems into the advanced equipment of modern cars will change the man-machine interface of cars considerably. Assistant systems are meant to provide the driver with more detailed information about the actual driving situation than a human is able to perceive and also give active support in critical situations as well as in tiring and monotonous driving tasks. The question arises then whether human judgement of traffic situations and the corresponding operations remains valid and meaningful if there is a considerable fraction of vehicles equipped with support systems that are able to carry out certain manoeuvres automatically.

Trust in the technical system can lead to a subjective overestimation of safety reserves and cause a more risky or too relaxed driving behaviour. The expected safety gain therefore may be compensated or even overcompensated by the driver.

Another point to be considered is the interpretation and anticipation of the behaviour of other traffic participants. Most drivers are able to predict actions of other drivers, because they usually observe other traffic participants carefully and know by experience that certain patterns of driving behaviour correlate with corresponding intentions and activities. This knowledge enables them to react appropriately and in due time. This mechanism may not work in the case of partly automated traffic.

After a discussion of the mentioned problems, it is proposed that more detailed investigations should be done by means of computer simulation. For this purpose cognitive capabilities that determine human driver behaviour must be described using methods that are applicable for setting up a computational model.

2 THE VEHICLE-DRIVER SYSTEM

Man-machine systems like aircraft and pilot, plant and operator or computer and user are often classified with respect to time behaviour and complexity of the technical systems, the kind of human activity (manual operation, process regulation or supervision) or the information flow between man and machine.

Car driving is a complex task which is performed in a dynamically changing environment where short time constants predominate. The extent to which the vehicle can be controlled by the driver is dependent on skill, experience and physiological or psychological factors. Driving skills, however, are usually limited to the well trained execution of everyday manoeuvres and do not include knowledge of underlying functional principles of vehicles.

Information about the environment is incomplete and imperfect due to the limitations of human perception and has to be updated continuously. The perceived information about the state of the environment that includes also the behaviour of other drivers is recognised as a situation associated with a choice of feasible actions. Each of these actions is then evaluated with respect to e.g. internal goals, consequences due to possible reactions of other drivers and constraints by safety considerations or traffic rules in a decision process. The desired operation is finally executed by the driver more or less skilfully according to his or her driving expertise.

Car drivers are not a homogeneous group, but exhibit a large variability with respect to skill or purpose of driving, which differentiates them from professional operators like pilots. Moreover they are to a large extent subject to emotional and motivational factors.

3 INTERACTION OF AUTOMATED AND NON-AUTOMATED VEHICLES

Driver assistant systems are a very recent development, available functions of present systems include navigation support, warning services but also active control of the vehicle. Active control includes speed control, distance regulation and lateral control (lane keeping). A long-term objective is to realise automated traffic on highways by "intelligent cars" driving in platoons at small distances. The human driver only monitors the functioning of the assistant system and resumes control whenever necessary, that is, under abnormal conditions. The idea is to increase highway capacity and make driving safer and more comfortable.

For aviation automated traffic using devices like autopilots is long reality. Therefore experiences of pilots are a valuable source of information when exploring the prospects of driver assistant systems in cars. Reported experiences indicate that although mental workload in normal conditions is reduced, there was a negative effect on flying skills, which are still or even more required to handle abnormal and critical situations. Moreover, it has been observed that reaction times to critical events were increased due to the requirement to monitor many functions simultaneously. Reduced responsiveness to critical events can also be induced by overreliance on an automated system (Ward, 1996). Similar effects of automation have been reported by plant operators.

These findings suggest that the impacts of automated systems should be investigated in an early stage with the objective to correct too optimistic expectancies or avoid costly developments. In the context of this investigation only automatic distance and velocity control will be considered as these are essential features for automated traffic and will be available in near future (Metzler, 1998).

Automatic speed control on highways is already a well established technique, its widespread use would ensure a rather even speed distribution and enhance the possibility to apply automatic distance regulation systems.

For automatic distance regulation commonly accepted safety margins must be observed. It has been noticed, however, that human drivers generally follow much closer than recommended by safety boards. The interaction with non-automated vehicles would probably be such that somebody takes advantage of a wide headway and moves in, forcing the assistant system to start an emergency braking or making the driver to resume control. The effect of this interaction are critical situations, at least disturbance of the traffic flow. Small gaps between vehicles driving at high speed, on the other hand, could render it impossible for a human driver to intervene at all and lead to severe accidents in case of system failure. This could make a driver feel uncomfortable and stressed. The regulation procedure of the assistant system must be therefore adapted to realistic individual driving styles.

The already mentioned loss of competence and driving skills can make a critical situation even more dangerous, because if drivers have not learned how to react properly, hazards become more likely.

The same effect could be created by the belief that the system will somehow cope with critical situations and that it is therefore not necessary to drive in a defensive way. This kind of overcompensation has been observed in the first period of Anti-Blocking systems (ABS). Situations that usually caused drivers to be cautious like black ice on the street were not considered dangerous anymore when the vehicle was equipped with ABS. This behaviour resulted in more accidents, and more severe accidents than before the application of ABS, a tendency that was even more accentuated by the fact that many drivers misunderstood the function of the system. As a conclusion the use of assistant systems should be included in the training of drivers, but this seems not to be practicable in near future.

Drivers usually can anticipate of other drivers' intentions and future actions from observed behaviour, because they suppose that other drivers behave more or less similar to themselves. There are several reasons why these estimates are not longer valid when automated systems are involved.

Since driver assistant systems generally do not, at least not yet, drive like a human driver, the conclusions drawn from experience especially about driving skills or intentions of other drivers in an automated vehicle are incorrect. Another point is that the automatic may exhibit incomprehensible or unexpected behaviour that provokes immediate intervention of the operator. That intervention of the human driver into the automated driving mode will change the driving behaviour very abruptly and can not be deduced by others from observed behaviour in advance. Even if drivers learn to differentiate between automatic and human drivers, behaviour is still unpredictable due to the diversity of different systems that can also induce different driver behaviour, e.g. depending on the degree of automation.

4 MODELLING OF HUMAN COGNITIVE CAPABILITIES

In the section above questions about the application of driver assistant systems and their impact on the individual driver and traffic safety have been raised and discussed, but designers, engineers and economists need more elaborate assertions if they want to develop practicable solutions. While reliable predictions about the behaviour of technical systems are readily available by means of computer simulation, man-machine systems are much more difficult to cope with.

One reason is that the behaviour of technical systems can be efficiently described by valid physical laws using relationships between a number of measurable properties of the system. Human behaviour, on the contrary, does not follow well defined rules and is influenced by a great number of parameters that are not easily quantifiable or not observable at all. Although with the development of Artificial Intelligence quantitative modelling methods that allowed the simulation of human problem solving behaviour became also popular in the field of cognitive psychology, the main objective of psychological models of human behaviour was describing and explaining empirical observations by formulating general concepts and theories.

Consequently the models of the human driver that were to be used in computer simulations have been traditionally developed from an engineering point of view. The main approach was to set human behaviour equal to the function of a control element using the familiar mathematical descriptions of control theory. Such models were used to investigate and optimise the technical system, but reliable predictions of the driver-vehicle-environment system as a whole for various and especially critical situations were not possible.

Both approaches have particular objectives, advantages and deficiencies. Engineering methods can handle dynamical systems and produce practicable results which lead to technical solutions, but they mostly neglect human adaptivity and variability. Modelling methods used by psychologists, on the other hand, describe and imitate the human cognition process by means of information processing using methods like artificial neural networks or production systems, but are generally not the best choice to make exact predictions or cope with dynamic problems. Drivers can rather be described as intelligent autonomous agents that perform individual planning and goal directed acting dependent on the perceived actual state of a dynamically changing environment. The research on intelligent autonomous agents in a multiagent environment is presently an active field of research within Artificial Intelligence. The emphasis there, however, is not humanlike but rational behaviour.

Therefore some effort must be spent to combine different modelling approaches for a quantitative model common in the engineering sciences. It is desirable to include theories of cognitive psychology about motivation and emotion to formulate a model hat describes the decision making and acting of an agent dependent on the actual situation. The features of the entire system involve the following topics from different research disciplines:

- vehicle dynamics
- human perception
- information processing of incomplete and imperfect data
- time-critical problem solving
- estimation of other agents' goals and intentions
- influence of emotional and motivational factors
- constrained decision making and planning
- motor action and action regulation

For the purpose of this investigation information processing is a key issue, therefore emphasis is laid on two items of a more comprehensive cognitive driver model and will be discussed in detail:

1) Identification and interpretation of traffic situations
2) Estimation of other drivers' current goals and intentions

4.1 Identification and Interpretation of Traffic Situations

Much more information about the state of the world is picked up by the driver than is actually necessary or relevant for the driving task. The term "situation" in the context of the proposed driver model describes a specific constellation of relevant state variables and is understood as a certain range within the state space. A situation is closely related with a choice of associated actions.

To identify a situation, quantities like position or velocity are assembled into a feature vector, which is mapped onto a number of prescribed discrete situational categories. Rekersbrink (1994) found that certain values of relative distance and time to collision are evaluated by drivers with respect to e.g. individual safety margins and can trigger the start of an action like braking or overtaking. These values together with current velocity and driving mode (following or overtaking) are therefore taken as discriminating factors of a situation.

As a first approach the driver model compares perceived state variables with declarative knowledge about traffic rules and individual safety margins. If the deviations between the identified situation and the internally desired state are above a certain margin, a drive to change the current situation towards the desired state is created. As several action alternatives may be able to transfer the present state into a more favourable one, the alternatives have to be evaluated to make a choice.

The task to identify a situation can be considered as a kind of pattern recognition, and therefore the application of artificial neural networks (ANNs) could be an alternative. Previous work shows that ANNs are well suited to model humanlike identification of a situation with subsequent estimation of action urgency (Jürgensohn et al., 1994; Jürgensohn and Willumeit, 1997; Jürgensohn, 1997).

4.2 Estimation of other drivers' current goals and intentions

It was already mentioned that anticipation is a prerequisite for competent driving. The anticipation process can be explained by means of an internal model about the mental state of other road users that is generated by the driver from observed actions and used to infer other drivers' current goals and intentions. The model evolves in a learning process and is therefore linked to individual expertise and experience

Specific difficulties to correlate preceding driving behaviour of a person with future actions are the uncertainty and incompleteness of perceived data on the one hand and the variability of people's driving strategies on the other hand. Own understanding of driving in various situations is very important for the interpretation of perceived behaviour of other drivers, as it is plausible to presuppose that other drivers behave rational and follow goals that are similar to one's own.

Examples for internal goals that can be found commonly with drivers are the wish to arrive at the destination as early as possible, to minimise the risk of an accident, or to take the challenge of mastering a powerful sport car. The final setting of the internal goal is dependent on various internal and external factors like the purpose of car driving (professional, leisure), weather conditions or the time of the day. Examples for internal factors influencing the driver among many others are motivation, driving skill, audacity or fatigue. Internal goals are not necessarily constant, but can be modified by the driver during a journey.

In the model it is assumed that other drivers develop intentions in a procedure that is analogous to one's own. The internal goal is set equivalent to the desired speed, safety considerations and traffic rules are modelled as constraints. Both the internal goal and the constraints are dependent on the car type and motivational factors. The problem of predicting the goal of another driver thus reduces into identifying influencing factors like the actual motivational state. Data available for this task comprises actual values of position, velocity, acceleration, the vehicle type and setting of blinker or flash of the object in question.

After having estimated the other driver's desired speed and individual safety margins, comparing the actual state with the supposed internal goal results in a prediction of the next actions that are most likely. This way a driver can evaluate benefit and risk of own action alternatives taking into account probable reactions of other drivers.

4.3 *Critical Considerations about Modelling Methods*

The parts of the driving task that are important in this context involve a process of continuous problem solving and decision making dependent on the actual state of the world as it is perceived by the driver.

Aasman (1995) developed and implemented a cognitive driver model using the production system Soar which takes advantage of psychologically motivated problem solving procedures and algorithms, but reported the already mentioned difficulties with dynamic environments and multiple goals that have to be handled simultaneously. An alternative method for modelling cognitive driver behaviour, fuzzy logic, has been applied by Wolter, et al. (1997). Both fuzzy theories and artificial neural networks share the drawback that realistic data has to be available in order to adjust fuzzy parameters or to train the neural network. Moreover, the parameters that are found do not represent any attributes of the model, so that neither a variation of model parameters nor the interpretation of the influence of model parameters on simulation results are possible.

On the other hand, as a priori knowledge about driving is available that can be readily expressed by means of IF-THEN rules, and rule based systems are a proved means to model human problem solving, a production system is chosen as a framework for setting up the driver model. For the purpose to be able to combine various modelling methods, like differential equations for the vehicle model or artificial neural networks for situation recognition, it was decided to use a specially designed program that is flexible enough to allow the incorporation of the respective procedures as well as the direct modification of parameters like motivational factors or traffic rules.

The question of the validity of the entire model arises when rules that describe driving behaviour are set up using a priori or heuristic knowledge. As the objective is to simulate scenarios that are not yet existent, the entire model cannot be strictly validated. It is reasonable to assume, however, that aggregation of individually valid components paying special attention to mutual interactions will result in an altogether valid model. Individual components of the model, particularly rules that reflect the influence of emotion and motivation or preferred driver actions in specific situations should be verified by test rides in a simulator for selected situations or by questionnaires.

5 CONCLUSION

Automation of traffic using driver assistant systems can give rise to safety problems, caused by possible misinterpretations of observed driving behaviour or increased difficulties to predict future actions of other drivers at all. To investigate this problem by means of computer simulation, modelling methods from different research disciplines have been assessed with respect to being employed in a cognitive driver model.

REFERENCES

Aasman, J. (1995). Modelling Driver Behaviour in Soar, *Proefschrift, Rijksuniversiteit Groningen*, Leidschendam: KPN Research

Jürgensohn, T. (1997). *Hybride Fahrermodelle,* Pro Universitate Verlag, Sinzheim

Jürgensohn, T., H.-C. Raupach and H.-P. Willumeit (1994). A Model of the Driver Based on Neural Networks, *SAE - Paper, No. 945082, Proceedings of the 25th FISITA Congress, Beijing,* pp.161-166.

Jürgensohn, T. and H.-P. Willumeit (1997). Modellierung von Mikro-Steuerstrategien mit Hilfe künstlicher neuronaler Netze, *Autoombiltechnische Zeitschrift*, **Volume No 6/97**, pp. 348-352

Metzler, H.-G. (1998). Assistenzsysteme in der Fahrzeugführung. In: *Wohin führen Unterstützungssysteme? Entscheidungshilfe und Assistenz in Mensch-Maschine-Systemen* (Hans-Peter Willumeit, Harald Kolrep. (Ed.)), pp.206-226, Pro Universitate Verlag, Sinzheim

Rekersbrink, A.(1994). *Verkehrsflußsimulation mit Hilfe der Fuzzy-Logic und einem Konzept potentieller Kollisionszeiten*, Schriftenreihe Institut für Verkehrswesen, Universität Karlsruhe

Ward, N.J. (1996). Interacions with Intelligent Transport Systems (ITS):effects of task automation and behavioural adaptation. *ITS Focus Workshop "Intelligent Transport Systems and Safety", February 29th 1996,* London

Wolter, T.-M., T. Jürgensohn and H.-P. Willumeit, (1997) Ein auf Fuzzy-Methoden basierendes Situations-Handlungsmodell des Fahrerverhaltens, *Automobiltechnische Zeitschrift,* **Volume No 3/97**, pp. 142-147

USE OF SPEECH COMMUNICATION AS AN INTERFACE OF A NAVIGATION SUPPORT SYSTEM FOR COASTAL SHIPS

**Junji Fukuto * Masayoshi Numano * Keiko Miyazaki *
Yasuyoshi Itoh * Yujiro Murayama ** Kazuo Matsuda ***
Norio Shimono ******

* *Ship Research Institute,
6-38-1, Shinkawa, Mitaka, Tokyo, 181-0004, JAPAN,
Email: fukuto@srimot.go.jp*
** *Yuge National College of Maritime Technology, JAPAN*
*** *Mitsubishi Heavy Industries, Ltd.*
**** *All Japan Coastal Tanker Association*

Abstract:
An advanced navigation support system (ANSS) was developed aiming at one-man bridge operation (OMBO). The ANSS designed based on active sailor's demands collected through a series of simulator experiments and interviews. The ANSS has many automatic navigation functions such as a TRACK PILOT, etc. All functions for OMBO are controlled with speech communication.
The first ship, which is equipped with the ANSS, started her service from September 1997. Through the observation of actual use of the ANSS, the effectiveness of the ANSS and its man machine interface are discussed.
In this paper, after a brief introduction of our research program, the design principles and functions of the ANSS are described. Then the functions of man-machine interface using speech communication and its implementation are described. Finally, we discuss the effectiveness of the ANSS and its man machine interface through the observation of actual use and the subjective evaluation of active users. *Copyright ©1998 IFAC*

Keywords: Navigation Systems, Speech Control, Man/Machine Interfaces

1. INTRODUCTION

A steady decrease in the number of sailors, especially young sailors, and their aging cause a big problem for keeping domestic marine transportation in Japan. The decrease of skilled sailors affects total transport volume and navigation safety directly. To overcome this situation, many efforts to improve working environment and to realize safety navigation with less crew have been made by several organizations. A navigation support system is one solution. Navigation Support System (NSS) is also called INS (Integrated Navigation System) or IBS (Integrated Bridge System) in maritime society.

In the past decade, many R&D programs for NSSs were carried out for contributing laborsaving and navigation safety. (Fuwa *et al.*, 1989) (Kasai and Kobayashi, 1993) Unfortunately, these systems have not been used widely by customers yet. One main technical reason is the lack of accurate and reliable sensors. The other is poor usability of them.

Nowadays, the performance of sensors has been improved such as Deferential Global Positioning

System (DGPS). The error of position data goes to within a few meters.

On the other hand, recent technology of speech recognition and speech synthesis makes it possible for operators to converse with computers effectively. The speech recognition system reached to the level of practical use with large vocabulary and speaker independence. (Watanabe *et al.*, 1994)

From 1993, All Japan Coastal Tanker Association and Ship Research Institute started to develop an advanced navigation support system (ANSS) for coastal tankers. The ANSS was developed aiming at one-man bridge operation (OMBO). The first three years were spent for defining the design concept and the requirements of the ANSS. Through simulator experiments and discussions with active sailors, the required functions to the ANSS were defined. Then Mitsubishi Heavy Industries Ltd. manufactured the ANSS based on our requirements. [1]

In 1997, the first coastal LPG tanker equipped with the ANSS was delivered. The effectiveness of the ANSS was evaluated through on board observation and interviews to active users.

In the next section, the design principles and functions of the ANSS are described. Then the functions of man machine interface using speech communication and its implementation are described. Finally, the effectiveness of the ANSS and its man machine interface is discussed.

2. ADVANCED NAVIGATION SUPPORT SYSTEM

2.1 *Simulator Experiments*

A series of simulator experiments were carried out to clarify the requirements to a NSS for OMBO. Two types of simulator experiments were held. One is a series of experiments in a narrow channel. The other is that of long-term navigation experiments in open sea.

The simulator experiments in a narrow channel were carried out in one of the most famous and difficult channels in Japan, "The Kurushima Channel". (Numano *et al.*, 1995) The difficulty of this channel is caused by strong tidal current, many steep bends of waterway and heavy traffic. In the simulator experiments, some advanced information which is expected from the ANSS, were given to a subject in addition to conventional navigation aids. This situation corresponds to the busy condition of watch work.

[1] This ANSS is now on the market named "Super Bridge X TM" from Mitsubishi Heavy Industries, Ltd.

The simulator experiments of long-term watch (Itoh *et al.*, 1995) was held in open sea with few encounter ships. It took 3 hours for each run. The information given to a subject was the same as in narrow channel. In these experiments, the demands for long-term OMBO were clarified and the fatigue of operator was evaluated.

It is confirmed that the fatigue through 3-hour watch did not affect watch work seriously from observation of experiments and interviews.

From the experiments of OMBO and interviews to active sailors, following demands were pointed out.

(1) **Support for ship handling**
Auto Pilot and Track Pilot.

(2) **Warning for grounding**
Warning for a breakaway from the planed route and a shoal.

(3) **Warning for collision danger and information support for collision avoidance maneuver**

(4) **Support for miscellaneous works**
Support for miscellaneous works not to disturb continuous watch.

(5) **Eye free support**
Eye free support no to disturb continuous watch.

(6) **Advice for matters that require attention**
Legal and empirical suggestion to the operator to avoid careless mistake.

(7) **Warning for the disorder of operator and its safety guard**
Warning for functional disorder of operator and its countermeasure.

2.2 *Functions and configuration of the ANSS*

Conventional crew team is mainly composed of three crews. The first is a captain as a decision-maker. The second is a helmsman who keeps his ship on ordered course. The third is a RADAR operator who gathers the information of other ships and obstacles.

For one-man bridge operation, the captain's role should be left to human and the other roles should be transferred to the ANSS. The task that computer can do better than human should be allocated to the ANSS such as information processing and ship control.

From the view point of the task allocation policy, the functions for the ANSS were designed as shown in Table 1.

Fig. 1 is a block diagram of the ANSS. The ANSS is mainly composed of following 4 parts, (1) Navigation Information Processor (NIP), (2)

Table 1. The ANSS functions for OMBO

- Support for ship state and position recognition
- Automatic ship maneuvering and speed control including Track Pilot
- Warning for grounding and collision with other ships
- Information support of other ships and collision avoidance maneuver
- Dead man alarm

Touch Panel Type Screen, (3) Speech Information Processor (SIP) and (4) I/O Interface. The NIP manages all functions listed in Table 1. Through the I/O interface, the NIP gets information from sensors such as DGPS, RADAR/ARPA, etc. and puts order to actuators. Touch Panel Type Screen is a main console of NIP. All information can be displayed on this screen and all actions can be taken from the touch panel. The SIP is a main interface between operator and the NIP. It has a speech recognition system and an artificial voice synthesizer. Fig. 2 is a picture of bridge with the ANSS. The SIP gets operator's voice as input from a microphone and outputs information with artificial voice.

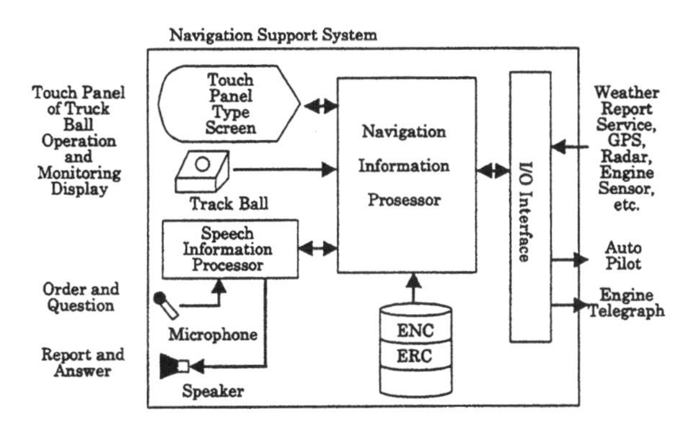

Fig. 1. Block Diagram of the NSS

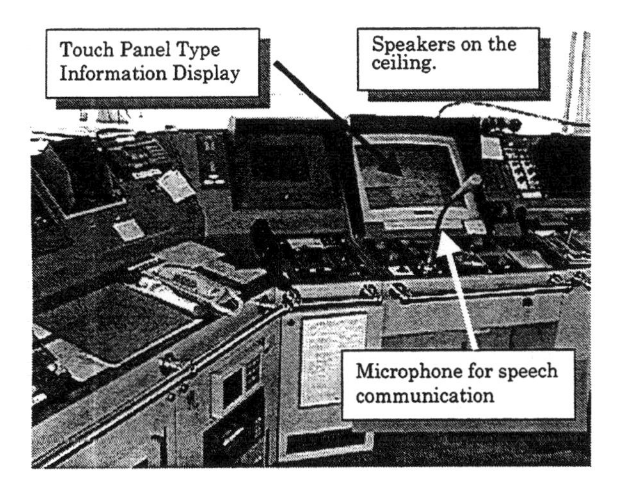

Fig. 2. Front view of a bridge and a console of NSS

3. MAN MACHINE INTERFACE FOR THE ANSS

3.1 *Demands for man machine interface*

In addition to automatic functions, three demands for the man machine interface of the ANSS are pointed out through simulator experiment. The first is that all supports should be given without disturbing continuous watch. Because of the lack of effective sensors for other ships, the support system for collision avoidance maneuver do not work effectively and the task remains for human operator even now. Therefore, active sailors request eye free operation. The second is that the ANSS should not need special training and attentive reading of manuals for daily watch operation. Basically, active sailors do not have enough training time for NSSs. To use advanced functions of NSS, they should learn to use the NSS in his job. At that time, they need to read manuals in details. This would be a barrier to use NSSs. Because of that, active sailors request easy learn interface or the interface close to the method which has been used, especially for daily use command. The third is quick and reliable response of the ANSS.

To meet the demands, the man machine interface for the ANSS was designed with positively use of speech communication. By using speech communication, eye free operation can be realized. The speech communication also matches to the second demand. Because they already use their voice to order commands.

3.2 *Functions of speech communication*

The speech communication is used in the following actions.

(1) Ordering a command to the ANSS
(2) Request information to the ANSS
(3) Answering the question
(4) Warning for grounding and collision danger
(5) Periodical navigation condition report

For example, an operator can order a command and ask a question to the ANSS such as "Change course to 190 degrees." or "Are there any collision danger? ". The order is carried out by the ANSS. The ANSS also answers the questions with artificial voice such as " There are two danger ships in front." The ANSS outputs warning message for grounding and collision danger based on the decision of the ANSS.

The ANSS reports the current status every 12 minutes (defined by administrator). This report is used not only for informing present state but also for monitoring the soundness of the ANSS.

The report is also used as a dead man alarm, which detect operator's disorder. The ANSS requests acknowledgment from operator after the report. When one does not response to the report, the NSS judges that some problem happens on the operator and warns that to the other crews. If no crew response to the warning, the ANSS slows down the ship automatically.

3.3 *Consideration of operation safety*

In spite of many merits of speech communication, it has some problems that must be taken into consideration, especially for keeping navigation safety.

The first is the quickness and reliability of speech recognition. The lack of enough quickness and reliability for command recognition may cause the deviation from proper maneuver and may lead the ship to an accident. To increase the reliability, following countermeasures are taken. The first is the use of the limited words for recognition. The SIP can only recognize the limited words and numbers. Every command for bridgework is tied to the predefined words. In the command sentence, the word order is also fixed. For example, when an operator wants to change the course to 330 degrees, one should order the command as "Course Three Thirty ". The SIP also uses personal voice data files for improving recognition performance. By using this way, the speed and the fault rate of speech recognition are improved enough for practical use.

The second is the restriction of communication channel of human. As human has only one channel for speech communication, it is difficult to listen to a message and to order a command simultaneously. Furthermore, when much information is coming in a short time, operator may not get important information within enough time for decision making.

To control coming message with voice, higher priority is given to the speech recognition process rather than the annunciation process. This priority enables to interrupt a long announcement and to order new command for next message or other action. The massage from the ANSS is also given the priority based on its importance. Then the operator can get important information and also skip the information by oneself if one does not think it is important.

The ANSS also has a touch panel type screen. All support information is displayed on the screen. Every action to the ANSS is also taken from hierarchical menu on the touch panel screen. In addition to the menu, some automatic pop-up buttons are installed on the screen. The pop-up buttons are made for frequent command such as the acknowledgment to a report. Operator can acknowledge a report with his voice or with the pop-up button.

From the safety point of view, the effective countermeasure for fault recognition of the command is also required. To eliminate the fault order by operator and speech recognition system, following procedure for the order of ship motion control is adopted.

(1) Voice order by operator
(2) Repeat the order by the ANSS and request the confirmation
(3) Confirmation by operator
(4) Execution of the order by the ANSS
(5) Report the change of the ship state.
(6) Acknowledge the report.
(7) Report the completion of the order
(8) Acknowledge the report.

In this procedure, the confirmation is requested for each order. At that time, the order is repeated with announcement. Any command is not activated without operator's confirmation. The ANSS also report the expected change of state, just after the execution of the order. These eight steps will make operator notice the error of the order.

4. EVALUATION

4.1 *On board observation*

The first ship, which is equipped with the ANSS, "Shin Propane Maru" is a 749GT LPG carrier with six crew member (Captain, 2 navigators, 2 engineers, 1 cook). This ship started her service from September 1997. The evaluation of the effectiveness of the ANSS was carried out through the on board observation and interviews to active users.

First, it is confirmed that the tasks are allocated properly to both the ANSS and an operator and that the operator works under proper workload.

To confirm the effectiveness of task allocation, the task analysis of the OMBO with the ANSS was carried out. At the observation, an observer recorded actions of operator and the ANSS as an event with time stamp. The bridgework categorized into 33 events for the advanced ship. The reliability of speech recognition and the utilization factor of the ANSS were also measured, through the observation.

The task analysis of the bridgework of a conventional ship with two crew members was also carried out to compare the workload of the advanced ship.

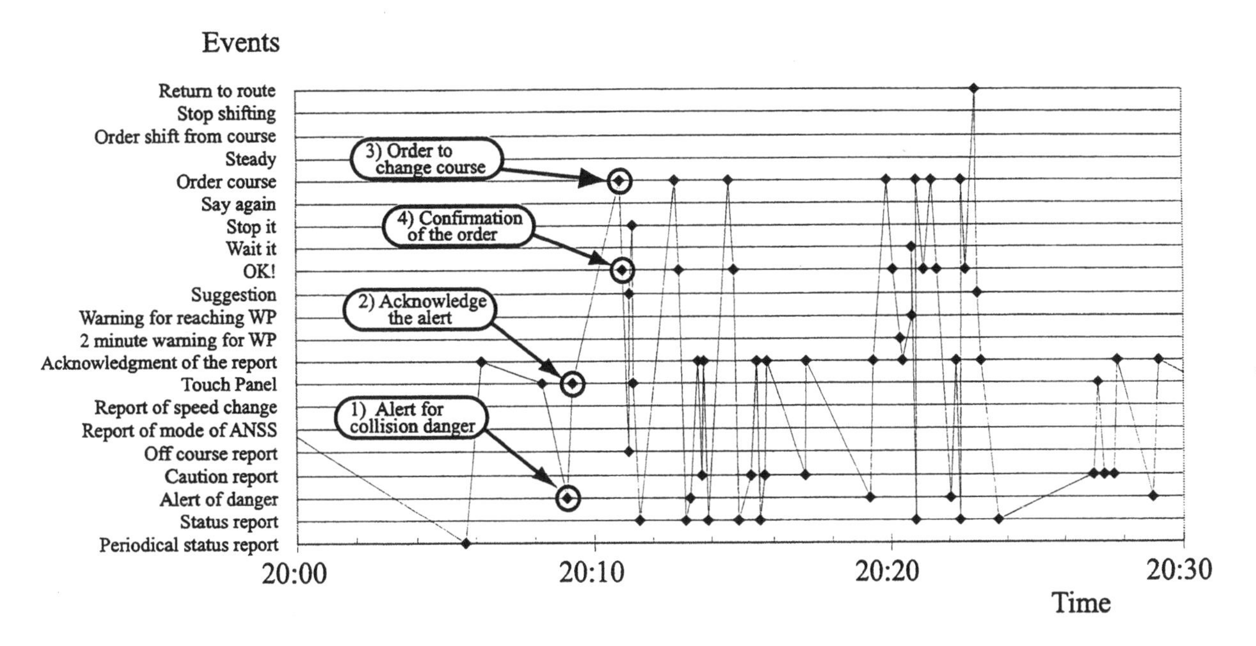

Fig. 3. Time History of events of advanced ship

4.1.1. *Results of on board observation* Fig. 3 is a part of time history of the recorded events. The events are plotted in ordinate and the time is plotted in abscissa. This time history expresses the procedure of collision avoidance maneuver. When the ship encounters a ship, first the ANSS alerted collision danger 1) and then an operator acknowledged· the report with touch panel 2). Next, the operator ordered to change course for avoiding collision 3). Then the ANSS repeated the order and the order was executed after operator's confirmation 4).

The observation was carried out from Mizushima to Kawasaki and the utilization factor of the ANSS was 100%. The route includes the Akashi Strait and the Uraga - Nakanose waterway, which are one of the most difficult waterways in the world. The rate of correct speech recognition was also measured at the same time. The rate was 92%. This measurement was carried out just after installing personal data files. After sufficient experience in the operator for speaking to the ANSS, the rate becomes almost 100%.

4.1.2. *Task comparison with the conventional watch work* To compare the tasks of the advanced ship with the tasks of the conventional ship, both events are categorized into 6 items, (1) Command & Confirmation, (2) Report, (3) Acknowledgment, (4) Question, (5) Manual Operation, (6) Touch Panel Operation. Confirmation and Touch Panel Operation are special item for the advanced ship. Manual operation of the conventional ship includes course change operation and checking residual distance and next destination. Manual operation of the ship with the ANSS includes direct steering with AutoPilot.

Fig. 4,5 is a bar graph of the number of events per hour. Fig. 4 is for the advanced ship and Fig. 5 is for the conventional ship. The gray bar is the data for all measured time, the white bar is for calm condition and the black bar is for congested sea condition.

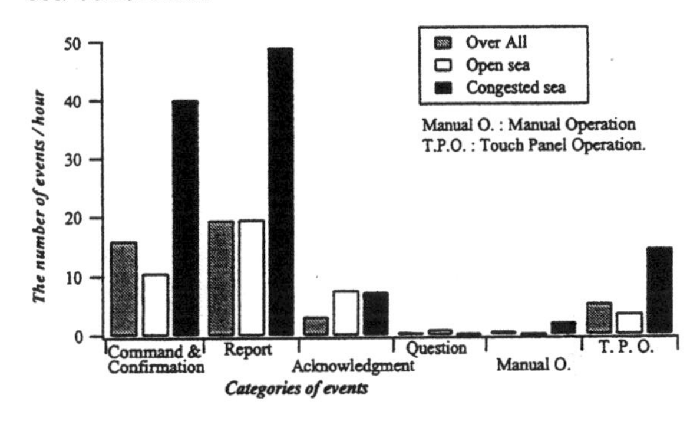

Fig. 4. Event frequency of an addvanced ship

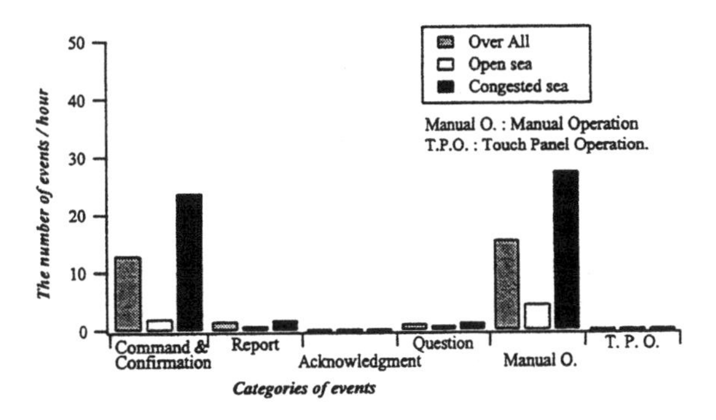

Fig. 5. Event frequency of a conventional ship

Distinct difference is appeared on "Command & Confirmation", "Report" and "Manual Operation". The number of "Command & Confirmation" and "Report" of the advanced ship is bigger than the number of the conventional ship. The number of manual operation of the advanced ship is smaller than that of the conventional ship.

As item "Command & Confirmation' includes command itself and confirmation, the substantial number of "Command " is a half of the number of that. Therefore, the number of "Command" of the advanced ship is almost same as of the conventional ship. From the number of "Command", it is confirmed that the support works effectively.

In the advanced ship, the number of "Manual Operation" is decreased remarkably. This means that the most of helmsman's tasks is transferred to the ANSS. Actually, in the observation, the voice command and automatic ship control functions were always used effectively and operators can do bridge work with appropriate margin and preparation time even in congested area.

4.2 *Subjective evaluation of the ANSS*

After crew was practiced the ANSS well, the subjective evaluation for each function and its human interface were taken. The functions of the ANSS are given a prize for its good usability and safety. (Fukuto *et al.*, 1998)

Speech communication is evaluated reliable and easy enough for practical use as human interface. Crews pointed out that they could use the ANSS without special training. They said it took about 1 week for mastering daily work by asking other crew through his actual watch. The use of touch panel screen compensates quick response. The combination of speech communication and touch screen is well accepted by active crew because of its eye free operation, quick operation and their trust for it.

Another merit is that speech communication produces the partnership between operator and the ANSS. This partnership helps to reduce the loneliness and the feeling of alienation of OMBO and encourage operator with friendly voice.

5. CONCLUSION

This paper describes an application of speech communication. The speech communication was applied to an advanced navigation support system (ANSS) to support one-man-bridge operation. The ANSS matches the demands of active sailors. The ANSS has some advanced functions under the requirements such as an ECDIS, a TRACK PILOT, etc.

The speech communication provides easy and simple interface of the ANSS without disturbing operator's continuous watch. With combining touch panel operation and certain confirmation for orders, the defects of speech communication are overcome.

The advanced functions and usable human interface of the ANSS gets eagerness for using it and trust for it. Actually, it is confirmed that active sailors accepted the ANSS and the navigation is carried out by one crew with this system. It is also observed that the rate of its use was nearly 100%.

Further research is needed to reduce operator's workload and errors with such a user-friendly support system. The improvement of sensors also needs to complete the ANSS, especially for other ship sensor. Now Automatic Identification System (AIS) has been announced by as a future communication system between ships and land stations. It is expected to be an effective information server for the ANSS.

Acknowledgments

Our special thanks go to Capt. Tanimoto and other crew of "Shin Propane Maru" for their help and valuable discussions.

6. REFERENCES

Fukuto, Junji, Masayoshi Numano, Keiko Miyazaki, Yasuyoshi Itoh, Yujiro Murayama, Kazuo Matsuda and Norio Shimono (1998). An advanced navigation support system for a coastal tanker aiming at one-man bridge operation. *Proceedings of IFAC Workshop on Control Applications in Marine Systems, CAMS'98 (to be published in Octorber 1998).*

Fuwa, Takeshi, Takeo Koyama, Kunihiko Tanaka and Junji Fukuto (1989). A knowledge-based system applied to an automatic ship navigation system. *Proceedings of IFAC Workshop on Expert systems and Signal Processing in Marine Automation, CAMS'89* pp. 45–57.

Itoh, Yasuyoshi, Kunihiko Tanaka, Yoshihiro Hirao, Yujiro Murayama, Yasuhiro Nakamura and Shigeki Ikeda (1995). Safety assessment of long-term operation using simulator (in japanese). *The Journal of Japan Institute of Navigation* **93**, 251–262.

Kasai, Hironao and Eiichi Kobayashi (1993). Maneuvering simulation approach to a ship's piloting expert system. *Proceedings of International Conference on Marine Simulation and Ship Maneuverability 93.*

Numano, Masayoshi, Nobuo Kiriya, Junji Fukuto and Kunihiko Tanaka (1995). Navigation support for narrow waterway (in japanese). *The Journal of Japan Institute of Navigation* **92**, 85–90.

Watanabe, Takao, Kaichirou Hatasaki, Ken'ichi Iso and Hiroaki Hattori (1994). Speech recognition (in japanese). *NEC Gihou (Technical Report).*

EXPERIMENTAL PLATFORM FOR SUPERVISION OF COMPLEX AUTOMATED SYSTEMS

B. Riera, G. Martel, M. Lambert et E. Cherifi

University of Valenciennes, LAMIH, Le mont Houy, BP 311, 59304 Valenciennes, FRANCE

ABSTRACT. Research in the field of supervision involves several different sciences like automatic control, human sciences and ergonomics. Progress will come, firstly with improvements of existing supervisory tools, and secondly with the integration of new algorithms. However, with regard to the strong human presence in the supervision loop, it is necessary to evaluate with human operators the supervisory tools which have been realized. Today, very often, studies in the field of supervision use process simulator and supervisor very specific with a realism which could be discussed. So, to solve this problem, an experimental platform dedicated to the study of supervision operators' activities with useful functions (simulation of feed-back controls, sequential command, simulation of failures, noise and so on) has been developed. This paper, after a synthesis about supervision, deals with the characteristics of the platform which includes a simulator, a supervisory software and a tool to analyse operator's activities.

Keywords : Supervision, Simulation, Man-Machine system

1. INTRODUCTION

Automation and data processing are two important factors of the development of industrial processes. Bigger these processes are, more complex it is for one or several human operators to supervise, to plan, to control, to command and more generally to understand the running and the behaviour of the whole system. In spite of progress in the field of automation, one can notice that supervisory tools for operators have not advanced a lot for thirty years. Displays, like P&ID diagrams, trends and alarms (triggered off when a fixed threshold is overtaken) are still the same. However, research in the field of supervision is very active. The study of human factors in the supervision is a very rich domain, at the intersection of several disciplines such as cognitive sciences, automatic control, and human sciences. Progress in the field of supervision will come from researchers of all these disciplines which have to work together in order to design new supervisory tools improving Man-Machine systems performance.

Because studied systems include human operators, it is necessary to analyse operator's behaviour using supervisory tools in realistic working conditions. However, often, performed studies, when they can not be realized on the real plant for obvious security, availability and cost reasons, use a specific experimental platform dedicated to a particular application. Consequently, a lot of time is spent to design the platform and to evaluate the results. In addition, realism of the platform can be discussed because simulated processes sometimes are too simple (few variables) and supervisory tools too light. Today, it exists several supervisory softwares running on personal computer, enough open to design important applications, and integrating all supervisory functions. In order to save a lot of time during the realization stage of an experimental protocol, we have developed an experimental platform integrating a numerical simulator, a supervisory software and a tool to analyse results coming from the experiments. The integration of these three functions is very simple, and consequently the time to develop a new experimental protocol and to analyse the results is reduced. The interest of the simulator is the functions which are directly included. These functions are really dedicated to the study of human activities. Indeed, the simulator has not be designed in order to solve complex equations for control-command applications for instance, but to represent dynamics taking into account running modes, failures, commands and so on. For that, the simulator runs in real time. In other words, dynamic evolution of variables respects an imposed time criteria. Data can be stored and after evaluated by the mean of the analysis tool. In addition, the simulator enables to simulate failures, breakdowns and noise on variables. At least, simulation of the control-command part (feed-back control and PLC) is possible. Hence, the simulator is like a virtual plant for the supervisory software. The experimental platform is very useful for the study of human factors in the field of supervision of automated processes. Analyse of the human perception of the control-command part, human errors, new kinds of information displays (ecological interfaces for instance), definition of information requirement of operators, support system for the training of operators, are some examples of the possible applications of the experimental platform

After recalls about supervision of automated processes, different possible fields of research, and the necessity of an experimental platform dedicated to supervisory experiments, the specifications and the concepts of the platform are developed.

2. SUPERVISION OF AUTOMATED PROCESSES

Supervision of automated processes is linked to the study of a particular family of Man-Machine Systems (MMS). Indeed, a MMS is defined as being a functional synthesis between a human being (or a group of human beings) and a technological system (Johannsen, 1982; Millot, 1988). Automated processes are dynamic MMS which could be divided into four levels (Sheridan, 1985) : sub-systems of production with their local command units, a stage of coordination of commands and the human supervisory team. Dealing without the hierarchy of the command, MMS requiring a "supervisory" control presents three components : an operative part, a control-command part and a "human" part.

Operative part, also called production system is the physical structure which performs the work. In the case of industrial production systems (it is to say: systems which give an added value to an input product), this structure includes sensors which enable to observe the system, actuators which command components and pre-actuators which supply necessary energy to actuators. Usually, production system are classified with regard to the production process : continuous or manufactured. In continuous processes, continuous flows of material and/or energy are processed. However, the complete process is more often a discontinuous sequence of continuous sub-processes.

Control-command part defines, from information coming from sensors, orders which are sent to actuators. It is possible to differentiate two kinds of automatic control-command. The first one concerns sequential PLC and the command is developed by the mean of logical equations, or Petri nets for instance. The second one concerns feed-back control (PI, PID, RST, and so on). It is the autonomy degree of the control-command part which defines MMS level of automation.

At least, the human part composed of human operators. In a very general way, they have to guarantee the security and the performances of the MMS. One can see the paradox of automation (Bainbridge, 1983). In one hand, during the normal running of an automated process, the control-command part involves an autonomy of the plant. Usually, operators manage the transitory and abnormal modes that the control-command part can not support. A serious problem comes from the variation of human workload induced by automation which have considerably modified human operator's work. One can note that when everything is going well, that is to say when there is a normal running of the process, the operator in front of his screen, far from the process, can damage his mental representation of the structure and the functions of the process. The associated risk is a drop of operator's performance when an abnormal running occurs. In addition, operators see the process through displays which propose P&ID diagrams, numerical values, trends, regulators and information coming from the automatic supervisory system (alarms and messages for instance). Consequently, the human capacity of decision depends partially on the efficiency and the performances of the supervisory tools. Today, these systems send a big quantity of information and particularly when a problem occurs (multiple alarms, "Christmas tree" effect and so on). This informational overload is at the source of numerous security problems.

The supervision of industrial processes includes a set of tasks aiming at controlling a process and at supervising its running (Millot, 1988). The control consists in acting on the process ; so, it is a top-down flow of information which acts on the lower levels (Benzian et al., 1994). On the contrary, the supervision is a bottom-up flow of information of which sources are the signals sent by the process. Control and supervisory tasks require high level of knowledge from the human operator and can be grouped into three classes (adapted from Rouse, 1983) :

- The control tasks and the supervisory tasks of the normal running of the process. The operator supervises the process and he tries to optimize the running by mean of adjusted tunings. Lejon (1991) points out that these tasks are going to disappear because of optimization algorithms.

- The transition tasks corresponding to a change of the running mode of the process (for instance, the start-up or the stop of the production system). In this case, the operator, most of the time, has to perform procedures already defined.

- The detection of failures tasks, the diagnosis tasks and the resumption tasks. The operator has in a first time to detect the presence of an anomaly by mean of alarms displayed or the trends of some variables. In a second time, the operator has to diagnose the state of the process. This consists in finding the initial cause of the failure but also the effects on the process. At least, the operator has to compensate or to correct the defect.

As a summary, operator's tasks consists in supervising process state and acting in order to maintain it as close as possible of its normal running point. To reach these objectives, an operator in a supervision room has to get at each instant a correct picture of the production system.

One can understand the importance of the man-machine interfaces in the supervision room. The man-machine interfaces must supply the right vision of the process state. From the information, which are presented on the screen, the operator must conjure up a correct structural, functional and behavioural model of the installation. However, it will be false to believe, that the quality of the interface can be a stopgap. Workers must be high motivated and efficient. A well defined work organization, that can allow the operator to maintain and improve his knowledge and a robust and stable process are necessary too. It is all of these parameters will guarantee a good performance of the global Man-Machine System.

3. RESEARCH IN THE FIELD OF SUPERVISION

Different points which have presented show the interest of researches about supervision. The improvement of dynamic MMS performance requires design of supervisory tools suited to operators. Simply, design of a supervisory tool necessitates to answer the two following questions : what information must be displayed ? and how to display information ?. If we consider that operators keep the choice of displaying views, research can be managed at two levels. The first one consists in improving existing supervisory tools. The second one concerns the design of new tools which could considerably modify the nature of the operator's work.

3.1 Improvement of existing tools

Improvement of existing tools requires supervisory displays better designed. Today, supervisory tools for operators are the following :

• disp^lays presenting all or parts of the process. Very often, they are designed from the lay-out of the plant and they copy the physical structure of the process,

• command displays enabling operator to act on the process,

• trends, which represent evolution of variables during the time,

• alarms.

As strange as it is, supervisory systems have really evolved in terms of power (number of variables, speed, etc.) but not in terms of function. Consequently, the difficult problem of the definition of the content of the displays is still existing. Indeed, what are necessary and sufficient information to enable an operator to supervise efficiently a process ? How to give the means to perceive the state of the operative and control-command parts ? Today, process is seen through views grouping variables. It is simply a filter. The way to filter information depends completely of the designer. More, it is possible, to aggregate and/or to synthesize variables and hence, give other information to operators. One can understand that too much information are not good. How to select and to group information is an interesting field of research. In addition, representation of information can be modified. Concerning that, these last years, new kinds of displays have been appeared aiming at facilitate the perception of process state and the understanding of the behaviour. Ecological interfaces like MDD (Mass Data Display) (Beuthel, 1985) or these which have been developed by Vicente (1995) or the PCED (Process Control Event Diagram) (Chung, 1997) are some examples. Even if these approaches do not solve all problems, they are interesting because new modes of representation have been proposed which could be used in the future. One can understand that this new generation of display has to be evaluated before being integrated on real site. Because of the human part, the evaluation stage is crucial. Very often, it is difficult on real site to manage a lot of experiments. Consequently, an experimental platform dedicated to supervision presents a big interest in this case.

3.2 Design of new supervisory tools

With regard to progress in automation and particularly in identification, one can think about new supervisory algorithms more efficient, using for instance analytical redundancy equations of the process. Integration of these algorithms inside a supervisory system can considerably modify human activities. Indeed, the a priori knowledge of a normal running model enables to generate more pertinent alarms based, not on a fixed threshold, but on the comparison with the normal value of the variable. More, a dynamic model could be used to generate synthetic information. Research performed at LAMIH about supervision of nuclear fuel reprocessing system is an example (Lambert, et al., 1997). Indeed, researchers have decided, firstly to design supervisory displays using ideas coming from the ecological school, and secondly to define high level information obtained from a dynamic model of the process. In the first case, Mass Data Display concepts for supervisory task, and casual graphs organised into a hierarchy have been used. In the second case, researchers have defined for each variable, qualitative information by the mean of a colour (from green to red) which are function of the gravity. Gravity is defined by the mean of the residue shape (it is to say : shape of the difference between the measure and the value of the model). More, shape of the residue is displayed in a symbolic way in each node (which represents a variable) in the casual graphs. Experiments performed with experienced operators have validated several proposed ideas. Indeed, the use of a normative model of the process has been accepted and appreciated by operators. The use of the symbolic information shape of residue and the colour code linked to the gravity permitted to operators an easier and an earlier detection of several abnormal running modes (slow drift for instance).

The use of an experimental platform enabling to analyse operator's activities is necessary in order to evaluate new supervisory tools.

4. AN EXPERIMENTAL PLATFORM DEDICATED TO SUPERVISION

Previously, we have shown necessity whatever the field of research about supervision of an evaluation stage. Only an experimental approach with operators can be thought of doing in order to test interest of new supervisory tools. Very often, performed studies use a specific platform. It is the case for instance for the GNP (Generic Nuclear Plant) developed by RISØ (Nielsen, 1991) and the Lee's pasteurization process (1994). Consequently, each time a new study has to be performed, a new platform has to be realized.

Hence, it seems necessary to design and to realize a platform enough open to be able to simulate different processes with different kinds of command. In addition, because our field of research concerns Man-Machine Interactions, it seems very interesting to simulate failures. Indeed, analysis of new supervisory tools presents an interest during the problematic stages : start-up, diagnosis, different running modes and so on.

4.1 Specification of the experimental platform

The main goal in the design of this platform is to create a virtual process completely autonomous which could run on Personal Computer. The platform is composed of a simulator, a supervisory software and an analysis software (see figure 1). The development of an experimental protocol requires the definition of the objectives and the choice of an experimental plan. This work necessitates to analyse the studied process in order to get one or several models. The design of Man-Machine Interfaces (MMI) and the realization are the last step. Problems come from the design of the simulator, the communication with MMI and at least the data analysis which is often performed with statistic tools or classical software like EXCEL. The objective of this platform is precisely to avoid difficulties of development which increase the time of development and analysis of results and to avoid the realization of specific platforms which can not be easily updated.

The three units (simulation, supervisory and analysis) are now going to presented.

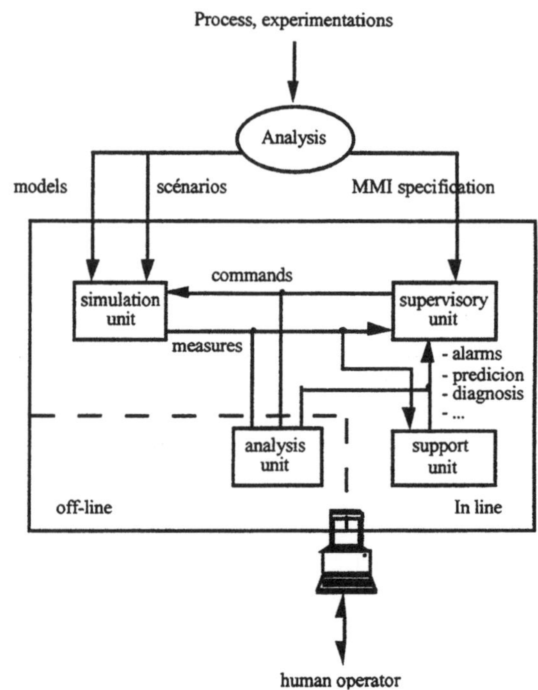

Fig. 1. General structure of the platform

4.2 The simulator

The heart of the platform is a simulator which has to be generic. It is to say : the simulator must be able to simulate different processes. It also has to be dynamic and it must react to external actions like setpoints or running modes change. The factor time has to be respected as well as possible. The simulator has to take into account the operative part but also the command part (sequential chart or feed-back control loops). In addition, the simulator has to integrate different functions linked to the objectives of the platform. To study the human activities, it is often important to place operator in situations which are not classical. So, it is important to integrate the possibility of generating noise and breakdowns on the command part, sensors, actuators, and so on.

Simulation of a process with the simulator is based on mathematical equations describing the behaviour of the variables. These equations can be temporal, recurrent or using numerical integration or a combination. The simulation is cadenced following a sampling period which is fixed with regard to the characteristics of the process. Of course, this sampling period modifies the accuracy of the results.

The description of the process is based on the declaration of the variables and the equations which are placed in files which are compiled during the initialisation phase by the simulator. The process is placed in an initial state with a specific file containing initial values of variables. After, equations are computed in order to have values of variables in real time. Data can be stored on hard disk during simulation with a specific sampling period.

The simulator can receive orders from the supervisory unit. The communication is performed with a specific protocol (DDE, DLL, TCP/IP, and son on).

The internal structure of the simulator is based around two units : variables management unit and updated variables unit.

variables management unit : it manages process variables referenced by its name and offers several functions enabling to read and to write variables, to fix thresholds, to add noise or failure on variables. Functions are internal to this unit and for this reason completely transparent for the user.

- For each variable, this unit manages the current value (at t time) and the previous values in a limit fixed in the description of the process. This function enables to manage in a simple way problems of time lag. Variables are updated by the updated variables unit. That can be blocked by the mean of a condition on the variable.

- It enables to take into account automatically low and high individual thresholds in order to limit value of variables. This is very useful to manage several non linearities.

- It is possible to add noise on variables. This noise can be either Gaussian or white and it evolves between a minimum and a maximum defined in the description files of the process. This is useful to simulate sensors.

- four kinds of failure can be automatically managed on each variable: variable at zero, variable at a constant value, offset and shift of the normal value. This is completely integrated in the simulator. Consequently, it is not necessary to take into account all the failures in the equations. At least, it is possible to get the value of the variable with or without the noise and the breakdown.

Updated variables unit: it computes at the time t+1, new values of variables with regard to equations. The order of the equations is very important. It is why the process can be described in several units (files) having the same format. For instance, one can decompose the process in control loops, actuators, external inputs, operative part and sensors.

So, the process is described with units having the following syntaxes : definition of constants, definition of variables and equations. Each equation can be conditioned by a logical equation. In the equations, all the classical mathematical functions can be used. In addition, one can use two specific functions in the equations : "time" and "dt". Function "time" gives the consumed time since beginning of simulation and "dt" is the sampling period. This simple way to write equations is interesting. Indeed, the up-date condition enables to simulate numerical regulators (RST type for instance). Conditions associated to each variable enable to simulate different running modes (automatic, manual and so on) and different kinds of non linearities.

4.3 The supervisor

The choice of the supervisory software is free. However, it is necessary to choose a software enough open with good performances in terms of animation and development of views but also in terms of communication. With the supervisory software, it will be easy to realize MMI on several computers, the real time database being managed by the supervisor and the communication between computers directly integrated.

4.4 The analysis tools

The results analysis stage of an experimental protocol is very heavy and demands a lot of time. Two kinds of data are principally analysed: objective data and subjective data. Objective data come from a "software spy" which stores all operator's actions. So, these data characterize physical operator's activities (change of setpoint, change of display and so on). Subjective data come from questionnaires or interviews. The analyst's work consists of determining operator's cognitive activities from these data. Tools like Mac-Shapa are useful during this stage (Sanderson et al., 1992). Results are often one or several files containing the set of physical and cognitive activities time dated for each scenario, each operator and each context. The next stage consists of putting in evidence the differences inter and intra operators. Statistical softwares or EXCEL are often used. It is at this level that our software is useful. It has been developed under ACCESS and it has been developed for non computer scientists and includes three functions: importation, data processing and data synthesis.

Notion of project and data importation : in order to facilitate data analysis, the tool proposes a project organization. So, it is possible to create, modify and delete projects. Data importation is easily managed by the mean of ACCESS assistant directly integrated in the tool. Imported data are stored as tables with defined format.

data processing : this operation consists of processing data by the mean of SQL requests. These enable to manage simple or complex data processing. It could be for instance interrogation of the table (example: selection of actions performed by one or several operators) or definition of indicators (example: consultation rate of one variable) or the research of cycle activity. For that, user has several possibilities. In addition of ACESS assistants, the tool proposes an assisted mode which enables to adjust SQL requests. Hence, it is possible to build requests "step by step" very simply with few knowledge about databases.

data synthesis: this operation consists in producing documents (states) or screens (forms) presenting the results of requests.

The tool has been used to analyse results coming from experiments performed with professional operators on a supervisory platform integrating new supervisory concepts. Results quickly obtained by non computer scientists have shown the interest of the tool.

5. CONCLUSION

The platform has been tested by the mean of several examples coming from automatic control. Particularly, a pasteurization process with about 150 variables and different running modes has been developed rapidly and is now used for laboratory experiments. A global approach to design a platform dedicated to supervisory studies seems interesting because its flexibility will enable to save time in development and in analysis of results.

6. REFERENCES

Bainbridge, L. (1983). "Ironies of automation", Automatica, vol 19, n°6, pp. 775-779, 1983.

Benzian, Z., J.L. Ermine, C.M. Falinower and B. Bergeon (1994). Modélisation des connaissances et spécification de systèmes de supervision de centrales thermiques. Journée d'Etude S3 (Sûreté, Surveillance, Supervision) du 8 décembre. Stratégies de conduite en présence de défaillances.

Beuthel, C. et al. (1995). "Advantages of mass-data-display in process S & C", *Proceeding of 6th IFAC symposium on Analysis, Design and Evaluation of Man-Machine Systems*. MIT Cambridge,USA, 1995.

Chung, P.W.H. et al. (1997). "Functional Modelling for hazard identification", Fith International workshop on functional modeling of complex technical systems, Paris, July 1997

Johannsen, G. (1982). "Man-machine systems - introduction and background.", *Proceeding of the firstIFAC Conference on Analysis, Design and Evaluation of Man-Machine Systems*, Baden-Baden, Federal Republic of Germany, p. xiii- xvii, 1982.

Lambert, M. (1997). "Conception d'une Interface de Nouvelle Génération pour la Supervision d'un Procédé de Retraitement de Combustibles Nucléaires", 6ème congrès français de génie des procédés - Paris, Cité des Sciences et de l'Industrie, 1997.

Lee, J. et al. (1994) "Trust, self-confidence and operator's adaptation to automation", Int. Journal of Human-Computer Studies 40, pp. 153-184, 1994.

Lejon, J.C. (1991) *L'évolution de la conduite sur S.N.C.C.* Edition Dunod, Collection point de repère.

Millot, P. (1988). *Supervision des procédés automatisés et ergonomie*. Hermés, Paris.

Nielsen, F.R. (1991). "Simulation of a PWR Power Plant for Process Control and Diagnosis", RisØ National Laboratory, Roskilde, Denmark, 1991.

Rouse, W.B. (1983). Models of human problem solving : detection, diagnosis and compensation for systems failures. Automatica, vol 19, n° 6.

Sanderson, P. et al. (1992). MacSHAPA and the enterprise of exploratory sequential data analysis (ESDA). International Journal of Human-Computer Studies, 41, 633-681.

Sheridan, T. (1985). "Forty-Five Years of Man-Machine Systems : History and Trends", *Proceeding of 2nd IFAC Conference Analysis, Design and Evaluation of Man-Machine Systems*. Varese Italy, 1985.

Vicente, K.J. (1995). "Ecological interface design: a research overview", *Proceeding of 6th IFAC symposium on Analysis, Design and Evaluation of Man-Machine Systems*. MIT Cambridge,USA, 1995.

SUPERVISING AUTOMATION

Peter A. Wieringa, Zhi Gang Wei, Kang Li, Henk G. Stassen

*Man-Machine Systems & Control, Faculty of Design, Engineering and Production,
Delft University of Technology*

Abstract: This papers deals with the limitations to human centered design of partly automated systems. The application domain is the process industry assuming continuous processes. It will be argued that it could be advantageous to assign some tasks, that could otherwise be assigned to an automatic system, to the human operator. But we will also show that the number of sub-systems, which have some kind of interaction to each other and that form a system, should be limited so that the not more than about 20 to 40 serious control actions per hour should be taken by the operator at all times. *Copyright © 1998 IFAC*

Keywords: Supervisory control, Alarm systems, Operators, Stress

1. INTRODUCTION

Systems are becoming more complex due to the many new links and dependencies among various domains. Our society becomes automated in many ways and the feeling that even the smallest error of an falling automatic control system may cause a disaster, exists. The mechanisms that underlie such feelings may be compared with the butterfly effect: small numerical errors in the initial conditions of a deterministic whether model can unfold into simulated catastrophes; a finding by the meteorologist Lorenz described in a book on chaos by Gleick (1988).

Technological systems such as chemical processes show also an increasing number of interactions between different domains and between plants. Sometimes very tied links are made between plants owned and operated by different companies. The agreements between the companies, among other facts, affect the operations of the plant and thus the way operators are supposed to supervise. At the same time process control is automated further putting the operator only occasionally in control of the plant; namely during start-up, shut-down and mall functioning. Our research is focused on the dilemma that the increase in complexity due the on the in the context of an increasing complexity.

Normally the tasks in a control room vary by nature and by number. In other words there are monitoring, administration, communication, set-point adjustment tasks etc. which differ by nature. At the other hand a number of for instance the monitoring tasks may be of the same nature e.g. reading the temperature of identical reactor cells, or pressures in vessels.

A tendency exists to create larger control centers which control a larger plant or several plants. The advantage being that functions that require little workload can be combined, concentrated and executed by a smaller number of operators. Furthermore, the concentration of various tasks increases the number of tasks with the same nature creating enough workload to be executed by one operator or specialist. Some of these tasks may be almost independent in time and space of the tasks executed during normal operating modes. Consequently, these tasks may be executed by specialists outside of the control room (e.g. optimizing a plant, maintenance, administration). Although the consequences of this reasoning can be seen in plants already the danger exists that the situation awareness of the plant operators that remain in the control room becomes less.

We investigated the relationship between on the one hand the number of sub-systems and strength of the interconnections and on the other hand the operator performance, mental load and perceived complexity. Furthermore, the effect on mental load and system performance was studied in a system that was partly automated and that suffered from a sudden loss of automation.

2. EXPERIMENTAL SETUP AND RESULTS

In the next section the artificial system and the experiments performed with that system will be described. In the next system the laboratory system is depicted.

2.1 Artificial system: System description and experimental set-up

The system consisted of a series of first order systems, called sub-systems, that where connected in a strictly forward sense (cascade connection). The number of input-signals and output-signals of the system equaled the number of sub-systems. The number of sub-systems for a specific experiment was either 4, 8, 12, or 16. The steady state gain of each sub-system was 1.0. For a specific experiment the coupling strength (gain) between the sub-systems was either weak (gain zero), medium (gain 0.25) or strong (gain 0.5). Set-point requests were generated to stimulate operator actions. The mean time between set-point request was either 6, 3, or 1.5 secs.

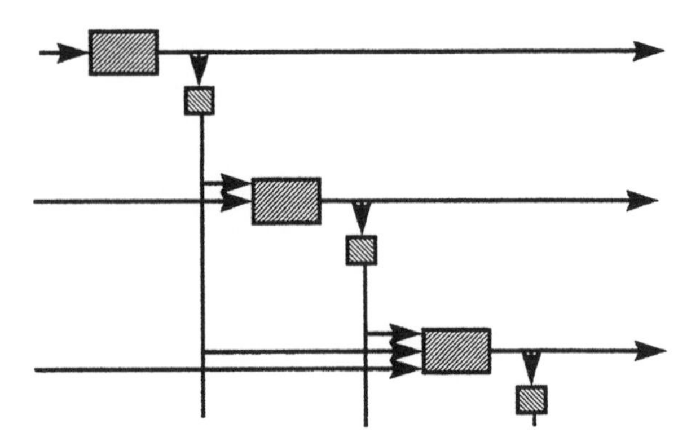

Fig. 1 Artificial System; example with 3 sub-systems.

The large boxes (//) are first order systems whereas the small boxes (\\) are connection gains.

A large display was provided showing all sub-systems in a regular matrix on the screen. The first sub-system, which output was connected to all other sub-systems, was positioned on the top left. The last sub-system, that received input signals from all other sub-systems appeared on the lower right corner. The operator could use a mouse or a keyboard to activate sliders on the screen or to create incremental steps or numerical values for the inputs to the sub-systems. Students of our department volunteered in the experiments and received a fee. Each experiment lasted 30 mins.

The performance was obtained from the error signal between the set-point request to a particular sub-system and the output of that sub-systems.

Results: Stassen, et al. (1993) used the artificial system to derive at an index for the system complexity. They found consistent results between the performance and the complexity index as long as the set-point request frequency was below 360 per hour, the number of sub-systems was about 8 (the did not use a configuration with 12 subsystems) and the gains of the interconnections was 0.5 times the steady state gain of the subsystems. For systems consisting of 12 sub-systems inconsistent results were obtained and the operators complained about the difficulty operating the system; even with low set-point request rates.

Wei, et al. (1998a) used the same system but with 12 sub-systems and a connecting gain of 0.25. They showed that the 12 subsystems could be controlled by the operators. Besides measuring the performance they used the so-called Rating Scale Mental Effort (Zijlstra, 1993) to obtain a subjective measure for the mental load.

Wei (1998b) also used the same system and performance criteria but equipped some of the sub-systems with an automatic control system to take care of the set-point changes. They used the task complexity (a.o. measures) to obtain an index for the Degree of Automation (DofA). The task complexity for each sub-system could be obtained by investigating the minimum number of action that should be taken to adjust a new set-point for a specific sub-system. This was done for each sub-system in a situation that all sub-systems were assumed manually controlled. The DofA was 0 if all sub-systems were manually controlled whereas the DofA was 1.0 in case all sub-systems were automatically controlled.
The experimental protocol included the falling (and recovery after some minutes) of the automatic control system for some of the sub-systems. It appeared that the operator could adequately take over the control of those sub-systems that were equipped with a falling automatic control system. In this case a temporary increase in mental load and a dip in performance occurred shortly after the automation failed but the values of these measures stayed within acceptable limits. This indicates that the operator awareness of the system state before the automation failed, thus at a higher degree of automation, was adequate to respond to the emerging situation.

2.2 *Laboratory experimental set-ups: System description and experimental setup.*

Wieringa and Li (1997) used the results by Stassen, et al. (1993) obtained for the artificial system on a somewhat more realistic system, namely a laboratory, experimental setup. The laboratory system consisted of heat exchangers with a controllable flow of cold water into a reservoir and hot water supplied via a separate tubing system. The temperatures of the inflow hot and cold water streams could not be controlled.

The task of the operators was to direct the water temperature in the reservoir to a certain value. Up to 5 sub-systems could be connected, depending on the experimental protocol. The sub-systems were connected such that the water temperature of one reservoir affects the valve setting for the cold water inflow to another reservoir. Wieringa and Li studied systems that were:

I. completely disconnected,
II. only connected in a cascade,
III. connected by weak feedback and feed-forward links, or
IV. connected by strong feedback and feed-forward links.

This classification was taken as an index for complexity so that configuration I had the lowest complexity index and system IV the highest.

The operators experienced different complexities of the system and difficulties in operating the system. A subjective rating scale was used to obtain these measures. The number of key strokes and the time needed to accomplish a set-point change were also recorded.

Results: Wieringa and Li (1997) showed in a preliminary study using this system that the perceived complexity and difficulty could be described by a linear function of the number of sub-systems in the system. The relationship shifted almost in parallel as a function of the way the sub-systems were interconnected, i.e. the index for complexity.

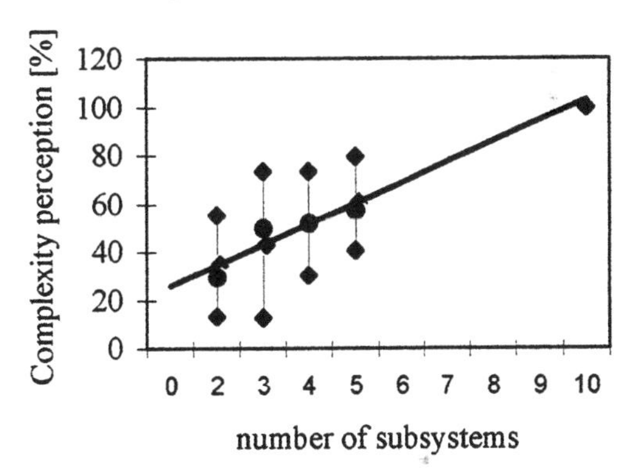

Fig. 2 Complexity perception [%] for a system configuration III. Mean (•) and 2×st.dev. ranges (•) are indicated (N=4). The solid line shows that a linear extrapolation reaches 100% when the system consist of 10 sub-systems.

The system with only disconnected sub-systems (I) was much easier to control than the other 3 configurations (II-IV). The system with configuration II and III scored equally well whereas the system with strong feedback and feed forward interconnections scored worse.

Although Wieringa and Li used at most 5 sub-systems extrapolation of their results predicts that the operators would have given a rating of “too complex” when the number of sub-systems would have been about 10 in case the sub-systems where linked (II to IV). Fig. 2 shows the example for system configuration III. The average time needed to perform this job was, after extrapolation as well, 30 min. (Fig. 3 shows the example for system configuration III where the extrapolated average operation time was 40 min.) This suggests that the operator was not willing to or could not spent more effort to adjust 10 set-points in about 30 to 40 min. or 15 to 20 set-point per hour.

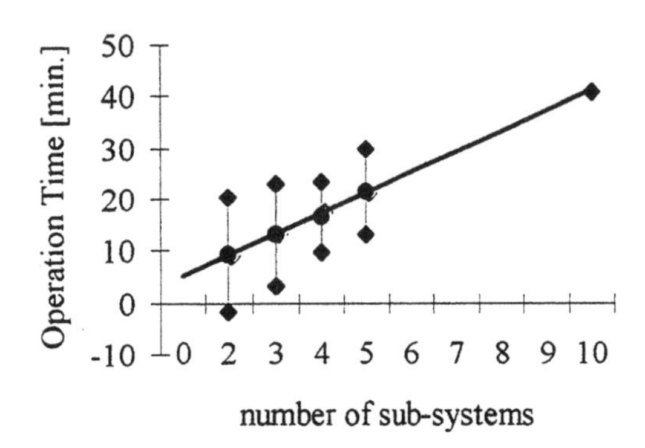

Fig. 3: Operation time as a function of the number of sub-systems for system configuration III. Mean (•) and 2×st.dev. ranges (⧫) are indicated (N=4). The solid line shows that a linear extrapolation reaches 40 min. when the system consist of 10 sub-systems.

3. DISCUSSION

The experiments show that the number of sub-systems that a human operator is able and willing to control is limited. Even if part of the system is automated one should take into account that the operator has to be able to control the system after an outage. In that case the workload may cause an unacceptable rise in workload and drop in performance.

Interconnections between sub-systems have a strong effect on the complexity perception. The gain of the interconnected systems should not be too high because otherwise the total system becomes too difficult to control. This is understandable because high interconnection gains between two sub-systems makes them less distinguishable from each other. Hence, in that case a new system is made with an order that equals the sum of the orders of the original two subsystems. The order of the assembled system may be higher than 3 which makes it too difficult for the human operator to control.

Wieringa and Li (1997) showed that for a system with relatively weak connections between the sub-systems the presence of feed-back or only strict feed-forward connections did not affect the experimental results differently. Stassen, et al. (1993) showed that the interconnecting gain affected the results a lot more. Therefore it seems that the magnitude of the gains of the links is more important than the configuration of the sub-systems.

3.1 Implications

It is well known that operators sometimes have to handle more than 2000 alarms per person per 24 hours with peak workload lying in the order of 900 per hour. Our experiments reveal that the number of sub-systems that can be operated upon should be restricted to about 20 per hour. How does those two numbers relate?

There are a few factors that need to be taken into account when bridging the gap between these two numbers:

- The operators used in the experiments were students and did not have a long training background and experience for operating the plant. Although the results showed that the subjects were at the end of their learning curves it is plausible that they may develop strategies when operating the simulated plants for a longer period of time. We therefor estimate that the number of subsystems that can be controlled by professional operators will be an order of 2 larger; say 40 per hour.
- The experimental protocol did not push the operators to go to the limits. However, for the first set of experiments using the artificial system, when 360 set-point requests per hour were given for 16 sub-systems, the subjects reported that this was too much and some panicked. More consistent results among the subjects were obtained when the number of sub-systems was less, namely 8. No mental load measurements were made at that time. The results suggest that repetition of alarms should cause less problems than the number of interconnected sub-systems.
- Not all alarms of the 900 per hour that are mentioned for the industrial plants are coming from separate sub-systems or different process variables. What we need to know is how many alarms required actions from the operator comparable to a change in set-point values. Often alarms have causal relations and are repetitive etc. Classification of the alarms into categories such as standing alarms, consequence alarms, repeating alarms, precursor alarms, and redundant alarms, reveal that the number can be reduced considerably. Examples of 80% repetitive alarms and 15% causal alarms have been reported to us (personal communication). Only 5% are real alarms that require considerable operator work. Furthermore, the repeating alarms are mostly coming from a

few subsystems. Hence, the number of serious alarms that should be dealt with by the operators are an order of 10 smaller. It is reasonable to assume that the number of serious alarms that the operators had to work on in this example lies in the order of 90 per hour. The rest causes nuisance and does probably not contribute to the fault diagnosis process.

This analysis shows that the workload of professional operators is probably much too high during periods of alarm flooding. It is necessary to reduce the number of alarms that require no serious attention and action from the operator. These findings have serious implications for the design of partly automated systems.

REFERENCES

Gleick, J. (1998). *Chaos, making a new science*. Penguin Books.

Muir, B.M. and N. Moray (1987). Operators' trust in relation to system faults. In : *Proc. of 1987 IEEE Conference on Systems, Man and Cybernetics*.

Stassen, H.G., J.H.M. Andriessen and P.A. Wieringa (1993). On the human perception of complex industrial processes. In: *Preprints of the 12th IFAC World Control Congress*, **6**, pp. 275-280.

Wei, Z.G., A.P. Macwan, P.A. Wieringa (1998a). A Quantitative Measure for the Degrees of Automation and its Relation to System Performance and Mental Load. (accepted for publication *The Human Factors Journal*).

Wei, Z.G. (1998b). Mental Load and Performance at different Automation Levels. *PhD thesis* Delft University of Technology, Delft, The Netherlands.

Wieringa, P.A., Li, K. (1997). Reducing Operator Perceived Complexity. *Proc. of the IEEE International Conference on Systems, Man, and Cybernetics*, **Vol 5(5),** p. 4498-4502.

Zijlstra, F.R.H. (1993). Efficiency at Work. *PhD thesis*, Delft University of Technology, Delft, The Netherlands.

DESIGNING METHOD OF HUMAN INTERFACE FOR SUPERVISORY CONTROL SYSTEM

Shinichiro Hori
Yujiro Shimizu

Takasago Research & Development Center, Mitsubishi Heavy Industries, LTD.
2-1-1 Shinhama, Arai-cho, Takasago, Hyogo, JAPAN

Abstract: In this paper, designing methods of human interface for supervisory control systems are proposed. For the design of hardware systems, especially the design of layout, human factors engineering was applied in terms of visual field and visual angle. On the other hand, for the design of software systems, especially the design and evaluation of a construction of screens, cognitive engineering was applied with the method of analyzing structure. *Copyright © 1998 IFAC*

Keywords: Human-machine interface, Human factors, Cognitive science, Supervisory control, Computer control system design, Structured analysis, Computer hardware, Computer software

1. INTRODUCTION

In these days, supervisory control (SVC) systems have been changed from the distributed one to the centralized one. The former is used by operators who operate machines or plants nearby them, and the latter is used by fewer operators who operate the objective plants in a SVC room. Accompanied with this change, the systems are requested to process or display much information, and so their human interface (HI) must be changed, too. The traditional typed HI has consisted of some analog meters, consoles, and graphic boards. However, the new typed HI has consisted of some CRT displays (CRTs) and a large-scaled screen (LSS), and it has been applied to many industrial plants recently (power plants, railway and traffic SVC systems, iron and steel plants, drainage pump plants, chemical plants, etc.). The new typed HI has some features in the following.

- It has much displaying area and can display much information than the previous one.
- It can be composed compactly by adopting centralized display and CRT operation. And so it needs less space than the previous one.
- LSS like a video wall system helps operators to monitor all around of the objective plants, and to share information between operators.
- It brings various styles of SVC with applying a multimedia technology or various types of displays.

Especially, these features are available for the SVC systems of some plants located in a wide area, or a large-scale plant which has much equipment or many instruments. However, it has not been clarified that how to select devices, information, or displaying methods, and how to design the optimum HI in terms of reliability and compactness. In this paper, the designing flow of HI was tried to clarify and the designing methods concerning with specifications and layouts of hardware in a SVC room and a construction of screens for SVC software was proposed.

2. DESIGNING FLOW OF SUPERVISORY CONTROL SYSTEM

To design HI for SVC systems, three main items must be investigated: displaying devices (specification, number, size, layout), displaying styles of information (quantity, media, size of character), and displaying methods (timing, position in a screen, method to change a screen). These items have some relations that can have influence with each other. For example, if displaying area is too small, many screens and a task of frequent changes of screens are required and it makes an operator's workload increase. On the other hand, if displaying area is large and there is too much information on a screen, it makes an operator's workload increase, too. So, it is necessary to arrange and clarify a designing procedure in consideration of independence and priority of each item. The designing flow diagram of HI for SVC systems is given in Fig. 1. HI can be designed by following this flow with considering some conditions :

- Features of a plant : operators' number and organization, space of a SVC room, quantity of information, system function for SVC, etc.
- Human factors : an ability to process information, range of visual field or visual angle, etc.
- Limitations of specification, cost of displaying devices

3. ACTUAL JOB TO DESIGN SUPERVISORY CONTROL SYSTEM

In this section, an actual job in each step in Fig. 1 is mentioned with examples of application to the SVC system for some plants in wide area. This system is assumed to be consisted of one LSS and some CRTs used by two operators (a chief-operator and an assistant-operator).

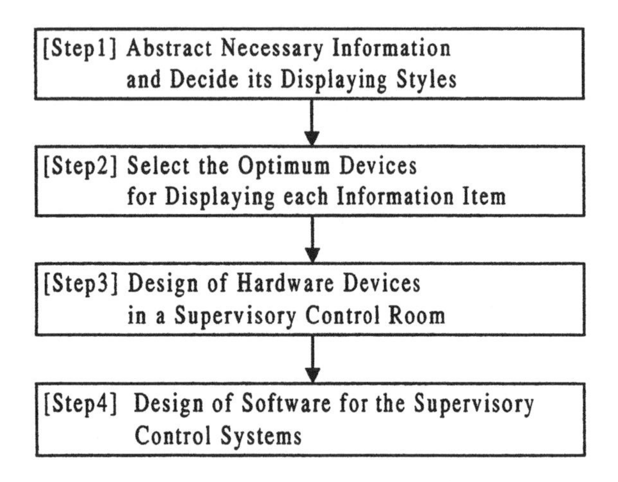

Fig.1. A flow diagram shows a procedure of designing human interface for supervisory control systems.

3.1 Abstract necessary information and decide its displaying styles

In the 1st step, to abstract necessary information for each operator, a task analysis method is adopted. A procedure of this method is explained in the following.

1) Each operator's task sequence is clarified in each operation mode (normal, trouble shooting, etc.).
2) Necessary information items are abstracted in each task.
3) A displaying style and media (text, graphics, image, sound, etc.) of each information are selected.

An example of the result of task analysis applied to the above SVC task is demonstrated in Table 1.

3.2 Select the optimum devices for displaying each information items

In the 2nd step, a displaying device for each information must be selected. CRT, liquid crystal display (LCD), and LSS are candidates of them. The selected devices for each information is shown in Table 2. LSS is suitable to monitor information with graphic or image styles, that uses wide area in a screen and is required to be shared between operators. LCD is suitable to monitor SVC or operation support information, that needs more fine resolution and less space on a desk top. CRT is available for all kinds of styles of information, and so it can be a back up device for another devices.

Table 1 An example of the result of task analysis applied to a SVC task. Necessary information and its displaying styles is decided in each operator's task.

[Operation Mode : Normal Operation]

Task Sequence	Operator		Information Item	Styles of Information
	Chief	Sub		
1.Monitoring data of environment of the plants	◎	O	rainfall, level	graphic
2.Confirm the status of each plant	O	◎	plant: start/stop gate : open/close	graphic,text ITV image
3.Predict status of the plants and Plan to operate plants	◎	O	current and predicted levels	trend graph
4.Check operation mode and Prepare to start	O	◎	start condition	table list
⋮	⋮	⋮	⋮	⋮

(◎ : Main person, O : Assistant Person in charge of each task)

Table 2 The result of selection of displaying devices for each information of SVC system

Item of Information	Displaying Devices		
	CRT	LCD	LSS
Graphic information of total basin	O	△	◎
Image information of ITV camera	O	△	◎
SVC information with trend graph or graphic style	O	◎	△
Operation support information with text or table style	◎	◎	△

(◎ : Suitable, O : Possible, △ : Unsuitable)

3.3 Design of hardware devices in a SVC room

In the 3rd step, design items : number, size, and layout of hardware devices in a SVC room must be decided considering space of a SVC room, limitation of human factors, necessary functions, and operators' organization.

Design layout of a large-scaled screen. At first, layout of LSS, one of main devices in a SVC room, is examined in view of human factors, especially a range of visual field and readable size of a character on a screen.

The relation between human ability to receive information and a range of visual field is examined by Sanders (1970) and the following results are obtained as a guideline.

- 0-30[degree] : Easy to see with little movements of eyes (the best range in the visual field).
- 30-80[degree] : Easy to see with natural movements of eyes.
- over 80[degree] : Necessary to move the head.

Shurtleff (1980) examined the readable size of a character on a screen of CRT and the following results are obtained as a guideline of visual angle.

- 0-14[minuet] : Difficult to recognize a character.
- 14-24[minuet] : General size of a character.
- over 24[minuet] : Easy to recognize, but less character can be displayed on a screen.

Applying these guidelines, a distance between LSS and operators' console can be determined on condition that size and resolution of LSS are limited. An example of the result of this designing method is given in Fig. 2. In this example, size of one screen is assumed as 70" and 3 patterns (1, 2, and 4 screens) are examined. Resolutions of all patterns are evened as 80characters x 60lines. As shown in Fig. 2, the optimum distances of 3 patterns are obtained as about 3, 5, and 7[m]. In view of the guideline of visual field, the lower limit is determined, and in view of the guideline of visual angle, the upper and lower limits are determined.

Design layout and construction of displays. Layout and construction of CRTs or LCDs are determined in consideration of kinds of information items, kinds of functions, and organization of operators.

The objective plants, mentioned in the top of this section, are assumed as two-men operation. As shown in Table 1, each operator has a different part in SVC tasks. For example, a chief-operator's tasks are to monitor total area, to predict status of plants or environments, to set up goals, and to plan operation. An assistant-operator's task is to check each plant's status in detail, and to operate them following the plan made by a chief-operator. So it is necessary to distribute information items and functions for each operator. And following this result, layout and a construction of CRTs should be determined. An example of the result of this approach is given in Table 3 and Fig. 3. In Table 3, two plans of display construction is shown. And in Fig. 3, an example of a SVC room layout with LSS (Middle pattern in Fig. 2) and LCDs (Plan A in Table 3) is shown.

Table 3 The result of an arrangement of displays for each operator

Operator	Displaying Information	Number of Displays A	Number of Displays B
Chief Operator	Graphic information of total basin	1	1
Chief Operator	Operation support information with text or table style	1	
Assistant Operator	Operation support information with text or table style	1	1
Assistant Operator	SVC information with trend graph or graphic style	1	1
Assistant Operator	Image information of ITV camera	1	Displaying on LSS

(A: plan of 5 displays, B: plan of 3 displays)

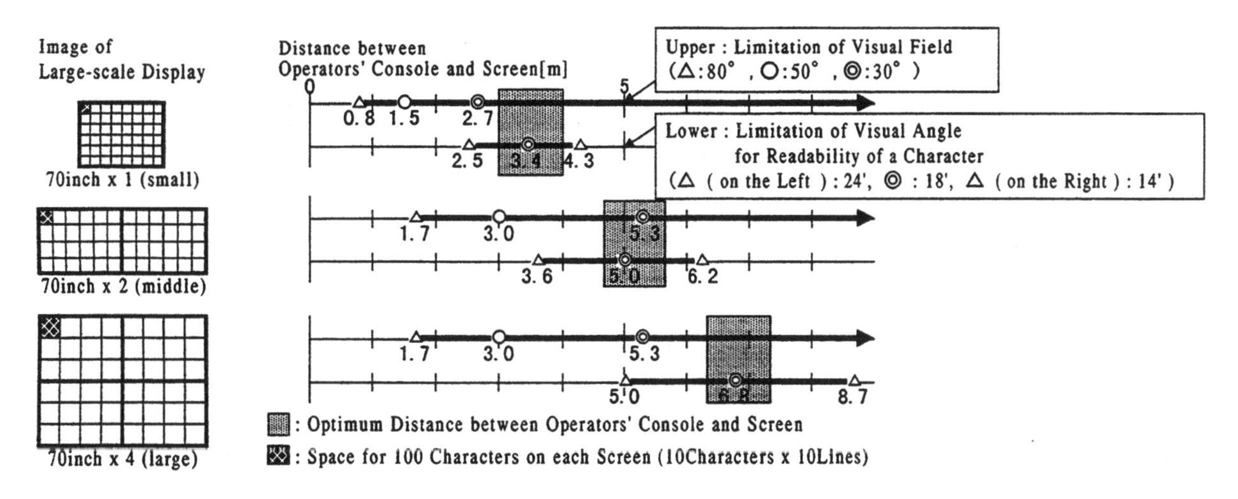

Fig.2. An example of a result of the designing method for a distance between operator's console and a large-scaled screen in a SVC room in view of visual human factors.

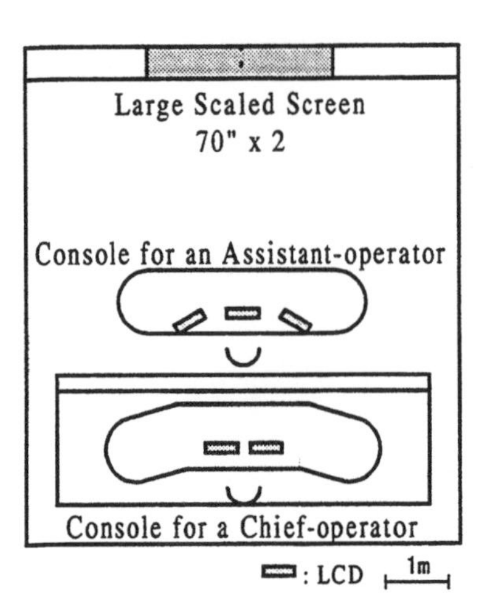

Fig. 3. An example of a SVC room layout with large scaled-screen and liquid crystal displays

3.4 Design of software for the supervisory control systems

In the 4th step, to design software for the SVC systems, it is necessary to investigate a construction of total screens for each displays (general flow of screens or links between relational screens), and a specification of each screen (quantity of information, size and color of buttons or characters, consistency, etc.).

For the latter, some methods to design and evaluate specifications of each screen are proposed in the field of the usability evaluation. For example, a walk-through method with a check-list for a specification of each screen in view of consistency, intelligibility, or clarity can be found in S. Ravden and G. Johnson (1989).

On the other hand, for the former, available methods never be obtained, and its design depends on the subjectivity of a designer. One of the main reasons is difficulty to visualize a construction of screens.

To design and evaluate a construction of screens, one of the analyzing methods of a problem structure, "DEMATAL" (Decision Making Trial Evaluation Laboratory) is adopted to visualize them. A flow diagram of an analyzing process of this method is shown in Fig. 4. A procedure of DEMATEL method is explained in the following.

1) A relation of each set of two screens is evaluated in view of frequency and direction of mutual movements by a system designer. And as the result of this evaluation, a matrix of direct relation is obtained as an initial data of the DEMATEL analysis. In this study, 4 levels are adopted as a relation between two screens.
2) A matrix of total relation including direct and indirect relations is obtained as a result data of the DEMATEL analysis. The indirect relation is defined as a relation between two screens those can move only indirect path from one to another.
3) Using values of D (sum of lines) and R (sum of rows) in the matrix of total relation, a levels of influence and a level of relation are defined. And these feature of each screen can be visualized as the oriented graphs on a 2-dimensional plain.

An example of the result of this method, applied on screens for the SVC system is shown in Fig. 5. With this visualized result, links of screens, general flow of screens, and groups of screens can be clarified. And some problems can be abstracted with attention to a separation of the same group, or a reverse flow to a general flow in the graph of "Before Improvement" in Fig. 5. After the improvement of these problems, screen groups became separated clearly, and the general flow of screens became well-arranged as shown in the chart of "After Improvement" in Fig. 5.

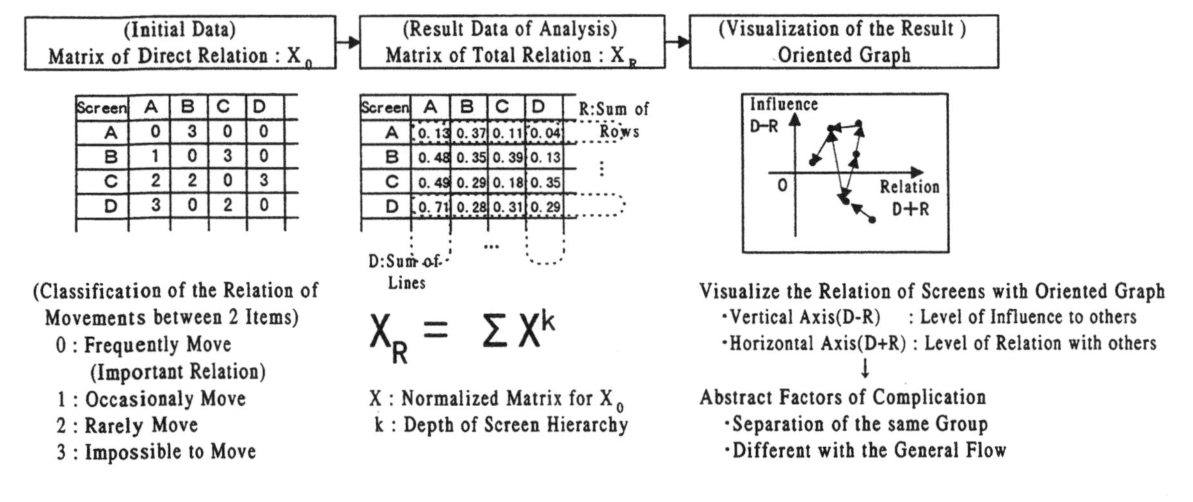

Fig. 4. A flow diagram shows a procedure of the DEMAETL analysis method, applied to visualize a construction of screens.

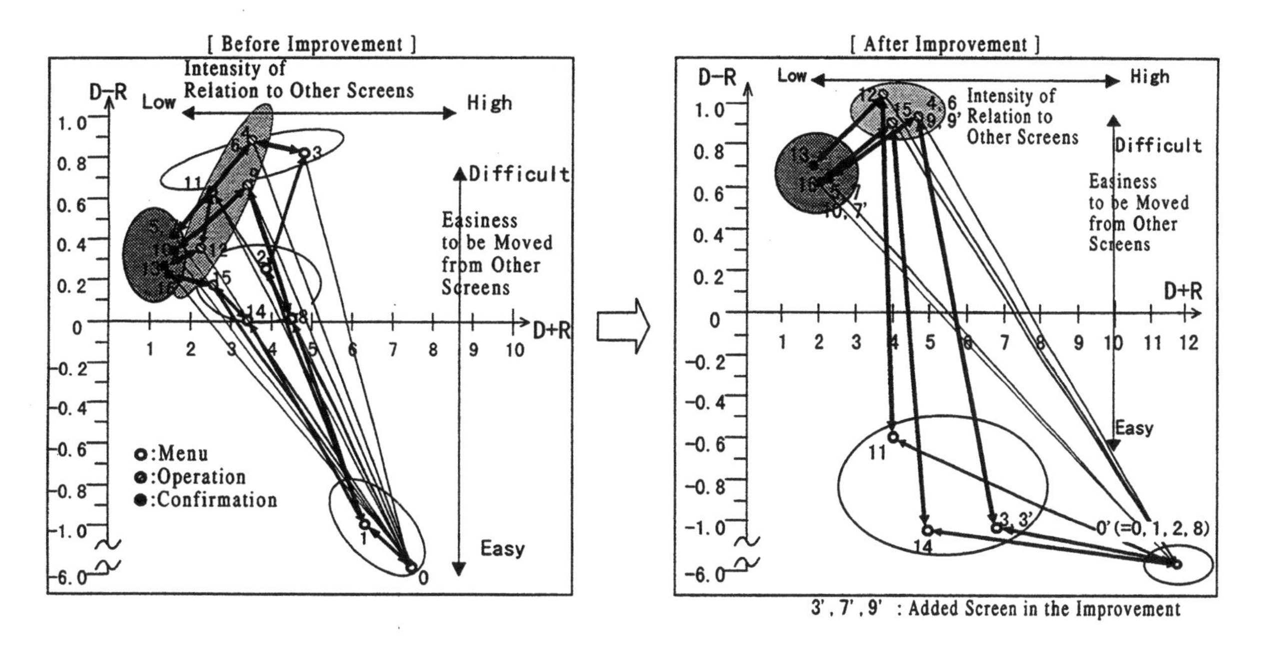

Fig.5. An example of the result of the DEMATEL analysis method, applied on screens for the SVC system.

Besides of above investigation, it must be studied that how to support operators' SVC tasks with software functions, those tasks are tempted to be imposed burdens by limitations of hardware designs. For example, if too much information but not enough numbers of CRTs, operators will be required to change screens frequently and this makes his workload increase. So it is available to urge to change to the optimum screen at the optimum timing with the supporting function of software. This is one of the subjects to study in the future. And an approach using expert system will be able to apply with expanding the approach, that can be found in S.Hori and Y.Shimizu (1994).

4. CONCLUSION

Designing methods of HI for the SVC systems are proposed. For the design of hardware systems, especially the design of layout, human factors engineering was applied in terms of visual field and visual angle. On the other hand, for the design of software systems, especially design and evaluation of a screen construction, cognitive engineering was applied with the method of analyzing a problem structure "DEMATEL". Father study will be continued to investigate the support function of software in the future.

REFERENCES

Sanders, A.F. (1970). Some aspects of the selective process in the functional visual field. *Ergonomics*, **Vol.13**, No.1, pp.101-117.

Shurtleff, D.A. (1980). *How to make displays legible.*, Human Interface Design, La Mirada, CA.

S. Ravden and G. Johnson (1989). *Evaluating usability of human-computer interfaces*, Ellis Horwood, Chichester.

S.Hori and Y.Shimizu (1994). Advanced study of human interface for drainage pump stations. *Japan Joint Automatic Control Conference*, **Vol.37**, pp.403-404.

ANALYSIS, DESIGN AND EVALUATION OF A 3D FLIGHT GUIDANCE DISPLAY

A. Helmetag, R. Kaufhold, P. M. Lenhart, M. Purpus

Prof. Dr.-Ing. W. Kubbat
Head of
Darmstadt University of Technology
Institute of Flight Mechanics and Control
Petersenstraße 30
D - 64287 Darmstadt

VDO Luftfahrtgeräte Werk GmbH
An der Sandelmühle 13
D - 60439 Frankfurt

Abstract: The analysis of today's accidents focused the work on the avoidance of 'Controlled Flight Into Terrain'. Analysis of safety concepts led to the design of the proposed Synthetic Vision System that will be described in detail. Since most information in the 3D-Displays is contained in a graphical way, it can intuitively be seized by the pilots. Especially in phases of high workload, the highly pre-processed information and its redundant presentation will significantly contribute to flight safety. For evaluation this system was integrated into the DLR test aircraft ATTAS and flown at Frankfurt airport in August 1997. *Copyright © 1998 IFAC*

Keywords: Aircraft Control, Flight Control, Graphic Display, Guidance Systems, Human Machine Interface, Man Machine Interface, Supervisory Control

1 INTRODUCTION

Before the analysis phase there was the idea to visualise the flight path and outside view for flight guidance purposes. Work on 'Tunnel in the sky displays' has already begun in the late 50th and was a basis for this development.

2 ANALYSIS

About 75% of all accidents in civil aviation are caused by human factors. A major origin of the problems lies in the aircraft cockpit instrumentation. The pilots have to scan many instruments, convert the information, and mentally work out what is usually called the spatial situation awareness. In conditions of high workload the pilot is often overloaded with information and can loose situation awareness. The loss of situation awareness results in inadequate behaviour or reaction. A typical situation is an approach where the aircraft is near to terrain and reaction times are reduced. In case of lost SA the risk of ground collision is very high. In the next chapter causes of CFIT accidents will be analysed.

2.1 CFIT Problem

'Controlled Flight into Terrain' has become the most important reason for accidents. Typical conditions of CFIT accidents can be seen if one looks at some statistical values. 87% of CFIT accidents are in Instrumental Meteorological Conditions (IMC). The available information about the pilots situation will be discussed later on. 90% of CFIT happen within 15nm from the airport, and 40% without significant terrain elevation. In these cases the pilot crashes with an operative aircraft into flat country. This indicates that there is an enormous lack of information. Analysing the visual information requirements was the next step to the design of a new HMI.

Visual Information Requirement. In order to design a flight guidance display it is necessary to collect information requirements. Beside typical values the focus was especially on visual information that is not available due to bad weather condition. In this work only flight guidance tasks were analysed which are not concerning the aircraft system itself like mode or engine status. For the definition of information requirements the flight was separated into several flight phases like taxi out, take off, climb, cruise, decent, initial approach, final approach, flare, roll out and taxi back. The descend for example was separated once more into approach planning, control of flight path, collision avoidance and system management. Visual information for approach planning includes traffic, weather, energy level, preview of the approach profile and finally preview of the runway.
To complete the requirements it was very important not only to define information that is required for the defined task but also to keep SA since SA is not necessary to fulfil the job. If the pilot follows the demanded procedures he probably will not have an accident even if he has no sufficient SA. But once the pilot makes a mistake the system cannot support him on finding the way back to the flight-path and continue the procedure. This is the main reason for the requirement to ensure SA in any flight phase. Mistakes are the only way to learn and the capability to learn is the advantage the pilot has compared to the automation. Thus it is very important to have an error tolerant system. This includes presentation of limits and reversibility of actions. Condensed: a **kind environment** (Rasmussen, 1987).
The information requirement analysis phase was finished with final pilot discussions. Information presentation defined in a symbology specification was the next step to do. For this purpose today's information presentation was analysed which will be mentioned in the next chapter.

Natural versus Coded Information. The pilots job is to acquire a lot of information and combine it to a complex mental model. It is a kind of multimedia puzzle in a time critical environment. For some parts of this puzzle the way of presentation is mentioned here and compared for Visual Meteorological Conditions (VMC) and IMC. The types of information are distinguished into natural and coded information (Figure 1) (Helmetag, et al.,1997).

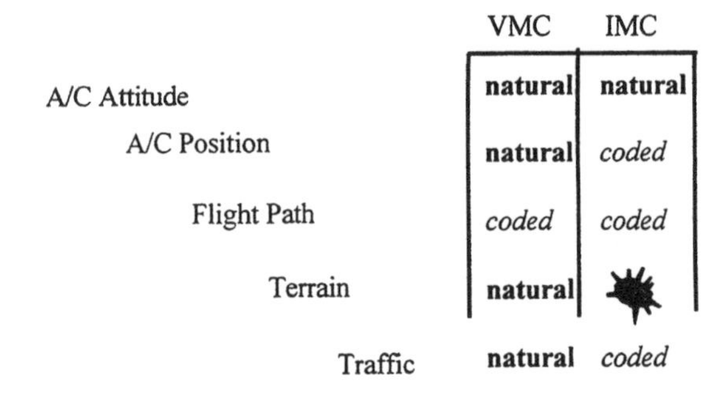

Fig. 1. Information Coding

Natural information means for example the altitude above ground if one looks out of the window compared to the coded one like a digital value indicating the radio altitude in feet. While natural information can be perceived intuitively and thus very fast, coded information has to be decoded and analysed.
If one regards the perception of aircraft attitude in both VMC and IMC one has a natural information in VMC by looking out of the window. In IMC there is the artificial horizon that is the only instrument with a natural pilot interface. In contrast to this, aircraft position and traffic are natural in VMC and coded in IMC. The pilot gets a digital information like altitude or vertical separation to the intruder.
Summary: It is very important to give the pilot the information to keep situation awareness in bad weather conditions. This has to be done in a natural and intuitive way to insure a fast perception and integration into his mental model.

2.2 *Safety Concepts*

A typical classification, as it is known from car constructors, is the separation into active and passive safety (Figure 2).
Active safety is pre-impact safety and passive safety is post-impact safety. Within passive safety we have to distinguish Minimising Post Collision Damage (MPCD) and Minimising Collision Damage (MCD).

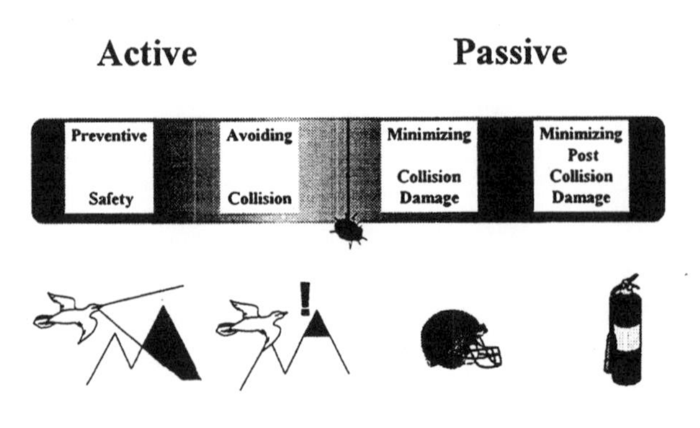

Fig. 2. Safety Concepts

MPCD is the last chance to reduce passenger injuries when the vehicle has crashed and is already burning. The installation of fire extinguishing systems is the final thing a constructor can do for the user.
MCD is a method situated between impact and MPCD. The vehicle is impacting but the chance for passengers to survive can be increased by the introduction of shock absorbers or airbag systems.
These concepts of post impact safety are very important, but a significant increase of safety can only be reached by active concepts.
In the last two decades safety increased tremendously by the introduction of Collision Avoidance Systems (CAS). In the seventies Ground Proximity Warning Systems (GPWS) and in the eighties the Traffic Alert and Collision Avoidance System (TCAS) were introduced. These airborne systems control the actual flight path and alert the pilot if there is the risk of an impact. Beside the alert like „Traffic" or „Terrain" these tools also generate a problem solving solution.
So these tools support a crew that has lost Situation Awareness (SA) and is no more on the right track or is intruded by an other aircraft.
The 3D Flight Guidance Display that is presented in this paper tries to increase safety even one step before the loss of situation awareness. **Preventive Safety Concepts** try to keep the pilot in the loop. He is supported to create and update his mental model with the required information in an intuitive way. If this can be reached a CAS system or passive concepts will never be vital.

3 DESIGN (SOFTWARE)

On 'Commercial of the shelf' Silicon Graphics Computers a software prototype was developed for simulation and flight test evaluations.

The 3D Flight Guidance Display proposed here, is based on a contact analogue synthetic vision system. On the one hand there is an „inside out view" of the actual situation called the Primary Flight Display (PFD), on the other a parallel projection including the same features called Navigation Display (ND) or moving map display. This artificial image of the outside world is extended by virtual elements. For example a spatial flight path predictor and the predetermined flight path are visualised on the display and permit the pilot to intuitively understand his actual situation and react with foresight.
This 3D view is superimposed by conventional 2D symbology like Basic T. They provide exact values for speed, heading and altitude. To avoid the different elements of the display from concealing each other, some of them like the tape scales are made transparent and are placed on the edge of the display.

3.1 Primary Flight Display

In the design of the PFD much attention has been paid to redundancy of information (Figure 3a,b). There is for example the altitude information during approach phase: The first clue to the present altitude is given by the size in which the grid appears which is rendered on the surface of the terrain. This allows to estimate the altitude above the ground at a first glance. When flying an ILS approach the glide slope channel provides the required altitude very precisely in a purely graphical form. Shortly before touchdown the shadow of the flight path predictor gets into the field of view giving another very accurate altitude reference.
In addition to these graphical hints a digital readout is integrated into the altitude scale.

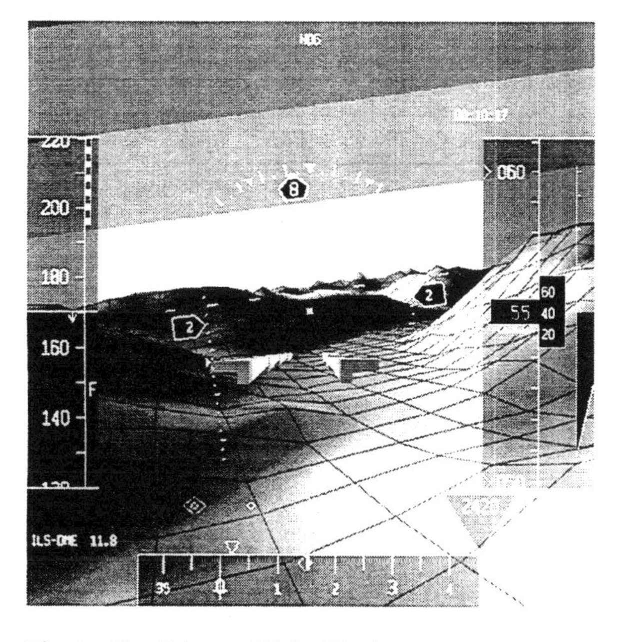

Fig. 3a. The Primary Flight Display

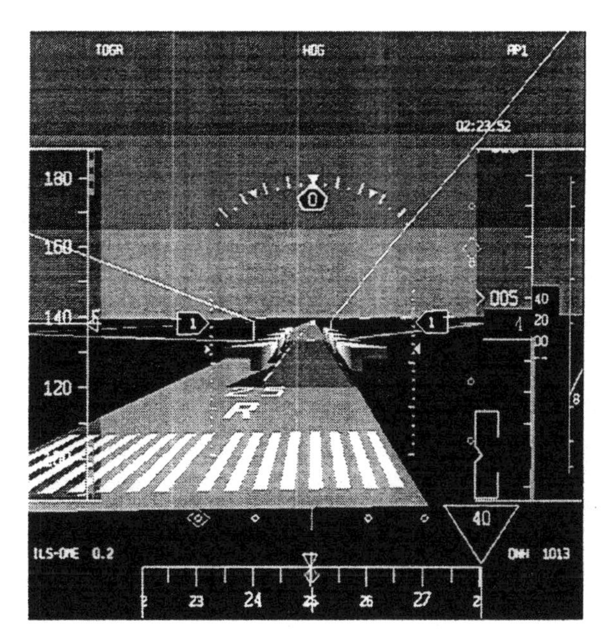

Fig. 3b. The Primary Flight Display

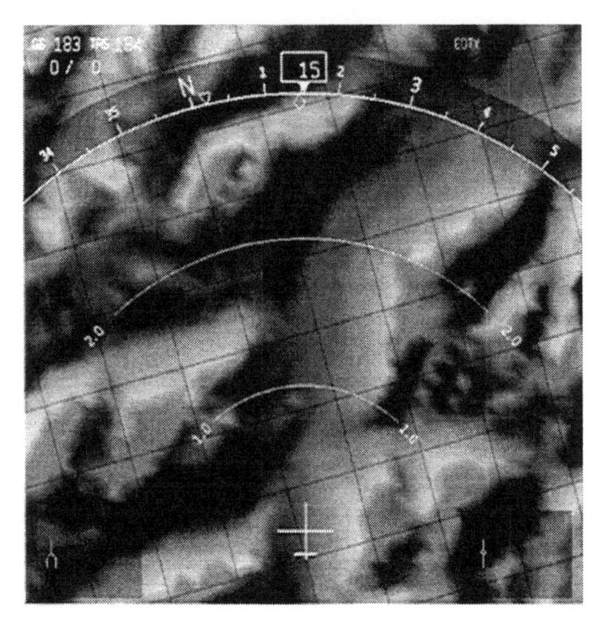

Fig. 4. The Navigation Display

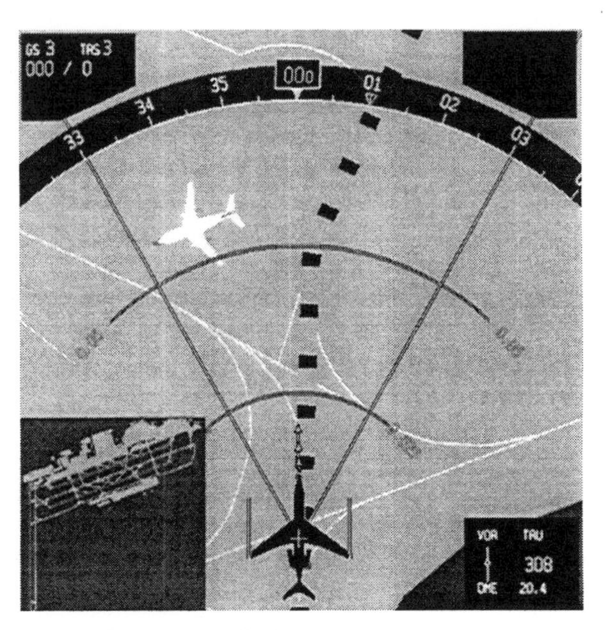

Fig. 5. The Taxi Symbology

Another example for redundant information is the speed indication. The exact value can be obtained from the scale with its digital readout. As an intuitively perceivable presentation the visual flow field (especially the grid) gives a sense of the actual speed. Furthermore, the predictor changes its length and colour according to the speed of the aircraft, thus giving a continuous feedback.
This redundancy of information helps to prevent the pilots from misinterpreting the presented data.

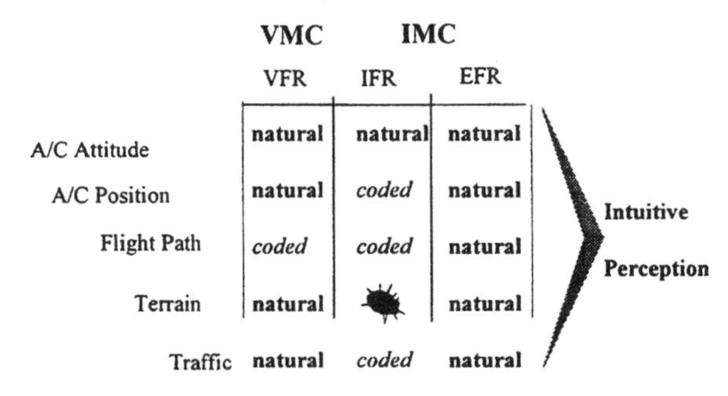

Fig. 6. Enhanced Situation Awareness

3.2 Navigation Display

In order to enable the pilots to develop a complete imagination about terrain, obstacles and adjacent air traffic, the PFD had to be complemented by a NAV Display (Figure 4).
The NAV Display format is computed from the same databases as the PFD, containing the necessary information about the terrain underneath the aircraft. Though using parallel projection it provides again a three-dimensional impression (relief-picture).
The new display combines a conventional navigation display with a map and virtual elements such as navigational aids, air traffic routes, commanded or preplanned flight path, taxi paths (Figure 5) and ground proximity warning symbology.
NAV display zooming is possible, allowing the pilot to observe in detail an area of his choice. The field of view may be selected with the pilots own aircraft at the centre or the lower edge of the display to give the pilot an impression of the terrain and traffic surrounding him.
One example of the integrity of the PFD and NAV Display is the feature of the PFD viewing area, appearing in the NAV Display. Thereby the objects visible on the PFD can clearly be associated with the ones on the NAV Display.

3.3 Prospect

Today visual separation in good weather condition is an important aid to increase airspace capacity; in the near future it will be possible to do the same in IMC if the aircraft is equipped with technology like this.
The 3D Display will greatly enhance the pilots awareness regarding their situation relative to terrain, obstacles and other aircraft. Since most information is contained in a graphical and natural way, it can intuitively be seized by the pilots (Figure 6).
Especially in phases of high workload, the highly pre-processed information and its redundant presentation will significantly contribute to flight safety.

4 FLIGHT TESTS EVALUATION

During the phase of development several tests and presentations have been done in a B 727 generic simulator. For these tests, pilots with varying experience have been selected to get new design ideas and to evaluate the design steps. In order to obtain more convincing results about the usefulness of a concept like this, flight tests have been done in 1996 together with VDO (Purpus, et al., 1997).

Fig. 7. The ATTAS Test Aircraft

Fig. 8. The Cockpit

In August 1997 a second campaign of flight tests was carried out at Frankfurt airport. Frankfurt was selected because this airport combines high traffic density, complex taxi environment and good infrastructure.

4.1 *Test Environment*

The tested 3D Flight Guidance Displays were generated on a ruggedised Silicon Graphics and visualised on a LCD. Test environment was the ATTAS test aircraft of DLR, a VFW 614. It is a 40 seater equipped with a simplex fly by wire control (Figure 7). The navigation was realised by a multi sensor solution including Inertial Navigation (INS), radio altimeter, ILS, D-GPS, data matching and barometric altimeter. Precision of navigation was in submeter range (Pfister, et al., 1997).
The Flight Guidance Display was installed in front of the conventional instruments on the left side of the cockpit while the safety pilot took the right seat (Figure 8). The Display hardware is a commercial of the shelf 13" LCD.

4.2 *Subjects and Task*

Twenty-five pilots of Deutsche Lufthansa, Deutsche Flugsicherung, Aero Lloyd, Luftfahrt Bundesamt, DASA and test pilots of Airbus Industry have flown the same task within 3 weeks. 75% of the subjects were captain with an average flight experience of 13.000 hours on Airbus and Boeing type aircraft. 90% had experience with glass cockpit and 80% knew Frankfurt airport and the environment.
The flight of about 90 minutes included taxi out, take off, climb, tunnel in the sky flights, low level flight under the ridge line of the Feldberg (2884ft), two ILS approaches at Hahn airport and the way back to Frankfurt. Each pilot was introduced into the test aircraft systems and briefed concerning the new symbology, flight path and questionnaire. In the debriefing pilots explained their impressions and filled in the questionnaire. Flight data were recorded and off-line analysed.

4.3 *Questionnaire and Results*

The evaluation of a complex man machine interface like this is very difficult and requires a high financial effort and man power. Since both are limited these flight tests were used to obtain an overall impression of advantages and problems of this 3D Flight Guidance System. 87 questions concerning 9 main topics were grouped and evaluated as an input for detailed research in the flight simulator.

Primary Flight Display. The overall impression of the PFD was good, resulting in an good to excellent acceptance for this kind of information presentation. The SA compared to conventional displays was evaluated as good to excellent. Legibility and colour coding of overlay symbology including α-blending contrast to the background were criticised rather bad. This was caused by poor brightness of the LCD in bright weather conditions.

Depiction of Terrain. Vertical and Horizontal aperture were fixed to 60°, terrain was rendered with an equidistant grid of about 300m and not oversized. While the vertical aperture was adequate the horizontal was a adequate for 70% and too small for 30%. Subjects mentioned that the actual altitude can be estimated within a range of 500ft. The presented terrain was a bit too abstract resulting in a terrain recognition rating of good to fair. It has to be mentioned that terrain avoidance was the objective and not terrain recognition. For CFIT avoidance purposes the depiction of terrain was noted excellent.

Comparison Outside View/Display. Orientation having the displays as only source of information was rated satisfactory to good. Correspondence of displays and real environment was rated the same. Answers have shown, that perception of dimension and presentation of FMS information like waypoints and selected flight path have to be enhanced in future.

Flights in the Tunnel. The tunnels were used as flight path directors supplemented by a flight path

predictor. Extrapolated position and altitude was presented for the next 2 to 8 seconds. No conventional crosspointer was shown.
On the one hand the ILS was presented as a linear tunnel in the sky approach on the other hand low level flight under the ridge line of Feldberg was guided by a curved tunnel for noise abatement reasons. In general the ILS tunnel was rated better than the low level tunnel concerning intercept, horizontal and vertical guidance. This may be caused by the dimension of the low level tunnel that was much smaller and resulted in unsatisfactory since it took a lot of effort to stay in. While pilots in command had the impression of flying inexact the safety pilot remarked that these were the most accurate ILS approaches he had monitored. This was certified by off-line data analysis.

Predictor. As mentioned above the predictor was used as a flight guidance aid. Pilots had no experience with flight prediction and had to get used to. Ratings for control of flight attitude was rated fair to excellent by 80% and poor to insufficient by 20%. Whereas 35% assessed the predictor as essential for tunnel in the sky flights and 55% as helpful.

Comparison Predictor/Flight Director. ATTAS was not equipped with an Autopilot system so no flight director values could be provided to the pilot. The comparison was exclusively based on experience with flight directors. Vertical as well as horizontal guidance comparison was balanced.

Warnings. Beside red coloured terrain above the own altitude over- and underspeed and high bank warnings were coded with different predictor colours. Terrain warning was rated excellent while over- and underspeed are ambiguous.

Nav Display. Terrain presentation in the navigation display was non-ambiguous for 80% of the pilots. Other questions concerned the symbology of navigation-aids, obstacle colour coding and the presented view segment of the PFD. During taxi phase runways, taxiways and selected taxilines were presented as well as an airport map. The taxiing from the runway to the park position was assessed easy (35%) to very easy (60%). The airport map was rated as essential to helpful by most of the pilots.

Subjective Stress Level. The level of stress of the eyes was judged low by 15%, acceptable by 55% and high by 25%. As mentioned above the brightness of the LCD was not sufficient for the good weather conditions. Stress of concentration compared to conventional displays was evaluated higher while 70% think that stress in the cockpit will decrease using these displays.

4.4 Flight Test Summary

Assuming the pilots comments the prevention of CFIT with this 3D Flight Guidance Display is excellent. First the pilot is always aware of the environment he is flying in, and during the low level flights the dangerous terrain and the escape route is visible. So the aim of an intuitive and easy to perceive display was reached. Taxiing symbology supports the pilot to reach his gate in an advantageous way, while today the pilot has no airborne support after touch down. The precision of navigation and database was very high. Comparing the external view of the taxi line and the synthetic one, pilots recognised a difference less than half a metre.

Outlook. An overall evaluation of the Flight guidance displays was reached in all flight phases from gate to gate. Subjective judges of these test pilots have to be evaluated in further more detailed tests. For this purpose a new generic flight simulator with an 180° by 40° collimated vision will be used at Darmstadt University.

5 SUMMARY

The next era of avionics has begun. 3D Flight Guidance Displays have reached a level of operability that is not only sufficient for flight tests but also for introduction into the next generation of aircraft or even for retrofit solutions. Required components like highly accurate navigation, databases and graphic hardware are feasible. Advantages are a tremendously increase of safety concerning CFIT accidents and in-flight collisions. Capacity problems in the air and even on the ground can be solved in order to reduce delays.

REFERENCES

Helmetag, A., R. Kaufhold, M. Purpus. (1997) 3D Flight Guidance Displays, CEAS, Free Flight Symposium, Amsterdam.

Pfister, J., J. Schiefele, L. May, H. Raabe, T. Hausmann (1997). A Concept for high accuracy Navigation and Databases in free flight Synthetic Vision Systems, CEAS, Free Flight Symposium, Amsterdam.

Purpus, M., C. Below, H.v. Viebahn (1997). Flight Tests of the 4D Flight Guidance Displays, SPIE, Enhanced and Synthetic Vision, **Volume 3088**.

Rasmussen, J. (1987). New Technologies and Human Error, Wiley & sons.

COCKPIT PROCEDURAL ADVISORY SYSTEM UTILIZING FLIGHT PHASE ESTIMATION

Keiji Tanaka, Kohei Funabiki and Koji Muraoka

National Aerospace Laboratory
7-44-1 Jindaiji-Higashi, Chofu, Tokyo 182-8522, Japan

Abstract: An advisory display against procedural deviations on the flight deck by using autonomously estimated flight phases was proposed. The system provided advisory information regarding control devices and switches critical in each flight phase. A piloted operational simulation yielded that the advisory system functioned appropriately and timely. Meanwhile, necessities of improving the flight phase transition after a missed approach and of integrating the display information with the existing warning system were also revealed. Potential implementation of a proposed system in a future warning system or a future flight management system was suggested.

Keywords: Aircraft operations, Alarm systems, Human-machine interface, Human error, Human factors

1. INTRODUCTION

Human-machine harmonization has been one of the critical themes for further improvement of aviation safety. Although cockpit automation in modern aircraft has drastically reduced the flight crew workload and improved flight safety, some of the recent aircraft accidents, such as China Airlines A300-600 crash at Nagoya in 1994, and American Airlines B757 crash near Cali in 1995, still demand further efforts to improve interactions between the flight crew and the advanced automatic flight system (AFS). Among primary causes of aircraft accidents related to human factors, deviations from nominal procedures have been most frequently mentioned. Such inappropriate crew action is often attributed to lack of so called Situation Awareness (SA). It is recognized that deficiencies in SA are occurring even in modern glass-cockpit environments. Owing to the fact that pilots in a glass-cockpit are getting more heavily dependent on electronic displays, display methodology to maintain SA draws attention. In this paper, SA is used as a "term to refer to the flight crew knowing and understanding the present and future status of the airplane and its systems, based on the airplane's state and flight path parameters (e.g., the airplane's position, speed, flight path, energy state, and position of the flight controls) and the status and behavior of the autoflight system relative to the operating environment (e.g., terrain, air traffic clearances, and other traffic)" (FAA, 1996).

Two approaches seem to be possible to maintain and improve SA. One is to maintain high SA by allowing appropriate evaluation, understanding and monitoring of the system at all times; the other is to compensate SA degradation by alarming human deficient evaluation, understanding and monitoring. An example of the former approach can be seen in a future cockpit display to enhance autopilot-mode awareness (Funabiki, 1997). The latter includes attempts to cope with degraded SA by alerting deviations from nominal procedures in the form of cautions or warnings, and this approach was investigated in this study.

Advisory messages in the modern flight deck are generally integrated and presented in electronic crew alerting systems located in the center instrument panel. In a previous study, a cockpit advisory system was developed to automate the crew's check procedures (Tanaka *et al.*, 1992). Results of a flight-simulator evaluation demonstrated that the system reduced crew workload in critical situations and enhanced crew situation awareness. It was also suggested that employing meta-level

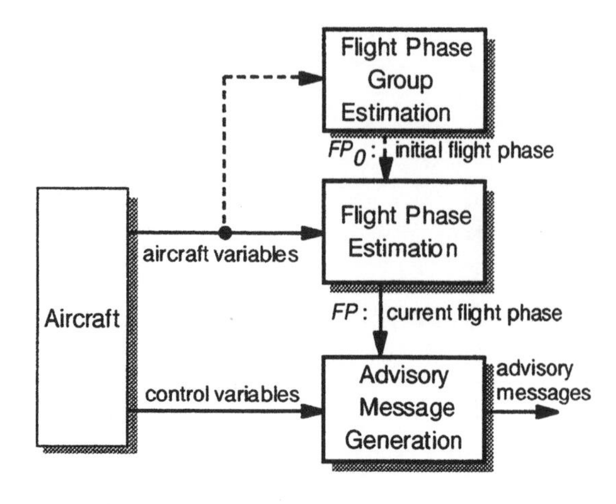

Figure 1. Block Diagram of Procedural Advisory Message Generation

knowledge of flight phases and aircraft conditions into the system was beneficial for presenting prioritized advisory information to ensure the flight crew's timely reactions. On the basis of the results, a new procedural advisory system to cope with doubtful operations critical to each specific flight phase was investigated and proposed in the present study.

2. MONITORING CREW ACTIONS

It is generally difficult to estimate whether a flight crew is in a condition of misconception of the situation, for there are no objective means to measure the crew's SA in real time. An alternative method is to monitor crew actions and to compare them with nominal actions. This monitoring by machine is theoretically possible by extracting what both crew members do during the crosschecking of their procedures. This method of monitoring is limited in a sense that it cannot generate alerts when misconception happens, but only after deviation of action happens. Although this delay is significant, the response of the method is expected to be sufficiently quicker than the usual warnings, which are activated by resulting changes of the aircraft motion.

Most of the human operations on the modern flight deck comprise tasks which are classified as the rule-based behavior of Rasmussen's taxonomy (Rasmussen, 1983). The tasks, considered as interactions between the crew members and the aircraft with AFS, should be treated as another kind of dynamic response of a feedback system (Tanaka *et al.*, 1986), if a human error is considered as a propagation of the chain of events in the human-machine system. The tasks form procedures which also form a chain. As summarized in aircraft operation manuals, a total flight is a combination of procedures of each flight phase (FP). At the end of the procedure in each FP, completion of the tasks is confirmed by crew members before proceeding

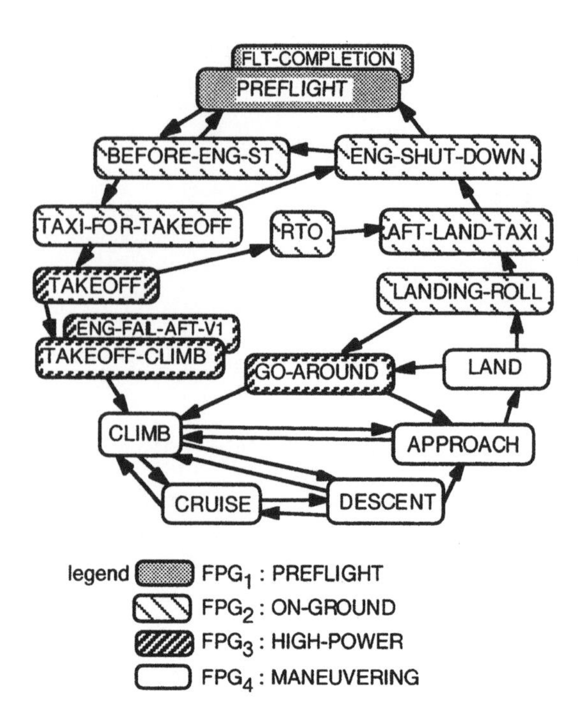

Figure 2. Possible Transitions of Flight Phases

to the next FP by using a checklist. In this meaning, the FP is a key parameter of the flight operation. To realize utilization of FPs in the system side, an algorithm to estimate the current FP becomes necessary. It is noted here that human controls cannot be used for flight-phase specification and estimation, because using the controls implies that all of the crew actions are assumed to be correct.

3. SYSTEM DESCRIPTION

3.1 Outline of the Procedural Advisory System

The proposed system continuously estimates current FP by monitoring the aircraft situation and by taking the flight progression into account. Then, the system compares the controls and switches manipulated by the crew members with allowable ranges of the values specified at each FP. If a deviation of the control positions or states of the switches is detected, the system generates a corresponding advisory message to be presented on the cockpit display. The conceptual block diagram of the message generation is indicated in Fig. 1. Detailed functions of the blocks in this figure are described below.

3.2 Flight Phase Estimation

The categorization of FPs in this study was arranged so as to agree with usual operation procedures from the flight crew's viewpoint. A set of FPs, including normal ones and their relationships, are illustrated in Fig. 2, which comprises 15 FPs as listed in Table 1. It

Table 1. Summary of Flight Phases

FP n	nominal range from	to
PREFLIGHT (FLT-COMPLETION)	power off	power on
BEFORE-ENG-ST	power on	door closed
TAXI-FOR-TAKEOFF	chock out	enter runway
TAKEOFF	set takeoff power	lift off
TAKEOFF-CLIMB (ENG-FAIL-AFT-V1)	lift off	altitude=460m
CLIMB	altitude=460m or climb after CRUISE or DESCENT	level off
CRUISE	level off	start climb or descent
DESCENT	descent after CRUISE or CLIMB	altitude=1520m
APPROACH	altitude=1520m	altitude=460m
LAND	altitude=460m	touch down
LANDING-ROLL	touch down	reach 41m/s
AFT-LAND-TAXI	reach 41m/s	arrive terminal
ENG-SHUT-DOWN	arrive terminal	engine stopped
RTO	start deceleration	full stop
GO-AROUND	set GA power	altitude=460m

is noted that in this figure, there is no transition path from APPROACH to GO-AROUND, because APPROACH in this study covers the altitude range from 1,520 m (5,000 ft) to as low as 460 m (1,500 ft), where a missed approach does not necessarily require GA (Go Around) power. Also, note that another non-normal case of Engine Failure after V_1 (Take-Off Decision Speed) is included in the TAKEOFF-CLIMB phase, for both require almost identical procedures.

A small number of variables were chosen for FP estimation; N_1 (Low Pressure Rotor Speed), Airspeed, Vertical Speed, Altitude, APU (Auxiliary Power Unit), and Parking Brake. Most of them were aircraft variables. The control variables, APU and Parking Brake, were used to simplify the estimation algorithm during the non-critical phases when the aircraft was fixed on the ground. It was assumed that the reliability of all the variables was high enough to neglect the possibility of signal source malfunctions.

The adopted algorithm of FP estimation utilized the limited possibilities of FP transitions given in Fig. 2. The algorithm tested conditions of the aircraft variables as follows: when the present FP is FP_m, which is one of the 15 FPs given in Table 1, and if the aircraft variables satisfy the condition to proceed from FP_m to FP_n, then FP_n becomes the new present FP. This process was repeated during the entire flight.

Another algorithm was prepared for resetting the system as well as for providing the initial condition. When the system was started, this second algorithm was activated initially. It estimates the initial present flight phase, FP_0 as much as possible by testing the present aircraft variables. When the algorithm cannot identify FP_0, it categorizes the situation as one of the flight phase groups (FPGs), denoted as FPG_k (k=1, ..., 4) in Fig. 2. FPGs are characterized by similar values of the aircraft variables among the FPs grouped in each FPG. After waiting for changes of the aircraft variables, they satisfy the condition to proceed from FPG_k to FP_l, and FP_l becomes the new present FP.

3.3 Generation of Procedural Advisory Messages

If a critical deviation of controls in a given estimated FP was detected, the corresponding advisory message to alert the flight crew was displayed on the screen. Existing warnings are activated when a signal exceeds a certain allowable range, while this method of generating advisory information modifies the allowable range according to FP. For example, existing warnings to lower gears are activated when the flap angles and the airspeed fall in certain ranges, while this method directly advises to lower gears if they are retracted when the aircraft enters into the LAND phase. Such direct monitoring enables a focus on critical items among various control variables, and decreases the chance of a misconception of warnings by minimizing the number of advisory messages for the corrective action.

4. OPERATIONAL SIMULATION

4.1 Setup

In order to evaluate the procedural advisory system in an operational environment, a piloted flight simulation which simulated an entire flight was conducted. This operational simulation was focused on the evaluation of the significance and timing of the advisory messages. A research simulator from the National Aerospace Laboratory (Funabiki *et al.*, 1995) was utilized for this purpose. Primary elements of the simulation are as follows:

Aircraft Model: A twin jet transport with a generic AFS model including Auto-Pilot, Auto-Throttle and FMS (Flight Management System) were implemented (Muraoka *et al.*, 1995).

Cockpit: As the simulator cockpit is equipped with three large CRTs, a set of PFDs (Primary Flight Displays) and NDs (Navigation Displays) for both crew members and upper and lower ECAM (Electronic Centralized Aircraft Monitor) displays were provided to crew members (Fig. 3). An FMS-CDU (Control Display Unit) was prepared in the aft part of the center pedestal. A full glass-cockpit operation of a transport airplane was thus realized. Though the flight simulator has a six-degrees-of-freedom motion system, it was used by fixing the base.

Advisory Display: A display window for advisory messages was opened next to the upper ECAM. The window displayed the estimated FP (or FPG), the previous FP, possible next FPs, the FPG, and advisory messages

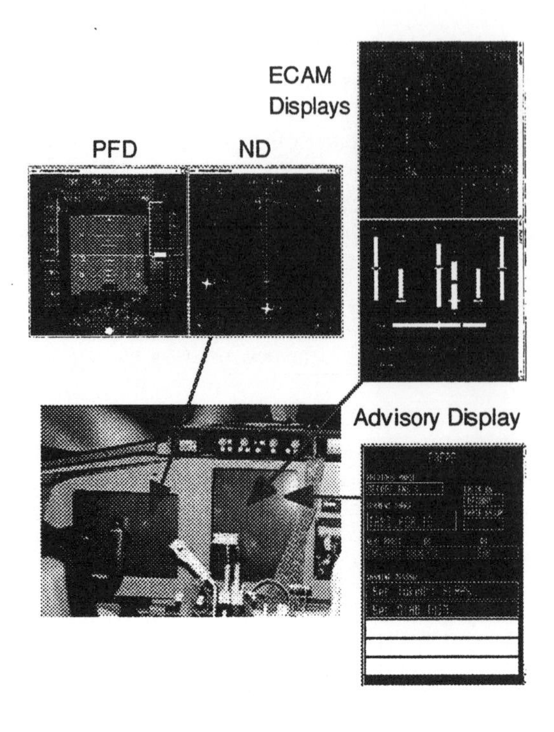

Figure 3. Displays of the Simulator Cockpit

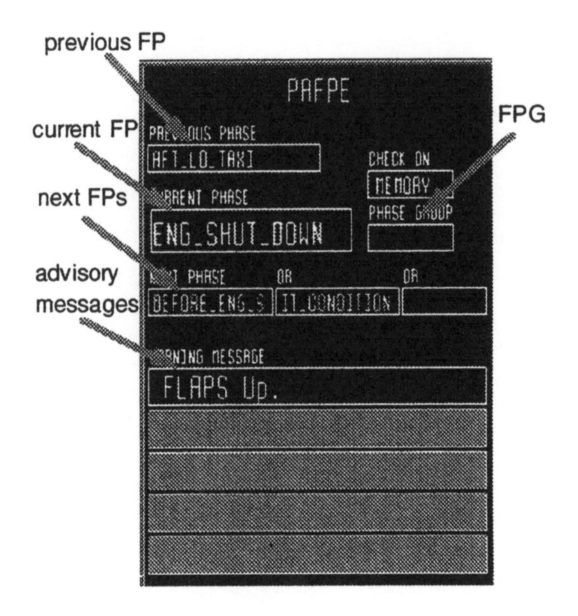

Figure 4. Advisory Display

(Fig. 4). Each advisory message was accompanied by an aural tone for two seconds to alert the crew members to take corrective actions.

4.2 Method

In the operational simulation, all the FPs beginning from Cockpit Preparation to Engine Shutdown including GA (Go Around) and RTO (Rejected TakeOff) were simulated. Basic functions and performance of the advisory system against various procedural deviations were evaluated by two research pilots. Each pilot performed PF (Pilot Flying) and PNF (Pilot Not Flying) alternately. Although they had no experience with line operations, a relatively short familiarization time was required before evaluating the advisory system so they were familiar with the simulated air route and the simulator set up.

Prior to the evaluation in the simulator cockpit, such information as the outline of the research, the purpose of the simulation, and functions of the advisory system were provided to the pilots. A pre-flight briefing was conducted before boarding. The two crew members and one of the experimenters were seated in the cockpit. Another experimenter was seated at the control desk of the simulator, and managed other simulated tasks such as ATC (Air Traffic Control) communications and the door control. As the present evaluation was concerned with crew operations, other conditions were simplified by assuming nominal ATC communications, ceiling and visibility unlimited and calm wind. Also, for simplification of the evaluation, usual cockpit warnings were not simulated by assuming that no malfunctions from the aircraft side happened.

The operational simulation was carried out as follows: The aircraft was initially located at a northern airport, and left for a southern airport. The distance between the two airports was about 300 km. A nominal SID (Standard Instrument Departure) for departing from the northern airport and STAR (Standard Instrument Rule) for arriving at the southern airport were prepared. Thus, for cases of automated flights, as soon as completing the take-off, the AFS was activated to start an automatic flight along the flight plan stored in the FMS beforehand; and as soon as capturing the ILS (Instrument Landing System) signals in the terminal area, the automatic landing was conducted. One operational simulation ended when all the engines were shut off at the assigned spot of the destination airport.

Three sets of scenarios for evaluation were prepared:
Scenario A: full automated flight with FMS;
Scenario B: full automated flight with FMS, including RTO and intentional deviations; and
Scenario C: minimum autopilot usage without FMS, including GA and intentional deviations.
In the latter two scenarios, intentional procedural deviations, e.g., leaving gears down after lift off, were requested so as to generate procedural advisory messages. After confirming the timing and contents of the advisory messages, pilots manipulated the controls or switches by following the messages on the screen.

It took about 90 min to complete one entire flight. After completing each flight, a post-flight briefing was held to collect pilot comments. Overall pilot ratings on the advisory system were scored after a series of evaluation flights.

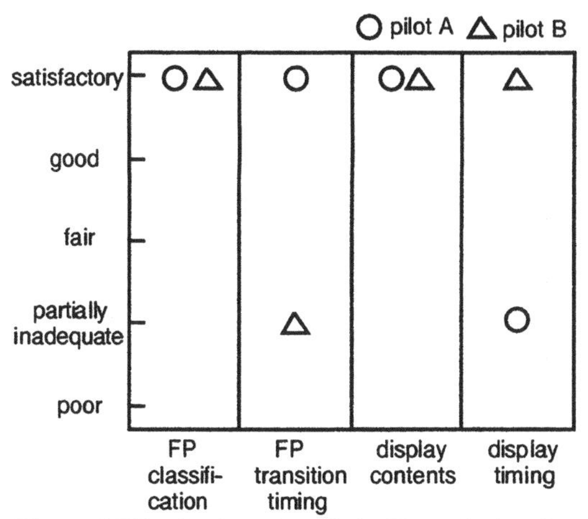

Figure 5 Pilot Ratings for Flight Phase Estimation and Advisory Display

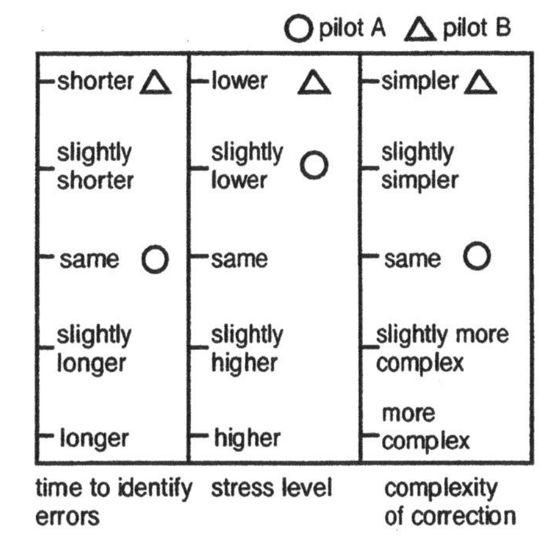

Figure 6 Workload Ratings Compared with Existing Warnings

5. RESULTS AND DISCUSSION

Seven flights in total were conducted before completing the pilot evaluation. The first four flights were devoted to familiarization and confirmation of the basic functions of the advisory system by repeating Scenario A. The last three flights, i.e., two Scenario B flights and one Scenario C flight, were allotted to evaluating the detailed functions of the advisory system. Although the contents of the tasks in Scenarios B and C were different, no significant differences were observed in utilization of the display.

Pilot ratings to the primary evaluation items are shown in Figs. 5 and 6. Other pilot comments were collected concerning FP estimation performance, display performance, workload, specific detailed display items, interface with other systems and future utilization of the system. The following discussions were made based on the ratings and the comments.

Performance of Flight Phase Estimation: Both pilots confirmed that the present phase classification was generally adequate. Regarding the timing of the FP transition, one transition was pointed out to be inadequate as shown in the rating given in the FP transition timing in Fig. 5. It was caused by a phenomenon that occurred as soon as reaching an altitude higher than the range of TAKEOFF-CLIMB after GO-AROUND. Then, FP proceeded to CLIMB, and again changed to APPROACH after level off at the planned altitude of the missed approach procedure. These transitions were too early for pilots to complete the missed approach procedure. As there are various patterns of missed approach flight profiles according to airports, flexibility of transitions should have been maintained by allowing any of the missed approach procedures. One method of modification would be to extend the GO-AROUND phase up to level off flight, and then to connect to the following APPROACH or CLIMB. Other timings of FP transitions were commented as adequate.

Display Performance: The pilots commented that the advisory information were useful in general and corrections after acknowledging the advisory messages were possible. These comments support adequateness and timeliness of the advisory information generated by the system. Meanwhile, as indicated by the rating in the display timing in Fig. 5, one of the pilots requested modification of the timing of providing the advisory message to lower the landing gears. The algorithm of generating the landing gear message was activated as soon as the FP becomes LAND after APPROACH. Although the usual gear horns were not provided in the present flight simulation, it was anticipated by the evaluation pilot that this advisory system may provide contradictory information to the existing gear warning logic, which usually employs the aircraft configurations such as extended flaps and reduced power for its activation. Moreover, landing gear lowering at high airspeed is prohibited due to structural reasons. The comment intended that all these conditions should be considered before presenting the advisory message for the landing gear. This inconsistency between the present FP based advisory information and usual warnings revealed the necessity of integrating the proposed FP based advisory information with the existing warning system.

Workload: Figure 6 shows ratings that the time to recognize the situation from the displayed messages was equal to or shorter than existing alarms, also that the stress levels while judging the situation and deciding the corrective actions were adequate. One pilot predicted stress may be increased by direct installation of the system together with other existing warnings, which would produce confusion among the sources of the warnings. This comment again implied the necessity of integration with the warning system. The complexity of the corrective procedure after noticing the advisory messages, including confirmation of situation, corrective action and confirmation of the correction, was proved to be equal to or simpler than the existing system. Namely, it was satisfactorily easy to recognize the situation with the message, owing to the fact that the system did not warn of any malfunctions but only procedural deviations.

Future Direction of Utilization: The evaluation results revealed the necessity of clarifying the consistency of the display information with existing warnings or existing operational limitations. Some advisory functions of the proposed system overlapped with existing warnings, such as those of flaps and landing gear, and there is potential inconsistency between them. However, it now appears to be advantageous to utilize the capability of generating advisory information earlier than current systems merited by the information of the progress of the flight. For example, in the proposed system, take-off configuration warnings were activated well before entering the runway. Also, utilization of estimated FPs can extend the alert inhibit functions, which are being adopted to recent aircraft, such as a Fire Warning Light and Fire Bell inhibition during the take-off phase.

Another potential application would be to integrate the proposed system into the FMS. The FMS realizes a planned flight with reference to the present aircraft position and altitude by sending commands to the autopilot. For this purpose, nominal sequential phase transitions from Preflight, Takeoff and so on to Approach have been incorporated into the FMS. It is anticipated that the FMS functions are going to be expanded to realize full coverage to ground operations and to off-nominal operations. The present FP estimation may provide a potential direction of enhancing the FMS functions.

6. CONCLUSION

A cockpit procedural advisory system was developed and evaluated in a flight operation environment. The advisory system has a unique feature of estimating the present FP by making use of FP progression in order to timely generate advisory information for flight crew actions. Piloted evaluation in the operational simulation yielded the following:

(1) It was confirmed that the present flight phase classification and timings of FP transitions were adequate. One exception was the FP transition after GO-AROUND, which should be designed to take missed approach profiles into account.

(2) Pilot comments support adequateness and timeliness of the advisory information generated by the system. However, it was anticipated that the advisory system may provide contradictory information to the existing warnings such as the landing gear warning, and the necessity of integration of the proposed FP based advisory information with the present warning system was emphasized.

(3) The pilot workload ratings regarding time to recognize the situation from the displayed messages, stress levels and corrective procedures were equivalent to or smaller than existing alarms.

The above results imply that the proposed system can cope with SA degradation. Meanwhile, the necessity of clarifying the consistency of the display information with existing warnings was revealed. Potential direction of implementing the proposed system in a future flexible warning system would be to integrate it into the existing warnings by making use of its early warning capability, and by applying estimated FPs to a flexible warning inhibit mechanism. Also, potential implementation into a future FMS would be to expand FMS coverage to ground and off-nominal operations by making use of the proposed FP estimation.

ACKNOWLEDGMENTS

This study has been conducted as a part of the Fundamental Research on Human Characteristics for Harmonizing Systems with Human Beings through Special Coordination Funds of the Science and Technology Agency of the Japanese Government. The authors would like to express their thanks to the committee members of the coordinated research for their valuable comments on this study, especially to the late Prof. Hidekatsu Tokumaru, and also to the pilots who participated in the evaluation and provided precious comments to the display system.

REFERENCES

FAA Human Factors Team (1996). The Interfaces Between Flightcrews and Modern Flight Deck Systems.

Funabiki, K. (1997). Tunnel-in-the-Sky Display Enhancing Autopilot Mode Awareness, *10th European Aerospace Conference "Free Flight"*.

Funabiki, K. and K. Tanaka (1995). A Flight Simulation for Human Error Study, AIAA-95-3410.

Muraoka, K., K. Funabiki, K. Tanaka, M. Nakamura. and Y. Terui (1995). Development of a Line Operational Simulator for Research Purposes, *Proceedings of Making it Real CEAS on Simulation Technologies.*

Rasmussen, J. (1983). Skills, Rules, and Knowledge; Signals, Signs, and Symbols, and Other Distinctions in Human Performance Models, *IEEE Trans.*, SMC-**13**, 257-266.

Tanaka, K. *et al.* (1992). A Flight Simulator Experiment of the Cockpit Advisory System, in Computer Applications in Ergonomics, *Occupational Safety and Health* (Mattila, M. and Karwowski, W. (Eds.)) , North-Holland, 63-70.

Tanaka, K. and K. Matsumoto (1986). A Hierarchical Model of Pilot's Procedural Behavior for Cockpit Workload Analysis, *Trans. Japan Society for Aeronautics and Space Sciences*, **28**, 230-239.

TUNNEL SIZE IN A TUNNEL-IN-THE-SKY DISPLAY: A CYBERNETIC ANALYSIS

Max Mulder *,[1] **J.A. (Bob) Mulder** *

* *Delft University of Technology, Faculty of Aerospace Engineering, P.O. Box 5058, 2600 GB Delft, The Netherlands.*

Abstract: The tunnel-in-the-sky display is a viable candidate to become the primary flight display of future cockpits. The tunnel size is one of the main tunnel display design variables, affecting pilot performance and workload. The paper describes the effects of varying the tunnel size on pilot behaviour from the perspective of cybernetics. After an analysis of the fundamental information-transfer characteristics of the tunnel display, an experiment was conducted to evaluate pilot behaviour both qualitatively as quantitatively. It is shown that the cybernetic approach can be used to obtain tunnel display design guidelines in a conceptual rather than empirical fashion. *Copyright ©1998 IFAC*

Keywords: Aircraft control, manual control, displays, information analysis, cybernetics, mathematical models.

1. THE TUNNEL-IN-THE-SKY DISPLAY

The volume of air transportation will show a considerable growth in the near future. New technologies are being developed with the dual objective of increasing the *efficiency* of air traffic management and enhancing flight *safety*. One of the expected measures is to increase the flexibility in air traffic control by allowing curved approach profiles. Flying these – inherently more complex – curved approaches increases the pilot task demand load and requires enhanced levels of situation awareness. Improving the presentation of information to the pilot by means of intuitive displays can alleviate these problems considerably (Oliver, 1990). A promising candidate to become the future primary flight display is the *tunnel-in-the-sky* display (Fig. 1), which shows a spatial analog of the planned trajectory. Previous research indicated that the tunnel display has certain advantages over current displays in both the pilot manual as supervisory task (Wilckens, 1973; Grunwald, 1984; Wickens, Haskell, & Harte, 1989; Theunissen, 1997).

[1] e-mail: m.mulder@lr.tudelft.nl

At the Delft University of Technology a research project was initiated to investigate the applicability of a tunnel display for the pilot manual control task. In contrast to other studies, its goal was not to compare the tunnel display with current displays in terms of pilot performance, situation awareness, and workload. Rather, the objective was to obtain an understanding of how pilots use the tunnel display as their main source of information in the aircraft guidance task (Mulder, 1995). A methodology has been developed, labelled the *cybernetic approach*, which allows substantial insight into the effects of varying display designs on pilot behaviour (Mulder, 1998). The paper will describe the application and merits of the approach in detail, at the hand of an investigation of the effects of the tunnel size design variable.

2. EFFECTS OF THE TUNNEL SIZE

The tunnel size is probably one of the main tunnel display design parameters. By constraining the aircraft airspace the tunnel size *commands* the level of path-following accuracy. Increasing the

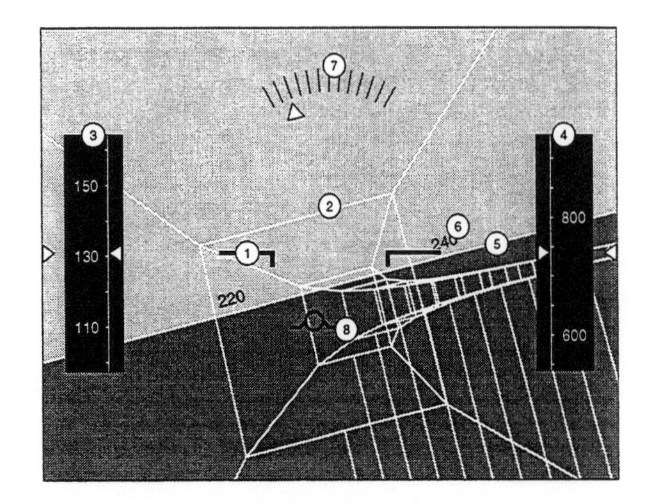

① aircraft reference symbol
② tunnel geometry (frames & poles)
③ airspeed [kts]
④ altitude [ft]
⑤ horizon line
⑥ heading angle indicator [deg]
⑦ bank angle indicator
⑧ flight-path vector

Fig. 1. The tunnel-in-the-sky display (left) and an explanation of its principal components (right).

tunnel size decreases this level, allowing pilots to decrease their control gains and to lower their level of attention allocated for the guidance task, yielding a more relaxed control strategy (Theunissen & Mulder, 1994). On the other hand, smaller tunnels command a higher level of trajectory-following precision, demanding more pilot attention for the guidance task. Hence, the tunnel size is an excellent example of the omnipresent design tradeoff between performance and workload.

2.1 *Pilot performance and workload*

Not surprisingly, the tunnel size parameter has played an important role in almost all investigations into perspective flight-path displays. In (Wilckens, 1973; Schattenmann & Wilckens, 1973) the effects of different tunnel sizes (range 20–400 [m]) on pilot performance and control activity were investigated in a task of following a straight trajectory. It was concluded that control activity increased for smaller tunnel sizes. The same holds for the performance, but only to a certain level: further reducing the size of the tunnel led to reduced stability. In a qualitative diagram (Fig. 2) the relation between the tunnel size and tracking accuracy is exemplified. The figure shows that the tracking accuracy is limited on the one hand by the size of the tunnel (large tunnel sizes) and on the other hand by the closed loop dynamic stability (small tunnel sizes). Hence, an *optimal* size is hypothesized, depending on the level of pilot adaptation with the tunnel display (Wilckens, 1973).

Grunwald (1984) investigated the use of two tunnel sizes (300 and 450 [ft], i.e. 91 and 137 [m]) in a task of flying curved approaches, in the presence of a position predictor symbol. Again, an increasing tracking performance and control activity for the smaller tunnel was reported. In an attempt to further examine the possible confounding effects of tunnel size and predictor symbology, the curved

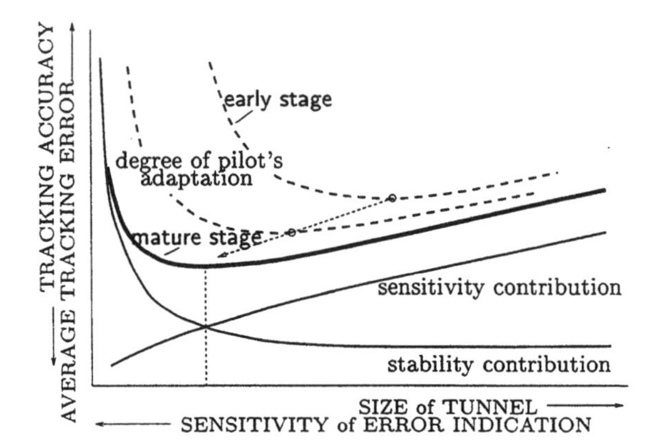

Fig. 2. Relation between tunnel size and tracking accuracy (cf. Wilckens, 1973).

trajectory following experiment of Grunwald was repeated by Theunissen (1995). The results of this experiment supported those of the earlier studies.

2.2 *Pre-experimental cue analysis*

A pre-experimental analysis was conducted to examine the effects of varying tunnel size on the characteristics of the fundamental optical sources (cues) of information in the tunnel display (Mulder, 1998). The analysis is limited to the lateral dimension and examines especially the effects of the tunnel size on the presentation of a lateral position error X_e and a heading angle error ψ_e (a zero-sideslip flight is assumed). In Figs. 3(a) and 3(b) the tunnel images are shown for tunnel sizes of 20 and 40 [m] respectively. Both figures show the situation of a zero and a non-zero lateral position error (X_e=+5 [m]), and a zero heading angle error.

2.2.1. *Cues for a position error* The lateral position error is conveyed by two categories of optical cues. First, consider the relative lateral displacements ϵ_{ij} and η_{ij} of the tunnel frames i and j, located at distances D_i and D_j from

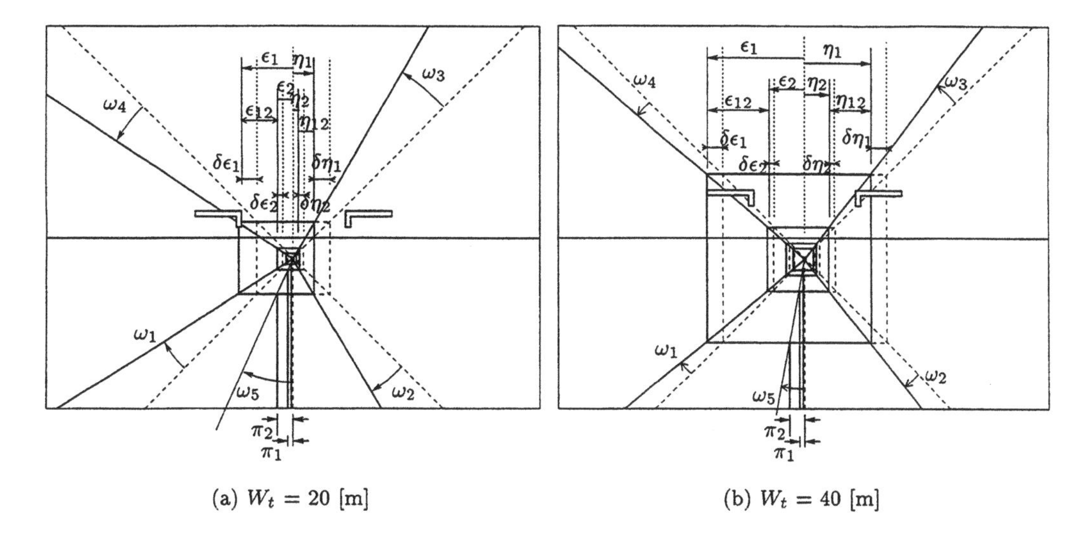

Fig. 3. Static optic cues for a lateral position error in straight tunnel sections. The dashed and continuous lines show the tunnel image for a zero and a non-zero (+5 [m]) lateral position error respectively. The heading angle error is zero.

the viewplane. Changes in these relative lateral displacements from the zero-error condition are a function of the lateral position error X_e only (Mulder, 1996):[2]

$$\epsilon_{ij} = +\kappa X_e \left(\frac{D_j - D_i}{D_i D_j} \right) \tag{1}$$

$$\eta_{ij} = -\kappa X_e \left(\frac{D_j - D_i}{D_i D_j} \right) \tag{2}$$

with κ a display constant dependent on the field-of-view of the perspective projection and the size of the screen. The displacement cues are *independent* of the tunnel size, an effect which can be seen from Fig. 3 when comparing the values of these cues ($\delta\epsilon_i$, $\delta\eta_i$) for both tunnel sizes. The *relative* displacements (ϵ_{12}, η_{12} etc.), on the other hand, *do* appear larger in relation to the size of the frames themselves, for smaller tunnels. Decreasing the tunnel size leads to a situation in which the first visible frame is smaller and seems to be positioned farther away. The subsequent frames are even smaller and appear considerably cluttered on the display. Because the relative displacements become difficult to perceive, they do not present very useful information. The relative displacements of the altitude poles (π_{ij}), however, remain visible and are unaffected by the tunnel size.
The second set of cues – the *splay angles* – are defined as the angles between the projections of the of the longitudinal lines connecting the tunnel frames, and the horizon. They are a function of *both* the lateral and vertical position error (Mulder, 1996). For a zero vertical position error, however, one obtains:

$$\begin{aligned} \omega_1 = \omega_4 &= -\frac{X_e}{W_t} \\ \omega_2 = \omega_3 &= +\frac{X_e}{W_t} \\ \omega_5 &= -2\frac{X_e}{W_t} \end{aligned} \tag{3}$$

with W_t the (square) tunnel width. In other words, the tunnel acts as a scaling factor – a *gain* – for the change in splay angle caused by a position error. This scaling effect was originally reported in (Wilckens, 1973) and led Theunissen (1995) to allocate the term *error gain* to the tunnel size. The 'virtual' line connecting the end-points of the altitude poles supporting the frames also yields a splay angle (ω_5), with a 'gain' that is twice as large as those of the other splay angles.

2.2.2. *Cues for a heading angle error* In straight tunnels, the *infinity point* is defined as the projection on the viewplane of an arbitrary point of the tunnel geometry at infinite distance D ahead (Mulder, 1998). The heading angle error ψ_e can be perceived from the lateral translation of the infinity point with respect to the center of the screen marked by the aircraft reference symbol, a translation which is *independent* of the tunnel size. Further, differentiating the splay angles yield a second potential cue for the heading angle error. E.g. for splay rates $\dot{\omega}_1$ and $\dot{\omega}_4$ one obtains:

$$\dot{\omega}_1 = \dot{\omega}_4 = -\frac{V_{tas}}{W_t}\psi_e \tag{4}$$

Hence, the heading angle error is also coded in the display with the *splay angle rates*, scaled by the aircraft velocity and the size of the tunnel.

[2] The same holds for the altitude pole displacements π_{ij}.

2.3 *Conclusive remarks*

The tunnel size affects mainly the presentation of the position error, by *scaling* the amplitude with which the position is presented on the display. The principal cues are probably the splay angles, especially for smaller tunnels. For larger tunnels, the relative lateral displacements of the frames and the altitude poles could be an alternative. It is clear that the tunnel size influences the perceivability of the visual cues which probably affects the pilot control strategy. It is difficult, however, to *isolate* these cues in order to evaluate their individual usability. The significance of a cue depends on the tunnel size, the forward velocity and probably also the control and observation strategy of an individual pilot (Mulder, 1998).

3. EXPERIMENT

3.1 *Objectives*

An experiment was conducted to investigate the effects of the tunnel size design parameter on pilot performance, control activity, and mental workload. The tunnel size commands the required level of path-following accuracy, and its functionality for this purpose has been shown in earlier investigations (Wilckens, 1973; Grunwald, 1984; Theunissen, 1997). These studies, however, showed the effects of the tunnel size on pilot behaviour using empirical performance data alone. The cybernetic approach, centred around a quantitative determination of the pilot control behaviour *adaptation* process (Mulder, 1998), allows a re-examination of the mere qualitative claims of (Wilckens, 1973) concerning closed loop stability and performance. Hence, the current experiment should throw a new light on how the tunnel size parameter affects pilot behaviour.

3.2 *Method*

Four subjects – all professional pilots – were instructed to control the lateral/longitudinal motion of an aircraft – the Cessna Citation 500, a small business jet – through the tunnel as accurately as possible. The trajectory was straight and the aircraft was disturbed by atmospheric turbulence. The experimental variables were the tunnel size W_t (4 levels: 80, 40, 20 and 10 [m]) and the aircraft velocity V_{tas} (3 levels: 50, 70 and 100 [m/s]), combined factorially, resulting in 12 conditions. The aircraft velocity was constant and the side-slip was zero. The aircraft variables in the vertical dimension were all fixed to their initial conditions. No flight-path vector was presented.

4. RESULTS

4.1 *The pilot questionnaire*

A pilot questionnaire revealed that for large tunnels pilots used the relative lateral displacements of the tunnel frames and the altitude poles. For smaller tunnels they tended to use the splay angles formed by the tunnel sides. These comments correspond well with the pre-experimental cue analysis. Pilot effort ratings showed a clear increase in mental workload for decreasing tunnels, especially for the smallest ones.

4.2 *Performance measures*

The 95% confidence limits of the empirical performance data are shown in Fig. 4. A statistical analysis (ANOVA) of the main performance variables revealed that, independent of the velocity conditions, a reduction in tunnel size yields increasing path-following performance (X_e), increases pilot control activity (δ_a, $\dot{\delta}_a$), and generally leads to more jerky control behaviour (increasing ϕ and $\dot{\phi}$). The heading angle error ψ_e depended on the aircraft velocity condition only. These effects were all significant at the p=0.01 level.

4.3 *Modelling results*

The model-based analysis supported the aforementioned results. A decreasing tunnel size increases the crossover frequency of the (outer loop) position error feedback considerably. This also leads to increasing bandwidths of the aircraft middle loop (heading angle error) and inner loop (roll angle) loop closures. However, whereas the stability margins of the latter two loop closures remain fairly constant, a decreasing tunnel size *does* lead to a significant reduction of the stability margin of the position error feedback. For a detailed analysis, including details on identification and pilot model validation issues, the reader is referred to (Mulder, 1998).

5. DISCUSSION

The translation of the experimental results to the in-flight situation seems rather obvious for the experiments examining the effects of tunnel size. The experimental findings reported in literature (Wilckens, 1973; Grunwald, 1984; Theunissen, 1995) are consistent in their statements concerning the tradeoff between performance and pilot workload. These findings, however, are all based solely on the measured performance data.

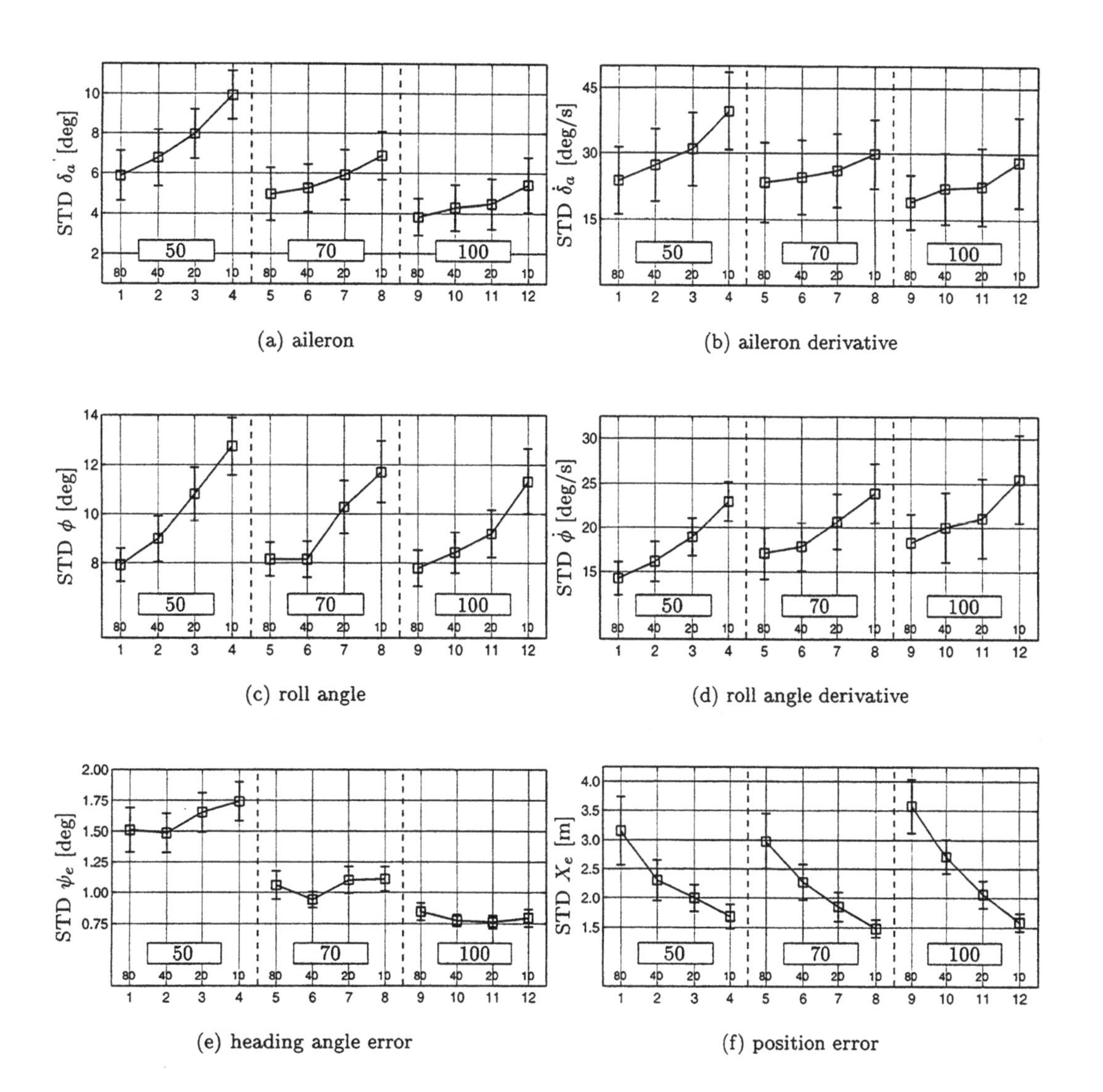

(a) aileron

(b) aileron derivative

(c) roll angle

(d) roll angle derivative

(e) heading angle error

(f) position error

Fig. 4. The 95% confidence limits of the experiment dependent measures (all subjects). In these figures the insets show the three velocity conditions (50/70/100 [m/s]). The numbers in the bottom show the tunnel widths (80/40/20/10 [m]), and those below each figure depict the experiment conditions.

The pilot workload has not been assessed explicitly, nor has there been any control-theoretic analysis of pilot control behaviour for the varying tunnel size conditions. Moreover, an important question remains unanswered, i.e. what are the *limits* in defining a particular tunnel size? Experiments confirm the statement that there is a direct relation between path-following accuracy and the tunnel size. The performance data (Fig. 4) suggest that the tunnel size can be decreased further to gain an even better performance, at the cost of some increase in control activity. The data state nothing, however, about any 'margins', or 'limits'. Although Wilckens (1973) did mention a deterioration of performance when decreasing the tunnel size beyond a certain level (Fig. 2), his approach did not allow any further analysis of that phenomenon. The cybernetic approach adopted here, developed in (Mulder, 1998) *does* allow such a control-theoretic analysis.

Consider Fig. 5(a) in which pilot performance (expressed in the STD of the position error X_e), is shown as a function of the position error feedback bandwidth ω_c^{out}. The figure shows that for the large tunnels performance is low and the crossover frequencies are smaller than for the smaller tunnels. Decreasing the tunnel size increases the bandwidth of the position error feedback considerably, yielding an increased performance. The figure clearly shows that the relation between performance and the bandwidth is non-linear and that a further reduction of tunnel size does not lead to an improved performance: the pilot/aircraft system becomes *saturated.* Fig. 5(b), showing the same performance data but now as a function of the position error feedback phase margin φ_m^{out}, marks an alarming decrease in closed loop stability for smaller tunnel sizes. Recall that a phase margin of 20 [deg] is generally considered as a lower bound, even in laboratory single-axis compensatory tracking tasks (McRuer & Jex, 1967). Hence, from a stability point of view, further decreasing the tunnel size is undesirable and should be avoided.

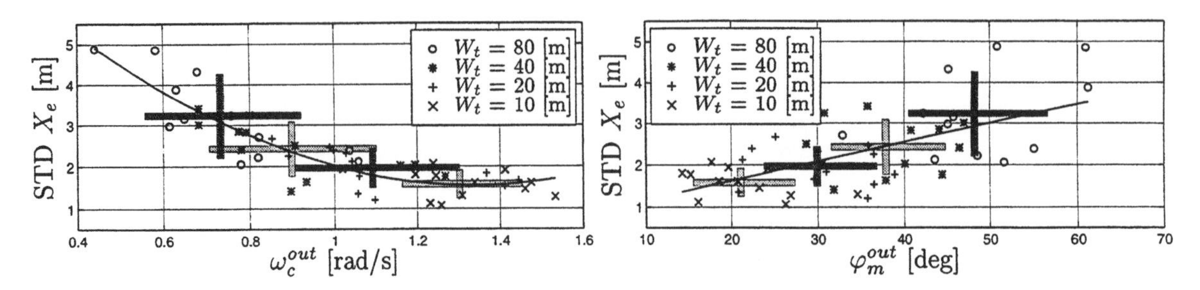

(a) performance vs. position error feedback bandwidth

(b) performance vs. position error feedback stability

Fig. 5. Path-following performance vs. position error feedback bandwidth and stability (all subjects, all conditions). The filled rectangles show the mean and standard deviation of the variables shown on the ordinate and abscissa, for the four tunnel sizes.

6. CONCLUSIONS

The experiment supports the findings in literature concerning the functionality of the tunnel size as a design parameter with which the level of path-following accuracy adopted by the pilot can be manipulated. One of the additional outcomes of the cybernetic approach propagated here is that it allows the introduction of stability requirements for the tunnel size design variable.

The notion of system stability should be considered as being equally important as system performance. A tunnel size that is too small could result in an ill-damped manual control situation which should be prevented from a performance, workload, and especially a *safety* point of view. In such critical tasks as landing an aircraft considerable – robustness – margins should exist with respect to the stability of the pilot/aircraft control system.

The cybernetic approach can be used to obtain guidelines concerning performance and stability requirements with a tunnel display in a *conceptual* rather than empirical manner.

REFERENCES

Grunwald, A. J. (1984). Tunnel Display for Four-Dimensional Fixed-Wing Aircraft Approaches. *Journal of Guidance and Control*, *7*(3), 369–377.

McRuer, D. T., & Jex, H. R. (1967). A Review of Quasi-Linear Pilot Models. *IEEE Transactions on Human Factors in Electronics*, *HFE-8*(3), 231–249.

Mulder, M. (1995). Towards a Control-Theoretic Model of Pilot Manual Control Behaviour with a Perspective Flight-Path Display. *Proceedings of the XIVth European Annual Conference on Human Decision Making and Manual Control, Delft, The Netherlands, June 14-16*, 1.2.1–1.2.13.

Mulder, M. (1996). Modelling Manual Control of Straight Trajectories with a Tunnel-in-the-Sky Display. *Proceedings of the XVth European Annual Conference on Human Decision Making and Manual Control, Soesterberg, The Netherlands, June 10-12*, 1.2.1–1.2.12.

Mulder, M. (1998). *Cybernetics of a Tunnel-in-the-Sky Display*. Ph.D. dissertation, Faculty of Aerospace Engineering, Delft University of Technology.

Oliver, J. G. (1990). Improving Situational Awareness Through the Use of Intuitive Pictorial Displays. *Society of Automotive Engineers, SAE Technical Paper 901829.*

Schattenmann, W., & Wilckens, V. (1973). *Vergleichende Simulatorstudien mit dem kontaktanalogen Kanal-Display und mit konventionellen Instrumentierungen* (Forschungsbericht No. FB 73-57). Deutsche Luft- und Raumfahrt. (in German)

Theunissen, E. (1995). Influence of Error Gain and Position Prediction on Tracking Performance and Control Activity with Perspective Flight Path Displays. *Air Traffic Control Quarterly*, *3*(2), 95–116.

Theunissen, E. (1997). *Integrated Design of a Man-Machine Interface for 4-D Navigation.* Ph.D. dissertation, Faculty of Electrical Engineering, Delft University of Technology.

Theunissen, E., & Mulder, M. (1994). Open and Closed Loop Control With a Perspective Tunnel-in-the-Sky Display. *Proceedings of the AIAA Flight Simulation Technologies Conference, Scottsdale (AZ), August 1-3*, 32–43.

Wickens, C. D., Haskell, I. D., & Harte, K. (1989). Ergonomic Design for Perspective Flight-Path Displays. *IEEE Control Systems Magazine*, *9*(4), 3–8.

Wilckens, V. (1973). Improvements in Pilot/Aircraft-Integration by Advanced Contact Analog Displays. *Proceedings of the Ninth Annual Conference on Manual Control*, 175–192.

CURVE INTERCEPTION WITH A TUNNEL-IN-THE-SKY DISPLAY: PILOT TIMING STRATEGIES

Max Mulder *,[1] **J.C. (Hans) van der Vaart** *

* *Delft University of Technology, Faculty of Aerospace Engineering, P.O. Box 5058, 2600 GB Delft, The Netherlands*

Abstract: The tunnel-in-the-sky display is a promising candidate to become the primary flight display of the next generation of cockpits. An investigation of the pilot's guidance task revealed that there are two subtasks: a regulating task and an anticipatory one. The paper describes the pilot decision making process in the anticipatory task of intercepting a curved section of the trajectory. It is hypothesized that timing the curve interception manoeuvre is determined by the handling characteristics of the aircraft and the nature of the available guidance information. An experiment was conducted to investigate two pilot timing strategies based on the pilot's estimate of either the *time* or the *distance* before the turn. The experimental data and the pilot opinions showed that the time-related strategy is preferred by pilots. Hence, the tunnel design should allow pilots to adopt the time-related strategy.

Keywords: Aircraft control, manual control, displays, information analysis, models, cybernetics, decision making, timing analysis.

1. INTRODUCTION

The concept of a tunnel-in-the-sky display, showing the reference trajectory to the pilot in a synthetic three-dimensional world (Fig. 1), is not new. Since the early 1950s it has been hypothesized that such a pictorial display could establish, in many ways, an important improvement in information-transfer to the pilot (Jones, Schrader, & Marshall, 1950). Its application remained impractical due to technical limitations, but several recent technological developments made the pictorial display concept practical.

Perspective displays in the cockpit have important consequences. In a conventional cockpit the pilot mentally reconstructs the aircraft's spatio-temporal situation from a number of *planar* displays. With a perspective flight-path display this information is presented in a *spatial* format, which is often stated to be highly compatible to pilots' internal representation (Haskell & Wickens, 1993; Mulder, 1994; Theunissen, 1997)).

At the Delft University of Technology, research is being conducted to investigate these implications on pilot manual control behaviour (Mulder, 1998). The issue here is *not* whether a perspective display might yield an 'improved' man-machine interface, but the impetus is rather on determining *how* pilots can control a complex dynamic system – the aircraft – along a space-constrained trajectory – the tunnel – with a perspective flight-path display as their principal source of information. The pilot's guidance task of following a trajectory was analyzed in (Mulder, 1995). It was concluded that, based on the properties of a curved approach trajectory, two sub-tasks can be distinguished: (i) a *regulating task* of following a trajectory of which the geometry is time-invariant, e.g. a straight or curved section of the path, and (ii) a task of fol-

[1] e-mail: m.mulder@lr.tudelft.nl

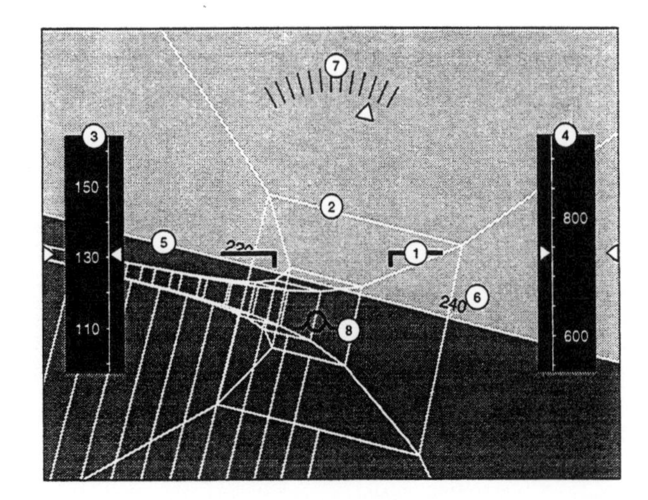

① aircraft reference symbol
② tunnel geometry (frames & poles)
③ airspeed [kts]
④ altitude [ft]
⑤ horizon line
⑥ heading angle indicator [deg]
⑦ bank angle indicator
⑧ flight-path vector

Fig. 1. The tunnel-in-the-sky display (left) and an explanation of its principal components (right).

lowing a trajectory of which the geometry changes in time, i.e. a task of controlling *transitions* between two reference conditions. This paper treats the second task, in particular the process of anticipating the transition between a straight section of the trajectory and a curved section.

2. CURVE INTERCEPTION

The interception of a curve is fundamentally different from an aircraft regulating task. Instead of maintaining an aircraft reference condition in the presence of – for instance – atmospheric disturbances, the pilot must control the aircraft from an initial aircraft reference state to another. Hence, other cognitive, perceptual and control-theoretical aspects play a role that are less important in pilot regulating tasks. The interception of a curve is essentially a matter of *decision-making behaviour* with the notion of *timing* and *control proportioning* as the main variables of interest.
Due to the inherent lags of the aircraft dynamics the pilot must initiate the curve transition manoeuvre well before the curve begins. But how far ahead? Pilots could for instance prefer a smooth manoeuvre that starts a relatively long time before the curve, or they could initiate a faster response just before the start of the curve. The objective of the research presented here is to obtain insight into the parameters underlying the pilot decision making behaviour in the curve interception manoeuvre.

3. RESULTS FROM A LITERATURE SURVEY

The perceptual mechanisms in approaching and following a bend in the trajectory are reported in (Mulder, 1998). The results of a literature survey showed that most – if not all – efforts in investigating human behaviour in this task originate in automobile driving research. Central in these investigations is the fact that the forward view of the road ahead conveys dual information, i.e. information necessary for guidance and information necessary for stabilization. Hence, a driver is generally modelled on two levels of control, consisting of anticipatory and compensatory components (Donges, 1978). The characteristics of the visual scene allow a driver to conduct the anticipatory actions at curve entrance. The phenomenon of *curve negotiation* (Shinar, McDowell, & Rockwell, 1977), i.e. the lateral eye movements preceding a bend in the road, reflect the driver's need to obtain information about the properties of the bend, probably its curvature. The anticipatory steering actions play a dominant role in the curve entrance phase (Godthelp, 1986). Further, experimental as well as theoretical evidence exists of the role of the *tangent point* – the inner-most position of the inner curb line of a curved road – for timing a curve-entrance manoeuvre (Riemersma, 1991; Land & Lee, 1994; Boer, 1996). In the aerospace literature, no previous studies on the aspect of *timing* a curve transition manoeuvre were found. Previous research on timing of the aircraft landing flare manoeuvre showed that information regarding time-to-contact (TTC, (Lee, 1974)) is a powerful cue (Pleijsant, Mulder, van der Vaart, & van Wieringen, 1996).

4. PRE-EXPERIMENTAL ANALYSIS

Based on the literature and a preliminary study of curve transition manoeuvres, it was hypothesized that such a manoeuvre depends on the dynamic characteristics of the aircraft on the one hand and on the nature of the presented information on the other (Mulder, 1998). These topics were the subject of a pre-experimental analysis.

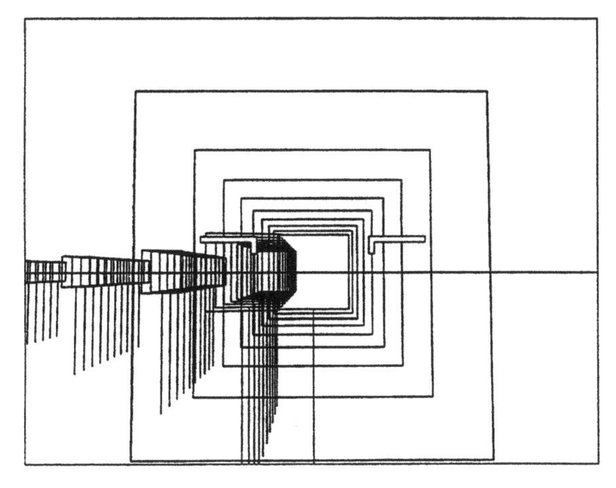

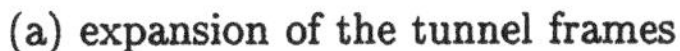

(a) expansion of the tunnel frames

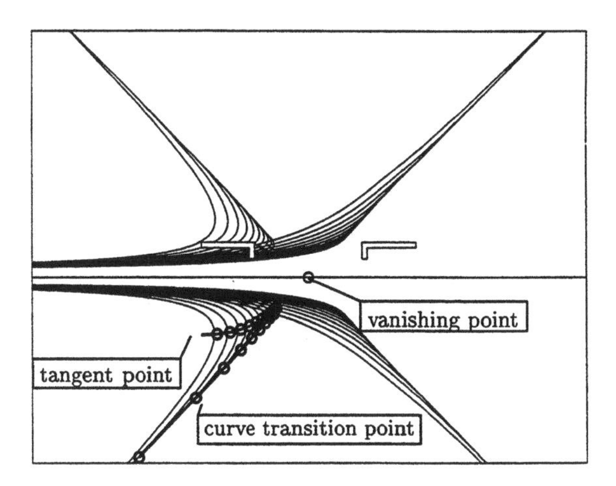

(b) motion of the tangent point

Fig. 2. The frozen images of a 3-D 'movie', showing the appearance of two basic elements of the tunnel display when approaching a curve. Figure left: only the frames of the tunnel presented. Figure right: only the lines between the tunnel frames presented. The tunnel images are generated with an interval of 1 second. The aircraft approaches the curve transition along a straight line with a velocity of 70 [m/s], starting at 700 [m] before the curve transition. In the bottom figure, the circles show the positions on the screen of the tangent point, the curve transition point and the vanishing point at identical time instances.

4.1 *Influence of the aircraft handling characteristics*

The investigation into the effects of the aircraft dynamics, conducted with a computer-based simulation, showed that especially the aircraft roll motion lag time constant is important (van Oorschot, 1997). It was concluded that although the pilot has considerable freedom in initiating and conducting the curve transition, a tradeoff exists between path-following accuracy and the magnitude of the roll-rates accompanying the manoeuvre (Mulder, 1998).

4.2 *Influence of the presented information*

A pre-experimental cue analysis of the presented information revealed that two types of information can serve as an onset cue for initiating the manoeuvre (Fig. 2). First, the expansion of the tunnel frames when approaching a curve contains time-to-contact information, which could allow pilots to obtain an estimate of the *time* before the start of the curved section (Lee, 1974). Second, and similar to the case of automobile driving, the motion of the tangent point could yield pilots an estimate of the *distance* to go to the turn. Hence, the information analysis led to the hypothesis of two timing strategies:

- Strategy I: a TTC-strategy using the expanding tunnel frames,
- Strategy II: a distance-related strategy using the tangent point in combination with the estimated distance to the start of the curve.

The TTC-strategy can expected to be superior to the second in that it is independent of the geometrical aspects of the tunnel trajectory, such as the tunnel size and the radius of the curve. In other words, the TTC-strategy would be *robust* over different tunnel geometries.

5. EXPERIMENT

An experiment was conducted to investigate the combined effects of aircraft dynamics and the nature of presented information on the pilot timing and manoeuvering characteristics of the curve transition task. Five subjects – all professional pilots – were instructed to control the lateral motion of an aircraft along a trajectory from a section that is straight into a curved section.

5.1 *Independent variables*

The independent variables can be categorized in terms of the characteristics of the aircraft and the tunnel display. To examine the effects of the *aircraft characteristics*, two aircraft types – the Cessna Citation, a business jet, and an Airbus A300, a large passenger transport – were used, representing a small (quickly responding) and large (sluggish) aircraft, respectively. Both aircraft types were simulated for a low and a and a high velocity, corresponding with the approach and cruise flight conditions, respectively. As far as the *tunnel display characteristics* are concerned,

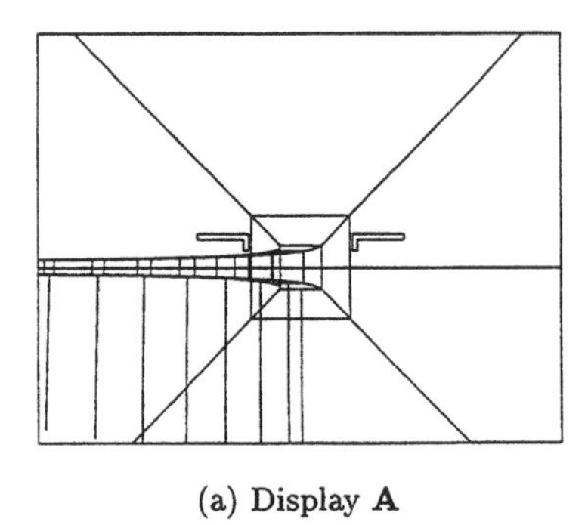

(a) Display **A**

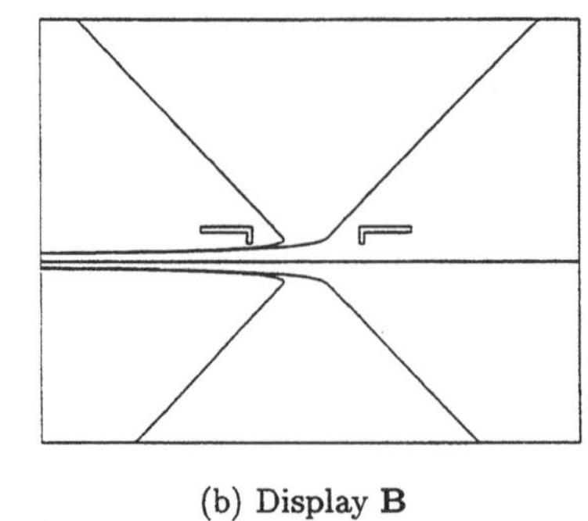

(b) Display **B**

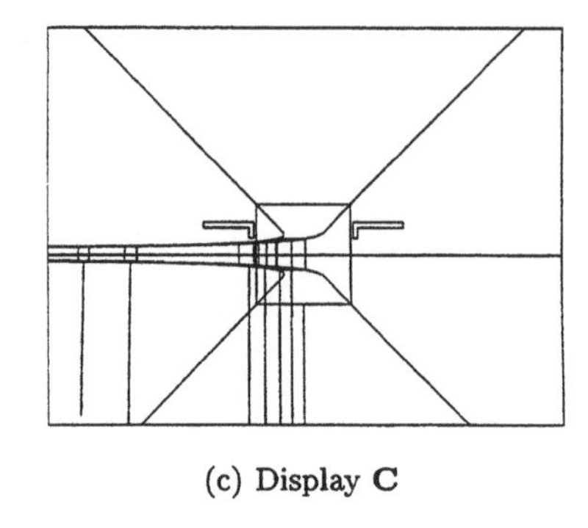

(c) Display **C**

Fig. 3. The three displays used in the curve transition experiment: Display **A**, the default tunnel; Display **B**, the tunnel without frames; Display **C**, the tunnel with frames placed at irregular distances.

three tunnel geometries were applied in the experiment (Fig. 3): (i) Display **A**, the baseline tunnel; (ii) Display **B**, the baseline tunnel without frames, and (iii) Display **C**, similar to **A**, except that the frames were positioned at random intermediate distances. Finally, two tunnel widths were applied (W = 20 and 40 [m]). A factorial design of all four independent variables yielded 24 experimental conditions.

5.2 *Procedure*

The *task* of the pilot was to initiate and conduct a *well-timed* and *well-proportioned* curve transition manoeuvre, just as he judged to be appropriate. No external disturbances acted on the vehicle. The main experimental hypothesis was that for displays **A** and **C** timing strategy I would be used, while with display **B** strategy II would be adopted by pilots. The experimental results consisted, first, of subjective data (a pilot questionnaire) and, second, an array of objective results categorized in data related to performance, timing, and manoeuvering. The trajectories to be flown consist of an initial straight section followed by a curve to the left. The curve radii were selected such that the aircraft yaw-rate (the derivative of the aircraft heading) corresponding with the curvature was fixed to 1.5 [deg/s]. Thus, for the low-velocity conditions the curve radius was substantially smaller than for the high-velocity conditions. As will become clear below, the relation between curve radius and aircraft velocity has consequences for the use of timing strategy II.

6. RESULTS & DISCUSSION

The effects of varying aircraft dynamics supported the findings of the pre-experimental analysis. First of all, the effects of the different aircraft dynamics did not interfere with the effects which could be attributed to the timing of the manoeuvre: although the curve transition was initiated earlier for the larger and heavier Airbus, the effects of manipulating the display were similar for both aircraft. Hence, the aircraft dynamics determine the freedom in pilot decision making behaviour merely in terms of a ***manoeuverability range***. Within this range, pilots can determine the moment they initiate the curve transition manoeuvre. A special algorithm was developed to determine this moment (Mulder, 1998). Figures 4(a) and 4(b) show the timing results of the experiment, expressed in either the *time-before-the-turn* T_{BT} (in [s]), or the *distance-before-the-turn* D_{BT} (in [m]). These figures illustrate clearly that the timing of the curve transition is not affected by the aircraft velocity for displays **A** and **C**. In fact, the timing with these displays was similar and consistent (p <0.01) for all experimental conditions. This provides strong evidence for the TTC-strategy (I) hypothesized for these displays. As far as *timing* is concerned, the results of display **B** show a less convincing but still sufficient argument for the use of a distance-related strategy, i.e. using the motion of the tangent point and the estimate of the distance to the curve transition. This can be seen when examining the significant effects of aircraft velocity and tunnel size on the timing data. Assume that the pilot initiates the manoeuvre when the position of the tangent point on the display exceeds a certain threshold. Then, (i) for equal curve radii (and thus equal velocity) a larger tunnel size would lead to an earlier response (T_{BT} larger) and (ii) for equal tunnel size a smaller radius (and thus smaller velocity) leads to the same effect. This is exactly what has been found in the experiment: Independent of tunnel size, the T_{BT} decreases for the higher velocity conditions (larger radius). Independent of aircraft velocity the T_{BT} increases with larger tunnel size. When investigating the effects together with those for the distances to the turn, the D_{BT}s, results show sufficient evidence for the distance-related strategy hypothesis, especially for the small tunnel size conditions. Because the tunnel width and the curve radius are variables characterizing the geometry of the tunnel trajectory, the inconsistency

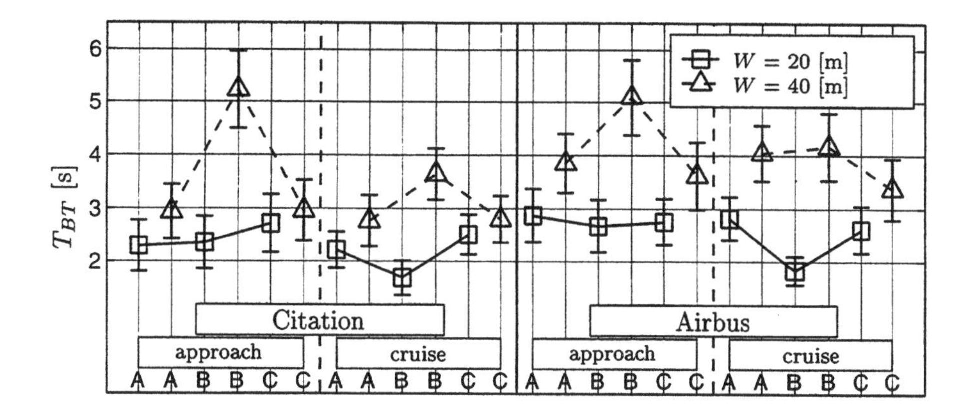

(a) the time-before-the-turn T_{BT}

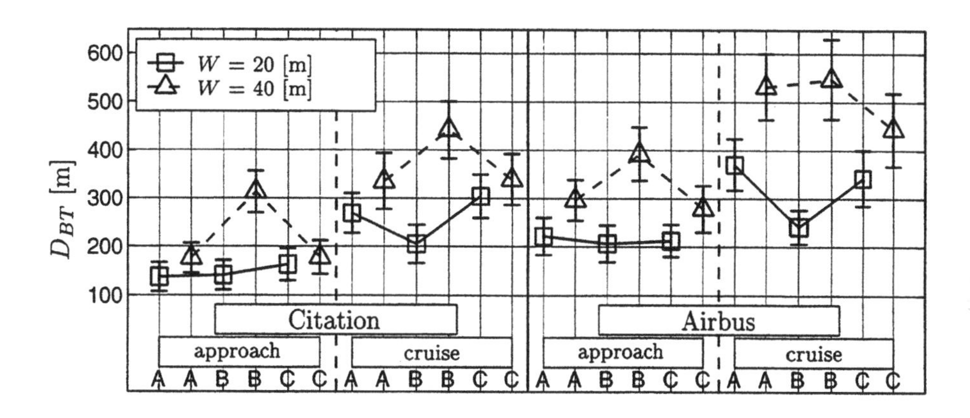

(b) the distance-before-the-turn D_{BT}

Fig. 4. The means and 95% confidence limits for the T_{BT} and D_{BT} variables (all subjects). The data are shown for both aircraft types (Citation and Airbus), for both velocities (approach=low and cruise=high), for both tunnel widths (20 and 40 [m]) and for all three displays (**A**, **B**, **C**).

of timing with display **B** illustrates the *vulnerability* of the distance-related timing strategy.
The pilot comments collected in a pilot questionnaire supported the finding of two different timing strategies. Pilots considered the task more difficult with display **B** than for displays **A** and **C**. Hence, the presence of tunnel frames near the curve transition allows pilots to adopt a – preferred – TTC-strategy.

7. CONCLUSION

The pilot decision making process in intercepting a curve is determined by the available visual guidance information. More specifically, when the expanding tunnel frames are present near the curve transition, pilots adopt a time-related strategy using TTC information. When no tunnel frames are presented and no TTC information is available, pilots are forced to adopt a sub-optimal distance-related strategy. The TTC-strategy is superior because it does not depend on the geometrical properties of the tunnel trajectory: it is a strategy which is *robust* over these variations. This study shows that, to allow pilots to use the TTC-strategy, a sufficient number of tunnel frames should be positioned near transitions in the trajectory to be followed.

ACKNOWLEDGEMENTS

The authors wish to thank Peter van Oorschot (M.Sc. student) for his contribution in the design, execution, and evaluation of the experiment.

REFERENCES

Boer, E. R. (1996). Tangent Point Oriented Curve Negotiation. *IEEE Proceedings of the Intelligent Vehicles '96 Symposium, Tokio, Sep. 19-20.*

Donges, E. (1978). A Two-Level Model of Driver Steering Behavior. *Human Factors, 20*(6), 691–707.

Godthelp, H. (1986). Vehicle Control During Curve Driving. *Human Factors, 28*(2), 211–221.

Haskell, I. D., & Wickens, C. D. (1993). Two- and Three-Dimensional Displays for Aviation: A Theoretical and Empirical Comparison. *The International Journal of Aviation Psychology, 3*(2), 87–109.

Jones, L. F., Schrader, H. J., & Marshall, J. N. (1950). Pictorial Display in Aircraft Navigation and Landing. *Proceedings of the I.R.E.*, 391–400.

Land, M. F., & Lee, D. N. (1994). Where we look when we steer. *Nature, 369*, 742–744.

Lee, D. N. (1974). Visual Information During Locomotion. In R. B. McLeod & H. Picks (Eds.), *Perception: Essays in Honor of J. J. Gibson*. Ithaca (NY): Cornell University Press.

Mulder, M. (1994). *Displays, Perception and Aircraft Control. A survey of theory and modelling of pilot behaviour with spatial instruments* (Report No. LR-762). Delft: Faculty of Aerospace Engineering, Delft University of Technology.

Mulder, M. (1995). Towards a Control-Theoretic Model of Pilot Manual Control Behaviour with a Perspective Flight-Path Display. *Proceedings of the XIVth European Annual Conference on Human Decision Making and Manual Control, Delft, The Netherlands, June 14-16*, 1.2.1–1.2.13.

Mulder, M. (1998). *Cybernetics of a Tunnel-in-the-Sky Display*. Ph.D. dissertation, Faculty of Aerospace Engineering, Delft University of Technology.

Pleijsant, J. M., Mulder, M., van der Vaart, J. C., & van Wieringen, P. C. W. (1996). Effects of Runway Outline and Ground Texture on the Perception of Time-to-Contact in Simulated Landings. *Proceedings of the fourth European Workshop on Ecological Psychology (EWEP-4), Zeist, The Netherlands, July 3-5*, 65–68.

Riemersma, J. B. J. (1991). Perception of Curve Characteristics. In A. G. Gale, M. H. Freeman, C. M. Haslegrave, P. Smith, & S. P. Taylor (Eds.), *Vision in Vehicles II* (pp. 163–170). Elsevier Science Publishers (North-Holland).

Shinar, D., McDowell, E. D., & Rockwell, T. H. (1977). Eye Movements in Curve Negotiation. *Human Factors, 19*(1), 63–71.

Theunissen, E. (1997). *Integrated Design of a Man-Machine Interface for 4-D Navigation*. Ph.D. dissertation, Faculty of Electrical Engineering, Delft University of Technology.

van Oorschot, P. W. J. (1997). *Simulaties voor Bochtinitiatie Tunnel-in-the-Sky*. Unpublished preliminary MSc. Thesis, Faculty of Aerospace Engineering, Delft University of Technology. (in Dutch)

INTEGRATED DESIGN OF SPATIAL NAVIGATION DISPLAYS

Eric Theunissen*, Henk Stassen‡ and Durk van Willigen*

**Faculty of Information Technology and Systems, Delft University of Technology, Mekelweg 4, 2628 CD Delft, The Netherlands*

‡Faculty of Design, Engineering and Production, Delft University of Technology, Mekelweg 2, 2628 CD Delft, The Netherlands

Abstract: Compared to current command displays, a spatially integrated presentation of the future trajectory has the potential to reduce the task demanding load for guidance and control while at the same time increasing spatial and navigational awareness. To allow a structured design of spatial navigation displays, this paper illustrates how the information contents of the visual cues can be described as properties of the optic flow field. The type of visual cues are related to the potential control strategies, and the results of experiments in which these control strategies are evaluated are discussed. *Copyright © 1998 IFAC*

Keywords: displays, navigation, human-machine interface, aircraft control, design.

1. INTRODUCTION

Due to the increasing number of aircraft passengers and the resulting increase in aircraft, bottlenecks in airspace capacity are beginning to emerge. Once the number of aircraft has reached a certain threshold, any further increase will cause unacceptable delays. Since the bottlenecks mainly occur in the vicinity of airports, this is also the place where a solution should start. The basic idea behind improvements is to abandon today's straight-in approaches and allow aircraft to intercept the final straight glidepath at predetermined locations. This provides air traffic control with more possibilities to manage the traffic flow, creating the opportunity to increase capacity. An additional advantage is that it becomes possible to avoid noise sensitive areas by using noise abatement procedures. This concept, however, increases the task demanding load of the pilot. In contrast to the current straight-in approach, pilots will have to fly curved approaches. Due to the more frequent changes in direction, it becomes harder to maintain an adequate level of spatial and navigational awareness. As a result, pilots will have to scan the navigation display more frequently. Because of the already high workload during the approach, an introduction of more complex approach trajectories is likely to reduce safety. By providing pilots with the information they need to accomplish their task more easily, while maintaining spatial and navigational awareness without the need to scan additional displays, it becomes possible to fly more complex approaches without a reduction in safety. It is highly unlikely that with all future developments, safety can be increased by extrapolating current concepts. New functionality and new technology cannot simply be layered onto existing concepts, because the current system complexities are already too high. Better human-machine interfaces may require a fundamentally new approach. Navigation, guidance, and control are not three independent tasks, but with the current flight director display the control task is isolated. Displays providing a spatially integrated presentation of the future trajectory, and thus presenting guidance *requirements* instead of control *commands* have advantages relative to current non-spatial displays. These advantages result from the fact that the pilot has to perform less integration of information and the fact that the more natural presentation requires less effort for interpretation and evaluation. Fig.1 shows a spatial navigation display during a flight test.

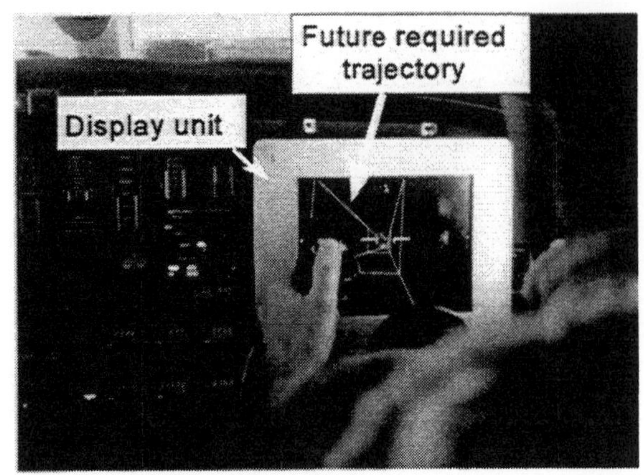

Fig. 1. Flight test of an experimental spatial navigation display.

2. AN INTEGRATED DESIGN APPROACH

A factor which has prevented the introduction of spatial displays for navigation and guidance is the complexity of their design. Fadden *et al.* (1987) indicated this problem by stating that *'while the promise of spatial displays is great, the cost of their development will be correspondingly large. The knowledge and skills which must be coordinated to ensure successful results is unprecedented. From the viewpoint of the designer, basic knowledge of how human beings perceive and process complex displays appears fragmented and largely unquantified'*. Hardly any detailed guidelines to the design of these types of displays exist which take specific human capabilities in the areas of perception, cognition, and control into account.

2.1 Design questions

The design of a spatial navigation display requires the specification of a frame of reference, a field of view, and a number of properties of the object representing the flightpath. Fig. 2 provides an overview of the different elements involved in the spatially integrated presentation of navigation data. In Fig. 2 *Static Synthetic Data* refers to data which describes abstractions of real-world objects. *Symbology specification* refers to data describing symbology which due to their specific representation have a particular meaning. Properties of the symbology such as position, orientation, color, and size can be used to convey information. *Dynamic Synthetic Data* refers to data which describes the geometry of objects according to a set of *representation rules* and a *forcing function*. An example is the representation of the required trajectory. Based on the *selection rules*, the *selection logic* controls which data is to be presented. The *transform rules* determine the dynamic properties of the objects to be presented such as position, orientation, size, color, and style. The *data transformation* applies the transform rules.

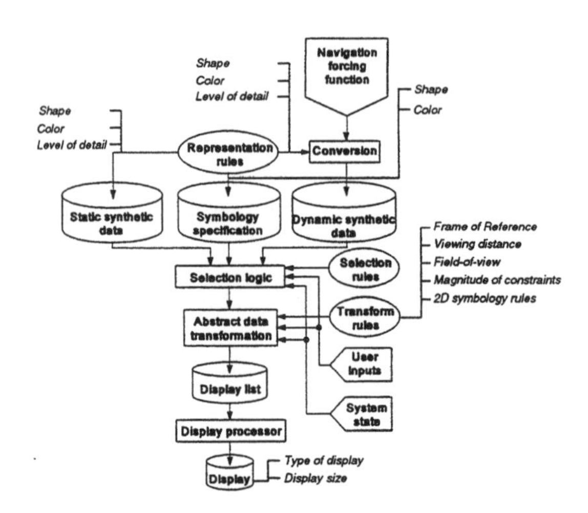

Fig. 2. Overview of the elements involved in the spatially integrated presentation of navigation data.

A proper selection of the design parameters requires an understanding of their relation with the magnitude of the visual cues conveyed by the display and the potential task strategies. Furthermore, tasks and measures are needed to evaluate the potential of a certain perspective flightpath display format for a range of control strategies. An approach is needed which supports a structured design process for perspective flightpath displays in which technical possibilities and human factors are truly integrated. To develop such an approach, the question on how to utilize the existing knowledge in the domains of perception, cognitive science, and control theory, has been addressed. To accomplish this, it was decided to translate specific design questions to these domains. In the remaining part of this paper, the approach used to describe the information contents of the visual cues and the resulting control strategies will be discussed.

2.2 Analysis of the visual cues

To benefit from research performed into perception and control of self motion, a model is needed which describes the relation between the data which must be observable and the magnitude of the visual cues which convey this data as a function of the display design parameters. For perspective flightpath displays the basis for this approach was developed by Wilckens (1973). Later, Warren (1982) derived similar equations to describe changes in the visual cues as properties of the optic flow field and introduced the concept of the splay angle. For a perspective flightpath display, the information content of the presentation is described by deriving a relation between the position and orientation errors of the aircraft and the resulting changes in the position and orientation of the perspectively presented trajectory. Fig. 3 shows the influence of a position error on the spatially integrated presentation of the flightpath.

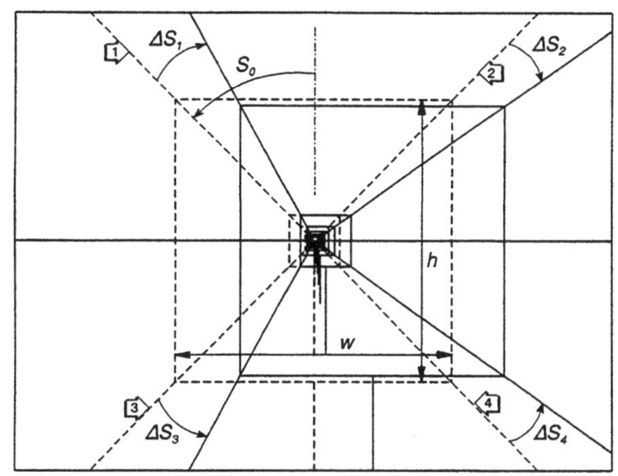

Fig. 3. Influence of a cross-track error on splay angle S_0. The dashed tunnel indicates the symmetrical reference condition and the solid tunnel is seen when the viewpoint is displaced to the left yielding a cross track error. Tunnel lines 1 to 4 rotate around the vanishing point as a result of the cross-track error. The change is splay angle is referred to as ΔS_i, in which the index i is used to identify the tunnel line.

The rate of change of the angles between the tunnel lines and the line perpendicular to the horizon provides visual cues which are also known as splay rate cues. Owen (1990) describes several studies which all indicate that splay rate is the functional variable for altitude control. A symmetrical object as the one depicted in Fig. 3 provides splay rate cues both for vertical and horizontal position control. It can be assumed that when presenting both the horizontal and the vertical constraints, splay rate is the functional variable for horizontal and vertical position control.

Besides cues obtained through changes in splay angle, the optic flow field provides additional information about the direction of motion relative to the trajectory. These directional cues allow the observer to distinguish between the orientation of the vehicle body axis and the direction of vehicle motion. Fig. 4 shows the optic flow field combined with a snapshot of the tunnel for a situation in which a directional error is present.

The dynamic presentation also conveys so-called temporal range information. This temporal range information can be used for the timing of anticipatory control actions, e.g. the moment to initiate a control action to enter a curve. In the situation of recti-linear motion, the dynamic perspective presentation of the trajectory allows the pilot to estimate the time until the center of projection reaches a certain reference point without knowing tunnel dimension or vehicle velocity. Lee (1976) refers to this phenomenon as time-to-contact (TTC). Kaiser and Mowafy (1993) showed that for rectilinear motion temporal range can already be extracted from the motion of a single point. They refer to this temporal range as time-to-passage (TTP).

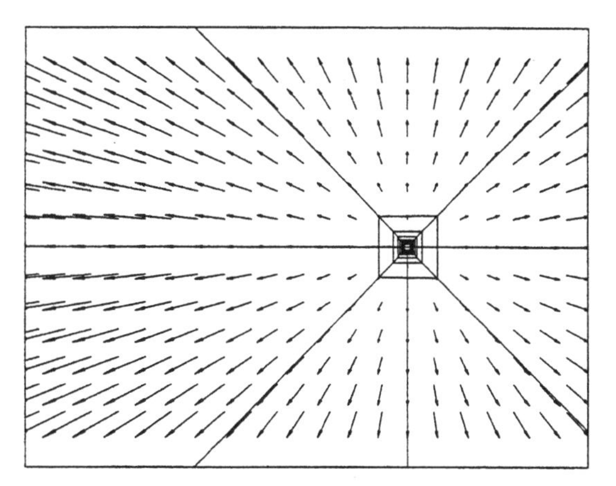

Fig. 4. Velocity streamer pattern for the situation in which the track of the aircraft is aligned with the tunnel axis and the aircraft flies with a crab angle of 10 degrees due to crosswind. The direction of flight can be perceived from the location of the center of optic outflow.

2.3 *Control strategies*

The presentation of the future forcing function and its constraints allows continuous compensatory, anticipatory (McRuer *et al.*, 1977), and error neglecting (Godthelp, 1984) control strategies to be applied. Continuous compensatory control is the strategy which is typically used with today's glideslope and localizer displays and flight directors. As indicated earlier in this paper, when splay rate is the functional variable for position control, an equal ratio increment in the variable should yield an equal interval improvement in tracking performance.

The presence of preview on changes in the required trajectory combined with information about the temporal range towards these changes provides pilots with the possibility to apply anticipatory control. The accuracy of the anticipatory control action is determined both by the accuracy with which the pilot can extract temporal range towards a change in the trajectory and the accuracy with which the magnitude of this change can be estimated.

When the task of the pilot is to remain within certain position constraints, instruments which force the pilot to continuously minimize position errors will unnecessarily increase task demanding load. In contrast, instruments which provide information about the current and future position margins that are available, allow the pilot to determine whether a certain position or orientation error requires a control action or not. The resulting tracking accuracy is determined by the threshold(s) the pilot uses when making a decision whether to intervene. Such an error-neglecting control strategy provides the pilot with the opportunity to make a trade-off between workload and tracking accuracy. With error-neglecting control there is a similarity with car driving in terms of control task (boundary control) and the type of visual cues (spatially integrated presention of constraints). When looking at the results from the time-to-line crossing (TLC) experiments performed by Godthelp (1984), this suggests that with this particular control-strategy, the pilot bases the moment of an error-correcting action on an estimate of the remaining time before the aircraft crosses one of the imaginary tunnel walls, the so-called time-to-wall crossing (TWC).

In the context of the Delft Program for Hybridized Instrumentation and Navigation Systems (DELPHINS), it has been investigated how the various design aspects influence the type and magnitude of the visual cues, what the consequences are for the translation of the data into useful information, and how useful the information is with respect to the ability to apply a particular control strategy. This analysis served to derive guidelines for the design aspects. To be able to investigate certain design questions in more detail, the concept has been implemented in a way which allows the manipulation of the design aspects for evaluation purposes. The following section discussed the results from experiments focusing on compensatory, anticipatory, and error-neglecting control strategies.

3. EVALUATION

For pilot-in-the-loop evaluation, tasks and measures are needed to quantitatively rate a certain design in terms of performance for the range of potential control strategies. Displays for guidance and control are typically compared with each other in terms of maximum tracking performance and control activity. With a well designed command display, maximum tracking performance is achieved when the pilot applies a continuous compensatory control strategy. Since perspective flightpath displays allow a mix of compensatory, anticipatory, and error-neglecting control strategies to be applied, tasks and measures should be used to assess and to rate the possibility of applying each of these different types of control strategies.

For evaluation purposes, the ability to apply a range of control strategy is a mixed blessing. A general problem when analyzing performance data obtained from pilot-in-the-loop studies with perspective flightpath displays is that as a result of the range of potential control strategies, more trade-offs between effort and tracking performance are possible. This can be compensated for by defining tasks in which the pilot is forced to apply a particular control strategy. Although an isolated control strategy might not always result in typical control behavior, it allows the potential of different display formats to convey task relevant information to be compared. Some examples of tasks and measures to rate compensatory-, anticipatory-, and error-neglecting control strategies will be discussed next.

3.1 Continuous compensatory control

Grunwald (1984) showed that with the addition of a position predictor, tracking performance with perspective flightpath displays can be improved. To investigate the combined effects of error gain and position prediction on tracking performance and control activity for a closed loop compensatory control strategy on both the straight and the curved segments, two experiments have been conducted. In these experiments, six pilots had to fly a curved approach trajectory, which presented them with quite a challenging tracking task. The experiments were performed in the moving base flight simulator of Delft University of Technology. Both experiments were a two factor (error gain and position prediction) repeated measures design.

Fig. 5 shows the lateral tracking performance as a function of splay gain and Fig. 6 shows the aileron control activity as a function of splay gain. The label FPV indicates a condition in which a flightpath vector was presented, and the label FPP indicates the condition in which a position predictor was presented. The results have been split into those for tracking of straight segments (indicated by the -S) and those for tracking of curved segments (indicated by the -C).

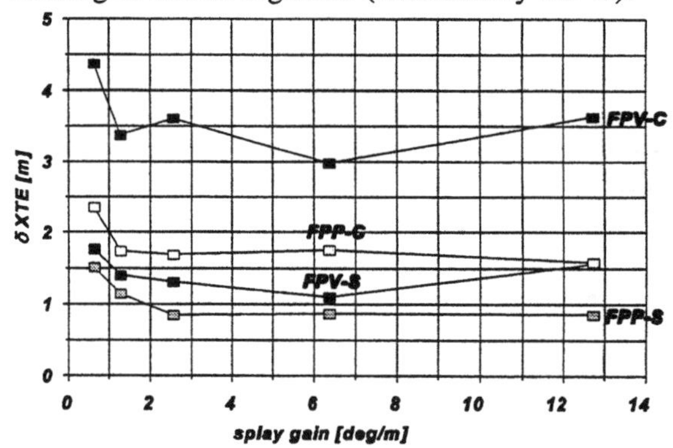

Fig. 5. Lateral tracking performance on straight and curved segments as a function of splay gain.

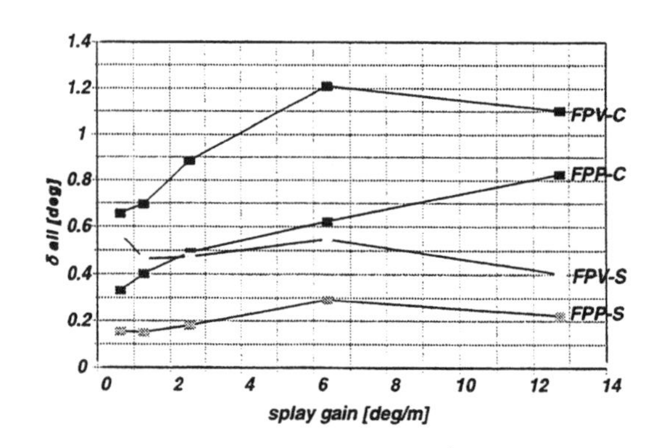

Fig. 6. Aileron control activity on straight and curved segments as a function of splay gain.

Fig. 5 illustrates that lateral tracking performance does not improve with a splay gain above approximately 2.5 deg/m. Fig. 6 shows that both on straight and curved segments the additional of a position predictor results in reduced control activity. Analysis of the data (Theunissen, 1997) led to the following conclusions:

1. The variation in the size of the tunnel shows the effect on tracking performance which is to be expected when changing the gain of the functional variable for position control, strengthening the hypothesis that with a perspective flightpath display the splay rate is a functional variable.
2. With a position predictor, the splay rate is no longer the functional variable for position control.

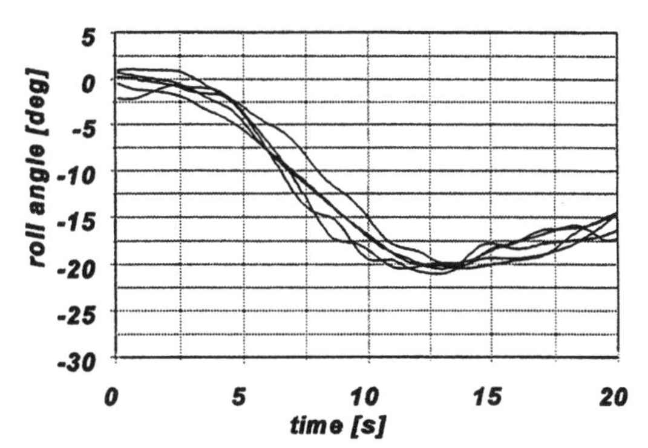

Fig. 8. Time histories of roll angle in the presence of a position predictor.

3.2 *Anticipatory control*

Each trajectory contained two curved segments, forcing the pilots to transition from a reference condition in which the wings are level to a reference condition in which the aircraft is banked to approximately 17 degrees for the commanded velocity and the specified turn radius. Theunissen (1997) illustrates that in an egocentric frame of reference, the magnitude of the visual cues conveying information about the rate of change of the direction of the future trajectory for a given turn radius makes it hard to accurately estimate the turn radius. Two parameters which can be used to rate the potential of a display format for anticipatory control are the deviation from a reference time which indicates the moment the open-loop control action should be initiated, and the deviation from a reference bank angle which is achieved as a result of the anticipatory control action. Fig. 7 shows a number of time histories of the bank angle during a change in the direction of the trajectory for a perspective flightpath display described in Theunissen (1997). Fig. 8 shows a number of time histories for the same part of the trajectory, but with the addition of a flightpath predictor in the display. As can be seen from Fig. 7, the basic tunnel does not provide adequate cues to support accurate timing and magnitude of the required anticipatory control action. Fig. 8 shows that the addition of the position predictor compensates for this deficiency. The results of the experiment clearly demonstrate that the integration of a position predictor allows the pilot to increase the accuracy of the anticipatory control actions.

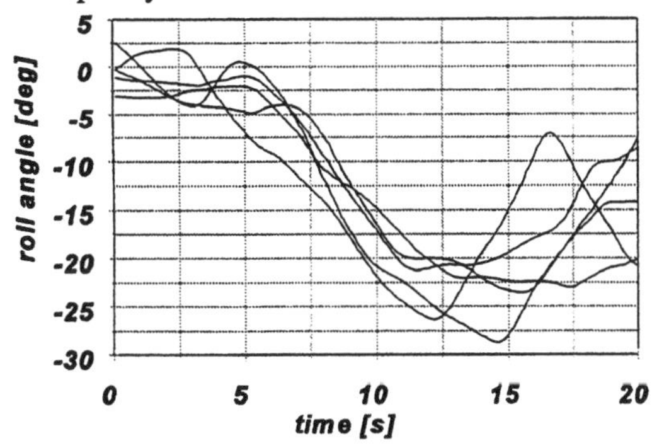

Fig. 7. Time histories of roll angle in the absence of a position predictor.

3.3 *Error neglecting control*

An experiment was performed to verify whether the moment the pilot initiates an error-corrective control actions is related to a prediction of the time-to-wall crossing (TWC), and, if so, whether the TWC can be approximated with a first or second-order model. In this experiment, four pilots had to fly a number of trajectories consisting of two straight and one curved segments. To reduce the need to apply many compensatory control actions, a relatively low position error gain of 0.42 deg/m (corresponding to a tunnel width of 135 m) was selected.

The task of the subjects was to try to minimize the number of control inputs while staying inside the tunnel, and only apply an error corrective action when the aircraft would otherwise violate the position constraints. In this way, the TWC measured at the moment a control action is initiated can be used as a measure to determine the ability to extract information about the constraints. Fig. 9 presents an estimate of the probability density function of TWC_{C2}, the 2nd order prediction of the time-to-wall crossing at the moment the pilot performs an error-corrective control action.

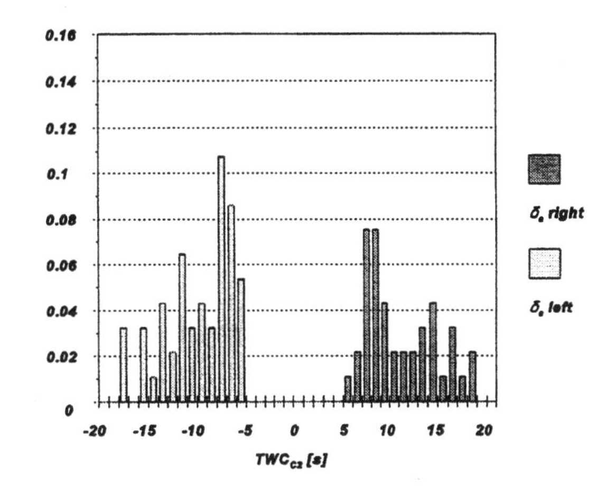

Fig. 9. Estimate of the probability density function of TWC_{C2}, the 2nd order prediction of the time-to-wall crossing at the moment the pilot performs an error-corrective control action.

Fig. 9 shows that with the type of perspective display

used in the experiment, a minimum temporal spacing of approximately 5 seconds was used. When forcing pilots to apply an error-neglecting control strategy, a distribution as depicted in Fig. 9 can be used to compare different display format for their potential to convey information about the margins towards the constraints. The experiment resulted in the following conclusions:

1. As a result of the multitude of visual cues conveying position and orientation information relative to the constraints, the perceptual response is not based on a single position or orientation error exceeding a certain threshold, but on a combination of these cues.
2. The spatially integrated presentation of guidance data allows pilots to extract information which allows them to make better than first-order estimates of the time when the aircraft would cross a tunnel wall, enabling them to apply an efficient error-neglecting control strategy.
3. Since pilots do not have to mentally integrate the values of position and angular errors and error rates and to verify whether the outcome exceeds a certain threshold which would be required for error-neglecting control with non-integrated displays, a perspective flightpath display reduces the task demanding load required for error neglecting control.
4. When comparing different display concepts for the guidance task, an analysis of the TWC variable provides more insight into the pilot's ability to utilize information about constraints.

4. CONCLUSIONS

Spatial navigation displays are superior to current 2-D command displays because the former provide information about the current and future guidance requirements rather than just plain control commands. The design of display formats providing a spatially integrated presentation of data is more complex as compared to conventional 2-D command displays. One of the requirements for a more structured design process is a model describing how the various design parameters influence the magnitude of the task related visual cues. Based on findings from research into perception and control of self motion and findings from research into control strategies, an analysis of the information contents of the visual cues conveyed by a perspective flightpath display has been used to provide more insight into the relation between the design parameters and the potential control strategies. Pilot-in-the-loop experiments have confirmed that it is possible to use the information to apply compensatory, anticipatory, and error-neglecting control strategies, and that the tracking performance can be influenced by the splay gain.

ACKNOWLEDGMENTS

The research into perspective flightpath displays is performed in the context of the Delft Program for Hybridized Instrumentation and Navigation Systems phase II (DELPHINS II), which is being sponsored by the Dutch Technology Foundation STW.

REFERENCES

Fadden, D.M., Braune, R. & Wiedemann, J. (1987). Spatial Displays as a Means to Increase Pilot Situational Awareness. In: *Spatial Displays and Spatial Instruments*, NASA CP10032, pp. 35-1 to 35-12.

Godthelp, H. (1984). *Studies on Human Vehicle Control*, PhD Thesis, Institute for Perception TNO, Soesterberg, The Netherlands.

Grunwald, A.J. (1984). Tunnel Display for Four-Dimensional Fixed-Wing Aircraft Approaches, *Journal of Guidance*, Vol. **7**, No. 3, pp. 369-377.

Kaiser, M.K., and Mowafy, L. (1993). Visual Information for Judging Temporal Range, *Proc. of Piloting Vertical Flight Aircraft*, pp. 4.23-4.27, San Francisco, CA.

Lee, D.N. (1976). A Theory of Visual Control of Braking based on Information about Time-to-Collision, *Perception*, Vol. **5**, pp. 437-459.

McRuer, D.T., Allen, R.W., Weir, D.H., Klein, R.H. (1977). New Results in Driver Steering Control Models, *Human Factors*, Vol. **19**, No. 4, pp. 381-397.

Owen, D.H. (1990). Perception & control of changes in self-motion: A functional approach to the study of information and skill. In: *The Perception and Control of Self Motion* (R. Warren and A.H. Wertheim (Eds.)), Hillsdale, N.J.

Theunissen, E. (1997). *Integrated Design of a Man-Machine Interface for 4-D Navigation.* ISBN 90-407-1406-1, Delft University Press, Delft, The Netherlands.

Warren, R. (1982). Optical transformation during movement: Review of the optical concomitants of egomotion, *Technical Report AFOSR-TR-82-1028*, Bolling AFB, DC: Air Force Office of Scientific Research (NTIS no. AD-A122 275).

Wilckens, V (1973). *Zur Lösung der Flugführungs-probleme vornehmlich bei der Nullsicht-Landung mit der echt-perspektivischen, bildhaft-quantitativen Kanal-Information.* Technical University of Berlin, Germany.

DESIGN OF A USER INTERFACE FOR A FUTURE FLIGHT MANAGEMENT SYSTEM

J. Marrenbach [(1)], M. Pauly [(2)] and K.-F. Kraiss [(1)]

(1) Aachen University of Technology, Department of Technical Computer Science, D-52074 Aachen, Germany. E-Mail: marrenbach@techinfo.rwth-aachen.de

(2) Dasa Navigation and Flight Guidance Systems GmbH, D-89070 Ulm, Germany. E-Mail: pauly@vs.dasa.de

Abstract: A Flight Management System (FMS) is a complex computer system that is used for flight planning in all modern commercial aircrafts. The FMS' human-computer interface needs substantial improvement to make the best use of the increasing number of features. This article outlines some ideas for a new user interface to replace today's Control and Display Units (CDUs). The alphanumerical flight plan editing is replaced by a graphical user interface. A software prototype of such a CDU has been created, using Statecharts for the definition of this interface. A development tool translates this graphical specification into C source code which can be integrated in the prototype. The software prototype is currently evaluated in experiments. *Copyright © 1998 IFAC*

Keywords: Flight Management System, Graphical User Interface, Statecharts, Direct Manipulation, Graphical Back Animation, Man Machine Interaction

1. INTRODUCTION

A Flight Management System (FMS) is an important part of the automatic flight guidance system of modern aircrafts. This complex system is meant to reduce the workload of the pilots, while at the same time conducting the flight in an economical way. This includes among other aspects low fuel consumption and meeting scheduled times of arrival.

Due to the increased complexity of cockpit instruments and gauges and to the higher degree of automation, it has become more and more difficult for the pilots to understand the connections and dependencies between the numerous cockpit systems. Analysis of accidents and training sessions performed in flight simulators show that system behaviour is not always as expected by the crew. Often enough, pilots are not fully aware of the functionality that is offered by FMS pages. Consequently, they avoid using these functions even if they could be of great help (Sarter and Woods, 1992; Sarter and Woods, 1994; Polson, *et. al.*,1994; Dornheim, 1996).

2. TODAY'S CONTROL AND DISPLAY UNIT

Today's Control and Display Units (CDUs) offer significant room for improvement. The CDU presents the man machine interface of the FMS. The growing number of features provided by the system has lead to a rather complex menu structure. This makes it more difficult for the pilots to access a function, because they need to memorise the provided functions and their respective locations in the menu system. Figure 1 depicts the User Interface of a current Control and Display Unit.

Also, stringent rules have to be obeyed when entering parameters into the system. Depending on what kind of value or position the pilot wants to enter, a particular format for the entry has to be used. These formats sometimes lack consistency; a more flexible way of entering data is highly desirable.

Fig. 1: Today´s Control and Display Unit

Problems are also caused by the fact that the users have to translate the flight plan into a string of characters. For a pilot, the flight plan is at least a lateral (if not a three dimensional) idea of how the flight will proceed and how individual waypoints are defined. For the FMS, this idea has to be converted into a textual description. This is usually not a difficult task, but it can introduce errors and it has in the past, with fatal results in some cases.

Finally, pilots complain about the small buttons of the alphanumeric keypad. This makes it difficult to enter text, especially when flying in turbulence. However, with only limited space available in the cockpit, a larger alphanumeric keypad is not likely to be used for an FMS (Gerlach, 1996).

3. AN ADVANCED CDU CONCEPT

In order to make using the CDU easier, especially when changes have to be entered quickly, such as during an approach or departure, the Department of Technical Computer Science (LTI) at RWTH Aachen is developing a graphical man machine interface for an Advanced Flight Management System (AFMS). This concept is carried out in close co-operation with NLR (National Aerospace Laboratory, Amsterdam, Netherlands). The project is funded by the European Union, and thirteen European organisations and companies are involved in it. A software demonstrator has been built by LTI and a corresponding hardware demonstrator by DASA NFS GmbH in Ulm, Germany. The prototype has the same physical size as the conventional system.

The most noticeable change in the user interface is the use of a graphical output device, as already used in today's navigation displays, for user inputs. The graphical, object-oriented user interface makes it possible to implement direct-manipulation aspects. Due to this approach only functions applicable for the selected object are presented to the pilot. Additionally, the function-oriented way is implemented as used in the current system, but only the selected functions, even more than one, are displayed at the same time. The combination of object-oriented and function-oriented user interaction will increase the situational awareness of the pilots. Less keys are necessary, which results not only in larger buttons, making it less likely to hit the wrong button by mistake, but also in more room for a larger display. This display contains navaids, airways etc. and the flight plan. Within the CDU, a touch pad is provided to control a cursor. Using the touch pad, the users can mark and select objects from the map and the flight plan (Marrenbach, *et. al.* 1997).

The graphical user interface relieves translating the pilot´s idea of a flight plan into the system language. There is, of course, still a translation process, but since the new system language of the ACDU (Advanced Control and Display Unit) and the pilot´s ideas have now a lot more in common, the translation is easier, giving less opportunities to introduce errors into the system flight plan. The important difference to today´s EFIS systems is that even though these systems do provide a graphical preview of a flight plan modification, the modification has to be entered the old way, with alphanumeric string input.

The object-oriented design of the user interface drastically reduces the number of functions from which the pilot has to make a selection. Once a graphical object has been selected, only a small number of functions can be reasonably used with this object. For example, there is no need to delete a VORTAC from the flight plan if it is not part of the flight plan. At any given time, so few functions make sense for a selected object that these functions can usually be associated with the CDU´s line select keys. Nevertheless, the traditional function-oriented approach has to be available, too, because a number of features is not related to a particular object on the map.

Figure 2 depicts the software prototype which is developed by the Department of Technical Computer Science.

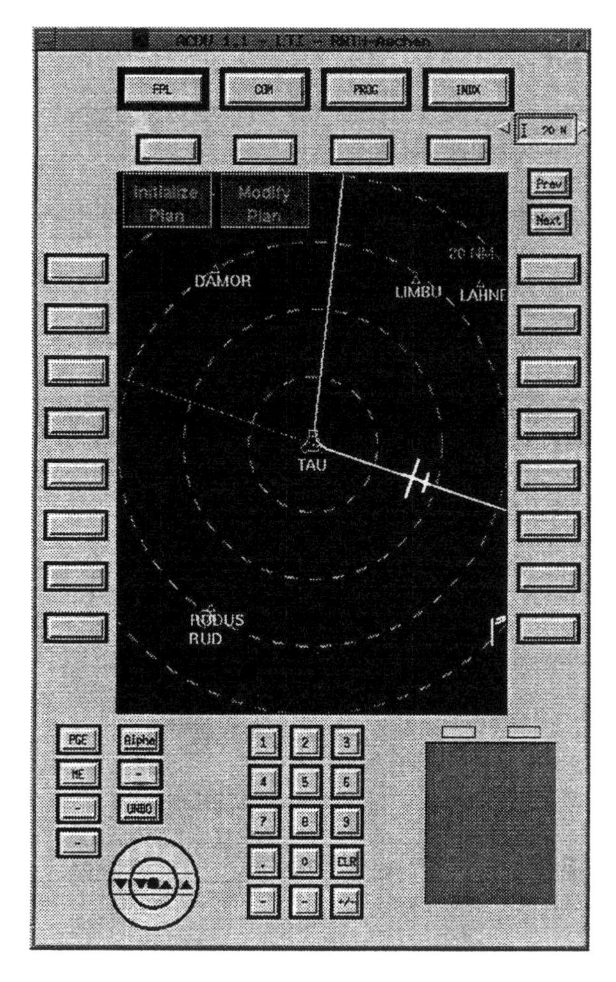

Fig 2: Advanced Control and Display Unit

Supplementary to the graphical representation of the flight plan (map-mode, plan-mode, vertical mode) an alphanumerical page is implemented, because it is much more easier for the user to gain an overview of the whole constraint list if it is presented on an alphanumerical page.

4. IMPLEMENTATION OF THE USER INTERFACE

The user interface was designed following the Seeheim Model which divides it into presentation component, dialog control and application interface (Green, 1985). The presentation component determines how information is presented to the user. It also accepts user inputs such as keystrokes and pointer movements, forming a stream of user events.

These events are further processed by the dialog control. Events may result in different actions depending on the current system mode, and a sequence of events may be required before a function can be executed. Dialog control performs some kind of syntactic analysis on the event stream, i.e. it knows how events can be used to phrase requests to the system. Statecharts, as introduced by Harel (1987), proved themselves to be very efficient in designing the dialog control. Development tools were used to translate the graphical statecharts into source code.

Finally, the application interface provides access to the FMS functions. These functions are invoked by the dialog control as soon as a complete transaction is recognised and confirmed.

Figure 3 depicts the three different components of the Seeheim Model as applied to the AFMS prototype.

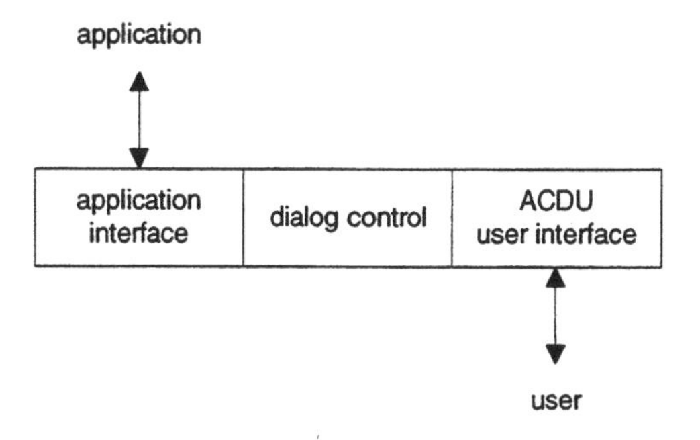

Fig. 3: AFMS structured according to the Seeheim Model

Harel's Statecharts are a powerful extension of finite automatons. Very briefly, they add hierarchy and orthogonality of states to the original concept of finite automatons. With hierarchy, a state may contain a set of substates, and whenever the parent state is active, exactly one of its substates is also active. Orthogonality allows a state to have substates that are all active whenever the parent state is active. With another level of hierarchy below these orthogonal states, independent parts of a state machine can be modelled very efficiently. However, until now the semantics of statecharts presents a number of problems, which means that some care is necessary to guarantee deterministic behaviour.

The prototype is modelled as a set of states interconnected by action transitions describing the reaction of the prototype due to user inputs. Using a connection between the ACDU prototype and a statechart modelling tool it is possible to visualise the interaction by highlighting state transitions within the graphical statechart representation. This functionality is called graphical back animation and allows a better validation of the user interface (Pauly, 1997).

Exemplarily, figure 4 shows the top level of the statecharts concerning the functionality of the ACDU.

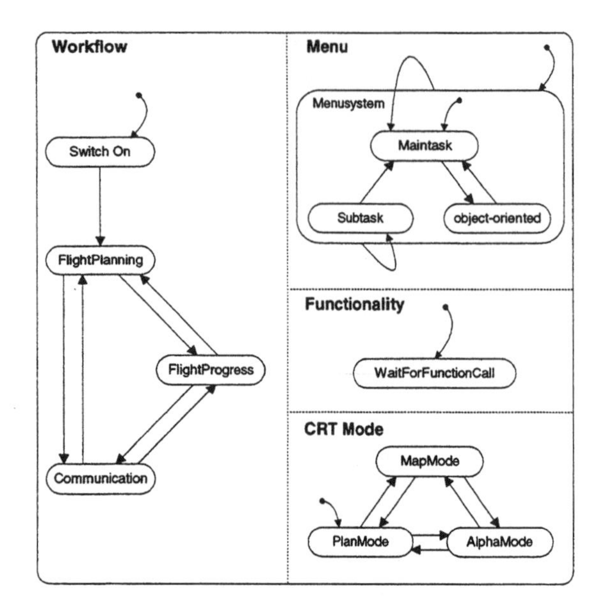

Fig. 4: Top level of the statecharts

5. RESULTS AND FUTURE WORK

First experiments with users and the new man machine interface showed that standard flight planning tasks can be performed with a significantly lower number of pilot actions. Therefore, these tasks can be completed in a shorter time, compared to conventional CDUs.

The following example is intended to describe how a graphic flight plan editing interaction between the pilot and the AFMS could look like. This example informs how the flight plan for a flight from Paris to Cologne (EDDK) is adapted to the expected vectored approach to Cologne airport when arriving from the west.

The example point out the following benifits using intuitive graphic interaction:

- objects and functions are closely related to each other which simplifies the menu structure and reduces memory load
- 'what you see is what you get' facilitates interaction and reduces input error
- alphanumeric input is minimised

The table shows a comparison between an interaction carried out on a commonly used alphanumeric CDU and the proposed graphic supported flightplanning based on the example.

The comparison applied to this experiments shows that if the intuitive graphic interaction is used, actions necessary to carry out the same task can be remarkably reduced, up to approximately 50%.

Table 1: Comparison between interaction on the alphanumerical CDU and the graphical ACDU

Function	Alphanumeric Concept	Actions	Graphic Concept	Actions
select appropriate page	press button *LEGS*	1	(perhaps)	(1)
create new *WPT1* as place / distance point on leg to *NOR* of active flight plan	write into scratchpad *NOR/-10/WPT1*	12	position cursor click *NOR* press DistLeg drag/click *-10*	4
insert *WPT1* before *NOR*	press line-select-key *NOR*	1		
create new *WPT2* as intersection of place bearing / place bearing	write into scratchpad *WPT1100/ILS32R169*	17	(position cursor click *WPT1*) press Radial drag/click *100*	2 + (2)
insert *WPT2* before *ILS32R*	press line-select-key *ILS32R*	1	position cursor click *ILS32R* press Radial drag/click *169*	4
delete/bypass old leg	press line-select-key *WPT1* press line-select-key *ILS32R* (delete legs)	1 1 (8)	position cursor click *WPT1* press *Direct-To* position cursor click new *WPT2* position cursor click *ILS32R*	7
calculate new flight plan	press *EXEC*	1	press *Execute*	1
		35 (41)		**18 (21)**

The reduced number of actions required to do a flight planning job will allow the pilots to make extensive use of FMS also during flight phases in which head-down times are to be minimised, e.g. the approach. This will help to fully exploit the capabilities of flight management systems during the whole time.

The AFMS prototype including the whole functionality, e.g. 4D flight planning and guidance and communication aspects, is already integrated in a flight simulation facility. Initial test runs with a simplified flight simulation program will be performed by some pilots. Due to this these results regarding usability, error rates, will become available later this year. The AFMS as designed and built by DASA NFS is called NFS-5000; it is currently being tested by Alitalia pilots in a DC-9 on regular airline flights to see if the system is suitable for the needs of a commercial carrier. Reports about this test are expected soon.

6. SUMMARY

This article outlines the goals of using Flight Management Systems, and it explains some of the difficulties that exist with today´s Control and Display Units. A graphical user interface is suggested as an alternative which provides a presentation that fits to the pilot's idea of a flight plan much more than a character string.

Technically, the graphical user interface is designed using the Seeheim Model, with statecharts used to define the dynamic behaviour. Apart from providing well defined reactions of the system, statecharts proved themselves to be very efficient when it comes to explaining the concepts of the user interface to other engineers.

The usability tests are still performed the final results are expected soon. However, the first reactions are very promising, and it could be demonstrated that pilots can edit their flight plans with significantly less CDU actions, compared to what it takes today to incorporate the same flight plan changes.

7. ACKNOWLEDGEMENT

This work is performed within the european Brite/EuRam III project No. BRPR-CT95-0044 - *AFMS Advanced Flight Management System*.

REFERENCES

Dornheim, M. (1996). Recovered FMC Memory Puts New Spin on Cali Accident. In: *Aviation Week & Space Technology*, 09/09/96, pp. 58-61.

Gerlach, M. (1996). *Schnittstellengestaltung für ein Cockpitassistenzsystem unter besonderer Berücksichtigung von Spracheingabe.* Fortschrittsberichte VDI, Reihe 12, No. 273, VDI Verlag, Düsseldorf.

Green, M. (1985). Design Notations and User Interface Management Systems. In: *User Interface Management Systems* (Pfaff (Ed.)), pp. 89-107. Springer-Verlag.

Harel, D. (1987). A Visual Formalism for Complex Systems. In: *Science of Computer Programming*, **8**, 1, pp. 231-274.

Marrenbach, J., M. Pauly and M. Gerlach (1997). Konzept zur ergonomischen Gestaltung der Benutzungsoberfläche eines zukünftigen Flight Management Systems. In: *Jahrbuch der Deutschen Gesellschaft für Luft- und Raumfahrt*, **I**, pp. 497-505.

Pauly, M. (1997). *Entwicklung und Implementierung der Benutzungsoberfläche zur Programmierung eines Flugwegrechners unter Berücksichtigung softwareergonomischer Aspekte.* Technical report no 97/76. Department of Technical Computer Science. Aachen.

Polson, P., S. Irving and J.E. Irving (1994). *Applications of Formal Models to Training and Use of the Control and Display Unit.* Institute of Cognitive Science, University of Colorado, Technical Report.

Sarter, N.B. and D.D. Woods (1992). Pilot Interaction with Cockpit Automation: Operational Experiences with the Flight Management System. *The International Journal of Aviation Psychology*, **2**, pp. 303-321.

Sarter, N.B. and D.D. Woods (1994). Pilot Interaction with Cockpit Automation II: An Experimental Study of Pilots´ Model and Awareness of the Flight Management System. *The International Journal of Aviation Psychology*, **4**, pp. 1-28.

PILOT'S PERCEPTION AND CONTROL OF AIRCRAFT MOTIONS

Ruud Hosman and Henk Stassen

Man-Machine Systems and Control, Faculty of Design, Engineering and Production, Delft University of Technology, Mekelweg 2, 2628 CD DELFT, The Netherlands

Abstract: For proper manual aircraft control, the pilot has to perceive the motion state of the aircraft. In this perception process both the visual and the vestibular system play an important role. To understand this perception process and its impact on pilot's control behavior a descriptive model was developed. The single channel information processor model was applied as the basic structure of the final model. Three groups of experiments were performed to refine the model structure and to define the majority of the model parameters. The model has been evaluated by measuring the control behavior in tracking tasks. *Copyright © 1998 IFAC*

Keyword: Human perception, Flight control, Manual control, Simulation.

1. INTRODUCTION.

Since the seventieth, civil air-transport pilot training is more and more accomplished on full flight simulators. Due to the development of digital computers, of visual displays and of motion systems, flight simulation has evolved to an almost full grown technology. However, due to the fundamental difference between the flight simulator and the real aircraft - the limited motion space of the simulator compared to that of the aircraft - there is an essential difference in motion stimulation of the pilot in flying a simulator or an aircraft, respectively. The positive influence of the vestibularly perceived motion stimuli on pilot's manual control behavior was recognized long ago and has been implemented in flight simulation by the application of six degrees of freedom motion systems. If transfer of training from the simulator to the aircraft has to be obtained, then a correct stimulation of the pilot is required. The present transformation of aircraft- to simulator motions by so-called washout filters, however, does not fulfill that role completely. An improvement can only be reached if a pilot's visual-vestibular perception and control of aircraft motions is understood, modeled and applied to the optimization of the washout filters.
At the Delft University of Technology a research project was executed to incorporate all available knowledge of the motion perception sensors - the visual and the vestibular system -, and their interaction in a descriptive model of pilot's control behavior. To reach that goal a single channel information processing model was adopted. Three groups of experiments were designed in order to obtain the missing information to complete the model for the present application. The experiments were performed to establish the dynamics of the visual attitude and rate detectors on the one hand, and the interaction of the visual and vestibular sensors in motion perception and control on the other. Based on the results, a descriptive model could be developed and adapted to the control behavior measured in tracking tasks with different combinations of visual and vestibular motion feedback.
In the next section manual aircraft control and the selection of the information processing model will be discussed. In the Sections 3 and 4 experiments will be described to evaluate the models of the visual and the vestibular systems and their interaction in the perception process. In Section 5 the tracking experiments and the results will be discussed. In section 6 the model will be described and evaluated. Finally a discussion and the conclusions will be presented in Section 7.

2. THE INFORMATION PROCESSING IN SKILL-BASED CONTROL.

The pilot's manual control task is configured as a nested control loop from which the inner loop is an attitude control and stabilization loop. The higher order dynamics of the aircraft attitude control may be approximated by a second order system. For this attitude control task, the control behavior of a well-trained pilot can best be described as skill-based behavior. To perform this skill-based task, the pilot has to process information from the presented attitude on the artificial horizon to the required control output. The information processing is considered to be performed by a single channel mechanism with multiple input due to the individual senses. A model of this single channel processor is presented in Fig. 1.

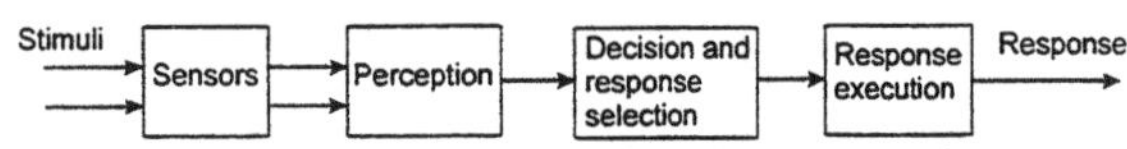

Figure 1. A model of the human information processing.

In this model four stages are considered: The sensors which convert the input stimuli to sensory output, the perception process, the decision and response selection, and the response execution. This basic model is well-known and documented in the literature (Wickens and Flach, 1988). Applying this model, raises not only the question which individual sensors contribute to the perception and control process, but also how the interaction of the sensory outputs in the perception process has to be modeled, and on what criteria the human operator selects his control output.
In the perception and control of vehicle motion, the visual and vestibular sensory output provide the most important information to the perception process. Due to the different dynamics of both the systems they contribute each in their specific way. Although accurate models of the visual and the vestibular (sub)systems were available, the interaction of the sensory outputs in the perception process had to be investigated. Furthermore, it was not known how changes in the perception process influenced the output selection.
Two groups of experiments were performed to investigate the speed and perception accuracy of motion stimuli and the interaction between the visual and the vestibular systems. A third group of experiments was directed to the influence of visual and vestibular feedback on pilot's control behavior in attitude tracking tasks.

3. ACCURACY AND SPEED OF ROLL ATTITUDE AND RATE PERCEPTION.

Due to the higher order dynamics of the aircraft attitude control, the pilot has to perceive the attitude as well as the angular rate. The visual perception of these two variables has been investigated in a stimulus-response experiment. The roll angle was presented to the subject on a central display (simulated artificial horizon), Fig. 2.a. The roll rate was presented on the central display and/or on two peripheral displays, Fig. 2b. Two TV monitors displayed a moveable checkerboard pattern controlled by the roll rate.

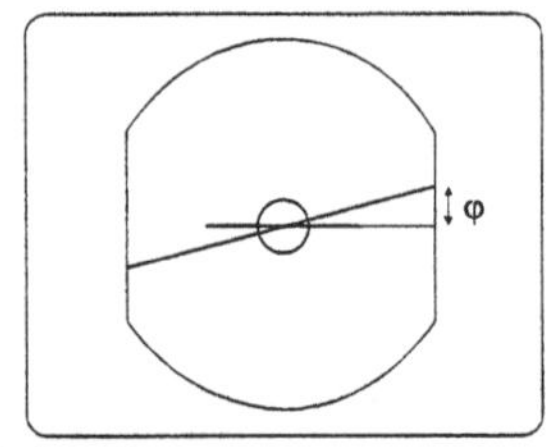

a. Central display format

b. Experimental station

Figure 2. Experimental station with the central display, the peripheral displays, and the keyboard.

The stimulus s_n, roll attitude or roll rate, was presented to the subject for a limited exposure time Δt_{exp}. After the exposition of the stimulus, the display was blanked, and the subject was required to give his response r_n by pressing the corresponding key on a keyboard, Fig. 2. Immediately after the response the remaining error $e_n = s_n - r_n$ was shown on the displays in the same way as the stimulus itself.
Approximately twenty discrete magnitudes of the stimulus were applied. Roll attitude was varied between ± 30 degrees, and roll rate was varied between ± 25 degrees/sec. Perception accuracy as expressed by the score parameter $S_c = \sigma_e^2 / \sigma_s^2$, the quotient of the variences of the error *e* and the stimulus *s*, and the mean response time $\overline{RT}$ were recorded. Both the parameters were determined as a function of Δt_{exp}.
The most important results of these experiments are shown in Figs 3 and 4. Roll attitude can be perceived

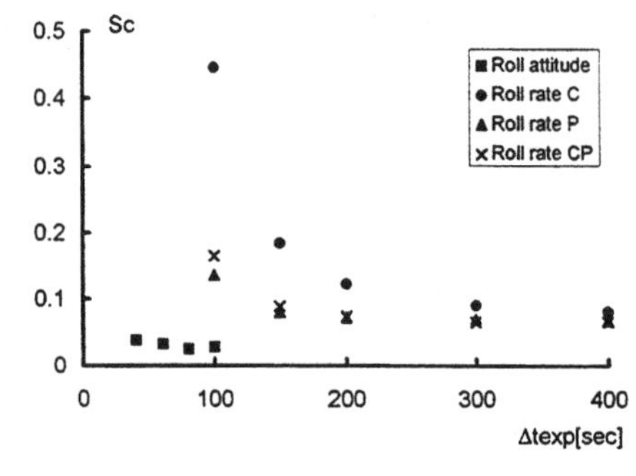

Figure 3. Score S_c for roll attitude and roll rate. perception. (Central display, C, Peripheral displays, P, and Central and Peripheral displays, CP)

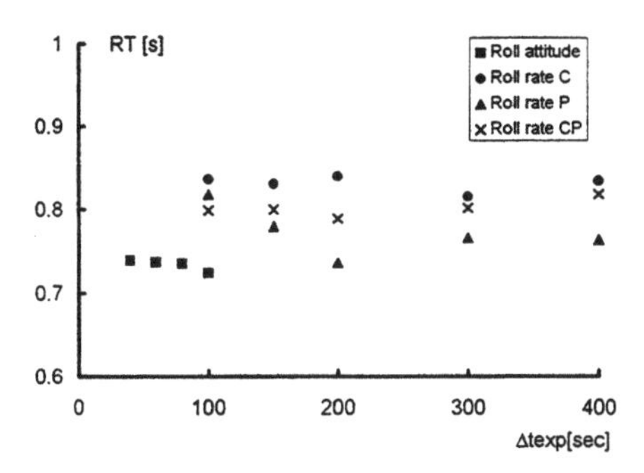

Figure 4. Mean response time for roll attitude and roll rate perception. (Central display, C, Peripheral displays, P, and Central and Peripheral displays, CP).

after an exposure time of 50 ms with a response time of approximately 0.73 s. Roll rate perception from the central display needs an exposure time of at least 300 ms to reach the final accuracy. The reaction time is approximately 0.83 s.

The results of the present experiment show that:

- The minimum required exposure time for accurate roll attitude perception is approximately 50 ms, for rate perception in the peripheral visual field 150 ms, and for roll rate perception in the central visual field 300 ms.
- The response time for rate perception in the peripheral visual field is 40 ms, and in the central visual field 100 ms longer than for attitude perception in the central visual field.
- Roll attitude perception is more accurate than roll rate perception.

Considering the experimental results and the model presented by van den Berg and van de Grind (1989), the models for visual perception of attitude and rate used in this paper are:

$$H_{C,att}(\omega) = Ke^{-j\omega\tau_{att}},$$

$$H_{C,rate}(\omega) = Ke^{-j\omega\tau_{C,rate}}, \qquad (1)$$

$$H_{P,rate}(\omega) = Ke^{-j\omega\tau_{P,rate}},$$

where $K \cong 1$, τ_{att}=0.05s is the delay for attitude perception, and $\tau_{C,rate}$= 0.15s and $\tau_{P,rate}$=0.09s are the time delays for central visual and peripheral visual rate perception, respectively.

4. VISUAL-VESTIBULAR INTERACTION IN THE PERCEPTION OF MOTION STIMULI.

By using models for the visual and vestibular system, the differences between the response of the sensors to a presented stimulus may be studied. Models for the visual system sensors has been presented in the previous section. Models for the vestibular system were derived by Fernandez and Goldberg (1971). For attitude perception and control the dynamics of the semi-circular canals are of direct importance. The semi-circular canals are sensitive for angular acceleration and the transfer function $H_{SCC}(\omega)$ describes the relation between the stimulus roll acceleration and the neural sensor output:

$$H_{SCC}(\omega) = \frac{K_{SCC}(1+j\omega\tau_L)}{(1+j\omega\tau_1)(1+j\omega\tau_2)}, \qquad (2)$$

where: K_{SCC} is the gain, τ_L is the neural lead term and τ_1 and τ_2 are time constants of the sensor's physical system.

In the stimulus-response experiment described in this section, the step response of a well damped second-order system has been used as the stimulus. In Fig. 5, the simulated sensory outputs to the step response stimulus (ω_o = 2 rad/sec, ζ = 0.7) is shown. From this figure it is clear that the semi-circular canal output leads, and the visual system output lags the stimulus angular rate.

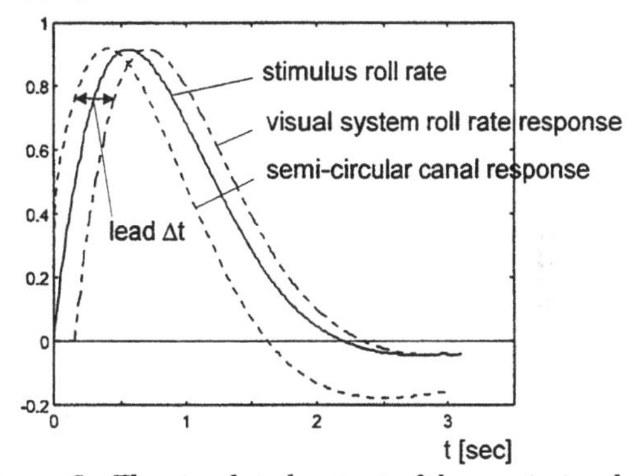

Figure 5. The simulated output of the semi-circular canals, the visual rate sensor, and the stimulus roll rate.

The vestibular system response leads the visual system response with a time lead Δt of approximately 300 ms. Due to the dynamics of semi-circular canals this lead depends on the time course of the stimulus. Hence, the lead depends on the natural frequency ω_o of the stimulus generating second order system, Fig. 6.

With a stimulus-response experiment the simulation results were verified. The subjects estimated the final magnitude of the stimulus and responded on the keyboard. In this experiment the stimuli were presented to the subjects with different combinations of the central display, peripheral visual displays, and a flight simulator motion system. The aim of the experiments was to verify the predicted change of the time lead as a function of the second-order system natural frequency ω_o. Three values of ω_o were applied, 0.65, 1 and 2 rad/sec.

Due to the limited roll excursions of the simulator 13 step magnitudes between ± 12 degrees were applied. The display configurations used in the experiment were Central display, C, Peripheral displays, P, simulator Motion, M, and the combination Central display and Motion, CM.

The accuracy of the estimation of the final magnitude of the stimulus turned out to be not significantly different due to the different display configurations.

From the results the mean reaction times $\overline{RT}$ as a function of display configuration and natural frequency of the second order system are of interest here. Significant differences ($\alpha < 0.01$) were found due to the display configurations, and the natural frequency ω_o. Based on the response times of the different configurations, the time lead Δt could be computed, Fig. 6.

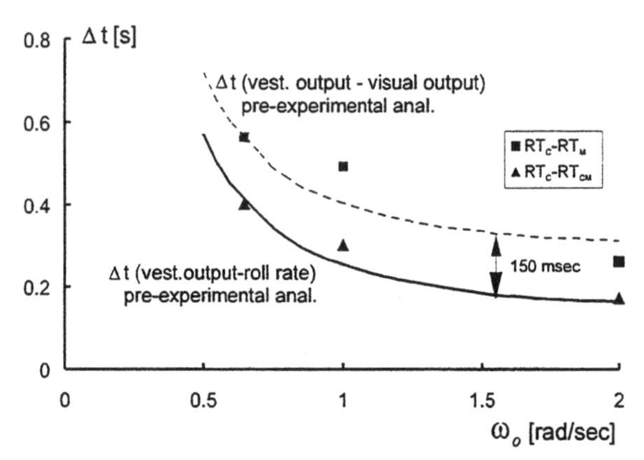

Figure 6. The time lead Δt based on the experimental results compared with the pre-experimental analysis.

The changes in the measured response times confirmed the results of the pre-experimental simulations. In Fig. 6 the time lead $\Delta t = RT_C\text{-}RT_M$ and $\Delta t = RT_C\text{-}RT_{CM}$ are compared with the time lead based on the pre-experimental analysis.

Considering the experimental results, the visual-vestibular interaction in the perception of motion stimuli was confirmed. The vestibular perception of motion stimuli is much faster than the visual perception. The models for visual rate perception (eq. 1) and the semi-circular canals (eq. 2) give a fair prediction of the experimental results.

5. THE INFLUENCE OF MOTION FEEDBACK ON CONTROL BEHAVIOR.

The third group of experiments consisted of a series of tracking experiments. In the literature tracking experiments investigating the influence of motion feedback on tracking performance has been described (Hosman, 1996). Most of these studies showed that the control of second or higher order systems (as vehicles) is improved if the motion output of the vehicle is fed back to the operator. To obtain a reliable data base on the influence of the perception process on the operators performance and control behavior an experiment was set up.

The aim of the experiment was to establish the influence of the visual and/or vestibular presentation of the tracking error and system response on subjects control behavior and performance. The Central display (C), Peripheral displays (P) and the simulator Motion system (M) were used to present the controlled variables to the subjects. Both the target following task and the disturbance task, Fig. 7, were incorporated in the experiment. In the target following task the subject minimizes the error $e(t)$ as presented on the central display while the peripheral displays and the motion system presents the system output to the subject. In the disturbance task the subject minimizes the system output; whereas all stimuli correspond with each other. All possible display configurations were applied, 4 for the target following task and 7 for the disturbance task, Table 1.

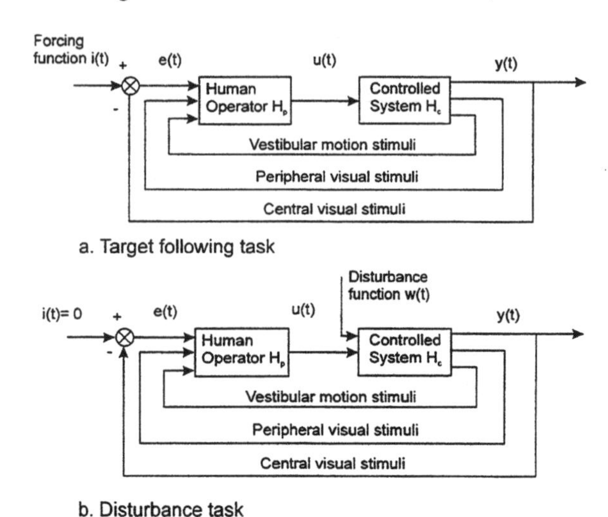

Figure 7. Configuration of the experimental tracking tasks with visual and vestibular feedback.

A double integrator, $K/(j\omega)^2$, was used as the controlled element which system forces the subject to generate lead to reach stable control. By applying different display configurations in both control tasks, the subject had to change his perception process and consequently the method he generates the required lead.

In both the target following task and the disturbance task, the tracking performance improved considerably when vestibular motion cues and/or peripheral visual cues were presented to the subjects, Table 1. The changes in control behavior due to motion feedback, however, were quite different. In the target following task the subjects decreased their gain, leading to a

Table 1. Tracking experiment results.

Display configuration	σ_e degree	ω_c rad/sec	φ_m degree
Target following task			
C	2.23	2.18	19
CP	1.63	2.27	32
CM	1.32	1.86	74
CPM	1.26	1.49	85
Disturbance task			
C	1.94	3.24	16
P	2.67	3.03	17
M	0.99	3.89	22
CP	1.64	3.29	20
CM	0.78	4.75	22
PM	1.00	4.19	24
CPM	0.70	5.03	21

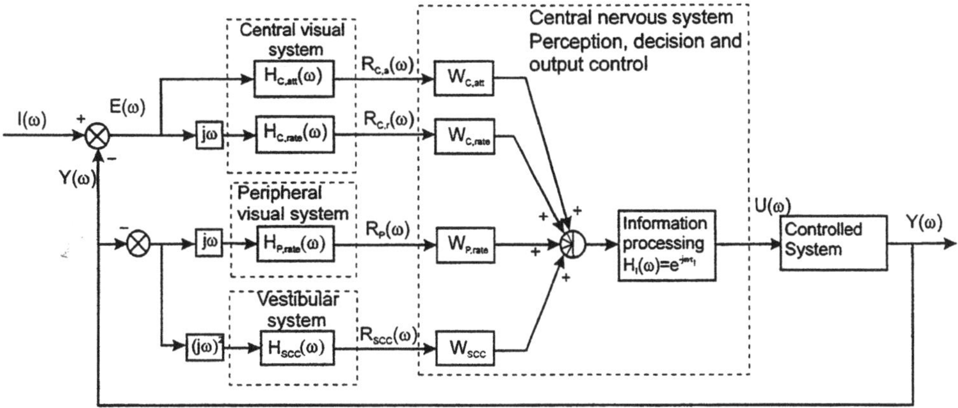

Figure 8. Block diagram of the descriptive pilot model in the target following task.

lower crossover frequency, but they were able to increase their phase margin considerably. In the disturbance task, subjects increased their lead and used this lead to increase the crossover frequency at approximately constant phase margin.

In the disturbance task both the single input configurations C, P and M and the combined input configurations CP, CM, PM and CPM have been applied. The frequency response of combined input configurations can not be derived by simply adding the frequency responses of the corresponding single input configurations. Considering the changes in control behavior in the disturbance task as a result of the different display configurations, Hosman (1996) showed that it is acceptable to assume that the contribution of the different sensors is weighted in the perception process by frequency independent weighing factors.

6. THE DESCRIPTIVE MODEL

Considering the model for information processing, Fig.1, and the results of the experiments described in the previous sections, it is now possible to extend this model. The aim is to obtain a model capable to describe the influence of the visual and vestibular perception of aircraft motions in order to determine its influence on pilot's control behavior. The final model is depicted in Fig. 8.

With the experiments described in Sections 3 and 4, the dynamics of the sensors and the interaction between the visual and the vestibular systems were evaluated. In the model the sensors, described by a transfer function, are placed in parallel and convert the stimuli, attitude, angular rate, and angular acceleration, to the sensory outputs $R_i(\omega)$.

In the perception and decision process the sensory outputs have to be converted to one output. Considering the results of the tracking experiment in Section 5, the model output is generated by a weighted sum of the sensory outputs. Each sensory output is weighted by its individual weighting factor W_i.

The time delay due to the information processing is incorporated next in the model. Finally, the neuromotor system transfers the commanded output to the actual output. Due to the high bandwidth of a well designed neuromotor manipulator system ($\omega_n >> \omega_c$), the information processing delay and the neuromotor manipulator system are modeled by one lumped time delay τ_I. Considering the effective time delay of the human operator in tracking tasks (McRuer and Jex, 1967), and the characteristics of the visual system, Hosman (1996) set $\tau_I = 0.2$ s.

With the model the influence of visual and vestibular motion feedback in both the target following task and the disturbance task can be described. The advantage of the model is that the only free model parameters are the sensory weighting coefficients W_i.

As shown in Fig. 7 there is an important difference in the feedback of the controlled system motion stimuli between the target following task and the disturbance task. The result is a difference in the describing function of the pilot model for both the tasks. If the pilot has to perform the control task with the central display and motion feedback, the model transfer function for the target following task is:

$$H_{CM_{Foll}}(\omega) = \frac{[K_{C,att}.H_{C,att}(\omega)+K_{C,rate}.H_{C,rate}(\omega).j\omega]}{[1+K_{SCC}.H_{SCC}(\omega).(j\omega)^2.K_I e^{-j\omega\tau_I}.H_c(\omega)]}.K_I e^{-j\omega\tau_I}, \tag{3}$$

and for the disturbance task:

$$H_{CM_{Dist}}(\omega) = \begin{array}{l}[K_{C,att}.H_{C,att}(\omega)+ \\ K_{C,rate}.H_{C,rate}(\omega).j\omega + \\ K_{SCC}.H_{SCC}(\omega).(j\omega)^2].K_I e^{-j\omega\tau_I}.\end{array} \tag{4}$$

By adjusting the weighting factors of the pilot model, the transfer functions could be adjusted to the measured subject frequency responses of the experiment described in Section 5, Fig. 9.

7. DISCUSSION AND CONCLUSIONS.

The results of the present study contribute to an improved understanding of a pilot's motion perception and its influence on pilot's control behavior. The experimental results turn out to be directly applicable to the proposed descriptive model.

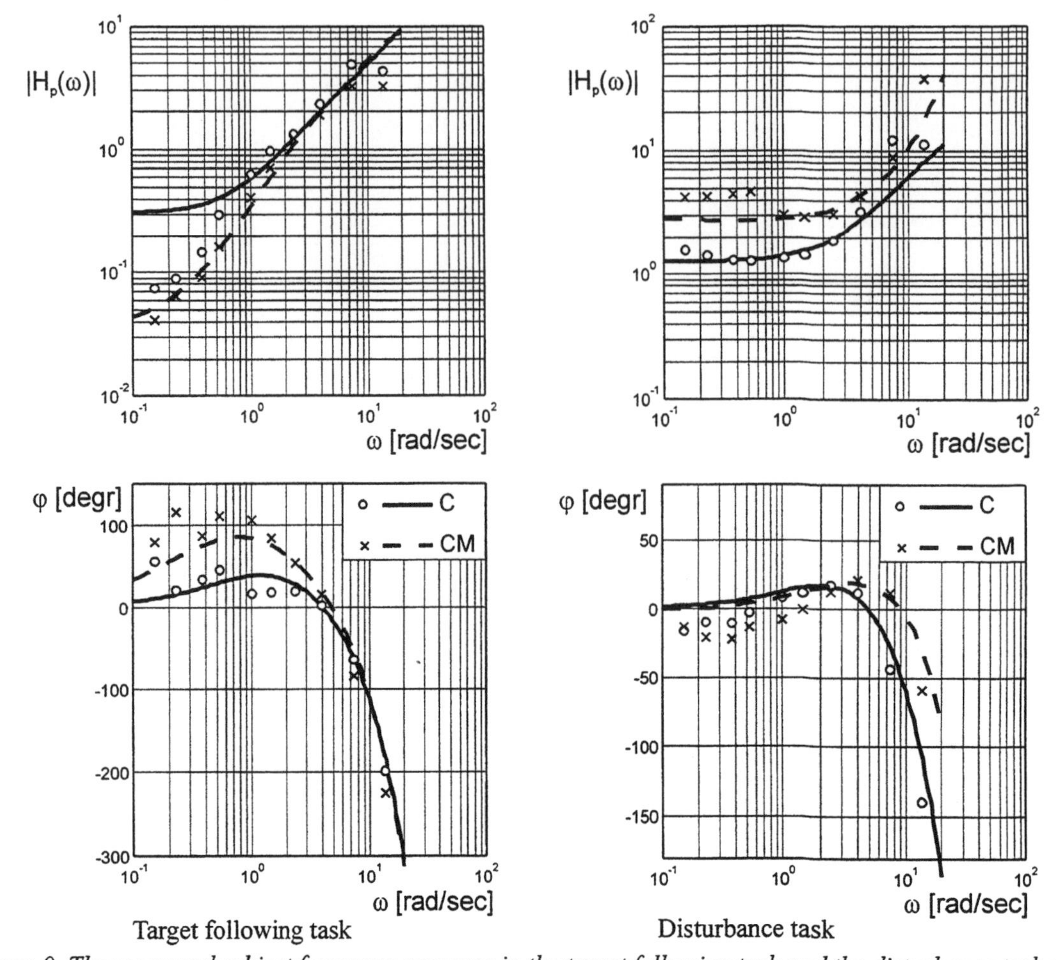

Figure 9. The measured subject frequency response in the target following task and the disturbance task and the matched descriptive model. Display configurations: Central display, C, and Central display with Motion, CM.

The model was easily fitted to the measured frequency response in the target following tasks and the disturbance tasks.

An important field of application of the obtained understanding is vehicle simulation in general and flight simulation in particular. The characteristic differences between real and simulated flight are time delays and a mismatch between the visual cues and the motion cues.

The proposed descriptive model provides the opportunity to study the effects of the above mentioned differences due to the simulation process, on pilot's motion perception and control behavior. For each individual sensor the effect of the simulation process on pilot's control behavior and performance can be studied. Inversely, the descriptive model can be applied to optimize the transformation of the aircraft motions to simulator motions.

The following conclusions can be drawn.

1. The results of the present study contribute to an improved understanding of a pilot's motion perception and its influence on a pilot's tracking behavior.
2. Based on a well-known model for information processing and additional experimental research, a descriptive model describing pilot's control behavior could be developed.
3. The application of the results of this study lies primarily in the field of flight simulation. The descriptive model has the capacity to evaluate the influence of the simulation time delays and washout filters on pilot's control behavior.

REFERENCES

Berg, A.V. van den, and W.A. van de Grind (1989). Reaction times to motion onsets and motion detection thresholds reflect properties of bilocal motion detectors. *Vision Research*, Vol.29, no. 9,pp 1261-1266.

Fernandez, C., and J.M. Goldberg (1971). Physiology of peripheral neurons innervating semicircular canals of the squirrel monkey: II Response to sinusoidal stimulation and dynamics of peripheral vestibular system. *Journal of Neurophysiology*, Vol. XXXIV, no. 4.

Hosman, R.J.A.W. (1996). Pilot's perception and control of aircraft motions. PhD Thesis. Delft University of Technology. Delft University Press.

McRuer, D.T., and H.R. Jex (1967). A Review of Quasi-Linear Pilot Models. *IEEE Transactions on Human Factors in Electronics*. Vol. HFE-8, no3, September 1967.

Wickens, C.D. and J. Flach (1988). Information Processing. In: E.L. Wiener and D.C. Nagel (Eds.). Human Factors in Aviation. Academic Press, N.Y.

ADVANCED AIR TRAFFIC MANAGEMENT IN THE AIRPORT TERMINAL AREA

Ralf Beyer

German Aerospace Center (DLR)
Institute of Flight Guidance
D-38108 Braunschweig
Lilienthalplatz 7
Federal Republic of Germany

Abstract: Air traffic is growing but limited resources of real-estate, environmental considerations and shrinking acceptability by the society make it difficult to extend certain airports to meet the demand. Therefore, a better utilisation of given airport resources is a way to achieve part or all of the required capacity enhancement. This requires a new automation philosophy, the development of intelligent air traffic planning, guidance and control systems, and a partial transfer and delegation of air traffic controller functions to these systems. Some fundamentals, an automation philosophy, and applications are presented. *Copyright © 1998 IFAC*

Keywords: air traffic control, automation, cognitive systems, human factors, man/ machine systems.

1. INTRODUCTION

Future air traffic is influenced by a number of quite different factors ranging from changes of societal values and growing air traffic demand to new technological developments and operational procedures of air traffic management (ATM) in the terminal area of airports. Operations research has shown that the operation of an airport can be regarded as a system of interacting processes which are often not optimised today with respect to airport capacity and to the more general requirements of energy conservation, protection of the environment, and the integration of air traffic into a multi-modal transport system. As the chances to extend existing airports in terms of real-estate for new buildings, runways and service areas are rather limited in many cases, advances towards a greater airport capacity are expected from a better utilisation of given airport resources. Major improvements of airport capacity are expected, therefore, from a more efficient design, integration and operation of these interacting processes employing intelligent technologies for situation monitoring and situation diagnosis on one hand and for plan generation, plan selection, and plan execution to modify the current situation on the other hand. This is particularly true for the control of the arriving, taxiing, and departing air traffic. Process optimisation in these areas is considered an efficient means for any progress towards a more efficient air traffic management in the airport terminal area to gain greater airport capacity. Accompanying effort is required for advances of operational procedures and as far as possible improvements of the airport infrastructure.

In this context the human operator will remain the responsible element in the teamwork of interacting processes of an airport for years to come. However, it is expected that more cognitive functions of the human operator can be transferred to supporting machines in the future. Next century needs of air traffic management in the extended terminal area of airports require this transfer of cognitive functions to machines in order to better support those working in air traffic management (ATM) and to achieve the required air traffic capacity enhancement at an airport. In the light of this perspective the role of the human operator in aviation and some hereto related fundamental considerations, automation philosophies, and applications of DLR's automation philosophy to future air traffic management at airports are presented.

2. ON THE ROLE OF THE HUMAN OPERATOR IN AVIATION

As more and more functions of the human operator and in particular cognitive functions can be implemented by machines the question today is not how much can be automated but what degree of automation is acceptable responsibly. For instance, the Concept Document for the future European Air Traffic Management System (EATMS) lists as one of the principles for its development (EUROCONTROL, 1992):

"The concept introduces new methods of task sharing between the automated system elements and the human operator. That development shall be carefully balanced in particular in relation to the responsibilities for traffic monitoring and decision making; this may set limits to the degree of automation. The man/machine interface is a key element of successful application of automation and of overall system performance."

A forecast of the potential future of the US Air Force (USAF, 1995) lists the item „people" including modelling of the human operator, training, education, man/machine interaction, and chemical intervention as one of its six identified essential capabilities:

- Global Awareness
- Dynamic Planning and Execution Control
- Global Mobility in War and Peace
- Projection of Lethal and Sublethal Power
- Space Operations
- People

This much improved role of „people" and associated developments of man/machine interactions and interfaces in comparison to the situation some years back cannot be overemphasised in times of severe cuts in military funding and rearrangements of development priorities.

3. FUNDAMENTAL CONSIDERATIONS

Human error will come, regardless of the quality of design and training. Suppressing human error is seen as a naive goal. A better approach would be to correct system ergonomics but keep the human in the loop (Amalberti, 1997). Air traffic controller tasks not only refer to the volume and activity of the air traffic, but also to the continuously changing set of tasks and procedures associated with air traffic control. For instance, Lenorovitz, Olason, Krois, and Tobey, 1991, presented a list of nine variables differentiating air traffic controllers and their various tasks. Jackson, 1989, identified elementary functions of air traffic controllers. Both investigations emphasise the cognitive functions of air traffic controllers which can be summarised as follows:

- monitoring — what is the state of air traffic?
- diagnosing — what happens and why?
- planning — what options are available?
- deciding — what option is optimum?
- implementing — what implements an option best?

The keywords here are „Attention" and „Situation Awareness" which Wickens defined as follows:

- Attention

A selective filter, that chooses and determines what information will be processed, and what activities will be carried out at any given time (Wickens, 1992).

- Situation Awareness

The continuous extraction of information about a dynamic system or environment, the integration of this information with previously acquired knowledge to form a coherent mental picture, and the use of that picture in directing further perception of, anticipation of, and attention to future events (Wickens, 1996).

Attention and situation awareness and their relation pertain to safety, design and training and have led to an order of priorities which is today the basis for a number of human-centred development projects:

- Provide the human operator with an awareness of the situation as complete as possible and direct his/her attention to the most urgent task at hand. Assist the operator in his/her work by appropriate support and assistant systems and by human-engineered man/machine interfaces.

- Provide machine support for situations where the workload of the human operator would exceed an acceptable level and for a reduction of task complexity to a degree which the human operator is able to manage.

4. AUTOMATION PHILOSOPHIES

The design of automation for ATM systems largely depends on the automation philosophy pursued. Current experiences show that an approach to automation on the basis of a provision of rather isolated ATM support functions most probably ends up in an equal number of different man/machine interfaces. On the other hand an approach relying on a more conversational style of interaction between technical and human agents in an automated ATM system requires an integrated and highly interactive man/machine interface. In the following, therefore, some characteristics of different automation philosophies are discussed.

4.1 General automation philosophies

The human operator is usually the weakest link in the total system, and technology itself may easily overload or underload the operator causing human errors. The traditional answer to this problem has been to provide various automated tools to aid the operator in the performance of system subtasks. Unfortunately, automation of this type may induce unique problems as summarised by Wickens, 1992: increased monitoring load, component proliferation, out-of-the-loop unfamiliarity, higher level operator error, and loss of co-operation between human operators.

A refinement of this tool-based approach is the design of closed-loop adaptive systems. This approach evolved from the difficulties associated with tool-based philosophies (i.e. increased monitoring load, component proliferation, etc.) while maintaining operator control. In closed-loop adaptive systems, the level of system automation is simply enhanced or reduced as a function of the inferred workload of the human operator (Rouse, 1988; Wickens and Kramer, 1985).

Another alternative, and perhaps the most sophisticated one, is to combine operational parameters (traffic, weather, and others) and measures of system performance in an attempt to model the operators attentional state. Modelling the operator's attentional state is a sophisticated technique that will discriminate overload from underload, and assess whether performance has failed because a primary task has become too difficult or because another activity has diverted the human operator's attention. Modelling depends upon a carefully crafted integration of system performance measures, operator behaviour, and physiological measures.

4.2 DLR's automation philosophy

DLR's automation philosophy assumes that the performance of cognitive machine functions should be at least comparable to if not better than that of the same functions performed by the human operator. This will enable machines first to support and later to assist the human operator efficiently. In this context three development phases are considered:

- Support tools

The human operator is able to delegate tasks to machines of known performance. These are tasks he is not interested in primarily and for which he knows that these tasks can be fulfilled equally well or better by a machine. A machine failure would have almost no impact on the safety of operations but might cause a predictable loss of operational performance. A delegation of this type of tasks to machines has already resulted in a reduction of operator workload and has contributed to a smooth and predictable work flow. A typical example is the Computer Oriented Metering, Planning, and Advisory System (COMPAS) developed by DLR and the German Air Navigation Services (DFS).

- Assistant systems

Assistant systems comprise more than a single support function. They make use of several elementary support functions and their interactions. They are able to infer new facts and to arrive at conclusions which are exchanged and assessed in a two-way communication with the human operator. This allows more than a delegation of tasks to a machine. It is the begin of a co-operative machine-based assistance where the skills of the human operator are augmented by the machine. Although the human operator still assumes full responsibility for most vital processes it is the begin of a shift of authority at least for non-safety related processes like plan generation and plan selection. However, the acceptance and the implementation of the selected plan would still be the responsibility of the human operator. The Cockpit Assistant System (CASSY) developed by the University of the German Armed Forces with support by DLR and the Cockpit Assistant for Military Aircraft (CAMA) developed by DLR and industry are typical examples.

- Autonomous systems

At a later stage assistant systems may become more mature and more adaptable to unexpected situations so that they are able to perform an increasing number of tasks on their own under the supervision of the human operator. At this stage of development the best chances to optimise air traffic management and control processes are expected. However, the introduction of autonomous systems will depend to a lower degree on their technical feasibility and performance but mostly on the comprehensibility of their decisions and actions, the predictability and reliability of operation, and their acceptance by the human operator (air traffic controllers and pilots), the users of the air transport system, and the society in general.

DLR's approach to automation for ATM results in two self-contained lanes of functionality: One is present in the human operator and the other is implemented by the machine. Fundamental aspects are:

- From a functional and procedural point of view the machine must be able to handle the tactical control of air traffic and the strategic management of the flow of air traffic automatically.

- From an operational point of view there would be a partnership between the human operator and the machine: The human operator decides what type of machine support suits his momentary and foreseeable needs best and what tasks he can delegate to the machine for the time being.

This means in short: Automate as much as technically feasible but utilise machine functions only as far as responsibly acceptable. By this approach the human operator will be provided with a suite of tools to select from according to his needs and with a chance to tune the tools in a way that their pace is matched to that of the human operator. Full authority is assumed by the human operator when new functions are introduced but this may change when the human operator gains better insight into the new machine functions and greater confidence in their performance and reliability.

5. APPLICATIONS

Pursuing its automation philosophy, DLR in co-operation with the German Air Navigation Services (DFS) and the Programme for Harmonised Air Traffic Management Research in EUROCONTROL (PHARE) has developed a number of support tools for the air traffic management in the terminal area of airports. These tools include:

- The Flow Monitor

Monitoring and analysing an air traffic situation is a fundamental process for any airport capacity enhancement plan. Knowledge is required about the air traffic demand, traffic flow and traffic delay and associated operational parameters like aircraft separations, runway direction in use, occupation of adjacent control sectors and so on. The Flow Monitor provides this basic knowledge automatically in form of on-line displays as well as in form of statistical representations for a period of interest. The Flow Monitor is in operation since 1992 at the airport of Frankfurt/Main.

- COMPAS

The Computer Oriented Metering, Planning and Advisory System (COMPAS) estimates arrival times, determines the required separation between aircraft of same or different weight classes, and suggests an optimum sequence of aircraft with regard to a smooth flow of the arriving traffic while making best use of the existing airport capacity. COMPAS has led to a smooth work flow in the management of the arriving air traffic at an airport. COMPAS is in operation since 1989 at the airport of Frankfurt/Main.

- The 4D Planner

Time-based arrival planning and position and time-based (4D) guidance of aircraft in the extended terminal area of an airport are expected to reduce the variability of arrivals and thus to allow a reduction of aircraft separation towards the minimum separation standards. Observations have shown that the statistical variation of the arrival separations can be approximated by a Gaussian distribution with a standard deviation of about 18 seconds. Flight tests at DLR with a laboratory version of a ground-based support tool to plan the arriving air traffic in space and time (4D Planner) have shown that the arrival error for a given instance of time can be reduced to about 5-8 seconds (1 σ). The 4D Planner is currently under development at DLR.

- PHARE Advanced Tools

The objective of the Programme for Harmonised Air Traffic Management Research in EUROCONTROL (PHARE) (EUROCONTROL, 1997) is to organise, co-ordinate and conduct studies and experiments aimed at providing and demonstrating the feasibility and merits of a future air-ground integrated air traffic management system in all phases of flight. DLR was in charge of a PHARE demonstration which focused on advanced tools for the management and control of arriving air traffic in the extended terminal area of an airport. From the results obtained in a realistic air traffic simulation with the inclusion of DLR's test aircraft it was concluded that the advanced tools have proven their suitability. The advanced tools showed the potential to improve air traffic throughput and quality of service at an airport. These improvements can be achieved at acceptable levels of air traffic controller workload. The advanced tools were well accepted by air traffic controllers from 7 European nations participating in the experiments.

6. CONCLUSIONS

Next century management of air traffic in the extended terminal area of airports requires a greater transfer of cognitive functions from air traffic controllers to machines in order to enhance traffic capacity and quality of service while maintaining achieved safety standards. The experimental investigations at DLR and the resulting systems already in operation at an airport have shown that machines can be designed to handle the tactical control of air traffic and the strategic management of air traffic flow automatically. A partnership between the human operator and the machines is envisaged where the human operator decides what type of machine support suits his/her momentary and foreseeable needs best and what tasks can be delegated to the machine for the time being. The successful installations of advanced monitoring and planning tools at a major European airport by DLR support this view and provide orientation for research in this area for the next century.

REFERENCES

Amalberti, R. (1997). *When Human Errors Serve Safety Goals*. CSERIAC Gateway, **VII**, No. 3, 1997.

EUROCONTROL (1992). *European Air Traffic Management System*. Issue 1.1. EUROCONTROL Agency, June 1992.

EUROCONTROL (1997). *Programme for Harmonised Air Traffic Management Research in EUROCONTROL*. EUROCONTROL Agency, May 1997.

Jackson, A. (1989). *The Role of the Controller in Future ATC Systems with Enhanced Information Processing Capability*. EUROCONTROL, EEC Report No. 224, June 1989.

Lenorovitz, D. R., S. C. Olason, P. A. Krois, and W. K. Tobey (1991). Customizing the ATC computer-human interface via the use of controller preference sets. In: *Proceedings of the Sixth International Symposium on Aviation Psychology* (R. S. Jensen (Ed.)), 454-459. Ohio State University, Columbus, Ohio, USA.

Rouse, W. B. (1988). *Adaptive aiding for human/ computer control*. Human Factors, **30**, 431-443.

USAF (1995). *New World Vistas - Air and Space Power for the 21st Century*. US Air Force Scientific Advisory Board.

Wickens, C. D. and A. Kramer (1985). Engineering psychology. *Annual Review of Psychology*, **36**, 307-348.

Wickens, C. D. (1992). *Engineering Psychology and Human Performance*. New York: Harper Collins.

Wickens, C. D. (1996). *Attention and Situation Awareness*. A NATO AGARD Workshop.

EXPERIMENTAL STUDY OF SITUATION-ADAPTIVE HUMAN-AUTOMATION COLLABORATION FOR TAKEOFF SAFETY

Makoto Itoh*, Yasuhiko Takae, Toshiyuki Inagaki*, Neville Moray#**

*Institute of Information Sciences and the Center for TARA,
University of Tsukuba, Tsukuba 305-8573 Japan
** Doctoral Program in Engineering, University of Tsukuba,
Tsukuba 305-8573 Japan
Department of Psychology, University of Surrey,
Guildford, Surrey GU2 5XH, United Kingdom

Abstract: This paper investigates usefulness of dynamic and flexible allocation of authority for decision and control during takeoff of an aircraft. Even though one of the authors' previous mathematical analysis claims that the authority should be exchanged between humans and automated systems depending on situations, it is not obvious whether human pilots are willing to accept the trading of authority. This paper gives an experimental study for the efficacy of such situation-adaptive autonomy. *Copyright ©1998 IFAC.*

Keywords: Aircraft operations, Automation, Decision making, Human-centered design, Human factors, Man/machine systems, Safety, Supervisory control

1. INTRODUCTION

Human-centered automation (Billings, 1991; Woods, 1989) is one of the central issues in the study of a human-machine system with a supervisory control configuration (Sheridan, 1992). It is often claimed that human supervisors must bear effective authority over automation (Woods, 1989). It is not obvious, however, whether the humans must have full authority at all times in every occasion. The humans in a large-complex system may face extremely difficult situations. Inagaki (1993) has thus proposed the concept of *situation-adaptive autonomy* (SAA). Mathematical analyses (Inagaki, 1993, 1997) have shown that computers may be given authority for decision and control when safety is a factor.

By taking, as an example, a rejected takeoff (RTO) problem of an aircraft, it has been proven in Inagaki (1997) that the authority of decision making during takeoff should be traded between humans and automated systems in a situation-adaptive manner. It is not clear, however, whether human pilots are willing to accept the trading of authority. This paper conducts an experiment to examine the usefulness of SAA for attaining takeoff safety. How humans feel toward SAA is analyzed from a viewpoint of human acceptance of SAA.

2. PROBABILITY MODEL

Consider the situation where an aircraft is taking off. Suppose an engine failure warning is given during the takeoff run. Human pilots have two alternatives: GO (continue the takeoff) and NO-GO (abort the takeoff). The aircraft can stop by the end of the runway if the human pilot initiates procedures to abort the takeoff before V_1. The aircraft, on the other hand, may cause an overrun if the takeoff is rejected after achieving V_1. The standard procedure for the decision, upon an engine failure, is: (i) If the airspeed is less than V_1, then NO-GO. (ii) If V_1 has already been achieved, then GO. Note here that V_1 should be regarded as the maximum speed by which the human pilot must *initiate* RTO *actions*, instead of just *deciding whether to go or not.*

Inagaki (1997) has proven the necessity of dynamic trading of authority for attaining takeoff safety. This section summarizes the results of his work.

Let L_{AS} denote the conditional expected risk when a decision is made by an automated system (*AS*). An engine failure warning can be false. There are two policies, which are taken by a human when he/she hesitates to say either that the warning is correct or that it is incorrect: (i) trustful policy (*TP*), in which the warning is trusted, and (ii) distrustful policy (*DP*), in which the warning is distrusted. Let L_{TP} or L_{DP} denote the conditional expected risk when the decision is made by a human pilot with TP or DP, respectively.

In Inagaki (1997), the RTO problem was analyzed by distinguishing some takeoff rolling phases (Fig. 1): (i) If an engine failure warning is given at an airspeed V_{EF} which is far below V_1 (Fig. 1(1)), then $L_{DP} \leq L_{TP} \leq L_{AS}$. This result suggests that *the takeoff should not be fully automated.* (ii) If a warning is given before but near V_1 (Fig. 1(2)), then $L_{DP} < L_{TP}$. No fixed order relation is found between L_{AS} and L_{TP}, or between L_{AS} and L_{DP}. (iii) If an engine failure warning is given almost at V_1, and no human pilot can initiate RTO actions by V_1 but AS can (Fig. 1(3)), then $L_{DP} < L_{TP}$. (iv) If neither a human pilot nor AS can initiate RTO actions by V_1, then $L_{AS} \leq L_{DP} \leq L_{TP}$, which implies that *it is inappropriate to assume that a human pilot must always bear full authority for decision and control.*

The above results (i) - (iv) suggest the necessity of SAA.

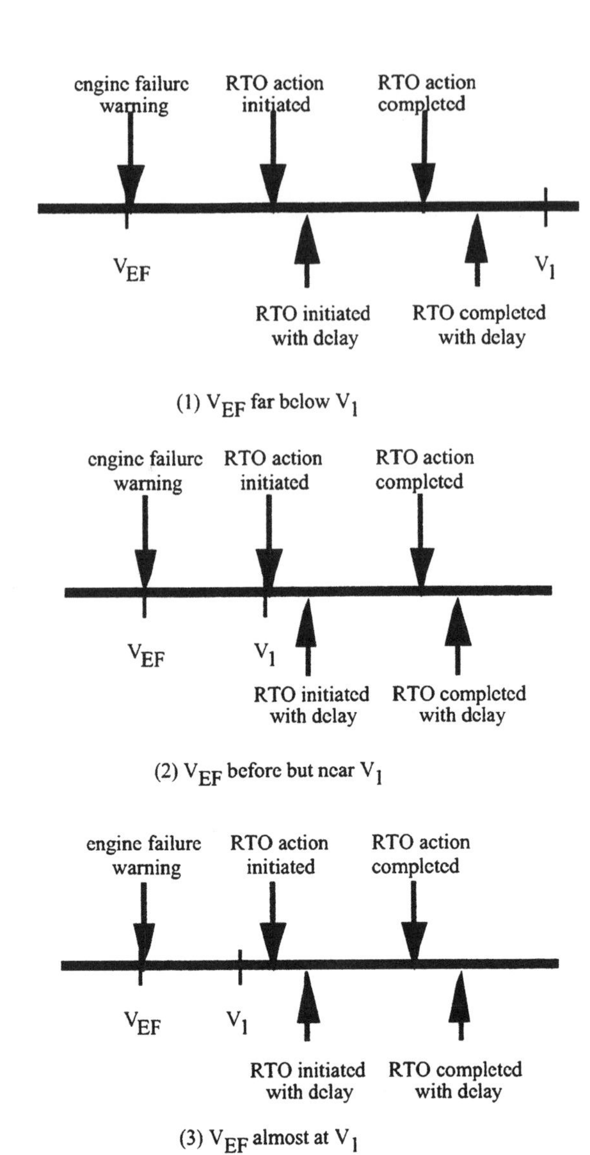

Fig. 1. Engine failure and rejected takeoff (after Inagaki, 1997)

This paper conducts an experiment to investigate the effectiveness of SAA for phases (ii) and (iii).

3. RESEARCH VEHICLE

A flight simulator has been developed for the RTO experiments. The flight simulator runs on a graphic workstation (SGI), and its interface is designed with VAPS (Virtual Prototypes, Inc.). Fig. 2 depicts the interface of the simulator, which consists of the outside view seen from the windshield (top), PFD (Primary Flight Display: left), and EICAS (Engine Instrument and Crew Alerting System: right). PFD gives: (a) the aircraft speed, (b) V_1, (c) V_R at which the liftoff procedure must be initiated, and (d) remaining length of the runway. Callouts are given by the computer at 80 knots, V_1, and V_R. EICAS gives parameter values

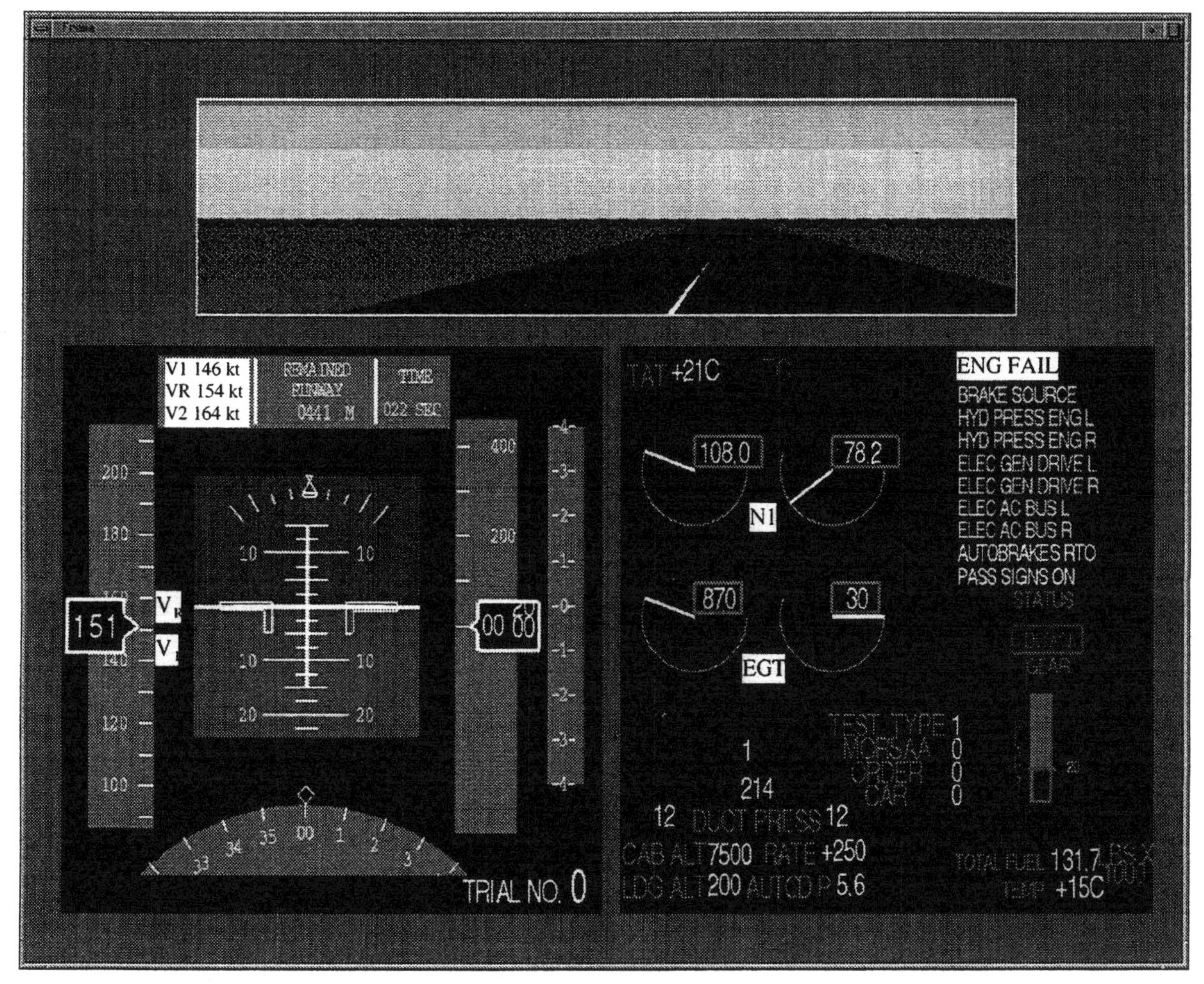

Fig. 2. Interface for RTO experiment

for each engine, such as N_1 (which indicates engine power in terms of rpm) and EGT (Exhaust Gas Temperature). Subjects are requested to lift off the aircraft safely or to reject the takeoff if necessary.

The simulated aircraft has two engines and an engine may fail during a takeoff run. A yaw movement is observed when an engine fails. A crosswind, however, may also produce a yaw movement. Thus, the yaw movement may not always indicate that an engine has failed. There are some cues to detect the engine failure: (1) an engine failure warning in red, which will be given at the top right of EICAS, (2) decline in engine parameter values, and (3) change in rate of acceleration of airspeed.

Subjects are told to control the aircraft with a control box which is shown in Fig. 3. The control box consists of a joystick, two levers, and some switches (including a trigger on the joystick).

The takeoff procedure under normal operating conditions with the simulated aircraft is as follows:

(1) Pull the trigger to start a takeoff run. The two engines are then set at full power.

(2) Maintain directional control by twisting (if necessary) the joystick for compensating a crosswind effect.

(3) Pull the trigger when he/she decides to "GO."

(4) Pull the stick when the airspeed has reached V_R.

The standard procedure for the GO/NO-GO decision,

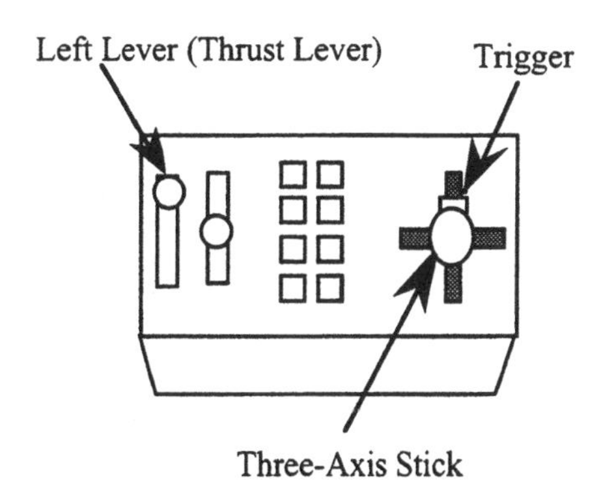

Fig. 3. Location of switches on the control box

upon an engine failure, is specified as follows:
(1) If an engine fails at a speed which is less than V_1, then pull the left lever to cut off the engines for rejecting the takeoff.
(2) If V_1 has already been achieved, then pull the trigger to continue the takeoff.

There are four types of outcomes for each trial.
(i) *Successful takeoff*: The aircraft attains height of 35 feet at the end of the runway.
(ii) *Successful RTO*: The takeoff is aborted and the aircraft stops on the runway with no damage.
(iii) *Marginal takeoff*: The altitude (screen height) achieved at the end of the runway is less than 35 feet.
(iv) *Overrun accident*: The aircraft can not stop safely by the end of the runway due to a late NO-GO decision.

SAA carries out the takeoff procedure if an engine fails at an airspeed above V_1. If SAA has the authority, a human pilot is not allowed to make NO-GO decision. SAA never aborts a takeoff once V_1 has been achieved. It could be said that SAA is *GO-Minded.* If there is no engine failure, SAA never flies the aircraft.

4. EXPERIMENT

4. 1 *Methods*

The experimental design is a 2 x 2 x 2 factorial design, mapping onto Mode of Control x Phase of Takeoff x Order of Condition. Mode of Control is Manual (M) or SAA. Phase of Takeoff refers to phase 2 (Fig. 1(2)) or phase 3 (Fig. 1(3)). In phase 2, an engine fails at some time point in the time interval $[T_{V1}-1.68, T_{V1}-0.84]$, where T_{V1} denotes the time point when V_1 is achieved. The time interval $[T_{V1}-0.84, T_{V1}]$ is defined as phase 3. These time intervals were set based on a preliminary experiment. Order of Condition refers to the fact that half the subjects received all the M trials before the SAA trials (M->SAA), and half the opposite (SAA->M). Mode of Control and Phase of Takeoff are within-subjects factors, and Order of Condition is a between-subjects factor.

Subjects were 12 volunteer undergraduate and graduate students who were paid ¥1,500 as a basis for participating a two-hour session. They could win additional bonuses (¥200 or ¥500) according to their performance. None of them had any prior experience with the takeoff-simulator. Participants were randomly assigned to one of Order of Condition.

Training: At the beginning, subjects received a written description on the purpose of the experiment. Explanations on technical terms, interface, and takeoff procedures were also given. Each subject received at least ten trials with an engine failure in phase 1 or after V_1 to learn how to manipulate takeoff procedures. Subjects were given a few trials without any engine failure. Each subject proceeded to the next stage if his or her actions were correct in more than 9 trials out of the last 10.

Data Collection: Every subject performed 30 trials under each control mode. Each set of trials consisted of 12 trials in phase 2, 12 trials in phase 3, and 6 dummy trials composed of two phase-1 trials, two after-V1 trials, and two trials with no engine failure. The parameters, such as V_1, V_R, V_{EF}, site of an engine failure, and direction of a crosswind, were randomized at every trial. At the end of each trial, subjects were requested to give a subjective rating of "self-confidence" (SC) to represent how confident they were in their ability to perform the task manually. An 11-point rating scale with "0" indicating "not at all" and "10" indicating "completely" was used. In addition to SC, subjects were requested to give subjective ratings of "trust" (T) in the automation and "reliance" (R) on SAA. Subjects gave subjective ratings after they were told whether there had been an overrun accident or a marginal takeoff, the value of root mean square error for the directional control of the aircraft, and whether SAA had took over the control of the aircraft.

Performance Measure: Several performance measures were recorded: (1) response time, (2) outcomes of trials, (3) Self-Confidence (SC) , (4) Trust (T), and (5) Reliance (R).

4.2 *Results and Discussion*

Before discussing effectiveness of SAA, it may be useful to examine whether the experimental condition, such as the definition of phases, was appropriate or not. The probability of how often SAA performed the takeoff procedure was evaluated. Table 1 gives the summary of the data. People could hardly initiate RTO actions

Table 1 Probability of takeovers by SAA

Order	Phase 2		Phase 3	
	mean	s.d.	mean	s.d.
M->SAA	0.486	0.281	0.945	0.101
SAA->M	0.514	0.244	0.972	0.043

before V_1 if an engine failure occurred in phase 3. On the other hand, subjects were able to reject the takeoff with a probability of about 50 percent if an engine failed in phase 2. The results indicate that the definition of the phases in this experiment meets the assumptions in the mathematical model.

Data on reaction time were not used in the analyses, because subjects pulled the trigger sometimes too late for recording the exact time of making "GO" decision.

As stated in section 3, each trial ends with one of the four types of outcomes. On the successful takeoffs, an ANOVA gave two significant main effects. Subjects lifted off the aircraft more successfully under SAA mode than under M mode ($F(1,10)=6.50$, $p<0.03$) (Fig. 4 (a)). The proportion of success in takeoff with an engine failure was higher in phase 3 than in phase 2 ($F(1,10)=595.56$, $p<0.01$). Fig. 4 (b) depicts a nearly significant interaction between Mode of Control and Phase of Takeoff ($F(1,10)=4.17$, $p<0.069$), which shows that subjects lifted off the aircraft more successfully under SAA mode than under M mode when an engine failed in phase 3.

An ANOVA on successful RTOs showed that the proportion of success in RTO with engine failure was higher in phase 2 than in phase 3 ($F(1,10)=46.96$, $p<0.001$) (Fig. 5(a)). The interaction between Mode of Control and Order of Condition was significant ($F(1,10)=14.96$, $p<0.01$) (Fig. 5(b)). The interaction

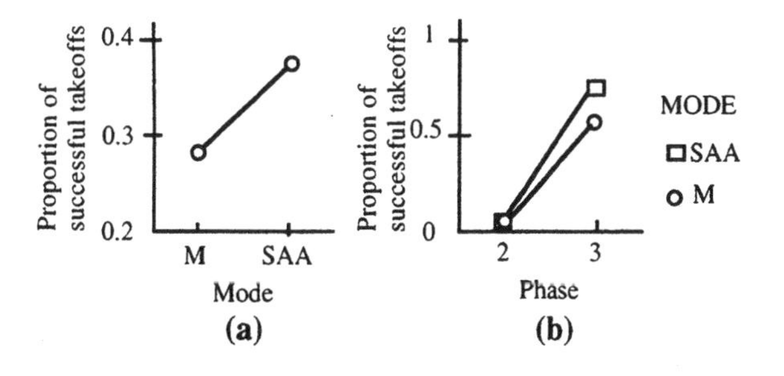

Fig. 4. Effects on successful takeoffs.

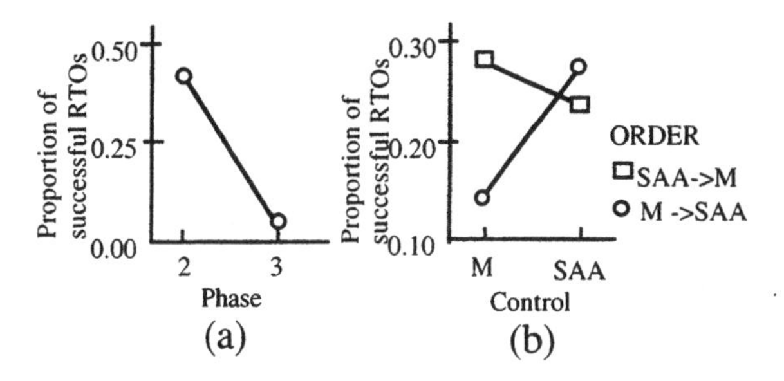

Fig. 5. Effects on successful RTOs.

can be interpreted as follows. As humans experience more takeoffs with an engine failure at a high speed, the proportion of success in RTO under M mode approaches to that under SAA mode. If people are not experienced at making the GO/NO-GO decision, on the other hand, an RTO is more successful under SAA mode than under M mode.

On the marginal takeoffs, an ANOVA showed that there was a significant main effect of Phase of Takeoff ($F(1,10)=25.18$, $p<0.001$), indicating that the proportion of marginal takeoffs with an engine failure was greater in phase 2 than in phase 3.

There were three overrun accidents under SAA mode, but there were many (52 accidents out of 288 trials) under M mode.

If the GO/NO-GO decision is made in a *too* GO-Minded manner, takeoffs with an engine failure tend to be continued even when they can be aborted. An ANOVA on the probability of continuing an takeoff with an engine failure in phase 2 did not give the significant main effect of Mode of Control. The result shows that SAA is *not* too GO-minded.

The above results illustrate efficacies of SAA under time-critical conditions. It is important to know here how people felt toward SAA and whether they were willing to accept it. An ANOVA on SC showed that the interaction between Mode of Control and Phase of Takeoff was significant (Fig. 6). Sheffe test showed that there was a significant difference between the modes when an engine failed in phase 2. The result implies that subjects lost SC to some extent under M mode when an engine failed in phase 2. The subjects were likely to hesitate in deciding which alternative to choose, and thus they sometimes caused overrun accidents. On the other hand, if subjects were supported by SAA, they did not lose SC even when an engine

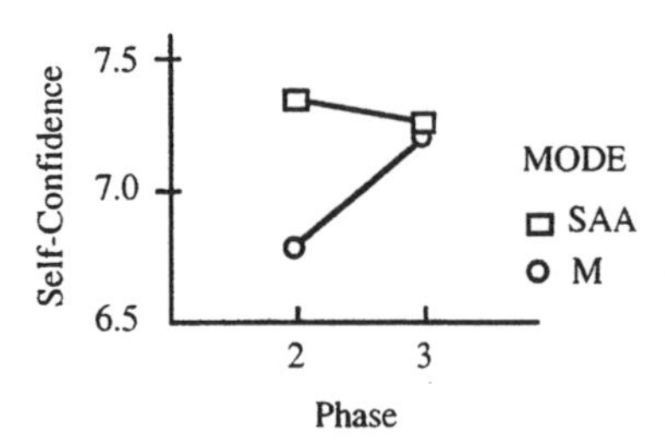

Fig. 6. Two-way interaction between Phase of Takeoff and Mode of Control on Self-Confidence

failed in phase 2, because SAA could avoid overrun accidents.

T and R were measured only when SAA had the authority to make GO/NO-GO decision. ANOVAs on them were not straightforward, because some subjects experienced few cases in which SAA took authority when an engine failed in phase 2. Thus, the data of T and R with an engine failure in phase 3 were analyzed. The mean value of T was 8.13 for group M->SAA, and 9.25 for group SAA->M. The mean values of R were 7.83 and 9.02 for groups M->SAA and SAA->M, respectively. ANOVAs showed no significant effect, which implies that Order of Condition affected neither T nor R.

The data of subjective ratings under SAA mode were broken down depending on whether or not decisions made by a subject and SAA were contradictory (Table 2). ANOVAs showed that SC and R were lower when the human and SAA did not have the same intention than when they did ($F(1,208)=11.20$, $p<0.001$; $F(1,208)=5.04$, $p<0.03$). There was no significant effect on T. The results on the ANOVAs can be interpreted as follows. A subject may lose his or her confidence when SAA does not accept his or her decision. The reliance on SAA may also decline if the decision of SAA conflicts with the subject's one, even when the subject thinks that SAA is trustworthy.

Table 2 Subective ratings under SAA mode

decisions (Man, Auto)	SC		T		R	
	mean	s.d.	mean	s.d.	mean	s.d.
conflicting (NO-GO,GO)	6.34	1.79	8.34	1.54	7.95	1.68
coincident (GO, GO)	7.28	1.80	8.68	1.24	8.49	1.49

5. CONCLUSION

This paper found that SAA was useful for attaining takeoff safety. As a whole, SAA in this experiment was trusted and accepted by subjects. However, SAA should be implemented with care, because human reliance on SAA may be lost if intentions of a human and SAA are contradictory even though SAA is reliable. Note that SAA takes only actions to continue the takeoff, and never takes action for NO-GO. It means V_1 has been already achieved at the time moment when SAA begins to control the aircraft. Thus, one reason for the decline of reliance is that humans may not understand that it is too late to reject the takeoff when the humans choose "NO-GO." SAA should therefore be more capable of expressing what actions are to be taken and why. How to design human-machine interface to give SAA the capability must be studied for making SAA acceptable to humans.

ACKNOWLEDGMENTS

This work has been partially supported by the Center for TARA at the University of Tsukuba, Grant-in-Aid for Scientific Research 07650454, 08650458, and 09650437 of the Japanese Ministry of Education, Science, Sports and Culture, and the First Toyota High-Tech Research Grant Program.

REFERENCES

Billings, C. E. (1991) Human-centered aircraft automation: A concept and guidelines, NASA TM-103885.

Inagaki, T. (1993). Situation-adaptive degree of automation for system safety. *Proc. ROMAN*, 231-236.

Inagaki, T. (1997). To go or not to go: Decision under time-criticality and situation-adaptive autonomy for takeoff safety. *Proc. IASTED Int'l Conf. Applied Modeling and Simulation*, Banff, 144-147.

Sheridan, T. B. (1992). *Telerobotics, automation, and human supervisory control.* MIT Press.

Woods, D. D. (1989). The effects of automation on humans' role: Experience from non-aviation industries. In: *Flight Deck Automation: Promises and Realities* (Norman and Orlady, Eds.), NASA CP-10036, 61-85.

VIRTUAL REALITY-BASED NAVIGATION OF A MOBILE ROBOT

M. Schmitt* K.-F. Kraiss*

* *Department of Technical Computer Science*
Aachen University of Technology (RWTH)
Ahornstr. 55, 52074 Aachen, Germany
Email: {schmitt, kraiss}@techinfo.rwth-aachen.de

Abstract: This paper focuses on the control and localization of mobile robots using a three-dimensional model of the operating environment. We provide a new concept for a control center which makes use of a virtual reality environment model to improve mission management, operator situation awareness, and access to robot sensor information. The model is built from the robot's sensor data together with photos taken by a swivel-mounted onboard camera, and the robot is controlled by a straightforward goal designation in this virtual environment. Furthermore, the model is used to determine the robot's position by comparing the virtual with the real view.

Keywords: Autonomous mobile robots, Virtual reality, Human-machine interface, Model-based control, Position location, Telerobotics, Remote control, Robot navigation

1. INTRODUCTION

The use of service robots for rationalization in processing and transportation tasks emerges to a significant field of innovation in the attendance sector. There is a great market potential for applications in logistics and cleaning, transport automation, and the security area. Most tasks that make use of mechanical actors need very special, individual solutions, whereas monitoring tasks should be performed by a universal system, capable of serving the purpose in different operating environments. This implies that the robot relies on its own skills only and does not make use of a special infrastructure. Therefore our research aims at developing concepts and systems that are able to perform exploration, stocktaking and documentation tasks without the need for installation of dedicated infrastructure.

However, until now most mobile robots do not operate fully autonomously. Most often, mobile systems operate semi-autonomously and need external help by an operator from time to time. Universally usable systems need at least a task selection or goal designation - in special cases even a manual control might be required.

This explains the demand for a control center as man-machine interface, that supports the operator in determining the goal, defining the tasks and accessing the gathered data and the results of a robot mission. The design of this control center is essential for the acceptance and the efficiency of a mobile robot service system.

Virtual Environments enable the efficient visualization of complex data. They allow operators to perceive and control complex systems in a natural fashion and are therefore predestinated for a control center. Virtual Reality tools have already shown to be useful in the context of direct teleoperation and supervisory control of robotic vehicles (Piguet *et al.*, 1995). Their application has been restricted to very special tasks like planetary or subsea exploration, though.

In the project *VERONA* we investigate the use of Virtual Environments for Robot Navigation in more general applications - beyond the scope of those special missions and beyond the visualization task. This includes the exploration of an unknown (indoor) environment, the construction of a textured three-dimensional model of the robot's operating environment, the application of the self-generated model for mobile robot localization, and a model-based database access. We introduce a control center that makes use of a Virtual Environment for all aspects of mobile robot application, fitting the requirements much better than two-dimensional maps commonly used for robot navigation.

Our approach for a mobile robot control center aims at the following improvements:

- The operator's survey over the operating environment is enhanced by a three-dimensional graphical man-machine interface. Navigation and task management of one or many robots by 'point & click' has obvious advantages as compared to handling abstract coordinates.
- Evaluation of robot navigation data is difficult without reference to the corresponding recording position. The assignment of sensor data to the correct position of a 3D model allows a very intuitive topological search that is much easier and more efficient.
- Building a virtual environment from shaded polygons is expensive, does mostly not look realistically enough, and is not applicable to large environment high resolution modeling. Hence the model skeleton is assembled from the platform sensor data and is subsequently textured with pictures taken by the onboard camera. This approach provides an economic visualization with improved realism.
- The model described above can be used for accurate localization by comparing the robot's current camera view with the corresponding snapshots stored at the corresponding position in the model, i.e., by comparing the real view with the 'virtual camera' view. This optical technique provides more accuracy than mechanical tracking systems or systems based on, e.g., ultrasonic distance measurement.

In the following sections we describe the concept of this control center and a possible way of creating the 3D model of an unknown operating environment. We then show how this model can be used to identify the robot's position.

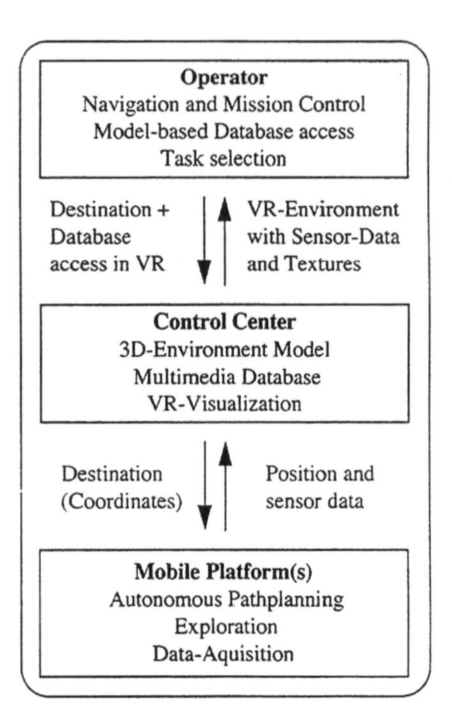

Fig. 1. System overview

2. VIRTUAL REALITY CONTROL CENTER

The main components of the control center as shown in figure 1 are a 3D environment model, a database and the interface for visualization and interaction. The control center gathers all information of the mobile platform and supports the operator by an intuitive graphical interface. All interaction between operator and mobile platform(s) is performed through this model-based interface.

The operator can move around in the virtual environment (VE) without the limitations of the real world and can take viewpoints that otherwise may not be reached (due to some kind of danger or physical laws) as, e.g., bird's-eye view. Views of areas are possible that cannot be provided by the onboard camera and the operator is not bound to the current position of the platform(s). This feature simplifies the designation of goals the robot should reach autonomously as well as manual robot control, that might be needed in special cases, as additional information (onboard camera plus, e.g., virtual sideview) can be provided.

All information is stored in a relational database. The robot's current data set as recorded in the real world is assigned to the corresponding position in the VE together with all information of previous exploration trips. Therefore, a combination of position, orientation and recording time is used as key index to ensure a correct correlation between the model and the recorded data. This spatial correlation enables a model-based database access and makes the job of, for example, finding a specific snapshot much easier as compared to manually scanning a video of a whole mission.

Determining the goal of one or even several robots and specifying an object to be inspected can be done very intuitively by pointing at it or marking it with the cursor in the VE. This is much easier than typing in coordinates. In addition to the latest sensor information displayed in the model, previously recorded data is accessible via pop-up menus, providing easy comparison of several exploration trips.

Platform and control center are connected via two radio links: a digital radio link (wireless ethernet) for all control and data exchange and an analog one for continuous transmission of the onboard camera's video. The pictures used for image-based navigation are digitized onboard and transmitted via TCP/IP to avoid artefacts. Live video transmission is needed in case of manual control only.

However, during our experiments a window containing the virtual view and a window with the live video from the onboard camera are displayed in parallel to verify navigation and model creation. The virtual camera can be attached to the robot's position. Viewpoint, viewangle and aspect ratio of the virtual camera are therefore matched to the real camera's parameters.

The mobile robot's tasks are autonomous collision avoidance, pathplanning and data acquisition, e.g., it collects all data from wheel encoders, ultrasonic sensors and camera pictures (with its recording time and position) and periodically transmits them to the control center.

VERONA's data flow can be divided into two phases, as depicted in figure 2: an exploration phase and an operation phase. During the exploration phase the environment model is built incrementally from the robot's sensor data, as outlined in the next section. During operation the model can then be used as interface for all interaction with the platform and for an image-based robot localization.

3. CREATING A MODEL OF THE ENVIRONMENT

During the initial exploration of an unknown environment, navigation of our mobile robot can rely on odometric and ultrasonic measurements only, since we do not allow the use of external markers and expect to have no a-priori knowledge about the environment. Wheel encoders determine the robot's position relative to its starting position and ultrasound sensors enable obstacle avoidance and distance measurements.

The aim of our research is to create the environment model without manual input of CAD data as usual in other approaches (Kosaka and Pan, 1995) (Lanser and Zierl, 1996), but to use the data supplied by the moving platform instead. Manual input of a model makes the system less flexible and raises new problems: most existing ground plans do not reflect the final measurements in a building due to later modifications and additional installations. Furthermore, the environment characteristics initially detected by the roving robot system itself are those most likely to be identified during the following trips. Navigating by comparing information from different sources is obviously more error sensitive than comparing those originating from the same system. Hence only those characteristics are to be recorded, which are noticeable for the robot and which are useful for creating the model and for subsequent navigation.

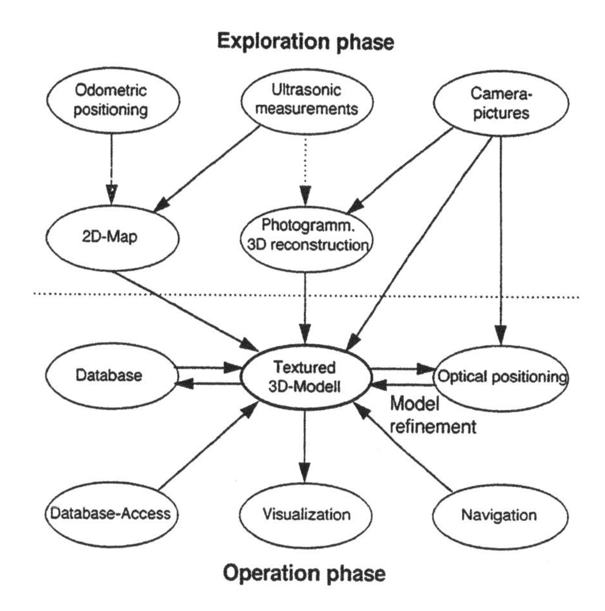

Fig. 2. Data flow at exploration and operation phases in VERONA

At the beginning, the robot is told to follow the walls during his exploration trips and to create a ground plan from the ultrasound distance measurements. We make use of the Recursive Line Fitting algorithm to extract the lines of walls and obstacles (Pauly, 1997). From this ground plan we then derive the third dimension by 'pulling up' the walls. Thus the skeleton of the 3D model is created. This results in a visualization as shown in figure 3.

However, for the monitoring tasks considered a more sophisticated and more detailed model of the operating environment is needed. Therefore the surrounding of the robot is being photographed during an exploration trip with the onboard camera and the pictures are mapped as textures onto the corresponding sections in the VE. To reduce distortions we currently allow only orthogonal views onto the walls during model generation. Figure 4 shows the expected visualization quality (for the time being created manually).

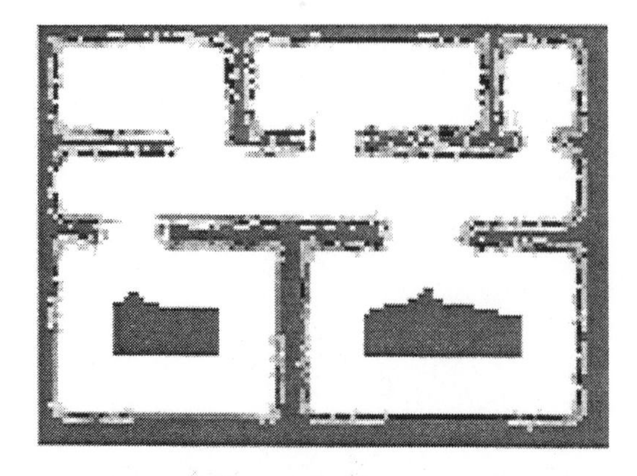

Fig. 3. Ultrasonic map and derived 3D visualization

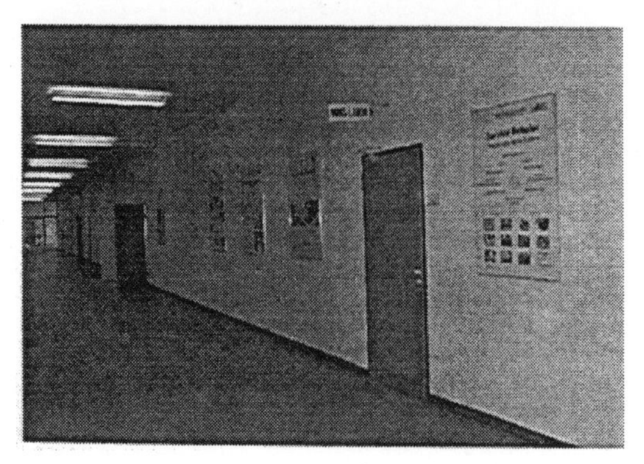

Fig. 4. Model view and real view of the robot's environment

When texturing the walls, an exact match of neighbouring pictures is required. The Cepstrum technique has shown to be able to efficiently and precisely detect the relative displacement of two pictures if the overlapping area is large enough (Lehmann *et al.*, 1995). The algorithm does not

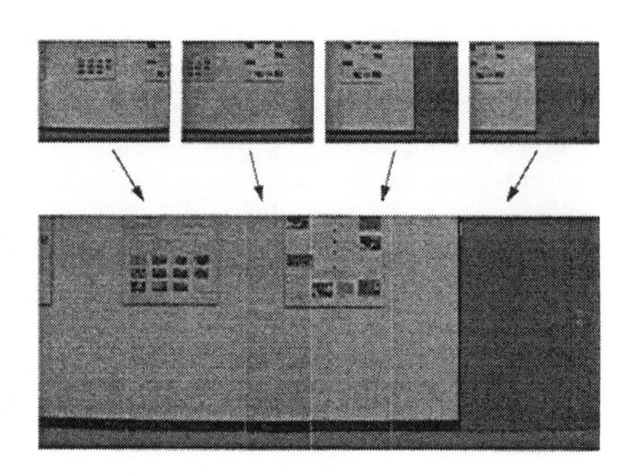

Fig. 5. Piecewise composition of a textured virtual environment from camera pictures

need a segmentation of the images and is insensitive to noise. However, it is only applicable to quite similar images. Figure 5 outlines the piecewise texturing with vertical white lines indicating the single composed segments. This piecewise assembly of pictures from the swivel-mounted onboard camera is done horizontally and vertically. Hereafter, floor and ceiling parts are cut off. Yet, this merging is only possible with flat surfaces to yield a photorealistic VE. Modelling more complex environments (e.g. rooms with furniture) requires the detection of structure (doors, obstacles, signs). To this end ultrasound distance measurement will be used in combination with photogrammetric techniques (Foehr, 1990).

4. MODEL-BASED NAVIGATION

During exploration of an unknown environment position determination relies on the robot wheel encoders only. However, subsequent missions can make use of the model described above for position correction. As positioning errors of the wheel encoders during model creation result in an inadequate model, techniques for a model refinement have to be investigated (Borenstein, 1996). Once the model is created, a position correction can be realized in the following way: When the robot and camera orientation is approximately known from the wheel encoders and the ultrasound distance measurements, the onboard camera image is compared to the corresponding view of a 'virtual camera', e.g., the section of the VE a real camera would see at the position the robot is assumed to be. The Cepstrum technique again is one possible procedure to identify the degree of overlap. The amount of displacement is calculated and the position can then be corrected accordingly (figure 6). Therefore, no recognition of single objects is needed, but the whole image, i.e., its position is used.

However, this method only provides the displacement for almost orthogonal views. Pictures taken

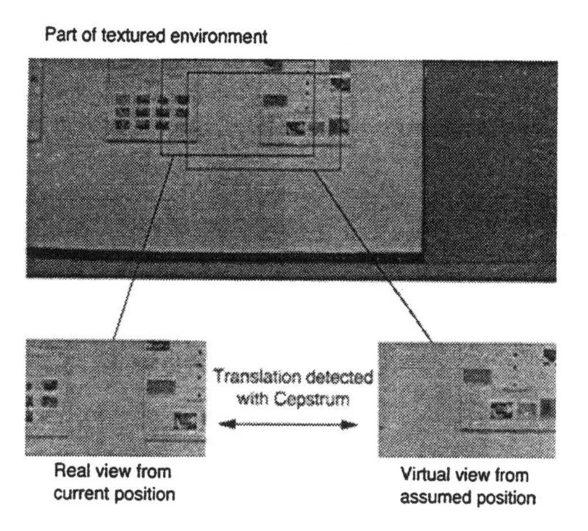

Fig. 6. Translation between virtual and real camera view

from arbitrary angles show image distortions and cannot be compared with this techniques. Nevertheless, this method is especially useful for navigation in long hallways, where ultrasound often fails to detect appropriate edges to be used for position estimation parallel to the walls. For position determination with pictures taken from arbitrary angles an edge detection based on the Hough transformation is applied to both the real and the virtual camera view, followed by a comparison of the vertically edges. The Hough transformation has already been used by (Kosaka and Pan, 1995) in this context. We extend this technique by applying it to several pairs of images from different perspectives to enable absolute positioning and to eliminate the dependency of position accuracy on heading. Bearings derived from these pictures in the VE are combined as shown in figure 7. The angle between two bearings is identical for all possible locations lying on a circular arc. The intersection of two arcs then specifies the location of the camera and the robot, respectively.

As the camera is swivel-mounted, this method is applicable even in environments with less structured walls (some contours are needed for Cepstrum application however). If positioning is not possible with the images taken on one side of the robot, a view in opposite direction may be examined. It is a remarkable advantage of this optical positioning that the achievable accuracy does not depend on heading, as is the case in other approaches (Lanser and Zierl, 1996).

Using the derived position, heading and additional images the model is gradually improved. As the model becomes more realistic the error of subsequent localization is continuously reduced. The described algorithms for image-based navigation are very computational intensive. Thus, they are not executed onboard the robot but at the control center residing on a powerful desktop workstation, as the results are not needed for a permanent robot control but for position correction once in a while.

Fig. 8. Mobile platform TAURO-II

5. THE MOBILE ROBOT TAURO-II

After gathering experience with our first mobile robot TAURO (Pauly and Kehr, 1995), (Pauly and Kraiss, 1995), a second prototype has been designed and realized at the Department of Technical Computer Science (Pauly, 1997). This platform - TAURO-II - is based on an electric wheelchair and is able to carry much more sensor equipment than our first robot. At this time two onboard computers, a CCD camera, 14 ultrasound sensors, two wheel encoders and several microcontrollers for controlling the hardware (motor and camera pan-tilt control) are installed. Currently, a manipulator (called MANUS), is being mounted, carrying a gripper and an additional camera. It will be used for simple mechanical actions (like activating switches) to widen the operational range accessible to the robot.

6. CONCLUSION

We introduced a concept for a control center based on a three-dimensional model of the operating environment, that provides new possibilities for mission and data management in the context of mobile robot control. Although virtual environments are already well known in the area of

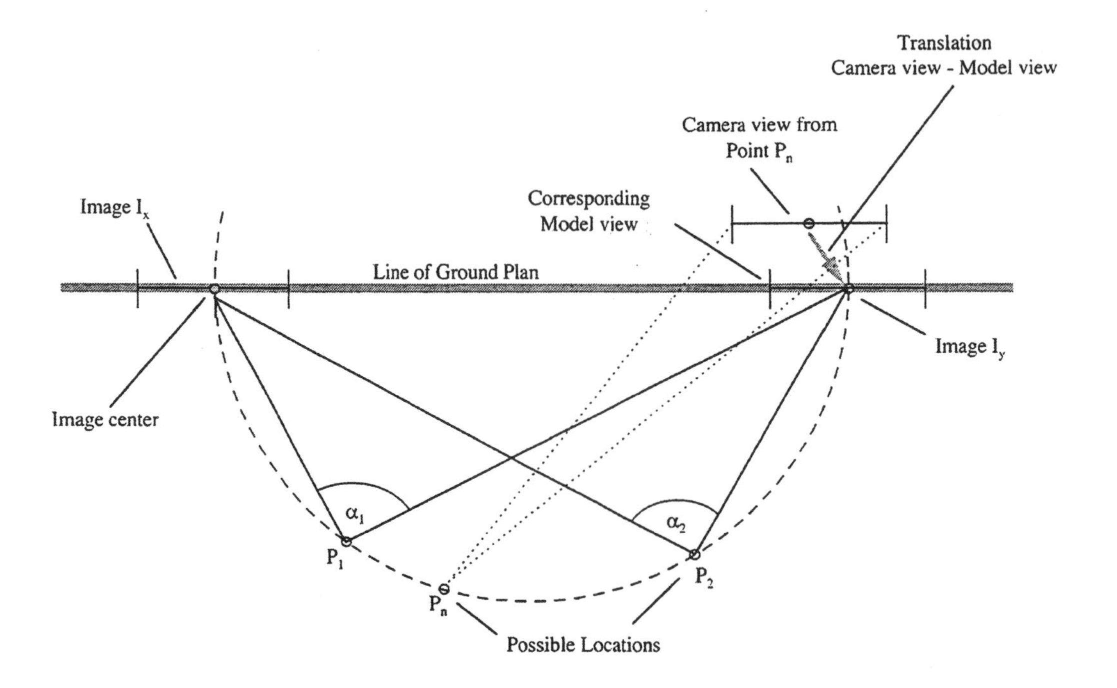

Fig. 7. Mobile robot localization by comparing real and virtual views from several bearings

static industrial robots for visualization and off-line path-planning of, e.g., a robot arm, they have scarcely been investigated for mobile robots. Remarkable improvements have been shown to be able regarding navigation, localization and data management. Most essentially a virtual environment can help to avoid the need for dedicated infrastructure installations.

A prototype of the control center has been implemented on a UNIX-Workstation (SGI) using the WorldToolKit (Sense8) for VR-Visualization and a relational database (ORACLE8). The image-based navigation has been tested within the Khoros toolkit (KHORAL Research Inc.), that is connected to the control center. Right now, the method of texturing the environment model and the achievable accuracy of the localization method we described are studied in detail. Further research is concerned with the detection of image structure and the modeling of more complex environments.

7. REFERENCES

Borenstein (1996). *Navigating Mobile Robots.* A K Peters Ltd. Wellesley.

Foehr, R. (1990). *Photogrammetrische Erfassung räumlicher Informationen aus Videobildern.* Vieweg, Braunschweig.

Kosaka, A. and J. Pan (1995). Purdue experiments in model-based vision for hallway navigation. In: *Proceedings of 'Workshop on Vision for Robots in IROS'95 Conference', Pittsburgh, PA,*. pp. 87–96.

Lanser, S. and C. Zierl (1996). Moral: Ein System zur videobasierten Objekterkennung im Kontext autonomer, mobiler Systeme. In: *Autonome Mobile Systeme 1996* (G. Schmidt and F. Freyberger, Eds.). Springer, Berlin. pp. 88–98.

Lehmann, T., C. Goerke and W. Schmitt (1995). Rotations- und Translationsbestimmung durch eine erweiterte Kepstrum-Technik. In: *Informatik aktuell: Mustererkennung 1995 - Verstehen akustischer und visueller Information* (G. Sagerer, S. Posch and F. Kummert, Eds.). Springer, Berlin. pp. 395–402.

Pauly, M. (1997). *Ferninspektion mit mobiler Sensorik.* Ph.D. Thesis, Department of Technical Computer Science, Aachen University of Technology, Logos, Berlin.

Pauly, M. and K.-F. Kraiss (1995). A concept for symbolic interaction with semi-autonomous mobile systems. In: *Proceedings of '6th IFAC/IFIP/IFORS/IEA Symposium on Analysis, Design and Evaluation of Man-Machine Systems'.* Boston.

Pauly, M. and M. Kehr (1995). Steuerung eines teilautonomen mobilen Systems mit einer ultraschallbasierten künstlichen 3D-Sicht. In: *Autonome Mobile Systeme 1996* (R. Dillmann, U. Rembold and T. Luth, Eds.). Springer, Berlin. pp. 289–298.

Piguet, L., T. Fong, B. Hine, P. Hontalas and E. Nygren (1995). Vevi: A virtual reality tool for robotic planetary explorations. In: *Proceedings of 'Virtual Reality World 95'.* Stuttgart, Germany.

DESIGN AND EVALUATION OF THE HUMAN-MACHINE INTERFACE FOR INSPECTION TASKS OF A TELEOPERATED SPACE MANIPULATOR

E.F.T. Buiël, M. de Beurs, A. van Lunteren and H.G. Stassen

Delft University of Technology, Faculty of Design, Engineering and Production, Department of Mechanical Engineering and Marine Technology, Man-Machine Systems and Control Group, Mekelweg 2, 2628 CD, Delft, the Netherlands. E-mail: e.buiel@wbmt.tudelft.nl

Abstract: The anthropomorphic European Robotic Arm (ERA) is a large telemanipulator that will be mounted on the new International Space Station. To the human operator, the execution of ERA's inspection tasks is difficult because of the lack of spatial information in the available camera pictures, the manipulator dynamics, and the potential presence of time delays in the control loop. The human-machine interface must assist the operator in order to avoid collisions between the manipulator and hazardous objects in its environment. The results of simulator experiments with three different configurations of ERA's human-machine interface show that automatic collision avoidance as well as a perspective distance visualisation in a graphical camera overlay are indispensable components of the interface. The automatic collision avoidance decreases the task completion time. Besides, it releases the operator from frequently correcting the commanded movements. The perspective graphical camera overlay substantially improves the operator's situational awareness. As a result, the automatic collision avoidance has to correct his control commands less often. *Copyright © 1998 IFAC*

Keywords: teleoperation, manual control, human-machine interface, human-centred design, evaluation

1. INTRODUCTION

The anthropomorphic, six-DOF European Robotic Arm (ERA) is a lightweight telemanipulator (Sheridan, 1992) that will be located on the new International Space Station (Dooling, 1995). Figure 1 shows the properties of the ERA[1] (Kampen *et al*, 1995; Boumans *et al*, 1996). The ERA is designed for inspection and maintenance tasks and for displacing objects, e.g. Orbital Replaceable Units and solar arrays. Recurrent object displacements will be controlled by an automatic controller. Inspection and maintenance tasks will be controlled manually. Research at the Man-Machine Systems and Control Group focuses on the design of the human-machine interface for the manually controlled ERA (Bos, 1991; Breedveld, 1996; Buiël, 1998). Generally, manual teleoperation of a large space manipulator like the ERA is difficult to the human operator. Firstly, the lack of spatial information in the available monoscopic camera pictures complicates manual control. Secondly, control performance suffers from the dynamics due to the flexibility of the lightweight limbs. Finally, when the

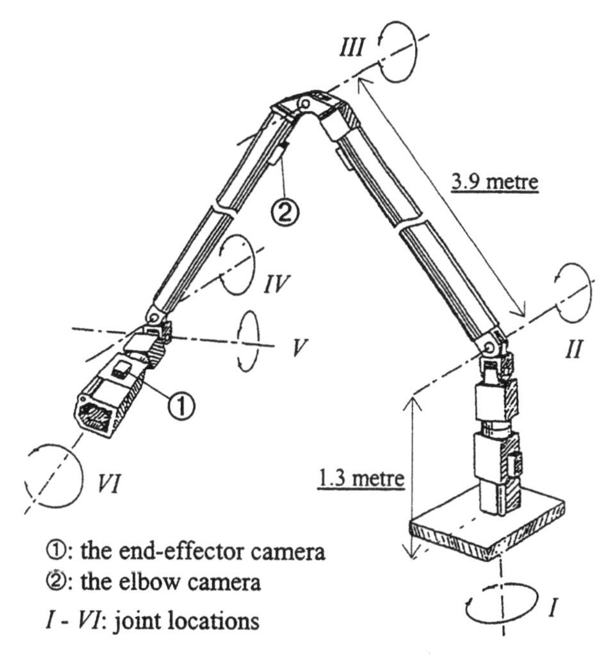

Figure 1 The European Robotic Arm

[1] Currently, Fokker Space B.V. in Leiden, the Netherlands, is developing the European Robotic Arm under contract of the European Space Agency (ESA).

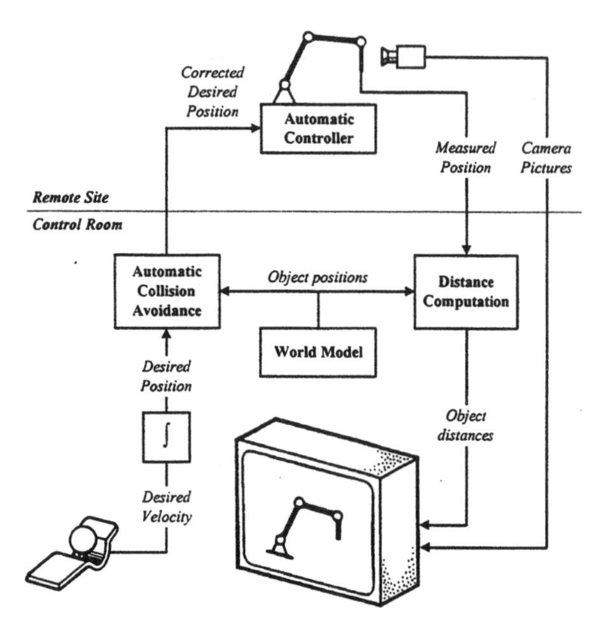

Figure 2 The control scheme

operator is located on Earth, time delays are introduced in the control loop as a result of the transmission of control commands from Earth to space, and the transmission of camera pictures from space to Earth. In the case of the ERA, the last-mentioned problem will not complicate the control, since the operator will be located in one of the modules of the International Space Station. This paper considers the design of a human-machine interface for ERA's inspection tasks and the evaluation of that interface by means of simulator experiments. Detailed information about the presented work can be found in the PhD thesis of Buiël (1998).

2. THE HUMAN-MACHINE INTERFACE FOR INSPECTION TASKS

2.1 Introduction

Inspection tasks of the ERA are performed by means of the pictures from two cameras: the end-effector camera and the elbow camera (Figure 1). The end-effector camera provides an accurate view of the inspected location. The elbow camera provides a global view. The operator uses the latter view to move the end-effector to a subsequent location that has to be inspected. During these intermediate displacements, the manipulator covers large distances. At that time, the operator has to avoid collisions between the manipulator limbs and the hazardous objects in their environment. Collision avoidance is rather difficult to the operator, since the monoscopic, black-and-white elbow camera picture provides only limited depth information. As a result, the distances between hazardous objects and the manipulator can hardly be estimated. The human-machine interface for ERA's inspection tasks must facilitate the avoidance of collisions by exploiting the available information on the relative positions of objects and manipulator limbs. Two information sources are available: the geometry of the International Space Station is registered in a database (the *world model*), whereas the manipulator position is measured by means of joint angle sensors.

2.2 The Control Scheme

Figure 2 shows the control scheme that has been adopted in the design of the human-machine interface for ERA's inspection tasks. The human operator uses a six-DOF Spaceball hand controller to define the desired translation and rotation velocities of the end-effector. A computer system in the control room first computes the accompanying setpoints for the joint angular velocities (inverse kinematics computation). Subsequently, it integrates these setpoints in order to obtain new setpoints for the joint angles (the manipulator position setpoint). Finally, it computes the resulting distances between the manipulator and hazardous objects by means of the world model. If the last-mentioned computation indicates that the manipulator is about to collide, the *automatic collision avoidance* corrects the direction of the commanded movement in order to avoid the collision. At all times, the operator can easily see when the manipulator is about to collide, since a perspective graphical camera overlay for the elbow camera picture, the *Raindrop Display*, visualises the actual distances between the manipulator and hazardous objects. These distances are computed by means of the measured manipulator position and the world model.

2.3 The Automatic Collision Avoidance

The automatic collision avoidance corrects the direction of the commanded manipulator movement when the operator intends to place the manipulator at a location where the distance between the manipulator and an object in its environment is smaller than the maximum amplitude of the manipulator vibrations that result from the limb deflections. Figure 3 visualises this correction. Firstly, the automatic collision avoidance computes the commanded velocity of the manipulator node at closest distance from the hazardous object, the *critical node*. Subsequently, it decomposes this velocity into two components: the velocity *parallel* to the surface of the object and the velocity *perpendicular* to that surface. The automatic collision avoidance zeros the latter velocity component and computes the joint angular velocities that are needed to move the critical node in the corrected direction. Finally, it obtains a new setpoint for the manipulator position by integrating the corrected joint angular velocity setpoints. As a result of this correction, the manipulator seems to glide along an invisible safety shield that protects the object.

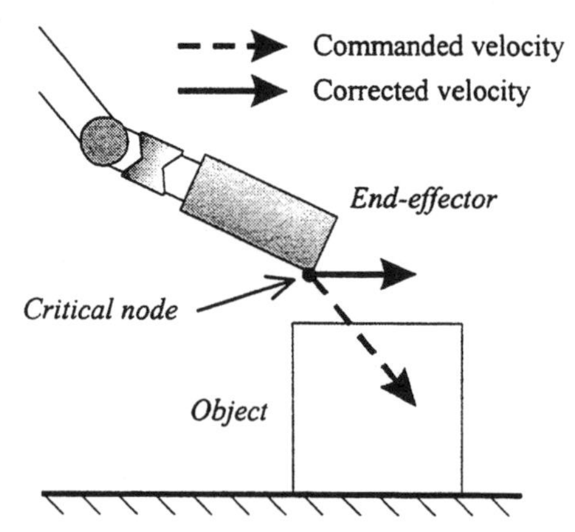

Figure 3 Automatic collision avoidance

Table 1 Two types of distance lines

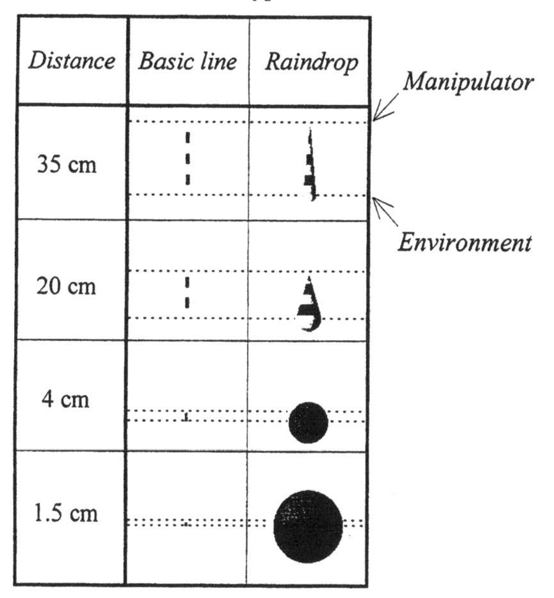

Distance	*Basic line*	*Raindrop*
35 cm		
20 cm		
4 cm		
1.5 cm		

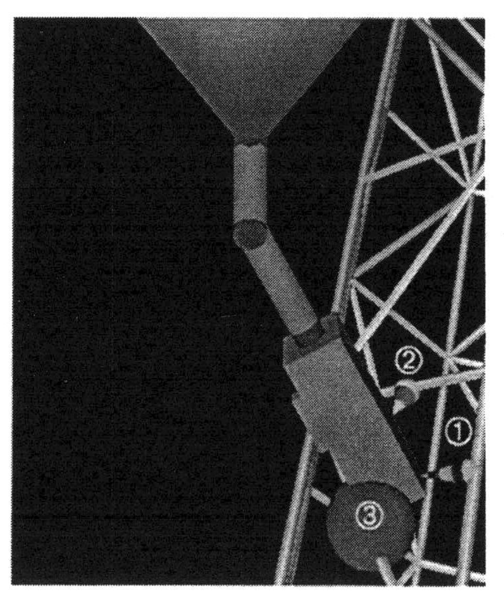

Figure 4 Elbow camera picture with raindrop-shaped distance lines (①, ② and ③)

2.4 The Raindrop Display

The Raindrop Display visualises the distances between the manipulator and the objects in its environment by means of distance lines (de Beurs, 1995; Buiël, 1996). A distance line visualises the distance between the manipulator and a hazardous object by connecting the manipulator-node and the object-node at closest distance (the second column of Table 1). Basically, two problems inherent to the shape of this line complicate its application in practice. Firstly, if both the environment-end and the manipulator-end of the line are (almost) in one line with the viewing direction of the elbow camera, the operator can hardly estimate the line's length. Secondly, the danger of a collision *increases* at the moment the length of the line *decreases*. This is a disadvantage from an ergonomic point-of-view: at the moment the danger grows, its display indicator becomes less eye-catching. In order to solve the two problems, the basic distance line has been turned into a raindrop (the third column of Table 1). The cross-section of the environment-end of the raindrop increases at the time when the indicated distance decreases. As a result, the operator observes that the raindrop expands just like a balloon that is blown up. Figure 4 shows a part of the elbow camera picture where three raindrops visualise the distances between the ERA and an inspected truss of the International Space Station. Note that the colours of the raindrops indicate the remaining distance as well. When a raindrop is coloured green (① in Figure 4), the remaining distance is large enough. When it is coloured yellow (②), the automatic collision avoidance is about to intervene. When it is coloured red (③), the automatic collision avoidance intervenes, since the remaining distance is about to become smaller than the maximum amplitude of the vibrations due to the limb flexibility.

2.5 Discussion

By itself, the *joint* application of a sophisticated perspective distance visualisation and automatic collision avoidance might sound rather controversial to the reader. After all, why should the human-machine interface still inform the operator about impending collisions if the automatic collision avoidance automatically 'guides the manipulator along hazardous objects'? The answer to this question is twofold. Firstly, the Raindrop Display can enlarge the operator's trust in the manipulator's control system, since it explains the corrections that are made by the automatic collision avoidance. Secondly, the display can be of essential importance in maintaining the operator's situational awareness. If the interface console does not show any information on the actual danger of collisions, the operator will never be fully aware of the actual state of the remote environment. This lack of situational awareness might effectuate misconceived control commands, e.g. when the operator overestimates the remaining distance between manipulator and hazardous objects.

Although the Raindrop Display can improve the operator's situational awareness, it probably *won't* diminish the need for the automatic collision avoidance. This tool seems to be indispensable when collisions due to slips of the operator's mind are to be avoided. Besides, the tool may well decrease the stress that is felt by the operator, since he knows that the system safety is guaranteed as much as possible. Therefore, the joint application of automatic collision avoidance and the Raindrop Display seems to be the right choice. In order to determine the actual value of both the operator aids in ERA's inspection tasks, two series of human-machine experiments have been performed.

3. HUMAN-MACHINE EXPERIMENTS WITH THE DEVELOPED INTERFACE

3.1 Simulation Facility

The experiments have been performed in the laboratory research simulator of Figure 5. In this simulator, subjects can practise a reference ERA inspection task: the inspection of the inside area of a truss of the International Space Station (Figure 6). In this task the operator uses the elbow camera picture to place the end-effector inside the truss. Subsequently, he points the

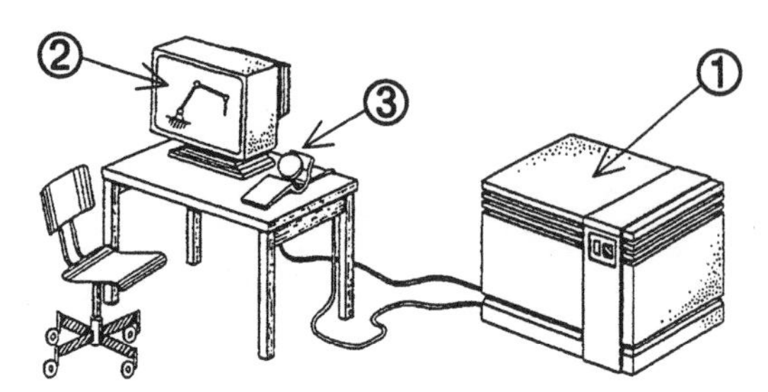

Figure 5 The simulation facility
(① graphical workstation, ② animated camera picture, ③ Spaceball hand controller)

end-effector tip in the direction of the location that has to be inspected. In the task simulation, a Silicon Graphics Indigo II workstation performs all input transformations, distance computations, and setpoint corrections, whereas it simulates the movements of the European Robotic Arm by means of a real-time mathematical model of the ERA dynamics (Breedveld, 1996). The ERA movements are visualised in a simplified elbow camera picture.

3.2 Experimental Design

The discussion at the end of the previous section hypothesises that both the automatic collision avoidance as well as the Raindrop Display are indispensable parts of the human-machine interface for ERA's inspection tasks. In the test of this hypothesis, experiments must show the consequences of removing the two tools.

In order to investigate the effects of removing the automatic collision avoidance (Experiment I in Table 2), six subjects have been trained to complete the simulated truss inspection task with the help of two human-machine interfaces: the Raindrop Interface and the Complete Interface. The *Raindrop Interface* visualises the impending collisions in the Raindrop Display, but does *not* avoid collisions automatically. In the task runs with this interface, subjects were asked to complete the task without causing collisions. Besides, they had to complete the task as fast as possible. The *Complete Interface* visualises the impending collisions in the Raindrop Display and automatically avoids them as well. Here, subjects were asked to complete the task as fast as possible.

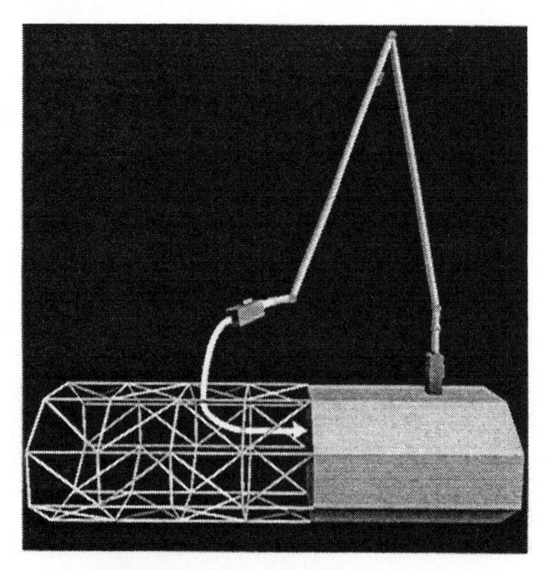

Figure 6 The truss inspection task

Table 2 The experimental design

Interface	Interface components: Raindrop Display	Automatic Collision Avoidance	
Raindrop Interface	X		*Experiment I*
Complete Interface	X	X	*Experiment I*
ACA Interface		X	*Experiment II*

In order to investigate the effects of removing the Raindrop Display (Experiment II in Table 2), another group of six subjects has been trained to complete the simulated truss inspection task with the help two interfaces: the Complete Interface and the Automatic Collision Avoidance (ACA) Interface. The *ACA Interface* avoids collisions automatically, but does *not* visualise the impending collisions in the Raindrop Display. In the task runs with both the evaluated interfaces, subjects were asked to complete the task as fast as possible.

The twelve tested subjects were students in Mechanical Engineering, without any experience in tasks like the simulated truss inspection task. They all had to complete 24 task runs with each of the evaluated interfaces. In each run, they had to move the end-effector to a different location inside the truss. At the start of the run, the elbow camera picture showed this desired location as well as the truss triangle that had to be passed in order to move the end-effector to that location. At the day before the measurement session, the subjects practised the experiment task for about four hours. With each of the tested interfaces, the subject performed at least 18 practise runs. The experimenter increased the number of training runs when the subject needed more time to attain stationary task performance in terms of run completion time and the number of collisions. At the end of each experiment, the subject was asked to choose his favourite interface in a written questionnaire. Detailed information on the experimental design is provided by Buiël (1998, Chapter 9).

3.3 Experiment I: The Evaluation of the Automatic Collision Avoidance

The results of the evaluation of the automatic collision avoidance showed that the Raindrop Interface enables the human operator to avoid collisions on his own accord. Five subjects never caused a collision during the 24 measurement runs with that interface, whereas the sixth subject caused one collision only. Yet, this observation did not turn the automatic collision avoidance into a superfluous tool. At the time when the subjects used the Raindrop Interface, they acted very carefully. In order to avoid collisions, they corrected the direction of the commanded movement much more often than when they used the Complete Interface. As a consequence, the manipulator generally operated at a relatively large distance from the inspected truss. This is illustrated by the boxplot for the collision avoidance time of Figure 7b. In the case of the Complete Interface, the collision avoidance time equals the time where the automatic collision avoidance corrects the direction of the commanded manipulator displacement. In the case of the Raindrop Interface, it equals the time where the

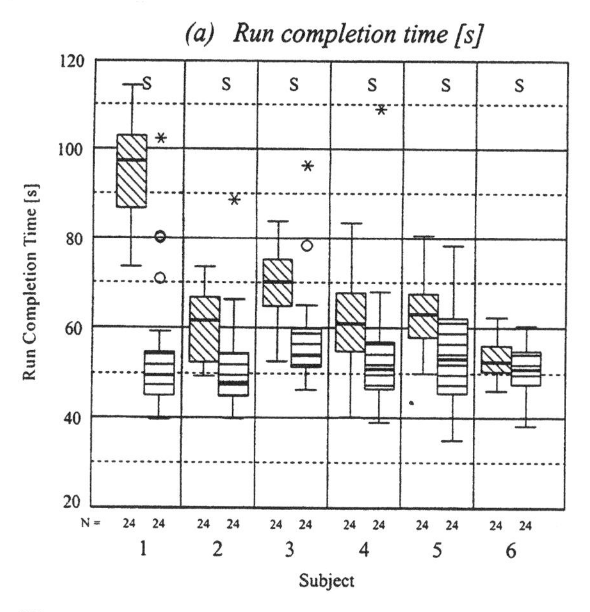

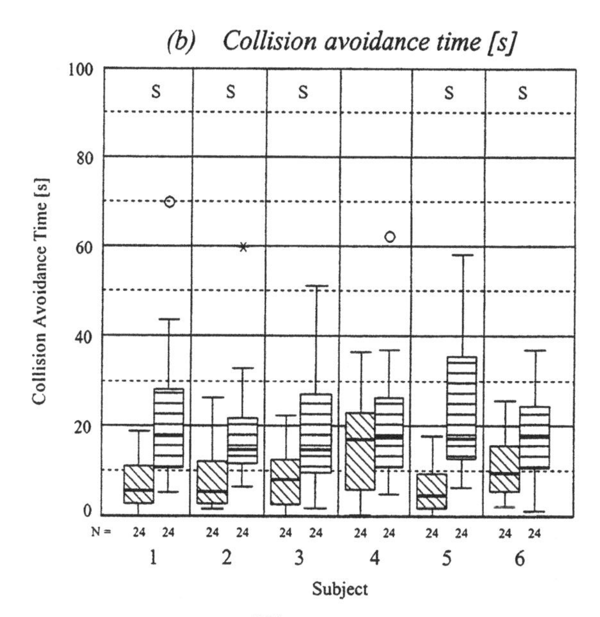

Figure 7 Results of Experiment I: a comparison between the Raindrop Interface (▧) and the Complete Interface (▤)

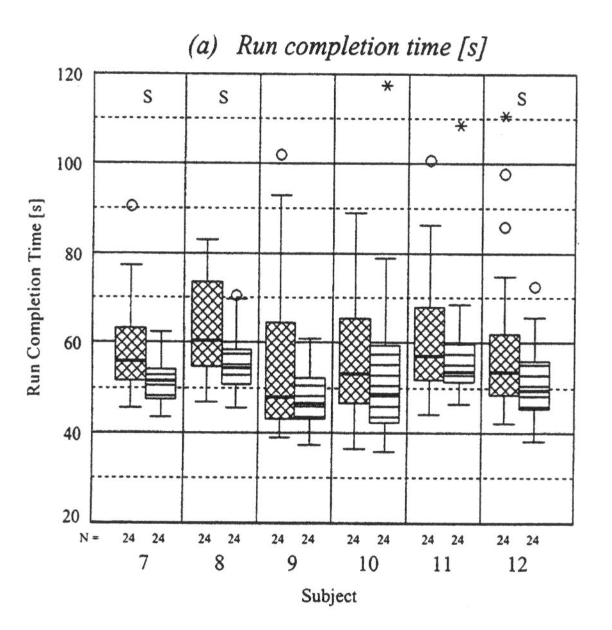

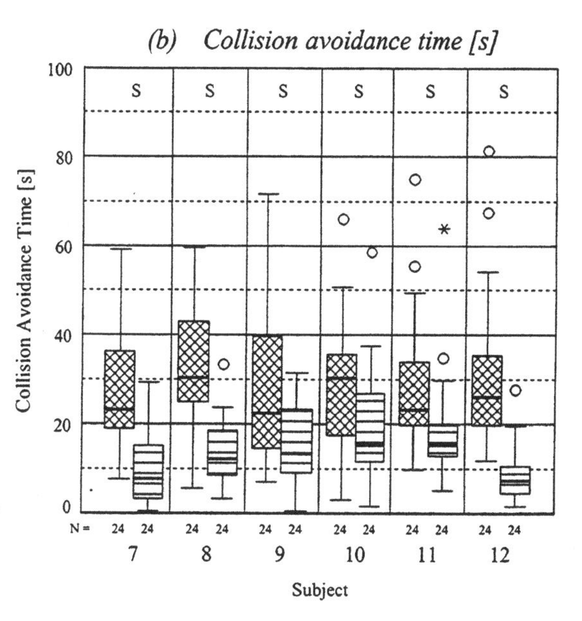

Figure 8 Results of Experiment II: a comparison between the ACA Interface (▩) and the Complete Interface (▤)

S	*Significant difference (Wilcoxon Signed-Ranks Test, $\alpha = 0.05$)*
*	*Extreme (more than 3 box-lengths from 75th percentile)*
o	*Outlier (more than 1.5 box-lengths from 75th percentile)*
	Largest normal observation
	75th percentile
	Median
	25th percentile
	Smallest normal observation
o	*Outlier (more than 1.5 box lengths from 25th percentile)*
*	*Extreme (more than 3 box-lengths from 25th percentile)*

Figure 9 Boxplot properties

commanded manipulator displacement would have been corrected if the automatic collision avoidance would have been activated. Averaged across subjects, the mean collision avoidance time is 19.9 seconds for the Complete Interface (36 % of the run completion time) and only 9.2 seconds for the Raindrop Interface (14 % of the run completion time). This difference is significant for five subjects.

As a result of their careful control strategy in the task runs with the Raindrop Interface, it took all subjects significantly more time to complete the measurement runs with this interface (Figure 7a). Averaged across subjects, the task completion time was 54 seconds for the Complete Interface, whereas the measurement runs with the Raindrop Interface were completed in 67 seconds. Note that the setpoints for the translation and rotation velocities of the end-effector cannot exceed a predefined maximum value in order to avoid large limb deflections. As a consequence, at least 35 seconds were needed to complete the simulated truss inspection task.

At the end of their experiment, four subjects mentioned

that they finally appreciated to complete the simulated truss inspection task by means of the Complete Interface, since this interface eliminates the danger of collisions on its own accord. The other subjects preferred the Raindrop Interface, since it challenged them to avoid collisions all by themselves. They said they were more satisfied when they had completed a task run with the Raindrop Interface without any collisions than when they had completed a task run with the Complete Interface in short time.

3.4 *Experiment II: The Evaluation of the Raindrop Display*

The results of the evaluation of the Raindrop Display showed that only 3 subjects completed the simulated truss inspection task significantly more quickly when they used the Complete Interface instead of the ACA Interface. Averaged across subjects, the task completion time was 53 seconds for the Complete Interface, whereas the task runs with the ACA Interface were completed in 59 seconds (Figure 8a).

Despite the relatively small completion time difference, all subjects said they by far preferred to complete the task by means of the Complete Interface, since this interface clearly visualises the distances between the manipulator and the tubes of the inspected truss. In the questionnaire, subjects said that due to the lack of distance information in the elbow camera picture, they had regularly over- or underestimated distances in the task runs with the ACA Interface. Besides, at times when they noticed that the automatic collision avoidance corrected their control commands in order to avoid a collision, they sometimes did not know the location of the avoided collision. In task runs with the Complete Interface, the colour changes in the Raindrop Display explained the interventions by the automatic collision avoidance. As a consequence, it was much more easy to the subjects to understand a noticed intervention.

The stated subjective opinions on the two interfaces confirm the lack of situational awareness of the human operator in task runs with the ACA Interface, just as this lack was theorised in the discussion at the end of Section 2. The boxplot for the collision avoidance time in Figure 8b illustrates the lack of situational awareness once more. Because of the difficulty of estimating distances in absence of the Raindrop Display, the average time where the automatic collision avoidance corrected the commanded manipulator movement was 29 seconds in the 24 task runs with the ACA Interface and 15 seconds in the 24 task runs with the Complete Interface. This difference was significant for all subjects.

4. CONCLUSIONS

To the human operator, the execution of teleoperated inspection tasks of a large space manipulator like the European Robotic Arm is difficult because of the lack of spatial information in the available camera pictures, the manipulator dynamics, and the potential presence of time delays in the control loop. The human-machine interface must assist the operator in order to avoid collisions between the manipulator and hazardous objects in its environment. The results of experiments with three different configurations of ERA's human-machine interface clearly show that a perspective distance visualisation in a graphical camera overlay as well as automatic collision avoidance are indispensable components of this interface. The automatic collision avoidance decreases the task completion time. Besides, it releases the operator from frequently correcting the direction of the commanded movement. The distance visualisation substantially improves the operator's situational awareness. As a result, the automatic collision avoidance has to correct his control commands less often. So, the joint application of automatic collision avoidance and a perspective distance visualisation is definitely the right choice.

ACKNOWLEDGEMENTS

This research has been supported by the Dutch Technology Foundation STW (Utrecht, the Netherlands).

LITERATURE

Beurs, M. de (1995). *Collision Avoidance Displays* (in Dutch). Report A-697. Delft UT, Dept. of Mechanical Engineering and Marine Technology, Man-Machine Systems Group, Delft, the Netherlands. 124 p.

Bos, J.F.T. (1991). *Man-Machine Aspects of Remotely Controlled Space Manipulators.* PhD Thesis. Delft UT, Dept. of Mechanical Engineering and Marine Technology, Man-Machine Systems Group, Delft, the Netherlands. ISBN 90-370-0056-8. 177 p.

Boumans, R., C. Heemskerk, E. Holweg, S. Kampen, M. van Lent and A. Pouw (1996). ERA: Baseline Capabilities and Future Perspectives. *4th ESA Workshop on Advanced Space Technologies for Robot Applications*, ESTEC, Noordwijk, the Netherlands. 5 p.

Breedveld, P. (1996). *The Design of a Man-Machine Interface for a Space Manipulator.* PhD Thesis. Delft UT, Dept. of Mechanical Engineering and Marine Technology, Man-Machine Systems Group, Delft, the Netherlands. ISBN 90-370-0147-5. 238 p.

Buiël, E.F.T. (1996). The Development of a Man-Machine Interface for Space Manipulator Displacement Tasks. *Proceedings 15th European Annual Conference on Human Decision Making and Manual Control*, Soesterberg, the Netherlands. 8 p.

Buiël, E.F.T. (1998). *Design and Evaluation of a Human-Machine Interface for a Teleoperated Space Manipulator.* PhD Thesis. Delft University Press, Delft, the Netherlands. ISBN 90-407-1687-0. 280 p.

Dooling, D. (1995). Research outpost beyond the sky. *IEEE Spectrum*, October 1995. pp. 28-33.

Kampen, S., W. Wandersloot, A. Thirkettle and R.H. Bentall (1995). The European Robotic Arm and its Role as part of the Russian Segment of the International Space Station Alpha. *46th International Astronautical Congress*, Oslo, Norway. 11 p.

Sheridan, T.B. (1992). *Telerobotics, Automation, and Human Supervisory Control.* The MIT Press, Cambridge, MA, USA. ISBN 0-262-19316-7. 415 p.

A STUDY ON DESIGN SUPPORT FOR CONSTRUCTING MACHINE-MAINTENANCE TRAINING SYSTEM BY USING VIRTUAL REALITY TECHNOLOGY

H. ISHII *, T. TEZUKA * and H. YOSHIKAWA *

*Graduate School of Energy Science, Kyoto University, Gokasho, Uji-shi, Kyoto-fu, JAPAN

Abstract. A design support system has been developed for constructing VR-based training environments for machine maintenance work without any expertise knowledge and programming effort on VR. Using the developed system, the users can easily construct various training environments under GUI environment. It was verified through some experiments that the developed system can reduce the working hours remarkably and that novice users who have no prior knowledge on the system could construct a training environment successfully after a few hours of tutorial on the system. In this paper, the system configuration and experimental results are described. *Copyright ©1998 IFAC*

Key Words. Virtual reality ; Design systems ; Training ; Petri-nets ; User interfaces

1. INTRODUCTION

Recently, Virtual Reality (VR) technology has emerged and been developing remarkably so that it becomes now possible for us to apply VR technology for various training purposes (Miwa *et al.*, 1995; Arai *et al.*, 1997). In fact, it was reported by NASA that VR-based training had been successfully used for on-the-ground pre-training of space shuttle staffs who were in charge of Hubble telescope repairing in outer space (Loftin and Kenny, 1995). The authors have developed a VR-based training environment in which trainees can disassemble a check valve used in the nuclear power plant (Yoshikawa *et al.*, 1997).

Compared with the training systems based on the real machines or real-size mockups, VR-based training system has a number of advantages: economical, safe for the trainee, no need of large space, and so on. However, from the author's past experience, there arises a serious problem for developing software systems for VR-based training. In fact, the workload of constructing the training environments is very large when a new training environment should be constructed for different kinds of training tasks. The important point is that expert knowledge and skills on computer programming are required to construct the VR-based training system. And it is very difficult for those who are not so familiar with the programming technology to construct the VR-based training system for various kinds of machine, which are necessary for practical and effective training.

In this study, the user support system has been developed for constructing a VR-based machine maintenance training environment, which has the following features:

- The users can construct a training environment without programming effort,
- The users can feed the necessary information through Graphical User Interface (GUI),
- The users can set the state transition of objects in the virtual environment by visual construction of Petri net model (Peterson, 1981),
- The constructed environment can be easily changed and reused, and
- The users can execute the training simulation by the same system.

There have been several systems developed for constructing the virtual environment under GUI environment (Fujii *et al.*, 1996), but those existing systems are designed only for constructing simple virtual environments in which objects cannot be manipulated just like in real world. No support system has been yet developed for constructing complicated virtual environments for training purpose, where various kinds of machines and equipments can be freely assembled and/or disassembled with input devices.

In this paper, the basic idea for supporting the construction is described in section 2, the system configuration is illustrated in section 3, and then the results to validate the system in section 4.

2. CONSTRUCTION OF VR-BASED TRAINING ENVIRONMENT

2.1. *Basic concept*

Firstly, it is necessary to define the term "object" for this training system in virtual environment;

"object" includes various parts of the machines to be assembled or disassembled, tools to be used for assembly/disassembly work, and both hands.

To make it possible for trainees to manipulate object models just like in real world, the interaction between objects and physical laws must be simulated; for example, if a pen is grasped by a hand, the pen must attach to the hand. It is possible to detect the contact of each object with others by using the surface model information. But it is very difficult to judge whether or not each object would contact with each other and then if contacted, to describe the movement of contacted objects. It takes a lot of computation time to do so even by high performance computer.

So in this study, the concept "state of objects" is introduced to virtual environment and the information on how an object in virtual environment moves in accordance to the trainee gestures is prepared. To put it concretely, the users prepare "state transition database" and "motion database". The state transition database gives information on what kind of states the object can take and how the states of objects change in virtual environment. And the motion database gives the information on how the object moves at each state.

In the training system, the movement of each object is decided based on the motion database according to the present state. And if an event (e. g. a collision of objects) happens, all the states of the concerned objects are changed based on the state transition database. For example in case of an action "grasp a nut on a table", the relationship between the state and its movement is described as shown in Fig. 1. In this system, a user interface is constructed by which the users can construct these two databases efficiently under GUI environment.

Concerning the state transition database especially, a new method has been developed to model the state transition of objects in virtual environment, by which the users can input the state transition by constructing Petri net visually. The detail of how to model the state transition of objects by Petri net was published in the authors' preceding paper (Yoshikawa *et al.*, 1997). And concerning the motion database, it is possible for users to make the motion database by selecting items and setting the related numerical values. To make it possible to construct various training environments, many kinds of variable items must be prepared. In this system, about 60 parameters can be set: on the surface of objects, on the movement of objects, and so on. The detail of these parameters will be explained in the next subsection.

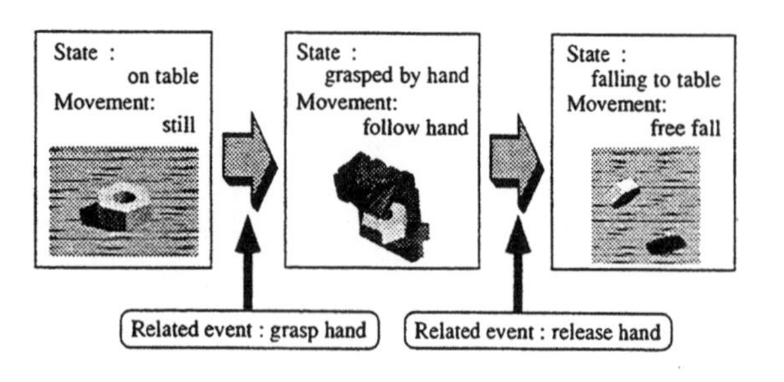

Fig. 1. Relationship between states and its movement.

But if the number of parameters to be set is so large, the user's work for setting those parameters becomes so time consuming that the efficiency of the system would not be expected. To cope with it, the system is equipped with "object template function" by which frequently used parameter set can be registered when it first appears and re-used repeatedly afterwards under GUI environment.

2.2. *Variable parameters*

The major parameters, which can be set in the developed system, are summarized in Table 1, with respect to types of the information, parameter numbers and the examples. By setting those parameters, various objects such as "open/close hand" , "falling object" and "flying balloon" can be rightly represented in virtual environment, by the way as will be explained below.

- Surface model of object
  In some cases, it is necessary for describing how an object will change its shape according to the states. For example, the shape of 'open hand' and that of 'grasping something' must be different. In the system, therefore, the file name of object's shape and texture must be set for each state.
- Initial position and direction after object changes its state
  In some cases, it is necessary for describing how an object will change its position and direction according to the states. For example, if a trainee grasps a pen, the position of the pen must change with the center of the hand. In the system, therefore, an initial position can be set for each state.
- Movement depends on time
  To simulate the movement of a motor, "free fall", and so on, it is necessary for the system to support to describe "movement depends on time". In the system, the movement of an object can be represented by time function and the user can select type of preset time function and set its coefficients.
- Limit of movement
  To simulate the interference of objects, it is necessary to limit the movement of objects. For example, in real world, an object can not

Table 1 Variable parameters

Information item	Number of parameters	Example
Surface model of objects	4	Texture (RGB format), Shape (DXF format)
Initial position and direction	7	Position (x, y, z), Direction (yaw, pitch, roll)
Movement depends on time	12	Sin., Cos. and linear function can be set.
Limit of movement	18	The limit of position and direction can be set.
Movement depends on gesture	12	Attach to the hand

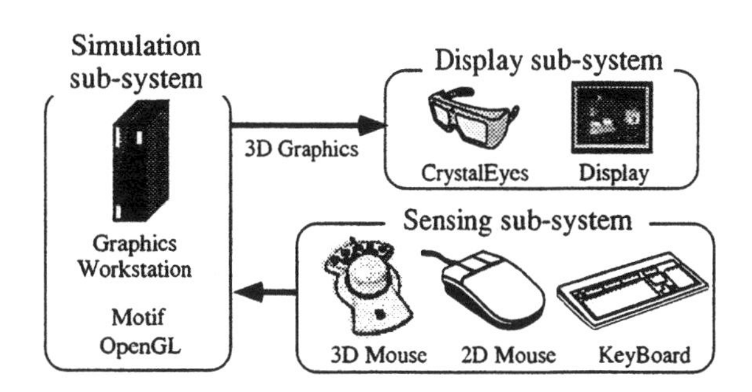

Fig. 2. System configuration.

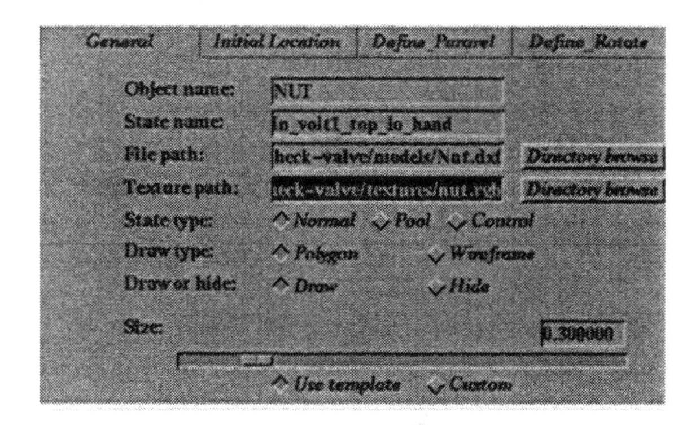

Fig. 3. An example of GUI for setting parameters for states.

penetrate into the other object. But to simulate this, the method of detecting the contact of objects by exact calculation is inappropriate because of large amount of computation time. In the system, the limit of movement can be set by numerical values.

- Movement according to trainee's gesture
  To make it possible for a trainee to manipulate objects with his hand, it is necessary for objects to move according to trainee's gesture. If trainee grasp a pen, for example, the pen must move by the same way as the trainee's hand. In the system, the relation between the movement of an object and trainee's gesture motion can be set freely by setting over 10 parameters.

Besides these parameters, some useful parameters can be set in the system; the sound can be set which is played when any event has occurred. Using this parameter, the collision of objects can be presented with real sound. Moreover, a simple shadow of an object can be drawn to make it easy for the trainee to manipulate objects in 3D world.

3. SYSTEM CONFIGURATION

3.1. *System configuration*

As shown in Fig. 2 of system configuration, the system consists of, (i) simulation sub-system, (ii) display sub-system, and (iii) sensing sub-system. The simulation sub-system corresponds to the graphic workstation, the display sub-system includes CrystalEyes and a display, and the sensing sub-system corresponds to a keyboard, a 2D and a 3D mouse. We have adopted the OpenGL library for rendering the 3D images, the Motif library for constructing GUI environment and the SGL library for stereo viewing with CrystalEyes.

For feeding the necessary information, the user manipulates the 2D and 3D mouse, with viewing 3D images on the display. The 2D mouse is mainly used for selecting information and inputting object's name, and the 3D mouse is used for pointing the 3D position in virtual environment.

For training, the trainee manipulates a pack and buttons of the 3D mouse with stereo viewing by CrystalEyes. Through the 3D mouse, the trainee can choose gestures such as grasp a hand, release a hand, drop objects, and so on.

3.2. *The interface of developed system*

Examples of the interface of the developed system are shown in Figs. 3 and 4, respectively, for setting parameters and constructing Petri net. The users feed the necessary information for constructing training environments through selecting toggle buttons and setting numerical values. And in this system, the users can input the state transition of objects in virtual environment by constructing Petri net visually. Besides, the buttons of the system is designed as shown in Fig. 5, in order that the users can easily imagine the function at first sight of the buttons.

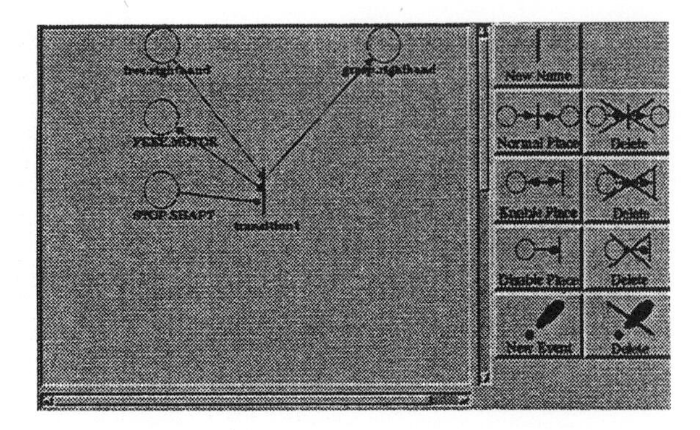

Fig. 4. An example of GUI for constructing Petri net.

Fig. 5. An example of icon buttons.

3.3. *Procedure of constructing the training environment*

The construction of a training environment is made by the following steps:

1. Preparation of materials for training,
   The materials necessary for training are the 3D surface model of objects, texture and sound. Those materials are prepared by using an appropriate application software such as CAD(Computer Aided Design).
2. Preparation of "states" of objects,
   Secondly, the possible states the objects will take and the motion database must be provided.
3. Construction of Petri net,
   By constructing Petri net visually, the state transition database is created.
4. Setting of initial state of objects,
   By marking tokens in Petri net, the initial state of objects is set.
5. Training execution.

4. SYSTEM VALIDATION EXPERIMENTS

4.1. *Construction of complicated training environment*

The workload for constructing a training environment was experimented by the developed system, for assembling/disassembling a check valve, as an example of constructing a complicated training environment. Then it was compared with that of the authors' previous study (Yoshikawa *et al.*, 1997).

4.1.1. *The target machine.* The structure of a check valve is shown in Fig. 6. For simplification, the maintenance place is limited to the lid of the valve, and the number of volts and nuts

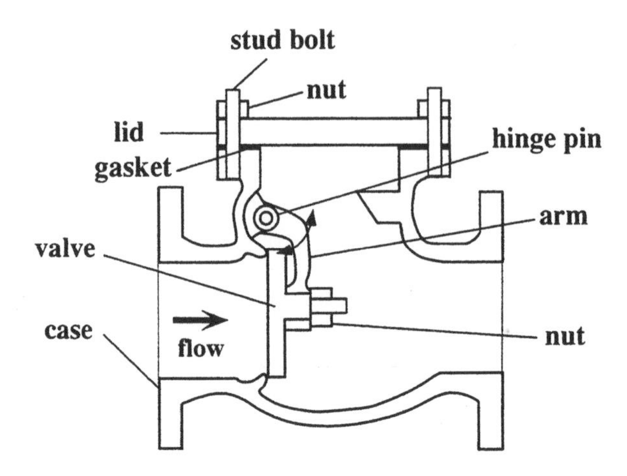

Fig. 6. Structure of check valve.

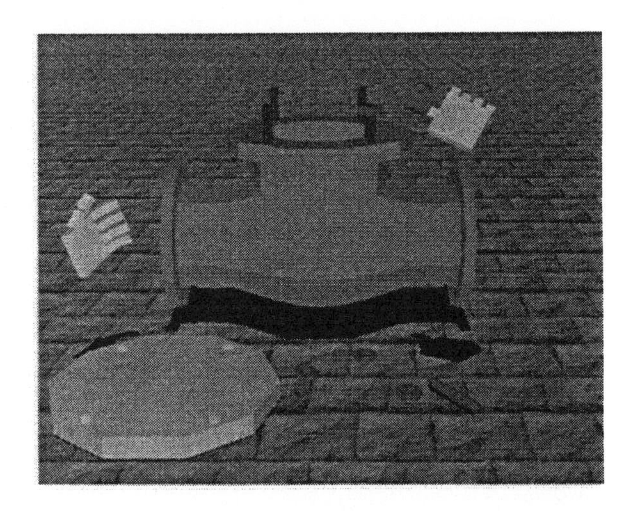

Fig. 7. An example scene of maintenance training of check valve.

is decreased from 16 to 4. The developed training environment consists of 14 objects: pen, nut, volt, spanner and so on. In this training environment, trainees can take various actions such as loose nuts with a spanner, check a mark on a valve with a pen, get off a lid of a valve, and so on. And the training environment is constructed by the same way, or as real as possible; for example, before a nut is loosen by a spanner, the trainee cannot loose the nut with his hand. To complete maintenance of the check valve, about 100 actions are needed. These conditions are almost the same as the previous study except for utilizing textures and sounds by the present support system.

4.1.2. *Results.* An example scene of the developed training environment is as shown in Fig. 7. In this experiment, it costs about 12 hours for a skilled user to construct this training environment. This is the time except for constructing 3D surface models, but compared with the previous study construction of almost same training environment (it costs about 4 months by two programmers), the efficiency of the workload is remarkably improved (see Table 2).

Table 2 Comparison of construction time

	previous work	with support
persons	2 (programmer)	1 (system user)
working time	4 months	12 hours

4.1.3. *Discussion.* The reasons for the remarkable improvement of efficiency by the developed system are as follows:

- In the previous work, it was necessary to invent the algorithm for various simulations, since the construction of training environments was made by coding programs. But using the developed system, the user did not need such kind of troublesome programming work.
- A large number of errors will always occur during the course of constructing complicated training environments. In coding programs, debugging is very difficult and time consuming. But using the developed system, errors can be corrected easily, because of the simple construction procedure.

4.2. *Construction capability by novice users*

To verify that even novice users can construct a training environment using the developed system, two students were asked to construct a simple training environment.

4.2.1. *Subject.* The number of subjects was two (subject A, B). Both of them had no experience of constructing virtual environments with coding programs. They were accustomed to using the 2D mouse and the keyboard, so they could input necessary information very easily but they used the developed system for the first time.

4.2.2. *Methods.* The surface models of the target machine and textures were prepared beforehand with other application. Prior to the experiment, guidance course of 30 minutes was allocated to the subjects, to explain how to use the developed system and show them the procedure to construct an example environment, in which a trainee can grasp a pen and release a pen with his right hand.

The time necessary for constructing the indicated environment was measured and the results of the construction were also recorded. Moreover, after the experiment, questionnaires were given to the subjects about the developed system.

4.2.3. *Target environment.* The target environment is such that it is composed of a motor, a rotating shaft and a fan, and a motor can start and stop its rotation by touching a switch. The number of prepared surface models is 4; motor, shaft, switch and base of switch. Necessary information for setting the construction of rotating a shaft are shown in Table 3 and Fig. 8. It was predicted that the operation of constructing Petri net would be easy, but that feeding the information for the movement of a shaft would be difficult.

Table 3 State information of rotating shaft

parameter	value
state name	rotate
surface model	shaft.dxf
texture	shaft.rgb
state type	normal
draw type	polygon
draw size	1.0
initial position	none
translation	disable
rotation	able
axis of rotation	local z-axis
mouse operation	disable

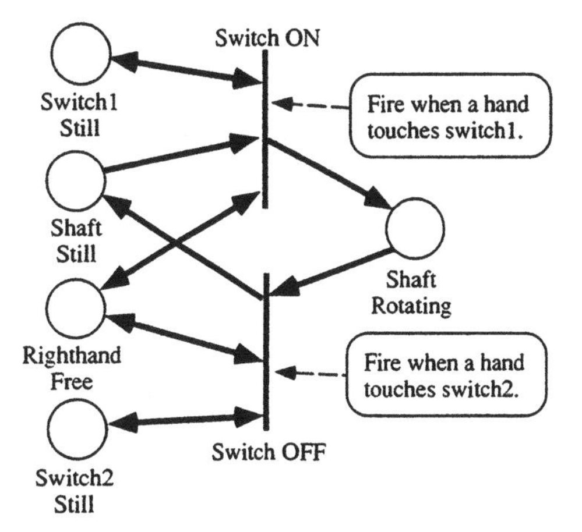

Fig. 8. Example of Petri net (state transition of rotating shaft).

4.2.4. *Results.* Both of the subjects could construct the objective training environment correctly. An example of the constructed training environment is shown in Fig. 9. The working time for the construction were 41 minutes(subject A) and 38 minutes(subject B). They could feed the necessary information for setting shaft movement, which was assumed difficult to feed. Subject B confused how to construct Petri net, but after trial and error, he could construct Petri net correctly. The results of the questionnaires from both subject are summarized bellow:

- At first I could not understand how each button works, but using the system for a few hours, I could understand how to use these buttons.

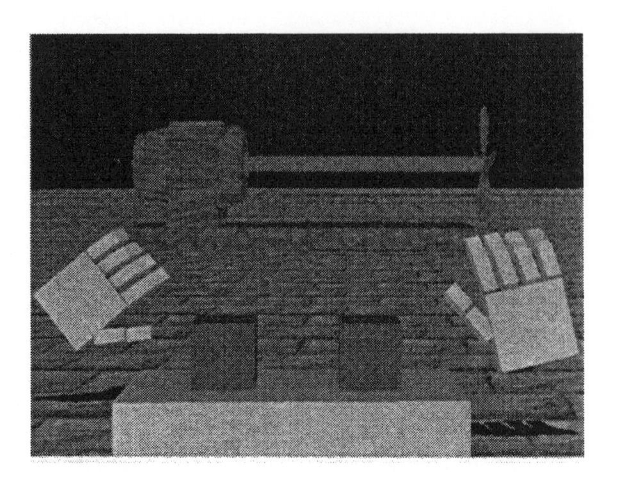

Fig. 9. A scene of training motor operation.

- With more familiarization with the system, I felt the motivation that I would like to construct more complex virtual environments.
- The positioning work of setting objects in 3D world was very difficult.
- If I would use the system for a longer time, I will be able to construct virtual environments more freely.

4.2.5. *Discussion.* It can be said from the above results that using the developed system, even novice users can construct VR-based training environments through trial and error if the environment is simple. Both of subjects have no experience of constructing virtual environments by coding programs, so if they try to construct virtual environments from the beginning, it will require much more time and larger workload because they need to study the basic knowledge of programming technique.

On the other hand, it was pointed out from the questionnaires that the interface for positioning the objects in 3D world should be improved. Since novice users who are not familiar with a 3D mouse are difficult to place objects correctly in VR space, alternative interface will be required for them, which will use a 2D mouse under GUI environment.

5. CONCLUSION

In this study, a support system was developed for constructing VR-based training environments without coding. The effectiveness of the system was confirmed through the validation experiments with respect to the reduction of workload and working hours and the familiarity with even novice users.

The developed system can be applied not only for training environments but also for generating other interactive virtual environments such as virtual show room, game, and so on. But in conjunction with the application of this system for actual training of maintenance personnel in the nuclear power plant, the following additional functions should be required:

- The function that offers alarms or advice when the trainee executes an improper task during "self-learning" course, and
- The function that offers, in virtual environment, the automatic performance of tasks to be executed next, for "demonstrative" teaching.

There is also a problem in the present system, that the more complex the machine for training becomes, the larger Petri net must be constructed. If Petri net becomes too large, it becomes difficult to understand the structure of the net and to update it. To deal with this problem, the application of colored Petri net is in consideration for future works.

6. REFERENCES

Arai, K., K. Abe and N. Kamizi (1997). Development of a simulator using virtual reality technology. *The Transaction of the Virtual Reality Society of Japan* **2**(4), 7–16.

Fujii, T., T. Yasuda, S. Yokoi and J. Toriwaki (1996). Visual simulation system for city planning with realtime interactive processing. *The Transaction of The Institute of Electrical Engineers of Japan* **116-C**(1), 36–42.

Loftin, R.B. and P.J. Kenny (1995). Training the Hubble space telescope flight team. *IEEE, Computer Graphics and Applications* **15**(5), 21–37.

Miwa, S., T. Ueda and S. Nishida (1995). A learing environment for maintenance of plants and equipments based on virtual reality. *The Transaction of The Institute of Electrical Engineers of Japan* **115-C**(2), 203–211.

Peterson, J.L. (1981). *Petri net theory and the modelling of systems.* Prentice-Hall,Inc.

Yoshikawa, H., T. Tezuka, K. Kashiwa and H. Ishii (1997). Simulation of machine-maintenance training in virtual environment. *Journal of the Atomic Energy Society of Japan* **39**(12), 72–83.

EVALUATION OF EMOTION BY USING PERIPHERAL SKIN TEMPERATURE: THE EFFECTIVENESS OF MEASUREMENT ON THE BACK OF FINGER

Tomio WATANABE, Masashi OKUBO, and Tsutomu KURODA

Faculty of Computer Science and System Engineering,
Okayama Prefectural University
111 Kuboki, Soja, Okayama, JAPAN 719-1197

Abstract: From the measurement of peripheral skin temperatures on the nose, temporalis, back of finger, and ankle simultaneously by thermistors and a thermography on the condition that peripheral blood vessels contract in smoking, it is demonstrated that the back of finger is most useful for the measurement area of skin temperature. The measurement of skin temperature on the back of finger in watching a scare video that is seized with fear is made for confirmation of the effectiveness, which is demonstrated from the result that the skin temperature on the back of finger for the scare dropped statistically in 18 subjects in comparison with that in watching a scenery video. *Copyright©1998 IFAC*

Keywords: Human Interface, Emotion, Skin Temperature, Physiological Measurement

1. INTRODUCTION

The purpose of this study is to develop a method to evaluate emotion and mental workload quantitatively by using physiological indices such as skin surface temperature closely related with the autonomic nervous system from the standpoint of the design and evaluation of human interface. We already reported that the facial skin temperatures decreased corresponding to the increase of the mental workload in which the memory task of number was imposed by displaying the different number of digits (Watanabe, et al., 1996a). The drop in the facial skin temperature as an index of emotional stress has been used for infants' emotional behavioral research (Mizukami, et. al.,1987,1990). They demonstrated that changes in the facial skin temperature of infants who were separated from their mothers were similar to those observed in adults during stress. We also proposed the measurement of cervical skin temperatures from the side view instead of the frontal view, because the deep breath disturbs the thermal emission of facial skin in the measurement from the frontal view by a thermography (Watanabe, et al., 1996b). The measurement of skin temperature in these analyzing areas, however, is not enough accurate for evaluating emotion from the stage of practical application.

The evaluation of emotion using skin temperature is based on the effects of the autonomic nervous activity on the blood circulation caused by the vessel blood contraction. The effects are remarkable in the periphery. In this present study, the peripheral skin temperatures on the nose, temporalis, back of finger, and ankle are measured simultaneously by some thermistors and a thermography on the condition that peripheral blood vessels are forced to contract in smoking, and it is demonstrated that the back of finger is most useful for the measurement area of skin temperature. The effectiveness of skin temperature measurement on the back of finger for the evaluation of emotion is made for confirmation on the basis of the physiological indices of skin temperatures, heart rate variability and respiration measured simultaneously during watching a scare video in which the emotion of a subject seized with fear is

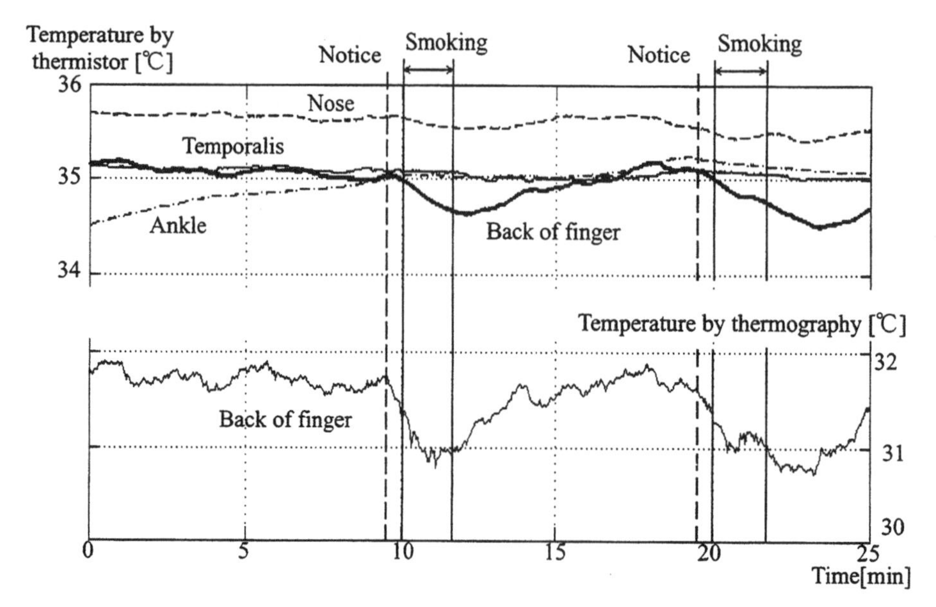

Fig.1. Representative time changes in skin temperature by thermistor and thermography in smoking.

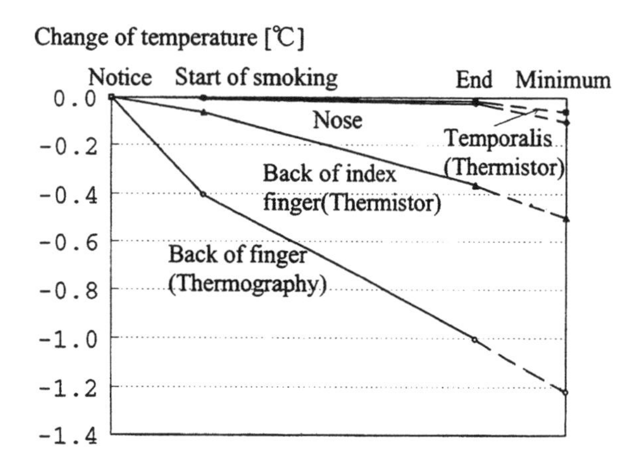

Fig.2. Changes of skin temperature on the basis of the advance notice of twice smoking in 4 subjects.

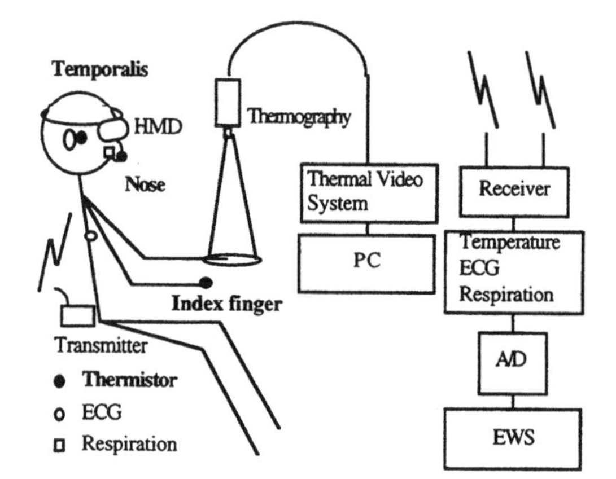

Fig.3. Setup for the experiment.

expected to be remarkably changed.

2. MEASUREMENT OF PERIPHERAL SKIN TEMPERATURE

On the condition that the vessel blood contraction in the periphery is caused by smoking, peripheral skin temperatures on the nose, temporalis, back of left index and little fingers, and ankle were measured simultaneously recorded by a DAT recorder (TEAC RD-130TE) through thermistors of a mutitelemeter system (NIHON KODEN WE-5000), and then digitized by a 12 bit analog-to-digital converter with 8 channels at a sampling rate of 1 kHz to store a computer. The thermal images of the back of left fingers were also simultaneously measured by a thermography system (AVIO TVS-8000), and then stored at a sampling period of 3 seconds in the built-in optical disk. The nail radix was selected as the analyzing area in the back of finger for the thermography because the temperature was noticeable in both height and change. The maximum temperature of 5x5 pixels in the analyzing area was measured, and the average of the values in the thumb, middle and third fingers was calculated. The subjects consist of 4 male student smokers. The subject was made a notice 30 seconds in advance of the start of smoking, and smoked two times at 10-minute intervals. Figure 1 shows the representative result. Figure 2 shows the changes of skin temperature on the basis of the advance notice of twice smoking in 4 subjects. Because the temperature change in the back of fingers is remarkably greater than that in the nose which has been selected as the analyzing area of facial skin temperature for the evaluation of mental workload (Genno, et al., 1995; Ishikawa, et al., 1996), the effec-

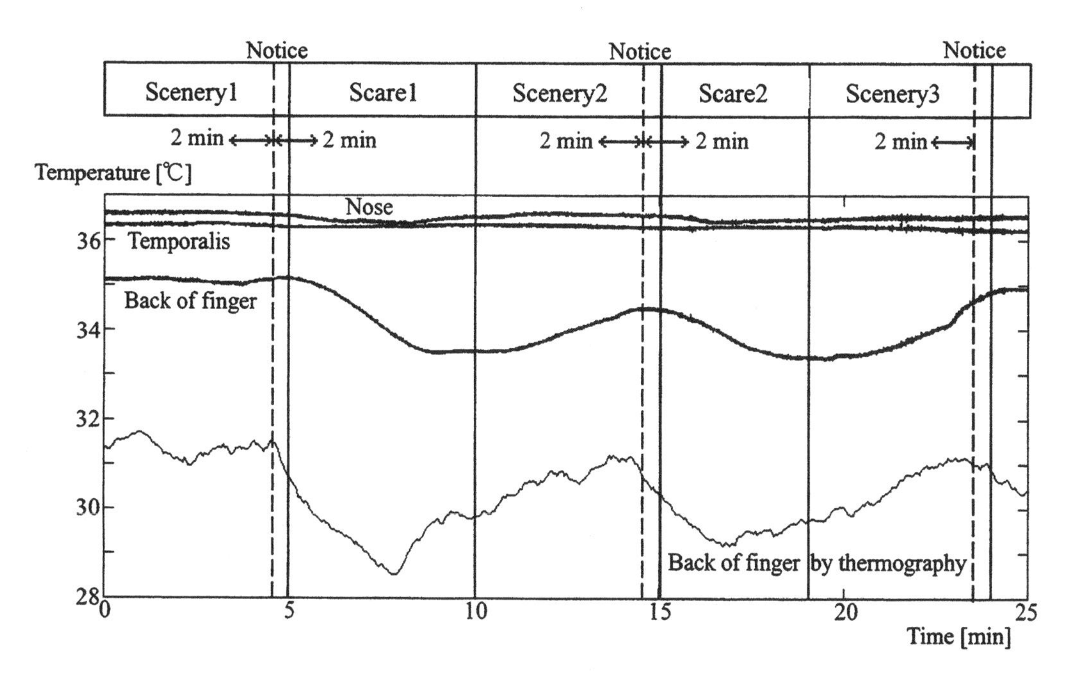

Fig.4. Representative time changes in skin temperature in watching the scenery and scare videos.

tiveness of the back of finger measurement is demonstrated. We can also find that the measurement by thermography is superior in dynamic sensitivity to that by thermistors, and that the skin temperatures on the back of fingers decrease from the advance notice of smoking. This drop in 30 seconds is estimated to be the change of emotion.

3. EVALUATION OF EMOTION IN WATCHING VIDEO USING HMD

Fear was taken note of as a fundamental emotion, because the emotion of fear was assumed to be evaluated by measuring the drop of skin temperature caused by the blood vessel contraction for fear. In concrete, watching a scare video was selected to be studied using a head mounted display (HMD; SONY Glasstron) which is effective in the change of emotion. The video was composed of the scenery video in which a dolphin is swimming in the calm background music and the scare video in which spiritual phenomena are introduced in the horrible background music alternately, inserted with the advance notice 30 seconds in advance of the scare video, in totally 29 minutes as follows; the scenery in 9.5 min, the advance notice of fear in 0.5 min, the scare in 5 min, the scenery in 4.5 min, the notice in 0.5 min, the scare in 4 min, the scenery 4.5 min and the only notice in 0.5 min. Eighteen subjects (9 males, 9 females) took part in the study. The subject's skin temperatures, heart beat-to-beat intervals, and respiration were measured simultaneously during his watching the video as shown in figure 3. The skin temperatures were measured in the nose, temporalis and the back of right index finger by the thermistors, and the back of left five fingers by the thermography. The heart beat-to-beat interval of cardiac cycle, or R-R interval (RRI), was measured with an accuracy of 1 ms by detecting the peaks of ECG. The time sequence of RRI was calculated in each 10 ms by using the cubic spline interpolation. Respiration data were measured on the average in each 10 ms by the thermistor picking-up in the nose. The number of respiration was calculated on the basis of the increase in exhalation and the decrease in inhalation.

Figure 4 shows an representative example of skin temperature measurements. The drop of temperature on the back of finger for the emotion of fear is remarkable in the same manner as that for the smoking. This tendency was shown in 12 (6 males, 6 females) out of 18 subjects. Figure 5 shows the mean and standard deviation in 18 subjects of the average drop of temperature on five back of fingers by thermography in 2 minutes before or after the advance notice of the scare on referring to figure 4. The drops for the fear were significantly greater than those for the scenery at the significant level of 5 % of t-test in every next pairs between them. These results demonstrate the effectiveness of skin temperature measurements on the back of finger.

The data of RRI were analyzable for 15 out of 18 subjects. Figure 6 shows an example of time series of RRI. There were no significant differences in the variability of RRI between 70 beats before and after the advance notice of the scare on referring to figure 6, though it is well known that the variability of RRI decreases in the state of attention and increases in the state of relaxation. Focusing on the decline of RRI time series after the advance notice of the scare, we calculated the angle of inclination approximated with a linear line using the least squares method. Figure 7 shows the result. The decline of RRI for the scare was greater than that for the scenery at the significant level of 5 %.

Figure 8 shows the mean and standard deviation in the measured 6 subjects of the number of respiration in the same analyzing time as the measurement of skin temperature, i.e. 2 minutes before or after the advance notice of the scare. The number of respiration for the scare is greater than that for the scenery. These results suggest that respiration so strongly influenced heart rate that the variability of RRI did not decrease in spite of the high stress for the scare (Brown, et al.,1993). We need to evaluate emotion on the basis of the relationships among skin temperature, respiration and RRI.

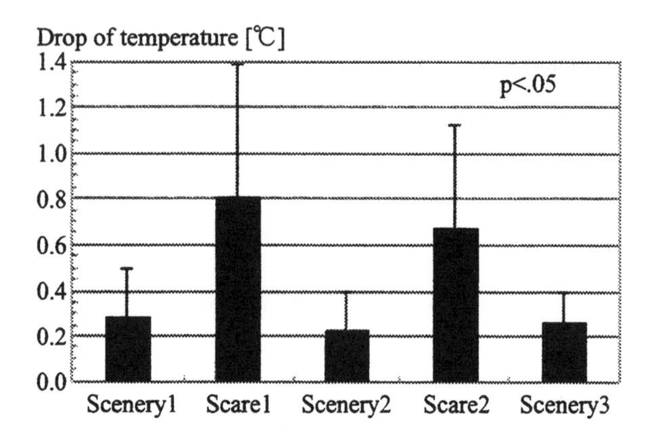

Fig.5. Mean and standard deviation in 18 subjects of the average drop of temperature on five back of fingers by thermography in 2 minutes before or after the advance notice of the scare on referring to figure 4.

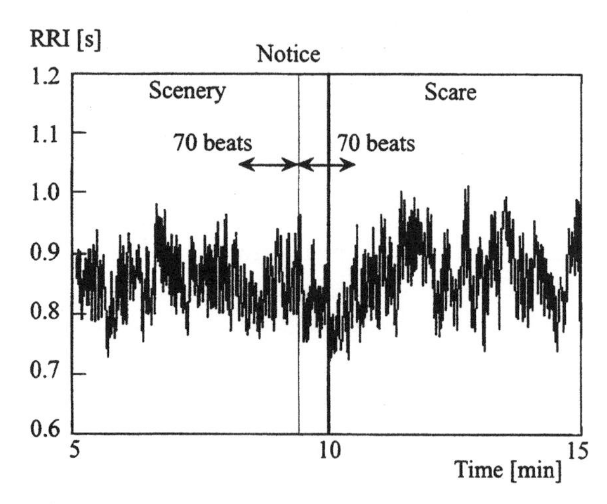

Fig.6. RRI before and after the advance notice of the scare.

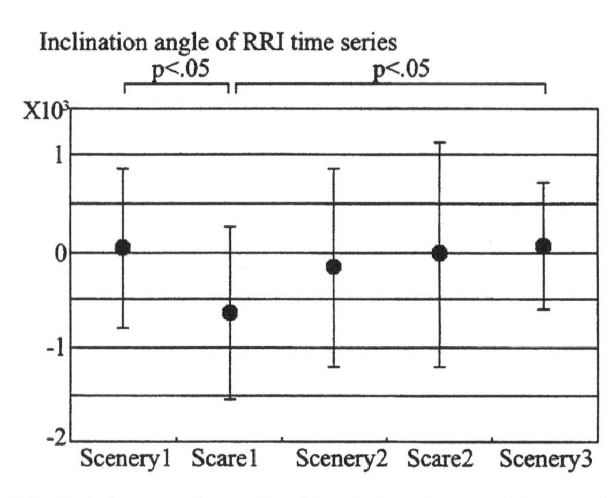

Fig.7. Mean and standard deviation in 15 subjects of the inclination angle of RRI time series in 70 beats before or after the advance notice of the scare on referring to figure 6.

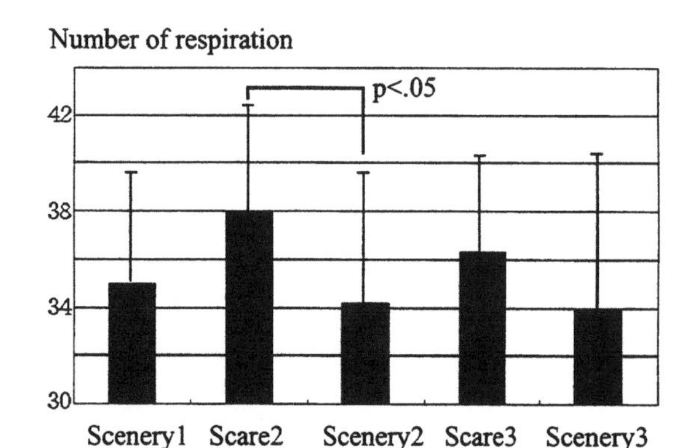

Fig.8. Mean and standard deviation in the measured 6 subjects of the number of respiration.

4. CONCLUSIONS

The effectiveness of measurement on the back of finger was demonstrated from the various peripheral measurements of skin surface temperature simultaneously by some thermistors and a thermography on the condition that the contraction of blood vessels was caused by smoking. It was also found that blood vessels contract to prepare for smoking because the skin temperature on the back of finger measured using a thermography dropped remarkably from the advance notice of smoking. The effectiveness of this measurement was confirmed from the evaluation of emotion in watching the scare and scenery video alternately where the drop of skin temperature for the scare was significantly greater than that for the scenery. The decline of heart beat-to-beat interval and the number of respiration for the scare were also greater than those for the scenery. These suggest the importance of the evaluation of emotion based on the relationships among physiological indices.

REFERENCES

Brown, T.E., L.A. Beightol, J. Koh and D.L. Eckberg (1993). Important influence of respiration on Human R-R interval power spectra is largely ignore. Journal of Applied Physiology, **75**, pp.2310-2317.

Gennno, H., K. Matumoto and K. Fukushima (1995). Evaluation of thermal sensation by using facial skin temperature. Trans. of the Society of Instrument and Control Engineers, **31**, pp.973-981.

Ishikawa, K., H. Genno, M. Osuga, T. Kurihara, Y. Nishio and M. Suzuki (1996). Evaluation of mental stress during a monotonous task using facial skin temperature. Proc. of the **12**th Symposium on Human Interface, pp.349-352.

Mizukami, K., N. Kobayashi, H. Iwata and T. Ishii (1987). Telethermography in infants' emotional behavioral research. The Lancet, **8549(ii)**, pp.38-39.

Mizukami, K., N. Kobayashi, T. Ishii and H. Iwata (1990). First selective attachment begins in early infancy: A study using telethermography. Infant Behavior and Development, **13**, pp.257-271.

Watanabe, T., M. Okubo and T. Kuroda (1996a). Evaluation of mental workload using facial skin temperature and heart rate variability. Proc. of the Japan. Society of Mechanical Engineers (JSME), **96-15**, pp.368-369.

Watanabe, T., T. Kuroda, M. Okubo, M. Hirose and Y. Nakagaki (1996b). Evaluation of mental workload using cervical skin temperature for evaluation of virtural environments. Proc. of the JSME, **96-45**, pp.177-178.

BASIC EXPERIMENTS FOR EVALUATING PECULIAR OCULAR CHARACTERISTICS CAUSED BY ARTIFICIAL BINOCULAR VISION

S. FUKUSHIMA*, D. MORIKAWA, H. FUJIYAMA*** and H. YOSHIKAWA*****

**Advanced Technology Research Laboratory, Matsushita Electric Works, Oaza Kadoma 1048, Kadoma-shi, Osaka, JAPAN*

***Graduate School of Engineering, Kyoto University, Yoshida Honmachi, Kyoto-shi Sakyou-ku, Kyoto, JAPAN*

****Graduate School of Energy Science, Kyoto University, Uji-shi Gokasho, Kyoto, JAPAN*

Abstract. By using Eye-Sensing Head Mounted Display (ES-HMD), basic laboratory experiments were conducted to evaluate convergent eye movement and pupil size under some specific visual conditions induced by artificial binocular vision; one experimental condition was that asymmetric luminous intensity for both eyes was illuminated. The result shows that each pupil size of both eyes in asymmetric condition is different from in symmetric condition, even if each eye is illuminated by the same luminous intensity for the both conditions. Since the experimental condition would never usually happen in real world, longtime exposure to light in asymmetric luminous condition may lead to visual fatigue. The other experimental condition was binocular vision generated by binocular parallax. It was found that (i) some subjects, who are called vergence normal, can adjust convergence according to the moving target with binocular parallax, but the other called vergence anomaly cannot, and (ii) pupil size in a vergence normal changes with the vergent eye movement induced by binocular parallax. These results show that a vergence anomaly, who is also called stereoanomaly, can be easily detected by using simple test image and observing vergent eye movement or pupil size. The obtained relation between pupil size and binocular parallax is also a new experimental proof of influence of vergent eye movement to the change of pupil size. *Copyright © 1998 IFAC*

Key Words. Visual environment; binocular vision; head mounted display; pupil; convergence

1. INTRODUCTION

It was recently reported in Japan that many children who had been watching a popular TV animation program suddenly complained of severe physical condition and had been hospitalized by ambulance (Takahashi, 1997). In fact, they had sudden spasm induced by visual stimulus which was composed of fast sensible flicker images.

The problem above may also be the case with virtual reality (VR) application that such kind of peculiar visual environment would threaten physical condition. A head-mounted display (HMD) is used as audio and visual output device in many VR applications, which promotes a sensation of immersion in virtual environment. However, since an HMD usually has two displays for both eyes, new inexperienced visual environment such as asymmetric visual condition or binocular vision would cause a serious problem of human visual function and result in visual fatigue or feel sick. For example, a past research pointed out that artificial stereoscopic view by binocular parallax would cause the conflict between convergence and accommodation, which would never happen in real world, and this is one of the causes of visual fatigue in artificial stereoscopic view (Wann and Mon-Williams, 1997). For wider spread of VR technology as intuitive human interface, it is significant to evaluate the effect of new visual environment such as artificial binocular vision given by VR technology from the physiological view point.

In order to obtain ocular physiological measures such as eye movement or pupillary dynamics while in artificial binocular vision, a new HMD, "Eye-Sensing HMD (ES-HMD)" was developed as a VR-based human interface shown in Fig. 1 (Fukushima *et al.*, 1996). What the authors call ES-HMD is an extension of HMD device which is added "eye-sensing" function to the conventional HMD. The specific feature of this ES-HMD is that it is equipped with CCD cameras for the both eyes of a user while his wearing the HMD and the supporting hardware and software for realtime processing to the monitored images of both eyes from the two CCD cameras to obtain various ocular characteristics in situ, such as eye movement, the size of pupil and eye-blinking. Specifications of display function and eye-sensing function, which represents valid area for capturing pupil, resolution of obtained pupil size and position, are shown in Table 1.

ES-HMD has the potential of not only a new-type interface of direct input-output device but also a basic measurement device to diagnose personal visual function from psycho-physiological aspect. Concerning the former type application of ES-HMD, the authors have been conducting on a laboratory experiment to realize a mutual adaptive interface for CAI by utilizing ES-HMD's eye gaze input function (Fukushima and Yoshikawa, 1997).

In this paper, for the latter area of psycho-physiological measurement device, the authors conducted on laboratory experiments with the following two experimental conditions to examine pupillary dynamics and vergent eye movement in artificial binocular vision; (i) one is under asymmetric luminous intensity of displays for both eyes, and (ii) the other is under the artificial 3 dimensional visual environment generated by binocular parallax.

Since an HMD usually has two displays for both eyes in light-bound environment, it is possible that luminous intensity of the two displays can be controlled differently from each other, which would never happen in real world. In fact, this would take place because of poor adjustment of the both displays or natural difference between the both images themselves such as stereoscopic images. The past research shows by means of subjective evaluation that such asymmetric light condition causes discomfort (Beldie and Kost, 1991).

In this paper, first, the pupillary response under the asymmetric luminous condition for both eyes will be measured for the following Experiment 1, while for the Experiment 2, the behaviors of pupil size and vergent eye movement while gazing at a moving target with binocular parallax will be measured, in order to examine the relationship between such ocular characteristics and binocular parallax.

The pupil size of human eye would tend to constrict when human looks at the target which is near to his eyes, and this phenomenon is called near reflex of pupil (Osaka *et al.*, 1993). It has been generally said that three elements of convergence, accommodation and constriction of pupil give rise simultaneously in near reflex, but the behavior of pupil is not yet well clarified in the case of disparity between convergence and accommodation, which is a typical condition of stereoscopic vision. It is expected to be examined in the Experiment 2.

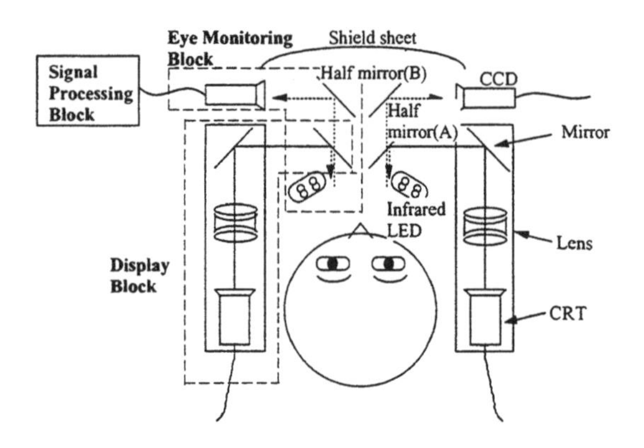

Fig. 1. Configuration of Eye-Sensing HMD.

Table 1 Specifications of Eye-Sensing HMD.

Display function	
Visual field size	48 deg (H) × 36 deg (V)
Vergence angle	0 deg. (fixed)
Focal length	1 m
Horizontal resolution	350 TVline
Eye-Sensing function	
Valid area	35.5 mm (H)× 26.5 mm (V)
Spatial resolution	0.0467 mm (H)× 0.109 mm (V)
Time resolution	16.7 msec

Since the situation of disparity between convergence and accommodation can be easily generated by ES-HMD with the accurate measurement on the change of eyeball rotation and pupil diameter, the use of ES-HMD for the study of eye movement and pupillary dynamics in both near reflex and vergent eye movement can offer new experimental environment, although it would be difficult to construct by using conventional apparatus and even if possible, experimental condition would be much limited.

2. METHOD

2.1. *Experiment 1*

Both eyes of a male subject TO wearing ES-HMD were exposed to uniform light generated by the displays embedded in the HMD. Intensity of illumination for each eye ranges from 0.74 [lux] to 8.26 [lux] by controlling luminous intensity of the both displays separately. He was placed in peculiar visual environment of asymmetric luminous intensity as shown in Fig. 2 for five minutes.

During a measurement, pupil diameter was measured and the subject is instructed to keep his eyes open as long as possible and fix his gaze point at a cross mark cursor at the center of the display in order to avoid near reflex of pupil caused by change of focal length.

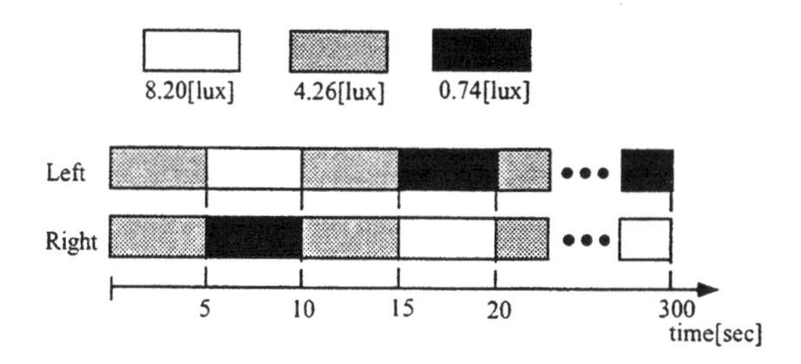

Fig. 2. Procedure of 5 minutes exposure to various luminous intensity for both eyes (Experiment 1).

Table 2 Luminous intensity of both displays in Experiment 1.

	Eye	Light intensity	
		1st half [lux]	last half [lux]
case 1	right	4.26	8.26
	left	4.26	8.26
case 2	right	4.26	0.74
	left	4.26	0.74
case 3	right	4.26	8.26
	left	4.26	0.74

Furthermore, short time light response of pupil of three subjects TO, TK and HF were also measured for ten seconds under the same visual condition as 5 minutes measurement mentioned above. As shown in Table 2, a measurement consists of the first half and the last half, in which each continues five seconds. At the transition from the first half to the last half, brightness of right and left displays changes simultaneously. Table 2 shows that case 3 is completely asymmetric luminous condition, which situation can rarely happen in real world, while the case 1 and 2 is similar condition as in real world.

All three subjects have normal phoria and no pupil anomaly. Measurement starts followed by 15 minutes' adaptation to darkness. Total time of measurement for one subject is within 15 minutes, which is too short to influence pupil size caused by mental load or arousal level.

2.2. *Experiment 2*

Another basic laboratory experiment was conducted on eight subjects (TO, YA, TK, SF, DM, HF, NI and TM, all male students who have normal visual function), to observe the changes of pupil size and eyeball rotation of their both eyes in case of watching a dynamically moving target in 3 dimensional space, which is also a typical visual situation in VR applications (Fujiyama *et al.*, 1997). The mechanism of binocular parallax is schematically illustrated in Fig. 3. The visual image used in the experiment includes a

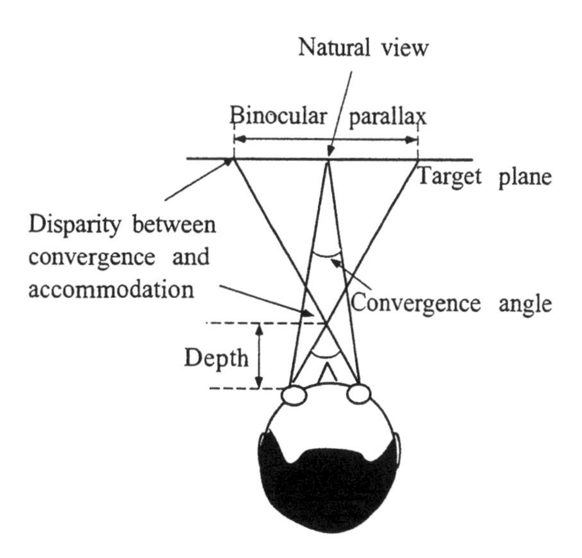

Fig. 3. Schematic of artificial binocular parallax.

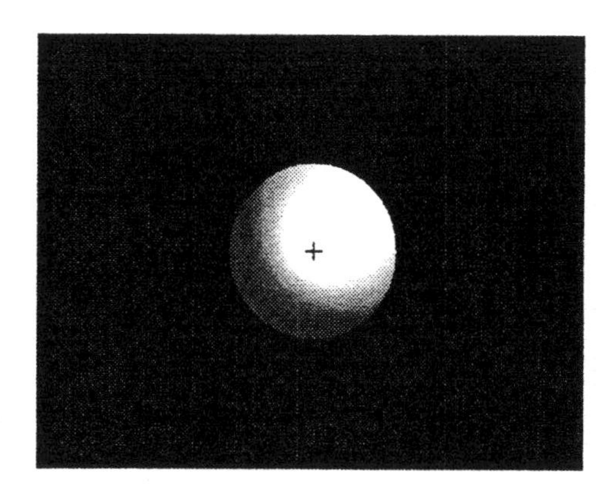

Fig. 4. Display image used in the Experiment 2.

monochrome shaded sphere object in empty black background as shown in Fig. 4. Black background can keep luminous distribution of the background constant and uniform even if eye movements occur. Two identical images of Fig. 4 with the controlled binocular disparity between the both were projected on the both eyes, and the subject was asked to try to fuse the both images into a single image by watching the cross mark (+) on the white ball. Tests were conducted to all subjects, with smooth change of the convergence angle from 0 to 16[*deg*] (this is controlled by binocular parallax) and then vise-versa, with cyclic periods between 2 and 16 seconds.

3. RESULTS

3.1. *Experiment 1*

After 5 minutes exposure to asymmetric luminous intensity condition, TO complained of visual fatigue. Pupillary response of TO to brightness change for ten seconds according to Table 2 are shown in Figs. 5, 6 and 7. All figures represent the pupil diameter of the left eye and those for

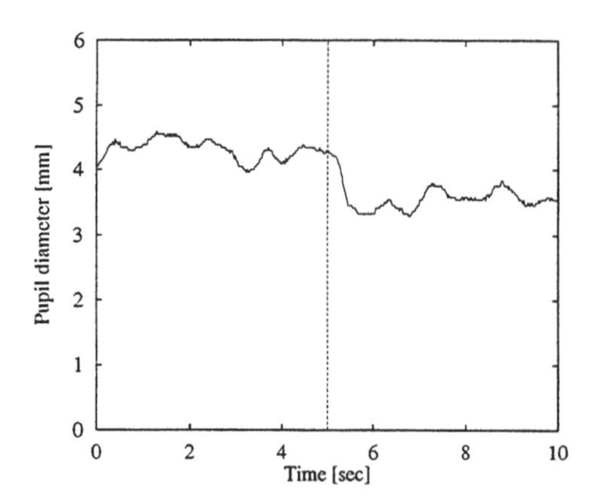

Fig. 5. Pupil diameter when luminous intensity of both displays changes from 4.26 [lux] to 8.26 [lux] symmetrically (case 1).

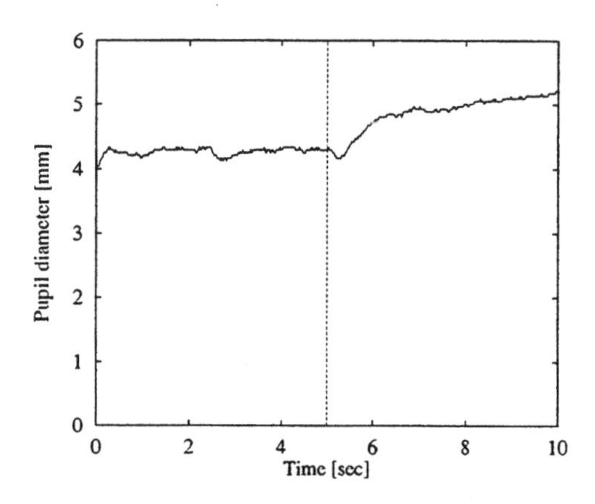

Fig. 6. Pupil diameter when luminous intensity of both displays changes from 4.26 [lux] to 0.74 [lux] symmetrically (case 2).

right eye are omitted because both pupils for a normal subject moves completely similarly.

As shown in Fig. 5, pupil fluctuates even at the constant luminous intensity in the first half five seconds and then constricts followed by the sudden increase of luminous intensity. In the same way, Fig. 6 shows that pupil dilates followed by the decrease of luminous intensity, however, whose time constant is longer than in constriction. On the other hand, Fig. 7 shows that pupil diameter has no tendency of step response as seen in Fig. 5 or Fig. 6 after the change of luminous intensity except instant small change which cannot be almost distinguished from normal fluctuation. This means imbalanced state of pupil size, because excessive light comes into the right eye, while the luminous intensity for the left eye is too small for the pupil size.

The asymmetric luminous condition can occur when using a head-mounted display which has two displays for each eye, or when presented 3 dimensional images which have luminous difference in the left and right images because of given binoc-

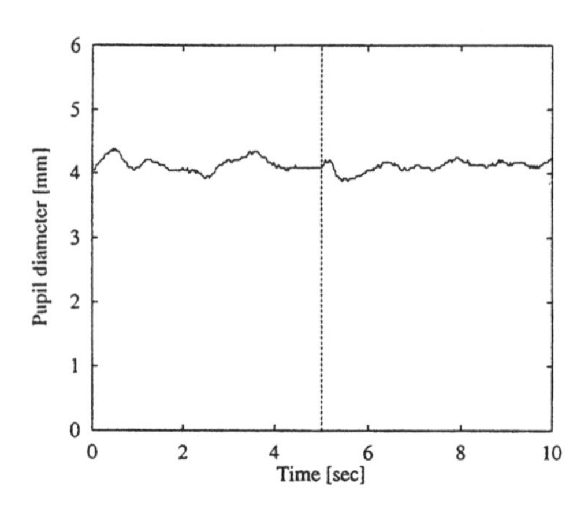

Fig. 7. Pupil diameter when luminous intensity of both displays changes asymmetrically (case 3).

ular disparity to each eye. The obtained results indicate that only five minutes exposure to such visual environment leads to visual fatigue and that one of the reasons of such fatigue may be caused by imbalanced state between pupil size and luminous intensity.

3.2. *Experiment 2*

Fig. 8 and Fig. 9 show both eyes movements of TO and NI, respectively, in the case of 8 [sec] of the period of the moving target. In these figures, 0 and 8 in horizontal axis indicates the time when no binocular parallax is given, while 4 and 12 correspond to the time when the maximum binocular parallax (about 16 [deg] in convergence angle) is given. Positive value in vertical axis indicates right direction of eye rotation, while negative for left direction.

Fig. 8 shows both eyes of TO, who is called vergence normal, can track the moving target and vergent eye movement appears. On the other hand, NI, who is called vergence anomaly and, furthermore, stereoanomaly, does not track the target by both eyes and conjugate eye movement, in which both eyes move in the same direction, appears instead of vergent eye movement. The fact agrees with his comment that double image of both eyes never fused.

Fig. 10 shows pupil size of TO and TK who is also vergence normal. In the same way, Fig. 11 shows pupil size of NI and HF who is also vergence anomaly. Horizontal axis of these two figures is completely same as in Fig. 8 or Fig. 9.

It is found from Fig. 10 that the pupil size of vergence normal would tend to change in accordance with the dynamic change of the moving target. The pupil constricts as binocular parallax becomes larger, while it dilates as binocular parallax becomes smaller. This phenomenon is simi-

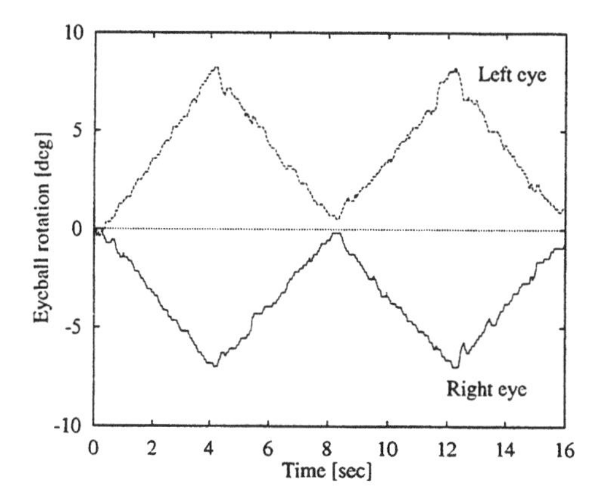

Fig. 8. Eyeball rotation of subject TO in the case of period 8 [sec].

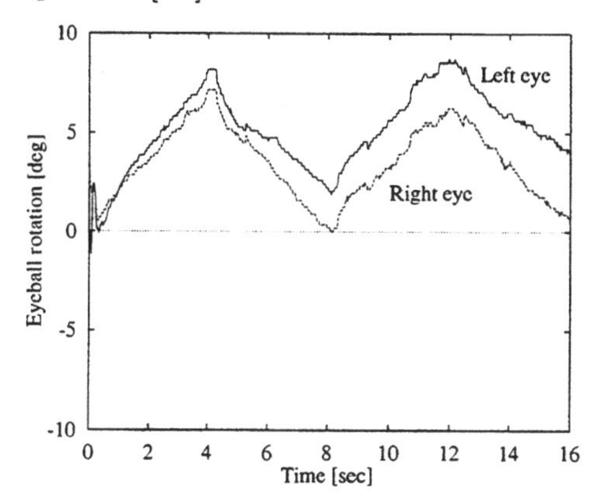

Fig. 9. Eyeball rotation of subject NI in the case of period 8 [sec].

lar to near reflex of pupil induced by vergent eye movement. On the other hand, such tendency of correlation of pupil with binocular parallax never appears in stereoanomaly. This is a new experimental proof of influence of vergent eye movement induced by binocular parallax to the change of pupil size.

From the result mentioned above, it is also found that a stereoanomaly subject could be easily diagnosed by using ES-HMD.

4. CONCLUSION

By using the developed ES-HMD, basic laboratory experiments were conducted to evaluate convergent eye movement and pupil size under some specific visual conditions induced by artificial binocular vision; one experimental condition was that asymmetric luminous intensity for both eyes was illuminated. The result shows that each pupil size of both eyes in asymmetric condition is different from in symmetric condition, even if each eye is illuminated by the same luminous intensity for the both conditions. Since the experimental condition would never usually happen in real world, longtime exposure to light in asymmetric luminous condition may lead to visual fatigue, but this remains to be evaluated for further conviction.

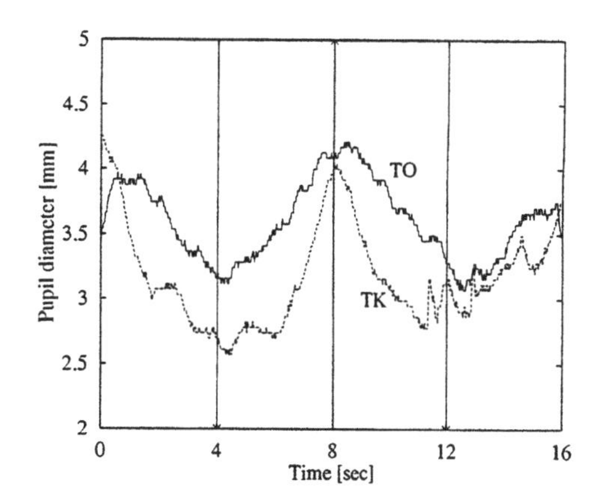

Fig. 10. Pupil diameter of TO and TK in the case of period 8 [sec].

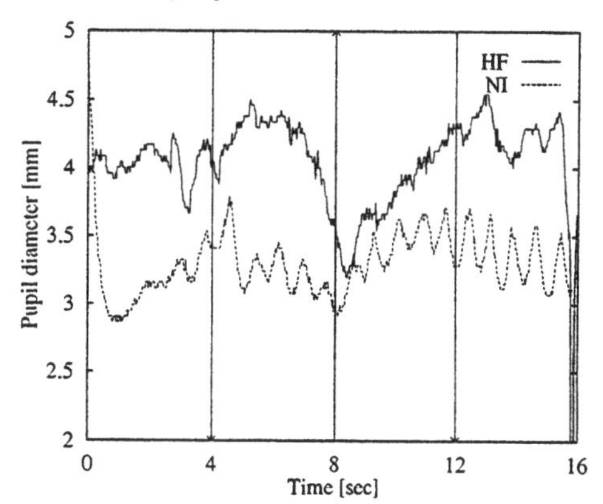

Fig. 11. Pupil diameter of HF and NI in the case of period 8 [sec].

The other experimental condition was binocular vision generated by binocular parallax. It was found that (i) some subjects can adjust convergence according to the moving target with binocular parallax, but a stereoanomaly cannot, and (ii) pupil size in a vergence normal changes with the vergent eye movement induced by binocular parallax. These results show that a stereoanomaly can be easily detected by using simple test image and observing vergent eye movement or pupil size. The obtained relation between pupil size and binocular parallax is also a new experimental proof of influence of vergent eye movement to the change of pupil size.

As the concluding remarks from the above results of basic experiments using ES-HMD, the authors found some peculiarities of human ocular characteristics caused by artificial binocular vision which would give rise adverse effects both mentally and physiologically, and therefore, have to be considered carefully for the various practical application of VR technology.

5. REFERENCES

Beldie, I. and B. Kost (1991). Luminance asymmetry in stereo TV images. *Proc. of SPIE* **1457**, 242–247.

Fujiyama, H., S. Fukushima and H. Yoshikawa (1997). An experimental analysis on gaze point history of stereoscopic vision by using Eye-Sensing HMD. *Proc. of 13th Symposium on Human Interface (in Japanese)* pp. 69–74.

Fukushima, S. and H. Yoshikawa (1997). Application of a newly developed Eye Sensing Head-Mounted-Display to a mutual adaptive CAI for plant-diagnosis. *Elsevier* **21B**, 225–228.

Fukushima, S., M. Takahashi and H. Yoshikawa (1996). A head mounted display with the function of measuring eye images. *Human Interface News and Reports (in Japanese)* **11**, 197–202.

Osaka, R., Y. Nakamizo and K. Koga (1993). *Experimental Psychology of Eye Movements (in Japanese).* Nagoya University Press.

Takahashi, T. (1997). For prevention Pocket Monster incident. *Asahi-Shinbun (in Japanese)* **29th Dec. morning**, 4.

Wann, J.P. and M. Mon-Williams (1997). Health issued with virtual reality displays: What we do know and what we don't. *Computer Graphics* **May**, 53–57.

COMPUTERISED PROCEDURE GENERATION

Professor Erik Hollnagel[1,2], Dr. Yuji Niwa[3], Mark Green[1]

(1) Institute for Energy Technology (IFE), Norway
firstname.lastname@hrp.no

(2) IKP/HMI, Linköping University, Sweden
eriho@ikp.liu.se

(3) Institute of Nuclear Safety Systems (INSS), Japan
niwa@inss.co.jp

Abstract. This paper reports from a project to develop a system for computerised procedure generation. The purposes of the system are to ensure procedures that are consistent and complete, and to facilitate the revision of procedures caused by changes to the plant. The approach makes a fundamental distinction between procedure presentation and procedure generation, and between the various types of knowledge needed to write a procedure. A system has been specified which generates a procedure to accomplish a specific goal, based on plant operations knowledge. The output is an intermediate procedure format, which can be further processed by a computerised procedure presentation system. The system is presently being implemented, and will be evaluated by professional nuclear power plant operators using a Steam Generator Tube Rupture scenario.

Keywords: Emergency procedures; human-machine interaction; task analysis, knowledge representation

1. INTRODUCTION

In the nuclear power plant (NPP) control room the functions related to procedures have been among the least affected by advances in information technology. This, however, does not mean that procedures have remained unchanged or unaffected by the onslaught of information technology. On the contrary, operating procedures must reflect the state of control room technology, for instance in terms of sensors, indicators, automatic control systems, safety systems, etc. In this sense the **content** of procedures has continually changed to reflect the developments in control room I&C and automation. But the **structure** of procedures has basically remained the same, as has their format, i.e., printed documents or handbooks.

One reason for this apparent lack of development is that procedures, and especially emergency operating procedures (EOP), represent a fallback position as something that must be guaranteed to work when all else fails. Information technology has therefore been applied very conservatively to procedures. Another reason is that procedures traditionally need to be independently verified and approved, hence must be separable from the control room equipment as such. A third reason is that procedures often contain elements that do not have a straightforward technological solution; a procedure needs operators to carry it out, and the steps described by the procedure cannot easily be automated. A further reason is that procedures do not only address activities of the control room, but also external functions that cannot easily be instrumented. Finally, procedures are produced towards the end of the design cycle, hence may be difficult to integrate in the actual control room.

2. ADVANTAGES OF COMPUTERISED PROCEDURE GENERATION

In order to fulfil their purpose, procedures must meet several goals, which appropriately can be described as technical and human factors goals respectively. The two main **technical goals** are:

1. that the procedures are **effective** in bringing about the desired changes to the process or the plant, implying that the correctness of the procedure can be verified; and

2. that the **production** or writing of the procedure process is explicit (transparent) so that decisions can be traced and so that the procedures are easy to amend and maintain,

   while the two **human factors** goals are:

3. that the procedures can be **carried out** by operators without any undue difficulties, i.e., the procedures must not induce a task overload; and

4. that the procedures are **presented** - hence structured - such that they are easy to understand and follow.

While the human factors goals can be met by computerised procedure presentation, cf. Niwa et al., 1996, the technical goals must be addressed at an earlier stage when the procedures are first written or generated. This problem has been addressed in a specific project to develop a system for computerised procedure generation, called the AG-EOP or the system for Automated Generation of Emergency Operating Procedures.

2.1.Automated Procedure Generation

Procedures are usually written by a team of people with specialised knowledge about the plant and the process, with operational experience, and possibly also with some knowledge or experience of how to write good and efficient procedures. Procedure writing clearly requires a combination of many skills that may be difficult to bring together in a single team, let alone in a single person. Automated procedure generation may therefore bring several advantages.

1. First, it ensures a **uniform format** or structure of the procedure, hence a better compliance with human factors criteria and procedure writing guidelines.

2. Second, the **contents** of the procedures will be consistent within and between procedures, leading to improved usability, in particular in relation to the level of detail in the descriptions of procedure steps. Automated procedure generation also requires that the necessary and sufficient information is available at the time when the procedures are generated, and this will ensure more uniform and more explicit demands to the operators.

3. Third, automated procedure generation improves the **process** of procedure generation, both in terms of efficiency and reliability. It also makes it easier to trace the development of the procedure (an audit trail), which may reduce the effort in updating (maintaining) or modifying procedures. Once the principles for procedure generation have been implemented and approved, it will require less effort to produce a revised version of it - in part or in whole. Specifically, whenever changes are made to the underlying plant knowledge the AG-EOP can be used to identify those procedures that will be affected by the changes, and efficiently generate new versions of them.

Automated procedure generation can be considered both as an off-line and an on-line function. The former is a way of producing **generic** procedures, while the latter is a way of producing **situation specific** procedures. Although both alternatives will improve the current situation, only off-line procedure generation is expected to be feasible in the short term. This can be achieved by developing a set of high-level requirements and use that to specify a corresponding method for automated procedure generation.

2.2.Required Procedure Knowledge

Computerised procedure generation requires explicit descriptions of the knowledge needed to operate the plant and the knowledge needed to generate the procedure. The **plant operations knowledge** is already encapsulated in existing procedures, as well as in the procedure designer's experience. It can also be specified by means of formal methods such as a Goals-Means Task Analysis (Hollnagel, 1993). The **procedure design knowledge** includes the procedure writing guidelines, although these only provide a high-level description of the procedure writing activity. Other aspects are the principles for unambiguous language, simple logic, clear

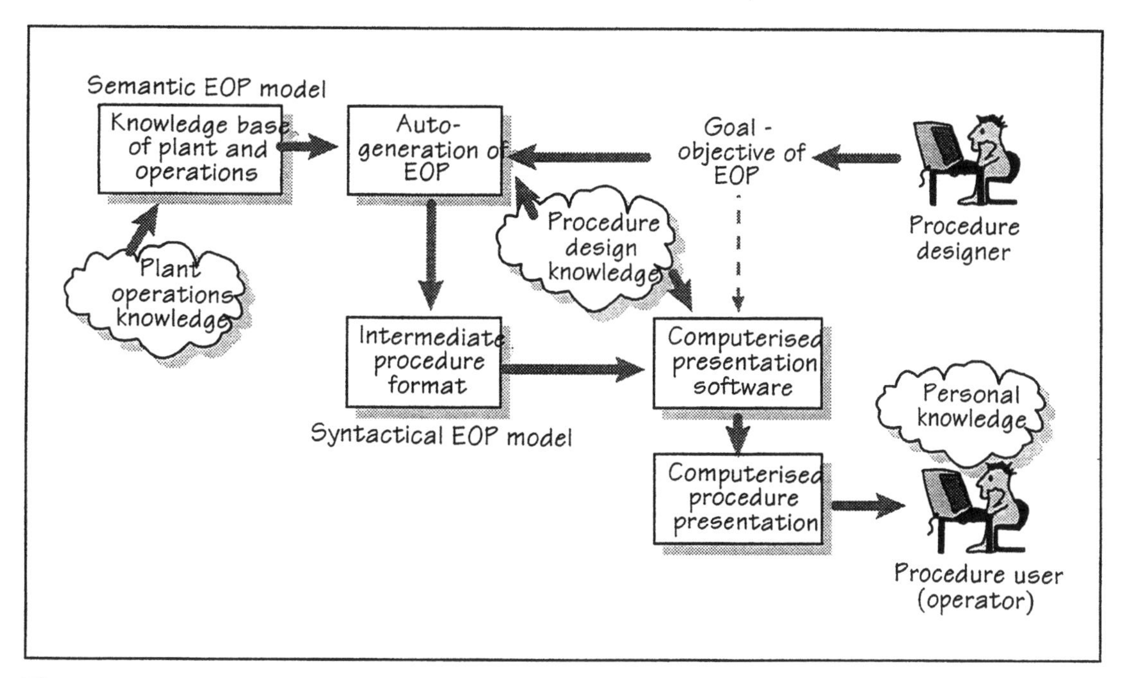

Figure 1: Knowledge required for procedure generation.

grammatical structure, etc., which are usually not explicitly represented or written down. The procedure design knowledge applies both to the functional structure or logic of the procedure and to the human factors aspects of procedure presentation. The relations between the various types of knowledge are illustrated in Figure 1, which also shows the flow of procedure generation. Since a procedure never can be a complete representation of the knowledge needed it is necessary that operators have **personal knowledge** of how to use the procedure, that they understand the basics of the process, that they are familiar with the control room and the equipment therein, etc. This knowledge unfortunately cannot be encapsulated in a computerised procedure generation system.

3. SYSTEM ARCHITECTURE

As shown by Figure 1, the AG-EOP requires the procedure designer to specify the goal or purpose of the procedure. This is usually expressed as a well-defined system state, such as "ruptured SG has been isolated". The AG-EOP uses the plant operations knowledge of the semantic model to generate a procedure that will achieve the stated goal. This results in an intermediate procedure format, or the syntactical model, which can be further edited. Whereas the semantic model contains the generalised knowledge of how to operate the plant, the syntactical model is a description of the specific steps needed to achieve a specific goal. In order to be used by the operator, the syntactical model must be transformed into the specified presentation format. This can be achieved by computerised procedure presentation software, as described by Niwa et al., 1996.

3.1. The Semantic Model

In order to support the automatic generation of procedures, it is necessary that the information needed by the procedure is represented in a knowledge base. For the purpose of the AG-EOP the underlying knowledge structure has been kept relatively simple, following the principles of a goals-means decomposition. The technique has been refined over a number of years, and has been applied both as a basis for the description of plant functions (Lind & Larsen, 1995), as a basis for display design (Rasmussen & Vicente, 1987), and as the basis for a task analysis method (Hollnagel, 1993). In relation to the AG-EOP it is especially interesting that the goals-means decomposition has been used as the basis for proposing a task description that can support a procedure. The representation used by Lind & Larsen (1995) is a generic frame structure with slots for the goal description as well as for various types of conditions. This is very similar to the formalism proposed for the Goals-Means Task Analysis (Hollnagel, 1993) which also is based on a generic frame concept. The semantic model of the AG-EOP was therefore defined to have the following properties:

IDENTIFIER:	The name of the procedure step
GOAL:	The purpose of the procedure step. This is essentially the same as the state that is obtained as a result of carrying out the procedure, or the main post-condition
PRE-CONDITION:	This is an ordered list of the conditions that must be met before the procedure step can be carried out, i.e., before the means can be activated. This may include Cautions and Notes, which in turn will become goals if they are not fulfilled at the start of the step
MEANS	The action or means that will bring about the goal or the post-condition.

Note that the main post-condition is not defined explicitly. It is taken for granted that it is identical to the goal description. It may, however, be convenient to include also the main post-condition as a separate attribute, although in terms of information this will be redundant. This resulting structure is quite general and can be applied for a number of purposes and domains. Although it, in principle, is all that is needed it is nevertheless convenient to add a few more categories, which basically are distinctions between different types of conditions.

EXECUTION-CONDITION:	The execution condition describes states that must remain true throughout the execution of the procedure step. Execution conditions are particularly important in the case of NPP EOPs, since very often an action can only be carried out given that a specific condition exists - e.g. that PRZ level has a certain value, or that subcooling is established and is maintained.
SIDE-EFFECT:	Whereas the goal is the main effect, any particular procedure step may also have one or more side-effects. These are the consequences of carrying out the step, even though they are not part of the main goal. E.g., a reduction in temperature also leads to a reduction in pressure. In special cases, such as success paths, the side effect may be used to achieve a goal that otherwise would not be reachable.

In practice some goals can be composite goals that describe two or more post-conditions. In such cases it is useful to be able to distinguish between different types of goal descriptions. This can be achieved in the following way, where the previous description of a goal is renamed **simple goal**.

SIMPLE GOAL:	The purpose of the procedure step. This is essentially the same as the state that is obtained as a result of carrying out the procedure, or the main post-condition
CONJUNCTIVE GOAL	A conjunctive goal describes two or more goals that both/all must be achieved in order for the conjunctive goal to be achieved. Either or both/all of the goals may be simple goals or disjunctive goals.
DISJUNCTIVE GOAL	A disjunctive goal describes two or more goals where the conjunctive goal is achieved if at least one of the goals is achieved. Either or both/all of the goals may be simple goals or conjunctive goals.

The proposed representation does not contain specific slots for e.g. warnings or cautions. This is because these should properly be considered goals or pre-conditions that are attached to a procedure step. In many cases the cautions actually define goals that are to be considered in parallel with the procedure steps proper.

3.2.AG-EOP Syntactical Model

The syntactical model corresponds to an intermediate procedure format that can be used as input for, e.g., computerised procedure presentation (Niwa et al., 1996). This format can be expressed by using a basic BNF notation, corresponding to a simple database structure. The syntactical model may need to include some information or fields that only serve the purpose of the presentation software. This information should be provided by the procedure writer, rather than by the semantic model. The differences between the intermediate representation and the final format - whether as a printed or computerised procedure - should, however, be kept as small as possible.

3.3.The AG-EOP In Action

In using the AG-EOP, the procedure writer specifies a goal that corresponds to a desired system state, for instance that the RCS has been depressurised. The AG-EOP then searches the semantic model, applying a set of interpretation rules, to produce an ordered set of frames or a search list. The order is given by the goals-means relations defined in the semantic model, but is not realised or brought to bear until a specific top-level goal has been defined. This first step has thus started the generation of the procedure step, and imposed an order of the task steps. The semantic model contains multiple possible sequences, corresponding to the goals-means links among the slots but none of these will be realised until a top-level goal is defined. The procedure, as the syntactical model, represents a transformation of the search list. The essential point is, however, that no changes are made to the order, hence to the semantics of the procedure. The transformation that produces the syntactical representation may require some principles to add morphological elements, such as brackets, etc. Notice also that information about, e.g., the responsible operator, procedure step name, etc., may have to be provided by the procedure writer during the editing of the search list.

In the resulting database, a procedure is a set of records, and each procedure step may be represented by several records. The number of records depends on the number of operations, conditions, and comments that are part of a procedure step. Some procedure steps can be very long and will require several display pages, while other steps may be short enough to fit on one display page. The database format may, for instance, include the following fields:

[***SequenceNumber***, ***OperatorId***, ***Operation***, ***Condition***, ***Comment***]

Of these fields, the ***SequenceNumber*** is not a part of the semantic model, but can be assigned to a record (or frame) in the syntactical model when it is created. Both the ***OperatorId*** and the ***Comment*** may be part of the semantic model, but may also be supplied by the procedure writer. The contents of the fields ***Operation*** and ***Condition*** are basically the same as in the instance of the semantic model. It may, however, be necessary to perform both a linguistic transformation and a reordering of the actions and (pre)conditions in order to produce a sequence that is easy to follow.

3.4. An Example

As an example, consider the specific instance of the semantic model that corresponds to achieving the goal of "RCS has been depressurised". The result of the search in the semantic model consists of six frames (the numbering of the frames is arbitrary). From this search list a representation can be produced that contains the pre-conditions and the actions ordered according to the way they were accessed by the inference rule. This produces the following structure:

Goal: [RCS has been depressurised] **(1)**
- Action: [Spray PRZR with maximum available spray] **(2)**
  - Pre-condition: [Normal PRZR spray is available] **(3)**
    - Action: [Check that normal PRZR spray is available]
  - Action: [Open PRZR spray valves]
  - Action: [Continue maximum spray]
- Action: [Close spray valves] **(4)**
  - Pre-condition: [RCS subcooling based on core exit TCs < 21^0 F (22^0 F for adverse conditions)]
  - **OR**
  - Pre-condition: [PRZR level > 19% (20% for adverse containment)]
  - **OR**
  - Pre-condition: [RCS pressure < ruptured SG(s) pressure **AND** PRZR level > 17% (18% for adverse containment)]
  - Action: [Close normal spray valves] **(5)**
  - Action: [Close auxiliary spray valves] **(6)**

This intermediate representation of the procedure is a simple transformation of the information in the frames that constitute the instance of the semantic model. It is therefore isomorphic to the semantic model. The next step is to transform this into a syntactical model.

In order to produce the syntactical model we must consider the format that is used by presentation the system. The BNF defines the elements of the syntactical model but may have several implementations or realisations, each of which will correspond to a specific database format. The syntactical model, in the nature of things, represents the syntax of a specific "language", the language being the procedure presentation format. As long as there is no universal format, there can be no universal syntactical model - and vice versa. Thus, a syntactical model for the generic Westinghouse format might not necessarily be adequate for other forms of computerised presentation.

As an example, a transformation of the six goals that corresponds to achieving the goal of "RCS has been depressurised" into the generic procedure format mentioned above, provides the following result.

Operation	Condition	Comment
Depressurise RCS		
Check that normal PRZR spray is available		
Spray PRZR with maximum available spray		Open PRZR spray valves
		Continue maximum spray
Close spray valves		Close normal spray valves
		Close auxiliary spray valves
	RCS subcooling based on core exit TCs < 21^0 F	(22^0 F for adverse conditions)
	OR	
	PRZR level > 19%	(20% for adverse containment)
	OR	
	RCS pressure < ruptured SG(s) pressure **AND** PRZR level > 17%	(18% for adverse containment)

(For the sake of the example, the fields of *SequenceNumber* and *OperatorId* have been left out.) This way of representing a procedure is close to being readable by itself, and can easily be used as the input for a computerised procedure presentation system. In fact, a similar representation, although generated manually, was the basis for the presentation system described by Niwa & Hollnagel, 1995.

4. CURRENT STATUS

The AG-EOP system has been developed over a period of several years in a collaboration between INSS, IFE, and Computer Software Development, Ltd. (Japan). The development has systematically applied the principles of cognitive systems engineering to develop a comprehensive set of software requirements, which have served as the basis for implementing a full prototype of the system (Hollnagel & Niwa, 1996). The test case that has been used during system development is the Steam Generator Tube Rupture scenario. When completed during the summer of 1998, the system will be evaluated by experience NPP operators, and the results will be used to refine the specifications.

5. REFERENCES

Hollnagel, E. (1993). *Human reliability analysis: Context and control.* London: Academic Press.

Hollnagel, E. & Niwa, Y. (1996). *A cognitive systems engineering approach to computerised procedure presentation.* In: Proceedings of Cognitive Systems Engineering in Process Control (CSEPC 96), November 12-15, Kyoto, Japan.

Lind, M. & Larsen, M. N. (1995). Planning and the intentionality of dynamic environments. In J.-M. Hoc, P. C. Cacciabue & E. Hollnagel (Eds.), *Expertise and technology: Cognition and human-computer interaction.* Hillsdale, N. J. Lawrence Erlbaum Associates.

Niwa, Y. & Hollnagel, E. (1995). *The design of computerised procedure presentation for nuclear power plants.* Paper presented at HCI International '95, July 9-14, Yokohama, Japan.

Niwa, Y., Hollnagel, E. & Green, M. (1996). Guidelines For Computerised Presentation Of Emergency Operating Procedures. *Nuclear Engineering and Design, 167,* 113-127.

Rasmussen, J. & Vicente, K. J. (1987). *Cognitive control of human activities and errors: Implications for ecological interface design* (Risø-M-2660). Roskilde, Denmark: Risø National laboratory.

USE OF NUCLEAR-REACTOR CONTROL ROOM SIMULATORS IN RESEARCH & DEVELOPMENT

Jacques Theureau

CNRS / Université Technologique de Compiègne, 60206 Compiègne, France

Abstract: Simulator studies are powerfull means to know, design and manage the complexity of nuclear-reactorcontrol, if they are correctly designed for that purpose. This contribution to an international state of the art precises the trends and novelties in the use of the results, in the theories and methodologies and in the construction of the simulated situations, i.e. in the conditions for an efficient use of the techniques of simulator design. *Copyright © 1998 IFAC*

Keywords: Complex systems, Design, Ergonomics, Modelling, Nuclear reactors, Power control, Simulators.

1. INTRODUCTION

The interest of the use of simulators in Research & Development concerning nuclear-reactor control rooms stems from the necessity we have today to design and manage their living, social and cultural complexity. For that purpose, we must know sufficiently the underlying dynamics of this complexity. The knowledge of the deviations it shows from what is prescribed by the management helps to set up the problem but not the solutions. What do we mean by "living, social and cultural complexity" in matters of nuclear-reactor control? We characterize this way the system made of the control room, including its diverse operators. In fact, if we consider the control room, the classical definition of complexity (many elements and many differentkinds of relations between them) is not sufficient. We need at least the Santa-Fe Institute definition: "systems with many different parts which, by a rather mysterious process of self-organization, become more ordered and more informed than systems which operate in approximate thermodynamic equilibrium with their surroundings". And it's itself not enough to take in account the presence of human actors who have the peculiarity to be autonomous, i.e. to have at every moment a subjective view of the whole system, including themselves, i.e. of what we can call the "situation at hand", and who interact at this moment with elements of this situation which have been shaped as relevant by their past interactions up to that moment. Such a living, social and cultural complex system can't be breacked up into simpler sub-systems to be studied apart from each other and aggregated afterwards to get the complex system, or, anyway, these breacking up and aggregation are drastically unsufficent. The complexity gives rise to important phenomena which can be missing in the simpler sub-systems studied separately.

Hence the three steps of the best use of simulator studies from the point of view of the scientific knowledge and management of the underlying dynamics of living, social and cultural complexity of the control room: (1) systematic studies of natural situations, that is of the real complexity to manage; (2) systematic studies of full-scale simulated situations, with simulators and scenarios designed especially, and operators chosen especially to match the characteristics proved essential in natural situations; (3) systematic studies of part-task simulated situations designed in the same way, where, thanks to their greater flexibility and lower cost, design alternatives can be tested. Still from this point of view, to proceed another way, performing only one or two of these steps, or performing them in the reverse order, can give only an illusory or inoperative knowledge, or at least a poor one, on which can be based only a poor management.

During each of these three steps, we need also, in the present state of the cognitive science, to practise both inductive and deductive method. Inductive methods proceed from data to concepts by descriptive generalization. Deductive methods proceed from an a priori mathematically organized view of the tasks to be performed to the concrete concepts describing the empirical systems. If we stuck to the first one, we take risks of getting pure clinical analysis, that is poor generalization. If we stuck to the second, we take risks of misplacing concreteness, Alfred North Whitehead's expression, that is of taking the a priori for the real, of finding in the real what we have put a priori in it. Concerning such a living, social and cultural complexity, in the present state of cognitive science, inductive methods mean process tracking methods, and deductive methods mean dynamical systems modeling.

These considerations shape the interpretative frame of a contribution we made recently to an "international state of the art review on the use of simulators in hazardous industries for purposes other than training", asked by the Human Factors Group of the Reliability Studies Department of the Direction of Studies and Research of EDF, the french electrical power public company (Theureau, *et al.,* 1997). In using this contribution here, we will put the emphasis, not on the techniques of simulator design, but on the epistemological conditions for an efficient use of them.

2. USE OF THE RESULTS

The results of the studies in full-scale simulators or in sufficiently rich and relevant part-task simulators, like those of studies in natural situations, are by construction, multi-uses: design of control rooms and of their organizations, devices and procedures (human-machine interfaces, paper or computer driven procedures, operation manuals); Probabilistic Human Reliability Studies (PHRA). Nethertheless, a few of these uses are dominant and increasing: (1) An increasing number of simulators studies aim at preventing potential negative effects of automation; (2) There exist more Verification & Validation studies than studies integrated in the design process, in spite of the possibilities open by part task simulators; (3) More interest exists in the improvement of training and certification; (4) More emphasis is put on qualitative aspects of Human Reliability Analysis than on its quantative aspects; (5) Still a poor interest is expressed, at least in the litterature, in testing the design of procedures, yet drastically changed and computerized in different ways all around the world since Three Miles Island events.

As the essential purpose of studies on nuclear power plant control room simulators is often to provide data for Probabilistic Human Reliability Analysis (PHRA), we will consider more thoroughly this point. Many studies continue to implement the conventional methodology initiated by A. Swain, which bypass the operators'cognitive activity. Different research and development teams tend to query the relevance of this conventional methodology in various ways. At VTT-Espoo, a recent objective of the psychological research group is to integrate cognitive analysis of control activity into the new stochastic dynamic model called the "marked point process" (Arjas and Holmberg, 1995). It matches well with the idea by which one should analyze the construction of the action and not model it in some way as a predefined sequence. At Westinghouse-Pittsburg, the human factors research group works upline of PHRA by implementing a checklist of cognitive task requirements produced from simulator tests (Roth, *et al.*, 1994). At OECD-Halden, Erik Hollnagel integrates a similar concern to a structured approach for contribution to PHRA, called CREAM (Cognitive Reliability and Error Analysis Method). The principle of the approach is to combine two interpretation methods, the first of which is a logical progression of the customary behaviourism of PHRA, and the second a logical progression of cognitivism, (Theureau, *et al.,* 1997). It is a clear recognition of the kind of complexity which is involved in nuclear power control.

3. THEORIES AND METHODOLOGIES

Much work is being done practically everywhere on innovation and development of data-collation and analysis methods. This work can be characterized by: (1) a reduction in the ambitions of cognitive simulation and a return to process-tracking methods; (2) a trend towards eclectism, that is the coexistence of heterogeneous or even contradictory theories and methods (which can be both an hindrance to research and a recognition of the complexity of the problem and of the limits of each of the theories and methods available), and search for theoretical and methodological complementarities; (3) a tendency to go beyond traditional cognitive psychology by means of the still-confused notion of situation awareness coming from human factors research in aircraft piloting; (4) a tendency to consider cognitive aspects of co-operation within the control crew, with distributed computerized information, and to develop the corresponding methods and theories; (5) an important issue only tackled in VTT-Espoo is the evolution of operators' competence and confidence in the automated systems, which requires longitudinal studies. All these trends or novelties are different ways to deal with the complexity of nuclear power control. We will insist here on points (1) and (3) which provoque the more discussions.

Just a few years ago, cognitive simulation—computer modelling of control activity, based on a symbolic representation of the task and considerations derived from experimental psychology—was the lode star for simulator studies. It still is today, but instead of expanding, this perspective vanishes. For example, the series of studies by (Roth, *et al.*, 1994) was developed within a broader project undertaken by the Nuclear Regulatory Commission (NRC) to study the performance of the control crew during simulated emergencies and develop a cognitive simulation of the cognitive activities involved. In a past series of studies, two variants of an ISLOCA (leak from the high pressure reactor coolant system to the low pressure residual heat removal system) were studied on a full-scale simulator. But generalization of the results of this study encountered many limitations: (1) solely ISLOCA incidents; (2) control crews made up of training staff and not actual operators; (3) only one crew for each ISLOCA variant; (4) control crews made up of two persons, and not the usual three to five. The simulated situation was far removed from a real full-scale situation as far as the composition of the control crew was concerned. Also, the cognitive simulation developed dealt only with "certain" of the cognitive activities engendered. It was therefore decided to develop a new, more extensive series of empirical studies with richer simulated situations. It was planned to at the same time develop the cognitive simulation, but on the NRC's recommendation it was decided to first focus on the empirical study because of the difficulties, cost, and time entailed. To our mind, this postponement implicitly reflects the relative failure of cognitive simulation in its current form, with respect to both knowledge of activities in complex dynamic systems and to design and management. We feel that at the moment use of this tool is of interest only if: (1) it develops in connection with a systematic analysis of activities and not on the basis of symbolic representation of the task and general considerations derived from experimental psychology; (2) it is restricted to modest objectives in both theoretical and practical terms. This is the case, for example, in collective activities in an emergency rescue service or in air-traffic control, as presented in (Pavard, 1994).

This limitation on the ambitions of cognitive simulation raises the problem of seeking out new channels for modeling, taking inspiration from the mathematical theory of dynamic systems, for example, together with the problem of replacing the paradigm of"man as an information-processing system" by a new paradigm of cognition. In any case, at the moment it is leading to a renewal of what some authors call "process-tracking methods". These process-tracking methods are related to the methods of French-language occupational ergonomics analysis, and more specifically to those of course-of-action analysis and their collective interlinking (Theureau and Jeffroy, 1994). This is not entirely fortuitous since, like course-of-action methods, they go back to (Newell and Simon, 1972) who, at the dawning of cognitive psychology and Artificial Intelligence as we know it today, developed a new fashion—in psychology in any case—for validation of theories and models which stresses systematic description of verbal protocols collated at the same time as the activity was in progress and which gives a secondary status to conventional experimentation and statistical analysis. The essential instruments of this new fashion for validation of theories and models are the problem-solving graph and computer simulation, the ancestors of process-tracking methods and cognitive simulations. The decline in the ambitions of cognitive simulations results in process-tracking methods being reinstated to a position they had lost since (Newell and Simon, 1972), except in certain French-language research in ergonomics. Let us look at another example of development of process-tracking methods: the "realistic" approach developed in studies on the VTT-Espoo full-scale simulator. As is explained in (Hukki and Norros, 1994), the approach is contextual (including the social situation), dynamic (acts are not considered as isolated events) and subject-centred (the operators' point of view is considered to be essential). The researchers speak of situation activity or socially constructed activity, or quite simply of activity in the sense of Vygotsky. Their fundamental concepts are "activity", "orientation", and "how to do things", associated with process-tracking methods .

The notion of situation awareness (SA) that came into being in aerospace studies is invading the nuclear field. It has become emblematic of the presence of man in highly automated technical systems. Still, situation awareness is unanimously considered to be a vague notion which has multiple definitions and gives rise to multiple complementary or alternative methods. A recent congress (Garland and Ensley, 1995) gives testimony to this. The definitions that follow are just two chosen from a multitude of others and felt to be exemplary of their radical theoretical heterogeneity: "the condition of the knowledge of the persons or the mental model of the situation around them", or "perception of environmental elements in a volume of time and space, understanding of their meaning, and projection of their condition into the near future" (Ensley); "dynamic cognitive coupling of an agent and a situation" (Flach). According to Meister, situation awareness is thus "a concept for aggregation rather than for analysis". According to Billings, it is "too clear, too holistic, and too attractive" a construct about which one might wonder if its utility compensates its complexity. For Charniss, situation awareness is a "default construct", i.e. we appreciate it most when it is absent: "when someone loses his situation awareness, the result is a crash". Some authors in the same congress stress the 'family resemblance' between the notion of situation awareness and that of work load, especially mental work load : same fuzziness, same practical necessity in the absence of better established notions, same measurement problem. In fact, the notion of situation awareness reflects both the incapacity of traditional cognitive psychology to answer the practical questions of control of complex dynamic systems, and the efforts to go beyond this traditional cognitive psychology, whereas no alternative has yet fully asserted itself. Its fuzziness is evidence of a scientific crisis that has not yet been resolved, but its very existence evidences the need to give the designers of complex dynamic systems if not criteria, then at least a principle concerning the relationships to be established between human operators and automatic systems: maintain the situation awareness of operators. This new principle results in a search for synthetic criteria to replace the usual analytical performancecriteria. It is therefore worth examining what can be done to clarify this notion.

In (Sarter and Woods, 1991), a preliminary clarification of the notion is made by showing that it cannot be the equivalent of "effective conscious knowledge", for "that would suggest that only the information in work memory could be considered to be 'aware'", and by considering that "any definition of situation awareness must refer to the information that is available or that can be activated, when it is relevant for evaluating a situation and dealing with it". If one agrees with these authors, the notion of situation awareness can be assimilated to that of potential actuality proposed in the course-of-action theory (Theureau and Jeffroy, 1994) as part of an human cognition paradigm alternative to that of "man as an information-processing system" most authors dealing with situation awareness continue to refer to. The definition of potential actuality is close to Flach's definition of situation awareness mentioned previously. Flach considers that the theoretical precision of situation awareness requires a " theory of the field of cognition" inspired by the mathematical theory of dynamic systems—although he wonders if such a theory is possible. It is precisely this theory that is the synthetic lode star of the course-of-action theory. The "potential actuality" at a given moment is considered to be co-produced by the situation and by the "agent's involvement in the situation" at that moment. This "agent's involvement in the situation" is the product not of the situation itself but of the entire course-of-action up until that point. The notion of "potential actuality" is thus built up in a way that is strictly the converse of the usual notion of situation awareness. Along this usual notion of situation awareness, what comes first is the "situation" independently of the person involved or "agent", whereas along the notion of potential actuality, what comes first is the "involvement in the situation" inherited from the past course-of-action, independently of the instantaneous "situation", with its characteristics of " orientation " or " confusion ".

This divergence between the current notion of situation awareness and that of potential actuality has important methodological consequences. If the situation does indeed come first, a method for documenting situation awareness which involves 'freezing' the simulator at certain times during the scenario and asking the operators to answer a questionnaire on the situation is legitimate. With the usual notion of situation awareness, it can be considered such a radical change to their "involvement in the situation" has little or no effect on the situation awareness they will express. If, on the other hand—as in the course-of-action theory—the "involvement in the situation" comes first, this sort of intrusion into the control activity is incapable of producing data reflecting the potential actuality. Potential actuality can only be reconstructed indirectly, through analysis of the control activity. For example, as part of the OECD-Halden international research and development programme, the SACRI (Situation Awareness Control Room Inventory) method was developed, which adapts to the nuclear power industry the SAGAT method (Situation Awareness Global Assessment Technique) developed in aeronautical studies. Both these methods are based on the principle of intrusion into the simulation that we have just criticized.

On the contrary, in (Roth, *et al.*, 1994), the question of situation awareness was re-examined following (Klein, 1995). While he does not specify the notion of "situation awareness", this author proposes to study it through the control activity, i.e. in the same way as one studies a "potential actuality". According to him, "Instead of studying the question of WHAT—what is the 'situation awareness' content of a person?—we can study the question of HOW—how does 'situation awareness' affect action? In this way, we can identify some important aspects of 'situation awareness'—those affecting judgement and decisions". Whence the stress on what these authors call 'process-tracking methods' which are very closely related to course-of-action analysis methods, as has already been said. Whence too a decision-support model said to be 'recognition-primed' which "models the way people make decisions in natural situations without having to compare options". The key to these decisions is " that people use their expertise to define situations and recognize typical courses of action that should be given consideration first " (ibidem). This decision-support model in fact implies a definition of "situation awareness" that brings it closer to what we call a "potential actuality". Klein's work is also among the essential theoretical and methodological referencesof the VTT-Espoo psychological research group referred to above.

4. CONSTRUCTION OF THE SIMULATIONS

Studies of human activity using simulators run into the problem of what it costs to conduct them, the problem of integrating them into the process of designing new systems, and the problem of relationships between the simulator and its scenarios and real situations. We observe: (1) More and more use of part task simulators, more and more rich and flexible and less and less expensive, due to the progress in computer techniques, in order to test alternative design options; (2) More use of usual training or certification simulated situations, with methodological cautions and limits and a consideration of training design issues; (3) More linkage of incidental/accidental simulator studies with retrospective incident/accident studies; (4) A growing (but still modest) interest for the study of natural, normally disturbed, situations, in order to insure a better relevance of the simulations and to know better the transfer made by the operators from the situations they usually live in to incidental/accidental situations; (5) A tendency to build simulation scenarios from theoretical hypotheses concerning control activity, and not only practical and empirical hypotheses. We will put emphasis on points (1) and (5) which exemplify the relation between scientific knowledge and design of complexity.

When people talk of simulators, they usually mean an ideal simulator, a "full-scale" one. The point of part-task simulators is to represent another ideal fulfilling another function. For example, at the NASA-AMES aerospace research centre, part-task simulation begins when pilots are not put in an exact replica of a real cockpit that reproduces the accelerations and movements of the actual aircraft. From this point of view, the HAMMLAB nuclear-reactor control room simulator of the international OECD-Halden programme is a part-task simulator. What is new is less the reality of part-task simulation (it might be said that traditional human-factors studies concern situations of this type) than the very notion of part-task simulation (as a simplification and reduction of full-scale simulation and not as a complication of psychological experimentation) and the fact that today's information-technology brings part-task simulation closer to full-scale simulation. Several considerations lead to studies being carried out on part-task simulators. The first two are the interconnected considerations of cost and integration into the design process: a part-task simulator costs less and is more quickly designed, transformed, or enhanced with new systems than a full-scale simulator. It therefore allows for easier comparison—from the point of view of control activity—of design alternatives for such new systems. The other considerations are of an ontological and epistemological nature, and imply two parallel trends.

The first trend arises explicitly or implicitly from a recognition of the living, social, and cultural complexity. For the supporters of such an ontology and epistemology, natural situations do not simply add complications to experimental situations. They add complexity and thus engender cognitive phenomena, some of which can be radically different. The resulting method for acquiring scientific knowledge of these cognitive phenomena starts from studies in natural or close-to-natural conditions (particularly when, as for certain emergency situations, it is absolutely necessary to use the simulator) in order to determine the cognitive phenomena involved. It works towards studies on part-task simulators intended to examine the cognitive phenomena more exhaustively and better validate them, but the pertinence and validity of these studies depend on the first studies. In the context of this recognition of the living, social, and cultural complexity, both full-scale and part-task simulators take on a scientific function instead of just a practical function or a role as ill-adapted substitutes for experimental situations in the laboratory. The researches of the Westinghouse-Pittsburg group, for example, are along these lines (Roth, *et al.* 1994).

The second trend arises out of an ontology (implicit) and epistemology (explicit) of "Lego" (internationally reputed children's building-block game) by which complexity is considered to be both capable of and having to be attained by putting together simple elements—or generic concepts of what is simple—produced by the laboratory situation studies. Part-task simulation is then thought out in relation to the ideal of laboratory experimentation. It is no longer thought out from the point of view of simulation. This is similar to traditional "human-factors" studies. The only difference between a part-task situation and a laboratory situation, from this point of view, is that because of the practical interests involved, researchers benefit from greater material resources than if they were to remain in their laboratory. A large number of studies on part-task simulators encountered in the literature result from this second trend. Their scientific interest is secundary relative to rigorous experimental procedures in the laboratory and field studies, full-scale simulator studies, or sufficiently rich part-task simulator studies developed from the simulator point of view. Nevertheless, their practical merits are not to be overlooked. They help demonstrate the interest of developing part-task simulators for integration of human factors into design processes. Their results can be re-interpreted in connection with an ontology and epistemology of complexity if one also has rigoutous studies in the natural situation or on full-scale simulator. At OECD-Halden, both points of view co-exist. In summing up 10 years of test and evaluation studies in (Folleso and Volden, 1993), it is considered that a high degree of realism was attained, to the detriment of systematic control of the experiments, and therefore suggested reducing realism in order to increase control, starting with the less realistic and more controlled studies in order to "demonstrate effects of vital aspects of the system", and then using more realistic situations to more broadly test the validity of their hypotheses. On the contrary, in (Kvalem, *et al.*, 1996), it is suggested to put "less stress on well controlled experimentation and more on 'simulated field studies' to analyze complexity" as a long-term prospect for the use of the HAMMLAB simulator.

It is commonplace to design the simulation scenarios in order to test practical and empirical hypotheses, such as the hypothesis of performance improvements due to a given system, or various organizational arrangements for the control crew. What is new is the trend to build scenarios from theoretical notions in order to test theoretical hypotheses regarding control activity, and not only practical and empirical hypotheses. This trend is seen in certain full-scale simulator studies and in most of part-task simulator studies, both in those that tend to stick close to the epistemological paradigm of Lego and those that—more or less implicitly, it must be said—consider part-task work from the simulator point of view, in relation with the paradigm of living, social, and cultural complexity. The series of studies by (Roth, *et al.*, 1994), for example, dealt with two variants of ISLOCA (Interfacing System Loss of Coolant Accident) and two variants of LHS (Loss of Heat Sink) with eleven complete crews of real operators for each event. The model of cognitive activities linked to operator behaviour in the emergency situations involved comprises two components: situation assessment and response planning. Situation assessment is similar to situation awareness (see above). Response planning corresponds to the decision to take a course-of-action, bearing in mind a particular situation assessment. The two ISLOCA variants were especially designed to be difficult from the point of view of situation assessment. The objective was to create situations in which the control crews would have to identify and isolate the breach without explicit guidance. The emergency procedures did indeed include ISLOCA procedures, but it was possible to create situations where the control crews could not find the ISLOCA procedure through the network of emergency procedures. The specific dynamics of the event led the operators to a LOCA (Loss of Coolant Accident) procedure. As for the two LHS variants, they were designed to be demanding in terms of both situation assessment and response planning.

5. CONCLUSION

Such trends in the use of the results, in the theories and methodologies and in the construction of simulated situations, leave room to a more efficient use of the techniques of simulator design in matter of knowledge, design and management of the complexity of nuclear-reactor control.

REFERENCES

Arjas E. & Holmberg J. (1995) Marked point process framework for living probabilistic safety assessment and risk follow-up, *Reliability and System Safety*, **49**, 57-73.

Folleso K. & Volden F.S. (1993) *Lessons learned on test and evaluation methods from test and evaluation activities performed at the OECD Halden reactor project*, Institutt for Energiteknikk, Halden, Norway.

Garland D.J. & Ensley M.R. (1995) *Experimental analysis and measurement of situation awareness*, Embry-Riddle Aeronautical University Press, Daytona Beach, USA.

Hukki K & Norros L. (1994) A method for analysis of nuclear power plant operator's decision making in simulated disturbance situations. In *XIII European Annual Conference on Human Decision Making and Manual Control* (Norros L. (Ed)), 200-217. VTT, Espoo, Finland.

Klein G. (1995) Studying situation awareness in the context of decision-making incidents. In *Experimental analysis and measurement of situation awareness, Proceedings of an international conference* (Garland D.J., Ensley M.R. (Eds)), 177-181. Embry-Riddle Aeronautical University Press, Daytona Beach, USA.

Kvalem J., Berg O., Fordestrommen N.T., Groven A.K., Hollnagel E., Pettersen F., Solie A.S., Stokke E., Sundling C.V. (1996) *HAMMLAB 2000: long term perspectives for use of HAMMLAB*, Institutt for Energiteknikk, Halden, Norway.

Newell A. and Simon H. (1972) *Human Problem Solving*, Prentice Hall, Englewoods Cliffs, USA.

Pavard B. (1994) *Systèmes coopératifs: de la modélisation à la conception*, Octares, Toulouse, France.

Roth E., Mumaw R. and Lewis P. (1994) *An empirical investigation of operator performance in cognitively demanding simulated emergencies*, NUREG/CR 62108, US NRC, Washington, USA.

Sarter N. & Woods D. (1991) Situation awareness: a critical but ill-defined phenomenon, *International Journal of aviation psychology*, **1 (1)**, 45-57.

Theureau J., Mosneron-Dupin F. and Schram J. (1997) *Contribution à un état de l'art international sur l'utilisation des simulateurs dans les industries à risque à des fins autres que de formation*, EDF/DER/ESF/HT-54/97/004/A, Clamart, France.

Theureau J. and Jeffroy F. (1994) *Ergonomie des situations informatisées: la conception centrée sur le cours d'action des utilisateurs*, Octares, Toulouse, France.

Woods D., Pople H. and Roth E. (1990) *The cognitive environment simulation as a tool for modeling human performance and reliability*, NUREG/CR 5213, US NRC, Washington.

LIFE CYCLE INTEGRITY MONITORING OF NUCLEAR PLANT WITH HUMAN MACHINE COOPERATION

Makoto Takahashi, Catur Diantono and Masaharu Kitamura

Department of Quantum Science and Energy Engineering
Graduate School of Engineering, Tohoku University
Aramaki-Aza-Aoba-01, Aoba-ku, Sendai, 980-8579 JAPAN
TEL&FAX: +81-22-217-7907
E-mail: makoto.takahashi@qse.tohoku.ac.jp

Abstract: Motivation, technical framework, and state-of-the-art of a research project called Life Cycle Integrity Monitoring are described in this paper. The project was established to significantly improve operational safety of large-scale artifacts by developing and introducing advanced techniques for efficiently supporting maintenance and repair activities. As the main technique of crucial importance, the life cycle information management system was proposed with component techniques of distributed artificial intelligence and advanced human-machine interface. Actual implementation of a prototype system, and observations obtained through preliminary evaluation of the prototype as well, are summarized and ongoing technical evolutions are briefly addressed. *Copyright © 1998 IFAC*

Keywords:Maintenance engineering, Human-machine interface,Distributed artificial intelligence,Intelligent instrumentation

1. INTRODUCTION

Operational safety of complex, large-scale artifacts like a nuclear power plant is ensured through careful execution of design, operation and maintenance activities. Among them, the issues related to maintenance activities have been studied to a lesser degree up to present. However, the effect of failures in maintenance activities on the operational safety is by no means less than design and operation failures. The causal breakdown of anomalies experienced in Japanese nuclear power plants during the period of 1990-1995 is given in table-1. Here, sixty-five anomalies, each resulted to unscheduled shutdown of the plant, were carefully analyzed and categorized to four groups according to the direct cause of the shutdown. The four categories considered are: design and construction (DC), operation and control (OC), maintenance and repair (MR), and others (OT). Just to make the analysis more informative, the shutdown events are also classified to two groups; manual shutdown (MS) and automatic shutdown (AS). Needless to say, the latter is more disturbing and undesirable to operators than the former.

Admittedly, the classification is not free from uncertainty and subjective bias. The sample plants are not stationary as the number of plants in operation was changing from 39 to 49 during the period. In spite of these factors, the table clearly indicates that the maintenance and repair activities are the most dominating contributor to the occurrence of anomalies among other activities. It would be reasonable to infer that the weight of maintenance and repair will become much higher in near future since the plants in service will become older while introduction of new plant will be less.

Another claim to place more emphasis on maintenance can be given from a different perspective, namely, more realistic view of human-machine interaction. In recent survey of supervisory activities of human operators (Hoc, et al., 1995), it is stressed that the diagnosis during operation should be considered as a part of more general activity called Dynamic Environment Supervision, where diagnostic decision-making is tightly coupled with subsequent maintenance and repair activities. Similar claim was made (Kitamura et al., 1996) in conjunction with design of intelligent monitoring system of nuclear power plant where fault

Table-1 Causal classification of the nuclear plant anomalies resulted to unscheduled shutdown during the period of 1990-1995 in Japan.

	DC	OC	MR	OT	**Total**
Manual Shutdown	15	0	26	7	48
Automatic Shutdown	5	6	4	2	17
Total	20	6	30	9	65

detection, diagnosis and remedial action synthesis should be treated as a mutually coupled activity rather than as a series of separate activities.

However, actual field practice is in the state that the maintenance activities are usually considered almost independently from the other operational activities. The research and development efforts/funds spent on maintenance improvement are smaller than their own right, nor they are properly organized.

One of the reasons of this lack of technical emphasis on maintenance can be attributed to the traditional belief in the concept of design-based safety. In other words, safety of complex machines has been expected to be attained through careful design in which all necessary countermeasures against envisioned failures are built in. This concept is fine as an ideal goal but not so in practice, since it is not possible for any human designer to envision all the possible modes of mechanical and human originated failures and to implement appropriate countermeasures against them.

Since a large portion of anomalies was introduced by certain defects and/or failures in maintenance activities as typically exemplified above, it is reasonable and necessary to take actions to reduce it. Based on this recognition, several research projects intending to stimulate evolution of maintenance activities were established in Japan. The major ones among them are; autonomous plant project (Kitamura, 1997) sponsored by Science and Technology Agency which includes projects for long-term degradation prediction and distributed maintenance robots(Asama, 1997), and intelligent maintenance system project sponsored by Ministry of International Trade and Industry (Okano, 1996). Both of the projects are aimed at developing key technologies for improving maintenance activities by utilizing advanced data management techniques and robotics. The present paper also deals with a relevant research project named life cycle integrity monitoring of plant systems. The project is in progress at Tohoku University under the sponsorship of Ministry of Education with the main emphasis of the study on life cycle information management (LIM), distributed artificial intelligence (DAI) and on human-machine interface (HMI).

2. FRAMEWORK OF LIFE CYCLE INTEGRITY MONITORING

2-1 Organization of Monitoring Techniques

The design concept and some of the essential key techniques are described in this chapter.

The information management throughout the plant life cycle is necessary since the nuclear plants are expected to be in operation for a long period of time (thirty to forty years or even longer) and most of important components are supposed to be used under the effect of repetitive maintenance activities. It would be clear that knowledge of the stage in the life cycle, and the records of preceding maintenance and repair activities as well, are crucially important for successful diagnosis and prognosis activities (Decortis, et al., 1991, Hoc, et al., 1995). The essential information to be handled includes the records of; scheduled inspection and testing, unscheduled (anomaly-based) inspection and testing, repairs of components and devices, etc. The symptoms of experienced anomalies are also regarded to be rich sources of information and thus need to be stored as a symptom database.

The usability of the LIM system is expanded by utilizing the DAI technology such as distributed local processors for maintenance-related information acquisition, processing, storage, retrieval and provision. The distributed function allocation enhances the on-site accessibility to the crucial information such as previous anomaly records, testing and repairing procedures, advisory messages, etc. More precisely, the LIM system consists of the distributed local LIM units covering various subsystems and components within the plant , and the central LIM unit dedicated to support supervisory and higher-level decision-making. The central LIM unit also takes the role of long-term data back-up since the memory size of the local database cannot be sufficiently large even taking into account the rapid growth in available memory size. Each distributed LIM unit, equipped with a local processor, local databases (LDBs), and HMI, is assigned to take care of integrity of a predefined set of components. The LDBs are used for storage of symptoms, activity records, maintenance procedural guides and related drawings. These knowledge and data are provided to local maintenance personnel upon request. Furthermore, the maintenance personnel can place a order to local support agents instantiated as dedicated relocatable sensors mounted on mobile robot arms to conduct supplementary measurements for better decision-making. The local support agents are also main constituents of the DAI. Another example of the local agents is a software robot, or a worm, that can carry out prespecifed functions related to monitoring, diagnosis and information retrieval in an autonomous manner.

The advances in maintenance information management naturally necessitate more careful consideration on HMI, since increase in the amount and diversity of information is useful if and only if the information is provided in a *systematic and coherent* manner. In other words, special considerations must be given to the design of the information display system assigned to the local as well as central LIM unit. A particular emphasis is needed to visualize functional implication and contribution of each component provided by each local LIM system to the overall plant functionality. Just giving localized information about the integrity of a particular component is not informative enough to the operators expected to make authorized decision-making.

The technical developments in this project are being carried out bearing these requirements in mind. The proposed framework of the present methodology is schematically illustrated in Fig. 1.

2-2 Functional Specifications

In this LIM system, several functional specifications are required to successfully conduct the expected missions. First, the time coverage of the system is enlarged significantly to allow access to the past data by specifying the date of concern, or by similarity-based search. It is a common experience of plant operators to occasionally observe signals behaving somewhat differently from everyday observation. It is desirable in such an occasion to have a tool to search previous records to find out a situation that provided similar signal behavior. The similarity can be defined in terms of various symptoms, i.e. time series record itself, statistical indices (amplitude probability density, auto and cross power spectral density, etc.), functional indices (response time transfer function, etc.). By finding the similar symptoms and by examining the operational and maintenance records related to them, the operator's decision-making about the cause of the anomaly and the procedure for recovery becomes far easier.

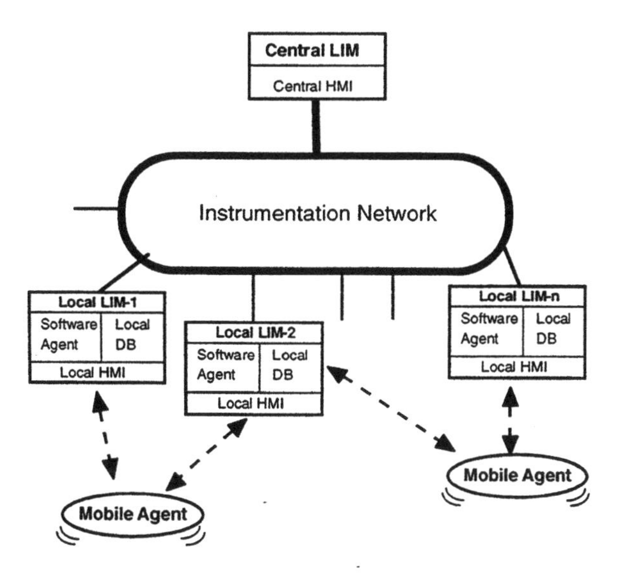

Fig. 1. The proposed framework of the present methodology

Second, search sensitivity needs to be set adjustable. Note that the adjustment is needed in on-line anomaly detection and in the database search as well. In detection phase, the sensitivity should be high enough to catch most of anomalies but should not be too high to respond to non-significant anomalies. This contradictory requirements can only be met through a kind of adaptation. In the database search phase, the threshold value of similarity needs to be adjusted to select the most appropriate candidates of the similar situations. This search sensitivity adjustment is regarded as another kind of adaptation.
Third, the mode of human-machine interaction should be carefully designed. In the present LIM system, the anomaly detection and similarity search tasks are primarily assigned to local processors and support agents. In other words, the LIM system should remain silent as far as the severity of anomaly exceeds a preset threshold of significance. The operators do not want to be disturbed by nuisance alarms in the detection phase. But at the same time, they do not want to miss any of potentially harmful anomalies. Therefore, the monitoring machine should be functioning in "latent" or "silent" mode in usual operation as already proposed in our previous report(Kitamura et al., 1996). Whenever significant anomaly is detected, however, the monitor must switch to "interactive" mode to inform the onset of anomaly to alert the operator, and provide the results of symptom database search upon request.

3. IMPLEMENTATION

The development of the LIM system is being conducted in a modular manner though overall framework is shared in common. In this section, technical features of essential subsystems already developed are described.

3.1 Local-Database

The symptom database system plays a central role in the local LIM database. The version-1 database system was developed and implemented on a personal computer with operating system Windows 95/NT and Borland OWL(Object Windows Library) and DBE(Database Engine). The time series data are actually categorized to three groups in terms of the sampling frequency; fast (100Hz), intermediate(10Hz) and slow (1Hz) sampling. The three data groups are memorized in the database at every specified time duration. The statistical and functional indices are computed and memorized with the same time duration.

It should be stressed that the sampling and storage of data are conducted in an event-driven manner also. Whenever the LIM system detects any deviation from normal signal behavior, it starts data sampling, processing and storage. The deviation is conventionally defined in terms of an average value of monitored signals. However, it can also be defined in terms of any subtle changes in the statistical and functional indices.

Although the storage of the raw time series data is demanding in terms of the memory size, the potential benefit of this policy is the capability of future evolution of the database to accommodate the novel techniques of signal processing coming up in future. This evolutionary nature is, we believe, another important issue for long-term monitoring system of long lasting artifacts. Since techniques of signal analysis are still rapidly developing, it is unrealistic to confine our scope within currently available views. The storage of raw data as the primary information source allows us maximum flexibility for future evolution of the LIM system.

The definition of similarity is yet another complex issue. The similarity between two time traces , for instance, can be measured by transforming each of the two traces to feature vectors Xi and Xj in a multi-dimensional Euclidian space and then defining an index of similarity as the norm of the vector dXij= [Xi - Xj] . This is the simplest definition and much more modifications are definitely needed to replicate diverse sense of similarity that human expert are employing. In addition to the time trace, the similarity needs to be defined for other measurements such as time domain correlation functions and frequency domain spectra. This issue, appropriate and context-dependent definition of similarity, is going to be studied in depth during the project.

3.2 LIM-HMI

The HMI of the LIM system is required to display information and data stored in local LIM database in a systematic and organized manner. The contents to be displayed are; time records, statistical and functional indices and so on together with text-like records and graphic data. As mentioned earlier, a special emphasis of the HMI is the accessibility to previous data which can be specified in terms of the similarity and of the time index. The local personnel can request review of data at any specified time, and of data with similarity as well.

Another important aspect considered in this HMI is, as addressed earlier, the embedding of the local information within the overall plant semantics. In other words, implication and potential impact of the particular component are designed to be visible to the local personnel. In the present design, this semantic representation is instantiated in a form of the "Structured Functional Display"(Takahashi, et al., 1997) where functional information provision is mapped on a plant overview display.

3.3 Local Agents

The local agents actually consist of two types, namely software agents and dedicated monitors mounted on mobile apparatus. The software agents are basically similar to software modules. But the former can travel through local processor network line to execute assigned tasks like signal processing and failure mode identification, and thus are more suited for long-term repetitive monitoring. We have already developed and tested a prototype of the traveling software agents as a worm-like system (Washio et al., 1995). In the present project, a new realization of agent as dedicated monitors on a mobile apparatus is attempted. Presently, an intelligent robot arm (Mitsubishi MOVEMASTER-EX) is employed as the mobile apparatus. A laser displacement sensor (KEYENCE LK-2000) capable of measuring distance and mechanical vibration is mounted on the robot arm as a multi-function mobile sensor. As the sensor allows remote and relocatable measurement of small-scale vibration, it can be used as a viable supplementary tool for incipient anomaly detection and characterization. A TV camera can replace the laser sensor to provide additional modality of sensing.

4. OBSERVATIONS

The experimental test bench of the project and the sample instantiation of the key techniques in an integrated manner are illustrated in Fig. 2. An extensive evaluation of the overall LIM system is scheduled to be conducted after completion of the full system development, most likely after eighteen months. Prior to that, the evaluation is carried out in an incremental manner with each subsystem. The evaluation of local LIM system was carried out with the tentative local symptom database and HMI with limited human-friendliness and multi-modality. But even at this phase, the design concept of this database system was strongly supported by experienced operators of nuclear plants. Further upgrading of the HMI need to be made by incorporating the functional separation into situation-dependent display modes, as typically proposed in (Johannsen, 1992) where presentation, dialogue, user model, application model and tasks to be performed

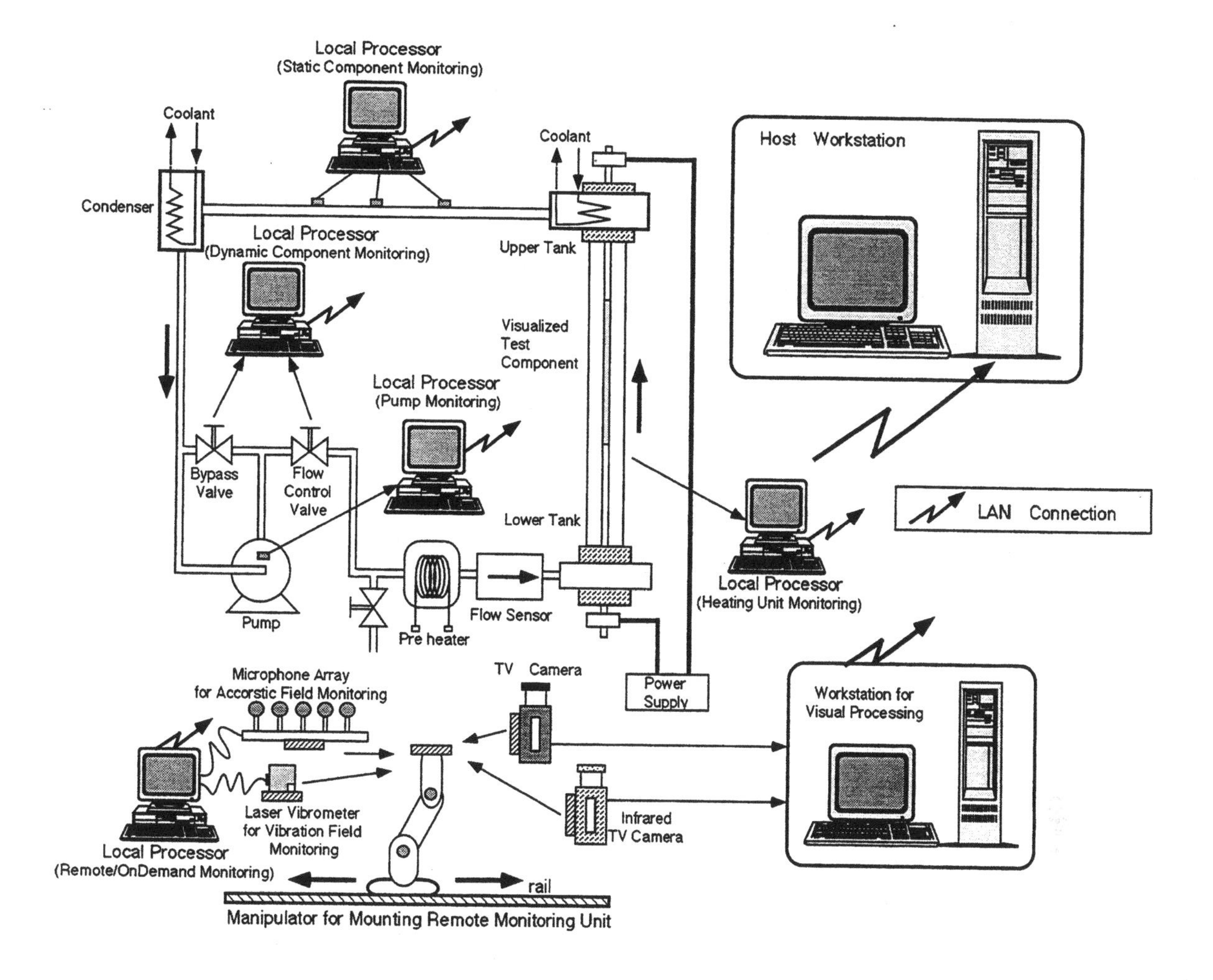

Fig.2 Experimental test bench of the proposed methodology

are suggested as the possible candidate modes. Detailed analysis and taxonomy of use knowledge requirements (Sturrock, et al., 1997) is another issue of importance being conducted under a close cooperation with experienced operators.

Other subsystems, e.g. maintenance record database, acoustic field monitors to be mounted on the robot arm, etc. are in the process of design and implementation. As a basic study toward effective operation of the mobile measurement system, a HMI for operation of the remote manipulator system was developed and tested. Operational performance of the human-machine total system was evaluated for various HMI design modifications and different levels of task complexity. Some of the experimental observations from the study can be found elsewhere (Takahashi, M., 1998). Though the results are promising as a whole, further improvements of the HMI are to be made by taking the experimental observations into consideration.

5. CONCLUDING REMARKS

The basic concept of, and methodological developments in, the research project called Life Cycle Integrity Monitoring are described in this paper. The project was established to significantly enhance operational safety of large-scale artifacts like a nuclear power plant by systematically introducing advanced techniques devoted to maintenance and repair activities. As the key technologies of crucial importance, the life cycle information management system was proposed with accompanying techniques of distributed artificial intelligence and advanced human-machine interface. The local information databases, software robots (worms) and mobile autonomous measurement systems are sample instantiation of the distributed artificial intelligence. The structured functional display is the example of the proposed HMI. Though the prototype systems are positively accepted by domain experts, further developments are necessary. Improvements of individual techniques and attempts to properly integrate the component techniques are being conducted in parallel.

ACKNOWLEDGMENTS

The authors would like to express their sincere thanks to G.Johannsen and E.Averbukh of University of Kassel and E.Hollnagel of OECD Halden Project for their stimulating discussions and comments. They also wish to thank Yuji Niwa of INSS Corp. for valuable comments on the issue of user requirements, and to members of AIR-IHAS Cross-over Research Group for informative discussions. This work was supported by Grant-in-aid for Scientific Research G(A)(2)-09308012.

REFERENCE

Asama,H.(1994). Task Oriented Evaluation System for Maintenance Robots, Proc. AIR'94, Specialists Meeting on Application of AI and Robotics to Nuclear Plants, Tokai, Japan, .11-20

Decortis,F., de Keyser, V., Cacciabue, P.C., Volta, G. (1991). The Temporal Dimension of Man-Machine Interaction. In: Human Computer Interaction and Complex Systems (G.R.S.Weir and J.L.Alty Ed.) 51-72, Academic Press.

Hoc,J-M., Amalberti,R., Boreham,N. (1995). Human Operator Expertise in Diagnosis, Decision-Making, and Time Management. In: Expertise and Technology—Cognition & Human-Computer Cooperation (J-M.Hoc, P.C.Cacciabue, E.Hollnagel Ed.) 19-42, Lawrence Erlbaum Assoc., Hillsdale, NJ.

Johannsen,G. (1992). Towards a New Quality of Automation in Complex Man-Machine Systems, Automatica, 28, 355-373.

Kitamura,M (1994). Keynote Speech, Proc. AIR'94, Specialists Meeting on Application of AI and Robotics to Nuclear Plants, Tokai, Japan, 3-8.

Kitamura, M., Furukawa, H., Kozma,R., Washio,T.(1996). Guiding Rules for Development of Intelligent Monitoring System of Nuclear Power Plants, Proc. SMORN VII; A Symposium on Nuclear Reactor Surveillance and Diagnosis 2,OECD, 493-501, (1996)

Okano,H., Kobayashi, M., Matsuda,K., Ikeda,I. (1996). Development of Intellectual Maintenance System,1996 Annual Meeting of Atomic Energy Soc.Japan, paper #D23 —in Japanese—

Sturrock, F., Kirwan,B., Baber, C.(1997). Using Knowledge Requirements To Define Interface Designs, paper presented at European Conference on Cognitive Science, April 9-11, 1997, Manchester. UK.

Takahashi, M., Takei, S. ,Kitamura,M.(1997). Multimodal Display for Enhanced Situation Awareness Based on Cognitive Diversity, Advances in Human Factors/Ergonomics 21B, 707-710.

Takahashi,M., Fukui,K., Kitamura,M.(1998). Preference-Based MMI for Complex Task Environments, paper to be presented in this conference.

Washio,T., Kitamura,M. (1995). Worm-Type Agents for Intelligent Operation of Large-Scale Man-Machine Systems, Advances in Human Factors/ Ergonomics 20A 925-930.

ANACONDAS : DATA ANALYSIS TO ASSESS WORK ACTIVITY IN SIMULATED CONTROL ROOMS. SETTING UP AN OBSERVATORY OF NUCLEAR POWER PLANT-OPERATION AT EDF

Geneviève FILIPPI *, Geneviève SALIOU *, Patrice PELLE **

*** EDF/DER**
**** EDF/ DSN**

Abstract : The use of simulators is essential to study human factors in emergency conditions of nuclear power plants. In order to ensure efficient coupling between technical arrangements—including procedures—and nuclear power plant operating crews in emergency conditions, EDF has set up an 'observatory': a monitoring system for the collection of simulator data. This paper focuses on a data-collection methodology called ANACONDAS. It will highlight the interest of qualitative models describing and explaining emergency operation on simulators. *Copyright.© 1998 IFAC.*

Keywords: Test and Evaluation, Nuclear power plants, Man-machine systems, Human-centered Design, Cognitive ergonomics, Simulators

I. A NUCLEAR POWER PLANT EMERGENCY -OPERATION OBSERVATORY

In order to ensure efficient coupling between technical arrangements - including procedures - and the operating crews of nuclear plants in emergency conditions, EDF is setting up a monitoring system for the collection of simulator data. This observatory allows the year by year assessment of the evolution of emergency operation. It combines qualitative analysis based on a few tests specifically dedicated to the assessment of emergency operation and quantitative analysis based on data collected during training session by instructors on a large number of tests.

This observatory has effects on the enhancement of procedures and training, and assessment of choices related to work organization.

This paper focuses on a data collection and analysis methodology for the intensive-analysis and will demonstrate the interest of qualitative models of emergency operation on simulators.

II. NEW ISSUES RAISED BY EMERGENCY OPERATION

Simulator test are often considered the equivalent of laboratory situations, where simplification of reality allows the reproduction of precise controlled conditions in order to compare and interpret results by statistical means. In this case, the use of simulators is profitable when the aim is to compare the performance of different crews through indicators such as error rates or the performance of actions important to safety, etc. (Kijima, 1993).

However, it is now necessary to study other issues such as the evolution of knowledge and competence, the role of procedures as a support for reasoning, the modalities for co-operation within the crew (Halbert & Meyer, 1995)

To approach these topics, it is not sufficient to use the quantitative methods based on performance indicators. The crew's activity on the simulator must be studied in depth in order to understand how the operators accord meaning to the situation and construct their point of view during the recovery from incidents. Questions linked to the collective aspects of the work should also be explored: how do the operators collaborate? Are the technical arrangements in the control room - including the procedures - of assistance or do they hinder this collaboration? How do the individual and collective work phases combine to produce an efficient collective activity?

Bearing this in mind, it is possible to obtain revealing results by considering simulator tests not as experimental laboratory protocols , where the aim is to simplify reality in order to control the variables, but as a specific kind of real world environments, in which certain characteristics must be considered in all their complexity.

III. DEFINING RELEVANT 'THEORETICAL OBJECTS'

To address these issues , at least in part, the study of the crew's activity on simulator tests must bring to light sufficiently rich, precise and valid data. As it is, the system composed by the agents and their environment is too complex to be directly appreciated. Consequently, the analysis of human activity supposes a reduction, i.e. the selection of the aspects chosen to be studied Yet this reduction actually should not kill off the essential phenomena that are to be apprehended e.g. (1) the dynamic nature of reasoning, (2) the importance of context in which actions take place, (3) the collective aspects of working activity.

(1) To start with, human activity is in fact global: it cannot be reduced to the sum of separate psychological functions such as planning, diagnosis, decision -making. Actions are not isolated but rather included in a continuum made up of other actions and communications, interpretations, perceptions which are linked over a period of time.

(2) Then, human activity is 'situated' (Suchman, 1987), i.e. the agent acts according to particular circumstances and his objectives at a given moment. Thus, in the simulated control room, activity cannot be reduced to a predetermined task of compliance with prescribed procedures, but on the contrary it is an active adaptation to the context and the social environment. It is worth noting that even if the simulator's social environment is restricted, operators are still constructing a context in order to act.

(3) In control rooms, it is difficult to delineate between individual and collective activity, because multiple forms of co-operation interact to affect each operators work (Heath & Luff, 1991; Filippi & Theureau, 1993)

Taking inspiration from the conceptualization of work analysis based on the study of course-of - action proposed by Theureau (1992), it is possible to describe how reasoning can be dynamically constructed. Activity is comprised of two interwoven parts: a part which can be observed by a third party and a part which can be reported and commented by the agent himself. The course-of-action represents a reduction of the activity to its meaningful part for the agent, in his relationship to the context and his time dynamic.

With this object of analysis, it is possible to report on individual activities and the interactions that an agent has with the others. To approach the collective aspect of working activities in the control room, it is fruitful to combine course-of-action analysis with other theoretical and methodological frameworks, for example those proposed by

- the recent development in ethnomethodology (Goodwin & Goodwin, 1991) and 'situated cognition' where the stress is laid on the joint construction of interaction by partners and the processes of mutual adjustment ;

- cognitive anthropology (Hutchins, 1995) which states that the pertinent level of analysis is not the individual activity, but the collective itself within whose framework cognition is socially distributed between agents and technical artifacts ;

IV. ANACONDA[1]: ANALYZING EMERGENCY OPERATION ON SIMULATORS

1. Data collection

The full-scale simulator replicates an actual control room of a given technological nuclear power plant. The physical simulation of the phenomena is as realistic as possible. Crews from power stations, consisting of a reactor operator, a turbine operator, a supervisor and an operations manager take part in the tests. In certain tests, auxiliary field operators also help to simulate the interactions between the control room and the field.

The scenarios are defined beforehand and enacted several times with different crews. These scenarios often comprise accumulated failures of varying degrees of complexity.

During the test, each operator is followed by an observer. Some of the observers have technical knowledge of the process (process expert), whereas others are trained in social sciences and in activity analysis (human factors experts). Two instructors control the test and play the part of the other departments of the power plant.

Two sorts of description of the activity. In order to better situate the context of the methodology. a

[1] Stands for ANAlyse de la CONDuite Accidenttelle in French.

distinction must be made between two sorts of description of activity:

• one - **intrinsic description** - represents an internal view of cognition from the agent's point of view : it addresses cognition, that is to say reasoning, actions and communications "here and now".

• the other -**extrinsic description**- allows for a characterization of the activity on the basis of its external factors, from the point of view of an observer, for example, the different aspects of the situation such as the state of the simulated process and its evolution during the test, the performances of the crews described by the execution or non-execution of certain important actions and the time taken to execute them, the level of training of the agents and their culture (for example, their previous experience outside the nuclear field).

This distinction between intrinsic and extrinsic is fundamental As in fact demonstrated by various studies, particularly French- language occupational ergonomics studies, the identification of actor's "abilities" outside his work activity leads to scientific limitations with negative practical consequences. To show an "error" is useless if the cognitive process of production of this error is not known. Accordingly, an analysis of the agents' reasoning must be carried out through an intrinsic description of the activity in the light of extrinsic descriptions.

Data gathered. The intrinsic description of the activity is based on the combination of data from observation during the test and data from induced verbalizations after the test. The methodology aims at restricting analysts' interpretations to take account of the phenomena inherent to the activity.

1. data on behavior during the test is gathered by video and audio recordings (2 cameras) of the actions and communications and completed by observers taking notes. Recordings are continuous in order to take account of the dynamics of activity.

2. verbalization data is gathered in self-confrontation interviews: the analyst presents the video tape of the test to the operator so that he may explain his actions and communications and give his interpretation of events. The purpose of these self-confrontation interviews is for the agent to reconstruct the course of his activity in its context.

Transcription of this observation data and verbalizations provides a script that records the temporal organization of the actions. It is the basis for analysis of the data.

The data for *the extrinsic description of the activity* is quite varied. It is mostly relative to :

1. information gathered during debriefing meetings with the crews who took part in the tests for a better knowledge of their training and their experience in the job,
2. the instructors' and process experts' judgments concerning how the scenarios were handled - in terms of technical performance and crew behavior - based on their observations of the tests.
3. all other sources of data, in particular those concerning written evidence, such as the computer log of the test, documents filled out by the operators during the test, etc..

2. Analysis of data

With a view to carrying out more in-depth studies of emergency operation, it is necessary to steer towards modeling the individual and collective work of the operators in the simulated control room. This implies analyzing operators' activity as a whole in order to both describe it's dynamic organization and explain this organization by reconstructing the underlying reasoning. This approach is consistent with a general trend to go beyond traditional cognitive psychology (Winnograd & Florès, 1986; Bannon, 1991; Carrol, 1997; Norros & Nukki,) and to consider the agent's point of view as preferable when it comes to interpreting his activity.

Analysis of emergency operation involves firstly taking account of how each agent construes the global organization of his activity, and secondly identifying regularities in several agents' activity , in other words, modeling the organization of this work activity. Let us consider these two steps of data analysis.

The construction of reduced accounts. First of all, it is a matter of dividing the ongoing course of activity into significant units for the agent and naming them by answering the question " from the agents point of view, what is this about?" These significant units reflect the fact that each action is not isolated but included in sets of temporally organized actions that are coherent for a given agent. Each script divided into significant units in this way produces a reduced account and provides a particular description of the scenario observed.

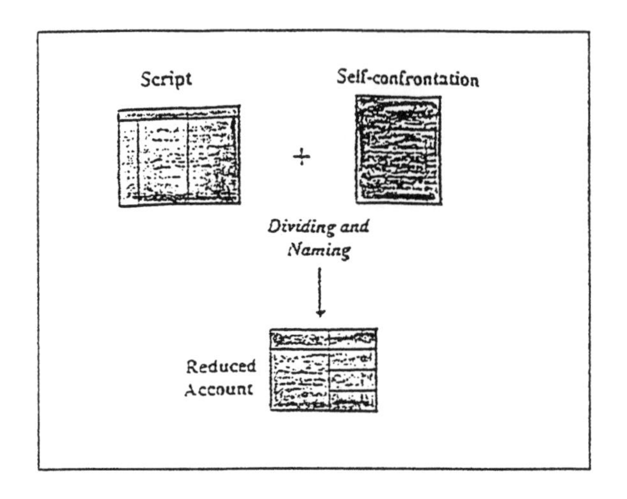

Figure 1 : The Construction of the Reduced Account

The reduced account thus produced represents different levels of activities' structure in which the participants are engaged. In the example in figure 2, the finest level is that of the elementary significant units which records the organization of actions and communications as they emerge step by step. The second level is that of the sequences in which groups of significant units are embedded. The last level is that of the macro-sequences expressing actor's engagement at a higher level. Each level of this structure can be read autonomously and renders a more or less abstract account of the script .

Figure 2 represents the reduced account of a short fragment of the transcription of the Supervisor's activity during a test. In this extract, the succession of meaningful units in the left-hand column are classified in sequences, the first relative to the monitoring of level II operating instruction and the others to the auxiliary feedwater supply loss (AFS) . The Supervisor applies reluctantly the level II operating instruction being aware that the situation requires the application of the level III instruction, because one of the steam generator (SG) is radioactive. However, he can not take this level III instruction until the turbine operator has finished isolating the SG and tick off the state of this SG in a specific box of a crew reference document. At the same time, the Supervisor has noticed that, shortly before and thanks to its procedure that the AFS is no longer functioning; he must at the same time solve this specific problem (macro-sequence "handling the AFS loss") and integrate it in the rest of the operation (sequence "ensuring the request for cooling is appropriate with the AFS loss ").

The Supervisor's course of action		
Elementary significant units (from the point of view of the Sup, what is happening here?)	**Sequences**	**Macro sequences**
1. Asks the Reactor Operator if he has requested a stabilization of the primary coolant temperature →no. it is cooling at 56°C/h	Applying a level II operating instruction while waiting to be directed towards a level III instruction	**Applying an instruction not adapted to the situation while waiting for the adequate ticking off of the steam generators**
2. Is surprised that there has been a cooling requested when there is no auxiliary feedwater supply	Ensuring that the request for cooling is appropriate with the auxiliary feedwater supply loss	**Handling the auxiliary feedwater supply loss**
3. Applies level II operating instruction sequence 4b [monitoring of Steam Generator]	Applying level II instruction while waiting to be directed towards a level III instruction	**Applying an instruction not adapted to the situation while waiting for the adequate ticking off of the steam generators**
4. Reactor Operator asks him if he too is_starting again sequence 4 → no, because he has a problem on the auxiliary feedwater supply		
5. Applies level II p.4b instruction [monitoring of SG] → nothing in this module orients him towards level III operating instruction		
6. Looks at page 4r	Looking for how to pass to level III instruction	
7. Operations manager asks him to apply sheets T204 and T205 to recover the auxiliary feedwater supply pumps	Deal with the AFS loss through application of recovery sheets	**Handling the auxiliary feedwater supply loss**
8. Takes sheets T 204 and T 205		

Figure 2: A Fragment of the Supervisor's Activity Reduced Account

A model of the global organization of activity. The second step in the analysis is modeling the organization of emergency operation during these scenarios. This model corresponds to generalization based on comparison of several reduced accounts. This process of comparison and generalization of reduced accounts reveals archetypal sequences reflecting regularities amongst agents and amongst crews in operation activity on simulators.

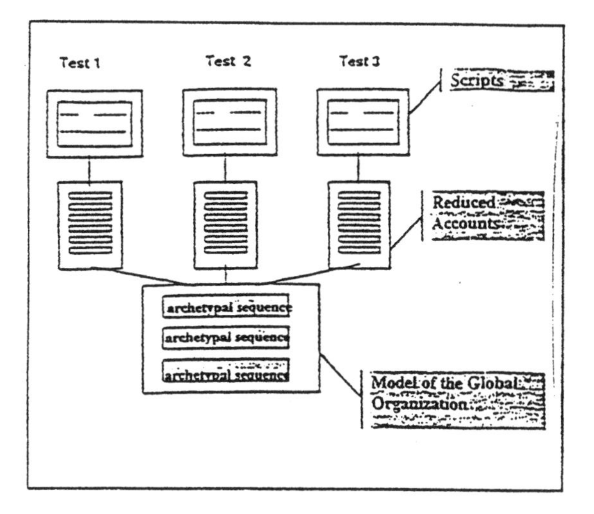

Figure 3: Modeling the global organization of activity

It is the identification of these archetypal sequences that makes it possible to compare the agents logic with the logic underlying the design of the technical setup. It thus becomes possible to examine procedure-design principles with respect to the logic of reasoning and consequently to detect any discrepancies.

V. WORK ACTIVITY LOGIC AND PROCEDURE-DESIGN LOGIC

In order to show the interest of such an approach, this paper will be illustrated by the topic of procedure use in emergency operation on simulators.

In incident or emergency situation, the operators use operation documents presented in the form of logic_diagrams. According to the answers given to different tests relative to the state of the process variables, the operators are supposed to follow step by step specific paths in the procedure where they are asked to carry out actions to bring the nuclear power unit towards off-position enabling normal condition recovery.

A new generation of procedures for emergency operation is in the course of being implanted in French nuclear power plants. These procedures are symptom based, i. e. it is a case of identifying the symptoms rather than diagnosing the event from which the incident originated as was the case of previous event-based procedures. This type of procedure has a very great advantage of dealing easily with multiple failures.

The underlying principle to procedure-design is to assist the operators in managing situations with which they are not used to dealing; even very complex situations that they can not analyze in real time. Kasbi & Lewkowitch (1990) and Dien (1993) have already shown that the monitoring of emergency procedure in simulated control room calls on a great deal of operator skills and know-how.

The tasks predefined by procedure designers do not claim to represent the steps of human reasoning. On the contrary, analysis of the managing of a simulated nuclear power unit in emergency condition highlights the fact that agents do not merely apply procedures, but above all, they solve problems showing their active engagement in the situation.

These problems appear in the reduced accounts with the naming of sequences and macro-sequences. In some cases, they can be directly linked to the process, for example, "understand and correct a drop in primary pressure", in other cases they are linked to the application of the procedure, for example "apply the level II procedure while waiting to be directed towards the level III procedure", or even linked to the activity of the others, for example, "monitor the Turbine Operator activity in his ticking off of the state of the Steam Generators", etc...

The systematic analysis of activity on simulators enables to emphasize the part played by initiative required for the procedures to be applied efficiently in the particular context of each test. It has been noted, thus, that less than half the significant units of a test (the significant units are not proportional in time) correspond to the strict application of the procedure. The other segments of the activity correspond to a permanent task of overall understanding of the situation in order to identify changes and monitor the progression of the test, ensure that the actions requested by the procedures match with the state of the process, interpret the procedures themselves, etc..

Furthermore, these operating documents are primarily designed to assist a 'single user' of procedures. A division of tasks is assigned to the operators and formal times of co-ordination between crew members are prescribed in the procedures. However, this pre-determined co-ordination is not sufficient: a specific job of articulation is implemented by monitoring the course of actions of the other crew members in order to make everyone's work easier. In this sense, operators must co-ordinate collaboration with the ongoing individual activity of others.

In other words, this important part of activity during which other things are carried out besides the strict application of procedures consists actually of building the context of their interpretation and supplying a framework, shared by all, for their application. This ability to put the prescribed tasks into context is founded on operators' know-how and experience.

Bringing to light the significant structures making up the activity of the operators who deal with scenarios enables to inform different situations such as :

1. the procedure makes up an assistance support for problem-solving.
2. the problems are treated independently from the application of instructions.
3. the procedures hinder the understanding of the situation.

The typology of these procedure-using situations and their detailed understanding allows for the enhancement of both their design and the training given to the operators.

VI. CONCLUSION

This type of methodology aiming at fine comprehensive analyses and modeling of the operating activity on simulator is fundamental for explaining the quantitative results derived from other methodologies or for supporting the results derived from more intuitive analyses (e.g. : experts judgment).

The approach described in this paper must be completed by a comparison of these analyses of emergency situations or simulated accidents with analyses derived from perturbed or even incidental situations on units which will enable a better understanding of the whole organization without being restricted to a reduced operating team (to 4 to 7 people) and a reduced simulation of the additional potential resources by the instructors.

REFERENCES

Bannon L, Robinson M. (1991). Questionning Representations, in *Proceedings of the European Conference on Computer-Supported Cooperative Work* , Amsterdam, The Netherlands.

Dien Y. (1993) Safety and Application of Procedures or How Do They Have to Use Operating Procedures in Nuclear Power Plants?; *HT-54/93-19A EDF*; Bad Homburg Workshop.

Filippi G., Theureau J. (1993) Analyzing cooperative work in an Urban Traffic Control Room for the Design of a Co-ordination Support System, *Proceedings of the Third European Conference on Computer Supported Cooperative Work,* G. de Michelis, Simone C, Schmidt K. (Eds) 13-17 September. Milan .

Goodwin C., Goodwin M. (1991). Formulating planes : seeing as a situated activity in Y. Engestrom and D Middleton (Eds) *Cognition and Communication at work.*

Halbert B., P, Meyer P. (1995). *Summary of lessons learned at the OECD Halden Reactor Project for the Design and Evaluation of Human-Machine Systems,* OECD Halden Reactor Project, Halden, Norway.

Heath C., Luff P. (1991). Collaboration and Control : Crisis management and multimedia technology in London Underground Lines Control Rooms *Computer Supported Co-operative Work - ECSCW ' 91,* **Vol 1,** n°s 1-2 .

Hutchins E. (1995)., *Cognition in the Wild.* MIT Press.

Kijima S. (1993). *Plant operator performance analysis using training simulators,* Japan IERE Council.

Suchman , L. (1987). *Plans and situated action : the problem of human/machine communication* ISL Xerox Palo Alto Research Center, Cambridge University Press.

Theureau J. (1992). *Le cours d'action : analyse sémio-logique. Essai d'une anthropologie cognitive située* (The course-of-action : semio-logical analysis. Essay on situated cognitive anthropology) Peter Lang, Berne.

Winograd T., Flores F. (1986). Understanding computers and Cognition: a new Foundation for Design, Norwood, NJ. Abbex.

A FREQUENCY-DOMAIN STOCHASTIC STABILITY CRITERION FOR HUMAN-CONTROLLED MULTI-INPUT MULTI-OUTPUT NONLINEAR SYSTEMS

Tscheh-Kyuhn Oh and Yun-Hyung Chung

Korea Institute of Nuclear Safety
P.O.Box 114, Yusung, Taejon 305-606, Korea

Abstract: A class of human controlled systems is modelled as multi-input multi-output (MIMO) randomly-sampled nonlinear control systems. The transfer function matrix of human controlled systems is defined and used to derive a criterion for the stability 'with probability 1' which can be considered as a stochastic analogue of the well-known circle criterion available for related classical deterministic feedback systems. In order to demonstrate the merits of the proposed criterion and the operator model, a completely worked out example is given. *Copyright ©1998 IFAC*

Key words: manual control, human controlled system, man in the loop, operator model, man-machine system, randomly-sampled control system, stochastic stability, stable 'with probability 1', non-linear control system, circle criterion, queueing theory.

1. INTRODUCTION

A lot of works has been devoted to sampled-data control systems with a randomly varying sampling period (Kuschner and Tobias 1969). This problem occurs, for instance, when the sampling is governed by a random process such as a sonar or radar echo. Newly emerging problems of human controlled systems from the field of nuclear power generation can be also classified as randomly sampled systems. Although Kuschner and Tobias (1969) have, already in the late sixties, referred to the modelling of the power plant operator as a significant applications field of the randomly sampled control system, existing criteria for the stability of such systems require, unfortunately, the explicit determination of a Lyapunov-function involving very tedious calculations. Therefore, no easy-to-use stability criterion is available for randomly sampled-data systems, as known to the authors.

The purpose of this paper is to formulate a stability-criterion in the frequency-domain which can be evaluated without determination of the Lyapunov-function of the involved system. Furthermore, a model of the operator in the emergency operation is proposed. Contrasting to existing models, this model does not require the precise internal representation of the plant to be controlled any more and it is strongly oriented to the queueing theory.

2. HUMAN OPERATOR MODEL AND RANDOM CONCEPT

The history of human operator models is only four decades old, but a great variety of models exist already.

At the Skill-Based-Behaviour-level (SBB-level) of the traditional three-level model of J. Rasmussen, many human operator models have been developed, mainly for manual control and detection tasks, such as the *Describing Function Model* (McRuer and Jex, 1967), and the *Optimal Control Model* (Kleinman et al., 1971) (See for the overview of this domain Stassen, H. G., G. Johannsen and N. Moray 1990).

As well acknowledged by control engineering community, the describing function method does not have a solid mathematical foundation to use as an engineering tool (Willems, 1971). Therefore, McRuer's Describing Function Model cannot address stability issues any more and it needs to be put on a sound theoretical foundation. On the other hand, Kleinman's Optimal Control Model requires that the operator should be so familiar with the processes to be controlled that he has to possess such a precise internal model that his accurate internal model of the processes can be utilized in the development of the human operator models.

However, it is an illusion to assume that the human operator has a precise internal model of the plant during an incident with which he is not familiar. Therefore, it is reasonable to develope the human operator model which necessitates only a rough internal model, if any.

Fuzzy logic controller: The so-called *cognitive controller* is based on a linguistic formulation of the non-mathematical knowledges from experts controlling the industrial processes. Very often, the interviewed expert can only say how the process input is to adjust in accordance with changing process outputs. In the course of this, the dynamic behaviour of system is not completely known. To circumvent the problems of precise internal representations, it is only natural to try to construct the human controller model on the basis of the concept of the fuzzy logic controller which is now quite popular in industrial processes, such as warm water plant, heat exchanger system, cement Kiln and so on. In other words, the designer of the fuzzy logic controller, models the behaviour of experts serving the process, instead of the process being served by the experts who have, possibly, only vague internal representations.

Network model: it is possible to create a network for human tasks and subtasks associated with their estimated completion time distributions and probabilities of successful completion. With a mathematical approach such as Monte Carlo Simulation, the overall completion time and probability of the successful completion of the job can be obtained (Baron *et al.*, 1982). However, the network models are totally specialized for dealing with procedural activities and discrete actions. The model is uncapable of dealing with the dynamic closed loop behaviours of human-machine system such as the control tasks which is one of the main tasks of an operator of Nuclear Power Plants (NPP).

The tasks assigned to the NPP-operator can be classified into state-continuous tasks such as the steam generator water-level control, the reactor coolant temperature control, and state discrete tasks which include, for instance, alarm handling and checking of equipment states. Although the state continuous tasks normally have a higher priority than discrete tasks, the performance of continuous tasks are unavoidably affected by state-discrete tasks. Hence, it is of great importance to quantitatively investigate to what extent a human operator can be interfered by the subsidiary state-discrete tasks without failing to meet the performance requirement for the main control tasks assigned to him.

Therefore, a need exists clearly for the stability criterion for the human-controlled systems on the basis of which the job of the human-operator can be designed in the early stage of man-machine system development. The early design-decision based on a stability-criterion with a solid theoretical foundation can save the time of useless trial and errors and the valuable development cost.

In this study, the control behaviours of NPP--operators will be modelled by means of an approach which is based on the modern control theories, such as the fuzzy system theory, and, in addition, is closely related to the queueing theory. This enables a symbiosis of several relevant features of existing models and the control theoretic model which is to be developed in the sequel.

2.1 Stochastic Task Duration

In the given accident situation, the primary state-continuous task of a turbine operator is the control of the water levels of four distinct steam generators. His secondary state-continuous task is the control of the steam generator pressure to depressurize and to cool down the reactor coolant system. However, he cannot pay attention simultaneously to all eight signals. In addition to state-continuous control tasks, he has to serve the environments such as the steam bypass system, the condenser system, the turbine system and so on. They impose dominantly state discrete tasks on him such as confirming component states and alarm handling which can involve not only Skill-Based-Behaviours (SBB) but also Rule--Based-Behaviours (RBB) of Rasmussen-model.

Hence, the operator can perform his control tasks only in a sequential manner so that merely a restricted time slice of the control period can be allocated to steam generator control tasks.

The time interval between k-th control activities and (k+1)-th control activities for a given steam generator is of a considerable magnitude of order because the operator has to accomplish the control tasks for three other steam generators as well as the environmental tasks. At the time $t=t_k$ the operator begins with the control task for SG_1 and at $t=t_k + \tau_{k,1}$ finishes with it, where $\tau_{k,1}$ is the completion time or the duration of the control task at the steam generator SG_1 in the k-th task cycle. He moves under the circumstance to the environment and performs the environmental task until $t=t_k + \tau_{k,1} + \zeta_{k,1}$, where $\zeta_{k,1}$ is the completion time or the duration of the environmental task associated with the steam generator SG_1 in k-th task cycle. At $t_{k+1} = t_k + (\tau_{k,1} + \zeta_{k,1}) + (\tau_{k,2} + \zeta_{k,2}) + (\tau_{k,3} + \zeta_{k,3}) + (\tau_{k,4} + \zeta_{k,4})$, the turbine operator finishes with the task associated with SG_4 and can begin with the task for SG_1 again. Also, the. task cycle is re-initiated. The duration of a task-cycle can be determined as follows:

$$T_k \equiv t_{k+1} - t_k = \sum_{i=1}^{4}(\tau_{k,i} + \zeta_{k,i}).$$

Since human behaviours are fairly irregular particularly in stressful emergency situations, the duration of a task-cycle can be considered as a random variable. Furthermore, the operator in the incidental conditions accomplishes, in a task cycle, a lot of tasks so that he cannot remember what he has performed at SG_1 a long time ago during the previous task cycle. Consequently, the behaviour of the operator in the k+1-th task-cycle is nearly independent of his behaviour in the k-th task cycle. This means that the sequence of the task durations $\{T_k,\ k \in \boldsymbol{N}\}$ can be assumed to be a *white stochastic process*.

2.2 Probability Density Function

The sequential process of steam generator water level control tasks is a special case of the

Markov process in which transitions are permitted only from state SG_i representing the control task at the steam generator i to the neighboring state ENV_i representing the associated environmental task or the subsequent state SG_{i+1}, as depicted in Fig. 1. Since the period of the task cycle $T(k)$ is a random variable, it must be specified by its stochastic characteristics, i.e., the state probability or the state Probability Density Function (PDF).

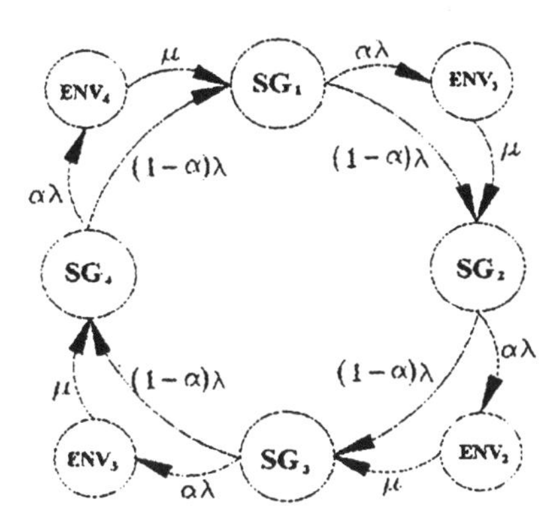

Fig. 1 A Task Network of the Turbine Operator as a Markov-Graph

Following the common notation of the queueing theory, let λ_i be the transition rate of the state SG_i at which the operator departs the state SG_i. Similarly, let μ_i be the transition rate of ENV_i at which the operator departs the environmental state ENV_i. The process will be a Markov process on the states SG_1, SG_2, SG_3, SG_4 and ENV_1, ENV_2, ENV_3, ENV_4, so that transitions are independent. The parameter a signifies the probability that the operator decides for the environmental task as the next job.

The fashion in which this four-stage process functions is that, upon finishing his task, the operator immediately moves from the fourth steam generator or from the neighbouring environment to the first steam generator. At least in the case illustrated in Fig. 1, PDF of the duration of the task cycle $T(k)$ can be determined by means of elementary calculations.

The environmental states consist of numerous sub-states so that their Markov-graph attains a remarkable size. Nevertheless, as a consequence of the well-known central limit theorem of the probability theory, the PDF of the duration of task cycle converges to the normal distribution especially in the case of many subtasks ($n>>1$), independently of the distribution of the duration of its subtasks. This implies that the busier the operator is, i.e., the more complex his task is, the more precisely his behaviour can be described by the standard Gaussian distribution (a consequence of the central limit theorem!).

3. RANDOMLY SAMPLED FEEDBACK CONTROL SYSTEM

Let us consider the non-linear sampled-data control system of Fig. 2. The sampler samples at a sequence of random time $\{ t(k), k \in \boldsymbol{N} \}$ and the holding intervals or sampling intervals $T(k)$ are defined by $T(k) = t(k+1) - t(k)$. The sequence of sampling intervals $\{T(k), k \in \boldsymbol{N}\}$ will be a stationary white stochastic process.

Let the [mxm]-matrix of transfer-functions $\underline{G}_{\infty}(s) \in \mathbf{C}^{m \times m}$ be given. Then, using established results of realization theory (Schwarz, 1971), one obtains this state space representation of the control system depicted in Fig. 2:

$$\boldsymbol{x}(k+1) = \boldsymbol{\Phi}[T(k)]\,\boldsymbol{x}(k) - \underline{H}[T(k)]\,\boldsymbol{\varphi}[\boldsymbol{y}(k)]$$
$$\boldsymbol{y}(k) = \underline{C}^t \boldsymbol{x}(k)$$
$$0 \le \varphi_i[y_i(k)] \cdot y_i(k) \le K_i \cdot y_i(k)^2, \quad \varphi_i[0]=0, \ i=1..m \tag{3.1}$$

where $\boldsymbol{\Phi}[T(k)] = e^{-\Lambda T(k)} \in \boldsymbol{R}^{n \times n}$,

$$\underline{H}[T(k)] = \underline{\Lambda}^{-1}(\underline{I} - e^{-\Lambda T(k)})^{-1}\underline{B} \in \boldsymbol{R}^{n \times m},$$
$$\underline{\Lambda} = Diag[\lambda_1, \lambda_2, \ldots, \lambda_n] \in \boldsymbol{R}^{n \times n},$$
$$\underline{K} = Diag[K_1, K_2, \ldots, K_m] \in \boldsymbol{R}^{m \times m},$$
$$\underline{C}^t \in \boldsymbol{R}^{m \times n}, \ , \ \underline{B} \in \boldsymbol{R}^{n \times m},$$
$$\boldsymbol{\varphi}[\boldsymbol{y}(k)] \in \boldsymbol{R}^m \quad \boldsymbol{y} \in \boldsymbol{R}^m.$$

The inequality; $0 \le \varphi_i[y_i(k)] \cdot y_i(k) \le K_i \cdot y_i(k)^2$, i = 1, „ ,m restricts the non-linearity $\varphi_i[y_i(k)]$ to a sector in the $[y, \varphi_i]$-plane, and this will be referred to as a non-linearity in the sector $[0, K_i]$. The positive real number K_i will be referred to as the upper bound of sector $[0, K_i]$.

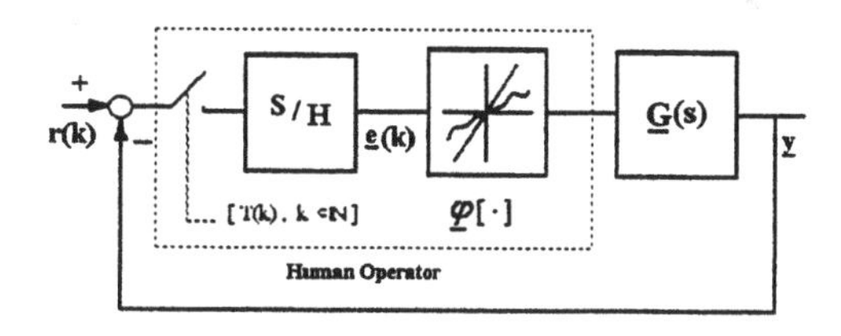

Fig. 2 Human Operator as Randomly Sampling Non-Linear Controller

In this paper, human-controlled systems will be viewed as non-linear systems with randomly varying sampling intervals. The non-linearities stem from the human as a controller in one part and from actuator characteristics in other part. In general, the human behaviour is hard to predict so that non-linearities can also be in the time variable form. In the case of the Single Input Single Output System(SISO), this fact can be taken into account by randomizing the sector-intervals K_i, as already demonstrated by Oh et al. (1997).

3. 1 *Fuzzy logic controller and sector bound non-linearity*

It is well-known that the fuzzy controllers can be viewed as a multi-level relay under certain assumptions (Kickert and Mamdani, 1978; Ray and Majumder,1984). As depicted in Fig. 3, a multi-level relay can satisfy the well-known sector-condition. Furthermore, because every interval-wise continuous function can arbitrarily be well approximated by a stair-case function, i.e., a multi-level relay, humans modelled as fuzzy controllers can meet the sector condition. In this fashion, the non-mathematical knowledges of field experts can be embedded into the set of sector bound non-linearities. This approach appeals more to the intuition than the mathematical representation of the plants to be controlled such as Kalman-filter and predictor.

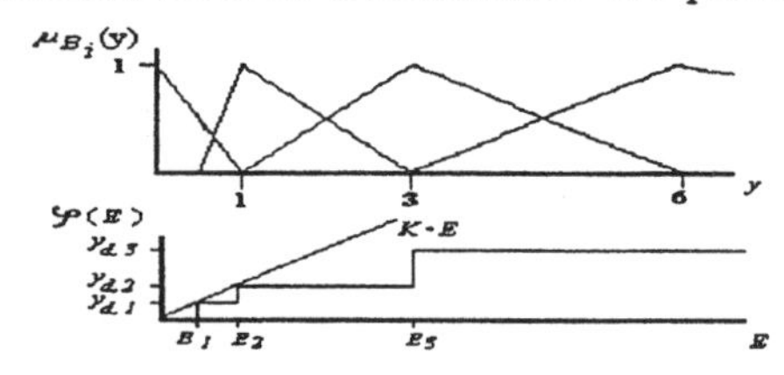

Fig. 3 Fuzzy logic Controller as a Multi-Level Relay

Nevertheless, even a fuzzy controller is a deterministic controller while a human as a controller shows a lot of irregularities. As mentioned afore, this fact can be handled by randomizing the sector-interval K_i, with i=1,, m.

4. MAIN RESULTS

The main objective of this paragraph is to build a bridge between the supermartingale theorem (Bucy, 1965; Kuschner, 1969) and the circle--criterion. This can be done by using an extension of Kalman-Yakubovitch-lemma (Szegö and Pearson Jr., 1964; Oh, 1980).

It is, therefore, fundamental to the development in the sequel to put the stochastic difference equation system (3.1) into the form on which Kalman-Yakubovitch-Lemma can be applied. So, we introduce these definitions:

$$\begin{aligned}
\Delta\underline{\Phi}(k) &= e^{-\Lambda T(k)} - E\{e^{-\Lambda T(k)}\} \\
&= \underline{\Phi}(k) - \underline{\Phi}_m. \\
\underline{\Phi}_m &= E\{e^{-\Lambda T(k)}\}, \\
\underline{H}_m &= \underline{\Lambda}^{-1}(\underline{I} - E\{e^{-\Lambda T(k)}\})\underline{B} \\
&= \underline{\Lambda}^{-1}(\underline{I} - \underline{\Phi}_m)\underline{B}, \\
\Delta\underline{H}(k) &= -\underline{\Lambda}^{-1}(e^{-\Lambda T(k)} - E\{e^{-\Lambda T(k)}\})\,\underline{B}, \\
&= -\underline{\Lambda}^{-1}\Delta\underline{\Phi}(k)\,\underline{B}
\end{aligned}$$

and rewrite (3.1) as follows:

$$\begin{aligned}
\underline{x}(k+1) &= \underline{\Phi}_m\underline{x}(k) - \underline{H}_m\underline{\varphi}[\underline{y}(k)] + \\
&\quad \Delta\underline{\Phi}(k)\,\underline{x}(k) - \Delta\underline{H}(k)\,\underline{\varphi}[\underline{y}(k)] \\
\underline{y}(k) &= \underline{C}^t\underline{x}(k),
\end{aligned}$$

$$0 \le \varphi_i[y_i(k)]\cdot y_i(k) \le K_i\cdot y_i(k)^2,\quad \varphi_i[0] = 0. \quad (4.1)$$

After these preparations, it is possible to give the stability criterion which is based on the early results in the unpublished dissertation of one of the authors (Oh, 1980).

Theorem 1 (Frequency-domain stability criterion)

Let $\{T(k),\ k\in N\}$ be the sampling process of the randomly sampled-data control system (4.1). If there exists a triple of real numbers $[\rho^2, \sigma^2, \delta]$ so that the following [nxn]-matrices are positive semi-definite:

1a)

$$\left[\frac{\frac{1-\rho^2}{1+\delta} - E\{\Delta\phi_i(k)\Delta\phi_j(k)\}}{\rho^2 - \phi_i\phi_j}\right] \ge \underline{O} \in R^{n\times n}$$

1b)

$$\left[\frac{\frac{\sigma^2}{1+\delta^{-1}} - \frac{E\{\Delta\phi_i(k)\Delta\phi_j(k)\}}{(1-\phi_i)(1-\phi_j)}}{\rho^2 - \phi_i\phi_j}\right] \ge \underline{O} \in R^{n\times n},$$

2a) all eigenvalues of the expectation of the [nxn]-dynamic-matrix; $\underline{\Phi}_m \in R^{n\times n}$ are inside the circle $\{z \in C \mid |z| = \rho\}$ so that it holds;

$$\phi_i^2 < \rho^2 < 1.0,\quad i = 1, 2, \ldots, n,$$

2b) for an arbitrary [mxm]-diagonal-matrix $\underline{Q} = Diag[Q_1, Q_2, \ldots, Q_m] \in R_+^{m\times m}$ this frequency domain condition is satisfied;

$$\bigwedge_{|z|=\rho} \frac{1}{2}[\underline{Q}\cdot\underline{G}(z)\cdot\underline{Q}^{-1} + \underline{Q}^{-1}\cdot\underline{G}(\bar{z})^t\cdot\underline{Q}] + \underline{K}^{-1} > \underline{0}$$

where $\underline{G}(z) = \underline{C}^t(z\underline{I} - \underline{\Phi}_m)^{-1}\underline{H}_m$,

$\underline{K} = Diag[K_1, K_2, \ldots, K_m] \in R_+^{m\times m}$ and

3) the m-dimensional vector-valued non-linearity $\underline{\varphi}[\underline{y}(k)] \in R^m$ satisfies the sector-conditions:

$$0 \le \varphi_i[y_i(k)]\cdot y_i(k) \le \frac{K_i}{1.0+\sigma^2}\,y_i^2(k),\quad \varphi_i[0] = 0,$$

then, the non-linear randomly sampled-data system (4.1) is asymptotically stable with probability one -wp1-, that is, for every positive real number ε, it holds that

$$\bigwedge_{\varepsilon\in R^+} \lim_{k\to\infty} P\{\sup_{m>k} \|\underline{x}(m)\| > \varepsilon\} = 0.$$

Proof: (See Oh, 1980).

The strength of the theorem 1 is in being independent of Lyapunov-function $V[\underline{x}(k)] = \underline{x}^t(k)\underline{L}\underline{x}(k)$ or the unknown [nxn]-matrix $\underline{L}$ except the triple of the real-valued parameters $[\rho^2, \sigma^2, \delta]$ which can be determined by means of simple graphical techniques as will be shown later. In addition, one of authors has already shown that SISO-system with complex-valued multiple poles can be handled with SISO-version of theorem 1. Further extension of theorem 1 is easy and will be omitted (Oh et al., 1997)

5. USABILITY DEMONSTRATION OF THE DERIVED STABILITY CRITERION

Now, it will be demonstrated that Theorem 1 represents a useful engineering tool. Indeed, examination of the condition 1a), 1b) and 2) proceeds fairly straightforward so that the application seems not to be difficult.

5.1 An Example

Let the transfer function matrix be given:

$$\underline{G}_\infty(s) = \begin{bmatrix} \frac{3}{4}, & 0 \\ \frac{1}{2}V, & \frac{3}{2} \end{bmatrix} * \frac{1}{(s+1)} + \begin{bmatrix} 0, & 0 \\ \frac{1}{2}V, & \frac{6}{2} \end{bmatrix} * \frac{1}{(s+2)}. \quad (5.1)$$

Furthermore, for the sake of the simplicity, the 'in the strict sense' stationary white sampling process $\{T(k),\ k\in N\}$ with the probability density function:

$$p_{T(k)} = 5 \cdot I_{[0.8,1.0]}(\omega) \equiv \begin{cases} 5 & 0.8 \leq \omega \leq 1.0 \\ 0 & \text{otherwise} \end{cases} \qquad (5.2)$$

is chosen.

In general, the explicit determination of a minimal realization of Multi-Input Multi-Output-system requires a time-consuming hard works (Schwarz, 1971). Fortunately, the transfer function matrix of the randomly-sampled control system can be obtained without the explicit knowledge of a minimal realization associated with (5.1), by using the so-called technique of 3-transformation of the Laplace-transform well--known in the traditional deterministic sampled--data control theory:

$$\mathfrak{Z}[\frac{1}{s+1}] := \frac{(z-1)}{z} \mathcal{Z}[\frac{1}{(s+1)s}] :=$$
$$\frac{(z-1)}{z}(\frac{z}{(z-1)} - \frac{z}{(z-e^{T(k)})}) = \frac{(1-e^{T(k)})}{(z-e^{T(k)})},$$
$$\mathfrak{Z}[\frac{1}{s+2}] := \frac{(z-1)}{z} \mathcal{Z}[\frac{1}{(s+2)s}] :=$$
$$\frac{1}{2}\frac{(z-1)}{z}(\frac{z}{(z-1)} - \frac{z}{(z-e^{2T(k)})}) = \frac{1}{2}\frac{(1-e^{2T(k)})}{(z-e^{2T(k)})}.$$

Replacing the random variables $e^{T(k)}$, $e^{2T(k)}$ by their expectations $E\{e^{T(k)}\}$, $E\{e^{2T(k)}\}$ the following is obtained (For the justification see Oh, 1980):

$$\underline{G}(z) = \begin{bmatrix} \frac{3}{4}, & 0 \\ \frac{V}{2}, & \frac{3}{2} \end{bmatrix} * \frac{1-E\{e^{-T(k)}\}}{z-E\{e^{-T(k)}\}} + \begin{bmatrix} 0, & 0 \\ \frac{V}{4}, & \frac{3}{2} \end{bmatrix} * \frac{1-E\{e^{-2T(k)}\}}{z-E\{e^{-2T(k)}\}}. \qquad (5.3)$$

Now, one can see clearly that the explicit knowledge of the minimal realization of $\underline{G}_{\infty}(s)$ is not necessary to evaluate condition 2.

Step 1a (Evaluation of Condition 1a): To examine the condition 1a (and 1b) the following variables are needed;

$$\phi_{mi} = E\{\phi_i(k)\}\ , \quad E\{\Delta\phi_i(k)\Delta\phi_j(k)\}$$

to build this matrix:

$$\left[\frac{\frac{1-\rho^2}{1-\delta} - E\{\Delta\phi_i(k)\Delta\phi_j(k)\}}{\rho^2 - \phi_{mi}\phi_{mj}} \right] \geq \underline{0}\ . \qquad (5.4)$$

The Sylvester-criterion can be used to test the positive-definiteness of the matrix given above. The dynamic matrix of the minimal realization associated with (5.1) can have identical elements stemming from various subsystems (Schwarz, 1971), matrix (5.4) contains identical rows and columns so that it's determinant vanishes. Since condition 1a does not require the strict positive-definiteness of (5.4) but merely semi-definiteness, identical columns and rows can be deleted except for the last. This means that matrices occurring in condition 1a and 1b can be determined by a simple inspection of poles in the transfer-function matrix of continuous systems in (5.1) (For details see Oh, 1980).

Now, let $F_2(\rho^2, \delta)$ be the determinant of the i-th principal minor of (5.4).. As an example, the graph of $F_2(\rho^2, \delta)$ is illustrated with various values of δ in Fig. 4. For the parameter value δ = 0.5, the largest value ρ^2 = 0.998 is chosen for which $F_2(\rho^2, \delta)$ remains positive.

Step 1b (Evaluation of Condition 1b): Taking the values chosen in the previous step; δ = 0.5, $\rho^2 = 0.998$, the following matrix is obtained for the examination of the condition 1b:

$$\left[\frac{\frac{\sigma^2}{1-\delta^{-1}} - \frac{E\{\Delta\phi_i(k)\Delta\phi_j(k)\}}{(1-\phi_{mi})(1-\phi_{mj})}}{\rho^2 - \phi_{mi}\phi_{mj}} \right] \geq \underline{0}\ .$$

In a similar way to the condition 1a, the determinant of the second principal minor $G_2(\delta, \sigma^2, \rho^2)$ of the corresponding matrix in condition 1b can be evaluated with various values of δ and ρ^2. For the chosen pair of parameter; {δ = 0.5, ρ^2 = 0.998 } the smallest value of σ^2 which does not make the function $G_2(\delta, \sigma^2, \rho^2)$ negative is taken: σ^2 = 0.070.

Step 2): Condition 2 resembles the classic circle criterion and, apparently, represents a stochastic analogue of the circle-criterion.

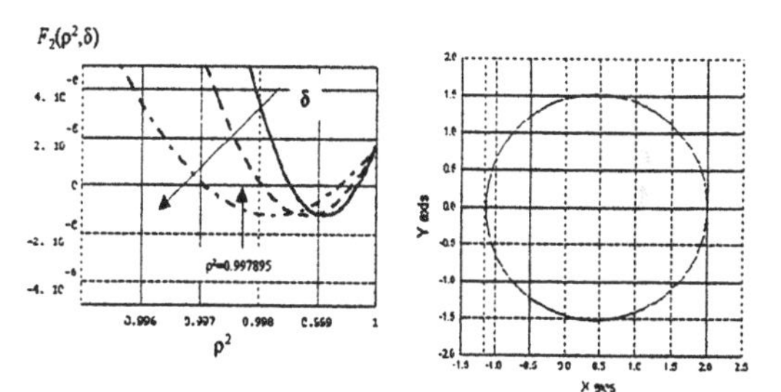

Fig. 4 Graph of $F_2(\rho^2, \delta)$

Fig. 5 Nyquist locus of $G_{22}(z)$.

The examination of this condition requires the investigation of the complex-valued function-matrix:

$$\underline{G}(z) = \underline{C}^T(z\underline{I} - \underline{\Phi}_m)^{-1}\underline{H}_m = \begin{bmatrix} \frac{3}{4}G_{11}(z), & 0 \\ \frac{V}{2}G_{21}(z), & \frac{3}{2}G_{22}(z) \end{bmatrix}$$

with

$$G_{11}(z) = \frac{1-E\{e^{-T(k)}\}}{z-E\{e^{-T(k)}\}},$$
$$G_{21}(z) = \frac{1-E\{e^{-T(k)}\}}{z-E\{e^{-T(k)}\}} + \frac{1}{2}\frac{1-E\{e^{-2T(k)}\}}{z-E\{e^{-2T(k)}\}}$$
$$G_{22}(z) = \frac{1-E\{e^{-T(k)}\}}{z-E\{e^{-T(k)}\}} + \frac{1-E\{e^{-2T(k)}\}}{z-E\{e^{-2T(k)}\}}.$$

According to the Sylvester-criterion, condition 2 is satisfied if these relations hold:

$$\frac{3}{4}Re\{G_{11}(z)\} + \frac{1}{K_1} > 0\ , \quad \frac{3}{2}Re\{G_{22}(z)\} + \frac{1}{K_2} > 0\ ,$$
$$(\frac{3}{4}Re\{G_{11}(z)\} + \frac{1}{K_1}) \cdot (\frac{3}{2}Re\{G_{22}(z)\} + \frac{1}{K_2}) > \| \frac{Q_2}{Q_1}\frac{V}{2}G_{21}(z) \|^2.$$

Choosing the pair of arbitrary positive numbers $[Q_1, Q_2]$ so that $Q_2/Q_1 \to 0$, one can see that condition 2 is satisfied, if these relations hold:

$$\frac{3}{4}Re\{G_{11}(z)\} + \frac{1}{K_1} > 0\ , \quad \frac{3}{2}Re\{G_{22}(z)\} + \frac{1}{K_2} > 0\ .$$

In the case of the control systems with a transfer function matrix of the triangle form, it is intuitively clear that the cross-coupling over

$V/2 \cdot G_{21}(z)$ can certainly affect the system dynamic but can hardly influence the asymptotic behaviour. The upper limit of the sector interval K_2 can be determined from Nyquist-locus depicted in Fig. 5: $K_2 = 2/(1.137*3)$ for $\rho^2 = 0.998$.

Since $3/4 \cdot G_{11}(z)$ is a system of the first order, the determination of K_1 is trivial and therefore will be omitted.

Step 3 (Evaluation of Condition 3): Now, the permitted upper limit of the sector interval $\underline{K}_{2\sigma} = \underline{K}_2 / (1+\sigma^2)$ can be obtained

As shown in step 1a, ρ^2 is located close to the neighbourhood of 1.0 and, therefore, the variance $\Delta\rho^2 = 1-\rho^2$ of the dynamic matrix is a very small number, in the case where the variance of the sampling process $E\{(T(k)-E\{T(k)\})^2\} \approx 0.00011$ is not substantially larger than in the demonstration.

Table 1 The permitted Upper Limit K_σ of the sector interval [1, K]

δ	σ^2	K_2	$K_{2\sigma}$
0.5	0.070	0.5863	0.5478

The variance δ^2 of the input matrix $\underline{H}[T(k)]$ is also quite a small number, as shown in Table 1. Because the transfer function matrix of the human-controlled system $\underline{G}(z) = \underline{C}'(z\underline{I} - \underline{\Phi}_m)^{-1} \cdot \underline{H}_m$ only needs to show a very small exponential stability degree $\Delta\rho^2 = 1-\rho^2 \approx 0.0$ the condition 2 in theorem 1 is not a hard requirement. It can now be seen that, in the case of the human-controlled system, the transfer function matrix $\underline{G}(z)$ formed out of the expectation of the system matrices has to show the degree of the exponential stability corresponding to the variance $\Delta\rho^2 = 1-\rho^2$ of the dynamic matrix $\underline{\Phi}[T(k)]$ and the upper limit of sector intervals K_i must be slightly reduced in accordance with the variance of the input matrix σ^2, namely, by the factor $1.0 / (1+\sigma^2)$.

Because the variance of the sampling process in many practical situations is small, the stability of the randomly-sampled or human-controlled system is determined mainly by the expectation of its system matrices $[\underline{C}', \underline{\Phi}_m, \underline{H}_m]$, equivalently, by the condition 2 in theorem 1.

This means that the human-controlled system can be treated as a deterministic control system with a constant sampling interval $T(k)$ = Const.. Consequently, the stability problem of human--controlled systems can be tackled with the results of the known deterministic control theory.

The pair $[\underline{\Phi}_m, \underline{H}_m]$ of the expectation of system-matrices depends upon the sampling process, i.e., the duration of the task-cycle $T(k)$. This cycle is partly determined by the subsidiary state-discrete tasks. Overloading NPP-operator with subsidiary tasks results in the unduly long sampling interval so that the stability of the human-controlled system can no longer be warranteed. However, this problem can be detected in the early stage of the man-machine system design by means of theorem 1 and can be resolved, for instance, by the timely reallocation of tasks of the operating crew.

6. CONCLUSION

In spite of the continuing efforts for the automation of the process industries, there exist some domains which will be reserved for humans even into the far future. The emergency operation during an incident in a nuclear power plant represents one of the domains reserved for the human operator. In contrast to the existing operator models (Kleinman, 1971; McRuer, 1967) whereby the human acts as a time-continuous controller with a precise internal representation of plants, the proposed operator model does not require the precise internal representation such as Kalman-filter.

The operators of NPP perform their tasks sequentially. Consequently, they inherently represent for time-discrete stochastic controllers. The above fact is reflected in the model by introducing a random sampling process. Furthermore, the stability-criterion presented in this paper gives a bright prospect and technical tool for quantitatively handling the random behaviour of human operators in the control loop.

REFERENCES

Baron, S. *et al.* (1982). An approach to Modelling Supervisory Control of Nuclear Power Plant *NUREG/CR--2988* N. R. C., Washington, DC

Bucy, R. C. (1965). Stability and positive Supermartingale *J. Differential Equations,* **Vol. 1**

Kickert, W. J. M. and E. H. Mamdani (1978) Analysis of a Fuzzy Logic Controller *Int. J. Fuzzy Sets and Systems* **Vol. 1 No. 1**

Kleinman, D. L., S. Baron and W. L. Levison (1971) A control theoretic approach to manned-vehicle system analysis *IEEE Trans. on Automatic Control,* **Vol. AC-16**

Kuschner, H. J. and L. Tobias (1969) On the Stability of Randomly Sampled Systems *IEEE Trans. Automatic Control* **Vol. AC-14, No. 4**

McRuer, D. T. and H. R. Jex (1967) A review of quasi-linear pilot models *IEEE Trans. Human Factors Electr.* **HFE-8**

Oh, Tscheh-Kyuhn (1980) Non-linear Control of Artificial Heart via Random Sampling *an Unpublished Dr.-Ing. Dissertation, in German* Technical Uni. Berlin, Berlin Germany

Oh, T. K., Y. C. Choi, and C. H. Chung (1997) A Frequency-Domain Stochastic Stability-Criterion for Human-Controlled Non-Linear Systems *Proc. of the 1997 IEEE 6-th Conf. Human Factors and Power Plants* June 8-13 1997 Orlando, Florida

Ray, K. S. and D. D. Majumder (1984) Application of Circle Criterion for Stability Analysis of Linear SISO and MIMO Systems Associated with Fuzzy Logic Controller *IEEE Trans. Systems, Man, and Cybernetics* **Vol. SMC-14, No. 2**

Schwarz, H. (1971) *Mehrfachregelungen* Springer--Verlag, Berlin, Heidelberg, New York

Stassen, H. G., G. Johannsen and N. Moray (1990) Internal Representation, Internal Model, Human Performance Model and Mental Work load *Automatica* **Vol. 26. No. 4**

Szegö, C. P. and J. B. Pearson Jr. (1964) On the Absolute Stability of Sampled Data System *IEEE Trans. On Automatic Control* **AC-9, No. 2**

Willems, Jan C. (1971) *Analysis of Feedback systems* M.I.T. Press , Cambridge, MA

DEVELOPMENT OF A PLANT NAVIGATION SYSTEM

Tsuneo NAKAGAWA[1)], Tomihiko FURUTA[1,†)], Ryuji KUBOTA[2)], Kouji IKEDA[2)]

1) Nuclear Power Engineering Corporation
3-17-1, Toranomon, Minato-ku, Tokyo 105-0001, JAPAN
†) present address: Toyo University
2) Hitachi Works, Hitachi, Ltd.
3-1-1, Saiwai-cho, Hitachi-shi, Ibaraki-ken 317-0073, JAPAN

Abstract : A prototype of a Plant Navigation System (PNS) has been developed to assist nuclear power plant (NPP) operators by automatically displaying the plant situation and plant operation procedures on a CRT screen when abnormalities occur. The operation procedures given in a symptom-oriented manual are expressed in a tree-type flowchart. The optimum operation procedure is selected automatically using built-in diagnostic logics based on the current status of the NPP. Copyright © 1998 IFAC

Keywords: Nuclear Power Plants, Operation Procedure, Navigation, Support System

1. INTRODUCTION

The development of automobile navigation systems which assist drivers of cars or similar systems has made remarkable progress. During operation, these navigation systems help to reduce human errors.

For nuclear power plants (NPPs), there are operator support systems which have already been developed and introduced. Some of these systems have been targeted to display only the measured physical process parameters on a plant block diagram or on trend charts. These systems only display the operational situation of NPPs. Recently, development of a system to display the operation manual on a CRT screen has been performed (Hollnagel and Niwa, 1996). A system to display the symptom-based emergency procedure guidelines (EPG) has been developed (Saijou *et al.*, 1998).

The aim of the present work is to develop a Plant Navigation System (PNS) which displays both the operation procedures and plant situation on a CRT screen. In NPP operation at abnormal evens, operation manuals (EPG) correspond to maps used in the automobile navigation systems. The PNS will have a similar function to that of the automobile navigation systems by showing suitable operations to be performed and progress of operations on the graphically displayed EPG.

2. SCOPE OF PNS

The target of the PNS is to support NPP operators under time pressure during operating conditions from the occurrence of scram to the prevention of degraded core. The PNS is intended to support a supervisor by showing the most suitable operation procedures and important plant parameters at abnormal events since he is responsible for judging the operation procedures and monitoring the plant parameters.

The PNS dose not force the supervisor to agree with its recommendation. This system is just a support system for the supervisor, and always shows the optimum procedures and plant information even if the supervisor select another procedure.

The PNS displays the symptom-based operation

manual. An object of the current version of PNS is a BWR type NPP.

3. DIAGRAMMTIC LANGUAGE SUITABLE FOR PNS DISPLAY

Different types of maps are used in automobile, ship and other navigation systems. However, the following are commonly shown on the map in the navigation systems:

a) The present position.
b) The goal position and the path to the goal position.
c) The start position and the path up to the present position.

In the case of NPPs, operation manuals correspond to the maps used in the automobile navigation systems. However, it is difficult to grasp the "start position", "present position" and "goal position", and to find a suitable path to the "goal position" from printed operation manuals. Diagrammatic languages such as flowcharts, therefore, should be adopted to draw a "map" and to make grasping the "positions" and "path" easier .

Flowcharts, which are frequently used for describing operation procedures, etc., were originally developed for computer algorithm. Therefore, the flowcharts are suitable for expressing procedures which have multiple paths. However, they are not suitable for expressing operation procedures which have hierarchical structure.

The tree-type flowchart (Yaku and Niki, 1978) which is a modified PAD (Problem Analysis Diagram) expresses, by adding arrows, the basic flow of operation procedures in the direction top to bottom and the hierarchical structure horizontally. This means that the tree-type flowchart uses a horizontally wide area of a CRT screen. Since it has various good points for displaying procedures on a CRT screen, the PNS has adopted the tree-type flowchart to show the operation procedures.

4. FUNCTIONS OF PNS

During normal operation, the PNS displays a block diagram of the plant, reactor water levels, reactor pressures, etc. to support supervisor's monitoring job.

When the reactor scram or the events leading to scram occur, the PNS starts navigation. Once a procedure introduction is identified, the PNS shows the optimum operation procedures and the paths to normal shutdown which is the goal point of the operation, by using the tree-type flowchart.

The symptom-based operation manual consists of many operation procedures. Operation procedures to be used are determined in accordance with the degree of progress of the abnormal event. In some cases, multiple procedures may be introduced simultaneously.

The important plant parameters are displayed on the important alarm tiles, plant block diagrams, trend charts and grid-type diagrams. The parameters to be shown are automatically selected according to the procedures currently displayed in the tree-type flowchart. For instance, when a trouble such as "low reactor water level" occurs, the plant block diagram describing the operational condition of the feed water system and the trend chart of the reactor water level are automatically displayed.

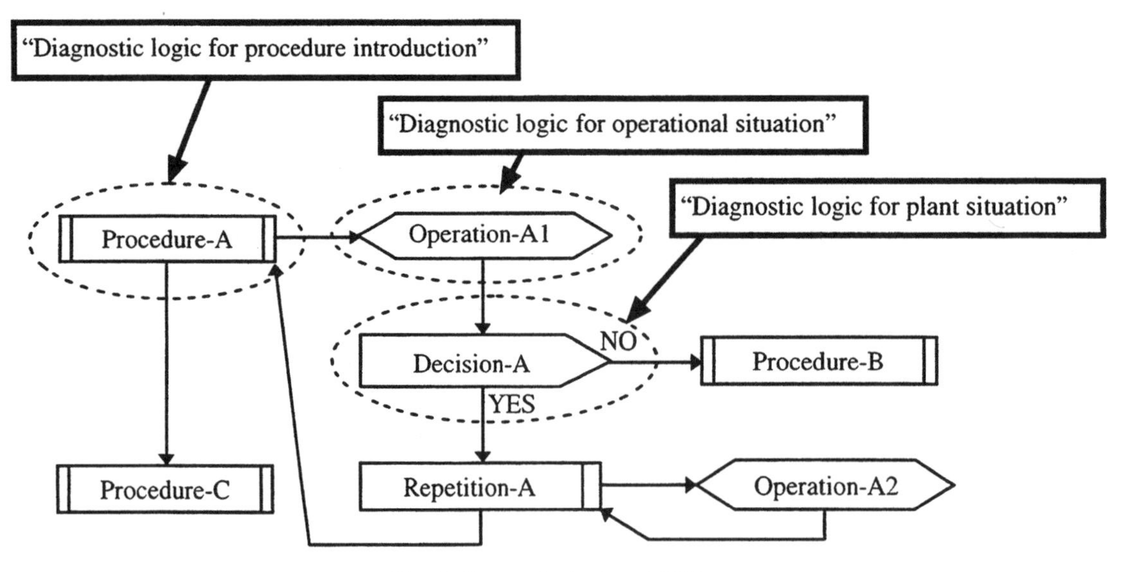

Fig.1 Diagnostic logic of PNS

5. STRUCTURE OF PNS

5.1 Diagnostic logic

In order to realize the above-mentioned functions, diagnostic logics are needed to compare the plant situation with the operation procedures. The diagnostic logic is constructed from the operation procedure by combining logical operations of information on ON or OFF of alarms, valves, pumps and limits of various plant parameters.

Elements of the operation procedures can be classified into three types in terms of their functions; judgment of procedure introduction, judgment of plant condition, and actual operations. Therefore, the following three types of diagnostic logic are introduced to the PNS corresponding to the three types of procedure elements.

a) Diagnostic logic for procedure introduction

Figure 1 shows an example of operation procedure displayed in the tree-type flowchart. A diagnostic logic for procedure introduction is defined at the element of “Procedure-A”, and compares plant parameters with its introduction conditions. Once the introduction conditions for the operation procedure are satisfied, all operations in the operation procedure become a target of navigation.

b) Diagnostic logic for plant situation

This is the diagnostic logic corresponding to judgment of the plant situation in the operation procedure. In the example shown in Fig.1, “Decision-A” has this type of diagnostic logic to decide the path of suitable operations.

c) Diagnostic logic for operational situation

This is the logic to judge whether the operation has been performed or not. Each actual operation such as “Operation-A1” in the operation procedure has this type of diagnostic logic, as shown in Fig. 1. This judgment is performed on the basis of the plant parameters which must be influenced by the operation.

Figure 2 shows an example of this type of diagnostic logic. This is the diagnostic logic which judges if an NPP injection system has worked with success. This diagnostic logic is based on the following conditions.

- An injection valve is open.
- A by-pass valve is close.
- Injection flow is enough .

If these 3 conditions are satisfied, it can be judged that the injection system has succeeded in injection into a reactor.

In the case where confirmation by the supervisor is needed, the PNS demands manual input of “confirmation”. The diagnostic logic for operational situation watches also the manual input from the supervisor.

5.2 Agents of PNS

Agents composition of the PNS is shown in Fig. 3. The PNS consists of several agents and a black board which preserves common data such as plant parameters, results of the diagnostic logics and the input data from the supervisor. Functions of each agent are described below.

a) Tree-type flowchart display agent

The tree-type flowchart display agent has two functions of "diagnostic" and "display of operation procedures". All the diagnostic logics for procedure introduction are active at the normal situation of the plant. When a procedure is introduced, diagnostic logics for plant situation and operational situation in the procedure become active in order to judge the current situation of the plant and the procedure; completed operations, uncompleted ones and an optimum path in the procedure. The results of their judgment are written on the black board.

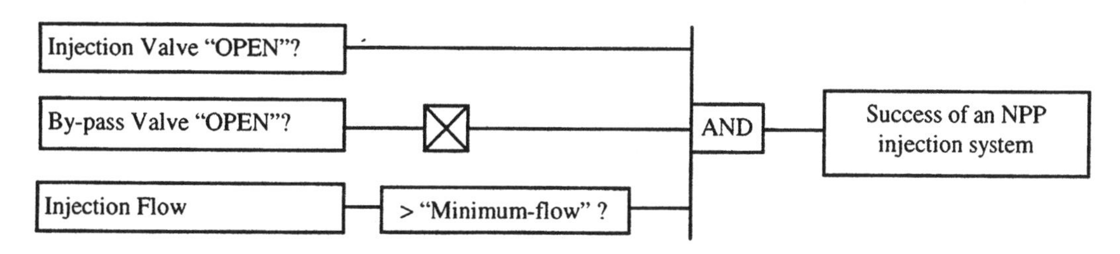

Fig.2 Example of diagnostic logic (NPP injection system)

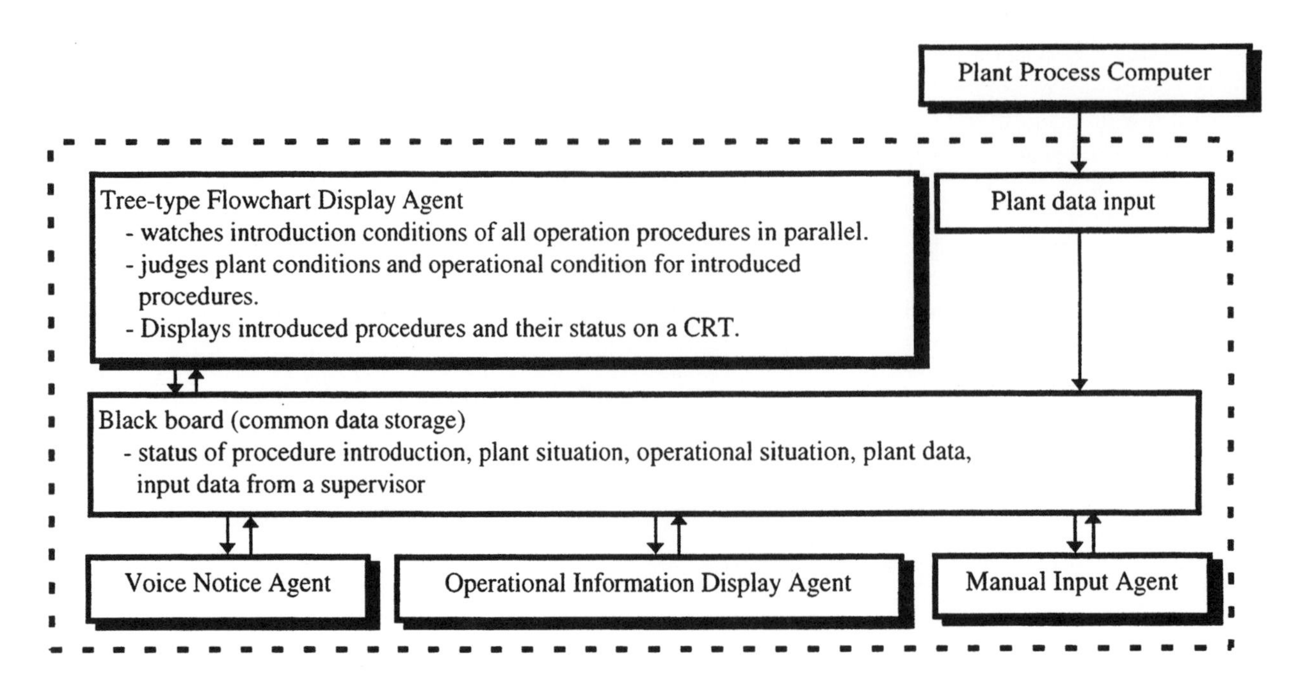

Fig.3 Agents composition of PNS

This agent displays the introduced procedures on a CRT, by using the tree-type flowchart and results of diagnoses. The current situation is explained by changing colors of tree-type flowchart elements.

- "green" : introduced procedures, completed operations or judgments.
- "pink" : operations not yet completed although the following operation has been completed.
- "red" : operation to be performed (present position).
- "white" : procedures or operations not yet introduced or not to be performed.

b) Operational information display agent

The operational information display agent shows plant parameters related to the introduced operation procedures. They are given in grid-type diagrams illustrating the relationship between two important plant parameters, trend charts and a plant block diagram as well as important alarm displays. The plant parameters to be shown are selected in advance for each operation procedure.

c) Manual input agent

The manual input agent carries out interactions with the supervisor. When a new operation procedure is introduced, this agent gives a dialog box to prompt the supervisor to confirm the introduction of the procedure. In other cases where automatic judgment is difficult by a computer, it demands also supervisor's decision. Furthermore, the supervisor can change the decision by the PNS through the manual input agent, if necessary. For example, the supervisor can introduce a procedure in advance of PNS's decision.

d) Voice notice agent

This agent watches plant parameters on the black board and announces important changes of plant situation by voice, for example, actuation of interlocks.

6. EXPECTED EFFECT OF PNS

Figure 4 shows effects of the PNS to cognitive processes of the supervisor.

When an abnormal event occurs, the supervisor has to collect information to make situation awareness, and to select the most suitable operation procedure. In such situation, all the information processing activities of Rasmussen's step ladder model (SLM) must be active. In the case where the PNS is installed on the supervisor's desk, the PNS displays automatically pertinent operation procedures together with related principal operational information (plant parameters). The supervisor need to ask operators only for other supplemental information. For this reason, only three of "detection", "observation" and "formulation of procedure" activities of SLM are important. Therefore, large reduction of his mental work load can be expected, and human errors at selection of procedures and observation of important plant parameters will be restrained.

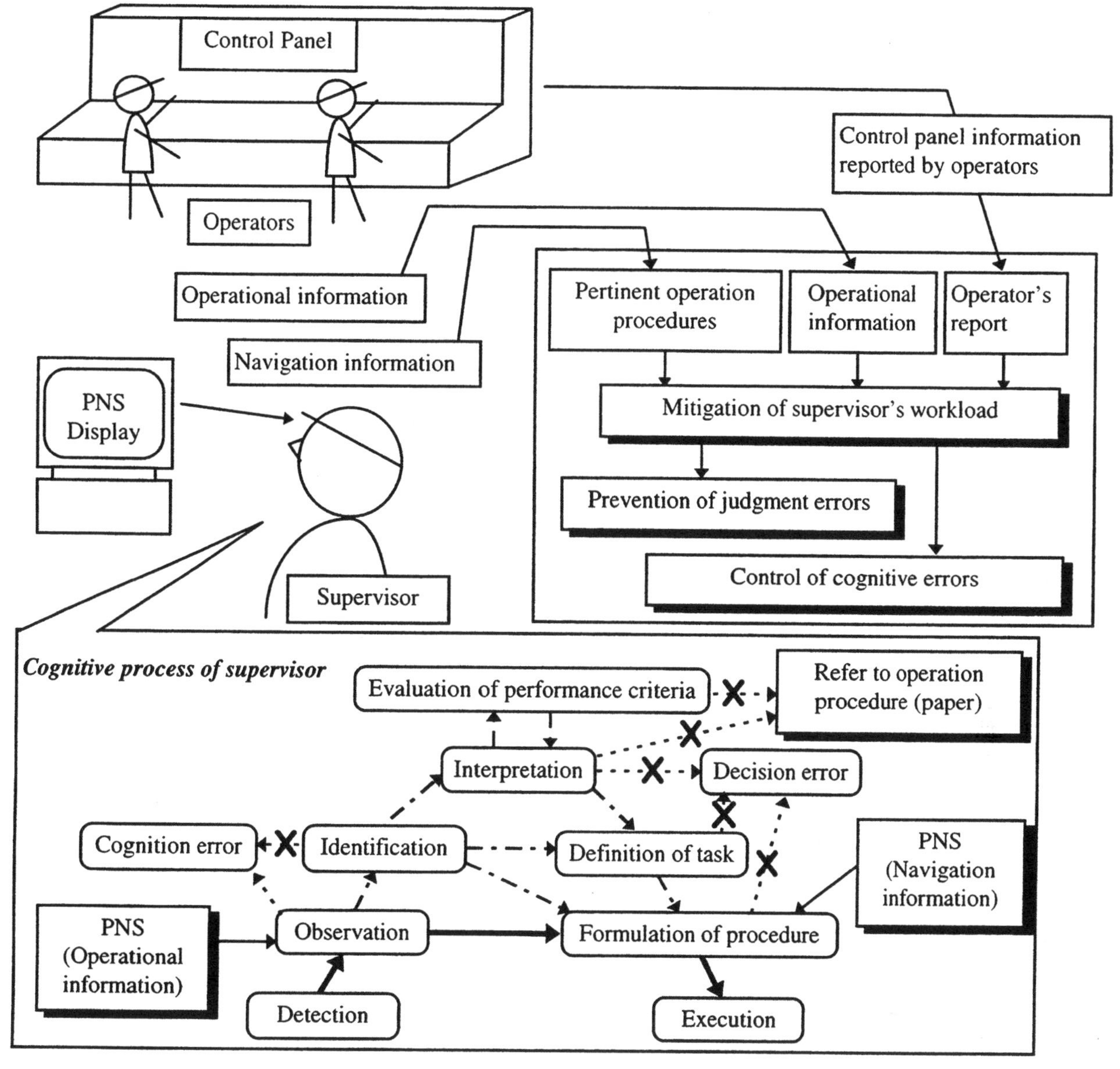

Fig.4 Cognitive process of supervisor (PNS introduced)

7. TRIAL MANUFACTURE OF PNS AND VERIFICATION TEST

A prototype of the PNS was constructed by using the real time expert system G2 (Gensys Corporation, 1992). An example of PNS displays is shown in Fig. 5. The selected procedures are displayed in the main part of screen. At the top of the screen, important alarms are located. Operational information is given on the right hand side in a trend chart, a grid-type diagram and a plant block diagram.

An experiment with 6 subjects was performed to investigate the usability of the PNS and valuable suggestions were obtained concerning the color of elements, trend charts, plant block diagrams, and so on.

8. CONCLUSION

The PNS to support a supervisor coping with abnormal events has been studied. The PNS is a system not to compel the supervisor to adopt its recommendation, but to show the most suitable procedures and the status of operation.

The prototype of PNS was constructed. It was confirmed that the PNS worked well as was expected.

It is scheduled to improve the prototype based on the experimental results. In addition, a new experiment in which the prototype of PNS will be connected with a plant simulator is also being planned.

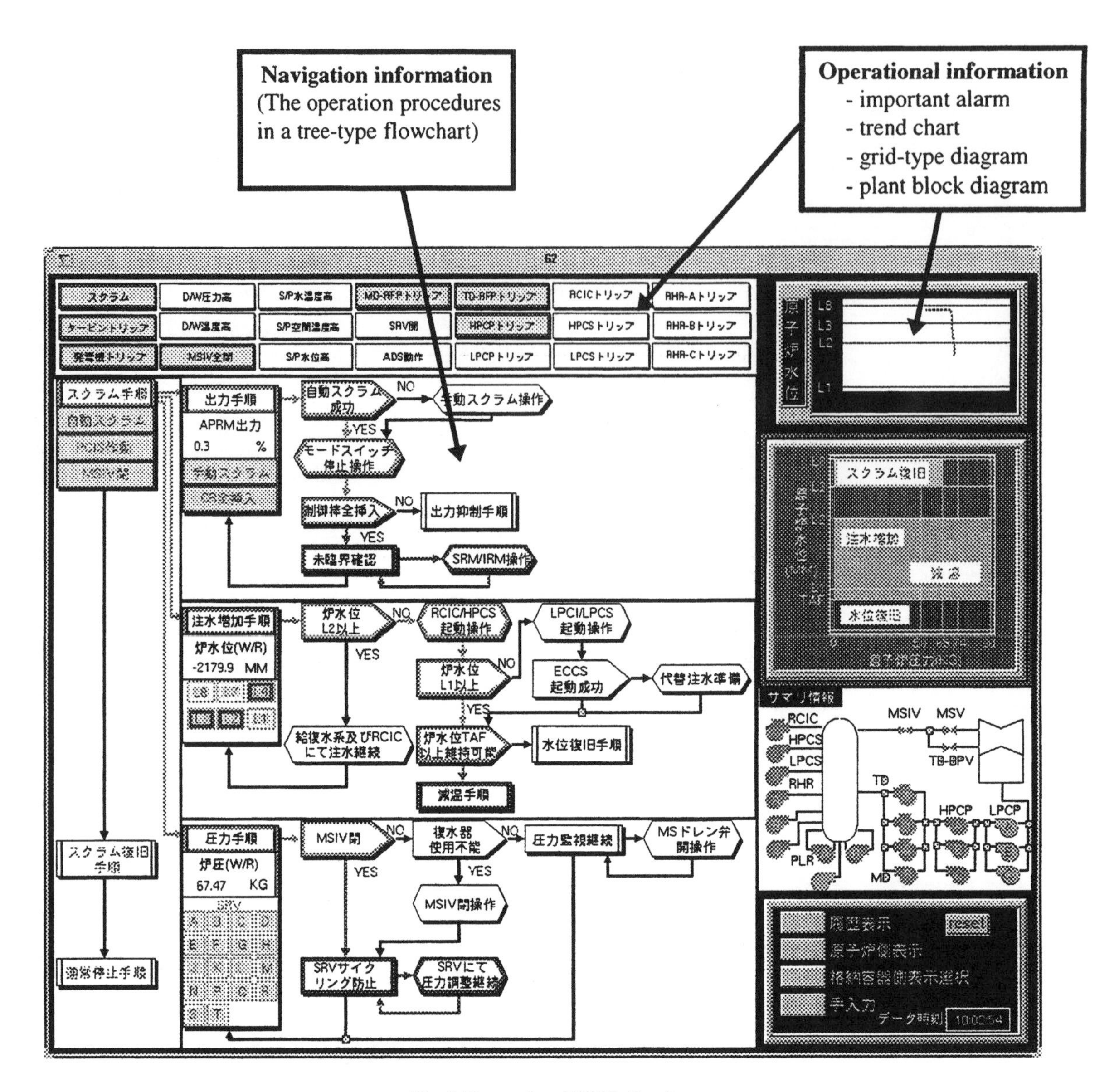

Fig.5 Example of PNS display

ACKNOWLEDGMENT

The present work was performed as a part of studies under a contract between the Ministry of International Trade and Industry (MITI) and Nuclear Power Engineering Corporation (NUPEC).

REFERENCES

Gensym Corporation (1992). *G2 Reference Manual.*

Hollnagel, E. and Niwa, Y. (1996). A Cognitive Systems Engineering Approach To Computerised Procedure Presentation, *Proc. Cognitive System Engineering in Process Control (CSEPC'96)*, Kyoto, Japan, Nov.1996, pp.286-291.

Saijou, N. *et al.*(1998). Development and Verification of Symptom Based Emergency Procedure Support System, *Journal of the Atomic Energy Society of Japan*, **40**, 225-234 [in Japanese]

Yaku, T. and Niki, K. (1978). Coding to Tree-type Flowchart, *AL-workshop in Institute of Electronics*, Information and Communication Engineers, AL-78-47 [in Japanese].

REALIZATION AND EVALUATION OF A NEW KIND OF SUPERVISORY SYSTEM

Manuel Lambert, Bernard Riera and Grégory Martel

Equipe coopération Homme-Machine et Informatique Industrielle
Laboratoire d'Automatique et Mécanique Industrielles et Humaines (LAMIH)
Université de Valenciennes et du Hainaut-Cambrésis
BP 311 Le Mont-Houy
59304 VALENCIENNES-Cedex
FRANCE
{Surname.Name}@ univ-valenciennes.fr
Under DRET convention: N° 93.34.098.00.470.75.01

Abstract: The development of a new kind of supervisory application has been carried out at the LAMIH of Valenciennes (France) in collaboration with the LAG ("Laboratory of Automation of Grenoble", France) and the LIA ("Laboratory of Advanced Computing") of Marcoule (France), a department of the CEA ("Atomic Energy Commission"). In fact, the LAMIH has realized supervisory interfaces which integrates original concepts and an advanced alarms filtering system developed by the LAG, based on a dynamical model of the process and on fuzzy reasonings. This design work has been finally concluded by an off-line evaluation on the site of the CEA at Marcoule by professional supervisory operators. *Copyright © 1998 IFAC*

Keywords: Design, realization, evaluation, supervision, continuous systems, dynamic models.

1. INTRODUCTION

Within a French project financed by the DRET (Direction of REsearch and Technology) laboratory, the LAMIH and the LAG have designed, realized and evaluated a supervisory application applied to a nuclear fuel reprocessing system supplied by the CEA, based on new concepts.

In fact, the supervision of the today production systems is more and more complex to realize not only because of the numerous number of variables to supervise but also because of the numerous interrelations existing between them, very difficult to interpret when the process is highly automated.

So, by reducing firstly problems of classical alarms filtering systems such as cascade phenomenons recalled by Riera et al. (1995), the challenge of the future years lies in the realization of support systems which let an active part to supervisory operators by supplying them tools and information which allow them to understand the working of their installation. To reach this major objective, this project proposes a research work following two axis : a work to improve existing tools thanks to the contribution of new concepts and the realization of a support tool dedicated to the diagnostic.

To be able to test these new concepts as regards to those already met in the industry, the realized supervisory application is composed in fact of two sub-applications. The first is classical, since this last proposes to the supervisory operators classical interfaces (synoptics) and classical tools (alarms filtering systems based on net thresholds). Besides, the second is original, since it proposes views using new concepts of presentation (Mass-Data-display concept for the monitoring tasks and causal graph concept for the diagnosis tasks) displaying high level information, i.e. information moreover symbolical than numerical.

In this sub-application, the classical alarms filtering system is replaced by an advanced one based on a dynamical model of the normal working of the process using fuzzy reasonings and developed by the LAG. Finally, a support system dedicated to the diagnostic which proposes diagnosis information (source of dysfunction and the propagation tracks) generated by an algorithm developed by the LAG, is also a new concept to test in this project.

Hence, the first paragraph briefly reminds today challenges of supervision, concepts (Mass-Data-Display and causal graphs) and the principles of the production workshop on which these concepts have been applied. Then, the second paragraph deals with the design steps related firstly to the monitoring views and secondly to the hierarchical diagnosis views. Finally, the more pertinent results of an evaluation realized in February 1997 with professional supervisory operators are presented.

2. PRESENTATION OF THE CONTEXT OF THE PROJECT

In general, supervision consists of commanding a process and monitoring its working (Millot, 1988). Hence, the supervisory system of an installation must collect, supervise and record important sources of data about the process working. The human operator (HO) performs two main classes of tasks. The monitoring tasks require a global vision of the process to allow an efficient supervision whereas the diagnosis tasks require a hierarchical vision to obtain views less or more detailed of the process.
In today supervisory systems, the HO has to his disposition synoptics (P&Id diagrams), trends views, alarms list, etc. which do not completely reply to their information requirements.

2.1 The MDD Interface and The Causal Graph

In function of the nature of tasks, two different kinds of representation have been employed: The Mass-Data-Display (MDD) for the monitoring tasks to the sense of Moray (1986) and the causal graph for the diagnosis tasks (cf. Fig. 1).

A MDD interface is characterized by the presentation on a single display of all the data allowing a HO to supervise a process (Beuthel, et al., 1995). Each variable must be symbolically represented and at least one of its intrinsical characteristics, such as its shape or its colour gives a picture of the temporal evolution of its state.

As for the causal graph concept, it is a graph whose nodes represent the pertinent variables and the oriented relations between these variables the causal links (Dziopa, 1996). This representation must theoretically facilitate cognitive reasonings coming into play in human diagnosis processes.

For instance, in the framework of a blast furnace, a process with a long response delay, studied by Samurçay and Hoc (1986) have shown that causal representation is more adapted to the diagnosis activities than topographical one.

Representation types / Characteristics	Mass-Data-Display	Causal graph
Principles	- Data Display in Mass with symbolical manner	- Causal representation of the process => Nodes = significant variables => Oriented relations =causes to effects links between the variables
Nature of Tasks	**Monitoring**	**Diagnosis**
Representations	working → Normal / Abnormal; 1 stick = 1 variable	(X) → (Y) The variable X influences the variable Y
Theoretical Interests	- Many variables represented by a single screen => exhaustive vision of process state - Visual attraction => facilitated detection	- Simple Description of the process working - Representation adapted to cognitive strategies developed by HO during diagnosis phases

Fig. 1. The MDD and causal graph representations

2.2 Principles Of the Workshop

This workshop is composed of two pulsated columns head to foot, the extraction column and the washing column and of their own supplying systems (cf. Fig. 2).

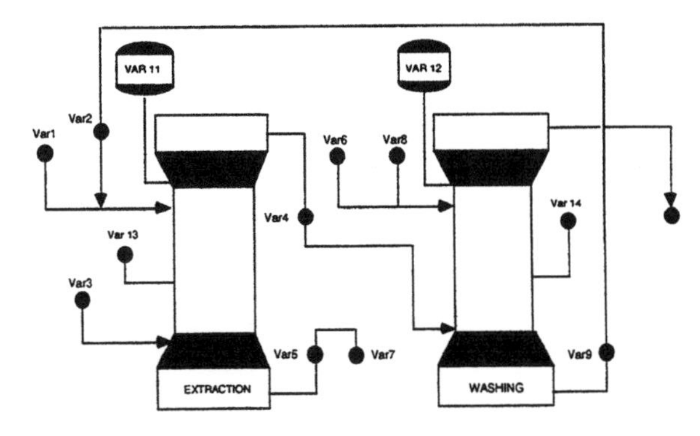

Fig. 2. The schematical representation of the workshop

These columns separate the couple uranium-plutonium from fission products, considered as impurities. The uranium, plutonium and fission products are originally dissolved in an acid solution. The extraction column receives this loaded solution (var.1) and extracts uranium and plutonium elements thanks to an organical solution (var.3). Within each column, a pulsation pressure (var.11, var.12) optimizes chemical exchanges.

Finally, at the base of the extraction column, the aqueous outflow contains only fission products (var.7) and at the top of the washing column the organical solution contains only the couple uranium-plutonium (var.10).

3. THE DESIGN OF SUPERVISORY INTERFACES

Two kinds of supervisory sub-applications associated with two kinds of alarms filtering systems have been elaborated in order to be able to evaluate the contribution of each concept.

3.1 *General organization of the application*

The first system, composed of synoptics and trends views is a classical supervisory system which is associated with a classical alarms filtering system. The second, the original supervisory system associated to the original alarms filtering system, is composed of the next views:

- Five views dedicated to the monitoring tasks based on the MDD concept : One general MDD view and four MDD views with a specifical numerical filter.
- Four causal views hierarchically organized and dedicated to the diagnosis tasks based on the causal graph concept.
- Four diagnosis views whose causal structure is the same than that of the four causal views, but which display the information generated by an support system to the diagnosis elaborated by the LAG. Thanks to these views, the HO can know the potential paths followed by the propagation of the dysfunction and the potential source of this dysfunction.
- A generical trends view, available for each variable which gives a picture in the time of its state.

3.2 *Design of Monitoring Views*

For the design of the MDD views (cf. Fig. 3), two different techniques of Functional Analysis (FA) have been applied to the studied production workshop.

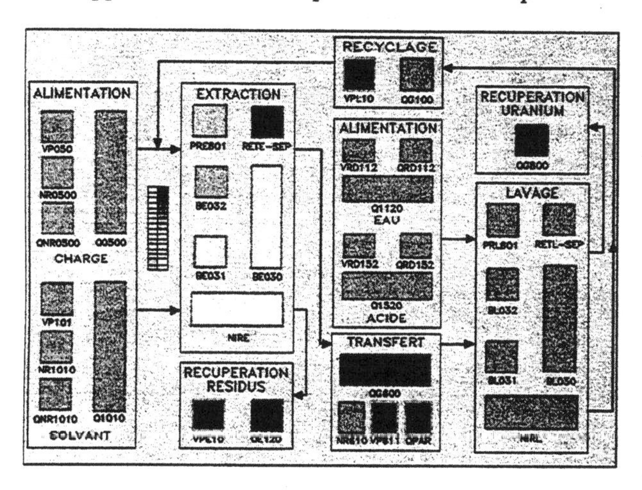

Fig. 3. The MDD view

The functional tree, see (Zwingelstein, 1995) allows to identify and to organize hierarchically the functions of the system. Then, the SADT model (Structured Analysis and Design Technique), see (Pierreval, 1990) allows to focus the study on the matter flows. Their combined application supplies both a structural and a functional model of the workshop under normal working.
As for the MDD symbol, each variable is represented by a colored quadrilateral whose two attributes, size and color, reflect respectively its significance within the function and its state (cf. Fig. 3). This symbol owns a range of colors with progressive contrasts (from green to red through the yellow) representative of the state of the residue (i.e. of the difference between the real value called measure and the theoretical measure called model).

These different values are generated by a numerical simulator of the workshop. The model value corresponds to a normal working whereas the measure to an abnormal working, i.e. when an abnormal event occurs at a given moment in the simulation of the process working.

So, the green color means that the variable is in a normal state (value of the residue null or near to null) whereas the red color translates a serious abnormal state. The panel of colors (30 colors) between these two colors allows one to translate, in a very gradual manner, the whole states from the normality to the abnormality. The MDD color as other high level information comes from an algorithm based on fuzzy reasonings developed by our collaborators of the LAG, see (Gentil, et al., 1997).

To elaborate this MDD color information, the state of a variable is established through the analysis of the instantaneous value of the residue and its evolution in the past (via its derivative).In fact, from this analysis, a table of nine qualitative primitive states has been elaborated by using fuzzy reasonings (cf. Fig. 4).

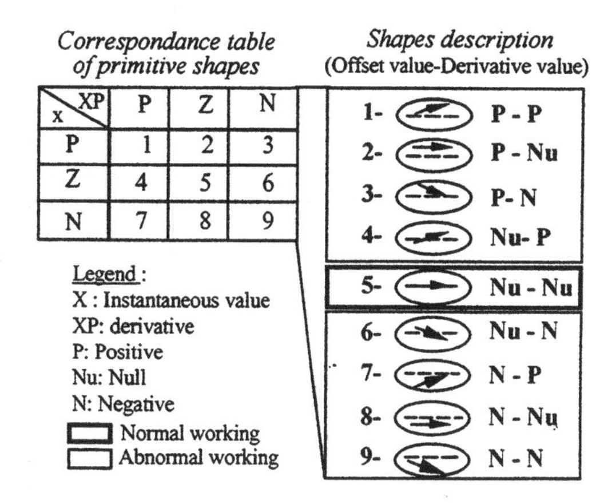

Correspondance table of primitive shapes

X \ XP	P	Z	N
P	1	2	3
Z	4	5	6
N	7	8	9

Legend :
X : Instantaneous value
XP: derivative
P: Positive
Nu: Null
N: Negative
Normal working
Abnormal working

Shapes description
(Offset value-Derivative value)

1- P - P
2- P - Nu
3- P- N
4- Nu- P
5- Nu - Nu
6- Nu - N
7- N - P
8- N - Nu
9- N - N

Fig. 4. The table of the qualitative primitive states

For instance, when a variable follows a normal behavior (value of the residue null or practically), the value of the membership function associated to the primitive state 5 tends towards 1 (cf. Fig. 4).

Besides, the states 2,3,4,6,7 and 8 correspond to suspect situations and the states 1 and 9 to serious situations.

3.3 *Design of diagnosis views based on causal graphs*

For the design of the diagnosis views, two points have been considered: Their general organization and their information contents.

Because of the informational density characterizing the complete causal graph of the workshop, an information filtering has been applied. To translate it at the level of the interface, a hierarchical organization following three levels has been chosen. The toppest level supplies a general view of the process (cf. Fig. 5), the intermediary level a median view and the inferior level two complementary views of the process, the extraction and washing views.

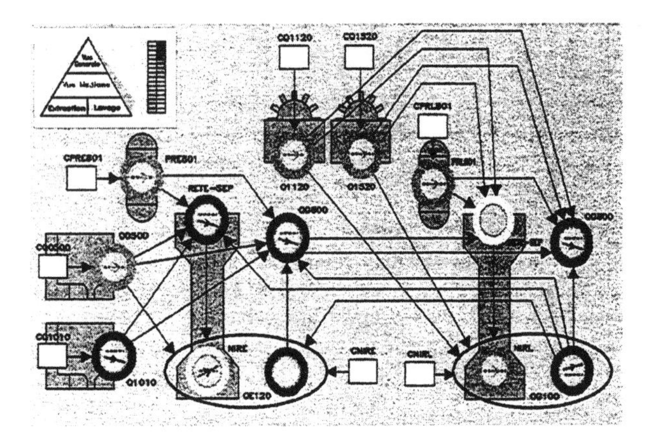

Fig. 5. The Diagnosis view of the toppest level, the general view

This pyramidal organization allows a gradual distribution of the information contents between different levels. These contents have been identified thanks to the use of the MFM (Multilevel Flow Modeling) model, created by Lind (1990). As for the representation, each variable is represented by a node composed of two entities. An exterior crown displays the MDD color already seen in the monitoring views and the interior disk a trend symbol representative of the shape of the residue also called "drift indicator". This symbol is selected within one of the nine qualitative primitive states of the figure 4.

4. THE EVALUATION

The evaluation of the supervisory system, developed on a Sun station has taken place at the CEA of Marcoule during the month of February 1997. The tested population was composed of seven professional supervisory operators, six experts and one novice.
Each operator has followed a learning stage in order to discover the functionalities of the supervisory application. Then, they have tested it through four simulations during which a dysfunction occurred at an unknown moment. The evaluation planning is described at the figure 6.

In fact, for each of these four simulations, new functionalities were available for the HO in order to evaluate the contribution of each new concept in the global efficiency. Hence, four different configurations called "contexts" have been defined.

Moreover, two different families of scenario have been elaborated: A first family composed of difficult dysfunctions to detect or to explain and a second family with less difficult dysfunctions.

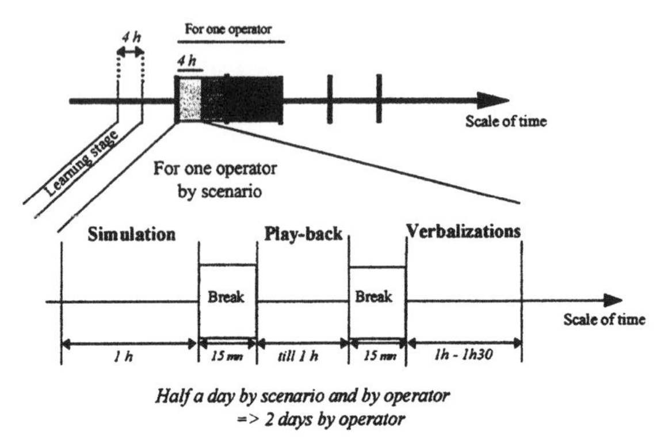

Fig. 6. The evaluation stages

The experts being divided into two groups of three operators, each group has tested a different family of scenarios. As for the novice, owing to his lack of experience, he has tested the supervisory application through the easier scenarios.

4.1 *The phases of activity*

This evaluation lies on two types of data: Objective data (actions of the HO and events of the simulation) recorded in a "spy" software file during each simulation and subjective data collected during the step of verbalizations following the simulation.

During this evaluation, two types of period have been studied: A period of monitoring which corresponds to the monitoring phase preceding the detection of a dysfunction by the HO and a period of diagnostic during which the HO searches for the source of this dysfunction.

4.2 *The analysis of objective data*

To organize and to treat the objective data, the LAMIH has developed a specifical tool called "Analysis". This tool is able to calculate different objective indicators from these database which give a quantitative picture of the evaluated concepts.

The indicators (for instance, proportion of consultation, duration of consultation, etc.) inherent to each type of views and variables (via the trends views) have been calculated by taking into account the next elements: The nature of the activity phase (monitoring or diagnostic), the expertise degree of HO (experts or novice) and the nature of scenarios (hard or easy).

The evaluation results concerning the monitoring and diagnosis phases are synthetically presented by describing the behavior of the HO during the simulation of the most complete context. In this context, the HO had the choice between all the views, classical (synoptics, trends) and new (MDD views, causal views) respectively associated to the classical and original alarms filtering systems. Moreover, a support system dedicated to the diagnostic was also available.

As for the views on which the detection has been realized, the presented results are a synthesis of the behavior of experts on the totality of the contexts in order to see its evolution.

To realize the synthetical representation of the figures 7 & 8, only the proportions of consultation above the arbitrary threshold of eight percents have been taken into account.

4.3 The monitoring phase in the most complete context

The general MDD view has met a large success during this phase (cf. Fig. 7). Five operators on the seven used it a lot, only the experts E1 and E6 have used it less. In all cases, the traditional synoptics have been deserted.

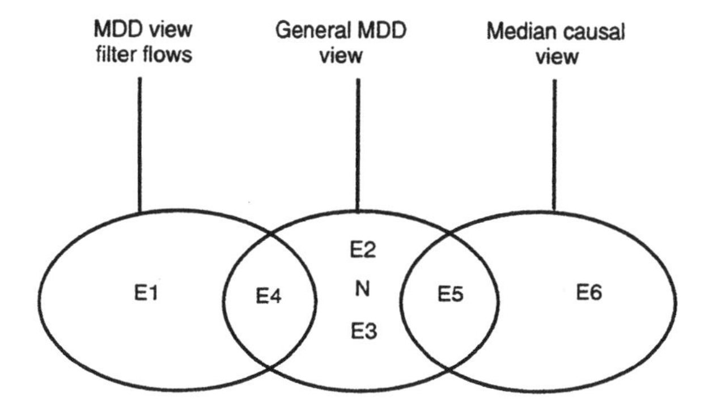

Fig. 7. The most consulted views

More precisely, three operators, the experts E2, E3 and the novice N use the general MDD view practically exclusively whereas two operators, the experts E4 and E5 have used it with an other view, either with the MDD view filter flows (E4) or with the median causal view (E5).

In an other hand, the expert E1 prefers to use only the MDD view filter flows and the expert E6 the median causal view whereas the median causal view has been originally designed for the diagnosis activities.

4.4 The detection views following the four contexts

On the totality of evaluation, experts have detected dysfunctions on four different kinds of views: on classical views (extraction and washing synoptics and two trends views) and on new views (general MDD view, MDD view filter flows and median causal view).

To summarize, new kinds of views, firstly the MDD views (13 detections i.e. practically 50 %) and secondly a good surprise since a view not designed for the detection, the median causal view (2 detections) have demonstrated its utility for the detection. The MDD views and the median causal view have been qualified by the operators as stay-up views i.e. views which allow the HO to know rapidly the state of the process.

On the totality of the evaluation, the HOs have mainly detected thanks to new views. Moreover, when they have the choice between classical views (synoptics) and these new views (MDD views and the Median Causal view), they have detected from these last one.

In an other hand, the classical trends views seem indispensable to detect because these kind of views allow the HO to know the state of the variable in the time and so to detect an abnormal change.

4.5 The diagnosis phase in the most complete context

Through the figure 8, it appears that:

- The expert E1 has gone on using a lot the MDD view filter flows.
- The expert E2 has deserted the general MDD view at the benefit of the views dedicated to the extraction part : the causal and diagnosis views.
- The expert E3 has moreover used the general MDD view but particularly the MDD view filter flows.
- The expert E4 has gone on using the general MDD view but has also used the causal views except the general view.
- The expert E5 has less used the general MDD and the median causal views at the benefit of the washing synoptic and the washing diagnosis views.
- The expert E6 has gone on using a lot the median causal view but this time has also used the views linked to the washing part (washing synoptic, washing causal and diagnosis views).

The use of the views whose structure is causal, is dependent of the nature of the dysfunction to diagnose. Indeed, it appears that for the experts (E2 and E4) whose dysfunction to diagnose has been located in the extraction part of the installation, they have more used the views dedicated to this part. In a same way, the other experts (E5 and E6) have moreover used the views dedicated to the washing part since, in their scenario, the dysfunction to diagnose has been located in the washing part.

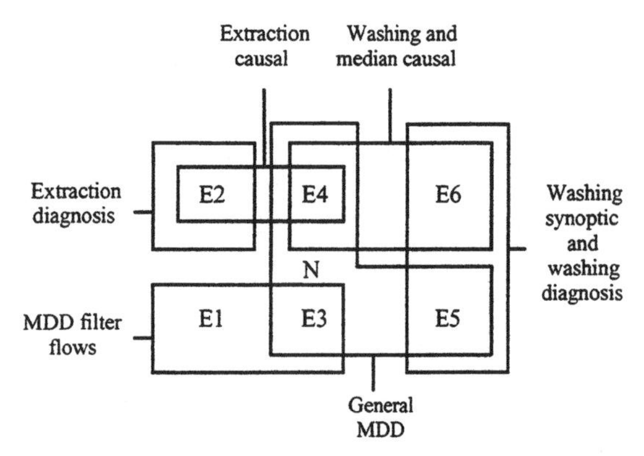

Fig. 8. The most consulted views

The success of causal views is mixed because only three experts (E2, E4, E6) have used a lot. This is also true for the support system dedicated to the diagnostic since only three experts (E2, E5, E6) have used it a lot. Moreover, the general causal view seems unuseful. According to the verbalizations, this view is too general. Besides, they consider the median causal view as a good view because it regroups all the useful information useful for the monitoring and for the diagnostic of the process working.

Whereas the views of MDD type have not been designed to fill the diagnosis tasks, these views have been used a lot by five operators. More precisely, the MDD view filter flows has been used by two experts (E1, E3) whereas the general MDD view by the experts (E3, E4, E5) and the novice. It is important to note that only two operators have felt the need to use the washing synoptic showing that this kind of views is not indispensable.

5. CONCLUSION

A design approach of a supervisory system based only on alarms condemns the HO to remain too much dependent on these alarms. So, an approach moreover centered on the HO is proposed. Indeed, the HO must be provided with tools which allow him to estimate at each moment the process state. For that, since the monitoring and the diagnosis tasks require specifical information, two distinct concepts have been applied: The MDD concept for the monitoring task and the causal graph concept for the diagnosis task.

So, dedicated to the monitoring, MDD views using symbolical information generated by an advanced alarms filtering system have been designed. In this view, the functional structure and the respect of the logical links between the functions performed by the system have to contribute to make more active the work of the HO by allowing him to detect as fast as possible the defaults. For the diagnosis task, the hierarchical views composed of causal graphs aim at allowing the HO to reflect more easily in terms of causes to effects relations and at perceiving the installation following different points of view.

All concepts presented in this paper have been tested by the means of an evaluation, led at the CEA of Marcoule with professional supervisory operators. The first results of these experiments are encouraging and show the interests to apply these new concepts in the supervisory system design to improve Man-Machine interactions.

REFERENCES

Beuthel, C., B. Boussoffara, P.Elzer, K. Zinser and A. Tißen (1995). Advantages of mass-data-display in process S & C. IFAC Analysis, Design and Evaluation of Man-Machine Systems, MIT Cambridge, 439-444.

Dziopa, P. (1996). Représentation multi-modèle pour la supervision de procédés industriels continus. Ph'D dissertation, Institut National Polytechnique de Grenoble, Grenoble.

Gentil, S., A. Evsukoff and J. Montmain (1997). Les systèmes flous pour l'aide en ligne aux opérateurs, Groupe Français de Génie des Procédés, Récents progrès en génie des procédés, Simulation des procédés et automatique, Paris, N°56, Vol. 11, 121-126.

Lind, M. (1990). Representing goals and functions of complex systems - an introduction to multilevel flow modeling, Technical report of Institute of Automatic Control Systems, Technical University of Denmark, Lyngby.

Millot, P. (1988). In: Supervision des procédés automatisés et ergonomie (Hermès, Ed.), Collection Traité des nouvelles technologies, Paris.

Moray, N. (1986). In: Handbook of Perception and Human Performance (K. Boff, L. Kaufman & J. Thomas, Ed.), Vol. 2, Chapter 40, pp. 40-6. Wiley, New York.

Pierreval, H. (1990).In: Les méthodes d'analyse et de conception des systèmes de production (Hermès, Ed.), Collection Technologies de pointe, Paris.

Riera B. et al. (1995). A proposal to define and to treat alarms in a supervision room. Sixth IFAC Symposium on Analysis, Design and Evaluation of Man Machine Systems, MIT Cambridge.

Samurçay, R. and J.-M Hoc (1996). Causal versus topographical support for diagnosis in a dynamic situation. Le Travail Humain, tome 59, N°1/1996, 45-68.

Zwingelstein, G. (1995). In: Diagnostic des défaillances; théorie pratique pour les systèmes industriels (Hermès, Ed.), pp. 90-91. Collection Diagnostic et Maintenance, Paris.

ACKNOWLEDGMENTS

The authors thank Christophe Louppe for his contribution in this project, and the professional operators of Marcoule for their participation to the evaluation stage.

COLLABORATIVE EMERGENCY OPERATION OF MAN-MACHINE SYSTEM IN NUCLEAR POWER PLANT

Yuji Niwa * **Masahiro Terabe** ** **Takashi Washio** ***

* *Institute of Nuclear Safety System Inc. 64 Sada, Mihama, Fukui 919-12, Japan*
** *Mitsubishi Research Institute, Inc., 3-6, Otemachi 2-Chome, Chiyoda-Ku, Tokyo 100, Japan*
*** *Institute for Scientific and Industrial Research, Osaka University, 8-1 Mihogaoka, Ibaraki, Osaka 567, Japan*

Abstract:
In this paper, the scientific investigations and analyses pointed out the necessity of the careful decision-making of NPP operators considering the critical parameters of NPP to attain the safety and those causalities amongst these parameters. Under the strong time pressure, it is virtually impossible to make optimal decision taking such causalities into account. As a result, the computerization of taking recovery actions is needed. Considering the requirement of such computerization, the autonomous system in collaboration with human (what we call the agent system) that is still effective even in such a unforeseen condition in NPP is proposed. *Copyright©1998 IFAC*

Keywords: Man/Machine systems, Nuclear plants, Artificial Intelligence, Agents

1. INTRODUCTION

Recent Advanced Information Technology (AIT) have enabled us to consider the automated recovery system of Nuclear Power Plant (NPP) during an accident. In the present situation, once an accident occurred in NPP, reactor is shut down automatically to maintain nuclear safety. As an application of traditional control theory, primary feedback control systems have been introduced into NPP. However, they only control each process, for example steam generator level, pressurizer pressure locally. Generally the automation of NPP remains at rather low level still now. However, AIT application alludes the possibility of automated recovery system during an accident. In this paper, automated recovery system as the application of AIT is proposed.

The plausible reasons the automation during an accident is so conservative has been quite acceptable, since feedback control system could not synthesized due to the strong non-linearity and complexity during an accidental stage. Hence the role of NPP operators are quite important currently. Traditionally, operators identify the event, i.e., they grasp what has occurred in NPP, when they encounter an accident. They try to recover NPP with referencing to Emergency Operating Procedure (EOP.) It specifies the operations to be taken in accordance with the event occurred in NPP. Whenever EOP is referred, the identification of the event is required. In this sense, such EOP is called Event-based EOP (EB-EOP.) Anticipated accidents can be settled with no problem by operators using EB-EOP unless they make errors. However all accidents that will occur in NPP cannot be expected necessarily in advance. Multiple failures and human errors may cause the loss of identification. Good example can be seen in the accident occurred in Three Mile Island

(TMI) No.2 unit. The experience of TMI accident emerged us the introduction of other EOP system based on a new approach. A new approach is focused on the NPP function. Corcoran (Corcoran,1981) defined the functions to attain the safety in NPP and called the (critical) safety function. The Safety Function (SF) is defined as the set of human/machine operations to avoid core damage. It has been concluded that the safety is guaranteed so long as NPP maintains all of the specified functions. The EOP to restore each function has been developed and it is called Safety Function based EOP (SF-EOP.) Since SF-EOP does not impose operators the event identification, operators can transfer from EB-EOP to SF-EOP when they fail to the event identification, when NPP never shows expected response and when Operators loose orientation in the recovery actions. The SFs are monitored continuously and even if one of them is lost, operators must restore it immediately even when the event has been identified and operators must taking actions for the restoration of violated function.

Although SF-EOP saved operators in the sense that operators are free from the event identification, the SF has been defined from expert judgments of NPP safety designer, mainly Probabilistic Safety Analyst. Therefore, it may be true that one of SFs is violated, it may cause core damage when left. However, it is worth while studying NPP is in safe state if and only if all SFs hold specified conditions. It should be guaranteed that EOP systems are applicable to all accident including unidentified ones. It is extremely difficult to verify the above subject. However, if SF is specified more objectively with minimum expert judgments that cannot be formulated, it is a great advantage to ensure this subject. The more serious problem is whether operators can take appropriate operations specified in SF-EOP in the upset status such that operators cannot realize what is going on in NPP with time pressure. In unforeseen status, the introduction of automated/computerized recovery system should be studied keenly.

Based on the above discussions, this paper concerns the following three main subjects. These are, 1. investigation on what objects such as SFs should be maintained during an accident in NPP and their structure, 2. architecture of automated accidental feedback recovery system taking their structure and how to collaborate with Human in NPP control room, and 3. design of Man-Machine System required for the collaboration human and machine (computer.)

The Safety Functions to avoid core damage (Corcoran, 1981)
Reactivity Control
Heat Removal from Core
Heat Removal from Reactor Coolant System
Prerssure Control of Reactor Coolant System
Inventory Control of Reactor Coolant System

* The Safety Functions for the integrity CV are not included in this Table

Fig. 1. Safety Functions given by Corcoran

2. THE RE-DEFINITION OF THE SAFETY FUNCTION AND THE ASSOCIATING SAFETY MEASURES

Five SFs were given by Corcoran to prevent core damage. They are given in figure 1.

It should be noted that SFs to prevent nuclear emission is not treated in this paper. The definition of the SF proposed by Corcoran was obscure for practical application. For example, the following understanding has been made for the SF, "Reactivity Control" in commercial NPP.:

The SF, "reactivity control" holds if:

$$[\text{Neutron flux is} \leq x(\%)] \cap [\text{Neutron flux level is stable} \cup \text{decreasing}] \quad (1)$$

Although Corcoran did not considered the interactions between each SF, it is natural to consider them so long as the SF is defined on the basis of plant parameters as above. In this paper, the Safety Function Designators (SFDs) are proposed to describe the SFs more objectively. Before starting the introduction of SFD, definitions of some key terminology must be clearly stated. An *accident* of NPP is the situation that its reactor core is threaten to be damaged by any failures of the plant safety components and operations. The following definition of *Safety Function Designator (SFD)* provides a basis of the evaluation:

Definition 1. (Safety Function Designator (SFD)). SFD is a set of quantitative NPP parameters and qualitative properties to specify objectively multi-purpose SFs to prevent core damage.

Taking this definition into account, five SFDs same as Corcoran gave was selected, since all SFDs are applicable to current EOP issues and correspond to current SFs in the sense that each SFD specifies corresponding SF respectively. On the basis of the definition in this paper, SFD can be formulated as follows:

$$SFD \stackrel{\text{def}}{=} [A, P, R, C, y], \quad (2)$$

where

A : a set of background assumptions of the SFD's model,

P : a set of parameters and variables in the SFD's model,

R : a set of constraints among parameters and variables in P,

C : a set of criteria to judge if the SFD is maintained or not,

y : the state variable that represents the behavior of SFD.

Hereafter called y critical safety parameter. The specific description of each SFD will give comprehensive idea to realize what the SFD is.

(1) SFD to maintain heat removal from the reactor core.
   C: The SFD is maintained if the maximum fuel clad temperature in the core is less than $1200°C$,
   T_f, Fuel clad temperature.
(2) SFD to maintain reactivity control of the reactor core.
   C: The SFD is maintained if the net reactivity is less than 1β in the normal operation mode and if the net reactivity is negative in the shutdown mode,
   ρ, Reactivity.
(3) SFD to control pressure of reactor cooling system (RCS).
   C: The SFD is maintained if the pressure of the pressurizer is larger than the saturation pressure of the coolant in the core,
   P_r, The pressure of a reactor or a pressurizer.
(4) SFD to maintain inventory control of reactor coolant system (RCS).
   C: The SFD is maintained if the entire fuel rods in the core are flooding,
   L, RCS level in a reactor.
(5) SFD to maintain heat removal from reactor coolant system (RCS).
   C: The SFD is maintained if the heat removal from RCS is larger or equal to the heat removal from the reactor core,
   Q_r, Heat removal *from* RCS.

The union of a set P and R gives the kinetic equations of NPP, i.e., NPP plant model that specifies each SFD. For example, the union (P,R) for reactivity control is represented as follows:

$$\rho(t) = \rho_{rod}(t) + \sum_{i=1}^{3} \rho_i, \tag{3}$$

where

ρ : reactivity (y of SFD_1),

ρ_1 : reactivity feedback due to Doppler effect,

ρ_2 : reactivity feedback from moderator,

ρ_3 : reactivity feedback from void,

ρ_{rod} : input reactivity by control rods.

The SFD specifies conditions required for nuclear safety objectively in the sense that the kinetics of the critical parameters based on NPP plant model can be considered in the SFD, where all plant parameters are formulated in the following state equations.The limitation of the model applied is manifested in a set C.

$$\frac{dX}{dt} = F(X,U), \tag{4}$$

$$Y = G(X,U), \tag{5}$$

where

t : time,

X : a state vector of the SFD's model,

U : a control vector of the SFD's model,

Y : an output (goal) vector of the SFD's.

Each element of the output vector Y is y in each SFD respectively and is measured or estimated by appropriate sensors. It also should be noted that all elements of state vector X are measurable in current commercial NPPs. Considering this relationship, Eq.(5) can be represented as a typical linear equation form, e.g., the inner product of constant vector and state vector. Equations (4) and (5) are called as a system process and a measurement process respectively. All elements of X at a time t are ensured to be estimated from Y of the time t through the measurement process Eq. (5) in our case. Accordingly, the state of the SFD at any time t, $X(t)$, is always predicted by solving Eq. (4) under a given sequence of operations for recovery to prevent core damage, U. It should be noted that a recovery actions during an accident are taken as a sequence of tasks.

$$U \stackrel{\text{def}}{=} [OP_1, OP_2, \cdots], \tag{6}$$

where

OP_n : an operation/task to maintain/recover SFD

Certain indices to evaluate the safety performance of U are required to maintain the SFD. Such indices can be defined and evaluated through the prediction process of the state vector X. Under a given U, the future trajectory of X starting from the present state is predicted in the model of the SFD. By applying the criterion C of the SFD, the judgment on the violation of the SFD can be conducted. In case that the SFD will be violated, the available time T_a at present until the violation will occur can be estimated. Also, the following safety margin

$$D = \int |\delta X|, \tag{7}$$

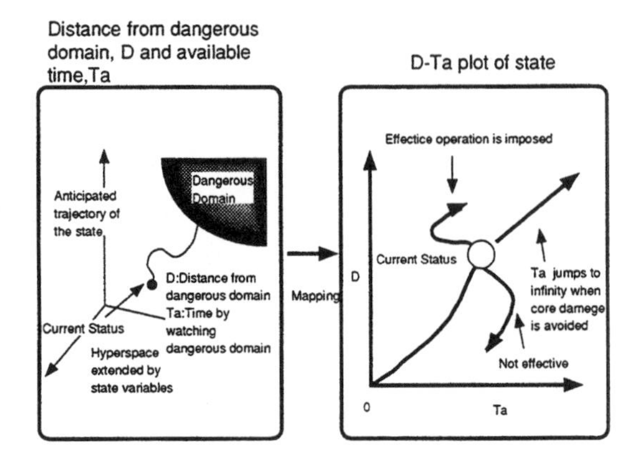

Fig. 2. The positions of $(X(0), U(t)(t \in [0, T_a]))$ on a $T_a - D$ plain

where

$|\delta X|$: the norm of the differential of the state vector X.

provides the distance between the present state and the state at the violation. For each combination of the present state $X(0)$ and the control sequence $U(t)$, its position can be plotted on a $T_a - D$ plain as depicted in fig. 2. The point closer to the origin (the left bottom corner) represents the more severe emergency, whereas the condition that the SFD holds and will hold in the future follows the infinite values of T_a. Therefore, the state point on a $T_a - D$ plain jumps to infinity by a successful emergency operation. Therefore, the above evaluation is required to each task in an operational sequence.

The last issue regarding SFD is to investigate the interaction amongst SFDs. Each SFD cannot be independent from the others whereas SFs form hierarchical structure and each SF is independent. The maintenance of some SFDs may support or prevent the maintenance of the other SFDs. Accordingly, the causal influences among those SFDs have been investigated through an analysis method named as *causal ordering* (Washio,1990), and the resultant dependency has been figured out as shown in fig. 3.

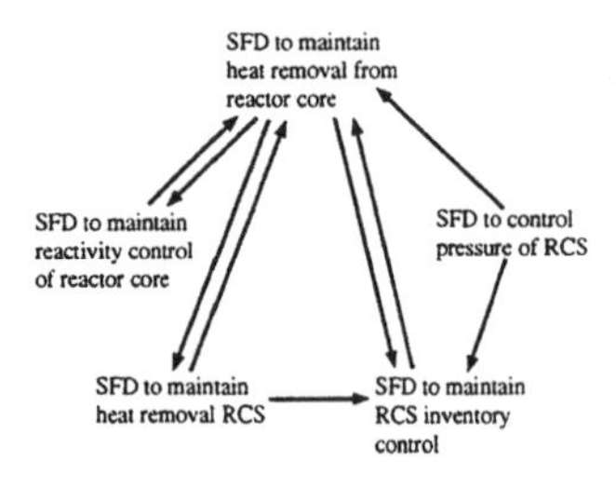

Fig. 3. The mutual dependency among SFDs.

The detail of the method is not explained here due to the limitation given space. However, it is clear that this information provides a basis to prioritize the SFDs to be maintained.

3. THE LEVEL OF COMPUTERIZED EOP AND PROPOSED SYSTEM

As explained in the earlier sections, NPP operators take recovery actions with referencing to EOP when they encounter an accident. EOPs as well as generic operating procedures have traditionally been provided in the form of documents. The transition to fully automated recovery system can be made in a number of steps, outlined below. In practice the question is not so much whether recovery actions should be computerized, but rather how far they should be computerized - or in other words, what the exact meaning of "computerized" or "automated" in an accidental recovery will be. Computerized EOP presentation (Niwa,et al.,1996) on GUI may be taken as a primary computerized EOP system. It is also important to grasp how advance the automated accidental recovery system that is proposed in this paper is and what research remains on this research. The most primary level is taken as 0, since it is the traditional, however most common form.

- Level-0 Procedures as documents.
- Level-1 Procedures as computerized documents.
- Level-2 Procedures as computer displays.
- Level-3 Computerized procedures with progress monitoring.
- Level-4 Automated procedures: Management by delegation.
- Level-5 Automated procedures: Management by consent.
- Level-6 Automated procedures: Management by exception.
- Level-7 Autonomous execution when required.

The role of NPP operators cannot be neglected, since full automation can be implemented only when all accidents can be analyzed previously. Therefore, it is a practical and preferable solution to study human and machine collaboration. In this sense, the automated recovery system proposed is the level-5 mentioned above. The various stages towards full automation and the changing role and responsibilities of the human operator have been conceptualized by e.g., (Sheridan,1982) which emerged the thoughts on collaboration with human and machine in this research.

4. THE AGENT SYSTEM

In this section, why we introduced a concept of agent to the operation support system, definition of agent, and the proposed agent system architecture are concerned. As mentioned in the section 2, there are quite complex causal relations between SFDs. Operators should recognize these relations whenever they operate in the emergent situation. Moreover, they should take not only available time but also executable operations at each moment into account. Since operators tend to confuse in an accident, especially they cannot realize what is going on, it is difficult for them to do these important tasks correctly. Therefore, we should make a system to take part in the part of these tasks. We designed the system under the concept of the agent. The definition of the agent or the agent system is *what acts autonomously*, in other words, something that acts intelligently. For example, Russell(Russell,et al.,1995) defined the agent as something that percepts and acts, and he also taken artificial intelligence as the research of rational agent. Our requirement to support system for NPP operators is just same as developing rational agents. In this paper, the terminology, *"rational"* is defined as *"acting (operating) to accomplish the goal (managing the emergent situation by operations) with the limited resources.*

The agent operation support system is proposed here. The agent architecture is depicted in fig.4. The agent system has two interfaces to interact/percept with its environment, i.e., NPP and operators. The machine (Agent)-machine (NPP) interface has instruments consisting of several sensors, transmitters, comparators indicators and so on together with automatic operation/control systems. On the other hand, the agent system has man-machine interface composed of seven windows on GUI as described in the next section. The agent system can show their result of information processing and suggestion to operators, and execute operation through these interfaces. The agent system is consisted of three main parts: a part of recognition, a part of resource management, and a part of decision making. The recognizing part grasps the situation of an accident from the information getting through the interface. The recognized situation is represented by not only NPP parameters but some induces such as time available and safety distance of each SFDs defined by Eq.(7). Time that is available for recovery and safety margin are calculated in the resource management part with process signals (plant parameters) sent from the recognizing part and the results are answered back to the recognizing part.

The resource management part manages two types of resources that are necessary for agent system to accomplish its target (mission) rationally. One of the resources is time resource, and the other is a subset of executable operations at each moment. It corresponds to safety components that are available in a given situation for recovery. The latter is called safety resource in this paper. The agent system refers to time available of SFD_1 (heat removal from reactor core) as index of time resources. The resource

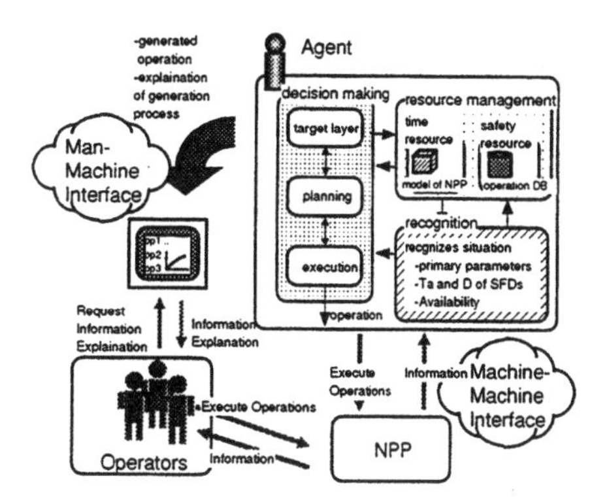

Fig. 4. The Agent System

management part has a simple kinetic model of NPP represented as lumped parameter system and estimates the index of time resource under the model with sensed parameters. This time resource is changing continuously with elapsed time. It will decrease if no operations for recovery are done and may increase if adequate operations are executed successfully. The agent system is able to utilize this time resource for taking recovery process, i.e., evaluating the situation, selecting the adequate operation sequences, and executing these operations. The agent system has database of all actions to be considered and a given situation changes its operability at each time in the database. These operations are grouped and allocated to their most effective SFDs in the database.

The decision making part has three layers: the target layer, the planning layer, and the action layer. The target layer generates the goals to be done of the man-machine system proposed in this paper. In this system, there are three main types of goals that are along with the degree of accident such as, "Goal A: Generating a adequate operation sequence", "Goal B: "Suggesting effective procedures",and "Goal C: "Preventing the damage of CV" Because of the limitation on number of pages given in this paper, mainly the explanation on the Goal A is written here. In the target layer, there are two targets. They are "avoid the core damage" and "make the available time longer". The former target is referred at first, and switches to latter target depending on the result whether the agent can generate appropriate operation sequence(s) , i.e., the target will change in the case that agent fails to generate operation sequence of avoiding core damage. At the operation generation layer, the agent system generates nominal operation sequence with referencing to the safety resources and estimates its effect, i.e., how long nominal operation sequence extend time available of SFD_1. If the available time is larger than one with no operation, the operation sequence is evaluated to be effective. Furthermore, the operation sequence makes the available time infinity, which means the operation sequence can avoid core damage in NPP, the agent shows the sequence as a

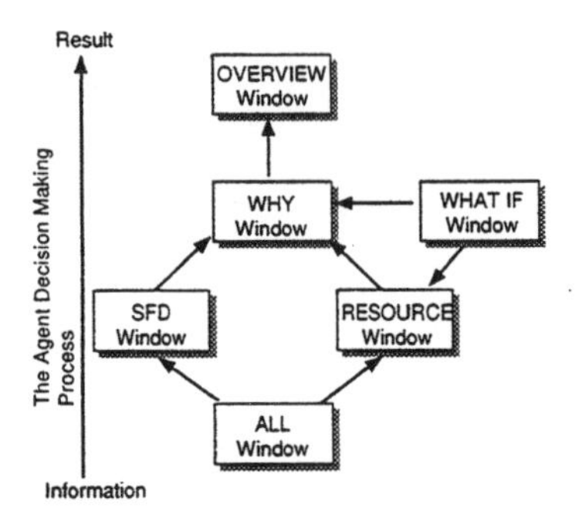

Fig. 5. Interface Window

successful generated result to the operators. At first, the agent searches operation sequence in the group of SFD_1. If there were no operations in the SFD_1, the agent searches the other SFD's operation group considering the safety structure discussed in the section 2. The group of operation sequence is given by an algorithm reflect operator's expert knowledge. Such knowledge is represented by bayesian network. Its application to this problem is left to the further study. The execution layer manages execution of the operations. The execution of generated operation sequence and confirmation of completing each operation in the sequence are done in this layer.

As mentioned in this section, the proposed agent system perceives and acts autonomously with the interface, and supports the operators intelligently by managing the resources.

5. MAN-MACHINE INTERFACE

Developing appropriate man-machine interface plays an important role in the man-machine collaborative work. Among the all factors of the interface, it has been focused how the agent could show its signal process clearly to the operators (so-called transparency) to help understanding. It should be noted that the trust in each other is critical in the collaborative work with human and machine. Under such principle, we compose the interface by seven windows as depicted in fig.5.

The composition of window and their layer comply with the agent information processing flow. Operators can understand the agent process from its information elements such as plant process and availability of the safety components in NPP to the final result by these windows.

The main window is overview window. The agent shows its final result of generated operation exactly on this window. It, of course, can be taken as an approval from the agent to operator. Operator can recognize the presented operation sequence and the effect at a glance by checking the D-Ta graph depicted in fig.2 on the window, and can decide whether operator accepts or rejects the presented operation sequence by the agent. Additionally, operators can make inquiry to the agent on its decision making process through "Why" window and "What If" window. For example, operators can obtain the answer to some question such as "why you suggest this operations?", and "how do you evaluate on this operations?" from the agent. Such communication would enhance understanding of the agent. "SFD" window is a kind of CAI window on SFD and its related operations. This window gives operators the explanation on each SFD and causal relations between SFDs. "Resource" window gives information on availability of all safety components, i.e., operability of all operation tasks and available time of each operation sequence. These information are all of information processed by/in the agent.

6. CONCLUSION

In this paper, at first, the more specific and objectively representation the SF is proposed as the SFD. The causal relations between SFDs are very complex so that taking them into account completely is heavy task for operators. If the difficulty is taken into account, the agent based operation support system is required. It generates the appropriate actions to avoid core damage and execute them automatically with operators approval. Moreover this system has been advanced to be more intelligent and autonomous by adding the resource management. Finally, the design principle of man-machine interface that enhances man-machine collaborative work is conjectured. The transparency of the decision making process is considered to be one of the most important factor.

REFERENCE

W.R.Corcoran et al.(1981). Critical Safety Functions. *Nuclear Technology*, **55** pp.690–712.

Y.Niwa, E. Hollnagel and M. Green (1996). Guidelines for computerized presentation of emergency operating procedures. *Nuclear Engineering and Design* , **167**, pp.114–127.

S.Russell and P. Norvig(1995). *Artificial Intelligence : A Modern Approach.*, PrenticeHall, New Jersy

T.B.Sheridan(1982). Supervisory control: Problem, theory and experiment for application to human-computer interaction in undersea remote systems. Dept of Mechanical Engineering, MIT.

T.Washio(1990). Derivation of Exogenously-Driven Causality Based on Physical Laws. *Journal of Japanese Society of Artif. Intell.* (in Japanese), **5** [4], pp.482–491.

ADAPTIVITY IN HUMAN-COMPUTER INTERFACE SYSTEMS: IDENTIFYING USER PROFILES FOR PERSONALISED SUPPORT

M. Dimitrova*, G. Johannsen, H. A. Nour Eldin***, J. Zaprianov*, M. Hubert*****

** Institute of Control and System Research, Bulgarian Academy of Sciences, PB 79, 1113 Sofia, Bulgaria; ** Laboratory for Systems Engineering and Human-Machine Systems (IMAT-MMS), University of Kassel, D-34109 Kassel, Germany; *** Group of Automatic Control and Technical Cybernetics atk, D-42097 Wuppertal, Germany*

Abstract: Adaptivity for user "personalised" support is viewed as complementary to systems for cognitive support and is related to the following aspects of user modelling: individual user profile, user style/state diagnosis, context of work and mental workload prediction. The paper reports the results of an empirical study revealing a particular kind of contextual influences on user behaviour, namely, the freedom to choose the set of operations to be performed from a larger set of tasks, valid for all users at the basic skill level of performance. Stability of patterns was found on the level of minimal reaction times (RTs), but not on the level of mean RTs. Neural networks successfully generalised the minimal RT profiles across contexts. *Copyright © 1998 IFAC*

Keywords: User interfaces, human-centered design, mental workload, interface, adaptation, monitoring, neural networks

1. INTRODUCTION

The human-computer interface plays a primary role in control of modern complex technological processes for efficient management of the information resources. One aspect of human-computer communication is the one-to-one dialogue on the task at hand. Distributed multimedia systems (DMS) have already been introduced in control rooms as intelligent means for timely and efficient communication between several users (Johannsen, 1997b; Gecsei, 1997; Kawai and Takaoka, 1994). This new role of multimedia systems requires a new design approach to the interface and new understanding of the "computer metaphor" in the eyes of the individual user. The functions of the computer as a negotiator have to be complemented and partially replaced by functions of a mediator, or perhaps, "moderator" of the dialogue between the user and his immediate interface. To accomplish this new role the interface should be able to accumulate, process, modify, retrieve and/or infer knowledge about the current user in an implicit, as well as explicit manner of interaction in order to achieve flexibility very much in the way people communicate with each other.

Important issues of implementing adaptivity in distributed multimedia systems are discussed in Gecsei (1997). Examples of current adaptive applications (mobile open-system technologies, teleconferencing, news-on-demand providing systems in Internet etc.) have been reviewed. These are grouped into 3 types of adaptivity of application - user-centered, system-centered and mixed, where "adaptation is based on observing parameters from both the system and the user" (Gecsei, 1997, p. 63). Several useful notions for future adaptive interface systems are given such as "personalisation" of the dialogue, design of "platforms" for architectural support of the adaptation, resource sharing, "user acceptance" and "user awareness" of the adaptation that have been viewed (under various but similar formulations) as "hot" research issues in human-computer interaction research by other authors as well (Johannsen, 1997a, 1997b; Johannsen *et al.*, 1997; Gecsei, 1997; Monostori et al., 1996; Kawai and Takaoka, 1994).

Design problems of adaptive interfaces have been surveyed in Norcio and Stanley (1989), an essential one of them being the need for adaptation to the individual rather than the average user. The adaptive interface system has to record user reactions to computer generated tasks and diagnose operators' skills, style of work and current psychological and/or physiological states. In this sense, the task on the computer counterpart can be set as pattern-based monitoring of the process of human-computer interaction (Rengaswamy and Venkatasubramanian, 1995; Dimitrova 1996).

"Style" of human-computer interaction has been revealed as a factor influencing mutual adaptation between operators and modern expert systems for decision support (Bushman *et al.*, 1993). In Stassen *et al.* (1990) three possible behaviours (resp. "styles of work") are outlined - of the "wise", "lazy" and "ambitious" supervisors. User profiles are essential when dialogue between more than one user or different classes of users has to be coordinated. In Dimitrova *et al.* (1997) it is proposed that, a) any set of job-relevant behaviours that can be retrieved by the computer can be represented as patterns of user styles of work; b) that the typology is *not* known *apriori,* and c) that knowledge about styles has to be retrieved in an implicit, rather than explicit manner in order to avoid frustration and additional workload on the users. Neural networks have been applied as appropriate tools for data extraction, classification of user cognitive style of work and recognition of individual users (Yoshikawa and Takahashi, 1994; Obaidat and Macchairolo, 1994; Dimitrova *et al.*, 1997).

The following section describes a framework for research on user-adaptive interfaces. Section three presents the results of an experiment designed especially to test the influence of context on patterns of data of individual users and an approach how to reveal stability across conditions. Section four gives the simulational results from neural network classification behaviour applied to the experimental data. Some ideas how to generalise the approach to higher levels of user cognitive behaviour are briefly discussed in section five.

2. CONCEPTUALISATION OF THE PROCESS OF HUMAN-COMPUTER INTERACTION IN USER-ADAPTIVE ENVIRONMENTS

It has early been pointed out that in the eyes of the user the system is the dialogue (Byrer and Jelassi, 1991). The same applies to current multi-user multi-purpose interface systems as well. The interface system manages and coordinates various applications of different purpose, functions and design. However, the user perceives the computer he is working with *holistically*, as a "personalised" entity. In the following conceptual scheme (Fig. 1) the computer is given as a metaphor of a partner in the dialogue so the various links representing its coordinating functions are omitted.

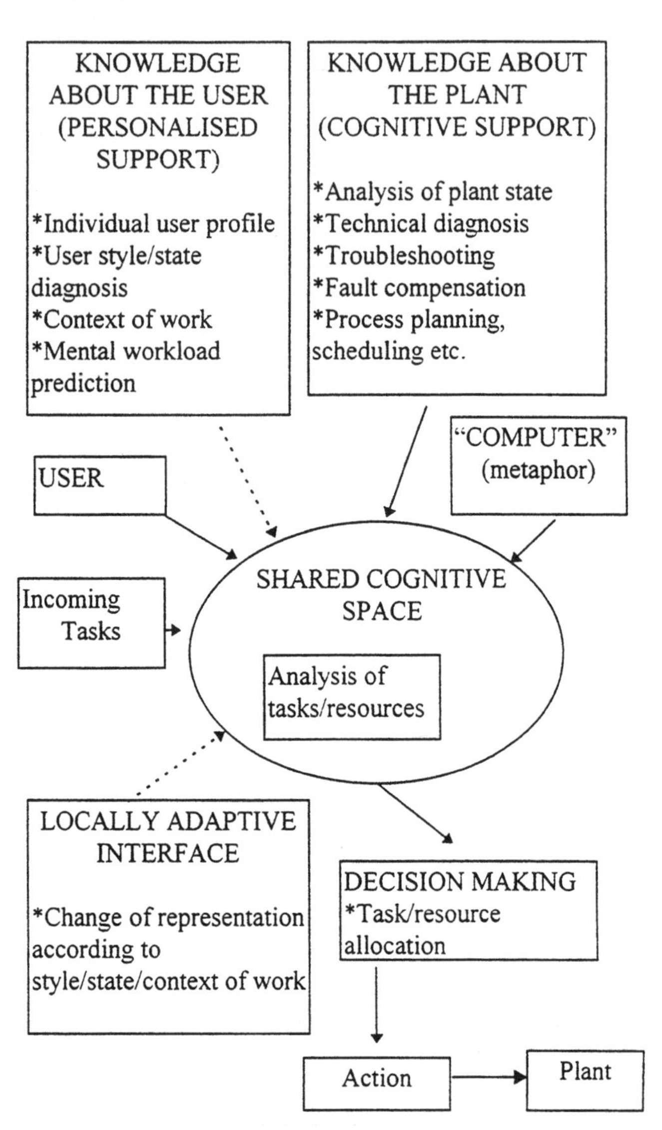

Fig. 1. Conceptualisation of the process of HCI.

The conceptual scheme of the process of adaptive human-computer interaction (HCI) summarises some of the types of knowledge that can be provided by the system as knowledge support for decision making and action in, for example, control rooms. The notion of the shared cognitive space can be viewed in the Cognitive Space Metaphor introduced by Johannsen (1997b) as important in revealing that human cognitive capacity is limited, so it is really important to apply processes of evaluation and selection of the information provided to the user at any particular moment of time. This is a major aspect of adaptation of computers to users that is essential for provision of **cognitive** support. Cognitive support is understood here as any application designed to assist the human in performing the job at hand. Research in HCI has been focused mainly in this direction, so it is given in the scheme by a solid line.

A second aspect of adaptation is the provision of **personal** support to the user. If the interface is capable of providing such kind of support, the interaction or the dialogue will be much more relaxed and "humane" and, hopefully, more efficient. Adaptivity for personal support is related to the following aspects of user modelling: individual user profile, user style/state diagnosis, context of work and mental workload prediction. The incorporation of these types of knowledge about the current user have not yet achieved the level of development on systems for cognitive support in current applications and is an area of intensive research. This is why it is given in the scheme by a dashed line. The assumption is that the adaptive interface should adopt complementary and compensatory strategies of providing cognitive support in order to avoid conflicts in task execution. The immediate implication of this assumption is that, at present, there is very little knowledge on how users *will* accept a system that is much more flexible and changeable than the standard unified interface systems. This is a new and exciting research issue that has recently revealed its importance in terms of "user satisfaction" and "system acceptance" (given by the second dashed line). Adaptivity should allow user awareness of the "personalisation process" and user involvement in management of the process of interaction (Gescei, 1997; Johannsen, 1997b).

What is new in this proposed approach to user-adaptive interface design is that the knowledge management process should perform on different levels in the hierarchy of the system for coordination of the dialogue between several users and knowledge-based systems, resp. in DMS. Systems for cognitive support should be available at a higher level and, respectively, should provide/distribute unified knowledge structures to all the users, whereas personalisation should take place only at the local level of the immediate interface. It is our view that if systems for cognitive and for personal support are separated as distinct design issues, flexibility of the dialogue will be achieved more easily and at a low cost.

3. INVESTIGATION OF LEVELS OF INTERACTION: THE ROLE OF CONTEXT

A locally adaptive interface has to be able to deal with both stable and changeable characteristics of the current user. Stability is important to provide a baseline of the set of possible behaviours of a given user or a class of users with similar profiles of cognitive behaviour. Dynamics of change is important for anticipation of future behaviour. It turns out that it is quite a difficult task to identify stable unique parameters that can be implicitly retrieved by the interface. Implicit retrieval is the information gathered from user actions such as mouse focusing and keyboard strokes. The individual profile of times (in msec) taken by a sequence of actions is used as data in our study.

Contextual influences are among the main factors causing instability of the retrieved patterns of behaviour. Context can be understood as influences from the working environment when either relevant or irrelevant information is being processed simultaneously with the task at hand. Another aspect of the influence of context is the "interiorized" context, i.e. the extent to which it *actually* changes performance even on simple reaction time tasks. One particular instance of contextual change is studied and presented below - user freedom to choose the sequences of actions to be performed. The purpose of the study is to identify stable patterns of data that are uninfluenced by the change of context from small samples of data.

3.1. The experiment

Subjects type in 5 words at a computer prompt. 7 users took part in the experiment. They had to type in the following words: Concept, Work, Design, Paper, Memory. In the first condition they can type in 3 of the words of their own choice (free condition). In the second condition they had to type in all of the words in the same order (normative condition). The word prompts were provided one after the other, immediately after the subject had entered the last letter of each word in both conditions. There were 9 trials in each condition. The interface records the choice and the reaction times. Each word which was typed in is considered a sequence of actions to be performed. Only the chosen words are dealt with further in the study. Table 1. presents the individual choices in the study.

Table 1. Individual choices of words to be typed in

Subject No	Concept	Work	Design	Paper	Memory
1	Yes	Yes	Yes		
2		Yes	Yes	Yes	
3	Yes		Yes		Yes
4	Yes	Yes		Yes	
5	Yes		Yes	Yes	
6		Yes	Yes	Yes	
7	Yes	Yes		Yes	

Since the choices do not match among users, the data is further processed for each individual word, i.e. within each column. The word Memory is left out because it was chosen only once.

3.2. Results

Table 2. represents the mean reaction times (RTs) for each keystroke in the group of subjects.

Table 2. Mean RT for each keystroke in the group of subjects

C	-O	-N	-C	-E	-P	-T
Normative	54.578	43.889	52.778	26.178	52.044	47.600
Free	49.511	37.933	36.356	24.756	40.067	25.378

D	-E	-S	-I	-G	-N
Normative	49.356	80.311	39.644	70.578	55.911
Free	48.933	61.200	32.689	58.422	40.444

P	-A	-P	-E	-R
Normative	54.578	43.889	52.778	26.178
Free	49.511	37.933	36.356	24.756

W	-O	-R	-K
Normative	59.511	51.378	68.933
Free	53.156	49.778	49.244

Statistically significant influence of change of context on reaction times was revealed for all subjects and all four remaining words. Table 3 summarises the level of statistical significance of the "Context" factor. There were also significant interactions between the three factors "Context", "RT" and "User". When subjects had to type in the preferred words in the context of a normative task, i.e. within a strictly followed sequence of words, it took them longer to type each individual word. This can cause difficulties in the attempts to find stable profiles when the influence of context is revealed at the very base level of cognitive performance. It seems quite an unexpected finding since it has long been assumed that skilled performance is most unlikely to be influenced by contextual change (see Rasmussen, 1986).

Table 3. Statistical significance of the "Context" factor

Word	F (1,40)	MSe	P
Concept	11.474	1294.682	0.002
Design	10.436	1254.096	0.002
Paper	21.473	192.538	<0.0005
Work	16.286	351.926	<0.0005

The present study reproduces the basic pattern of results of a preliminary study briefly reported in Ossikovska *et al.* (1996). The difference is in the group of subjects and in the new interface system for data recording. The present group consists of users who are experienced with computers but *not* computer professionals, whereas the previous group consisted of 5 computer professionals. In both cases the influence of context is present at the level of the individual words.

Stability of patterns can be found, however, on the level of minimal RTs for each subject in either condition. Fig. 2. exemplifies the shift of patterns towards each other when minimal RTs are considered in contrast with the mean times for an individual user/word.

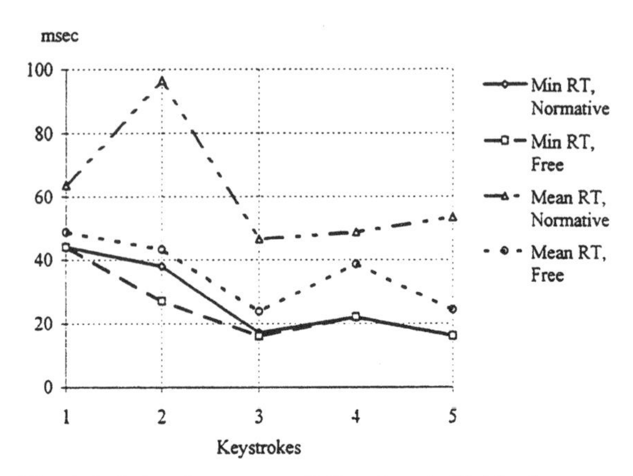

Fig. 2. Minimal vs. mean RTs of an individual user.

The statistical analysis reliably confirmed that the influence of context can be eliminated on the level of minimal RTs for individual users. A repeated measures ANOVA with factors "Context" x "Keystroke" did not reveal a main effect of the "Context" factor, nor any interaction between the factors. The results for the "Context" factor are given in Table 4.

Table 4. No main effect of the "Context" factor over minimal RTs

Word	F (1,4)	MSe	P
Concept	4.151	25.058	0.111
Design	2.691	31.400	0.176
Paper	3.137	34.713	0.151
Work	6.030	43.783	0.070

The identification of the level of stability of patterns is necessary for the interface system to be able to acquire means for implicit monitoring of the current state of the user as complementary means to the explicit dialogue especially in situations when the user is experiencing time pressure and fast decisions have to be made. Also, it can serve as a diagnostic tool for stress monitoring and "personalisation" of the dialogue depending on the individual stress resistance.

4. NEURAL NETWORK PERFORMANCE ON USER CLASSIFICATION

Neural networks were employed for recognition of individual users from the identified profiles of minimal reaction times. Four mini-networks were designed for each word, i.e. for each column of Table 1 (except for the last). The basic architecture is given in Fig. 3. A feedforward network with three layers - input, hidden and output employing the backpropagation learning algorithm with a nonlinear activation transformation function (tanh) on all layers except the input layer was used (McClelland and Rumelhart, 1988; Sorsa and Koivo, 1993).

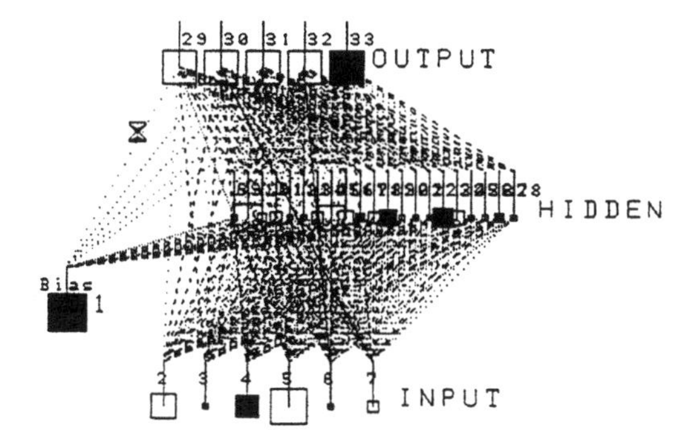

Fig. 3. Architecture of the employed neural network.

The aim is to test the ability of the network to generalise across contexts. The teaching pattern set was taken from the free context and performance was tested on both the studied and new patterns taken from the normative context. The network was able to successfully complete the task in the range of 10000 cycles, the lowest being for the word Design (5559 cycles).

5. PROFILES VS. STYLE OF COGNITIVE PERFORMANCE

We have observed that the data gathered by the interface system is characterised by similarity of patterns of different users. This observation applies not only to profiles drawn from skill-based behaviour, but to data from higher-level computer generated tasks, such as from the following list: reactions to extremal situations, visual tracking tasks, spatial (left-right movement) coordination, visual spatial orientation, visual pattern recognition, attention, visual working memory capacity, personality traits (extroversion/ introversion), simple motor reaction tasks and complex motor reaction tasks. Subject's scores on these tasks were used for testing neural network performance on individual user recognition in Dimitrova *et al.* (1997). Here, principal component analysis (PCA) is performed on the set of mean scores of 16 subjects to illustrate that it is difficult to identify clear-cut classes of users based on the pattern of results on the given tasks. Their results are plotted in Fig. 4 with the first two components of the PCA which explain about 51% of the total variance of the original data.

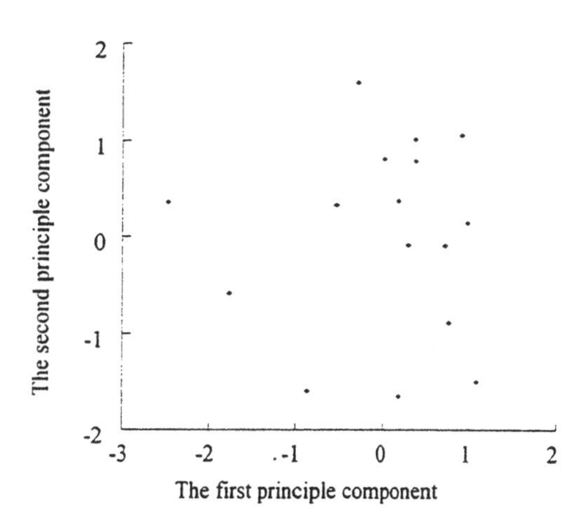

Fig. 4. Scores of 16 subjects on the 10 tasks presented with the first and second principal components.

It is interesting to note that the first revealed component can be labelled "extremal/undisturbed performance", and the second - "visual-motor tracking/working memory capacity" (see Baddeley, 1990). Further investigation is needed on the relation between general tendencies in human cognition, on the one hand, and the individual differences, on the other, and their appropriate account in interface adaptation. Yet any individual "profile" can be considered an aspect of the specific style of cognitive behaviour of the current user as his/her stable characteristic. This is why it is suggested that the interface retrieves **profiles** from the skill-based level and records **style** at higher cognitive levels.

6. CONCLUSIONS AND DIRECTIONS FOR FUTURE RESEARCH

Locally adaptive human-computer interface systems have to be able to adapt to the individual style or preferences of interaction, i.e. to provide personalised support via incorporation of implicit monitoring of the current user. This task is viewed as complementary to the more general approach for providing cognitive support based on knowledge of human cognitive strengths and limitations.

The paper reports the results of an empirical study revealing a particular kind of contextual influences on user behaviour, namely, the freedom to choose the set of operations to be performed from a larger set of tasks valid for all users at the basic skill level of performance. Also, it reveals a variety of preferences that can be accounted for in team composition. Stability of patterns was found on the level of minimal RTs, but not on the level of mean RTs. Neural networks successfully generalised the minimal RT profiles across contexts.

The study is viewed as a first step towards implicit monitoring of user behaviour. Further development of means for implicit monitoring are needed in representing the amount of change from a baseline for stress and workload prediction.

REFERENCES

Baddeley, A. (1990). *Human memory: Theory and Practice.* Lawrence Erlbaum Associates Ltd., Hove.

Bushman, J.B., C.M. Mitchell and P. Jones (1993). ALLY: An operator's associate for cooperative supervisory control systems. *IEEE Transactions on Systems Man and Cybernetics*, **23**, 111-128.

Byrer, J.K. and T.M. Jelassi (1991). The impact of language theories on DSS dialog. *International Journal of Operational Research,* **50**, 113-126.

Dimitrova, M. (1996). A compound neural network for user identification in adaptive human-computer interfaces. *Complex Control Systems*, **1**, 44-52.

Dimitrova, M., M. Hubert, D. Boyadjiev, A. Nabout, J. Zaprianov and H.A. Nour Eldin (1997). Neural networks for classification and recognition of individual users in human-computer interface. *Proceedings of the IEEE International Symposium on Intelligent Control (ISIC'97)*, Istanbul, Turkey, pp. 101-106.

Gecsei, J. (1997). Adaptation in distributed multimedia systems. *IEEE Multimedia*, **2**, 58-66.

Johannsen, G. (1997a). Conceptual design of multi-human machine interfaces. *Control Eng. Practice*, **5**, 349-361.

Johannsen, G. (1997b). Cooperative human-machine interfaces for plant-wide control and communication. In: *Annual Reviews in Control* (J.J. Gertler, Ed.), pp. 159-170. Pergamon, Elsevier Science, Oxford.

Johannsen, G., S. Ali and R. Van Paassen (1997). Intelligent human-machine Systems. In: *Methods and Applications of Intelligent Control* (S.G. Tzafestas, Ed.), pp. 329-356. Kluwer Academic Publishers, Dordrecht, The Netherlands.

Kawai, K and H. Takaoka (1994). Human interface design for highly automated power generating plants. In: *IFAC Integrated Systems Engineering* (G. Johannsen, Ed.), pp. 215-220. Baden-Baden, Germany.

McClelland, J.L. and D.E. Rumelhart (1988). *Explorations in Parallel Distributed Processing: A Handbook of Models, Programs, and Exercises*, pp. 11-44. MIT Press, Boston, MA.

Monostori, L., H. Van Brussel and Westkampfer (1996). Machine learning approaches to manufacturing. *Annals of CIRP*, **45,** 675-712.

Norcio, A.F. and J. Stanley (1989). Adaptive human-computer interfaces: A literature survey and perspective. *IEEE Transactions on Systems Man and Cybernetics*, **19**, 199-408.

Obaidat, M.S. and Macchairolo (1994). A multilayer neural network system for computer access security. *IEEE Transactions on Systems, Man and Cybernetics*, **24**, 806-813.

Ossikovska, S., I. Nikolaev and M. Dimitrova (1996). Reliability assessment of hybrid (human-machine) control systems. *Complex Control Systems*, **1**, 34-43.

Rasmussen, J. (1986). *Information Processing and Human-Machine Interaction*, New York, North-Holland.

Rengaswamy, R. and V. Venkatasubramanian (1995). A syntactic pattern-recognition approach for process monitoring and fault diagnosis. *Engineering Applications of Artificial Intelligence*, **8**, 35-53.

Stassen H.G., G. Johannsen and N. Moray (1990). Internal representation, internal model, human performance model and mental workload. *Automatica*, **26**, 811-820.

Sorsa T. and H.H. Koivo (1993) Application of artificial neural networks in process fault diagnosis. *Automatica*, **29**, 843-849.

Yoshikawa, H. and M. Takahashi (1994). Conceptual Design of Mutual Adaptive Interface. In: *IFAC Integrated Systems Engineering* (G. Johannsen, Ed.), pp. 221-226. Baden-Baden, Germany.

AUTONOMOUS GRAPHICS GENERATION SYSTEM FOR SUPPORTING HUMAN SELF-EXPRESSION

Takashi Miyatake* Katsunori Shimohara**

* *NTT Business Communications Headquarters 9-1 Konan 1-Chome Minato-ku, Tokyo 108-8019 Japan*
** *NTT Computer Science Laboratories, 2-4 Hikaridai Seika-ChoSohrakugun Kyoto 619-0237 Japan*

Abstract: It is very important for multimedia to truly take root in society because of its great potential to create affluence and improve the quality of life. We believe this requires an approach not only appealing to the rational side of human behavior but also to the emotional side, from which springs our desire for self-expression.
Specifically, we intend to make a system (called the "Autonomous Graphics Generation System") for generating CG by using an individual's biotic information for self-expression derived from the interaction between the individual and a computer. In this paper, we propose the concept of this system and introduce a prototype system for generating CG by using brain waves. *Copyright ©1998 IFAC*

Keywords: Multimedia, Man/Machine interaction, Computer graphics, Communication environment, Bionics

1. INTRODUCTION

It is very important for multimedia to truly take root in society because of its great potential to create affluence and improve the quality of life. For this to happen, it is important to find ways not only for people to search for and read information but also to make people want to generate and transmit information. We believe that this requires a technique to enable people to communicate with each other by using multimedia as simply as they use a telephone. In our study, we aim to develop a system and a technique to help people generate computer graphics (CG) that also satisfies their desire for self-expression. Concretely speaking, we want to make a system (called the "Autonomous Graphics Generation System") for generating artistic CG using an individual's biotic information (EEG, EMG, ECG, blood pressure, body temperature, etc.) as human self-expression information, which in turn reflects human mental information (Takashi Miyatake, 1997). We expect the system to popularize the use of multimedia communications. In this paper, we propose the concept of this system and introduce a prototype system for generating CG by using brain waves.

2. MOTIVATION

It is important for a multimedia environment to enable people to generate and transmit information that arouses creativity and imagination to nurture even more creative future communications. We believe the development of such a multimedia environment has to be approached from not only the rational side of human behavior but also from the emotional side, from which springs our desire for self-expression. Therefore, we are confident that creative communication through multimedia will become widespread if people can easily generate information that satisfies their desire for self-expression and arouses their creativity and imagination.

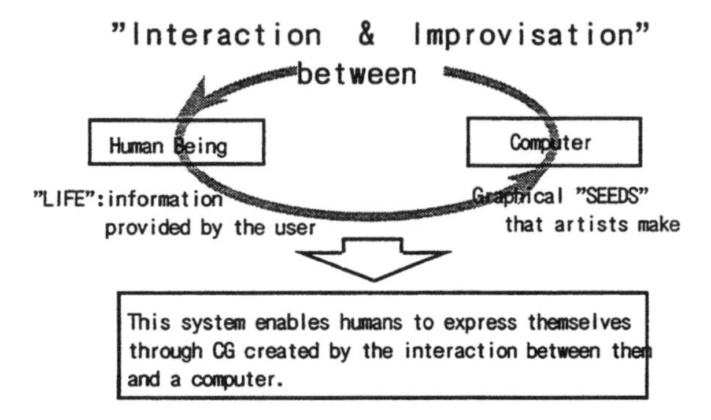

Fig. 1. The concept of the "Autonomous Graphics Generation System"(1)

Human biotic information (EEG, EMG, ECG, blood pressure, body temperature, etc.) is related not only to human physical states but also to human mental states. It always changes even in a single individual. Accordingly, CG generated by human biotic information should be considered human self-expression information that a person expresses only at one given time.

The study of CG generation by biotic information will be actively pursued in the future by using "Interactive Computer Graph ics" based on Artificial Life. For example, in "Interactive Plant Growing " (Sommerer Christa, 1993), CG are generated using electric signals from a person and a plant. However, studies on "Interactive Computer Graphics" have been almost exclusively approached from the viewpoint of art.

Therefore, our research is aimed not at professionals, who can easily express what they want to express, but rather at ordinary people who lack the skills or artistic imagination to express CG. Therefore we are developing a technology that lets people autonomously generate original and professional level CG from their biotic information. We expect that a technique for creating such CG would enable people to arouse sensitivity and imagination in others.

3. CONCEPT OF AUTONOMOUS GRAPHICS GENERATION SYSTEM

The concept of the proposed system is to generate CG by the improvisational interaction between a human being and a computer that can realize autonomous and creative functions (Fig. 1).

In other words, it is to generate original CG peculiar to an individual by giving "LIFE", in the form of user-provided information, to artistic CG "SEEDS", in the form of a CG program. The system uses the biotic information as the individual's "LIFE" information. Biotic information is greatly affected by human mental states, for instance, joy, anger, sadness and

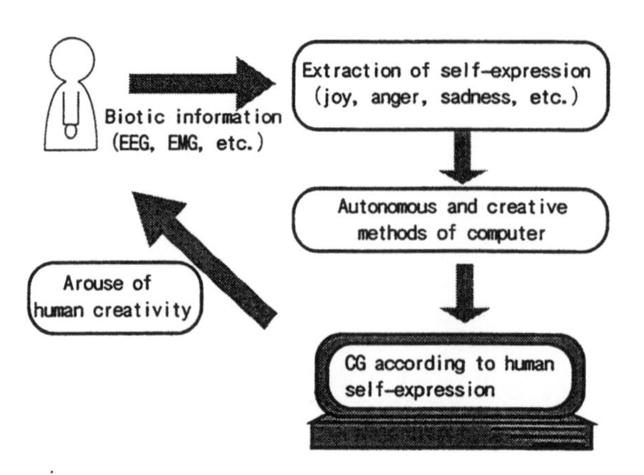

Fig. 2. The concept of the "Autonomous Graphics Generation System"(2)

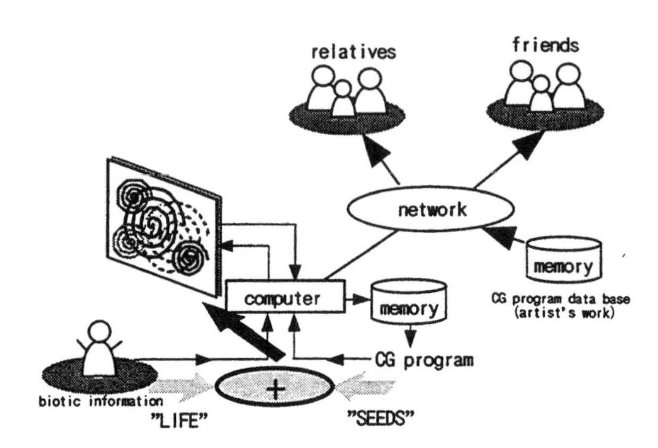

Fig. 3. CG KARAOKE system

happiness. We believe that using such information enables a person to generate original CG as self-expression without the need for special skills in creating CG (Fig. 2).

Fig. 3 shows a system to realize multimedia communication by using such a technique. The concept is a CG version of a KARAOKE system. A system user accesses a CG program data base to express CG through networks and selects the CG program in which he/she is interested. The interaction between his/her own biotic information and the CG program enables him/her to generate artistic CG based on his/her own information. The generated CG can be transmitted to the user's relatives, friends and others. We expect that such an approach would arouse the desire for creativity and imagination and realize a new type of multimedia communication.

4. PROTOTYPE SYSTEM

The first step of this study was to make a prototype system for investigating the possibility of CG generation from biotic information (Fig. 4). That system was designed as a platform for studying various problems and methods. The function of each processor in the prototype system is explained as follows.

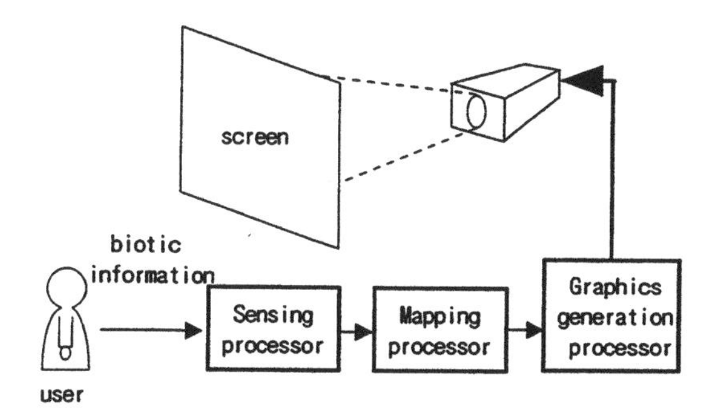

Fig. 4. The prototype system of the "Autonomous Graphics Generation System"

The sensing processor senses various biotic information and converts the analogue signals to digital signals with an Analogue-Digital converter.

The mapping processor maps biotic information processed by the sensing processor to the parameters needed for producing CG (CG parameters).

The graphics generation processor executes CG programs with those parameters and displays CG.

The prototype system uses brain waves, which contain rich information about the user's sensitivity and sensibility and may express the natural state of the human mind.

4.1 *Brain Waves*

We measured brain waves by EEG in the prototype system. EEG measurements are distinguished by frequency: δ waves (0.5Hz $\sim$ 4Hz), θ waves (4Hz $\sim$ 8Hz), α waves (8Hz $\sim$ 13Hz) and β waves (13Hz $\sim$ 40Hz). We do not use EEG as simply time-sequential signals but also pay attention to the frequency responses of EEG. In this prototype system, α waves and β waves are used because we regard EEG as human self-expression information. It is said that α waves emerge strongly if a person is relaxed.

Musha (Toshimitsu Musha, 1996) studied the relationship between EEG and four mental states: anger, sadness, joy and relaxation. Yoshida (Yoshida, 1995) studied the relationship between frequency-fluctuations of EEG and relaxation. Maekawa (Tadao Maekawa, 1995) studied the relationship between the power spectra of α waves and relaxation.

Furthermore, the power spectra of EEG have been studied when a person is in various states. Concretely, the following two states of EEG spectra have been studied experimentally (Takashi Miyatake, 1998).

(1) While listening to quiet classical music
(2) While playing a video game (a racing game)

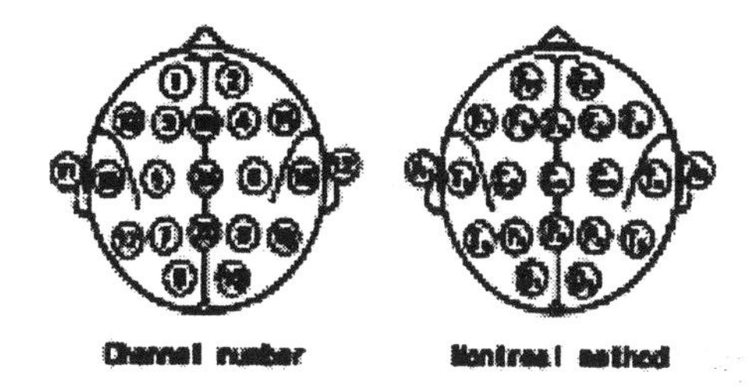

Fig. 5. Mapping between Montreal method and channel number of a electrode

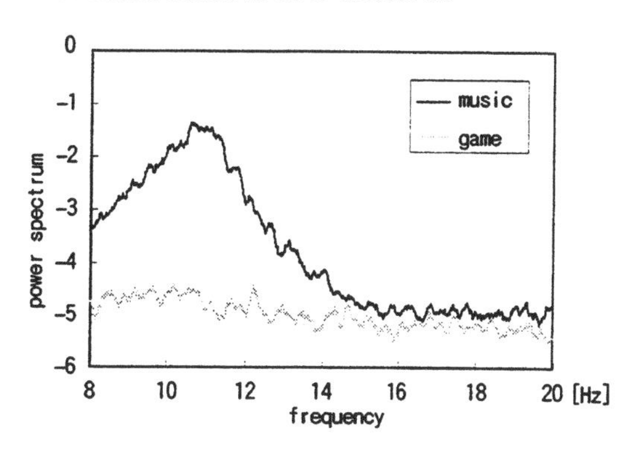

Fig. 6. Power spectra of each state

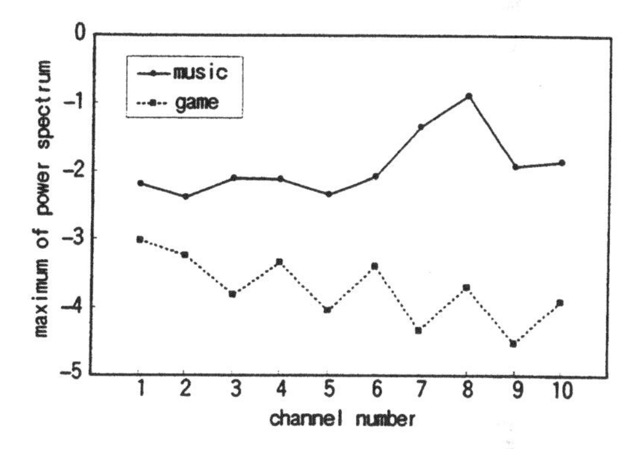

Fig. 7. Maximum power spectra of each channel

The positions of each experiment were taken at a sampling rate of 150 Hz, 10 electrodes(10 points on the brain) according to the Montreal method (Fig. 5) for five minutes and seven times. The number of subject is one.

The results are indicated below. Fig. 6 shows power spectra of a 7th electrode (P_3) in each state. The state of listening to classical music has stronger power spectra in the α waves band than does the state of playing a video game. The difference in the α wave bands is shown in the frequency of maximum power spectra and in the sum of the power spectra in 10-channel electrodes (Figs.7 and 8). In Fig. 8, power spectra of the state of listening to music is smaller because the absolute sum of power spectra is calculated.

The subject said that he had more relaxation in the state of listening to music. These results indicate that α waves have a relationship to human mental states and may be considered one of the indexes for discrimination of human mental

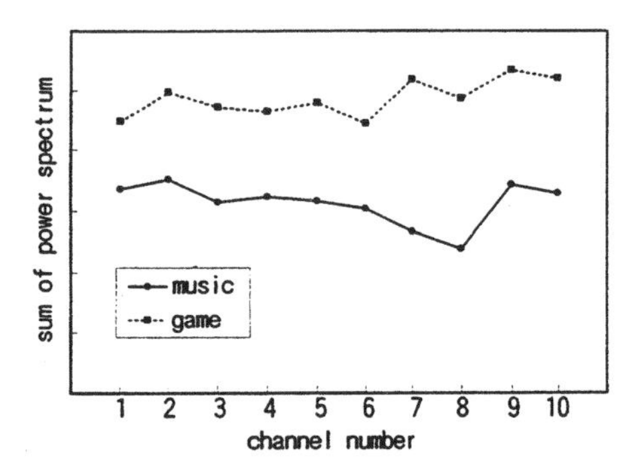

Fig. 8. Sum of power spectra of each channel

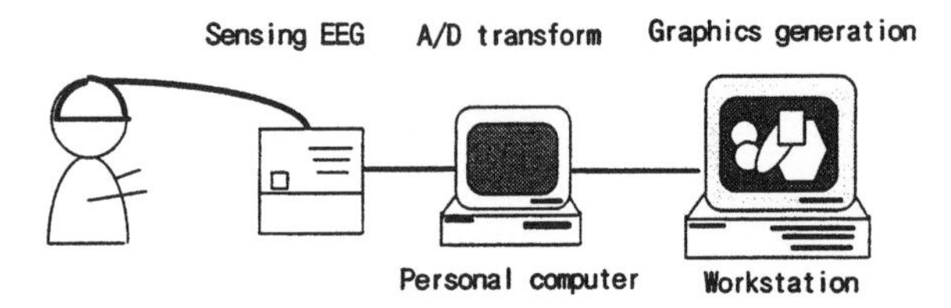

Fig. 9. The sensing processor

states. We are thus confident that it is effective for this prototype system to use α waves.

4.2 *Sensing processor*

Fig. 9 shows the sensing processor of this prototype system. In this processor, EEG is sensed, converted to digital signals with an analogue-digital converter, transmitted to a workstation and processed by various methods in the mapping processor. The sampling rate of EEG is 100 Hz. The number of electrodes is 10.

4.3 *Mapping processor*

The mapping processor calculates "strength of α wave spectra" and maps it along with time-sequential EEG data to the CG parameters. "Strength of α wave spectra" consists of two states: "alpha state" and "not alpha state". "Strength of α wave spectra" is decided by the following method.

(1) Frequency responses are calculated from 512 sample data of EEG by FFT (Fast Fourier Transform).
(2) The rate of the α wave band (8 $Hz \sim 13$ Hz) in the range from 8 Hz to 20 Hz is calculated in the 9th electrode (O_1) and the 10th electrode (O_2) (Fig. 10). These are averaged.
(3) If "alpha rate" is more than 65%, "strength of α wave spectra" is "alpha state". Otherwise "strength of α wave spectra" is "not alpha state".
(4) Return to 1 and repeat this algorithm.

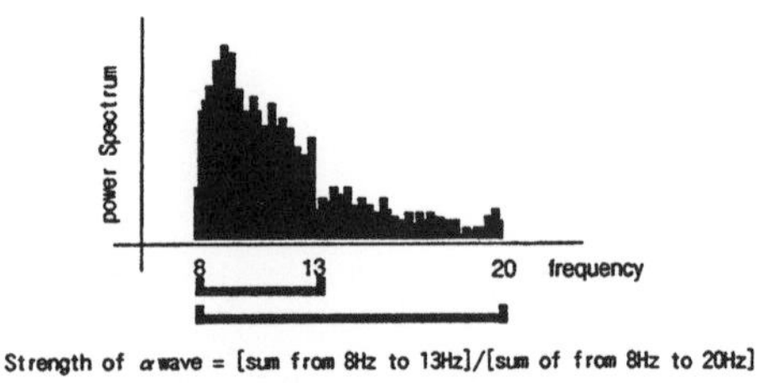

Fig. 10. Calculation of "Strength of α wave"

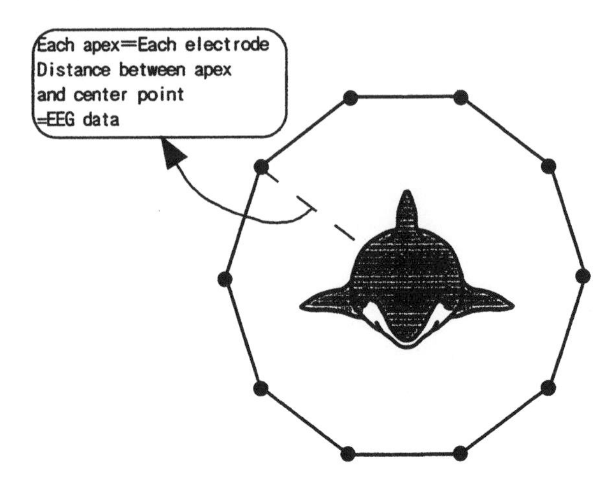

Fig. 11. The method of mapping EEG to CG (1)

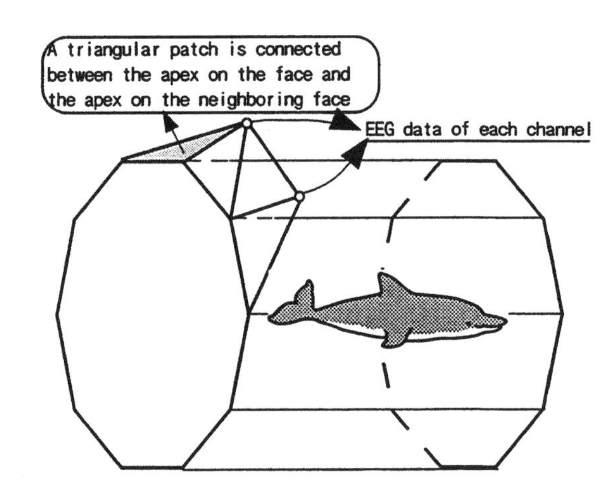

Fig. 12. The method of mapping EEG to CG (2)

Using this method, this processor maps "strength of α wave spectra" and time-sequential EEG data to the CG parameters. The processor displays a dolphin swimming and jumping in a tunnel made by time-sequential EEG data. A mapping method is indicated below.

(1) The 10 channels form 10 vectors in a radial configuration around the dolphin (Fig. 11). The face is arranged according to time. Triangular patches connect the face with a neighboring face (Fig. 12).
(2) The dolphin's direction of motion is determined by mapping eight channels to six directions and two rotations of the motion and then summing the eight vectors (Fig. 13).
(3) When there is a sudden change in data, the dolphin jumps.

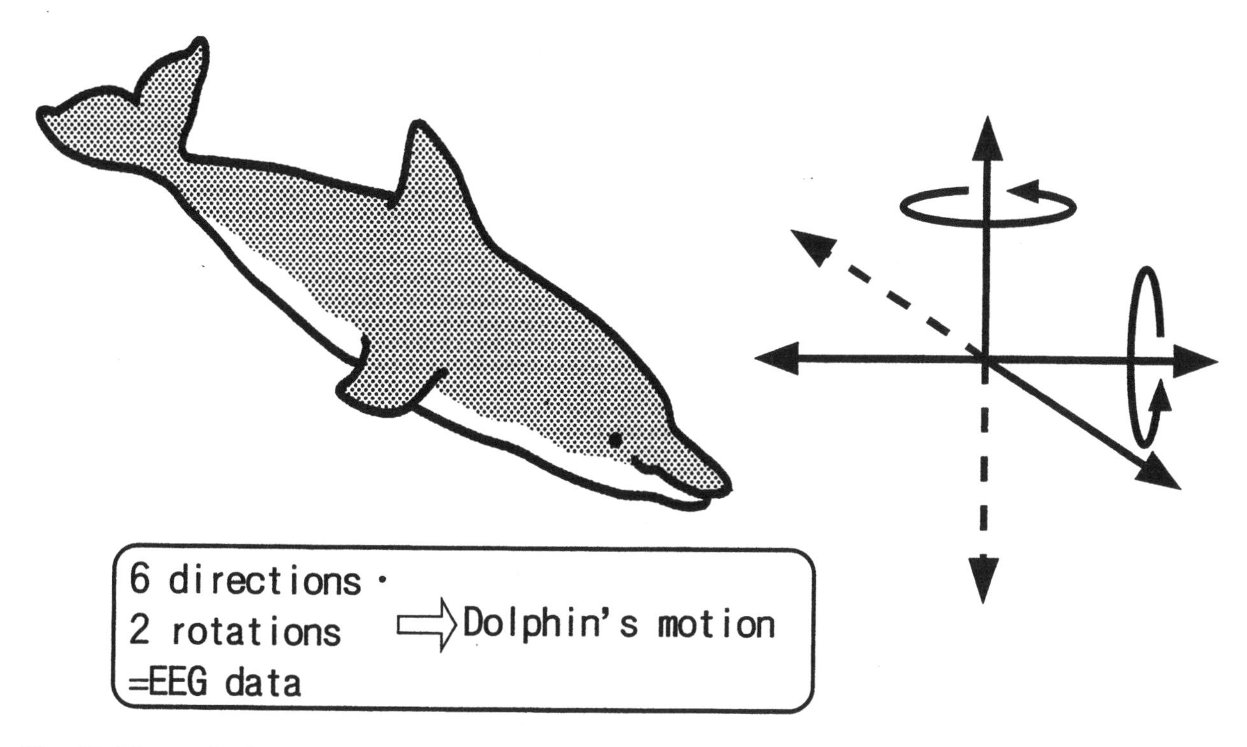

Fig. 13. The method of mapping EEG to CG (3)

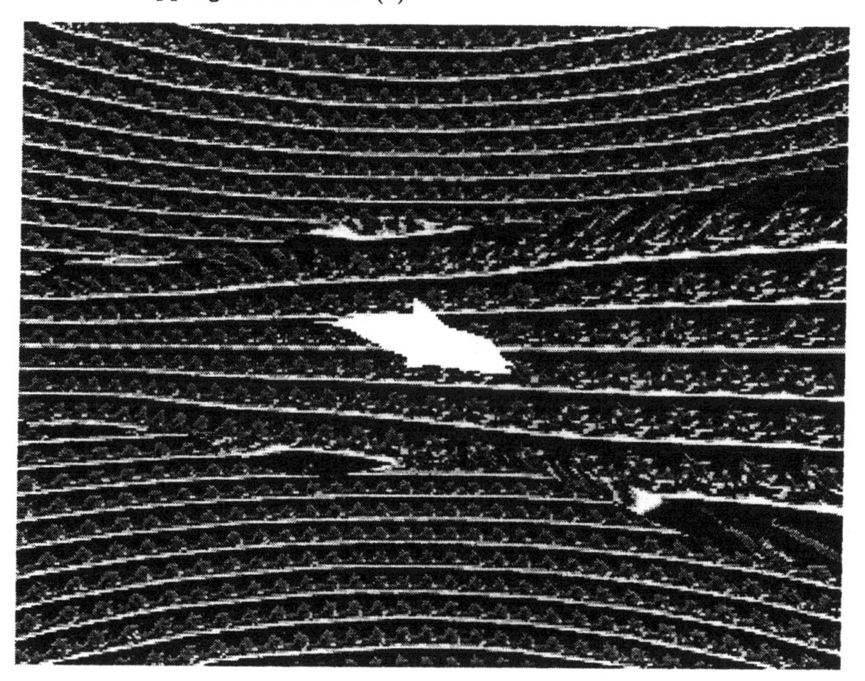

Fig. 14. CG generated in the graphics generation system

(4) The texture of the background is affected by the strength of the α wave. It toggles between two textures.

4.4 *Graphics generation processor*

The graphics generation processor displays generated CG (Fig.14).

5. CONCLUSIONS

We propose an Autonomous Graphics Generation System as a way to advance multimedia environments. We aim to develop a technique for the system to satisfy the user's desire for self-expression and arouse the user's imagination and creativity. We expect that the user will gain sensibility and sensitivity through the system. To realize this aim, the system allows improvisational interaction between a human being and a computer to be expressed through CG. Specifically, we are trying to develop a system to generate CG from biotic information, which is connected to human mental states. In this paper, we have introduced a prototype system using brain waves as the first step in realizing the Autonomous Graphics Generation System. The system generates CG by mapping "strength of α wave spectra" and time-sequential EEG data to a dolphin's motion and textures.

In the future, we will investigate the various human mental information included in biotic information, improve the Autonomous Graphics Generation System by using human mental information, and arouse the desire of human self-expression. Accordingly, we will study signal processing methods that enable us to extract mental information from EEG. We will also study the extraction of mental information from a large amount of biotic information. Moreover, we intend to study graphics generation methods that use interaction with autonomous and emergent methods based on artificial life.

6. REFERENCES

Sommerer Christa, Mignonneau Laurent (1993). Interactive plant growing. *ACM Siggraph Visual Proceedings.* pp. 164–165.

Tadao Maekawa, et al. (1995). Physiological and psychological assessment on texture amenity. *Proceedings of the 11th Symposium on Human Interface.* pp. 711–718.

Takashi Miyatake, Katsunori Shimohara (1997). Autonomous graphics generation system for supporting kansei expression. *Technical Report of IEICE MVE97-95.* pp. 17–21.

Takashi Miyatake, Katsunori Shimohara (1998). Discrimination of human states using brain waves. *The 1998 IEICE GENERAL CONFERENCE.*

Toshimitsu Musha, et al. (1996). Emotion spectrum analysis from eegs. *BPES'96.* pp. 413–416.

Yoshida, Tomoyuki (1995). The evaluation of emotion using the measurement of frequency-fluctuation of brain waves. *Journal JSME.* **98**, 71–74.

AN HCI-ENRICHED MODEL FOR SUPPORTING HUMAN-MACHINE SYSTEMS DESIGN AND EVALUATION

Christophe Kolski, Pierre Loslever

LAMIH - URA CNRS 1775
University of Valenciennes, Le Mont Houy - B.P. 311
59304 Valenciennes Cedex, FRANCE
E-mail: kolski@univ-valenciennes.fr

Abstract: The main development models from Software Engineering are too general and largely insufficient in term of interactive systems development. One solution to this problem is the proposal to use so-called HCI-enriched models. Such a model, called ∇ (pronounced "nabla"), is presented in this paper. During a questionnaire-based evaluation, this model has been compared with four classical models (Waterfall, V, Spiral and Incremental models) and five other HCI-enriched models. The first results of this evaluation, led with 43 subjects are discussed in the paper. *Copyright ©1998 IFAC*

Keywords: methodology, human-machine interfaces, evaluation, design, software engineering

1. INTRODUCTION

System development methods (OMT, Merise, UML, SASD...) are, in most cases, based implicitly or explicitly on development models (particularly the Waterfall and Spiral models and their variants, sometimes the V and incremental ones): these models precise globally the different stages to follow for developing the system (see Boehm, 1981 or Sommerville, 1994). For instance, the figures 1 shows the well-known Waterfall model used in hundreds thousands of industrial projects. The main development models from Software Engineering are too general and largely insufficient in terms of interactive systems development, particularly for applications in which human errors inherent in using the human-computer interface(s) can lead to safety problems, and/or ecological and/or economic consequences (Rasmussen, 1986). A solution to this problem is a proposal to use so-called HCI-enriched models (i.e. "Human-Computer Interaction"-enriched models). Such models consist of combining the usual stages found in the main models (project orientation, specification, global design, detailed design, coding, unitary tests, integration, validation...), with stages concerning more or less explicitly the interactive aspects of design and evaluation.

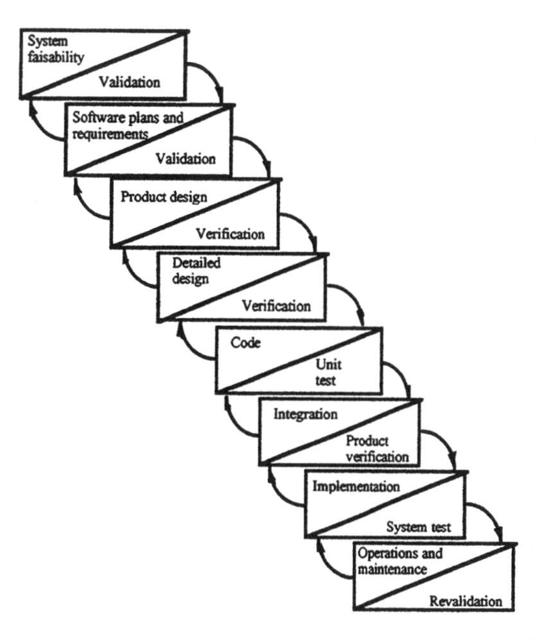

Fig. 1. Waterfall model (Boehm, 1981)

The notion of HCI-enriched model is not new. Several discussions and ideas concerning this notion exist in currently available literature (see for instance Hartson and Hix, 1989 ; Bass and Coutaz, 1991 ; Nielsen, 1992 ; Curtis and Hefley, 1994 or Collins,

1995). But no real and new model considered as a "Software Engineering model" has really been proposed. In this paper, a model called ∇ (pronounced "nabla") model, is presented. A questionnaire-based comparison of this model with four classical models and five other HCI-enriched models will also be explained. The first results are discussed.

2. GLOBAL PRESENTATION OF THE ∇ MODEL

The HCI-enriched ∇ model has been proposed by Kolski (1997 ; 1998). It concerns interactive system development. The first version is shown in figure 2. Its look is inspired by the well-known V model. Its objective is to locate the steps necessary to develop an interactive system, by distinguishing the HCI strictly speaking (left part of the model) from the application module(s) eventually accessible from the HCI (right part). One of its outstanding characteristics is to locate stages -that do not exist in the standard models - where the human factors have to be considered by the development team.

The first step is quite common in Software Engineering, and marks the beginning of the project by giving an orientation to the work, ie, realising objectives, project organization, constraints, and so on. Then the model emphasizes the importance of the analysis of the entire human-machine system during the project; this analysis deals more particularly with the system, the human tasks and the user(s). Modelling must be slanted towards:

- A real model corresponding to the current (existing or virtual) human-machine system, with its constraints, its strengths and weaknesses. Three cases can be considered. When the task consists of studying an already existing human-machine system to ultimately end with a new one, the modelling is of course carried out from the existing system. When the task consists of creating a new human-machine system from other already existing systems, the modelling is based on a synthesis of the date issued from each analysis. When there is no previously existing human-machine system and when the system is to be designed entirely from scratch, this model needs to be designed (Stammers *et al.*, 1990).

- A reference model corresponding to those of a human-machine system considered as ideal, by considering all the points of view and requirements of the different human partners concerned with the planned human-machine system. This model must in particular list a set of criteria which must be abided by. The nature of these criteria can be extremely wide-ranging (safety of human beings, of the facilities, of the environment, production, software ergonomy, energy-savings...), following the considered application field.

By comparing progressively the two models during the analysis of the human-machine system, and by reaching compromises aimed at satisfying a maximum of criteria, the data must be sufficiently relevant for the specification of an interactive system. This system is, in turn, adapted to the users' informational requirements, and to the requirements in terms of cooperation modes between the user(s) and the application modules. Then, the task consists of specifying the HCI on the one hand, and specifying the identified application modules on the other hand. This set of specifications will have to be evaluated and validated from a socio-ergonomic point of view, so as to verify the relevance of the new solutions being integrated into the targeted human-machine system; indeed, in most cases this includes several human beings, inter-connected software and hardware packages. We will have to consider the collective aspects of the work, aspects which are generally neglected by development teams (Rasmussen *et al.*, 1991; Zorola *et al.*, 1995).

After the specification of the HCI and application modules, and in order to reach the coding stage, the preliminary and detailed design stages, respectively associated links in the V model with integration tests and unitary tests, are carried out in the usual way. In relation to the design stages, it is important to ergonomically evaluate and validate the components of the interactive system. With regard to this remark, it is necessary to note the importance of evaluation methods; some of them are useable at this level. Many evaluation methods already exist in current literature (see for instance Wilson and Corlett, 1996; Kolski, 1997). Note also that, as indicated in the ∇ model, the chaining-process between the stages is propitious to a prototyping approach (Boehm *et al.*, 1984; Lichter *et al.*, 1994).

As is the case in each existing model, the acknowledgment stage has been located. In order to insist on the problematic approach of an interactive system, we have chosen to split this stage up by symbolically distinguishing it into an HCI-oriented acknowledgment and an application modules-oriented one. These two stages should be minimized if the complete interactive system is to conform to the data issued from the modelling of the human-machine system, and if each solution has been effectively evaluated and validated.

Finally this cycle ends with another stage, quite common in Software Engineering: the exploitation and maintenance stage. Note that this stage could return to the orientation stage, in a loop. Indeed this could aim at the evaluation of the interactive system being used, together with improvement. Thus, our approach would consist of proceeding towards an analysis of the human-machine system, leading to new specifications concerning the HCI and/or the application modules. Should these specifications be validated, the rest of the cycle could be viewed among the principles described previously. In conclusion, this development model could also be considered an evaluation and improvement model.

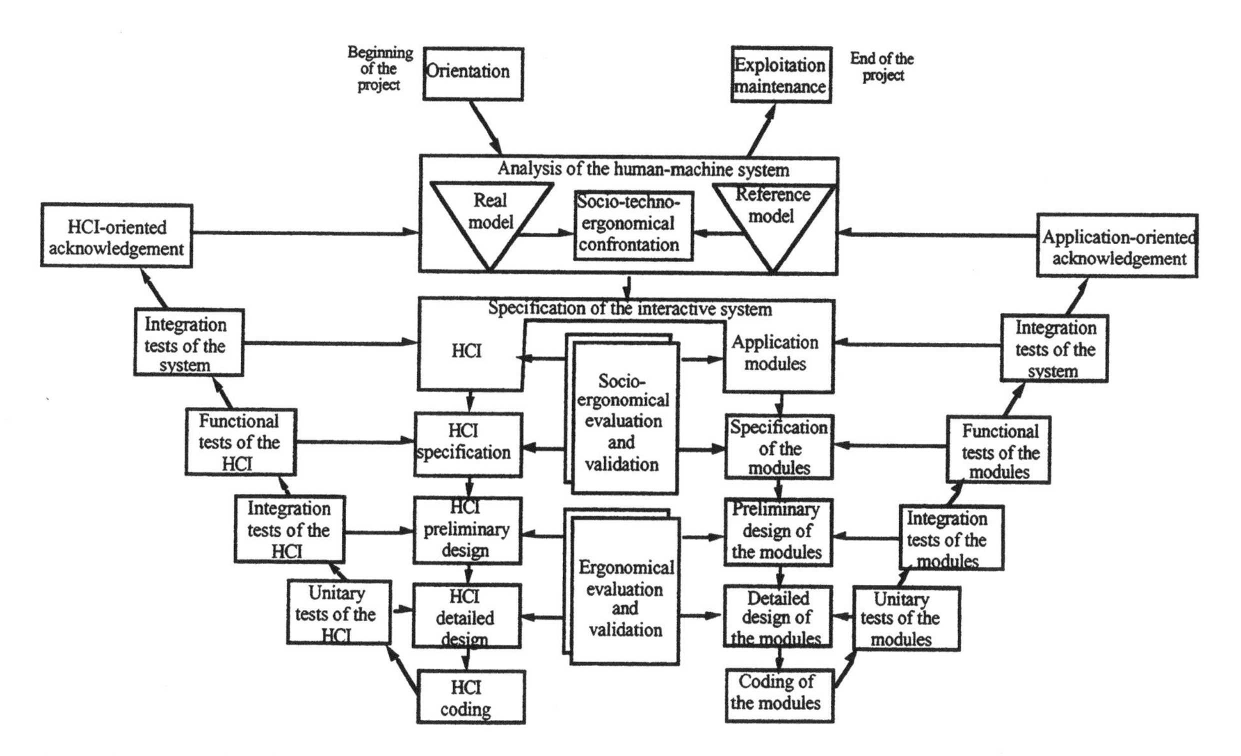

Fig. 2. ∇ model (Kolski, 1997; 1998)

Contrary to many of the main existing development models, the ∇ model has advantages, such as:

- importance being given to the human-machine system analysis and design,
- separation of interactive aspects from the application,
- importance being given to the system evaluation.

Such a model must provide interactive system developers with new ideas, and incite the developers to take into account human factors. This is why a questionnaire-based evaluation has been made, with the goal of comapring its impact with those provided by the classical models, and also by other HCI-enriched models.

3. QUESTIONNAIRE-BASED EVALUATION

Specific questionnaires have been given to 43 subjects. These subjects were academic students (with french diploma globally equivalent to Masters Degree), with primary experience in industrial and/or academic projects. These students came from three education types : (1) 6 ergonomists (5 academic years), (2) 15 computer scientists, with a specialization in human-computer interaction (5 academic years), (3) 22 computer scientists, without specialization (4 academic years). The subjects knew well the stages associated with the Waterfall, Spiral and V models; therefore they were able to understand each model.

For each question in the questionnaire, the subject had to answer with a value (if the answer was given without hesitation) or an interval in the contrary case (see figure 3). The subject could also comment on the answer given.

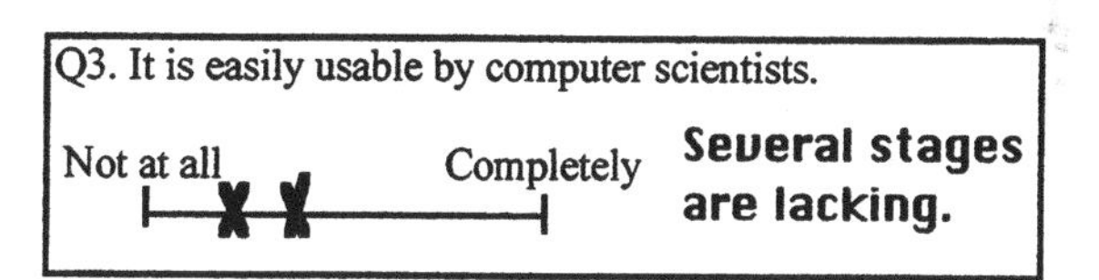

Fig. 3. Example of subject's answer

The questionnaire begins with a preliminary question: "QP : You feel concerned by the taking into account of human factors (requirements analysis, analysis of the human tasks the users have to perform, analysis of the users' characteristics, software ergonomics). It is a part of your job."

Ten models were successively displayed in the questionnaire without explanation (because a model must be self-explanatory). These consisted of four classical models (Spiral, Waterfall, V and Incremental models) and six HCI-enriched models: Produser model (James, 1991), the models of Valentin *et al.* (1993), Collins (1995), Curtis and Hefley (1994), the Star model (Hartson and Hix, 1989) and the ∇ (nabla) model. As examples, the model explained by Collins in his book is shown in figure 4, the model proposed by Curtis and Hefley in a paper is shown in figure 5; the Star model is shown in figure 6. Due to a lack of space the other models are not shown, but they can be easily found in currently available literature.

For each model, the subject had to answer the following questions:

"Q1. This model takes well into account the stages in classical Software Engineering.
Q2. It is sufficiently explicit concerning the development of the software interactive part.

Q3. It can be used easily by computer scientists.
Q4. It can be used easily by ergonomists.
Q5. It can be used by mixed teams composed of computer scientists and ergonomists.
Q6. It encourages teams to take into account human factors, and to integrate design ergonomics as early as in the first stages (specification and design stages).
Q7. It encourages teams to take into account the human factors. However the ergonomics can intervene rather after the software realization, for correction and improvement.
Q8. It allows a prototyping approach.
Q9. It insists on the stage of analysis and modelling of the complete human-machine system in which the human-machine interface is envisaged.
Q10. It allows identification of the difference between the HCI development and the application development.
Q11. It simplifies the conduct of the project when the target software must be interactive; therefore it is useful to the project conductor.
Q12. It must be taught during software Engineering lectures.
Q13. It must be taught during the lectures concerning interactive software design and evaluation."

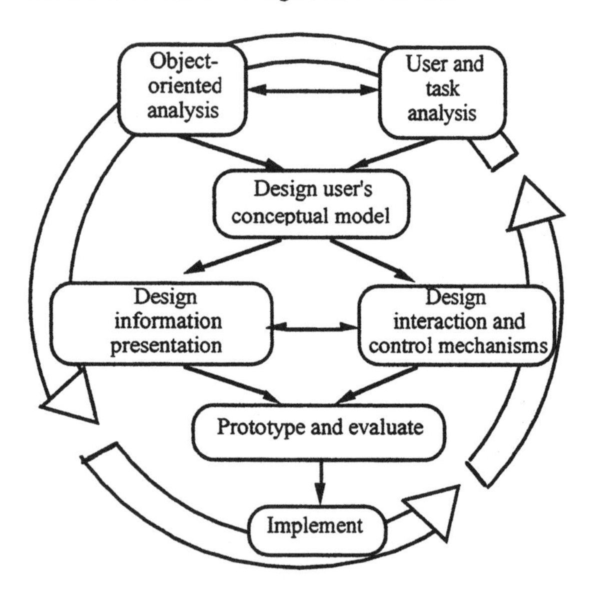

Fig. 4. Collins' model (1995)

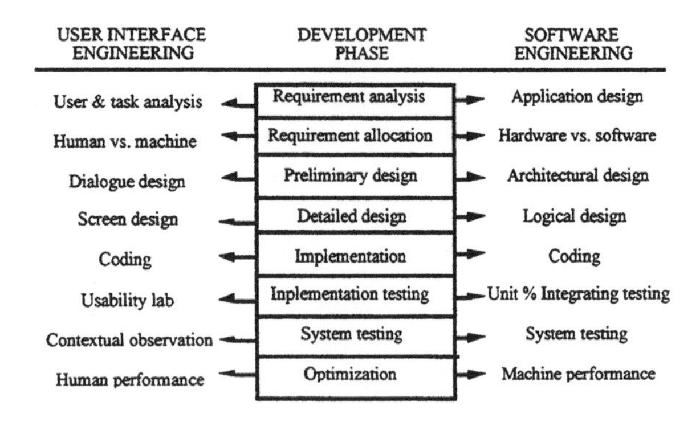

Fig. 5. Curtis and Hefley' model (1994)

All the data were analysed. Initial results are given below.

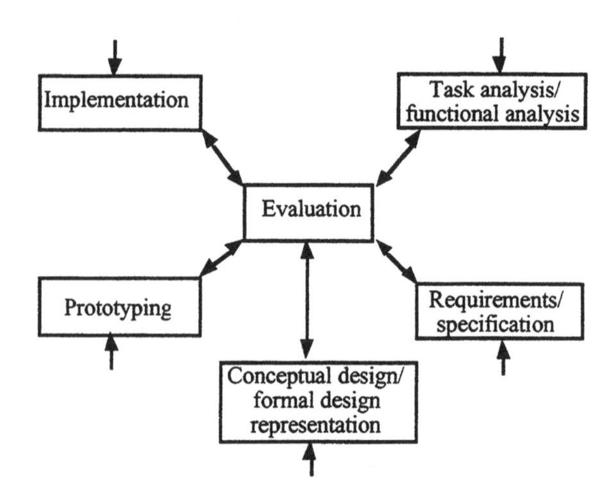

Fig. 6. Star life cycle (Hartson and Hix, 1989)

4. INITIAL RESULTS

With the presence of a multivariate aspect and responses through intervals, the statistical analysis is initially exploratory and global.

When observing the data roughly, one point becomes evident: only 10% of the responses are given with an interval; in fact few individuals do answer using an interval. Moreover, the use of intervals decreases with time (intervals are given for the first questions only). Thus, for this present and global analysis, the interval width is not considered (for the 10% of answers with a non-zero width, the values are summarized through the mean). A second point is that approximately 75% of individuals use the right part of the segment (from 2.5 to 5 cm) with the preliminary question ie. they feel rather involved by the human factors in software design.

The next stage of the exploratory analysis aims to show the most discriminant questions, their relationships and the influence of the model and the education type. To achieve this aim, the principal component analysis (PCA) is used. The data are considered through a table where the columns correspond to the 13 questions and the rows the the 430 combinations (s,i,m), *s* indicating the education type (s=1 (ergonomist), s=2 (computer scientist specialized in human-computer interaction) or s=3 (computer scientist)), *i* the individual within the education type and *m* the model (m= Waterfall, V, ..., nabla).

PCA yields two primary axes with very high relative inertia comapred with the next axes (46% and 12% vs. 7%, 6%, etc.). Thus we only focus on the factor plane results (58%). The variables with the highest contribution to the position of the first axis are respectively Q13, Q2 and Q11. The first axis highlights rather high correlations between all 13 questions. Observations of the 430 combinations cloud showed high intra and inter-individual differences. Nevertheless, it was possible to see, roughly, subclouds according to some model levels.

To assess the model factor effect on the questions, the 10 averaged profiles were computed and considered with a passive status in the PCA (they are projected as illustrative rows on the the first factor plane without participating in the positioning of the axes).

Such a procedure shows that models are organized into a hierarchy, figure 6; the highest (the best) responses being on the right side of the first factor axis and the lowest on the left side.

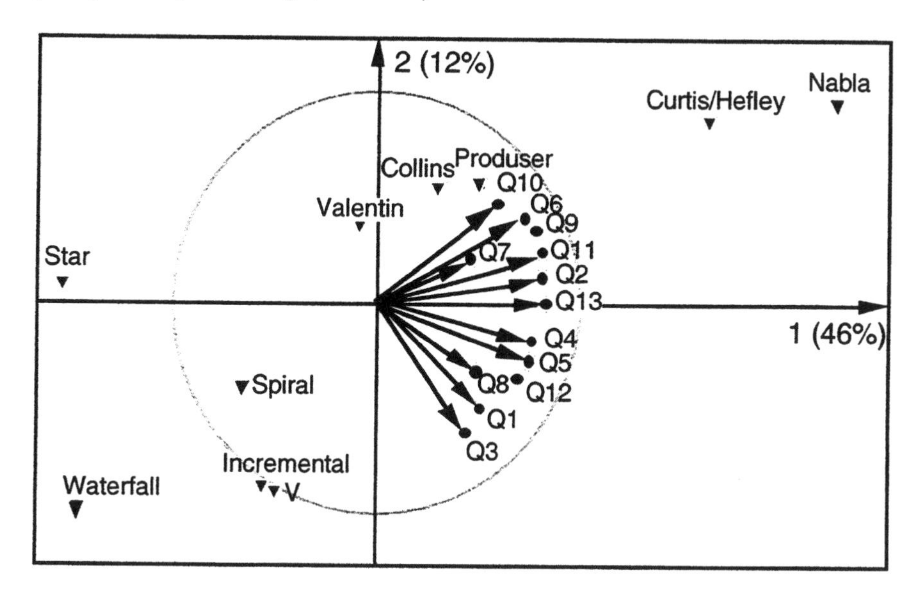

Fig. 7. Principal Component Analysis (PCA) of the questionnaire. Are represented on the first factor plane the 13 questions and 10 averaged points corresponding to the 10 models (from an integration over individuals and education types).

To go further with the questionnaire analysis, a confirmatory statistical approach based on the one way analysis of variance can be used. Most results are confirmed and more particularly the high influence of the model factor onto the most discriminant questions (Q13, Q2 and Q11) and a poor effect of the education type factor. For instance, the box plots of figures 7 and 8 show the influence of the model factor and the absence of influence of the education type factor on question 13.

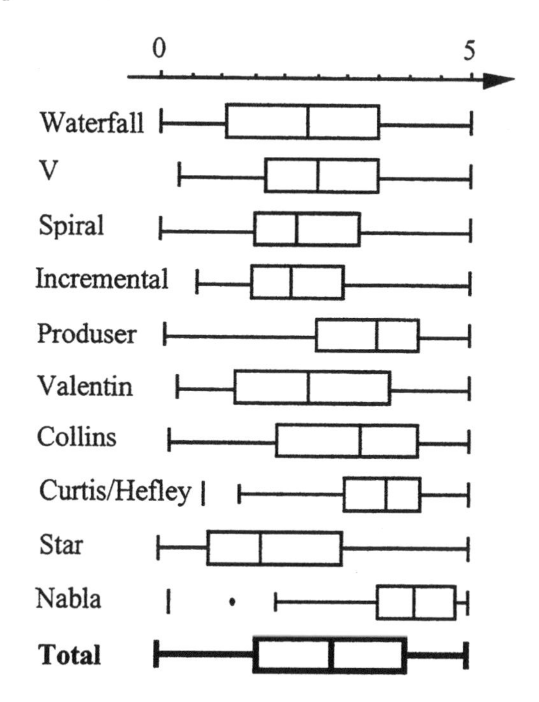

Fig. 8. Box plots of question 13 for the ten models

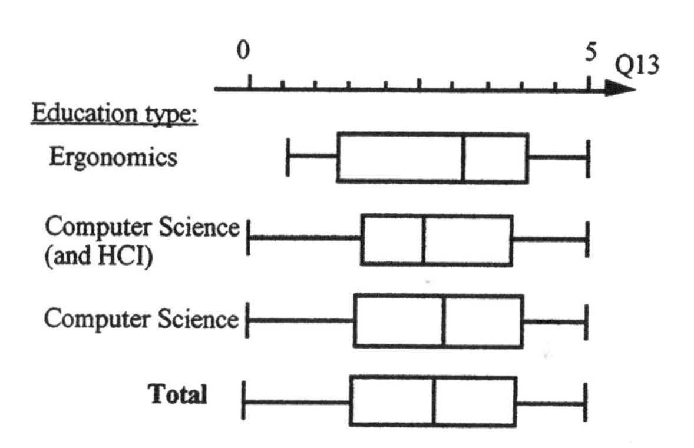

Fig. 9. Box plots of question 13 for the three education types

This first analysis shows clearly that the subjects can differentiate between the models. For interactive system development, the two most preferred are the ∇ (nabla) model and that of Curtis and Hefley, which are both HCI-enriched models. There is a high coherence between all the questions. The most discriminant one is question 13 (about the necessity to teach the model during lectures concerning interactive software design and evaluation).

5. CONCLUSION AND PERSPECTIVES

The HCI-enriched ∇ model has been explained in the paper. A first questionnaire-based evaluation shows a global positive impact with potential developers : computer scientists (specialized or not in human-machine interaction), as well as ergonomists. The statistical analysis can be enriched when studying more carefully the interval width and the connections between the variables.

This model could become a new theoretical and methodological framework for those researchers, concerned with the development of interactive systems.

Several research perspectives can also be envisaged. The most important of these would be validation of the model during real and/or simulated industrial projects and refining it over the next few years. Such validation is currently planned. Another perspective would be to improve existing development methods (SASD, UML, MERISE, and so on) by basing them on the stages proposed by the ∇ model.

ACKNOWLEDGEMENTS

The authors thank Joe Galway for considerable assistance in the translation of this paper.

REFERENCES

Bass, L., and Coutaz, J. (1991). *Developing software for the user interface*. Addison-Wesley.

Boehm, B.W. (1981). *Software Engineering Economics*. Englewoods Cliffs N.J. Prentice Hall.

Boehm, B.W., Gray, T.E, and Seewaldt, T. (1984). Prototyping versus specifying: a multiproject experiment. *IEEE transactions on Software Engineering*, 10 **(3)**, May.

Collins, D. (1995). *Designing object-oriented user interfaces*. The Benjamin/Cummings Publishing Company, Inc, Redwood City.

Curtis, B., and Hefley, B. (1994). A WIMP no more, the maturing of user interface engineering. *Interactions*, January, pp. 22-34.

Hartson, H.R., and Hix, D. (1989). Towards empirically derived methodologies and tools for human-computer interface development. *International Journal of Man-Machine Studies*, **31**, pp. 477-494.

James, M.G. (1991). PRODUSER: PROcess for Developing USER Interfaces. In *Taking Software Design Seriously*, J. Karat (Ed.), Academic Press.

Kolski, C. (1997). *Interfaces homme-machine, application aux systèmes industriels complexes*. Editions HERMES, Paris, january.

Kolski, C. (1998). A call for answers around the proposition of an HCI-enriched model. *Software Engineering Notes*, ACM Press, to appear.

Nielsen, J. (1993). *Usability engineering*. Academic Press.

Lichter, H., Schneider-Hufschmidt, M., and Zullighoven, H. (1994). Prototyping in industrial software projects, bridging the gap between theory and practice. *IEEE Transactions on Software Engineering*, 20 **(11)**, pp. 825-832, november.

Rasmussen, J. (1986). *Information processing and human-machine interaction, an approach to cognitive engineering*. Elsevier Science Publishing.

Rasmussen, J., Brehmer, B., and Leplat, J. (Eds.) (1991). *Distributed decision making*. Chichester, J. Wiley.

Sommerville, I. (1994). *Software engineering*. Fourth edition, Addison-Wesley.

Stammers, R. B., Carey M. S., and Astley, J. A. (1990). Task analysis. In *Evaluation of human work. A Practical Ergonomics Methodology*, Wilson J. R. and Corlett E. N. (Eds.), Taylor & Francis, London.

Valentin, A., Vallery, G., and Lucongsang, R. (1993). *L'évaluation ergonomique des logiciels, une démarche itérative de conception*. Montrouge, ANACT.

Wilson, J.R. and Corlett, E.N. (eds.) (1996) *Evaluation of human works : a practical ergonomics methodology*, 2nd edition. Taylor & Francis, London.

Zorola-Villarreal, R., Pavard, B., and Bastide, R. (1995). SIM-COOP: A tool to analyse and predict cooperation in complex environments, a case study: the introduction of a datalink between controllers and pilots. *Proccedings Fifth International Conference on Human-Machine Interaction and artificial Intelligence in Aerospace, HMI-AI-AS 95*, Toulouse, France, september.

Facilitate Cooperation of Humans and Machines from the viewpoint of Multi-Agent Systems

T. Yoshida*, T. Teduka* and S. Nishida*

**Department of Systems and Human Science, Graduate School of Engineering Science, Osaka University, 1-3 Machikaneyama, Toyonaka, Osaka 560-8531, Japan*

Abstract. This paper proposes a new approach for facilitating the cooperation of humans and machines from the viewpoint of Multi-Agent Systems (MAS). When a global problem is divided into subproblems, humans can take responsibility for *ill-structured* parts by utilizing their flexible and intuitive judgment. In contrast, agents are suitable for dealing with *well-structured* parts by exploiting their abundant computational resources and capabilities. An interface between humans and machines is provided by extending the cooperation method in (Yoshida *et al.*, 1998) to enable the flexible interaction among them. *Copyright© 1998 IFAC*

Key Words. Agents, Co-operation, Co-ordination, Design, Intelligent knowledge based systems, Interaction mechanisms, Interfaces, Man/machine interaction, Man/machine systems, Satelites

1. Introduction

It is widely recognized that the appropriate cooperation of humans and machines is important to solve large-scale and complex problems. The importance of collaboration among people with different backgrounds is also recognized and various researches have been carried out in the field of C-SCW (Stefik *et al.*, 1987; Winograd, 1989). As for the integration and collaboration among computational resources, various researches have been carried out in the framework of distributed systems to overcome the complexity of the problem to be dealt with as well as the total system (Bond and Gasser, 1988; Gasser and Huhns, 1989). In addition, it is natural to model large scale systems as distributed systems by mapping the distribution of resources and processes onto subsystems.

This paper proposes a new approach for facilitating the cooperation of humans and machines from the viewpoint of Multi-Agent Systems (MAS) by extending our previous research (Yoshida *et al.*, 1998). The framework of MAS is suitable for explicitly dealing with the interaction among subsystems since it can be naturally modeled as the communication among agents (Genesereth and Fikes, 1990; Finin *et al.*, 1997). Although it is difficult to utilize the strength and/or characteristics of MAS purely in terms of optimization compared with a monolithic system, it is possible to exploit its distributed nature by utilizing the respective advantages and/or strength of humans and agents [1] as a total system. For instance, when a global problem is divided into subproblems, humans can take responsibility for *ill-structured* parts by utilizing their flexible and intuitive judgment. In contrast, agents are suitable for dealing with *well-structured* parts by exploiting their abundant computational resources and capabilities. The communication from agents can also provide humans with various and useful information to support their decision making. Thus, it is expected that our approach enables to improve the effectiveness of human-machine systems as a whole within the framework of MAS.

In our approach both humans and machines cooperate each other on terms of equality toward global problem solving in MAS. They take responsible for different subproblems and communicate the results of respective problem solving each other. An interface between humans and machines is provided by extending the cooperation method in (Yoshida *et al.*, 1998) to enable the flexible interaction among them. The communication contents from agents to humans are converted and visualized through the interface so that they are intuitively understandable for humans, and vice versa for those from humans to agents. Through the interface humans and agents cooperate each other by utilizing the visualized information.

MAS has been also utilized for engineering design by exploiting its concurrent and collaborative characteristics (Lander, 1997; Darr and Birmingham, 1994). This paper focuses on facilitating the

[1] Hereafter the word "agent" refers to the computational agent.

capability of human-machine systems as a whole by exploiting their respective strength through an appropriate interface between them. Since the interface tries to explain why the communication content through the interface can be considered as design rationales for their respective problem solving (Shum, 1996; McKerlie and MacLean, 1994).

This paper is organized as follows. Section 2 describes our framework of MAS. The details of cooperation methods are explained in Sect. 3. Section 4 describes the experiments for the satellite design problem to investigate the effectiveness of our approach.

2. Framework of Multi-Agent Systems

2.1. *Framework of Problem Solving*

In our approach humans and agents carry out respective local problem solving based on the provided representation scheme toward the global problem solving by MAS as a whole. Local problem solving is carried out opportunistically by setting the appropriate parameter values in the scheme within the allocated amount of resource (e.g., CPU time). The system as a whole tries to maximize some measure on the quality of solutions while satisfying all the specified constraints on solutions.

Our framework of problem solving in MAS is based on the following assumptions:

i. All the necessary decomposition of global problem solving is already carried out and appropriately allocated to agents and humans.
ii. Each agent and human undertakes the problem solving on the decomposed subproblem and checks the constraint satisfaction on it.
iii. No single agent nor human can accomplish global problem solving sorely by him/herself.
iv. Agents know how the state is transformed via the application of operators.

2.2. *Problem Solving in Agents and Humans*

Agents construct their solutions in the conventional state space formalism via the application of *operators* to transit states. Agents apply the operator to the internal state to increase the quality of solutions within the fixed number of reasoning steps. The state to be expanded is chosen in best-first manner by applying the value function v to states to estimate their quality and by choosing the one with the best value. *Features* are defined to extract the important aspects of the state in problem solving. The value of states is defined as:

$$v(state) = g(\sum_k w_k \cdot f_k(state)) \quad (1)$$

g stands for a function, w_k for a weight, f_k for a feature. Sigmoid function is used for g in this research. w_k denotes the contribution of f_k for v.

In contrast, humans are allowed to carry out their local problem solving based on their intuitive decision making and not restricted to the application of pre-encoded operators. Their intuitive decision making is expected to be suitable for *ill-structured* subproblems which are difficult to handle for agents.

3. Cooperation Method

3.1. *Form of Cooperation*

We propose that both humans and agents cooperate each other *on terms of equality* toward global problem solving within the framework of MAS, as shown in Fig 1. Agents carry out problem solving on *well-structured* subproblems, which are easy to describe and encode necessary knowledge and thus are suitable for exploiting their abundant computation capabilities. In contrast, it is difficult to describe and provide appropriate knowledge for *ill-structured* subproblems (notably known as the bottleneck of knowledge acquisition) and thus are hard for agents to deal with. In that case humans can participate in the global problem solving and take responsibility for ill-structured and/or nebulous subproblems with their ability of flexible decision making. Thus, in our approach it is possible to exploit the advantages of both humans and machines appropriately and *uniformly* within the framework of MAS toward improving the effectiveness of human-machine systems as a whole.

3.2. *Cooperation between Agents*

We propose a new cooperation method with which agents utilize the "comments" from peers as the feedback on their behavior. Cooperation is carried out between agent A and B as follows (see Fig. 2):

i. A carries out local problem solving to construct its proposal and passes it to B.
ii. B incorporates the communicated proposal into the initial state in his problem solving. Then he[2] carries out local problem solving to modify it so that it can satisfy the constraints managed by him.

[2] Hereafter masculine gender is used to denote agents.

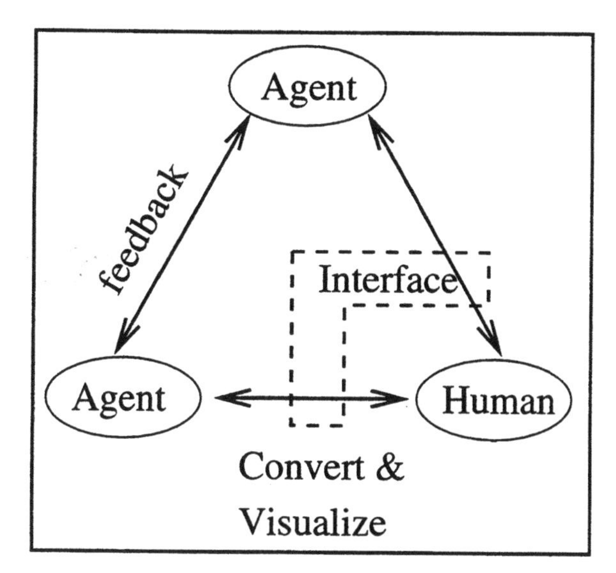

Fig. 1. Cooperation of Humans and Agents in MAS

iii. B returns the result (this is called as a *comment*) back to A as the feedback.
iv. A compares his proposal and the comment to find out which parts are agreed and disagreed. He also interprets from his viewpoints what kind of models B would prefer from the refinements made in it.
v. After the comparison, A carries out conflict resolution to construct the initial state in the next cycle of problem solving.

Comments for one agent represent how other agents like or dislike his solution and would modify it, and also show what would be brought about *to himself* as the result of his decision making. Thus, the communicated information can be utilized as the feedback to modify the behavior of A and motivates him to make concession or compromise according to his own necessity. In addition, this method is simple since it is not necessary to explicitly deal with the internal mental states in other agents. For instance, agents do not need to take care of why their proposals are refined in comments, or what kind of strategies other agents have taken to give comments. Here, one "cycle" of problem solving in agents consists of the above steps. One "epoch" of global problem solving is carried out by repeating the problem solving cycles.

The initial state in problem solving greatly affects its process and final result, especially in the case of opportunistic problem solving with limited reasoning steps and multi-peaked search space.[3] Conflict resolution to construct the initial state in step v. is carried out by each agent based on the following algorithm:

[3] Most realistic problems have this characteristics since "satisfactory" solutions are enough in most cases and the "best" one is rarely sought.

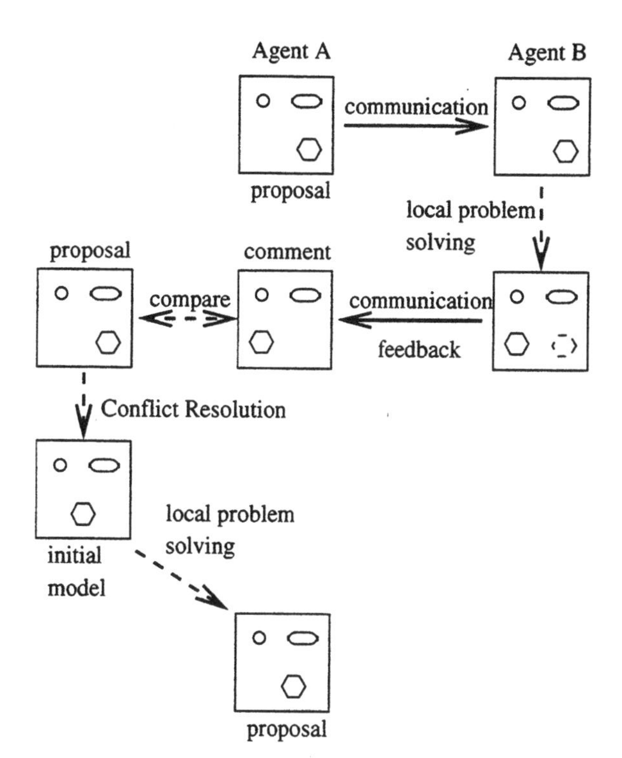

Fig. 2. The proposal from A is modified by B into the comment on it and given back to A as the feedback.

Do the modifications in comments affect the constraint or goal managed by him?

- No: accept and incorporate them into the initial state since they do not matter.
- Yes: analyze the history of his own proposals. Have the conflicts continued more than the pre-defined number?
  - Yes: make concession and accept them.
  - No: Is it possible to take weighted average between the parameter values?
    - Yes: take the weighted average.
    - No (exclusive cases): stick to individual opinions and reject them.

Our conflict resolution algorithm has the following properties toward the convergence of solutions and problem solving processes by agents in MAS:

a. eventual incorporation of opinions by others: taking weighted average and making concession when the conflict continues lead to eventual incorporation.
b. constraint satisfaction: since opinions by others reflect the constraint satisfaction managed by others, the solution constructed in step i. comes to satisfy all constraints toward feasible ones by incorporating them into the initial state in step v.
c. escape from local optima: conflicts can be conceived as that search is trapped into local optima. Change of initial state greatly alters the search region in problem solving, which can help to escape from local optima by leaping over their basin.

3.3. *Cooperation between Humans and Agents*

We propose an interface which visualizes the interaction between humans and agents in MAS to facilitate cooperation. The cooperation method based on the *metaphor of explanation* in the case of fully computational agents (Yoshida *et al.*, 1998) is extended in the interface to convert the communication between humans and agents. For instance, humans carry out problem solving graphically with physical presentation of solutions (e.g., designed products) to construct their proposals. Then the proposal is converted into the parameter values with the representation scheme through the interface and communicated to agents since they are handy to carry out constraint satisfaction and value iteration with (1) in agents. In contrast, the proposals constructed by agents within the parametric scheme are converted into the *physical* form so that it is easy for humans to *intuitively* understand or grasp the content of information.

The transition of constructed proposals by humans and agents is shown to them to illustrate and to let them recognize how their interactions affect the process and direction of problem solving in MAS as a whole. Their mutual responses to the communicated information also indirectly represent their characteristics such as roles in MAS, which can be utilized for determining effective responses each other. The interface is also utilized to *show the difference* in opinions between humans and agents, which plays the role of *explaining* the agreement and disagreement among them toward cooperation. The communication content is also *annotated* as in (Yoshida *et al.*, 1998) to indicate the degree of importance in the communication content for humans and agents, respectively.

Since the granularity of computation greatly varies for humans and agents, it is allowed that the number of internal reasoning steps which are carried out between interactions is different for them. Thus, in our approach agents do not need to wait for the reply by humans to exploit their abundant computational capabilities and respond only when humans communicate their response.

With our method it is expected that mutual understanding between humans and agents is facilitated by converting the information into the understandable form for both parties through the interface. Especially, humans can flexibly modify their responses to agents while they cooperatively carry out local problem solving in MAS. For instance, they can articulate nebulous conceptions into concrete forms during local problem solving in collaboration with agents. They can also improve and refine the ad hoc knowledge encoded in agents based on the feedback or responses from

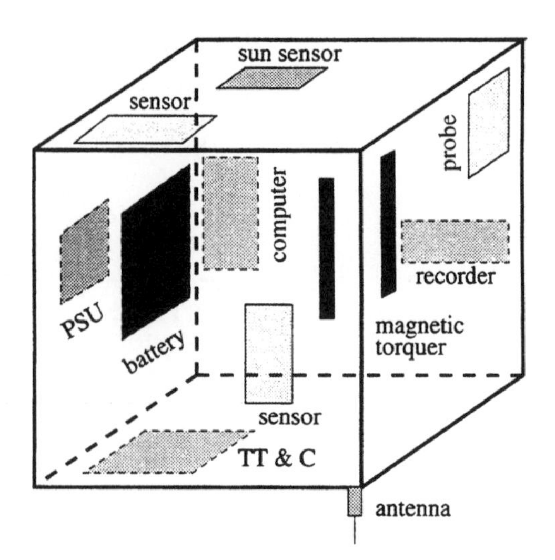

Fig. 3. Example of Satellite

agents. Furthermore, they can carry out flexible decision making with respect to multi-goal optimization or pareto optimality in cooperation with agents.

4. Experiments

Design of micro satellites is taken as the application domain to examine the effectiveness of our approach. [4] Since satellite design is complicated, usually a satellite is decomposed into various subsystems such as structure, thermal, electricity, communication, etc to make the design process easier and manageable (Wertz and Larson, 1991). Typical constraints imposed on these subsystems are shown in Table 1. Satellite design is carried out by disposing the specified mission devices onto the panels of satellite. Auxiliary devices might also be selected and disposed if they are necessary to satisfy constraints and to achieve the mission. An example of satellite is shown in Fig. 3.

Table 1 Constraints on subsystems for satellite design.

subsystem	constraint
structure	COG of satellite is within 5cm of the Center
thermal	temperature of each panel is within the tolerance level of the devices on the panel
communication	link margin is more than 3 dB

Agents have been implemented to carry out the design for structure and thermal subsystems. Features in Table 2 are used to estimate the quality of internal states in (1) to control search in the

[4] Micro satellites around 50 kg are often used for pilot research and much more simpler than commercial ones.

Table 2 Features in Satellite Design

Agent	Feature
all agents	mass of satellite
	number of modifications
	number of auxiliary devices
Structure	deviation of COG (Center of Gravity)
	natural frequency of satellites
Thermal	deviation in normalized temperature

Table 3 Operators in Satellite Design

Agent	Operator
Structure	disposition of devices
	addition of balance weight
	change of thickness of panels
Thermal	disposition of devices
	addition of heat pipes
	change of surface material

best-first manner. Operators in Table 3 are internally used by agents for local problem solving to modify and improve the quality of satellites.

Communication subsystem is designed by a human in the experiments as an extension of our previous approach. To design communication subsystem it is necessary to select the type of antennas and to change the length of antennas so that the link margin satisfies the constraint in Table 1. However, currently only ad hoc knowledge is available for the communication subsystem and usually its design is carried out *heuristically* by experts. Thus, it is reasonable to rely on a human for this subsystem and the form of cooperation among humans and agents for satellite design naturally fits into our framework of MAS. The agent for communication subsystem is also implemented for comparison.

We plan to examine how the cooperation between humans and agents is facilitated through the interface and the effect of visualization to facilitate cooperation. Experiments are analyzed in terms of the quality of problem solving in MAS, i.e., the quality of designed satellites by the system, and the process of problem solving, i.e., how satellites are modified through interactions between humans and agents during problem solving. The MAS with the participation of human designer for communication subsystem is compared with that with all computational agent, i.e., with the agent for communication subsystem. We are also interested in how the knowledge for design is improved or refined during problem solving with the participation of humans. Knowledge encoded in agents beforehand can be refined by humans based on the behavior of agents via the modification of operators and features. In contrast, knowledge in human designers can be refined when they come to acquire the appropriate domain heuristics to carry out local design. These will indicate the effectiveness of our approach for facilitating the cooperation of humans and machines from the viewpoint of MAS.

5. Conclusion

This paper has proposed a new approach for facilitating the cooperation of humans and machines from the viewpoint of Multi-Agent Systems (MAS) by treating them as agents toward improving the effectiveness of human-machine systems as a whole. In our approach it is possible to exploit the distributed nature of MAS by utilizing the respective advantages and/or strength of humans and machines as a total system. An interface between humans and machines is proposed by extending the cooperation method in (Yoshida *et al.*, 1998) to enable flexible interaction among them. The communication contents from machines are converted and visualized through the interface so that they are intuitively understandable for humans, and vice versa for those from humans to machines. We plan to examine the effectiveness of our approach through the experiments on satellite design by MAS and to refine the proposed methods. Scaling up of the system in terms of the number of agents and the complexity of design problem are also the future directions to extend our approach.

6. REFERENCES

Bond, A.H. and Gasser, L., Eds.) (1988). *Readings in Distributed Artificial Intelligence.* Morgan Kauffman Publishers, Inc.. San Mateo.

Darr, T.P. and W.P. Birmingham (1994). Automated Design for Concurrent Enginering. *IEEE Expert* **9**(5), 35–42.

Finin, T., Y. Labrou and J. Mayfield (1997). KQML as an Agent Communication Language. In: *Software Agents* (J. Bradshaw, Ed.). MIT Press. Cambridge.

Gasser, L. and Huhns, M.N., Eds.) (1989). *Distributed Artificial Intelligence Volume 2.* Morgan Kauffman Publishers, Inc.. San Mateo.

Genesereth, M.R. and R. Fikes (1990). *Knowledge Interchange Format Version 3.0 Reference Manual.* technical report logic-90-4 ed.. Computer Science Department, Stanford University.

Lander, S.E. (1997). Issues in Multiagent Design Systems. *IEEE Expert* **12**(2), 18–26. special issues in AI in Design.

McKerlie, D. and A. MacLean (1994). Reasoning with Design Rationale: Practical Experience with Design Space Analysis. *Design Studies* **15**(2), 214–226.

Shum, S. B. (1996). Design Argumentation as Design Rationale. In: *The Encyclopedia of Computer Science and Technology* (A. Kent and J.G. Williams, Eds.). Marcel Dekker, Inc.. New York.

Stefik, M., G. Foster, D.G. Bobrow, K. Kahn, S. Lanning and L. Suchman (1987). Beyond the Chalkboard: Computer-Support for Collaboration and Problem Solving in Meeting. *Communications of the ACM* **30**(1), 32–47.

Wertz, J.R. and J.L. Larson (1991). *Space Mission Analysis and Design.* Kluwer Academic.

Winograd, T., Ed.) (1989). *Groupware: The next wave or just another advertising slogan?*. Proceedings of IEEE Computer Society International Conference (COMPCON).

Yoshida, T., K. Hori and S. Nakasuka (1998). A Cooperation Method via Metaphor of Explanation. *IEICE Trans. EA* **81**(4), 576–585.

TRUST, SELF-CONFIDENCE AND AUTHORITY IN HUMAN-MACHINE SYSTEMS

T. Inagaki*, N. Moray , and M. Itoh***

* *Institute of Information Sciences and the Center for TARA, University of Tsukuba, Tsukuba 305-8573 Japan*
** *Department of Psychology, University of Surrey, Guildford, Surrey GU2 5XH, United Kingdom*

Abstract: This paper reports an experimental examination on trading of authority between the human and the automation. The effect of reliability of automation, magnitude of fault, human trust in automation, and human self-confidence in their ability to perform manual control on human acceptance of situation-adaptive autonomy (SAA) are investigated to show the effectiveness of SAA under time-critical conditions. *Copyright ©1998 IFAC.*

Keywords: Human supervisory control, Human-centered automation, Reliability, Safety

1. INTRODUCTION

How tasks should be allocated to the human and the computer in the supervisory control is one of the central issues in research on human-machine systems. When task allocation is discussed, a distinction must be made between sharing control and trading control (Sheridan, 1992). Sharing control discusses which tasks should be assigned to the human and which to the computer under multitask environment. Trading control, on the other hand, investigates cases where either the human or the computer turns over control to the other, and discusses when control should be handed over and when it should be seized back. There are many studies on sharing control, but little research has been done on trading control.

Trading control relates closely to the issue of authority and responsibility. In human-centered automation, it is often said that "the human must be at the locus of control," or "the human must be maintained as the final authority over the automation" (Woods, 1989; Billings, 1991). These statements are natural and reasonable, because it is not realistic to assume that machine intelligence is perfect, although it is often highly capable. However, there is still a room for discussion on whether the human must be at the locus of control at all times in every occasion. The operating environment may change with time, and the new environment may not be easy for the human.

In controlled flight into terrain (CFIT) accidents of aircraft, for instance, humans often show a loss of awareness and thus are late in decision and control. Even in cases where humans responded quickly, they may fail to complete all the necessary procedures to avoid accidents. In the crash of B757 near Cali in 1995, pilots responded immediately to the ground proximity warning alert by performing the terrain avoidance maneuver, but they failed to stow the speed brake which they had extended some time before with the (previous) intention of descending (Dornheim, 1996). The CFIT accident could have been avoided, if there had been a mechanism which retracts the speed brake automatically if it has not yet been stowed when the maximum thrust is applied: Using the speed brake and the maximum thrust at the same time is highly contradictory.

Table 1 Levels of autonomy (Sheridan, 1992)

1. The computer offers no assistance, human must do it all.
2. The computer offers a complete set of action alternatives, and
3. narrows the selection down to a few, or
4. suggests one, and
5. executes that suggestion if the human approves, or
6. allows the human a restricted time to veto before automatic execution, or
7. executes automatically, then necessarily informs human, or
8. informs him after execution only if he asks, or
9. informs him after execution if it, the computer, decides to.
10. The computer decides everything and acts autonomously, ignoring the human.

The issue of authority can be discussed in terms of the level of automation (LOA). Table 1 gives the scale of LOA suggested by Sheridan (1992). If it is assumed that the human must be at the locus of control and that the computer (or automation) should be subordinate to the human, then LOA must always stay within the range of 1 through 5, because the human may not be in full authority with LOA at 6 or higher. However, some accidents, such as the crash near Cali, suggest the need of high LOA for system safety.

By using mathematical models, Inagaki (1991, 1993, 1995) has proposed the concept of situation-adaptive autonomy (SAA) where authority is traded flexibly between humans and the computer. More precisely, SAA is a mechanism in which (1) LOA is altered dynamically depending on the situation, and (2) LOA at level 6 or higher may be adopted when judged necessary for attaining system safety. It has been proven that conditions exist for which the computer should be allowed to perform autonomous safety-control actions even when humans have given the computer no explicit directive to do so. However the mathematical proof is not always enough to show the efficiency of SAA. There are crucial factors, such as human trust in automation, self-confidence in his/her ability, which are not fully incorporated in mathematical models.

Humans working with highly automated systems often suffer negative consequences of automation, such as the out-of-the-loop performance problem. Endsley (1995) has investigated the impact of different LOAs on human performance, and has shown that humans are likely to lose situation awareness in highly automated environments. Sarter and Woods (1995) argue that humans may exhibit inferior failure detection performance, because the automation is sometimes strong enough to hide abnormalities occurring in the controlled process or in the automation itself. Moreover, an automation-induced surprise may happen if feedback of information is poor (Norman, 1990; Wickens, 1994).

The human's use of automation is also an important issue. Lee and Moray (1992, 1994) have investigated the impact of automation failure on human trust in automation, and have shown that reliance on automation is affected by human self-confidence and trust in automation. Parasuraman, Molloy, and Singh (1993) has argued that humans sometimes place too much trust in the automation, and that complacency potential is determined by a human trust in, reliance on, and confidence in automation. Riley (1994) has proposed and investigated a structural model for representing interrelations among various factors which may determine human's use of automation.

This paper reports an experimental examination on trading control of authority between the human and the automation. In a process control simulator, the effect on human acceptance of situation-adaptive autonomy (SAA) of reliability of automation, magnitude of fault in the controlled process, human trust in automation, and human self-confidence in their ability to perform manual control are investigated. It is shown that SAA is effective for fault-management under time-critical conditions. SAA has exhibited superior performance in safety-control to the conventional framework in which humans are invested with full authority for decision and control at all times in every situation.

2. SCARLETT

As a vehicle for research on human-automation interaction, Inagaki, Moray, and Itoh (1997) have developed SCARLETT (Supervisory Control And Responses to LEaks: TARA at Tsukuba), a simulated central heating system to maintain the temperature of the apartment complex shown at the bottom right of Fig. 1. The operator is given two tasks. One is to control manually the rate of flow and temperature of the fluid on the primary side of the heat exchanger so that the temperature at the apartment complex can be made as close as possible to the desired set-point. To do this, the operator must manipulate the pumps, heaters, and valves appropriately.

The other is a fault-management task. The operator must take appropriate actions when abnormalities occur in the controlled process. Even though SCARLETT can simulate various types of malfunctions, the current experiment has investigated

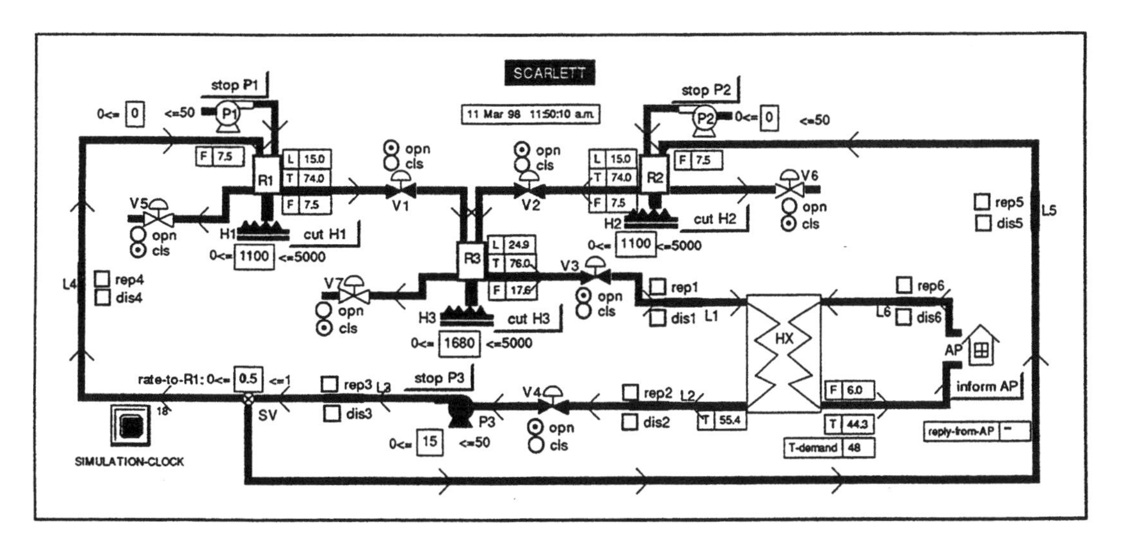

Fig. 1. Control panel of SCARLETT. Buttons for sending repair crew or for vetoing automation are adjacent to the site of faults. Heaters and pumps can be controlled by typing in values in the boxes below the components. Requested and actual temperature of the apartment are shown to the lower right of the heat exchanger.

cases with pipe faults which can occur at any site among L1,..., L6. The pipe faults are classified into two classes: (i) A leak where a portion of flow quantity is lost, and (ii) a break where the flow quantity is lost completely at the site of the break. Leaks can be fixed by sending the repair crew. In cases of breaks, the operator must shut down the whole process immediately.

The accident is defined as one of the following events: (1) A reservoir dries out while heated, (2) a pump runs with no incoming fluid, (3) the fluid coming out of the pipe floods the floor, or (4) the temperature of fluid to the apartment complex is too high or too low. Delay in fixing a leak entails loss of fluid and energy, and may cause failure in maintaining the temperature at the desired set-point. If the fault was a pipe break, delay or failure in shutting down the process may cause an accident.

3. EXPERIMENTAL DESIGN

The experiment has a 2 x 2 x 3 x 2 factorial design, mapping onto (Mode of Control for fault management) x (Type of Fault) x (Level of Reliability of computer message) x (Order of Condition). The first two are within-subject factors, and the latter two between-subject factors.

Upon detecting a leak or a break, the computer gives a red light which flashes at the site of the leak or the break, and gives suggestions for fault management.

Two Control Modes are distinguished for the fault management: Manual (M) and Situation-Adaptive Autonomy (SAA) modes. In M mode, the computer gives a suggestion, such as, "Leak! Send repair crew," or "Pipe break! Shut down the process." In this mode the computer performs no action without explicit commands by the human. In SAA mode, on the other hand, the computer may take control actions when it think it appropriate to do so, even if it receives no directive by the human. Computer message in SAA mode may be: "Pipe break! I will shut down the process in 15 seconds. You can veto with the Disagree button," where LOA = 6 (see, Table 1). If the operator fails to give the computer any directive within 15 seconds, the computer shuts down the process. If LOA is set at 7, the computer may shut down the process without telling anything to the human, if it thinks it appropriate. After shutting down the process, the computer may say "I have shut down the process because the accident was about to occur." SAA thus changes LOA depending on the situation.

There are two Fault Types, Leaks and Breaks. The observable phenomena differ depending on the type or the location of the fault, and the operating condition (such as the pump speed). Without having sound knowledge of the process, it is not easy for the human to identify and locate the fault.

Reliability of computer message is defined as the percentage of trials on which fault diagnosis by the computer will be correct. Three levels of reliability are distinguished: 100%, 90 %, and 70 %.

Unreliability of fault diagnosis is due to misdiagnosis in the sense that the computer would tell the human that a leak had occurred when in fact a break had occurred (and vice versa). Neither false alarms nor lack of alarms have been investigated in the current experiment.
The Order of Condition refers to the fact that half the subjects received all the trials under M mode before receiving any trial under SAA mode, and half the opposite.

Thirty graduate and undergraduate students of the University of Tsukuba Computer Science Program participated in the experiment. Each subject was paid 3,000 yen for participating in the experiment which lasted three days. The experiment consists of the Guidance, Training, Pre-Test, Data Collection, and Post-Test sessions.

The following data are taken as performance measures: (1) RMS error around the target temperature, (2) subjective rating of trust in automation, (3) subjective rating of self-confidence in the ability to control the process manually, (4) time to detect faults, (5) magnitude of fluid loss, (6) erroneous judgement on validity of computer messages, (7) inappropriate action for sending repair crew or for shutting down the process unnecessarily, (8) number of accidents.

4. RESULTS AND DISCUSSION

4.1 Probability of Accidents.

An ANOVA shows a significant main effect of Reliability of computer message ($F(2,24) = 9.825$, $p = 0.0008$). As reliability of fault diagnosis by the computer increases, the probability of accidents decreases. The ANOVA shows another significant main effect of Mode of Control ($F(1,2) = 438.86$, $p = 0.0023$). The control mode effect shows that there are far more accidents overall in M mode than in SAA mode, which justifies, as expected, the use of automation to reduce accidents (Fig. 2).

4.2 Probability of False Shutdowns.

An ANOVA shows a significant main effect of Reliability of computer message ($F(2,24) = 25.56$, $p < 0.0001$). The probability of false shutdowns decreases as reliability of fault diagnosis by the computer increases. There is also an interaction of Reliability with Mode of Control ($F(2,24) = 4.800$, $p = 0.018$), which proves that the human should be retained as controllers to avoid false shutdowns (Fig. 3).

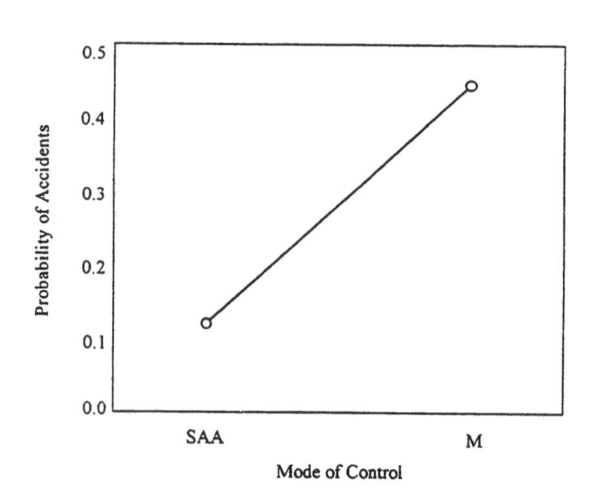

Fig. 2. Mode of control affecting the probability of accidents

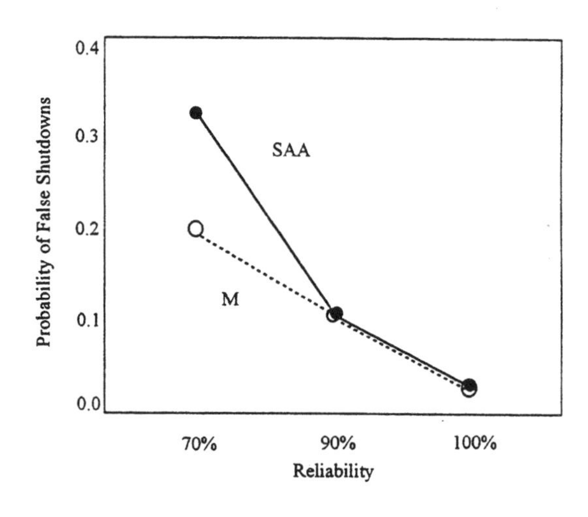

Fig. 3. Two-way interaction between reliability of computer message and mode of control affecting the probability of false shutdowns

4.3 Magnitude of Fluid Loss.

There is a significant main effect of Type of Fault ($F(1,2) = 2160.22$, $p < 0.001$), which is straightforward. An ANOVA also shows a three-way interaction of Reliability x Mode of Control x Type of Fault ($F(2,23) = 4.630$, $p = 0.020$) (Fig. 4), which implies that for small faults, and especially when the information provided by the computer is reliable, humans are more efficient. For severe problems, SAA is more effective in managing fluid loss during faults, whether or not the displayed information is reliable.

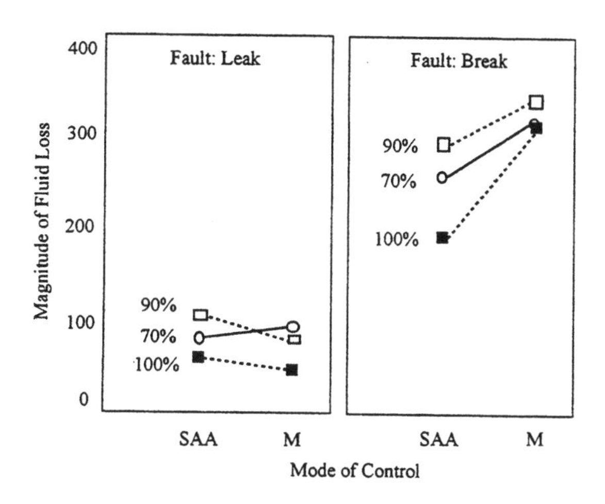

Fig. 4. Three-way interaction of reliability of computer message, mode of control, and type of fault affecting the magnitude of fluid loss

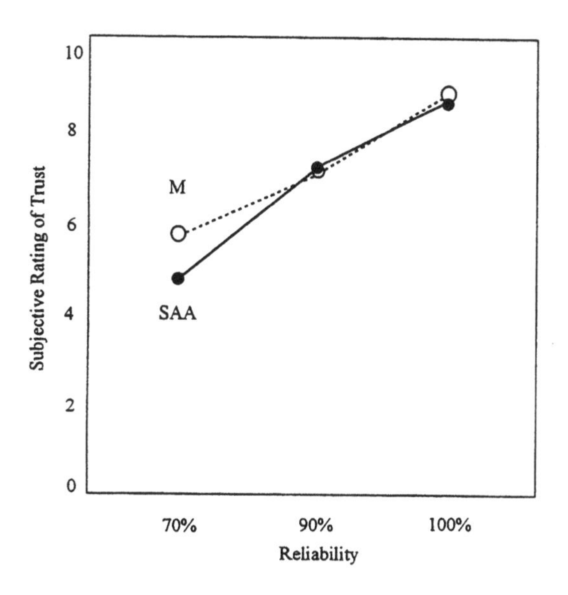

Fig. 5. Two-way interaction of mode of control and reliability of computer message affecting subjective rating of trust in automation

4.4 Trust in Automation.

An ANOVA shows a highly significant main effect of Reliability on subjective ratings of trust ($F(2,24) = 25.803$, $p < 0.0001$). Subjects trusted less as Reliability of computer messages declines. Fig. 5 illustrates a marginally significant interaction of Reliability by Mode of Control ($F(2,24) = 3.183$, $p = 0.062$), implying that while trust is not affected by Mode of Control when Reliability of computer message is high, trust in automation with high LOA is damaged when fault diagnosis by the computer is less reliable. Interviews with subjects have shown that experiencing an erroneous computer message did not always degrade the subjective ratings of trust in automation. An interpretation for this can be: If a subject is not very confident of his/her ability to deal with difficult situations without any aid of the computer, then he/she may be forced to rely on the computer to some extent, which may be enhanced under high time-criticality.

4.5 Self-Confidence.

An ANOVA has given no significant main effect of Reliability. The ANOVA show a nearly significant main effect of Mode of Control, and another marginally significant main effect of Type of Fault. However these are not important in terms of magnitude.

4.6 Automation-Induced Surprise.

The feelings toward highly autonomous safety control actions differ from subject to subject. Some accepted them willingly, but some did not. Those who accepted Level 7 autonomy said that they were happy with the fact that accidents were prevented from their occurring. Those who did not like Level 7 autonomy said that they could not be satisfied with the situation where they were not informed by the computer what it would do next, which suggests the need for a new LOA which comes between 6 and 7 (say, 6.5). Level 6.5 is defined as: "The computer executes automatically after telling human what it will do. No veto is allowed." Level 6.5 autonomy may be effective to avoid surprises induced by automated safety-control actions performed in case of emergency.

5. CONCLUSION

It is often believed (or assumed) that the human must be at the locus of control. The results of the current experiment prove the role of the human as a decision maker. However, the results also show the need of highly autonomous mechanisms for system safety, and effectiveness of the high level of automation is situation-dependent. The issue of authority is thus not simple, and dynamic trading of authority between human and automation must be sought.

Inagaki (1993, 1995) has proven mathematically that the effectiveness of high LOA is governed by various factors, such as dynamics of a controlled process, reliability of alarms, probabilities of human errors, cost of and recoverability from those errors. Not all the factors have been investigated in the current experiment, and thus some more extensive experiments are needed.

The automation does not exist in all-or-nothing fashion (Scerbo, 1996). As Sheridan (1995) suggests, further discussions need to be made on what is human-centered automation. The issue of authority must be investigated in relation to the human-centered automation.

ACKNOWLEDGMENTS

This work has been partially supported by the Center for TARA (Tsukuba Advanced Research Alliance) at the University of Tsukuba, Grants-in-Aid for Scientific Research 07650454, 08650458, and 09650437 of the Japanese Ministry of Education, Science, Sports and Culture, and the First Toyota High-Tech Research Grant Program. The authors express their thanks to Dr. Takeshi Matsuoka and Mr. Junji Fukuto, Ship Research Institute, Ministry of Transportation, for their help in modelling the heat exchanger of SCARLETT. Thanks are extended to Dr. Kazuo Monta, Visiting Professor at the Center for TARA, University of Tsukuba, for his valuable help and discussions.

REFERENCES

Billings, C.E. (1991). *Human-Centered Aircraft Automation: A Concept and Guidelines.* NASA TM-103885.

Dornheim, M. (1996). Recovered FMC memory puts new spin on Cali accident. *Aviation Week & Space Technology*, September 9, 58-61.

Endsley, M.R. and E.O. Kiris (1995). The out-of- the-loop performance problem and level of control in automation. *Human Factors*, 37 (2), 381-394.

Inagaki, T. and G. Johannsen (1991). Human-computer interaction and cooperation for supervisory control of large-complex systems. In: *Computer Aided System Theory* (Pichler, Moreno Diaz (Ed)), LNCS 585, Springer-Verlag, 281-294.

Inagaki, T. (1993). Situation-adaptive degree of automation for system safety. *Proc. IEEE ROMAN*, 231-236.

Inagaki, T.(1995). Situation-adaptive responsibility allocation for human-centered automation. *Trans. SICE of Japan*, 31(3), 292-298.

Inagaki, T., N. Moray, and M. Itoh (1997). Trust and time-criticality: Their effects on the situation-adaptive autonomy, *Proc.AIR-IHAS*, 93-103.

Lee, J.D. and N. Moray (1992). Trust, control strategies and allocation of function in human-machine systems, *Ergonomics*, 35, 1243-1270.

Lee, J.D. and N. Moray (1994). Trust, self confidence and operators' adaptation to automation. *Int. J. Human-Computer Studies*, 40, 153-184.

Parasuraman, R., R. Molloy, and I.L. Singh (1993). Performance consequences of automation-induced complacency *Int. J. Aviation Psychology*, 3, 1-23.

Riley, V. (1994). *Human Use of Automation*. Doctoral dissertation, Univ. Minnesota.

Sarter, N. and D.D. Woods (1995). Autonomy, authority, and observability: Properties of advanced automation and their impact on human-machine coordination. *Proc. IFAC MMS*, 149-152.

Scerbo, M.W. (1996). Theoretical perspectives on adaptive automation. In: *Automation and Human Performance* (Parasuraman and Mouloua (Ed)), LEA, 37-63.

Sheridan, T. (1992). *Telerobotics, Automation, and Human Supervisory Control*. MIT Press.

Sheridan, T.B. (1995). Human-centered automation: Oxymoron or common sense?, *Proc. IEEE SMC Conference*, 1-6.

Wickens, C.D. (1994). Designing for situation awareness and trust in automation. *Proc. IFAC Integrated Systems Engineering*, 77-82.

Woods, D. (1989). The effects of automation on human's role: Experience from non-aviation industries. In: *Flight Deck Automation: Promises and Realities* (Norman and Orlady (Ed)), NASA CP-10036, 61-85.

AN OPTIMIZATION STUDY ON MAN-MACHINE ASSIGNMENT FOR MULTIFUNCTION WORKFORCE UTILIZATION

Paulo Ghinato, Susumu Fujii and Hiroshi Morita

Graduate School of Science and Technology - Kobe University
1-1 Rokkodai-cho, Nada-ku, Kobe 657, Japan

Abstract: *This paper deals with the problem of multifunction worker assignment in U-shaped production lines. The authors present the problem as a multiobjective optimization problem where three objectives are pursued in a preemptive fashion. Moreover, a heuristic solution method based on the space filling curve technique and designed to solve the problem in three distinct phases is proposed. Considering the difficulty to obtain an optimal solution for a three-objective optimization problem, the heuristic procedure performs satisfactorily well as the computational experiments show.*

Key words: Heuristic searches, Lean manufacturing, Man/Machine systems, Mathematical programming, Multimachine, Multiobjective optimization

1. INTRODUCTION

New paradigms of production emerged in response to an increasingly competitive market have set new standards for production operations and management. In lean production system – a worldwide acclaimed production paradigm – continuous incremental improvements are systematically carried out by optimization procedures which focus on the complete elimination of wastes. The reduction of man-hour is one of such procedures which targets the labor excess as a means of cost reduction.

The reduction of the number of workers is only one step towards the full flexibilization of the workforce. What really makes a flexible workforce possible is a combination of an appropriate machinery layout (U-shaped layout), a multifunction labor, and a continuous evaluation and periodic revisions of the standard operations routine (Monden,.1983).

The man-machine assignment problem, which is basically a matter of defining the best possible allocation of workers to machines, plays a major role in the realization of the man-hour reduction.

However, the problem of allocation of multifunction workers to machines in a U-shaped line can not be appropriately represented by the existing man-machine assignment formulations nor can it be solved by typical solution techniques. It is necessary to develop a specific formulation and an effective method to solve it.

Considering the complex nature of the man-machine allocation problem in U-shaped production lines (briefly described as production lines in which the entrance of materials is located close to the exit of finished products), in the next section we propose a mathematical formulation which represents more accurately the relevant factors and conditions found in the actual assignment problem. In section 3 a heuristic solution method is proposed, and its applicability and performance are discussed and evaluated in section 4. Conclusions and final remarks are made in section 5.

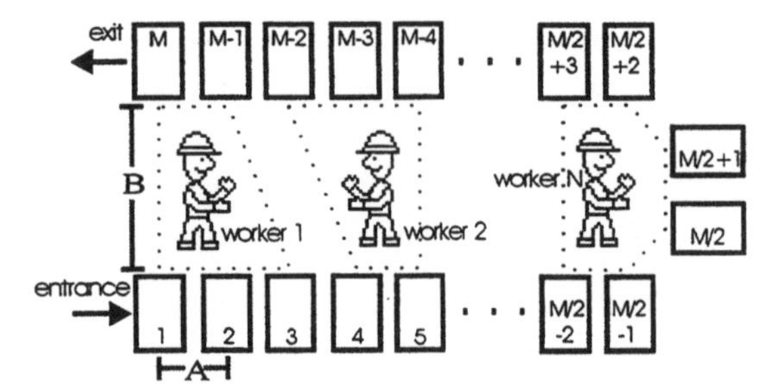

Fig. 1. A typical U-shaped layout with multifunction workers

2. A FORMULATION FOR THE MULTIFUNCTION WORKER ASSIGNMENT PROBLEM

Extensive bibliographical reviews on the man-machine assignment problem have shown that the objectives are, typically, to maximize production or to minimize costs. Moreover, the general formulation of the assignment problems invariably portrays the problem as a single-objective optimization programming (Baybars, 1986; Ghinato et at., 1998a; Stecke and Aronson, 1985).

In Lean production systems, however, the objective of minimizing the total cost of production unfolds itself into many different operational objectives. Therefore, the typical single-objective formulation cannot accurately represent the critical problem of man-machine assignment in lean production systems.

Recently, some researches have been considering the man-machine assignment problem in Just-In-Time production systems (see Miltenburg and Wijngaard, 1994; Ohno and Nakade, 1997). However, it is surprising that despite recognizing that the cycle time (the time interval between two successive outputs) is determined by the external demand, some formulations show the minimization of cycle time as the only objective to be pursued.

It seems reasonable to believe that, under the principle of "complete elimination of wastes", which is the lean production's hallmark, the objective of the man-machine assignment problem cannot be minimizing the cycle time. Any initiative to reduce the cycle time of any production line without the increase of market demand will result in waste of early overproduction – producing ahead of schedule – and additional secondary wastes as well.

It must be acknowledged that, although the reduction on the number of workers is of crucial importance, the workload balance (the balance of machine operation difficulty) has a significant impact on the workforce productivity, especially in the long-run. As Stecke and Aronson (1985) appropriately remarked, "if one operator [...] has a heavier workload than another, dissatisfaction might result."

In this paper the authors contend that, after setting the minimum number of workers, the minimization of workload imbalance should be immediately pursued. As a third objective, it is advocated that the imbalance of routine times should be minimized.

In dealing with multiple objectives some basic questions may arise: given a finite number of relevant objectives, which one or ones should be selected? Moreover, if more than one is used, what are the relations among them?
Basically, we can distinguish three possibilities:

(1) Cardinal ranking of objectives with fixed tradeoffs;
(2) Ordinal ranking of objectives with no finite tradeoffs;
(3) No ranking of objectives at all. Efficient solution methods.

The ordinal ranking without tradeoff between the objectives can be seen as a special case of (1), where the highest ranking objective is infinitely more important than the second which again is infinitely more important than the third and so forth. This concept is called preemptive priorities (Eiselt et al., 1987). First, the highest ranking objective is optimized, then the value of the objective function is fixed at this level and the method proceeds to optimize the second objective, fixes its value and continue in the same way until a unique solution or no solution is found (Eiselt et al., 1987).

The disadvantage with this procedure is that if the optimal solution, obtained by using the first objective, is unique then there is no possibility for the second or lower ranking objectives to be used. Even if alternative optimal solutions with respect to the first objective function exist, and the second objective function could be used to select the best alternative among them, chances are minimal that the third ranking or even lower ranking objectives will be used at all (Eiselt et al., 1987).

Although some authors believe that the lexicographical ordering/ranking (2) has been more widely adopted than it deserves to be and is rarely appropriate because of its oversimplified approach (Keeney and Raifa, 1976), most decision makers can easily apply it and, in fact, ranking is a concept that seems inherent to much of decision making (Ignizio, 1982). And it is exactly because of its simplicity and identity with the actual decision making process that the lexicographical ranking is suggested as the approach to solve the assignment problem.

Therefore, in an appropriate formulation for the man-machine assignment problem, we have the first objective function aiming at minimizing the number of workers. The second objective is to minimize the total workload deviation. And the third objective is to minimize the difference between routine times (routine time deviation). As for the constraints, constraint (4) guarantees that each machine is assigned to one and only one worker while constraint (5) implies that each worker will operate at least one machine. The zone constraint (6) is an important condition in our problem, nevertheless it is difficult to represent it mathematically. Zone constraint means to assign machines in such a way that the walking route of workers are not crossed. Constraint (7) sets an upper bound to the cycle time. And finally, in constraint (8), x_{ij} is 1 if worker i operates machine j and 0 otherwise.

Some assumptions for this formulation:

(1) The cycle time is fixed and determined by the demand of the succeeding production stage;
(2) Machines have similar sizes and, therefore, the machines in each row are lined up at regular distances from one another;
(3) Workers move at the same pace; The walking time between two machines is proportional to the distance between them;
(4) The time each worker takes to perform the manual operation on each machine is standardized and does not vary significantly from worker to worker. It varies only from machine to machine;
(5) There is only one kind of workpiece flowing through the production line, so the processing time and manual operation time on each machine are constant for a given period;
(6) Transference time of workpiece from machine k to machine k+1 is assumed to be nil. That is, the transference is performed automatically or by the workers while moving between the machines without interference on their walking times;
(7) Workers are 100% multifunctional skilled, i.e., any worker can operate any machine
(8) The model is deterministic, i.e., all parameters are assumed to be known with certainty.

Our formulation takes advantage of the fact that, in practical cases, the number of workers N is a small positive integer, typically between 1 and 20.

Therefore, the problem of man-machine assignment, as described above, can be represented as in the formulation below:

$$\text{Minimize } N \quad (1)$$

$$\text{Minimize } \sum_{i=1}^{N}\left|S_i - \frac{S}{N}\right| \quad (2)$$

$$\text{Minimize } \sum_{i=1}^{N}\left|T_i - \frac{\sum_{j\in\hat{M}} l_j + D}{N}\right| \quad (3)$$

Subject to

$$\sum_{i=1}^{N} x_{ij} = 1 \quad \text{for } j = 1,\ldots, M \quad (4)$$

$$\sum_{j=1}^{M} x_{ij} \geq 1 \quad \text{for } i = 1,\ldots, N \quad (5)$$

Zone constraint – Machines assigned to each worker are located in such a way that workers' walking routes are not crossed (6)

$$\max_{j\in\hat{M}}(l_j + p_j) \vee \max_{\forall i}(L_i + R_i) \leq CT \quad (7)$$

$$x_{ij} = 0 \text{ or } 1 \quad \forall\, i, j \quad (8)$$

Where:
$a \vee b = \max\{a, b\}$
N: number of workers available
$L_i = \sum_{j\in\hat{M}} l_j x_{ij}$: sum of operation times of machines assigned to worker i
R_i : shortest of the total walking times in one cycle among all possible routes of worker i, for $\forall\, i$
$T_i = L_i + R_i$: the routine time of worker i, for $\forall\, i$

Given:
CT : cycle time determined by the demand
M : number of machines in the production unit
$\hat{M} = \{1,2,\ldots, M\}$ set of all machines in the production line with $\hat{M} = \bigcup_{i=1}^{N} \hat{M}_i$, $\hat{M}_i \cap \hat{M}_f = \varnothing$ for $i \neq f$, $\forall\, i, f$, where $\hat{M}_i$ and $\hat{M}_f$ represent the set of machines assigned to workers i and f, respectively
p_j : processing time of machine j (the time required for machine j to transform the part/workpiece)
l_j : manual operation time for machine j (the time required to unload and load the jth machine)
$D = (M - 2)a + 2b$: the shortest walking time encompassing all machines in the production line
v_j : workload needed to be processed at machine j

$S_i = \sum_{j=1}^{M} v_j x_{ij}$ for $i = 1,\ldots, N$: workload assigned to worker i

$S = \sum_{i=1}^{N} S_i = \sum_{j=1}^{M} v_j$: total workload

3. DEVELOPING A HEURISTIC SOLUTION METHOD

In this section we present a modified version of a solution method developed by Ghinato et al. (1998c). The original method was devised for solving the assignment problem in U-shaped lines when there is only one objective, which is to minimize the number of workers under a given cycle time. Later, the algorithm was improved to solve the problem when two objectives were to be pursued: to minimize the number of workers and minimize the imbalance of workload distribution among workers (Ghinato et al., 1998e) (for details of these algorithms see Ghinato et al., 1998b, c, e).

The procedure assigns machines to each worker following the route of a spacefilling curve. A spacefilling curve is a continuous mapping which is an optimal tour through a set of n points, where n can be arbitrarily large. A useful property of a spacefilling curve is that it tends to visit all the points in a region once it has entered that region (Bartholdi, 1997).

Recent studies have shown that the spacefilling curve technique can be a useful alternative to solve problems such as the traveling salesman problem, the location of items in a storage rack, the facility layout problem, and special kinds of man-machine assignment problems (Kim et al., 1996).

In our problem, the basic idea is to generate a curve that visits all machines along the U-shaped layout starting from one of the four corners (machine 1, $M/2$, $M/2+1$, or M) of the production line. The curve visits the machines moving forward from a certain machine to one of its adjacent machines in a given routing pattern, which ensures that the zone constraint is respected throughout the assignment process. It should be noted here that the zone constraint – the key feature of the assignment problem in U-shaped layouts – has been the fundamental basis to build the proposed solution procedure.

The original solution method was developed upon a specific spacefilling curve which has a squared shape corresponding to the pattern SF01 in figure 2. Although there exist several patterns of spacefilling curve which satisfy the zone constraint (see some examples in fig. 2) and can be effectively used in the solution method, the algorithm proposed for solving the three-objective optimization problem utilizes a combination of two different patterns (SF1 and SF3).

The procedure attempts to satisfy the three objectives in three distinct phases. In the first phase, the procedure searches for the minimum N by assigning as many machines as possible to each worker. The assignment of machines to the first worker begins with machine 1, that is, using SF1 the proposed algorithm enters the U-shaped layout through the machine 1 and moves forward through machines M, M-1, 2, 3, M-2, M-3,..., $M/2+3$, $M/2+2$, $M/2-1$, $M/2$, $M/2+1$. The last machine visited by SF1 is always either M/2 or M/2 +1, depending on the number of machines (M) in the line.

Hence, the assignment of the first worker begins with testing whether the operation time of machine 1 is less than the specified cycle time. I.e., is the routine time of worker 1 less than the cycle time? If it is less, then the procedure moves to the next machine along the curve and tests whether the new routine time is still less than the cycle time and so on, until no machine can be added to the worker's routine. Then, the assignment of worker 2 begins.

Notice that the route and its respective walking time change each time a new machine is added. In case the routine time exceeds the cycle time the last machine added must be rejected, but the procedure continues to move forward along the curve neglecting the machines located in the same row of the *obstacle*'s (a machine that can not be included in the routine because it violates the time constraint). The assignment of worker i is terminated when the next obstacle is found. Then, there is no feasible assignment left to be tested and the assignment of the next worker begins from that first obstacle. This procedure goes on until all machines are assigned to the workers.

The second phase starts after the minimal number of workers has been found. N is then incorporated in the second phase as a constraint, and the objective becomes to minimize the workload imbalance. The procedure iteratively compares the values for the expression $|S_i - S/N|$ (from the second objective function) before and after assigning the machine j. If this value increases, then the machine j is not assigned (it is left for the next worker) and another machine which satisfies this condition is sought in the neighborhood.

In this phase, the spacefilling curve utilized is still SF1, however, the assignment begins in the last machine of the first phase, either M/2 or M/2+1. This procedure allows the assignment process to attack major imbalances resulted from the first phase which tends to allocate the least machines to the last worker. In case the procedure does not find a feasible solution at the end of phase 2, that is, when the machine 1 or M is reached, phase 2 restarts the allocation process by relaxing the test condition, i.e., the initial value of the expression $|S_i - S/N|$ is increased by one unit. Phase 2 ends when a feasible solution is found, usually after a couple of iterations.

Phase 3 starts at machine 1 by using the pattern SF3 depicted in figure 2. These changes in the pattern of the assignment sequence are likely to increase the number of feasible alternatives reachable in each phase since different orders of assignment are explored and tested. Again, the solution obtained in the previous phase, along with the result from phase 1, is transformed into another constraint. Phase 3 then begins with machine 1, attempting to assign the machines along the path SF3 to each one of the N workers under a certain workload distribution (determined by phase2). In this last phase, the objective is to minimize the imbalance of routine times which is pursued by iteratively verifying whether the expression $|L_i - (\Sigma l_j + D)/N|$ (from the third objective function) can be minimized each time a new machine is assigned, just like it was done in the previous phase with the workload distribution.

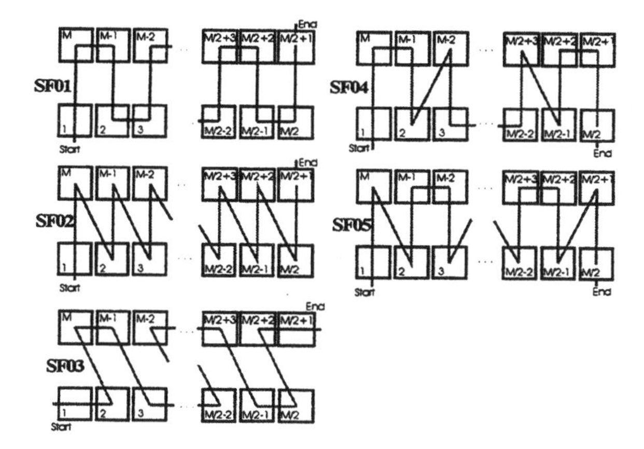

Fig. 2. Some Patterns of Space Filling Curve Applicable in U-shaped Layouts

Details of the assignment procedure as well as the generation of the space filling curve utilized in phase 1 and phase 2 are presented in Ghinato et al. (1998b, c, e).

4. COMPUTATIONAL EXPERIMENTS

Although satisfying the objectives (1), (2) and (3) is important for the utilization of multifunction workforce, it cannot be denied that the simple ordinal ranking without tradeoffs between objective functions may turn ineffective the existence of the third or even the second objective function.

In a previous work (Ghinato et al., 1998d), these authors have made an effort to show the significance of the three objectives of the proposed formulation and developed a computational experiment in order to demonstrate the applicability of the preemptive multiobjective approach to solve this problem.

In that paper, Ghinato et al. (1998d) applied a modified version of an exact solution method developed for problems with 2 men and m machines (Ghinato et al., 1997) to a series of 20 problems. The use of an exact algorithm allowed the authors to thoroughly analyze the weight and effect of each objective function on the final/optimal result, concluding that the preemptive approach is an effective solution method.

The heuristic method proposed in the previous section is applicable to problems of any dimension (n workers x m machines), however in order to evaluate its optimality, i.e., how close the obtained solution can be from the optimal one, the heuristic was applied to the same set of 20 problems solved by the exact algorithm in Ghinato et al. (1998d).

In order to enable the use of the exact method, guaranteeing N=2, it was assumed that $\sum_{j \in \hat{M}} l_j + D > CT$, therefore at least two workers should be allocated to the production line.

Therefore, the exact and the heuristic methods were applied to 20 problems with M=12, generated by randomly varying the machine operation times and machine workloads. The values for each parameter were obtained through a pseudo-random generator from the uniform distribution. The operation time ranges [3-10] and the workload index (operation difficulty index) is in the range [5-10]. It was also assumed that $\max_{j \in \hat{M}} (l_j + p_j) \leq CT$, hence the processing time had no effect on the assignment.

The cycle time was set to 60 time units and the walking times a and b were assumed proportional to the distances A and B, respectively (see fig. 1). Therefore, a and b were set to 1 and 2, respectively.

Table 1. Performance of the heuristic method

Objectives	Method	
	EXACT	HEURISTIC
1st	100%	100%
1st + 2nd	100%	50%
1st + 2nd + 3rd	100%	30%

Table 1 shows the performance of the heuristic method compared to the results obtained by the exact algorithm. Obviously, for all 20 problems there exists an optimal solution that satisfies the three objectives simultaneously, and that is what the column "exact" represents. The column "heuristic" shows how often the heuristic found the optimal value (the same value found by the exact method) for the objectives (1), (1)+(2), and (1)+(2)+(3), progressively.

The heuristic found N = 2 for all 20 problems, although only half – 10 problems – of these results satisfy the second objective. The optimality of the solution decreases even further when the third objective is added, that is, only in 6 problems the heuristic found the optimal solution.

It should be noted, however, that the analysis of the results revealed that in the 10 problems where the value for the second objective is not optimal, the result is this: in 9 times the value was the second best and once the result was the third best (from an average of 12 feasible alternatives for each problem). The same analysis, applied to the third objective, indicates that although only 30% of the solutions were optimal (satisfying preemptively the three objectives), in the remainder 70% of the solutions the value obtained for the third objective was the best or the second best among the feasible alternatives in 62% of the problems.

These results clearly indicate that the heuristic method proposed in the previous section can be successfully applied to solve this particular class of assignment problem.

5. CONCLUDING REMARKS

In this brief study, it was concluded that the current strategies for man-hour reduction in lean production systems concentrate excessively on the time issue and overlook a crucial aspect of the conditions found in many real production environments: the machines in a production line are generally different and have, therefore, different workloads or levels of difficulty of operation.

Finding a balanced distribution of workload among the workers should be one of the objectives of any effective workforce optimization technique. It is believed that an unbalanced workload distribution can have very negative effects – notably in the long-run – on the workforce willingness to accept and properly perform multifunction activities. In the same way, a balance among the routine times of each worker will produce positive effects on workers' morale and flexibility to accept enlarged responsibilities.

Therefore, a binary integer multiobjective optimization formulation was proposed as an alternative approach for this particular man-machine assignment problem. Moreover, a heuristic based on the space filling curve technique was developed and its performance evaluated against the results obtained by an exact procedure.

The computational experiments carried out in section 4 demonstrated that:

(1) a lexicographic order for the objective functions, despite some disadvantages discussed earlier, can appropriately represent the actual behavior of the decision maker when deciding upon the assignment of men to machines, and be an effective approach to solve the assignment problem.

(2) the special features of this man-machine assignment problem, combined with the robustness of the spacefilling curve technique, made possible that the proposed heuristic solution method be applied with relative success.

Further study will investigate the effects of different combinations of space filling curve patterns in the three optimization phases.

REFERENCES

Bartholdi, J. (1997). *Spacefilling Curves and Fractals*, http://homer.isye.gatech.edu/people/faculty/John_Bartholdi/mow.html.

Baybars, I. (1986). A Survey of Exact Algorithms for the Simple Assembly Line Balancing Problem, *Management Science*, **32**, 909-932.

Eiselt, H.A., G. Pederzoli, and C. L. Sandblom (1987). *Continuous Optimization Models*. de Gruyter & Co., Germany.

Ghinato, P., S. Fujii and H. Morita (1997). A Basic Approach to the Multifunction Workers Assignment Problem in U-shaped Production Lines. *Proceedings of the 3rd Int'l Congress of Industrial Eng.*, Brazil, CD-ROM.

Ghinato, P., S. Fujii and H. Morita (1998a). A Basic Study on the Multifunction Worker Assignment Problem in U-shaped Production Lines. *Memoirs of the Grad. School of Science & Technology of Kobe University*, Kobe, Japan, 143-150.

Ghinato, P., S. Fujii and H. Morita (1998b). An Analysis of the Performance of a Solution Method Based on the Spacefilling Curve Technique for the Man-Machine Assignment Problem. *Proceedings of the Third MSOM Conference*, Seattle (to appear in June).

Ghinato, P., S. Fujii and H. Morita (1998c). Two Effective Methods for Workforce Reduction Under Lean Production Principles, *Proceedings of the 8th Int'l Conf. on Flexible Autom. & Flexible Manufng.*, Portland (to appear in July).

Ghinato, P., S. Fujii and H. Morita (1998d). A Study on the Worker Assignment Problem Under Lean Production Principles. *Proc. of the 1998 Japan-USA Symp. on Flexible Autom.* Ohtsu (to appear in July).

Ghinato, P., S. Fujii and H. Morita (1998e). The Allocation of Multifunction Workers in U-shaped Production Lines: A Multiobjective Approach, *Proceedings of the 1998 Pacific Conference on Manufacturing*, Brisbane (to appear in August).

Ignizio, J. P. (1982). *Linear Programming in Single- & Multiple-Objective Systems*. Prentice-Hall, New Jersy.

Keeney, R. L. and H. Raiffa (1976). *Decisions with Multiple Objectives: Preferences and Value Tradeoffs*. John Willey & Sons, USA.

Kim, C., S. S. Kim and B. L. Foote (1996). Assignment Problems in Single-Row and Double-Row Machine Layout During Slow and Peak Periods, *Computer Industrial Engineering*, **30**, 411-422.

Miltenburg, G. J. and J. Wijngaard (1994). The U-line Line Balancing Problem. *Management Science*, **40**, 1378-1388.

Monden, Y. (1983). *Toyota Production System: Practical Approach to Production Management*, Industrial Engineering and Management Press, Norcross.

Ohno, K. and K. Nakade (1997). Analysis and optimization of a U-shaped production line. *Journal of the Operations Research Society of Japan*, **40**, 90-104.

Stecke, K. E. and J. E. Aronson (1985). Review of operator/machine interference models. *International Journal of Production Research*, **23**, 129-151.

PREFERENCE-BASED MMI FOR COMPLEX TASK ENVIRONMENTS

Makoto Takahashi, Koutarou Fukui and Masaharu Kitamura

Dept. of Quantum Science and Energy Engineering,
Graduate School of Engineering,Tohoku University
Aramaki-aza-Aoba01,Aoba-ku,Sendai,Miyagi,980-8579 JAPAN
Tel:+81-22-217-7907/Fax:+81-22-217-7900
E-mail: makoto.takahashi@qse.tohoku.ac.jp

Abstract: The concept of preference-based MMI has been proposed to realize more favorable and acceptable interface design. The positive and negative effects of user-driven customization have been discussed and basic requirements for effective customization have been identified. The laboratory experiments to evaluate basic aspects of user-driven customization are described, in which four subjects performed two kinds of tasks under different conditions. The selectable options were generally favorably accepted by all the users but the positive effect to the performance is still to be studied further in depth. *Copyright © 1998 IFAC*

Keywords:Man/machine interfaces, Manipulation task, Robot arms, Performance evaluation

1. INTRODUCTION

The importance of Man-Machine Interface(MMI) for a complex system such as a modern aircraft and a nuclear power plant has been widely recognized to realize enhanced reliability and safety. It is also pointed out that the MMI for operating remote manipulators for constructing Space Station (Sheridan,1994) or dealing with poisonous/radioactive material is critical issue, where the flexible mix of human and automatic control is being introduced. For both cases, the expected role of MMI is to transfer information on the complex task environment to the operator without distortion and failure and to transfer the intension of operators to the objective system. In addition to this basic requirement often called as transparency, the function of intelligent operation support based on the rapidly evolving computer capability is now being studied extensively with the aim of reducing mental workload and human error rate, especially in case of emergency. The design issue of intelligent support with MMI is closely related to the concept of level of automation (Sheridan,1992) and many problems remain unsolved concerning the task allocation, final authority and context dependency (Inagaki,et.al.,1996; Sheridan,1997). Another issue is the essential importance of diversity in the procedure of complex problem-solving (Takahashi,et.al. 1997). The central idea of cognitive diversity is to provide multiple information sources, each corresponding to an independent problem solving scheme.

In the present study, a concept of preference-based MMI (PBMMI) has been introduced to the design of MMI for complex system, with the emphasis on taking each user's diversity of preference into account. The conventional framework for designing MMI mainly utilized the experimental methods using mock-ups and analysis of human information processing behavior with the aid of cognitive psychology. Although the experimental methodology can provides important information for MMI design, it is often the case that the assumed users for analysis tend to be averaged and the diversity of each users preference is only partially captured. One possible solution for dealing with user's diversity is the *adaptive interface*, which is characterized by the ability to dynamically modify the contents and the form reflecting users' level of expertise and/or level of workload. One of the authors conducted an experimental study of adaptive interface, in which the internal cognitive state of the users has been estimated based on the physiological measures (Takahashi,.et.al.,1996). Although the concept of adaptive interface seems attractive for its potential for situation dependent user-computer interaction, it is quite difficult to realize positive effect of dynamic adaptation because of the difficulty of user state estimation and also because of the possible confusion caused by the inconsistency of the reaction of the interface. In the present stage of interface design, the authors believe that the requirement for dealing with diversity of the users are commonly recognized but the system driven dynamic adaptation still remain in the stage of fundamental study.

As for the interface design of relatively simple system

such as a home electric device or a word processing software on PC, the static adaptation, namely customization, has been widely accepted for dealing with variety of users. The customization is defined here as the adjustment of system function performed by each user based on the user's preference or level of expertise. The important point is that such an adjustment is allowed for end users. On the contrary, such adjustments are not allowed for operators in many cases of interface for complex systems. It is not straightforward to apply the concept of the customization to the interface of complex system because such systems are often operated by a team of operating personnels, where the customization made by each individual may cause a negative effect to others. In this regard, the concept of Preference-based MMI(PBMMI) is introduced and basic requirements for user-driven customization have been studied. In chapter 2, the concept of PBMMI is defined and both positive and negative aspects of user-driven customization are examined. In chapter 3, the laboratory experiments for manipulation task are outlined, which aimed at investigate basic requirements for user-driven customization. The overall conclusions derived through the study are summarized in chapter 4.

2. PREFERENCE-BASED MMI (PBMMI)

It is quite important to choose which aspects of preference should be focused for realizing effective customization. In our previous work (Washio, et.al. 1996), experimental evaluation of MMI using analytic hierarchy process(AHP) and Fuzzy integral has been proposed to quantitatively evaluate the goodness of MMI from multiple perspectives. Along with the process of developing a model of human decision-making, the existence of hidden or unstated requirements for MMI has been realized, which can not be described within the framework of conventional human model and theory of cognition. The development of preference-based MMI aims at modifying conventional MMI by capturing the hidden or unstated factors towards more favorable and acceptable interface design stating with the consideration of users' preference. In this chapter, the expected effects of user-driven customization for PBMMI is examined to identify the requirements to realize PBMMI which can enhance the user acceptance and reliability.

2.1 Positive Aspects of User-driven Customization

In the stage of interface design for operation of complex system, it is a common practice that the task analysis is performed to specify required functions. Then, by using mock-ups and example users, the subjective evaluation is performed to determine the design details. In each step of this design specification, the design with higher average score is generally selected. The specified interface design is desirable in the sense that it is good on the average. Taking diverse nature of human users into account, the variety of human preference should be considered as one of the factors for design specification. If only one fixed type of interface design is allowed technologically, the above mentioned design procedure can be accepted as optimal one. It has become more importance, however, to consider each user's acceptability and preference for interface design along with the progress of the computer technology, which provides us with extended freedom in terms of design flexibility.

The diversity of users' focus is another issue concerning interface design. In the case of operation of complex system, a team of operating personnel is responsible for operation. Although operators are usually trained well enough, it is plausible in the case of emergency that all the operators' are stuck in the same cognitive bias and cannot make correct decisions. As the operator trained more effectively, the possibility of losing diversity of viewpoint may increase. The introduction of effective customization for the interfaces of complex system, which provides enhanced variety of viewpoints, is expected to increase not only users' acceptability but also reliability and safety of total system by reducing possibility of human errors.

2.2 Negative Aspects of User-driven Customization

Typical examples of negative opinion are summarized as follows.

Possible confusion caused by enhanced freedom. In the case of *adaptive interface*, the adaptation by the system may cause serious confusion because the users cannot build a consistent mental model of the interfaces. Compared with this type of dynamic adaptive interface, the proposed PBMMI based on the user-driven customization will cause less confusion since the customization is performed by the users themselves. It is suggested, however, by the so-called *mode confusion* in the cockpit interface that the possibility of confusion still remains even when the selection of mode is executed by the users themselves. It has been pointed out in (Landauer,1995) that the allowed diversity in the interface design for computer has negative effects on the productivity and would cause serious problems. The increase in the number of selectable options simply results in the increased time required for selection process and the possibility of selecting inappropriate option cannot be ignored. In other words, the ordinary users are not capable of selecting appropriate options based on their limited experiences.

In the case of complex systems, where operation is performed by the team of crews, customization by one specific user is not necessarily favored by other members. Instead, it is quite possible that such selfish customization may have negative effects on others and may cause serious confusion.

Inherent nature of expert. The interview with experienced operators of nuclear power plant indicated that operators tend to perform their tasks within the capacity of given interface environments. In other words, they have been trained quite effectively to manage the situation based on the specific type of interface and if they feel any inadequacy of interface, they tend to attribute it to the lack of their own skills. Thus, it is often the case that actual operators tend to accept the given interface and they have few requests on the improvement of it. This tendency is closely related to their pride of their own expertise. Once experts become accustomed to the system with limited usability, they tend to be reluctant to accept revised system even if the usability is effectively improved. This tendency is also related to the conservatism inherent nature of expert, i.e. reluctancy to accept modification or new functionality. An example is the fact that operators of advanced boiling water reactor(ABWR), where user-driven modification of the display organization is partially allowed, are not in favor of utilizing the freedom. This fact can be interpreted as the reluctancy of expert to accept modification. Another example is also about the interface of nuclear power plant. The interface of ABWR is considered as the most advanced type of interface in terms of human factors aspects. Increased number of CRT displays and large size display for information sharing have been introduced with the touch screen capability for easy operation. Although this improved interface should result in lower operator workload and less possibility of human error, it is not favorably accepted by some of operators. This fact implies that even if the performance of interface has been improved in general, it is not always the case that it is favored by all of the operators.

2.3 Basic Requirements for User-driven Customization

Based on these arguments, the search for design guidelines which maximize the positive effects while suppressing the negative ones is worthwhile to be studied.The basic requirements for favorable user-driven customization are discussed in the following.

First of all, the range of freedom in customization should be constrained in order to avoid undesirable cognitive resource allocation to selection process. Excessive number of selection candidates cause confusion and is misleading to inappropriate choice. Each content of selection candidate should be provided after careful validation. Second, it is necessary for users to learn how they should utilize customization on specific situation during the course of prior training. It is quite difficult even for experts to perform appropriate customization if the selection is to be done ad hoc. Users should experience all of the possible candidate of customization to during training and should realize advantage and disadvantage of them to develope individual guideline for selection in advance. If some customization candidates have not been selected by anyone, they must be eliminated to reduce number of selections. This process of familiarization to the interface with customizing function may contribute to reduce the reluctancy of experts described earlier. Third, the distinction between the adjustable and fixed part of the interface should be explicitly defined in advance based on the task analysis . For example, customization should not be applied to the large screen display. The personal terminal for individual operator is one possible part of interface compatible with customization. In the case of complex system, even the personal terminal must be shared by other crew if they work in shifts. In such cases, the interface system should identify the user on duty by some method such as the Bar-code or the IC card. If individual terminal is capable of identify each user, individually customized interface can be provided for each user at any of the terminals in the operation room whichever terminal they use.

The user-driven customization with careful consideration on the above aspects may well be accepted by the experts with pride of their own skill.

3. EXPERIMENT

The laboratory experiments to evaluate basic aspects of user-driven customization are described in this chapter. The three issues shown below are examined through the experiments;

1. *Do users have different preferences?*
2. *Are users willing to utilize provided options ?*
3. *Does considering preferences lead to better performance?*

3.1 Task Domain and Experimental setups

The remote operation of robot manipulator has been adopted as the typical task domain in which operation is performed through computer generated interface. As for the intelligent support for operation of manipulator, automatization using task primitives and introduction of haptic feedback have been studied extensively (Bejczy,1994). In the practical situations such as Space Shuttle Mission, however, the manipulator is usually operated by the mission specialists with high level skills obtained through years of training. In our research group, experimental study for evaluating user skill was performed using simple mechanical manipulator

(Kitamura,1989). It was shown from the results of the experiments that operator's performance is closely related to the following two factors; (1)Capability of spatial perception and (2)Capability of spatial maneuvering. It was also shown that as the indicator of operator performance, the following experimental measures are applicable; (a)Task completion time, (b)Casewise variation and (c)Performance degradation caused by change in working environment. In contrast with this previous experiments mainly aiming to evaluate operator's inherent performance, experiments in the present study focus on evaluating the effects of introducing additional options to interfaces. Thus it should be noted here that the results of the present experiments must be interpreted in terms of inseparable two factors of inherent nature of subjects and the appropriateness of MMI with increased range of freedom.

Fig.1 shows the experimental setup. The MMI for manipulation is divided into two parts; the display of task environments for providing visual information, and the manipulation device to actually control the movements of the robot arms. Several selectable options for display and manipulation are prepared in the experiments. For display of visual information, a camera can be controlled through computer by using same device as the manipulation. Shooting angle (pan and tilt) and zoom ratio can be adjusted. For control device for manipulation, one out of two devices (joystick, mouse) was allowed to be selected.

Two types of task have been adopted to be performed by subjects; One is marking task, in which subjects are asked to touch a target point located at the center of a concentric circles. As for the marking task, two type of instruction have been given to subjects; One is "Perform the task as fast as possible" and the other is "Perform the task as precise as possible". The other is peg-in-hole task, in which subjects are ask to pull out a peg and insert it into other hole. As for the peg-in-hole task, the difficulty is controlled by the clearance. The marking task is assumed to be related to the spatial perception capability and the peg-in-hole task is assumed to be related to both of spatial perception and maneuvering capability.

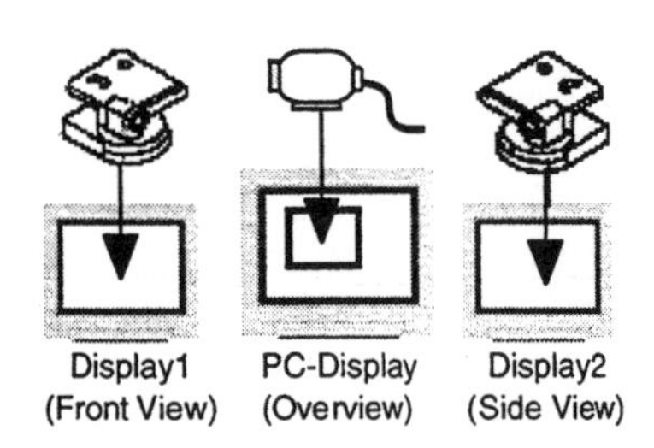

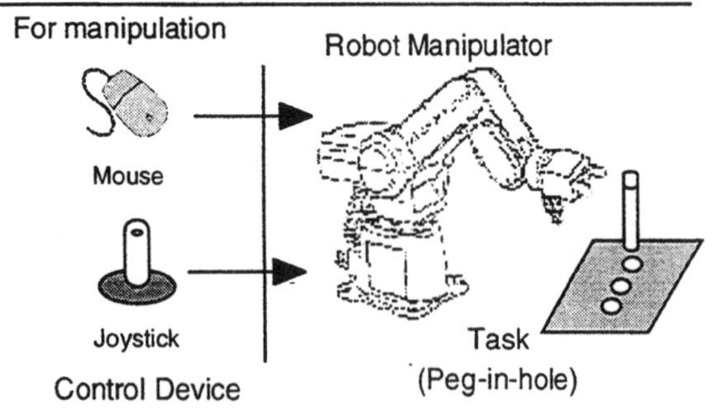

Fig.1 Experimental setups

Four subjects participated in the experiments. They are graduate students in school of engineering and are novice in this type of operation task. Experimental conditions actually performed are listed in Table.1. Each subject performed five trials for each experimental condition.

3.2 Results and discussions

Fig.2 - 4 show performance measures for each subject. Fig.2 shows medians of the task completion time (Tc) for variety of task conditions. Fig.3 shows the differences between best Tc and worst Tc, which can be considered as an index of stability of task execution. Fig.4 shows the differences of the Tc for different difficulty level (high and low), which is expected to reflect the robustness of each subject's performance to the change in the working conditions. Based on these results and observations during experiments, the inherent nature of each subject has been characterized as follows;

Subject A: Cautious, complaisant, amendable to change in environment, high learning capability
Subject B: Cautious,complaisant,self-directed to find new strategy
Subject C: Fast, better robustness to change in environment, stable manipulation capability
Subject D: Fast, relatively low spatial perception capability, unstable manipulation capability

Table.1 Experimental Conditions

Task	Difficulty	Instruction	Device	Camera
Marking	Low	Accuracy	Joystick	Allow control
			Mouse	Allow control
			Selectable	Allow control
			Selectable	No control
	Low	Speed	Joystick	Allow control
			Mouse	Allow control
			Selectable	Allow control
			Selectable	No control
	High	Speed	Joystick	Allow control
			Mouse	Allow control
			Selectable	Allow control
			Selectable	No control
Peg-in-hole	Low	Speed	Joystick	Allow control
			Mouse	Allow control
			Selectable	Allow control
			Selectable	No control
	High	Speed	Joystick	Allow control
			Mouse	Allow control
			Selectable	Allow control
			Selectable	No control

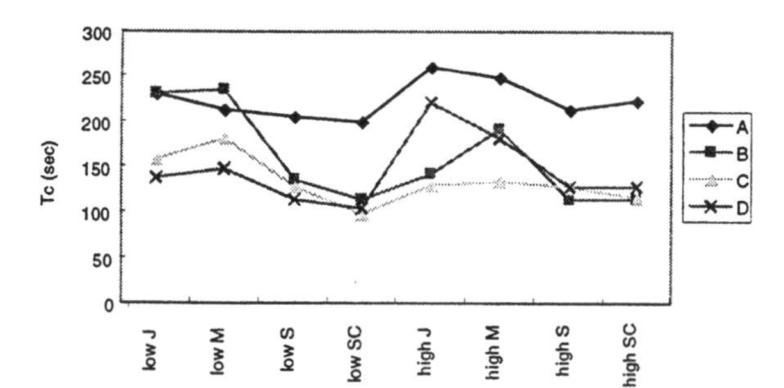

Fig.2 Medians of the task completion time (Tc)
Peg-in-hole task (high / low: difficulty
J:Joystick M:Mouse S:Selectable
SC:Selectable and Camera fixed)

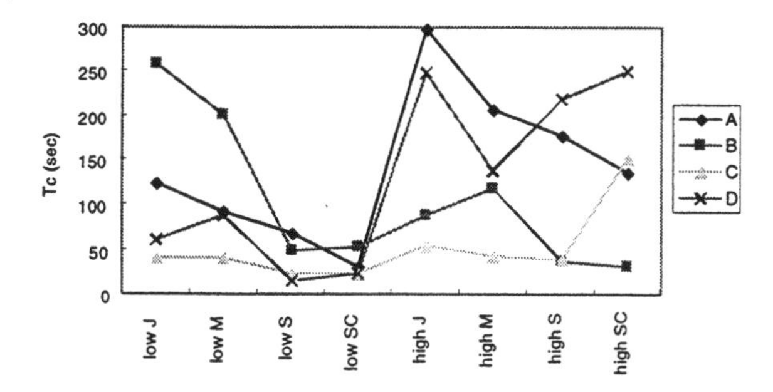

Fig.3 Diferences of Best Tc and worst Tc
Peg-in-hole task

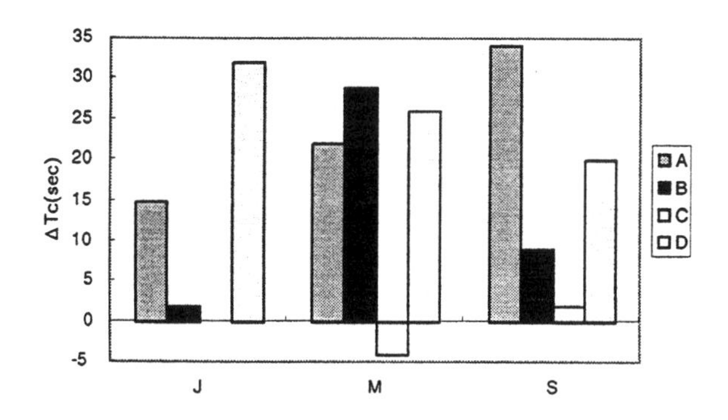

Fig.4 Change of Tc when difficulty increased
Median(high difficulty) - Median(low difficulty)

Although these observed differences may be interpreted as one of the reason to require PBMMI, the point here is whether the provided freedom in the interface has actually been utilized and lead to better performance. Based on the average Tc and robustness to the conditional change, the performance of subject C may be the best among all.

Table.2 summarizes the selected devices when each subject performed best and worst in terms of Tc. The devices with best Tc differ for each subjects, which implies the existence of differences of preferences for

Table.2 Selected devices in best and worst Tc cases

Subject	Task	Difficulty	Instruction	Best Tc Device	Worst Tc Device
A	Marking	low	Ac	selectable(j2m)	joystick
		low	Sp	selectable(m)	joystick
		high	Sp	mouse	selectable(m)
	Peg In Hole	low	Sp	selectable(j)	joystick
		high	Sp	selectable(m)	joystick
B	Marking	low	Ac	selectable(j)	mouse
		low	Sp	selectable(j)	mouse
		high	Sp	selectable(j)	mouse
	Peg In Hole	low	Sp	selectable(j)	mouse
		high	Sp	selectable(j)	mouse
C	Marking	low	Ac	selectable(j)	mouse
		low	Sp	selectable(j)	mouse
		high	Sp	selectable(j)	mouse
	Peg In Hole	low	Sp	selectable(j)	mouse
		high	Sp	selectable(m)	joystick
D	Marking	low	Ac	selectable(j)	mouse
		low	Sp	joystick	mouse
		high	Sp	selectable(j)	mouse
	Peg In Hole	low	Sp	selectable(j)	mouse
		high	Sp	selectable(j)	joystick

selectable(m): mouse selected when selection is allowed
selectable(j):joystick selected when selection is allowed
selectable(j2m):device changed from joystick to mouse when selection is allowed
Sp:Subject instructed to perform task as fast as possible.
Ac:Subject instructed to perform task as precise as possible.

operating devices. The best Tc was generally observed when subject can select devices. However, some question remains to be cleared. It could be the case that subjects just selected favored devices based on their prior experiences of using both devices. The other issue is the consistency of selection. Only subject A changed device during the trial (joystick to mouse) when accuracy of operation was required based on the given instruction. Subject B and D were reluctant to change device once they decided joystick.

Table.3 summarizes the results of subjective opinion on the preferred device obtained by the questionnaire answered by each subject after the all experiments. Though favored device is different for each subjects, the preference is consistent with the performance results in terms of Tc. It should be noted here that subject C, the best operator, thought suitable device was different according to task requirements. As for the control of camera, all subjects preferred the conditions when control of camera was allowed. Although some differences have been observed in the frequency of camera control (while subject B moves camera quite frequently, subject A and C controlled camera only when necessary), allowance of camera control has been accepted favorably by all subjects.

The result of the experiments can be summarized as follows; Subjects showed the different preferences in the selection of control device. Although the selection may have been influenced by the task requirements, the existence of variety of preference was obviously observed. Whether subjects were willing to utilize provided options remains unanswered as far as the present experiments are concerned. But the fact that selection was favorably performed by the best subject seemed to supports the effectiveness in accommodating preferences.

Further study is definitely required to evaluate the

Table.3 Answers to questionnaire

Subject	Which device do you prefer?	Is camera controll effective?
A	Mouse	yes
B	Joystick	yes
C	Mouse for precise operation Joystick for rough operation	yes
D	Joystick	yes

effectiveness of PBMMI. One factors to be considered is the learning effect. As all subjects were novice and handled the manipulator for the first time in their life, they are regarded to be still in the learning phase during the experiments. Although the process of skill development with the selectable option was our major concern, the behaviors of same subjects after learning phase should also be examined to validate the effectiveness of PBMMI for expert users. The other factor to be evaluated is the effects of range of selections. Too many possible options may cause confusion in selection process by users. The optimal number may depend on the level of expertise of users. Finally, the task domain should not be confined only to the remote manipulation task to ensure the generality of the proposed framework. The PBMMI for nuclear power plant has also been developed and the cognitive experiments to evaluate the effectiveness of it are now underway.

4. CONCLUSION

The concept of preference-based MMI has been proposed which aims at modifying conventional MMI by taking the users' preferences more extensively into consideration towards more favorable and acceptable interface design stating with the consideration of users' preference. The positive and negative effects of user-driven customization have been discussed and basic requirements for effective customization have been identified. The laboratory experiments to evaluate basic aspects of user-driven customization are described, in which four subjects performed two kinds of tasks under different conditions. The selectable options were generally favorably accepted by all the users but the positive effect to the performance is still to be studied further in depth. Although the present study needs additional experiments in variety of domains to confirm the effectiveness of PBMMI, the present results demonstrated the basic advantage of incorporating the freedom of user-driven customization.

ACKNOWLEDGMENTS

The authors would like to thank to Dr.Yuji Niwa of INSS Corp. for his valuable discussions.
This work was supported by Grant-in-aid for Scientific Research G(A)(2)-09308012.

REFERENCES

Bejczy,A.K.(1994).Toward Advanced Teleoperation in Space. In: Teleoperation in Space(S.B.Skaar and Carl F.Rouff Ed.),107-138,American Institute of Aeronautics and Astronautics,Inc.,Washington,DC.

Inagaki,T. and Itoh, M. (1996). Trust,autonomy, and authority in human-machine systems:Situation-adaptive coordination for system safety. Proc. of CSEPC 96,176-183.

Kitamura,M.,Takahashi,M. and Sugiyama,K.(1989) Evaluation of Operator Performance in Manipulator Handling,Proc. of ANS Third Topical Meeting on Robotics and Remote Systems, 10-5-1/7.

Landauer,T.K. (1995) The Trouble with Computers, The MIT Press, Cambridge.

Sheridan,T.B.(1994). Human Enhancement and Limitation in Teleoperation. In:Teleoperation in Space(S.B.Skaar and Carl F.Rouff Ed.),43-86,American Institute of Aeronautics and Astronautics,Inc.,Washington,DC.

Sheridan,T.B.(1992). Telerobotics,Automation and Supervisory Control,The MIT Press,Cambrdige.

Sheridan,T.B. (1997) Eight Ultimate Problems of Human-Robot Communication, Proc. of ROMAN97,9-14.

Takahashi,M.,Takei, S. and Kitamura,M.(1997). Multimodal display for enhanced situation awareness based on cognitive diversity,Advances in Human Factors/Ergonomics , Design of Computing Systems,21B, 707-710.

Takahashi,M. et.al., (1996). Validating Physiological Measures for the Cognitive Behavior Analysis Using Neural Network,Proc. of NPIC & HMIT '96,2, 1147-1153.

Washio,T. and Kitamura, M.(1996). Identification of Hidden Factors in Subjective Evaluation of Man-Machine Interface, Proc. of CSEPC96,172-175.

IMPROVEMENT OF DEPTH PERCEPTION AND EYE-HAND COORDINATION IN LAPAROSCOPIC SURGERY

Paul Breedveld, Ton van Lunteren, Henk G. Stassen

Delft University of Technology,
Fac. of Design, Engineering and Production,
Dept. of Mechanical Engineering,
Man-Machine Systems and Control Group,
Mekelweg 2, 2628 CD, Delft, the Netherlands. [1]

Abstract: Laparoscopic surgery is a minimally invasive surgical technique which is performed by using an endoscopic camera and long and slender instruments that are inserted through small incisions in the patient's abdominal wall. The indirect way of observing and manipulating complicates the surgeon's *depth perception* and disorders the surgeon's *eye-hand coordination.* This paper describes a concept of an endoscope positioning system that compensates *misorientations* between the surgeon's hand movements and the movements on the monitor by using a 90° endoscope. *Movement parallax* is realized by means of an endoscope positioning robot which is controlled by the surgeon's head movements. *Copyright ©1998 IFAC*

Keywords: Man/machine systems, manual control, human perception, medical applications, positioning systems.

1. INTRODUCTION

Laparoscopic surgery is a minimally invasive surgical technique which is performed by using an endoscopic camera and long and slender instruments that are inserted through small incisions in the skin of the patient's abdomen, Fig. 1. The endoscope is held by a camera assistant, and the surgeon performs the operation *indirectly* by spatially manipulating the instruments and by observing the camera pictures on a monitor (Cuschieri *et al.*, 1992, Frank *et al.*, 1997). Compared with conventional open abdominal surgery which is carried out through a large incision in the skin, the advantages of laparoscopic surgery are that it reduces the damage of the body and in principle also the risk of infection and the recovery time. Disadvantages are that the operation protocols are still far from optimal and that it is difficult for the surgeon to map pre-operative information from X-rays, Ultrasound

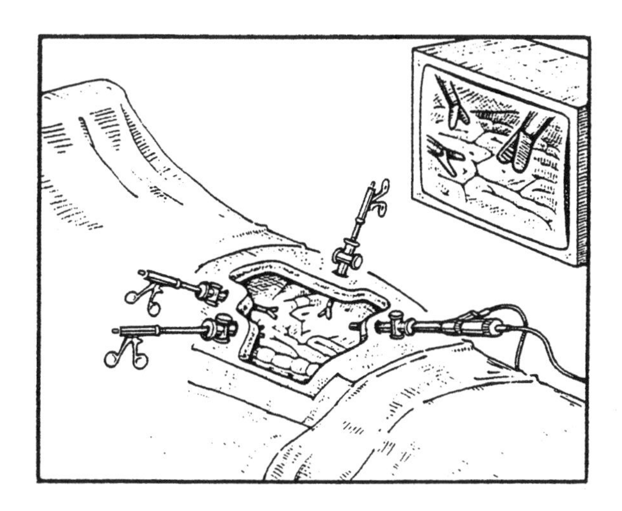

Fig. 1. Transparent view of the abdomen with an endoscope (right) and three laparoscopic instruments.

or MRI images onto the endoscopic camera pictures. Other difficulties concern the indirect way of observing and manipulating. They complicate the surgeon's *observation* and *manipulation* activities and disorder the surgeon's *eye-hand coordination* (Breedveld, 1997, 1998).

[1] This paper was written as a part of the MISIT research program on Minimally Invasive Surgery and Intervention Techniques. The program is part of DIOC-9: the Delft Interfaculty Research Center on Medical Engineering.

In order to find solutions for these difficulties, a research program on minimally invasive surgery techniques was initiated at the Delft University of Technology in cooperation with the Academic Medical Center in Amsterdam and several other hospitals in the Netherlands. The program is called MISIT, which is short for Minimally Invasive Surgery and Intervention Techniques. This paper describes the first results of a MISIT project that focuses on eye-hand coordination. Sections 2 and 3 give an overview of negative effects on the surgeon's depth perception and eye-hand coordination, and Sections 4, 5 and 6 describe a new method to reduce these negative effects. Section 7 ends with the conclusions.

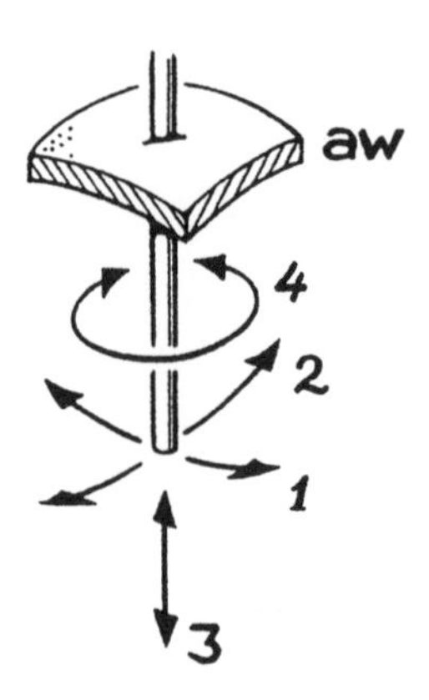

Fig. 2. The four degrees of freedom of a standard endoscope. *(aw) Abdominal wall.*

2. DEPTH PERCEPTION

The surgeon's *visual cortex* uses the camera picture on the monitor to determine the spatial position of the surgical instruments with respect to the anatomic structures. One of the largest problems in this observation process concerns the perception of distances and movements *perpendicular* to the image on the retina. A human can use three depth information sources to determine such distances and movements (Gibson, 1979, Regan *et al.*, 1979, Cuschieri, 1996, Voorhorst, 1998): *visuomotor cues, pictorial information* and *parallax.*

(1) Visuomotor cues concern the movements of the eyeballs and the eyelenses to focus an object. *Accommodation* is the adjustment of the eyelens to focus on an object, and *convergence* is the horizontal and inward rotation of the two eyes to point them to the object.
(2) Pictorial information concerns the pictorial cues in the retinal image, that give information about distances and movements perpendicular to that image. An example of such a cue is: *'an object touches a surface when it touches its shadow on the surface'*. This cue is very helpful for accurate spatial positioning tasks.
(3) Parallax concerns changes in the mutual positions of objects in the retinal image when the viewpoint of the eye changes. Two kinds of parallax can be distinguished: *stereovision* and *movement parallax.* Stereovision concerns the disparity between the two pictures seen by the left eye and the right eye caused by the distance between the two eyes. Objects with a different distance to the observer are shifted with respect to each other in the two pictures. The size of the shift gives information about their spatial position. Movement parallax concerns shifts in the picture seen by one eye when the observer moves the head. The head movement causes the visible objects to shift with respect to each other, and also this shift gives information about their spatial position. *Motion parallax* refers to a shift in the retinal image not caused by the observer's head movement, but by an external influence, for example the movement of a camera when the observer watches a camera picture.

In normal life, a human can use all these depth information sources to perform a spatial manipulation task. In conventional laparoscopic surgery, however, many sources are not available:

(1) When the surgeon looks at the monitor, the eyelenses focus on the *surface* of the television screen, and not at the visible objects 'behind' the screen. This makes the information coming from visuomotor cues useless for depth perception.
(2) In all standard endoscopes, the light source is located at the tip, forming a ring of light around the lens. This is advantageous for the brightness of the picture, but disadvantageous for the surgeon's depth perception, since the endoscopic camera pictures contain in principle no shadows.
(3) Conventional endoscopes are monocular and controlled by a camera assistant. The surgeon is thus not able to use stereovision and movement parallax as depth information sources. When the assistant moves the endoscope, motion parallax is present to some extent, but the amount of information is limited since the endoscope movements are limited. The incision point acts like a spherical joint that limits the degrees of freedom (DOFs) of the endoscope from six to four, Fig. 2. This makes it impossible to observe the anatomic structure from different sides while keeping the viewpoint in focus. Many laparoscopic surgeons experience this as a handicap.

Especially the absence of shadows, stereovision and movement parallax makes it difficult for a surgeon to determine spatial distances and movements accurately. The surgeon needs an intensive and time-consuming training period to improve his/her sensitivity for the depth information sources that are still present. For example: the surgeon learns to compensate for the absence of shadows by replacing the cue *'the instrument touches the tissue when it touches its shadow on the tissue'* by the cue *'the instrument touches the tissue when the tissue starts to deform'*.

In the laparoscopic literature, a number of aids has been found to improve the surgeon's depth perception (Breedveld, 1998). Shadows can be introduced by using *illumination cannulas* (Schurr *et al.*, 1996). These are trocars with light bundles integrated into their shaft that can be used as additional light sources. Stereovision can be introduced by using a *stereo-endoscope* (Cuschieri *et al.*, 1992, Frank *et al.*, 1997). Comparisons between monocular and stereo-endoscopes have shown, however, that the advantage of using stereovision as the *only* depth information source is only very small (Breedveld, 1998). Voorhorst (1998) describes a number of prototype *flexible endoscopes* that can in principle be used to observe the anatomic structure from aside while keeping the viewpoint in focus. This offers the possibility to add movement parallax to the control of the endoscope.

3. EYE-HAND COORDINATION

The *motor cortex* uses the spatial information from the visual cortex to determine the desired hand movements and to generate stimulation signals for the muscles in the upper- and forearms of the surgeon. The signals are generated by an *internal representation* that describes which muscles have to be stimulated and how far they have to contract to realize the desired hand movement with respect to the retinal image (Breedveld, 1997, 1998). The internal representation has been constructed and improved after years of everyday life training and experience in normal situations like in open abdominal surgery, where the surgeon observes and manipulates the surgical instruments *directly*. In the case of laparoscopic surgery, however, the situation is unnatural due to the *indirect* way of observing and manipulating. This leads to confusing discrepancies between the internal representation and the real situation causing the surgeon to have a disturbed eye-hand coordination (Wade, 1996).

When being introduced to the laparoscopic surgery technique, a resident surgeon associates the tips of the laparoscopic instruments on the monitor with his/her hands. When the resident uses the internal representation to determine the muscle stimulation signals, however, he/she notices that the instrument tips on the monitor do not move according to the expectation. This is caused by the following effects:

- The endoscope and the monitor show a *magnified view* on the laparoscopic instruments. Consequently, the instrument tips in the retinal image move faster than expected from the internal representation. The surgeon has to compensate this effect by scaling the hand movements mentally.
- The instrument incision points *mirror* the surgeon's hand movements. Consequently, the instrument tips in the retinal image move in the opposite direction than expected from the internal representation. The surgeon has to compensate this effect by mirroring the hand movements mentally.
- The endoscope, which is controlled by a camera assistant, is inserted into the abdominal wall at a location and at an angle where it does not hamper the surgeon and the operation team in their activities. The camera's line-of-sight is thus usually different from the surgeon's natural line-of-sight when he/she could look directly in the abdomen. Since the endoscopic camera pictures show a rotated view on the scene, the instrument tips in the retinal image move in a different direction than expected from the internal representation. The surgeon has to compensate this *misorientation* by rotating the hand movements mentally.

Especially the misorientation is very confusing for the surgeon. An effect similar to a 90° misorientation occurs in Window-based software when the mouse on the tabletop is rotated 90°. Left/right mouse movements then result in up/down cursor movements, which is very confusing. The surgeon needs an intensive and time-consuming training period to adapt the internal representation for eye-hand coordination in everyday life to laparoscopic situations.

4. COMPENSATION OF MISORIENTATIONS

The eye-hand coordination of the surgeon can be strongly improved if the misorientation is compensated automatically. Initially, it seemed to be most straightforward to compensate the misorientation in the endoscopic camera picture by rotating the picture *in software* such that its line-of-sight corresponds with the surgeon's natural line-of-sight when he/she could look directly in the abdomen. This approach was followed by Bajura *et al.* (1996) who used a 3D ultrasound scanner to project an ultrasound image on the patient's abdomen. The ultrasound scan was recorded, filtered, and transformed into a 3-dimensional graphical image. The observer wore a Helmet Mounted Display and the image was animated on-line on the two television screens before the observer's eyes. The observer's head position was measured, and the image was

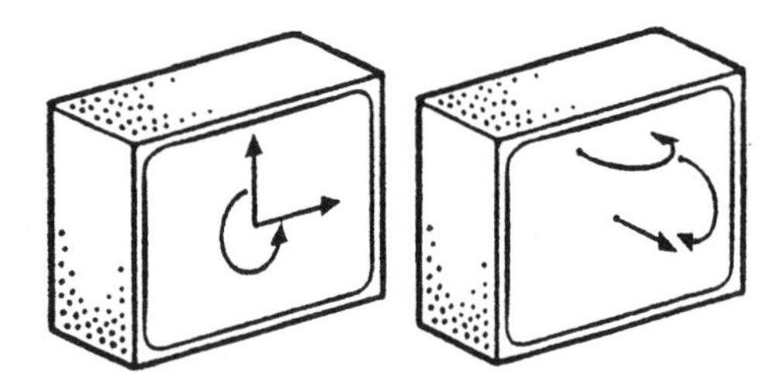

Fig. 3. Translations and rotations *parallel* to the screen (left) and *perpendicular* to the screen (right).

rotated such that its orientation matched the observer's line-of-sight when looking at the patient's abdomen.

A 3D ultrasound scanner generates a *3-dimensional* image that can be easily rotated in software to obtain the correct orientation with respect to the observer. An endoscopic camera, however, generates a *2-dimensional* picture of the operative area. This limits the freedom of rotation strongly. Rotations *parallel* to the television screen, Fig. 3, can be performed without problems in software. Rotations *perpendicular* to the television screen, however, cannot be performed in software since they cause the visible objects to *shift* with respect to each other. This can only be achieved by a change in the camera's line-of-sight.

Since the endoscopic camera picture cannot be freely rotated in software, it was decided to rotate the picture *in hardware* by rotating the endoscopic camera such that its line-of-sight corresponds with the surgeon's natural line-of-sight when he/she could look directly in the abdomen. With a standard endoscope, however, this results in a pose of the camera assistant that hampers the surgeon in the activities, Fig. 4. This is caused by the fact that the camera's line-of-sight is *in line with* the endoscope shaft. In order to solve this problem, it was decided to replace the standard endoscope by a *90° endoscope* in which the camera's line-of-sight is *perpendicular to* the endoscope shaft, Fig. 5. The endoscope can now be inserted at a location where it does not hamper the surgeon in the activities, and the endoscope tip can be rotated such that the camera's line-of-sight corresponds with the surgeon's natural line-of-sight when he/she could look directly in the abdomen. The 90 ° endoscope offers also the possibility to *look around* in the abdomen by rotating the endoscope around its shaft. This makes it more easy for the surgeon to find and to identify anatomic structures.

5. INTRODUCTION OF MOVEMENT PARALLAX

Besides compensating misorientations, the 90 ° endoscope can also be used to observe the anatomic structure from different sides *while keeping the viewpoint in focus.* This is achieved by translating the endoscope sideward and by simultaneously rotating it around its shaft such that the viewpoint remains centred in the camera picture, Fig. 6. *Movement parallax* can in principle be realized if the camera assistant is replaced by an *endoscope positioning robot* which is controlled by the surgeon's *head movements*, Fig. 7.

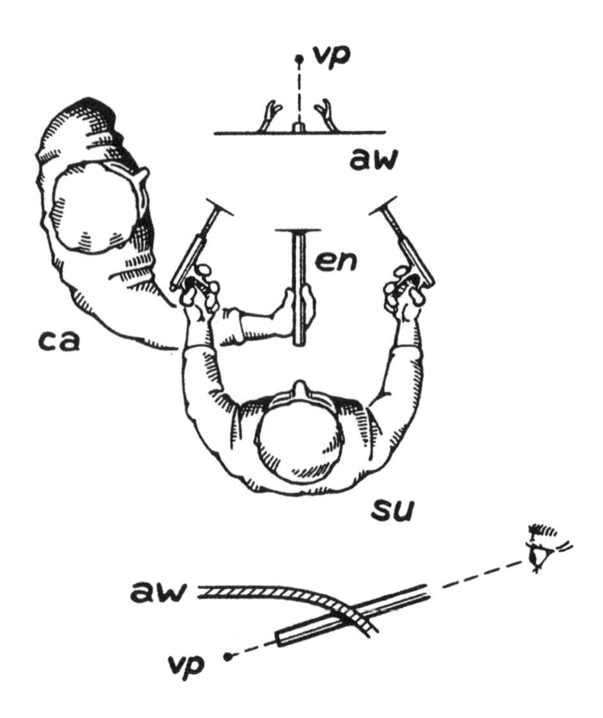

Fig. 4. Compensation of misorientations with a standard endoscope. *Top: top view of a section of the abdominal wall (aw), the surgeon (su), the camera assistant (ca), the endoscope (en) and the viewpoint (vp). Bottom: side view of the surgeon's eye, the abdominal wall, and the endoscope.*

Examples of head-controlled endoscope positioning systems can be found in Finlay & Ornstein (1995) and in Voorhorst (1998). The systems measure the surgeon's head movements and transform them into movements of a device that holds a standard endoscope. This enables the surgeon to position the endoscope in an intuitive way without the need for a camera assistant. The systems in Finlay & Ornstein (1995) and in Voorhorst (1998) have been designed for use in combination with a *standard* endoscope. As a result, it is not possible to observe the anatomic structure from aside while keeping the viewpoint in focus. Endoscope positioning robots for 90 ° endoscopes have not been found in the laparoscopic literature yet.

In order to realize movement parallax without misorientations, the endoscope incision point must be chosen such that a sideward translation of the

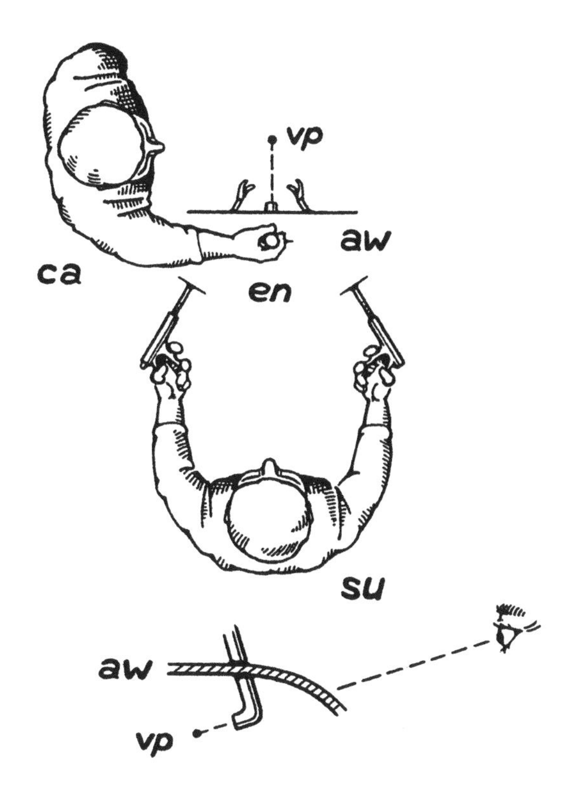

Fig. 5. Compensation of misorientations with a 90 ° endoscope. *Top: top view of a section of the abdominal wall (aw), the surgeon (su), the camera assistant (ca), the endoscope (en) and the viewpoint (vp). Bottom: side view of the surgeon's eye, the abdominal wall, and the endoscope.*

surgeon's head results in a sideward translation of the endoscope tip in a direction perpendicular to an imaginary vertical plane through the surgeon's head and the viewpoint on the organ to be operated, Fig. 8. The tip translation must be accompanied by a rotation around the shaft to keep the viewpoint in focus. Since the tip of the 90 ° endoscope has only four DOFs, the endoscope incision point must be located on the *intersection* between the vertical plane and the patient's abdominal wall. The incision point of the 90 ° endoscope is thus *restricted* to a line on the patient's abdomen.

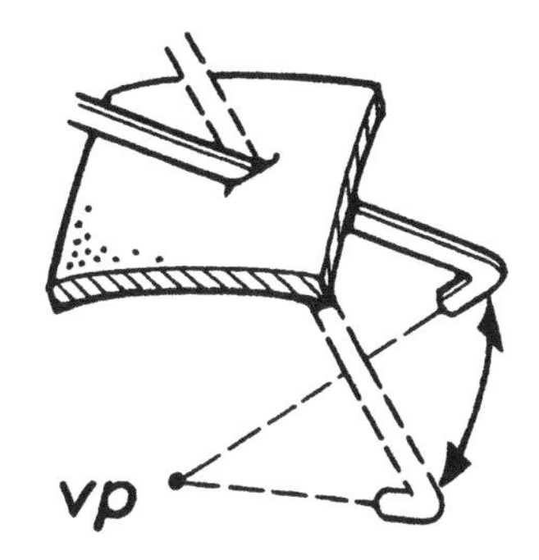

Fig. 6. Use of a 90 ° endoscope to observe the anatomic structure from aside such that the viewpoint *remains centred* in the camera picture. *(vp) Viewpoint.*

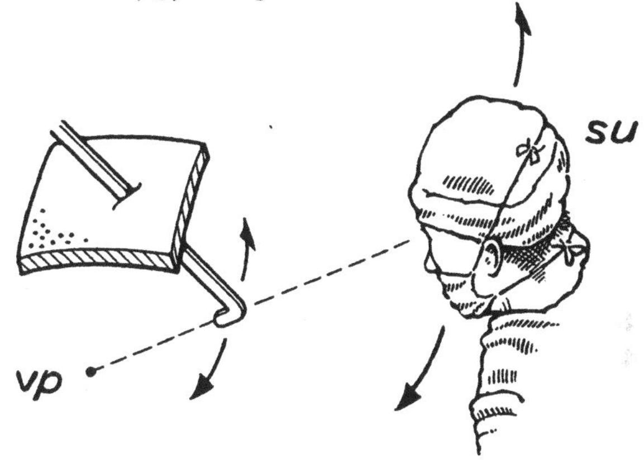

Fig. 7. Use of a 90 ° endoscope to obtain movement parallax. *(su) Surgeon, (vp) viewpoint.*

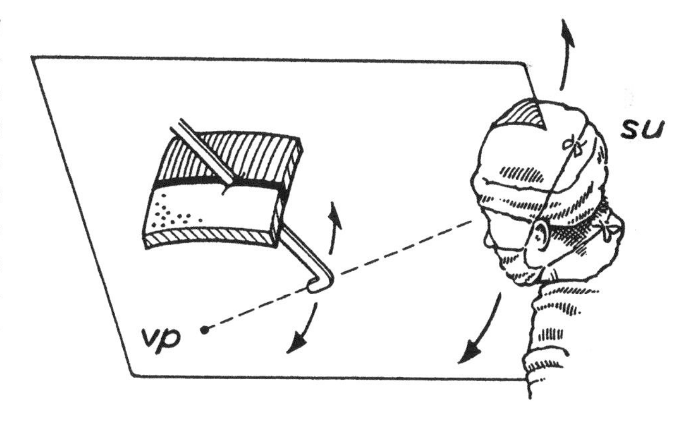

Fig. 8. Location of incision points of the 90 ° endoscope to realize movement parallax without misorientations. *(su) Surgeon, (vp) viewpoint.*

6. FURTHER IMPROVEMENTS

In order to give the surgeon more flexibility in choosing the endoscope incision point, the number of DOFs of the endoscope tip can be enlarged. This can be done by replacing the 90 ° endoscope by a *flexible endoscope* with a tip that can be rotated around an axis *perpendicular* to the endoscope shaft. This results in the endoscope tip to have five DOFs, Fig. 9: a sideward translation perpendicular to the shaft (1), a forward/backward translation perpendicular to the shaft (2), an up/down translation parallel to the shaft (3), a rotation around the shaft (4), and an additional rotation perpendicular to the shaft (5).

The two rotational DOFs (4) and (5) can be used to rotate the camera picture *perpendicular* to the television screen. Rotations *parallel* to the television screen can be realized by adding a third rotational DOF to the system. This DOF can be implemented *in software* since it does not cause the visible objects to shift with respect to each other, Fig. 3.

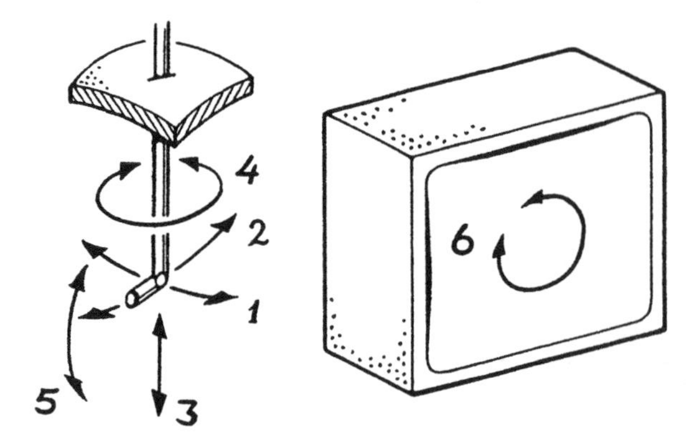

Fig. 9. The five degrees of freedom of the flexible endoscope, and the sixth degree of freedom on the television screen.

The six translational and rotational DOFs enable the surgeon to translate and rotate the camera picture *freely* in 3-dimensions, independent of the location of the incision point. The incision point is thus not restricted anymore to a line on the patient's abdomen. It can be chosen with a large amount of freedom without losing the ability to compensate misorientations and to realize movement parallax. Due to the addition of two extra rotational DOFs, movement parallax is not restricted anymore to *horizontal* head movements. The endoscope can in principle be controlled such that it reacts also on *vertical* head movements so that the anatomic structure can be observed from aside and from above.

7. CONCLUDING REMARKS

The paper describes a concept of a new endoscope positioning system that improves the depth perception and eye-hand coordination of the laparoscopic surgeon. The system compensates misorientations by using a 90° endoscope that can be inserted in the abdominal wall at a location where it does not hamper the surgeon in the activities. The endoscope tip is rotated such that its line-of-side corresponds with the surgeon's line-of-sight when he/she could look directly in the abdomen. Movement parallax can be realized if the camera assistant is replaced by an endoscope positioning robot which is controlled by the surgeon's head movements. The endoscope incision point can be chosen freely if the 90° endoscope is replaced by a flexible endoscope and if a third rotational DOF is added in software.

In theory, the system offers a convenient solution for two of the most confusing problems of the laparoscopic surgeon. A prototype of the system will be developed within the MISIT project at the Delft University of Technology. The development will be followed by a field evaluation with laparoscopic surgeons.

REFERENCES

Bajura, M., Fuchs, H., Ohbuchi, R. (1996). Merging Virtual Objects with the Real World: Seeing Ultrasound Imagery Within the Patient. In: *Computer Integrated Surgery - Technology & Clinical Applications* (Taylor R H, Lavallée S, Burdea G C, Mösges R (Eds.)), MIT Press, ISBN 0-262-20097-X, pp. 245-254.

Breedveld, P. (1997). Observation, Manipulation, and Eye-Hand Coordination Problems in Minimally Invasive Surgery. *Proc. 16th European Annual Conf. on Human Decision Making and Manual Control*, Dec. 9-10, Kassel Germany, 13 p.

Breedveld, P. (1998). *Observation, Manipulation and Eye-Hand Coordination in Minimally Invasive Surgery - Overview of Negative Effects, Experiments and Supporting Aids*, Report N-510, Delft University of Technology, Fac. of Design, Engineering and Production, Man-Machine Systems and Control Group, 49 p.

Cuschieri, A., Buess, G., Périssat, J. (1992). *Operative Manual of Endoscopic Surgery*, Springer-Verlag Berlin Heidelberg, ISBN 0-387-53486-5, 363 p.

Cuschieri, A. (1996). Visual Display Technology for Endoscopic Surgery. *Minimally Invasive Therapy & Allied Technologies*, Vol. 5, pp. 427-434.

Finlay, P. A., Ornstein, M. H. (1995). Controlling the Movement of a Surgical Laparoscope. *IEEE Engineering in Medicine and Biology Magazine*, May/June 1995, pp. 289-291.

Frank, T. G., Hanna, G. B., Cuschieri, A. (1997). Technological Aspects of Minimal Access Surgery. *Proc. Instn. Mech. Engrs.*, Vol. 211, Part H, pp. 129-144.

Gibson, J. J. (1979). *The Ecological Approach to Visual Perception*, Houghton Mifflin, Boston, USA, ISBN 0-395-27049-9, 332 p.

Regan, D., Beverley, K., Cynader, M. (1979) The Visual Perception of Motion in Depth. *Scientific American*, July 1979, pp. 136-151.

Schurr, M. O., Buess, G., Kunert, W., Flemming, E., *et al.* (1996). Human Sense of Vision: A Guide to Future Endoscopic Imaging Systems. *Minimally Invasive Therapy & Allied Technologies*, Vol. 5, pp. 410-418.

Voorhorst, F. A. (1998). *Affording Action - Implementing Perception-Action Coupling for Endoscopy*, PhD Thesis, Delft University of Technology, Fac. of Design, Engineering and Production, Lab. of Form Theory, Delft, the Netherlands, 204 p.

Wade, N. J. (1996). Frames of Reference in Vision. *Minimally Invasive Therapy & Allied Technologies*, Vol. 5, pp. 435-439.

IDENTIFICATION OF HUMAN OPERATOR CONTROL BEHAVIOUR IN MULTIPLE-LOOP TRACKING TASKS

M.M. (René) van Paassen *[,1] **Max Mulder** *[,2]

* *Delft University of Technology,*
Faculty of Aerospace Engineering,
P.O. Box 5058, 2600 GB Delft, The Netherlands

Abstract: Methods for identifying human control behaviour in compensatory and pursuit tracking tasks have been used extensively in the past. These methods are still very valuable, for example to study the effect of experiment conditions on control behaviour. In studies at the Faculty of Aerospace Engineering these techniques were used for studying control behaviour in multiple-loop tasks, in which multiple transfer functions for the operator were to be estimated. New in these studies is that, in these multi-loop tasks, analytical expressions were derived for the bias and variance of the estimates. This paper revises the technique, putting an emphasis on experimental set-up with modern equipment, the choice of test signals and the calculation of bias and variance of the estimates. *Copyright ©1998 IFAC.*

Keywords: Manual control, mathematical model, describing function, identification

1. INTRODUCTION

The combination of the quasi-linear pilot models and the crossover model theorem has become a well-established paradigm for describing and predicting human control behaviour in single-axis compensatory tracking tasks.

Methods for identifying the human control behaviour in compensatory tracking tasks have been known since the early applications of (Krendel and McRuer, 1960). For the identification of this control behaviour, the human operator is described as a linear describing function and a remnant noise. In combination with the quasi-linear pilot models, identification of human operator control behaviour proved to be an essential tool for understanding and describing manual control in many applications (McRuer and Jex, 1967).

Early users of these methods were limited in their experimental design to the then-existing techniques for recording and processing signals, and necessarily had to devote much attention to the experiment and signal processing hardware (McRuer *et al.*, 1965). This makes the older literature less accessible to today's readers.

Whereas identification of human control behaviour in a single-loop task is relatively clear-cut, the model identification and in particular the model validation becomes more intricate in multi-loop situations. First of all, there is the problem of determining the particular feedback loops that the operator will close: the model structure itself cannot be identified from experimental data (Stapleford *et al.*, 1969). Another problem that was only properly realized in the mid-1970s was the difficulty of identifying systems operating in closed loop. The rather elaborate identification techniques for these closed loop multi-loop applications were not available until the pioneer work of (Stapleford *et al.*, 1967; Stapleford *et al.*, 1969),

[1] M.M.vanPaassen@LR.TUDelft.nl
[2] M.Mulder@LR.TUDelft.nl

and it lasted until the late 1970s until they were formalised mathematically (van Lunteren, 1979).

As a result, not many multi-loop operator models have been described in literature. Moreover, in those cases where a multi-loop model *has* been published, the model identification and validation efforts have not been fully elaborated (Teper, 1972; Weir *et al.*, 1972; Weir and McRuer, 1972).

Although the method can now be called "old", it is still very relevant. It can for example be used to investigate human control behaviour in human-computer interaction. The increased use of simulation in training also requires a renewed focus on the manual control behaviour; the effect of simulation fidelity on the operator's behaviour must be assessed, which requires identification of the human operator's control behaviour. Some other research problems that require identification of the control behaviour of the human operator are:

- Investigation of the role of multi-modal feedback.
- Identification of human operator haptic (proprioceptive) control (van Paassen, 1994).
- Investigation of vehicle control (aircraft, automobile), in which usually more than one loop is closed and more than one channel is controlled simultaneously (Mulder, 1998).
- Investigation and evaluation of augmented systems, i.e. study of the interim between fully manual and automated control.
- Identification of the perception & action cycles in the paradigm of active psychophysics, i.e. the identification of the operator's use of (a subset of) visual cues in multi-cue displays (Flach, 1991; Mulder, 1998).

For most applications it is necessary to identify multiple feedback loops. As argued above, the methodology in multi-loop simulations is not well covered. In addition, the early papers about the method are less accessible due to the necessary emphasis on the technical difficulties of the measurement, test signal generation and signal processing.

In this paper we will revise the methodology and explain the approach to identification in multiple-loop tracking tasks, including the determination of bias and variance of the estimate. In Section 2 we will discuss the design of an experiment and the evaluation of the results for a single loop identification. Section 3 describes the multi-loop case. Some example results are summarised in Section 4 and a retrospective of the method and its findings is given in Section 5.

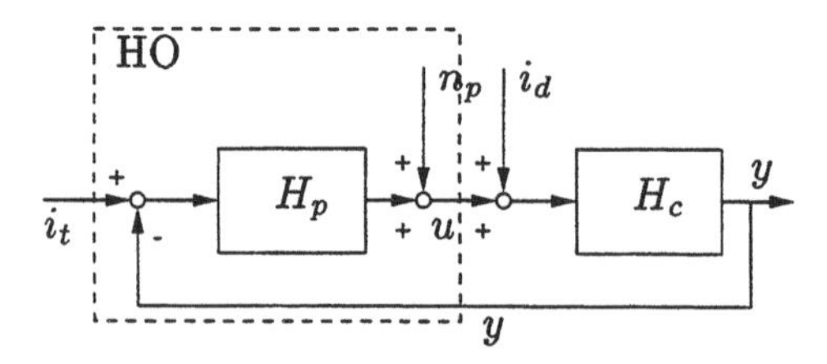

Fig. 1. Human Operator model for single loop control. The HO is represented by a linear control element H_p and an additional remnant noise n_p.

2. SINGLE LOOP HO IDENTIFICATION

2.1 *Introduction*

The emphasis of this paper is on the description of the identification of a HO's control behaviour in multi-loop control situations. We will describe the single-loop identification here, as a reminder, and as an introduction to multiple-loop identification.

The human operator is considered to be a continuous control element (H_p) in a closed loop with the controlled system (H_c), see Figure 1. The control behaviour by a HO is not entirely reproducible. To account for variation in control performance, a noise input, generated by the HO, is introduced in the model.

A test signal is needed to excite the combined system and obtain responses from the HO. This signal can be present at the input of the HO (i_t), so that a target following or tracking task is obtained, or at the input of the controlled system (i_d), resulting in a disturbance compensation task. Note that the human operator will try to adjust his control behaviour to optimize the overall task performance. Therefore the HO control behaviour can depend on properties of the test signal. The crossover regression phenomenon (McRuer *et al.*, 1965; McRuer and Jex, 1967) is one example of adaptation of the control behaviour in response to properties of the disturbance signal.

It has become common practice to apply a test signal that consists of a sum of sinusoidal functions. A describing transfer function of the HO is obtained that contains estimates at the frequencies of those sinusoidal functions. There are a number of things to consider in the design of such an experiment:

- Choose a reasonable recording period, in the order of one to several minutes. Whereas a short period limits the lowest frequency of the describing function, a long period will lead to HO fatigue, and possible changes in the control behaviour over the experiment.
- Choose the frequencies so that the sine waves fit an integer number of times in the recording period. The response to the test signal

will show up at distinct frequencies in the Fourier transformation of the measured signals. All signals at other frequencies are due to "pilot noise" inputs. It is best to check the test signal beforehand with an FFT.

- Choose the starting phase for each sine signal "randomly", but check that the test signal does not contain any unfortunate large peaks. Re-run the random generator if necessary.
- Add a run-in period before the recording period. At the start of a recording, the HO will display a transient behaviour. Recording should start after this transient has died out.
- Choose the frequencies in the test signal so that the test signal appears random to the HO. The trick is to have no sine signal be an harmonic of another signal. Use prime numbers (or combinations thereof) to choose the number of waves in the recordig period.
- Take into account that the manipulator dynamics and the display dynamics could very well influence the HO control behaviour. In most studies the manipulator position is taken as the input signal of the controlled system. This means that the HO model that is identified *includes* the manipulator. At least document these dynamics, as well as the physical properties of manipulator and display.
- Allow the subjects to practice, and start measuring after performance has reached a constant level.
- Consider measuring multiple runs for one configuration, and averaging the data to improve the signal-to-noise ratio.

2.2 *Data processing*

Consider Figure 1. For the following discussion a compensatory task will be assumed, so that $i_t = 0$. The symbol $y(t)$ represents the output of the controlled system as a function of time and $u(t)$ represents the output signal of the pilot. After Fourier-transformation this results in the signals $Y(\nu)$ and $U(\nu)$, where ν is the Fourier frequency. Assuming that the pilot behaves as a linear control element with an added noise input $n(t)$, the following holds for a certain realisation ζ of an experiment (McRuer and Jex, 1967; van Lunteren, 1979):

$$U(\nu;\zeta) = -H_p(\nu)Y(\nu;\zeta) + N(\nu;\zeta) \qquad (1)$$

Due to the choice of the test signal, as a relatively small sum of sine functions that have an integer number of periods in the test signal, the HO response to the test signal is only found in a distinct number of Fourier frequencies. Most of the frequencies ν will not contain any components of the test signal, but instead contain only the noise introduced by the HO. At a frequency ν_a at which the test signal contains a sine component, and assuming that the noise signal is relatively small compared to the test signal at that frequency, the HO transfer function can be estimated as:

$$\hat{H}_p(\nu_a;\zeta) = -\frac{U(\nu_a;\zeta)}{Y(\nu_a;\zeta)} \qquad (2)$$

Note that this estimate is calculated as if the considered system, the HO in this case, were tested *in an open loop*. Because the experiment involves closed loop identification, the estimate is biased, which can be verified by expressing the signals used in this estimate, $Y(\nu_a;\zeta)$ and $U(\nu_a;\zeta)$ in the test signal $I(\nu_a)$, the "true" HO transfer function $H_p(\nu)$ and the transfer function of the controlled system $H_c(\nu)$:

$$Y(\nu;\zeta) = \frac{(I(\nu_a) + N(\nu_a;\zeta))\,H_c(\nu_a)}{1 + H_p(\nu_a)H_c(\nu_a)}, \qquad (3)$$

$$U(\nu;\zeta) = \frac{I(\nu_a)H_c(\nu_a)H_p(\nu_a) + N(\nu_a;\zeta)}{1 + H_p(\nu_a)H_c(\nu_a)} \qquad (4)$$

And then substituting these in Eq. 2:

$$E\left\{\hat{H}_p(\nu_a;\zeta)\right\} = H_p(\nu_a) - \left(H_p(\nu_a) + \frac{1}{H_c(\nu_a)}\right) E\left\{\frac{N(\nu;\zeta)}{I(\nu_a) + N(\nu_a;\zeta)}\right\} \qquad (5)$$

The second term in this equation is the bias of the estimate. Taking x as the ratio between the intensity of the test signal at ν_a and of the noise at ν_a, $x(\nu_a) = |I(\nu_a)|^2/\sigma_N(\nu_a)^2$, and assuming that the Fourier coefficients of the $N(\nu;\zeta)$ have a normal distribution, the latter term in Eq. 5 can be calculated as (van Lunteren, 1979):

$$bias\left(\hat{H}_p(\nu_a;\zeta)\right) = \left(H_p(\nu_a) + \frac{1}{H_c(\nu_a)}\right) e^{-x} \qquad (6)$$

In a similar manner the variance of the estimate can be calculated (ν, ζ terms are dropped for brevity):

$$\begin{aligned}\mathrm{var}\left(\hat{H}_p\right) &= E\left\{\left(\hat{H}_p - E\left\{\hat{H}_p\right\}\right)^2\right\} \\ &= \left(H_p + \frac{1}{H_c}\right)^2 E\left\{\frac{N^2}{(I+N)^2}\right\} \\ &\quad - \left(bias\left(\hat{H}_p\right)\right)^2\end{aligned} \qquad (7)$$

The first term on the right-hand side in Eq. 7 can only be calculated when one accepts that there is a small probability ϵ that the variance is under-estimated. Then the following approximate solution may be found (van Lunteren, 1979):

$$E\left\{\frac{N^2}{(I+N)^2}\right\} \approxeq e^{-x+\delta} + e^{-x+\delta} - 1 + x\int_\delta^x \frac{e^{p-x}}{p}dp + xe^{-x}\int_\delta^\infty \frac{e^{-p}}{p}dp \qquad (8)$$

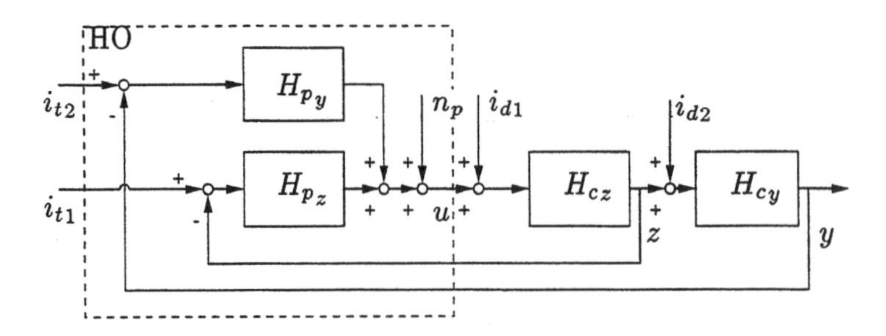

Fig. 2. Parallel Human Operator model for double loop control. The pilot is represented by two linear control elements placed in parallel to each other, H_{p_y} and H_{p_z}, and an additional remnant noise n_p.

With $\epsilon = 2e^{-x}\sinh(\delta)$. In Section 3 it will appear that the formulas of Eq. 6 and 8 will again be used in the determination of bias and variance for the multi-loop case. In practice, the biased nature of the estimate is not a problem; our experience showed that in cases where the bias is significant, so is the variance of the estimate, and these estimates, usually for the lowest frequencies of the test signal, are useless anyhow.

For the calculation of the bias – and the variance – the true transfer function of the HO (H_p) and the ratio between the test signal and the intensity of the HO's injected noise ($x = |I(\nu_a)|^2/\sigma_N(\nu_a)^2$) should be known. For lack of better the estimated transfer function is usually used. Thanks to the properties of the test signals an estimate for the standard deviation of the injected noise can be obtained from the measured data themselves. The Fourier frequencies neighbouring a frequency with a test signal component only contain the HO's injected noise. Assuming that the level of injected noise at neighbouring frequencies does not differ too much from the noise level at the considered frequency ν_a, one can make an estimate for the standard deviation of the noise from these data, e.g. with the three surrounding frequencies on each side:

$$s_N(\nu_a;\zeta)^2 = \frac{1}{6}\sum_{\substack{j=f_a-3\\ j\neq 0}}^{j=f_a+3}\left|\frac{U(j;\zeta)}{1+H_p(\nu_a)H_c(\nu_a)}\right|^2 \qquad (9)$$

3. MULTI-LOOP HO IDENTIFICATION

3.1 *Introduction*

In many control situations, multiple loops are closed instead of one. For instance, in (Mulder, 1998), the pilot control behaviour with a tunnel-in-the-sky display is studied. For aircraft control, it is assumed that the HO first closes the inner control loops, to control aircraft pitch and roll attitude, and then closes the "outer" loop, for control of flight path and position.

The model of the HO in that case is one of a set of transfer functions in series or in parallel. For a HO model consisting of multiple transfer functions in series, the bias and variance of the estimate can not be determined analytically (Mulder, 1998). We will limit ourselves here to a model in which the two transfer functions that model the HO are placed in parallel to each other, H_{p_y} and H_{p_z} in Figure 2.

To enable the simultaneous identification of these transfer functions, multiple test signals have to be applied, at appropriate places in the control loop. Again a disturbance task will be considered, with test signals i_{d1} and i_{d2} applied at the inputs of the partial systems H_{cz} and H_{cy}, and with the target inputs equal to zero, $i_{t1} = 0$ and $i_{t2} = 0$.

At any of the frequencies at which one of the test signals contains a sine component, an equation with two unknown transfer functions is obtained. For example, considering one frequency at which the test signal i_{d1} contains a sinusoidal component (Figure 2):

$$U_1 = -H_{p_z}Z_1 + -H_{p_y}Y_1 + N_1 \qquad (10)$$

Here U_1, Z_1, and Y_1 denote the Fourier coefficients at frequency ν of the corresponding measured signals. For the solution of the two unknowns, H_{p_z} and H_{p_y}, at least two – independent – equations are needed. A second equation is obtained by taking the Fourier coefficients for u, y and z at the neighbouring frequencies at which the test signal i_{d_2} contains sine components, and interpolating these to the considered frequency, so that a second set of data, $\tilde{U}_2$, $\tilde{Y}_2$ and $\tilde{Z}_2$ is obtained. Thus a set of two equations with the two unknowns, H_{p_x} and H_{p_y} are obtained:

$$\begin{pmatrix} U_1 \\ \tilde{U}_2 \end{pmatrix} = -\begin{pmatrix} Z_1 & Y_1 \\ \tilde{Z}_2 & \tilde{Y}_2 \end{pmatrix}\begin{pmatrix} \hat{H}_{p_z} \\ \hat{H}_{p_y} \end{pmatrix} \qquad (11)$$

Solving this set of equations leads to expressions for the estimated frequency responses, valid for the frequencies at which test signal i_{d_1} contains a sine component:

$$\hat{H}_{p_z} = \frac{U_1\tilde{Y}_2 - Y_1\tilde{U}_2}{\tilde{Z}_2Y_1 - \tilde{Y}_2Z_1} \qquad (12)$$

$$\hat{H}_{p_y} = \frac{Z_1\tilde{U}_2 - U_1\tilde{Z}_2}{\tilde{Z}_2Y_1 - \tilde{Y}_2Z_1} \qquad (13)$$

One can see that, as in the single loop case, the estimation of the transfer functions is relatively straightforward. The determination of bias and variance of the estimates is more complicated. However, following the principle of expressing the terms in the right-hand side of the estimation equation in the test signal (I_1 and $\tilde{I}_2$), the transfer functions of the controlled system (H_{cy}, H_{cz}) and HO (H_{p_y}, H_{p_z}), and in the noise injected by the HO (N_1 and $\tilde{N}_2$), and then substituting these in the estimation equations 12 and 13, one can produce expressions for the estimated describing functions in terms of the true HO transfer functions and the input. For the estimated transfer functions $\hat{H}_{p_y}$ and $\hat{H}_{p_z}$ one then obtains (van Paassen, 1994):

$$\hat{H}_{p_y} = H_{p_y} - H_{p_y}\left(\frac{N_1}{I_1 + N_1}\right) + \frac{N_1\tilde{N}_2}{H_{c_y}\tilde{I}_2\left(I_1 + N_1\right)} - \frac{\tilde{N}_2}{H_{c_y}\tilde{I}_2} \quad (14)$$

$$\hat{H}_{p_z} = H_{p_z} - \left(H_{p_z} + \frac{1}{H_{c_z}}\right)\left(\frac{N_1}{I_1 + N_1}\right) + \frac{N_1\tilde{N}_2}{\tilde{I}_2\left(I_1 + N_1\right)} - \frac{\tilde{N}_2}{\tilde{I}_2} \quad (15)$$

As for the single loop case, these equations form the basis for determining bias and variance of the estimates, similar to Eq. 5 and 7. For example, bias and the variance of the estimate $\hat{H}_{p_z}$ can be expressed as:

$$\begin{aligned} \text{bias}\left(\hat{H}_{p_z}\right) &= E\left\{\hat{H}_{p_z} - H_{p_z}\right\} \\ &= -E\left\{\frac{N_1\tilde{N}_2}{\tilde{I}_2\left(I_1 + N_1\right)}\right\} \\ &- E\left\{\frac{N_1\left(H_{p_z}H_{c_z} + 1\right)}{H_{c_z}\left(I_1 + N_1\right)}\right\} + E\left\{\frac{\tilde{N}_2}{\tilde{I}_2}\right\} \end{aligned} \quad (16)$$

$$\begin{aligned} \text{var}\left(\hat{H}_{p_z}\right) &= E\left\{\left(\hat{H}_{p_z} - E\left\{\hat{H}_{p_z}\right\}\right)^2\right\} \\ &= E\Big\{\frac{2N_1^2\tilde{N}_2^2\left(\hat{H}_{p_z}H_{c_z} + 1\right)}{H_{c_z}\tilde{I}_2\left(I_1 + N_1\right)^2} + \frac{N_1^2\tilde{N}_2^2}{\tilde{I}_2^2\left(I_1 + N_1\right)^2} \\ &+ \frac{N_1^2\left(\hat{H}_{p_z}H_{c_z} + 1\right)^2}{H_{c_z}^2\left(I_1 + N_1\right)^2} - \frac{2N_1\tilde{N}_2\left(\hat{H}_{p_z}H_{c_z} + 1\right)}{H_{c_z}\tilde{I}_2\left(I_1 + N_1\right)} \\ &- \frac{2N_1\tilde{N}_2^2}{\tilde{I}_2^2\left(I_1 + N_1\right)} + \frac{\tilde{N}_2^2}{\tilde{I}_2^2}\Big\} - \text{bias}\left\{\hat{H}_{p_z}\right\}^2 \end{aligned} \quad (17)$$

The fourth and fifth term in Eq. 17 vanish since the means of N_1 and $\tilde{N}_2$ are assumed to be zero, and N_1 and $\tilde{N}_2$ are uncorrelated. By applying the formulas from Eq. 8 and 6, the contributions of the remaining terms can be calculated.

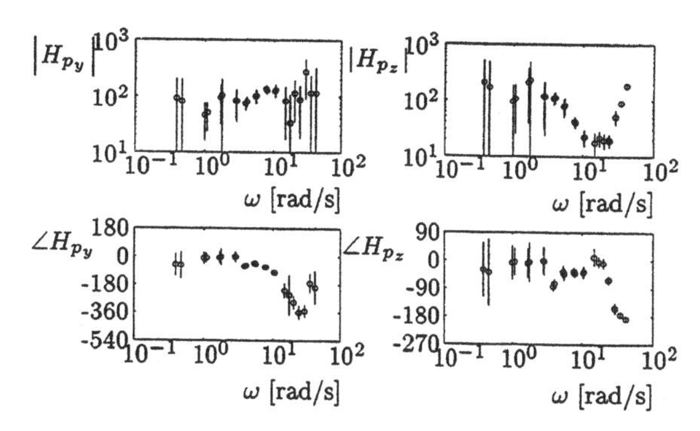

Fig. 3. Identification results of the HO's response to display error and to stick position. From (van Paassen, 1994).

4. EXAMPLE RESULTS

This method has been applied to the identification of HO control behaviour in a multi-loop control task (Mulder, 1998). Up to 3 describing functions could be identified within one experiment, producing reasonable values for the variance and bias of the estimated transfer functions.

In another study a closer look has been taken at the HO. In this study a side stick driven by a hydraulic motor has been used as the manipulator. The motor drive allowed the generation of an additional force test input, i_{d_1} in Figure 2. In this manner the HO's neuromuscular system can be tested. The "inner" loop, formed by H_{p_z} and H_{c_z}, represented the combination of the HO's neuromuscular system and the stick. Note that in conventional set-ups, the identified describing function (such as H_p from Figure 1) represents the dynamics of the HO control behaviour, of his perceptory system, of his neuromuscular system and of the manipulator used in the experiment, all lumped together.

The two describing functions H_{p_z} and H_{p_y} represent the proprioceptive feedback loop closed by the HO, respectively the visual feedback loop. An example result of the identification is given in Figure 3. The vertical bars on the symbols depict the standard deviation of the estimate.

The two rightmost plots give the proprioceptive feedback transfer function, H_{p_z}, i.e. the moment exerted on the stick in response to the stick position. It can be seen that the gain, and thus the stiffness of the HO's arm, is fairly high at lower frequencies. From other experiments (van Paassen, 1994) and literature (Winters and Stark, 1985) it is known that the mechanical stiffness of the arm is fairly low, the high stiffness at low frequencies must therefore have been generated by feedback loops in the neuromuscular system itself.

To create a representative control situation, a linear aircraft roll equation was used for the controlled system H_{c_y}. The moment exerted by the HO in response to the presentation on the display (the visual channel) is given in the two plots on the left side of Figure 3.

5. CONCLUSIONS

The method is an efficient means of estimating HO control behaviour. In the past it has been used as an instrument for studying HO manual control in continuous control tasks. It is now available as an instrument for testing experimental conditions in which one might expect a change in the control behaviour of the HO. With the advancement of modern control laboratory equipment, the troubles that plagued early practitioners no longer apply. It is now possible to design and perform an experiment, as well as process the data, on a PC with a good data aquisition card.

This paper gives a summary of the key elements of the technique. We try to warn for some of the more common pitfalls, and give a recipe for setting up a experiment and choosing the test signals. For a useful application of the method, the statistical validity of the results must be known. An overview is given of the determination of bias and variance of the estimate, both for the single loop case, as well as for the double loop case. The approach followed can be extended to a HO model with more parallel loops.

Many control task concern the simultaneous closure of multiple loops by the HO. It was already described how to apply the method in these cases. Here we present the determination of the bias and variance of the estimate for a certain class of multi-loop experiments.

6. REFERENCES

Flach, J. M. (1991). Control with an Eye for Perception: Precursors to an Active Psychophysics. In: *Visually Guided Control of Movement* (W. W. Johnson and M. K. Kaiser, Eds.). pp. 121–149. Number NASA CP-3118. NASA Ames Research Center.

Krendel, E. S. and D. T. McRuer (1960). A Servomechanics Approach to Skill Development. *Journal of the Franklin Institute* **269**(1), 24–42.

McRuer, D. T. and H. R. Jex (1967). A review of quasi-linear pilot models. *IEEE Transactions on Human Factors in Electronics* **8**, 231–249.

McRuer, D. T., D. Graham, E. S. Krendel and W. Reisener Jr. (1965). Human Pilot Dynamics in Compensatory Systems. Theory, Models, and Experiments with Controlled Element and Forcing Function Variations. Technical Report AFFDL-TR-65-15. Air Force Flight Dynamics Laboratory. Wright-Patterson AFB (OH).

Mulder, M. (1998). Cybernetics of a Tunnel-in-the-Sky Display. Ph.D dissertation. Faculty of Aerospace Engineering, Delft University of Technology.

Stapleford, R. L., D. T. McRuer and R. Magdaleno (1967). Pilot Describing Function Measurements in a Multiloop Task. *IEEE Transactions on Human Factors in Electronics* **HFE-8**(2), 113–125.

Stapleford, R. L., S. J. Craig and J. A. Tennant (1969). Measurement of Pilot Describing Functions in Single-Controller Multiloop Tasks. NASA Contractor Report NASA CR-1238. Washington (D.C.).

Teper, G. L. (1972). An Effective Technique for Extracting Pilot Model Parameter Values from Multi-feedback, Single-input Tracking Tasks. *Proceedings of the Eighth Annual Conference on Manual Control* **Report AFFDL-TR-72-92**, 23–33.

van Lunteren, A. (1979). Identification of Human Operator Describing Function Models with One or Two Inputs in Closed Loop Systems. PhD thesis. Faculty of Marine Technology and Mechanical Engineering, Delft University of Technology.

van Paassen, M. M. (1994). Biophysics in Aircraft Control - a Model of the Neuromuscular System of the Pilot's Arm. PhD thesis. Faculty of Aerospace Engineering, Delft University of Technology.

Weir, D. H. and D. T. McRuer (1972). Pilot Dynamics for Instrument Approach Tasks: Full Panel Multiloop and Flight Director Operations. NASA Contractor Report NASA CR-2019. Washington (D.C.).

Weir, D. H., R. K. Heffley and R.F. Ringland (1972). Simulation Investigation of Driver/Vehicle Performance in a Highway Gust Environment. *Proceedings of the Eighth Annual Conference on Manual Control* **Report AFFDL-TR-72-92**, 449–465.

Winters, J. M. and L. W. Stark (1985). Analysis of fundamental human movement patterns through the use of in-depth antagonistic muscle models. *IEEE Transactions on Biomedical engineering.*

MODELLING HUMAN CONTROL BEHAVIOUR WITH CONTEXT-DEPENDENT MARKOV-SWITCHING MULTIPLE MODELS

Roderick Murray-Smith

Dept. of Mathematical Modelling, Technical University of Denmark, DK-2800 Lyngby, Denmark. rod@imm.dtu.dk

Abstract: A probabilistic model of human control behaviour is described. It assumes that human behaviour can be represented by switching among a number of relatively simple behaviours. The model structure is closely related to the Hidden Markov Models (HMMs) commonly used for speech recognition. An HMM with context-dependent transition functions switching between linear control laws is identified from experimental data. The applicability of the approach is demonstrated in a pitch control task for a simplified helicopter model. *Copyright ©1998 IFAC*

Keywords: Multiple model systems, Hidden Markov Models, human–machine interface, learning algorithms, Human–machine interaction, flight control, simulation.

1. INTRODUCTION

If the intentions, goals and preferences, (the human 'state') and the accompanying skills of the human operator were known, the human–machine interaction problem would be to coordinate, adapt, and configure the automatic control system to ensure satisfactory performance of the full human–machine control system. Unfortunately, the states of the human are usually not known – at best they can be estimated. This paper describes an approach to the simultaneous estimation of human states and behaviour models – recognising operators' goals and 'modes' of behaviour from their actions.

The goal of this work is to develop approaches which could be used to estimate and predict operator skills, such that we would be able to learn individual preferences and expectations, and detect characteristic features and types of operators. In systems design, actual performance, workload and performance limitations for a given task could be better understood before construction of a prototype. Because of the complexity of human behaviour, and the richness of sensing and state, no conceivable model will be able to predict exactly what the human will do. In this paper we will use a probabilistic framework for the representation of human control behaviour, as this provides a common framework for describing the uncertainty in both the human and technical aspects of our system and allows us to develop models which for the given task behave statistically as a human would.

The need for human control models in systems design has long been known, but it was often impossible to identify and represent the complexity of human behaviour in a particular task at a reasonable cost. Improvements in computing power and learning algorithms have now made it feasible to implement complex operator models that can learn and represent high-level aspects of behaviour such as tasks and goals, as well as being able to discriminate between different human operators and various levels of operator performance and preferences.

Most classical approaches to modelling human manual control behaviour (quasi-linear, optimal control and sampled data models) are mainly applicable to low-level manual control tasks involving skilled operators. More complex tasks, higher-level information processing and inexperienced operators are typically not covered by such models. The more flexible model

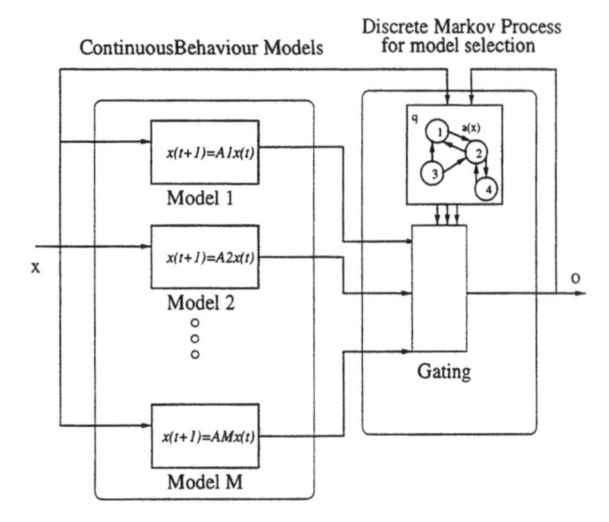

Fig. 1. The model structure – a discrete Markov process switches between the continuous state processes. The discrete process transitions are dependnet on the continuous state.

described in this paper assumes the operator is in one of a finite number of human 'states'. Each of these hypotheses has an associated behaviour which can be described in terms of a probability model. The learning algorithms used allow us to identify both the parameters of the individual behaviour models, and the switching functions simultaneously. We thus have a standard probabilistic framework for the interpretation of a time series of human action.

1.1 *Multiple-model representations of human control behaviour*

The multiple-model interpretation of human control action, instead of having a single complex model, prefers to describe control action by switching between a number of simple behaviours (see Johansen and Murray-Smith (1997) for a review of the multiple model approach to modelling). In experiments, it becomes clear that human control action often goes through rapid changes of behaviour, due to, for example, changes in the human's perception of the situation, goal changes, attention lapse, or change in the effective dynamics of the controlled system. [1]

This paper examines a model (as shown in Figure 1), which switches between a number of linear controllers. The transitions between models are instant, and do not involve blending of behaviours. The model switching is probabilistic but conditioned on the state/input vector, so it supports a spectrum of models from purely stochastic switching to purely deterministic switching, depending on its parameterisation. See Meila and Jordan (1997) and Bengio and Frasconi (1996) for further details. This gating or switching element can be viewed as a pattern recognition system which chooses the next model state given the 'pattern' of the current continuous state-vector. Its probabilistic nature takes into account both variations in human behaviour and measurement errors. It uses many of the tools common to speech recognition technology, and much of the framework is taken from the excellent review article by Rabiner (1989).

Related approaches to modelling human actions exist, but usually use discrete actions, not the mixture of continuous and discontinuous control used in this paper, and constant transition probabilities as opposed to the state-dependent transition functions used here. Yang et al. (1994) applied HMM's to learn human skills, and continued the work in Nechyba and Xu (1998). Pentland and Liu (1995) outline possible applications of HMM's to modelling driver control behaviour and prediction of immediate intentions.

2. MODEL STRUCTURE, INFERENCE AND LEARNING ALGORITHM

2.1 *Model structure*

We have a model $\mathcal{M} = \{A_i(x; w_i), f_i(x; \theta_i), \Sigma_i, \pi\}$, $i = 1, .., N_m$, with an observable continuous state x and a hidden discrete state q, which can be in one of N_m states. $A_i(x; w_i)$ is the state-transition matrix from discrete state q, dependent on continuous state x, the entries of which $a_{ij} = P(q_{t+1} = j | q_t = 1)$ are the transition probabilities at time t. This is effectively a pattern recognition system, mapping regions of the state-space to a transition probability distribution. In this paper, the transition probabilities are represented by a multinomial logit (or 'softmax') function (equation (6)), with parameters w.

The N_m submodels $b(o_t, i)$ define the emission probabilities – i.e. the probability of observing o_t given discrete state $q_t = i$, and continuous state x. In the continuous control action case examined here, one could use a mixture model density function, but in this paper we will only use a single component in each mixture, i.e. a linear model with Gaussian noise (mean $\mu_i = f_i(x; \theta_i) = \theta_i x$, variance Σ_i^2). The estimate of the initial discrete state distribution $\pi = \{\pi_i\}$, where $\pi_i = P(q_1 = i)$.

2.2 *Inference*

As reviewed in Rabiner (1989), there are several inference problems:

(1) *Evaluation:* Probability of model $\mathcal{M}$ given observed output time series $O = \{o_1, o_2, ..., o_T\}$?

[1] These ideas are not new; (Sheridan and Ferrell, 1974, Chapter 15) gives an interesting review of the reasons for using intermittent representations, but points out the disadvantage that at that time, such models were created by tedious 'cut and try' methods – now, however, available algorithms relieve the human of extensive parameter tuning.

(2) *Decoding:* Probability of hidden state i at time t given observed time series O?
(3) *Estimation:* What are the parameters most likely to have generated the output time series?

2.2.1. *Evaluation* To evaluate how well the model $\mathcal{M}$ matches the time series O we need to evaluate $P(O|\mathcal{M})$, which involves finding the probability of all possible paths through the hidden states, $P(O|\mathcal{M}) = \sum_{allQ} P(O|Q)P(Q|\mathcal{M})$, but this quickly becomes intractable, so to perform the calculations efficiently we use the standard Baum-Welch forward-backward procedure. Define $\alpha_t(i) = P(O_{1,t}, q_t = i|\mathcal{M})$ as the probability of a partial sequence, which allows us to recursively generate the probability of the whole sequence,

$$\alpha_1(i) = \pi_i b(u_1, i),$$
$$\alpha_{t+1}(j) = b(o_{t+1}, j)\sum_{i=1}^{N_m} \alpha_t(i)a_{ij},$$

giving us the probability of the whole sequence O, $P(O|\mathcal{M}) = \sum_{i=1}^{N_m} \alpha_T(i)$. There is an accompanying backward phase, which will be useful for the decoding step in the next section. Here we define $\beta_t(i) = P(O_{t+1,T}|q_T = i, \mathcal{M})$ and we now have a backward recursion:

$$\beta_T(i) = 1, \beta_t(i) = \sum_{j=1}^{N_m} \beta_{t+1}(i)b(o_{t+1}, j)a_{ij},$$

such that $P(O|\mathcal{M}) = \sum_{i=1}^{N_m} \pi_i b(o_1, i)\beta_1(i)$.[2]

2.2.2. *Decoding – which state are we in?* We wish to find out at each point in time the probability of the various behaviour hypotheses.[3] In other words, estimate $\gamma_t(i) = P(q_t = i|O, \mathcal{M})$. This can be expressed in terms of the forward-backward variables found in the previous section:

$$\gamma_t(i) = \frac{\alpha_t(i)\beta_t(i)}{\sum_{j=1}^{N_m} \alpha_t(j)\beta_t(j)}, \quad (1)$$

which is a probability measure which sums to one over all behaviours.

2.2.3. *Estimation* Given the ability to evaluate model quality and algorithms for decoding the hidden states, we can move on to the more difficult problem of parameter estimation. This problem cannot be solved analytically and there is no easy way of finding the optimal global solution, but we shall use the standard EM approach to local maximisation of $P(O|\mathcal{M})$.

We define $\xi_t(i, j)$ to be the probability of switching from state i at time t to state j at time $t + 1$,

$$\xi_t(i, j) = P(q_t = i, q_{t+1} = j|O, \mathcal{M}), \quad (2)$$

which can again use the results from the forward-backward stage, such that

$$\xi_t(i, j) = \frac{\alpha_t(i)a(x)_{ij}b_j(o_{t+1})\beta_{t+1}(j)}{P(O|\mathcal{M})}. \quad (3)$$

By summing $\gamma_t(i)$ for $t_{1,T}$ we have the expected number of samples where behaviour i was active, and similarly by summing $\xi_t(i, j)$ for $t_{1,T}$ we have the expected number of transitions from behaviour i to behaviour j. This leads us to reestimation formulae for the parameters to maximise likelihood:

Local Models The estimation stage for the local model parameters reduces to weighted linear optimisation, where the weighting function for each data point is provided by $\gamma_t(i)$, the probability of model i generating o_t. In the case of a single linear model,

$$\mu_i = \arg\min_{\theta_i} \sum_{t=1}^{T} \gamma_t(i)||o_t - f_i(x, \theta_i)||^2 \quad (4)$$

$$\Sigma_i^2 = \frac{\sum_{t=1}^{T} \gamma_i||o_t - f_i(x, \theta_i)||^2}{n_o \sum_{t=1}^{t=T} \gamma_t(i)}, \quad (5)$$

where n_o is the dimension of the observation vector o_t.

Transition functions As $\xi_t(i, j)$ represents the current estimate of the probability of changing from state i to j at time t, we use this as a target for the transition function (normalised by the probability of being in state i at time t). In this example we use a simple 'softmax' representation of the transition function,

$$a_{ij}(x) = \frac{\exp(w_{ij}x)}{\sum_k \exp(w_{ik}x)}, \quad (6)$$

but the same approach is valid for more complex networks or other representations, such as belief networks (see Jensen (1996) for background).[4]

To maximise the likelihood of the model the parameters of the transition functions w are adapted in N_m independent optimisation problems using a conjugate-gradient algorithm to bring $a_{ij}(x, w) \approx \frac{\xi_t(i,j)}{\gamma_t(j)}$. As we cannot guarantee that global optima are found, we are effectively performing a Generalised EM step. See Meila and Jordan (1997) for further discussion.

[2] To avoid implementation problems, we normalise the α's and β's such that $\alpha_t(j) = \frac{1}{N_t}\alpha_t(j)$, where the normalising constant $N_t = \sum_{i=1}^{N_m} \alpha_t(i)$. The β's are also scaled using the same N_t, $\beta_t(j) = \frac{1}{N_t}\beta_t(j)$.

[3] If we knew this, our problems would be trivial – we are thus viewing these variables as 'missing data' which have to be estimated.

[4] Note that in real applications we would often use different x state-vectors for transitions from different states, and certain transitions could be excluded from the model structure in advance. The x for transition functions may also be transformed in some way.

2.3 *Role of inference in modelling human behaviour*

The algorithms for inference described above have very concrete uses in a human modelling application. We take the continuous state x to be the input/state information the human bases his or her control on. That control action u is the observable sequence O referred to in the previous section.

2.3.1. *Evaluation for Classification* There are many applications where we would wish to classify a series of human control actions. The class chosen could be to estimate which of a number of known individual users performed them (possibly for security or insurance purposes), or to compare the behaviour to a number of *types* of user (e.g. beginner, average user, expert, tired performance). This could be useful in training operators in simulated environments, when classifiers which quantify the style of behaviour could be used to guide and document the results of a training programme. A further example is to differentiate between types of behaviour of a given human operator (e.g. normal behaviour, tired or inattentive behaviour, aggressive behaviour). The approach used is to collect data O_i for each class of control action, and estimate accompanying models $\mathcal{M}_i$. We then select the model with the maximum $P(O|\mathcal{M}_i)$. This is not an explicitly discriminative approach, and if classification is the ultimate aim of modelling, it may be worth using discriminative approaches.

2.3.2. *Decoding for segmentation of the time-series* The use of standard inference with the model, and EM to iteratively optimise the parameters, automatically gives us a segmentation of the human control time-series into sub-behaviours. The γ_t's provide us with the probability that the human was in the given state at time t. This is an attempt to infer human 'intentions' or 'sub-goals'. This information can then be used to improve the interaction in a human–machine control system, and can provide context information to human–machine interfaces. Multiple-Model Adaptive Control (MMAC) is an analogue method used in control applications, e.g. Schott and Bequette (1997).

2.3.3. *Estimation for modelling* Given the segmentation of the data provided by the γ_t's, the estimation stage provides us with local models corresponding to system behaviour in each hypothesis. Again, this information, with the γ's can be used to improve co-operation in a human–machine system – we have an estimate of the human's 'hidden state', which can be viewed as current intentions or a current mode of behaviour. We also have how the human usually behaves in this state – the local model associated with that state.

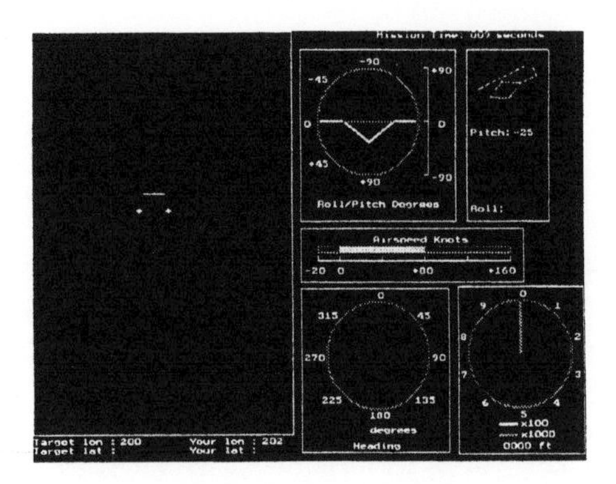

Fig. 2. The screen display used in the experiment. The human operator has to track the reference horizontal line with his/her own pitch indicator (double line)

3. PITCH CONTROL EXAMPLE

The modelling task used to illustrate the approach is that of pitch control in a simplified helicopter model, the 'flying brick' (see Bradley (1996)), which is basically a point mass steered by a force acting at a distance – a crude representation of a rotor disc. No other aerodynamic forces are included. This paper used the model in pitch control mode, where roll and yaw are always zero. In this experiment no attempt was made to control velocities, or position – only pitch was important. The relevant equations of motion for the pitch angle θ, given a control input u are thus:

$$\dot{\theta} = \frac{-l_h mg(1+\theta)\sin(u)}{I_{xx}},$$

where $l_h = 1.454\text{m}$, $m = 4313\text{kg}$, $g = 9.81\text{ms}^{-2}$, $I_{xx} = 2767\text{kgm}^2$.

3.1 *Experiment design*

The task was to track a changing pitch reference value, as indicated on screen (see Figure 2). The system was implemented on a standard PC and monitor. The sampling time was 0.05s which was then resampled at 0.1s for use in modelling. For actuation we used a centre-sprung games joystick with an 8 bit accuracy (CH products flightstick pro).

As in any empirical modelling task it is important to provide sufficient excitation in the experiment to be able to identify the parameters of interest, and to avoid numerical problems. The target trajectory used included large occasional random step changes, followed by frequent small step changes,and occasional ramp-like changes, as shown in Figure 4. This allows us to study the reaction to step changes of different sizes, and tracking a moving target. Note that these have no 'preview' effect, i.e. the human does not know what the reference signal is about to do, and cannot use feedforward control.

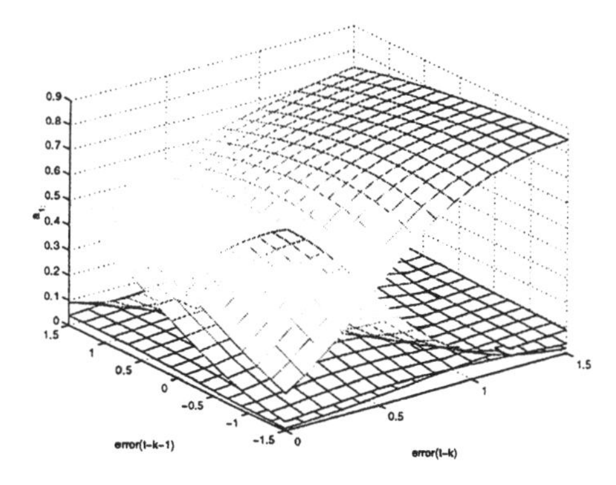

Fig. 3. A state-dependent transition function from a single state to 4 other states. Each curve indicates the probability of transition from this state for varying x. At any x the curves sum to one.

3.2 *Modelling results*

We identified simple multiple-model systems, with 3 models where each model was a second order discrete-time controller, with a pure k-delay element $u_t = \theta_i[e_{t-k}e_{t-k-1}u_{t-1}u_{t-2}]$. The transition functions were softmax functions scheduled on k-step delayed values of $\dot{e}$ and $||e||$ (e.g. see Figure 3).

3.2.1. *Segmentation of time series* An immediate question is whether we can recognise a 'sensible' segmentation of the time series from Figure 4 by plotting the γ's, as in Figure 5. Initially, $\gamma_t(i) \approx 1/N_m$, but as learning progresses we see the segmentation improve, and find a correlation between larger errors and probability of different hypotheses. Note that successful decoding does not necessarily mean that the model has learned the transition functions adequately.

3.2.2. *Parameter estimates* The parameters should look 'plausible' given available prior knowledge of the problem. This is relevant for both the local models and the transition functions. We could clearly see 'surge-like' behaviour (as discussed in Sheridan and Ferrell (1974)) with low gain models around small error regimes and high gain parameters in high error regimes. In some runs, one of the models would occasionally take on negative gains, corresponding to moments when the human went the 'wrong' way after a large change in reference value. In this example, the transition functions from each of the models were quite similar.

3.2.3. *Simulation results* Simulation of model behaviour in a closed-loop with the controlled system is probably the most interesting test of behaviour. We would hope to find typical features observed in the human reproduced in the simulation. Figure 6 shows such a test.

3.2.4. *Classification results* The classification experiment, involved 4 models trained on 4 different humans (all male researchers, new to the task), with 2 runs of 90s each, and tested on 3 new runs of 90s. The model with the largest $P(O|\mathcal{M})$ was selected for each run. Classification was 100% accurate on training runs, and for classification on new data 10 from 12 runs were classified correctly.

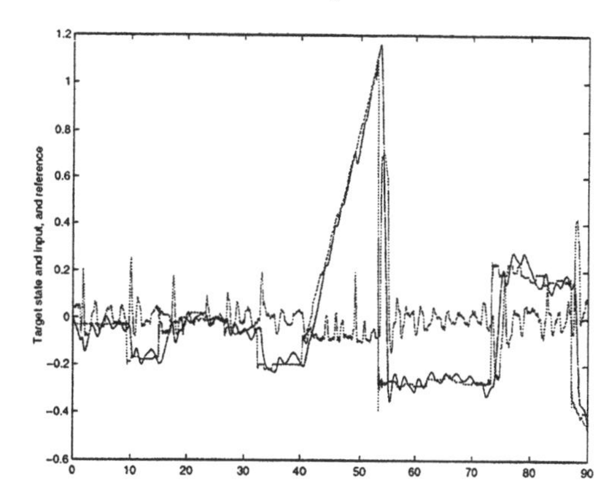

Fig. 4. A typical time-series recording from an experiment. The human control input indicates strongly that some form of intermittent model is needed.

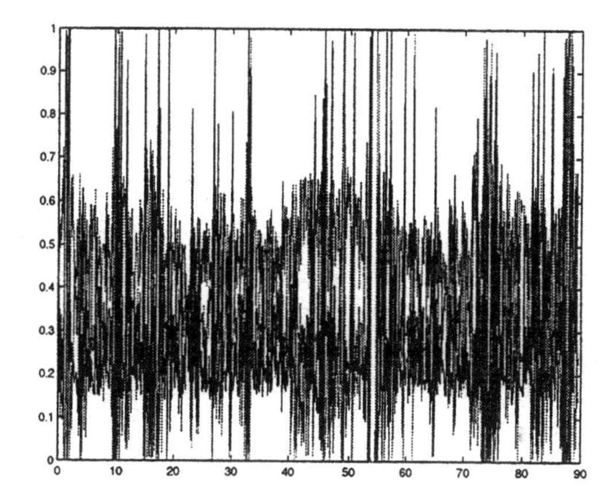

Fig. 5. The γ's segment the time-series – here we see some models are a more likely explanation of the data given large errors, while others compete in lower error regimes.

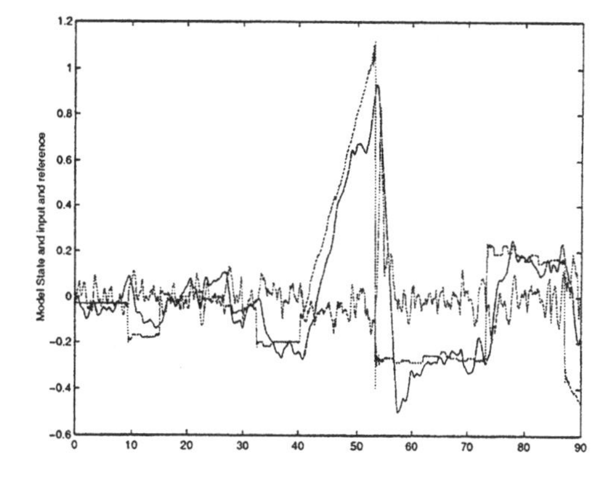

Fig. 6. The model's behaviour on the time-series shown in Figure 4. Note that the model should not be exactly the same as the human, but should be qualititatively similar. This run shows mismatch in the noise dynamics which a longer-tailed distribution would fit better.

3.3 *Relating model structure to a given application*

This example was not particularly complex, but selecting the granularity (how detailed the individual behaviour models should be) of the model is often a far from a trivial matter. In principle the engineer will examine the task, estimate the number of behaviours believed to be applicable, and the transition probabilities between them (if we can rule out certain transitions from the start, the learning task is eased significantly). [5] There are a number of sources of uncertainty in this model: Individual behaviours will vary, and the transitions between behaviours will also vary. We are trying to absorb much of the complexity of the model into the switching component. The context-dependent transition functions potentially provide us with arbitrarily powerful representations of transition uncertainty, conditioned on the measurable states.

4. CONCLUSIONS

The multiple model framework was introduced as a potentially powerful approach to modelling human control behaviour. The framework identifies a model, and estimates the current human 'state', and can be used to better coordinate human and machine control behaviour. It was illustrated in a simple pitch tracking task. The methods used were able to identify representative and meaningful models from experimental data, and were able to classify which human generated a given experimental behaviour, in a manner similar to speech recognition systems. For simplicity, these experiments examined a low-level manual control task, but we believe that the approach has more relative potential as a model of higher-level control behaviours and multivariable problems.

The models created are *generative* models. Even if we only wish to produce classifiers which recognise a given behaviour or human state, the use of generative models tends to make the approach less susceptible to minor changes in system configuration than *feature-based* classification methods would be, as well as allowing a more principled approach to the engineering task.

The modular nature of the approach means that the basic model can be extended incrementally to improve sub-models representing given behaviours, or to provide a more sophisticated behaviour switching logic. Bayesian networks could, for example, be used as a representation of the transition probabilities, or for individual behaviours. In fact, we could use a range of different approaches (e.g. fuzzy, classical control, Bayesian networks, neural networks) in a single model, given that they can be interpreted as providing a probabilistic output. We thus have the potential to integrate the uncertainty related to 'hard' engineering aspects with those of the 'soft' human aspects within a single framework.

[5] Note that the behaviours chosen need not be purely 'sensible' control actions, but can also include 'noise' behaviours which can be switched in rapidly, as well as 'human error' behaviours which are characteristic behaviours, but do not fulfill the human's stated objectives.

ACKNOWLEDGEMENTS

The author gratefully acknowledges the support of Marie Curie TMR grant FMBICT961369. Daimler-Benz research supported earlier stages of this work, and Marina Meila and Mike Jordan at M.I.T. introduced me to many of the techniques. The simulator was developed by David Murray-Smith and Graham Dudgeon at Glasgow University.

References

Y. Bengio and P. Frasconi. Input–Output HMM's for sequence processing. *IEEE Transactions on Neural Networks*, 7(5):1231–1248, 1996.

Roy Bradley. The flying brick exposed: Non-linear control of a basic helicopter model. Technical report, Dept. of Mathematics, Glasgow Caledonian University, 1996.

Finn V. Jensen. *An introduction to Bayesian Networks*. UCL Press, London, 1996.

T. A. Johansen and R. Murray-Smith. The operating regime approach to nonlinear modelling and control. In R. Murray-Smith and T. A. Johansen (Eds.), editors, *Multiple Model Approaches to Modelling and Control*, chapter 1, pages 3–72. Taylor and Francis, London, 1997.

M. Meila and M. I. Jordan. Markov mixtures of experts. In R. Murray-Smith and T. A. Johansen, editors, *Multiple Model Approaches to Modelling and Control*, chapter 5, pages 145–166. Taylor and Francis, London, 1997.

Michael Nechyba and Yangsheng Xu. On discontinuous human control strategies. Submitted to Proc. IEEE Int. Conf. on Robotics and Automation, 1998.

Alex Pentland and Andrew Liu. Toward augmented control systems. In *Intelligent Vehicles Symposium*, Detroit, 1995.

R. L. Rabiner. A tutorial on Hidden Markov models and selected applications in speech recognition. *Proceedings of IEEE*, 77(2):257–286, 1989.

K. D. Schott and B. W. Bequette. Multiple model adaptive control. In R. Murray-Smith and T. A. Johansen, editors, *Multiple Model Approaches to Modelling and Control*, chapter 11, pages 269–292. Taylor and Francis, London, 1997.

T. B. Sheridan and W. R. Ferrell. *Man-Machine Systems*. MIT Press, 1974.

Jie Yang, Yangsheng Xu, and Chiou S. Chen. Hidden Markov Model approach to skill learning and its application to telerobotics. *IEEE Transactions on Robotics and Automation*, 10(5), 1994.

PLANT OPERATOR MODEL FOR HUMAN INTERFACE EVALUATION

Teiji Kitajima*[1] Yoshihiko Nakayama*[2] and Hirokazu Nishitani

Nara Institute of Science and Technology,
8916-5 Takayama, Ikoma 630-0101, Japan
*(Present Address: *1 Toyohashi University of Technology; *2 Denso Corporation)*

Abstract: A simulation environment of plant operation, which consists of an operator model, a human-machine interface (HMI) model and a plant model was developd. When the operator model recognizes an abnormal situation, the model generates a countermeasure program in its short-term memory based on both the plant state information acquired from the plant model through the HMI model and the knowledge bases installed into the operator model. A comparison between the operator model's behavior and the human operator's behavior under the same malfunction was used to extract the characteristics of the human operator and to obtain information on human errors. *Copyright ©1998 IFAC*

Key Words: Plants, Chemical Industry, Human-machine interface, Human error, Human factors

1. INTRODUCTION

The usability testing of the plant control system is effective to discover human interface problems in the supervisory control scheme (Nishitani *et al.*, 1996, 1997a, 1997b). From these test results, improvements for many subsystems were proposed such as operational panels, alarm systems, training systems, operator support systems, personnel organization and so on, Each improvement should be examined experimentally. However, direct comparison by a human subject between with and without proposed improvements is difficult, because of the human ability to familiarize himself. For this reason, we developed an artificial operator model, which can supervise a boiler simulator under specified situations. We aimed at an operator model handling eight malfunctions, which are used in the basic course of operator training. The operator model was validated by examining its behavior under each malfunction. The human operator's behavior was also examined under the same malfunction to identify the characteristics of human errors.

2. SIMULATION ENVIRONMENT FOR PLANT OPERATION

Figure 1 illustrates the simulation environment of plant operation. The upper figure shows the actual plant operation environment. The operator supervises the objective plant through the operator console. The lower figure is the developed simulation environment. The plant is replaced by a plant model. For the plant model, a training simulator is used. The operator is replaced by the operator model, which is a human substitute. The operator console is replaced by the human-machine interface (HMI) model, where process data in the plant model is converted to the information for the console display. The operator model judges the plant state based on the information from the HMI model and the knowledge bases (KB) installed in the operator model. These models were installed on workstations connected via a local area network (LAN). The operator model and the HMI model were implemented with a tool for real-time applications, G2.

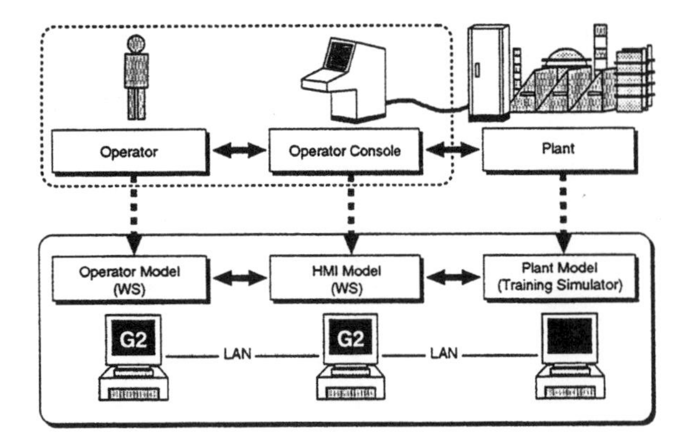

Fig. 1. Simulation environment of plant operation

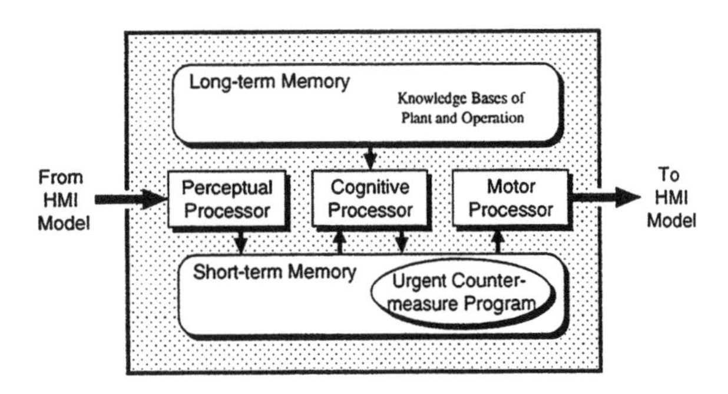

Fig. 2. Structure of operator model

3. OPERATOR MODEL

There have been some human models. For example, a human model processor (HMP) was proposed by Card *et al.* (1983) and an operator model in the nuclear power plant was developed by Takano *et al.* (1994). With reference to these human models, we developed a simplified operator model shown in Fig. 2. The model is composed of three processors and two memories.

The perceptual processor simulates the operator's awareness of panel information, i.e., whether the operator notices panel information. The cognitive processor simulates the operator's recognition of the current plant state. At the moment of alarm, an urgent countermeasure program is created based on the knowledge bases installed in the long-term memory. The created program is regarded as a temporal sequential program. The motor processor actuates the action intention given by the program. The short-term memory is a working memory that shares information with three processors. The urgent countermeasure program is also stored in this memory. The long-term memory includes knowledge bases of plant behavior and countermeasures.

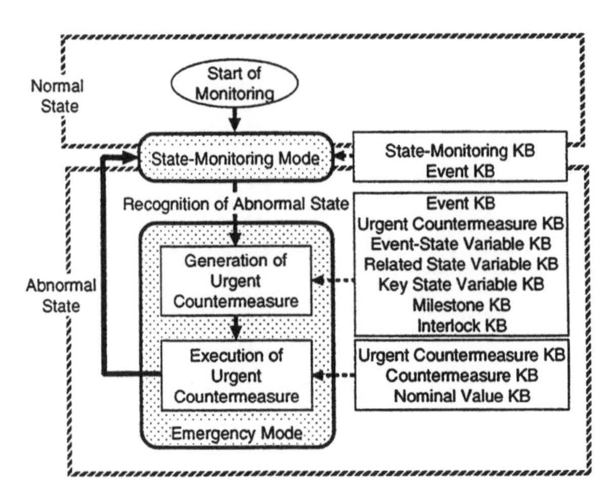

Fig. 3. Modes of cognitive processor

3.1. *Cognitive Processor*

The cognitive processor has two modes, a state-monitoring mode and an emergency mode as shown in Fig. 3. There are two knowledge bases for the state monitoring mode and eight knowledge bases for the emergency mode. Under normal plant state, the cognitive processor continues state monitoring. The cognitive processor monitors the plant state with information acquired by the perceptual processor and rules in knowledge bases in the long-term memory. Once an abnormal state is recognized, the cognitive processor switches on the emergency mode. In the emergency mode, an urgent countermeasure program is generated instantaneously and then the program is executed.

3.2. *Urgent Countermeasure Program*

Figure 4 illustrates an example of an urgent countermeasure program created with an event as a start, which suggests an abnormal plant state. This program is generated under the following assumptions:

1. The operator will predict the future plant state transition.
2. The operator will remember the corresponding state variable related to an event.
3. The operator will grasp the situation and make a plan based on key state variable information in an emergency.
4. The operator will correspond to the abnormal situation in the discrete event mode.
5. The operator will continue operations, if possible.

In the countermeasure program of Fig. 4, four basic components are included to represent a sequence of predictable future events. The event, state variable, milestone and interlock are indicated with a circle, a rectangle, a black point and a cross, respectively. In executing the program,

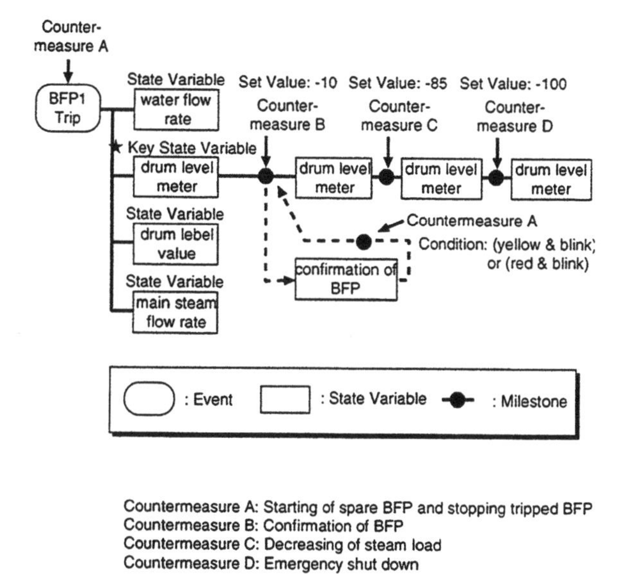

Fig. 4. Urgent countermeasure program generated by operator model

the countermeasure below the basic component is activated by the corresponding event. This program represented by the flow graph is generated by looking at the main issue in an emergency. The program is represented by the predictable event sequence with actions of the countermeasures. This program is effective until the plant state returns to the normal state or until it is terminated.

The generation procedure of an urgent countermeasure program is shown in Fig. 5. At step 1, the starting event is fixed. At step 2, a state variable related to the event is determined. At step 3, all state variables related to the state variable are enumerated. At step 4, a key state variable is selected. At step 5, milestones are determined for the key state variable. At step 6, the existence of interlock is examined. At step 7, the sequence of following events from interlock activation is added. At each step of program generation, appropriate knowledge bases in the long-term memory are referred to.

3.3. *Motor processor*

The motor processor has three functions. The first function is to send the intentions created by the program in the cognitive processor to the HMI model. There are three intentions:

set: the manipulation of panel elements.
get: the data acquisition of panel elements.
see: the monitoring of the plant state.

The second function of the motor processor is to control the operator model gaze point. For set intention, the gaze point is the window for the

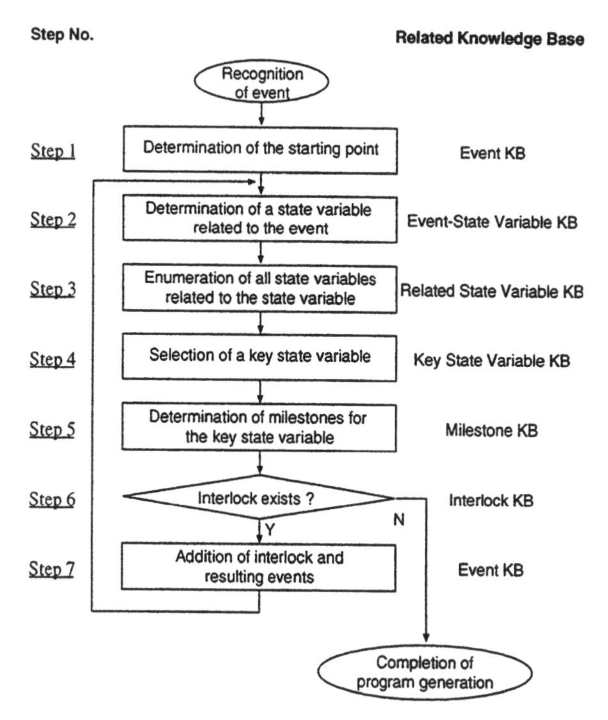

Fig. 5. Generation of urgent countermeasure program

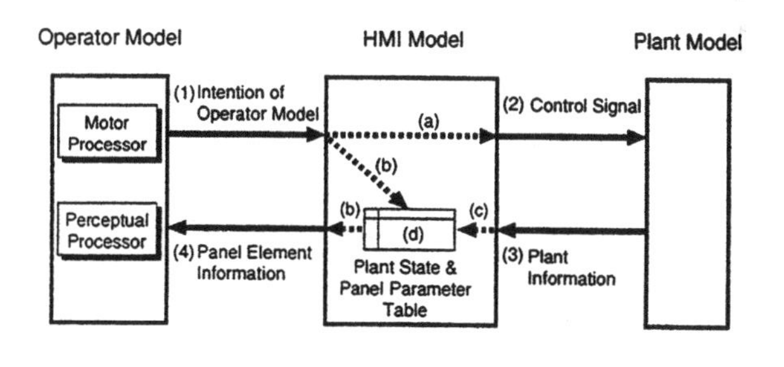

Fig. 6. HMI model

input; for get intention, the gaze point is the coordinate on the screen; for see intention, the gaze point is not controlled. The third function is to tune the time properties of the human behavior. In our model, all time properties are tuned in the motor processor for simplicity. Cycle time for set, get and see intentions are selected as 1.0 second, 0.1 seconds, and 0.1 seconds, respectively.

4. HMI MODEL

The HMI model simulates the interface between the operator model and the plant model. This model corresponds to the operator console. Figure 6 illustrates the mechanism of the HMI model.

The HMI model has a table to manage the plant state and the panel parameter information. The table is called the plant state and panel param-

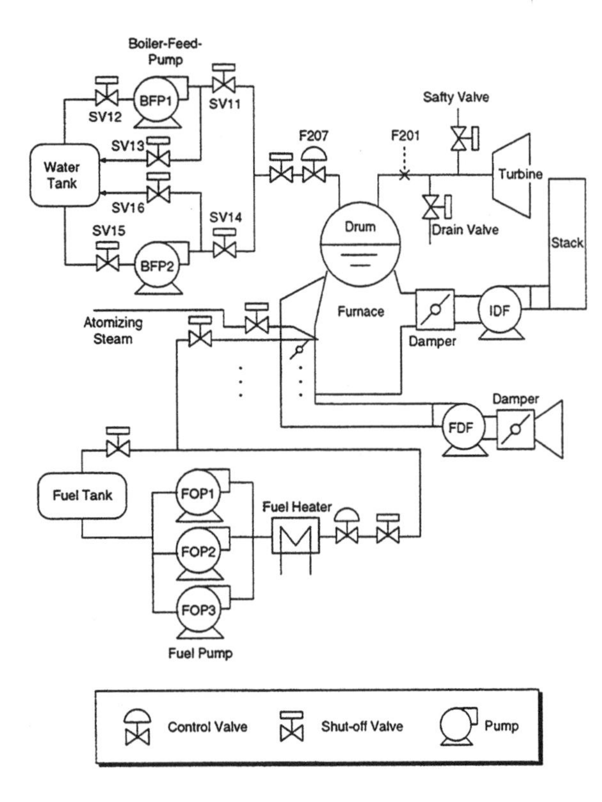

Fig. 7. PFD of a boiler plant simulator

eter table. The HMI model transmits the panel element information to the perceptual processor based on the get and see intentions given by the operator model. The HMI model repeats acquisition of plant information from the plant model at a cycle time of 0.1 seconds and manages the plant state and panel parameter table. When the intention received from the operator model is the get intention or the see intention, the HMI model sends a set of data of the specified element on the table to the operator model. If the intention is the set intention, the HMI model sends a control signal to the plant model. Figure 6 also shows these information flows through the HMI model.

5. EXPERIMENTAL STUDIES

We used a boiler plant training simulator to verify the performances of the operator model. Figure 7 illustrates the process flow diagram (PFD) of the boiler plant. The plant model was used to develop an operator model and a HMI model. Experiments of the operator model were performed under eight malfunctions, which are used in the basic course of simulator training (Yokogawa, 1991).

An experiment under a boiler-feed-pump trip is discussed in this section. Both plant behavior and correspondence in a BFP trip is summarized in Fig. 8. The left hand side of the flow chart shows the occurring phenomena in the plant, which are represented by a sequence of events. The right hand side of the flow chart shows the roughly described standard operation procedure for correspondence.

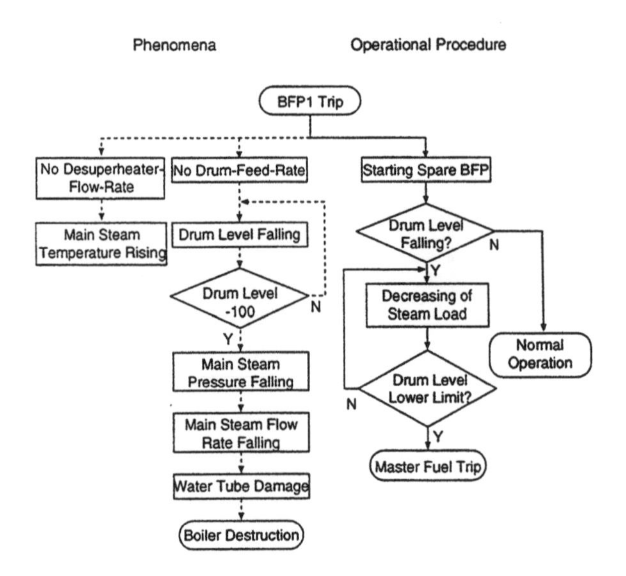

Fig. 8. Plant behavior and correspondence in a BFP trip

We can read this graph as follows: The operator will directly recognize the BFP1 trip by perceiving the blinking of the corresponding icon on the panel or the operator will notice an abnormal state by another piece of information such as the no flow rate of the boiler feed water. Once the operator recognizes the BFP1, he/she must start a spare BFP as soon as possible. Furthermore, he/she must adjust the steam load to prevent burning without water in the drum.

5.1. *Operation Result by Operator Model*

The urgent countermeasure program generated by the operator model was the same as shown in Fig. 4. The solid line represents the program when the operator model recognized the blinking BFP1 icon. The dotted line represents when the operator model recognized the no flow rate of drum feed water without noticing the BFP1 trip. The generating process of this program is summarized as follows:

1. The operator model detected the status change of the BFP1 icon by recognizing the color change to yellow and blinking of the icon. According to the "Event K" the operator model generated an urgent countermeasure program beginning with the initial event, "BFP1 trip".
2. According to the "Event and Related State Variable KB", the operator model found the related variable "water flow rate" and added this variable to the program.

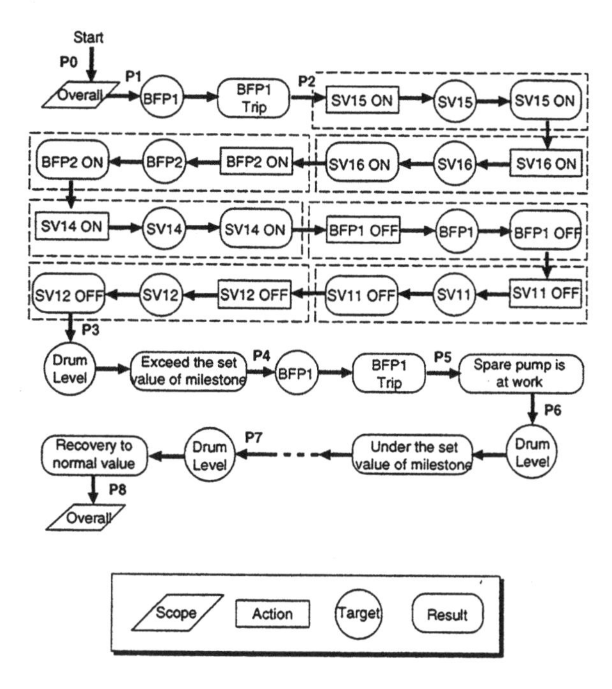

Fig. 9. Operational sequence by the operator model

3. According to "Related State Variable KB", three state variables, "drum level meter", "drum level value" and "main steam flow rate", related to "BFP1 trip" were added to the program.
4. According to the "Key State Variable KB", "drum level meter" was selected as a key state variable among four related state variables.
5. According to the "Milestone KB", three milestones connection with "drum level meter" were added to the program.

Representation of cognitive behavior is useful to analyze the operator's performance. A graphic representation of the operational procedure was developed to analyze the usability testing results of the plant operation control system (Nishitani *et al.*, 1997a, 1997b). Figure 9 shows the observed operational sequence by executing the urgent countermeasure program. The operation is represented by a sequence of steps after fixing the scope of attention. Each action by the operator model was confirmed by examining the corresponding target state on the screen. This is represented by a cycle of action, target, and result. Each cycle is shown by a dotted-line box.

In the state-monitoring mode, the operator model was watching all over the graphic panel for overview. This period corresponds to the period P0–P1 in Fig. 9. Then, a BFP1 trip occurred and an abnormal situation appeared. The operator model recognized the change of BFP1 element in the period P1–P2 and switched to the emergency mode to generate an urgent countermeasure program. This program is executed as a sequential control program. According to the generated program, "Countermeasure A" attached to the component "BFP1 trip", was executed. This appeared in the period P2–P3. After this step, the following components were scanned. Since scanned components were related state variables such as "water flow rate", "drum level value" and "main steam flow rate", these components were ignored and the pointer moved to the next component. When the component "drum level meter" was scanned, the following components of this variable were scanned because it was a key variable. Since this key state variable had the milestone "Set Value: -10", the operator model kept examining the "drum level meter" until its value gets to the set value. When "drum level meter" became -10, "Countermeasure B" was executed. This "Countermeasure B" was to confirm the execution of "Countermeasure A". This corresponds to P4–P5 period. Since "Countermeasure A" was already executed, this "Countermeasure B" was neglected (P5–P6). Then monitoring of "drum level meter" was repeated (P6–P7). After P7, the "drum level meter" recovered to the normal state. Consequently, the mode was switched to the state-monitoring mode. In this experiment, "Countermeasure C" & "Countermeasure D" were not activated.

From above results, the operator model behaved as we had planned. In this experiment, the operator model found visual indication of the BFP1 icon at first. In those cases when the operator model overlooked the change of the BFP1 icon, the operator model detected an abnormal state by noticing the decreasing water flow rate (F207 = 0), could generate an urgent countermeasure program, and succeeded in recovering it to a normal state. This shows the flexibility of the operator model based on knowledge bases.

5.2. *Operation Result by Human Operator*

We adapted the same malfunction of the BFP1 trip for a human operator, who mastered the basic operations of the boiler training simulator. The observed operational sequence by the human operator shown in Fig. 10, was analyzed by using logging data and protocol data.

The human operator continued monitoring the whole plant without persistence in specific elements during P0–P1. Then he found a trip of the BFP1, and started the countermeasure for the malfunction (P1–P2). The human operator could execute the standard operation, but he took extra actions and made some slips. The characteristics of human behavior, observed in this experiment, are summarized as follows:

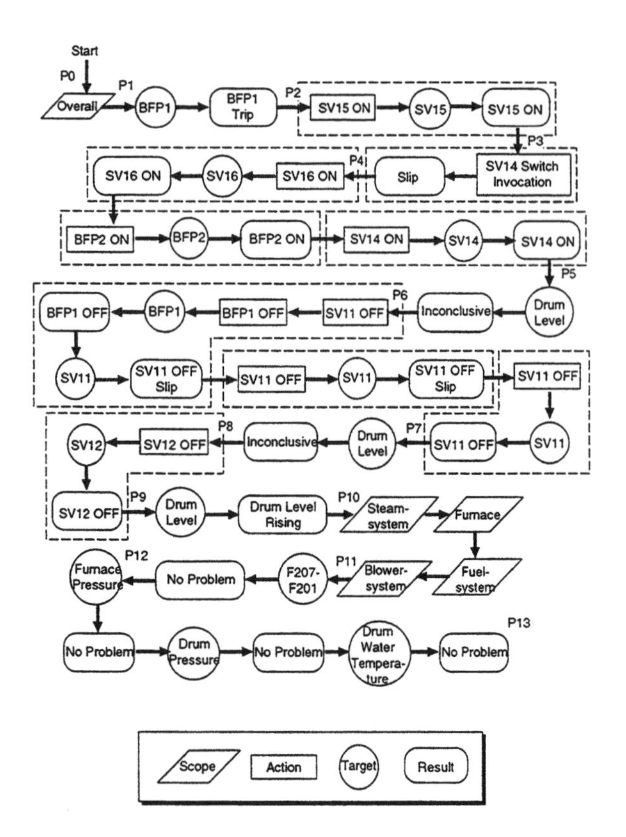

Fig. 10. Operational sequence by a human operator

P3–P4: He made errors in a hurry. He mistook SV14 for SV16,

P5–P6: He could not conclude that the drum level was starting to recover.

P6–P7: He failed twice in closing SV11 to stop the tripped BFP1.

P7–P8: He confirmed the value of drum level while he was stopping BFP1.

P11–P12: He reached a conclusion that the plant recovered to the normal state by multiple confirmations. In this case, two pieces of evidence were checked: (1) The difference of two flow rates, "F207−F201" became 0.; (2) The drum level recovered to the normal value.

By analyzing experimental results, we can get information on human errors such as occurring places and occurring frequency. This information can be applied to the improvement of human interface such as operational panels. Furthermore, elapsed time between the operator model and the human operator was compared. Due to human errors, the human operator took much more time than the operator model to get back to the normal state. From this case study, the human is fallible but flexible.

6. CONCLUSION

A simulation environment with the plant model, operator model and HMI model was developed. The operator model consists of the perceptual processor, the cognitive processor and the motor processor, and the long-term and short-term memories. This operator model can execute a standard operational sequence with an urgent countermeasure program created by knowledge bases installed in the long-term memory, although there remains difficulty of knowledge base maintenance. The operator model is useful in analyzing human errors and in considering countermeasures to cope with these human errors. The analyzing results help our judgment for full automation or for the development of human support systems. Moreover, the simulation environment with the operator model and the HMI model will enable various improvements in human interface elements to be studied experimentally.

Acknowledgement: The authors gratefully acknowledge partial financial support by the Grant-in-Aid for Scientific Research (#08305032) from the Japanese Ministry of Education, Science and Culture.

REFERENCES

Card S.K., T.P. Moran and A. Newell (1983). *The Psychology of Human-Computer Interaction*, Lawrence Erlbaum Associates, London.

Nishitani, H., T. Fujiwara, T. Kurooka, T. Kitajima and M. Fukuda (1996). A Preparatory Experiment for Usability-Test of Industrial Operation Control System, *Journal of the Society for Plant Human Factors of Japan*, **Vol.1, No.1**, pp.26–35.

Nishitani, H., T. Kurooka and T. Kitajima (1997a). Evaluation Method of Human Interface in CRT Operation Using Protocol Analysis, *Journal of the Society for Plant Human Factors of Japan*, **Vol.2, No.1**, pp.16–25.

Nishitani, H., T. Kurooka, T. Kitajima and C. Satoh (1997b). Experimental Method for Usability Test of Industrial Plant Operation System, In *"Advances in Human Factors/Ergonomics, 21B; Design of Computing Systems: Social and Ergonomic Considerations"*, edited by M. J. Smith, G Salvendy, and R. J. Koubek, Elsevier, New York, pp.625–628.

Takano, K., K. Sasou, Y. Yoshimura, S. Iwai, and Y. Sekimoto (1994). Behavior Simulation of Operation Team in Nuclear Power Plant—Development of An Individual Operator Model, *Human Factors Research Center Report*, **No. S93001**, Central Research Institute of Electric Power Industry, Tokyo.

Yokogawa Corporation (1991). Operation Manual of Boiler Simulator, TE34A1V91-21, Tokyo.

OPERATOR'S MENTAL IMAGES ON TEAMWORK PLANT OPERATION

Yoshihiko Nojiri*, Takehisa Kohda, Atsushi Kawamoto**, Koichi Inoue** and Hideo Noda*****

**Marine Technical College, Ministry of Transport*
***Graduate School of Engineering, Kyoto University*
****Institute for Sea Training, Ministry of Transport*
E-mail: nojiri@eng.mtc.ac.jp

Abstract: The interaction between/among operators must be considered in research domains such as a task analysis or human modeling of a teamwork operation. This paper shows that the interaction between/among operators can be represented by three kinds of "Mental Images" and operator's cognitive activity can be analyzed by using these mental images. Furthermore, it also shows (1) how mental images are formed in an operator's mind and (2) what kind of action would be taken when he/she senses a discrepancy between mental images. Finally, an illustrative example of an airplane accident shows how to make use of mental images in human error analysis. *Copyright © 1998 IFAC*

Keyword: Human error, Safety analysis, Man/machine interaction, Communications, Cognitive science

1. INTRODUCTION

The proportion of human-related accidents to the whole tends to increase in number recently, particularly in the area of advanced system operation such as nuclear power plant operation and airplane/ship navigation. It suggests that the safety operation can never be achieved only by high reliability of the system. In such a domain, we must also take account of reliability of operator's cognitive process. There is a little but important difference between single-operator's cognitive process and multi-operator's one in a teamwork. One of the most characteristic differences is the existence of the interaction between/among operators mainly through their verbal communications (Shepherd, 1989). An inadequate interaction would sometimes cause a dreadful misunderstanding that leads to a disaster, while a smart and good timing interaction would bring a marvelous ability to correct one's errors (Foushee and Helmreich, 1988). This paper describes the basic framework to represent an interaction between/among operators such as verbal communication. It also proposes that the method can be applied to the area of accident analysis and human modeling in teamwork.

2. PLANT OPERATION CHARACTERISTICS

In many advanced technological systems such as nuclear power plants and airplanes/ships, operators spend most of their time in monitoring system parameters ("Monitoring Mode"). During monitoring mode, they take no actual control action on the system. Instead, they are watching current system states (for

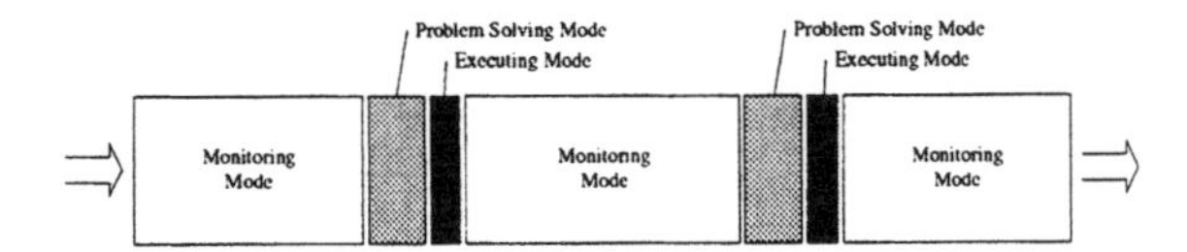

Fig.1 Operation Cycle

example, temperature, pressure, location, etc.) and simultaneously trying to compare these real states with the ideal ones in their mind under the current condition. Once they find a gap or discrepancy between these two states, they shift to "Problem Solving Mode" to remove or decrease the gap.

In this mode they diagnose the situation and formulate strategies to achieve the goal. After the strategy composition is completed, they in turn go to "Executing Mode" and put an actual control into the system. Soon after, they return to the monitoring mode. This is called "Operation Cycle" of plant operation (Itoh, 1997) (Fig.1).

An operation cycle as stated above can be applied not only in a steady state, but also in a transient state of a plant. In a transient state such as NPP's start-up/shut-down mode and airplane's takeoff/landing mode, operators have to take control actions on the system continuously. In this case, it can be said that the operation cycle goes around and around in a very short cycle. In this meaning, it is very important to understand the mechanism when, why and how they shift among each mode.

3. MENTAL IMAGES

In a monitoring mode it is of great importance for operators to get or have exact system state information since it will become a starting point of following activities. In a field such as a plant operation that requires cooperative works, tasks have to be shared among each member to keep their performance higher level as a whole. For this purpose, various information about the system should be shared with each operator and one operator should understand the other members. The team can not perform counter-measure correctly if shared information is inconsistent among them. Namely, it is required for reliable team operation that each member has to know not only correct information from the system but also his partners' mind.

Consider a team of two members, operator A and operator B. Fig.2 shows how to describe recognition state in an operator's mind about the system and his partner. The term "mental image" is defined as a recognized information about operator's environment (systems and other operators). Here,
RS(S) represents a real state of the system S.

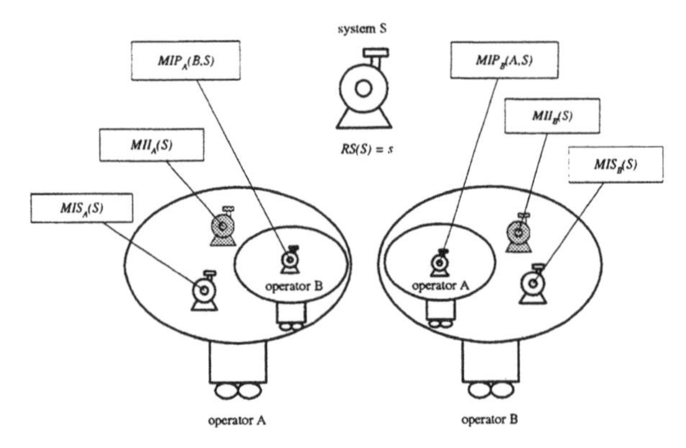

Fig.2 Mental Image

$MII_A(S)$ represents an ideal state of the system S that operator A is thinking under the current condition.
$MIS_A(S)$ represents a real state of the system S that operator A is thinking.
$MIP_A(B, S)$ represents operator A's mental image about what operator B is thinking about the system S.

In the same manner, three mental images of Operator B are represented as $MII_B(S)$, $MIS_B(S)$ and $MIP_B(A, S)$. With the use of these representations, it is possible to distinguish information about the system from that about partners.

4. PRODUCTION PROCESS OF MENTAL IMAGES IN MONITORING MODE

This section discusses the process how three kinds of mental images are produced.

4.1 Production Process of MII(S) (Fig.3-(1))

Monitoring is an essential task that is imposed on an operator to compare perceptive information from the system (temperature, pressure, etc.) with the ideal state of the system he is thinking or expecting. Therefore, the operator must already have *MII(S)* when he is working in the monitoring mode. Nevertheless once the system state or task goal is changed, he is obliged to produce new *MII(S)* through,
(1) cognitive activity based on his experience or knowledge,
(2) instruction from other team member(s) and
(3) instruction from a supporting system.

4.2 Production Process of MIS(S) (Fig.3-(2))

MIS(S) in an operator's mind is produced through,
(1) direct perceptive information from the system,
(2) communication with partner(s) such as reports and
(3) inference based on other related information.

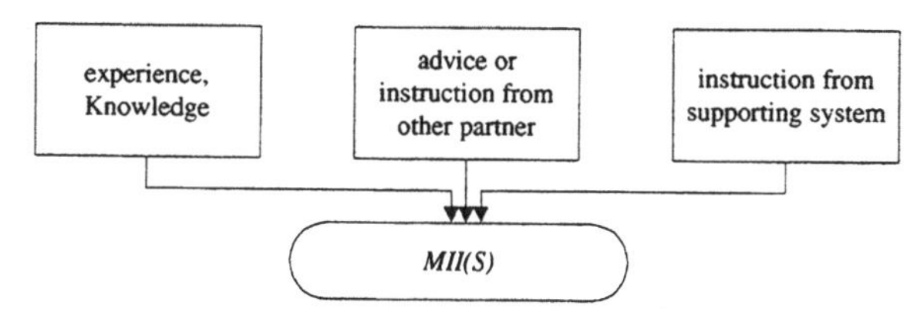

Fig.3 - (1) Production Process of Mental Image (MII)

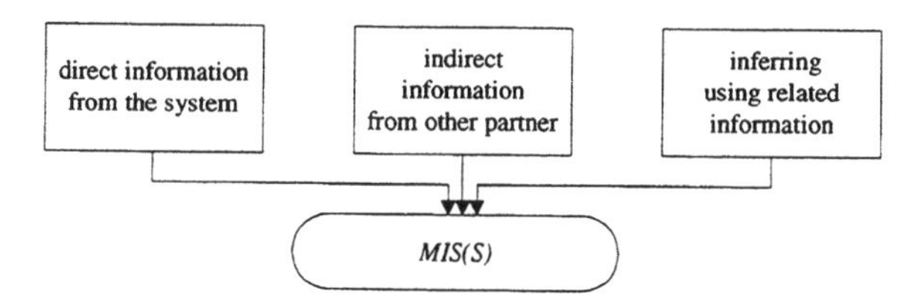

Fig.3 - (2) Production Process of Mental Image (MIS)

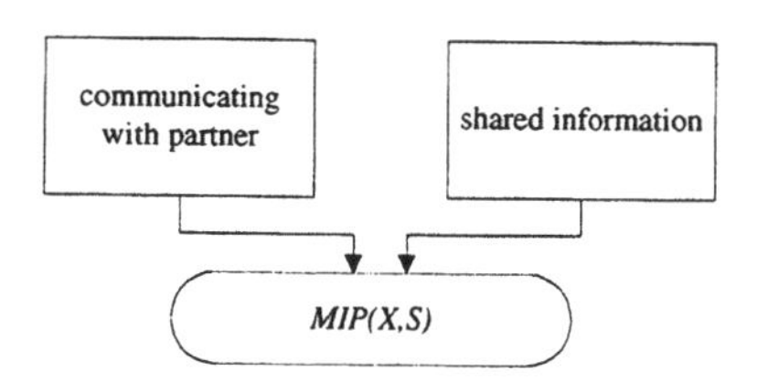

Fig.3 - (3) Production Process of Mental Image (MIP)

Since direct perceptive information usually comes through a man-machine interface, its design plays a crucial role in forming a correct *MIS(S)*. Misunderstanding or mishearing is commonly associated with verbal communication, so its style or method to transfer information through communication must be discussed. The last one, inference would be possible only if an operator had deep knowledge about the system function (so-called mental model of the system). Some researchers have developed or are developing an interface that enables operators to understand the system function easily in real time (Vicente and Rasmussen, 1992; Lindo, 1993).

4.3 Production Process of MIP(X, S) (Fig.3-(3))

Almost *MIP(X, S)* is produced through,

(1) communication with other operator(s) such as a report and an answer back and

(2) common information source such as a large-scale graphic display, which are simultaneously available to plural operators.

Teamwork operation requires mutual understanding. One operator must grasp his partner's intention and thinking about the system. If they fail to do this, timing errors, wrong object errors, omission errors, commission errors, etc. may occur in some operation. Well organized relationship in a relatively large scale teamwork is generally maintained mainly by verbal

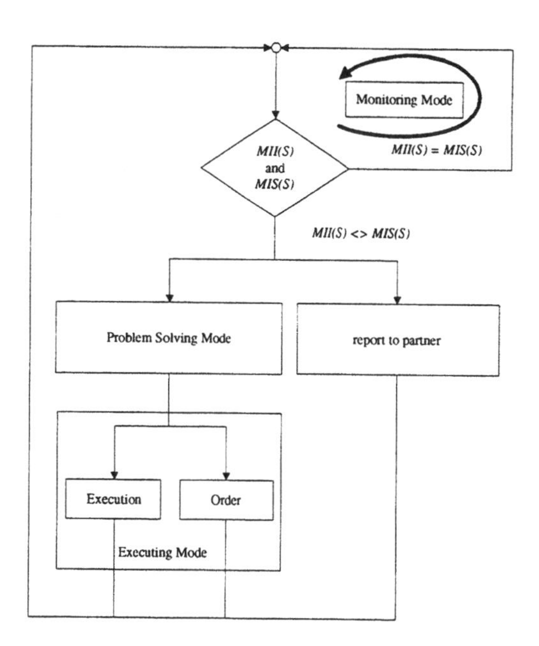

Fig.4 - (1) Discrepancy between MII and MIS

communication, though a so-called perfect-pair can get mutual understanding without any physical communication in case of doing simple task.

5. CONFLICT BETWEEN MENTAL IMAGES IN MONITORING MODE

No action is necessary when these three mental images are the same. Once an operator finds discrepancy among them, he starts an action to remove it. This section discusses operator's action when he finds discrepancy between *MIS(S)* and *MII(S)* or between *MIS(S)* and *MIP(X, S)*.

5.1 Discrepancy between MIS(S) and MII(S)

Since this means a deviation from an ideal system state, an operator shifts his mode to the problem-solving mode. In the first place he sets a new goal to get rid of the present discrepancy and formulates a strategy to achieve the goal. Then he shifts to the executing mode to do his strategy. In this process, a hierarchical problem-solving action like SRK stepladder model (Rasmussen, 1986) could be applied. Namely, if the operator knows an action to achieve his current goal, he immediately does it. If he knows no direct actions but has a plan to achieve the goal, he tries to unfold lower-level sub-goal(s) and replace his current goal with a new goal. If he knows neither actions nor plans, he tries to modify his current goal based on his knowledge or information on the system. If he can not achieve his current goal by himself, he tries to obtain help from the other members or supporting systems (Kohda, *et al.*, 1996) (Fig.4-(1)).

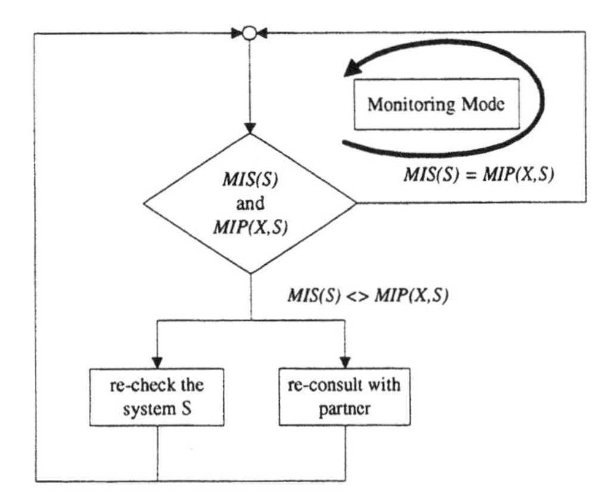

Fig.4 - (2) Discrepancy between MIS and MIP

5.2 Discrepancy between MIS(S) and MIP(X, S)

This inconsistency is caused by inadequate mutual understanding or communication. Since an operator knows that there must be only one true state of the system, he starts suspecting himself or his partner depending on his conviction. If he had stronger conviction than his partner did, he would try to confirm information from the partner; namely, he would give a question to the partner. Otherwise he would try to check the system again by himself (Fig.4-(2)).

Thus, a departure from the monitoring mode originates from an occurrence of discrepancy between/among mental images.

6. OPERATOR MODEL WITH MENTAL IMAGES

This section shows one of applications of the framework using mental images. To design an operator model as a team member, interactions not only between an operator and a system but also between an operator and his partner(s) should be taken into account. A mind state inside an operator must be changed by the influence of information from the system and his partner(s) simultaneously. Namely, an operator model should receive information through communication with his partner and understand his partner's mind. Fig.5 shows an example of operator models that take account of operators' interaction by using *MIP*.

7. EXAMPLE OF AIRPLANE ACCIDENT

Finally, the following shows how the framework using mental images is applied to an accident analysis. An accident happened in 1966. A Boeing-707 of Indian Airline (B707), flying at several miles to Geneva airport, crashed into the top of Mt. Mont Blanc (M.B.) in dense fog soon after contacting with a controller of the airport. The accident killed all passengers and crews

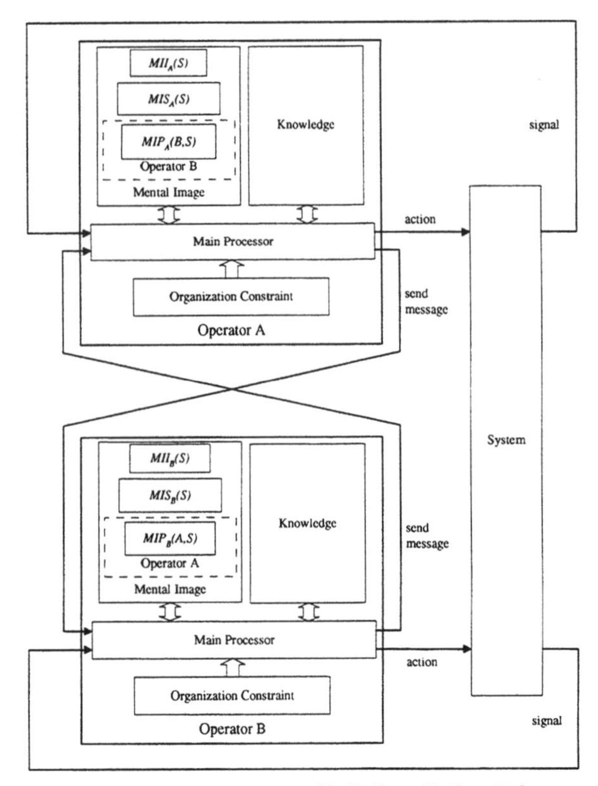

Fig.5 Example of Operator Model

(117 persons). According to the voice record between the pilot and the controller, the cause was undoubtedly due to their inadequate communications. The following are their conversation at that time.

(1) Several minutes before the accident, the pilot of the B707 reported or asked to the controller (COM-1),

"I think we are passing abeam Mont Blanc now."

(2) The controller answered (COM-2),

"I understand."

(3) The controller checked his radar and confirmed that the position of the B707 was at 5 miles straight to M.B. He gave the pilot the exact position to correct his mind like this (COM-3).

"You have 5 miles to the Mont Blanc."

(4) The pilot answered back (COM-4).

"I see."

The controller failed to correct the pilot mind about his position. The pilot thought that B707's position was not at 5 miles to M.B. but 5 miles abeam M.B. Fig.6-(1) shows the process of producing their mental images (*MIS*, *MII*, and *MIP*) according to the above-mentioned conversation. The initial states of each mental image are,

(Pilot)

$MII_P(B707)$ = {5 miles abeam M.B.}

$MIS_P(B707)$ = {abeam M.B.}

$MIP_P(C, B707)$ = { ? }

(Controller)

$MII_C(B707)$ = {5 miles abeam M.B.}

$MIS_C(B707)$ = { ? }

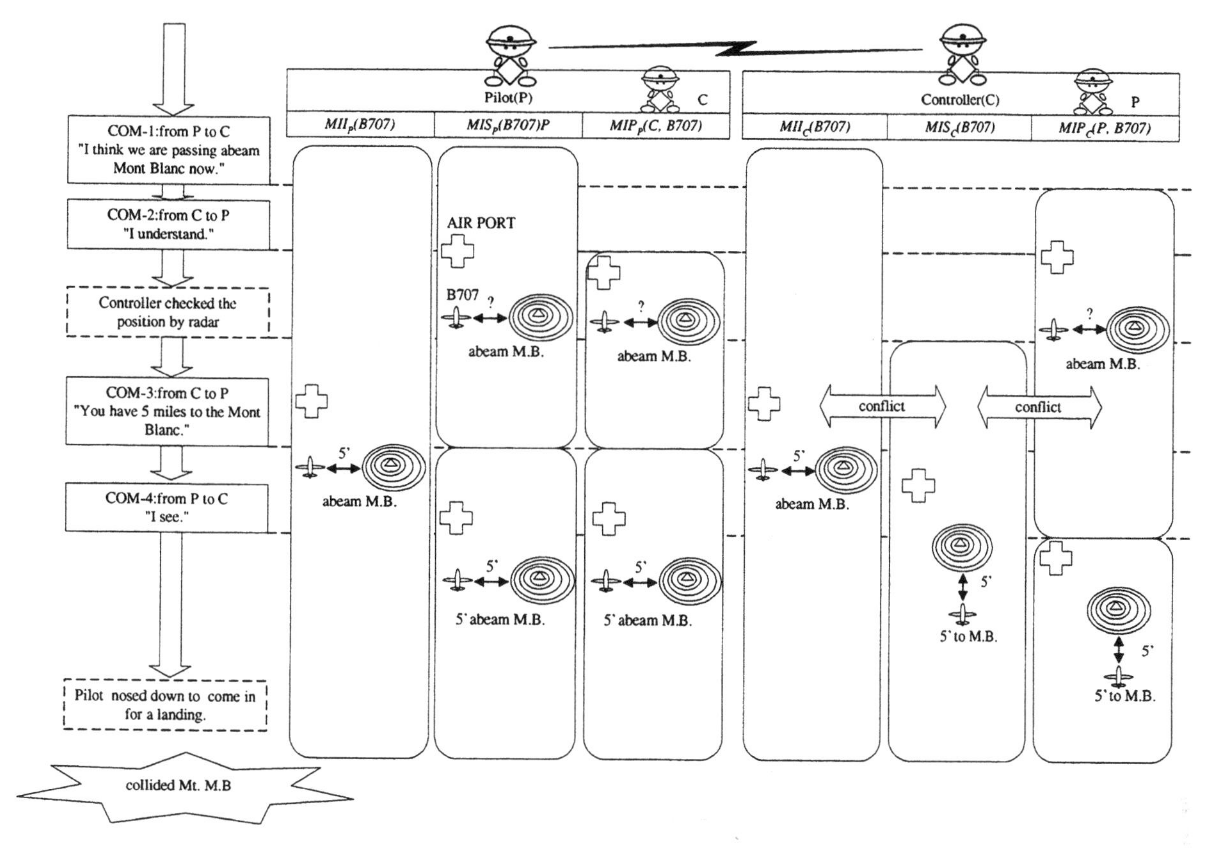

Fig.6 - (1) Boeing-707 Accident at Mt. Mont Blanc in 1966

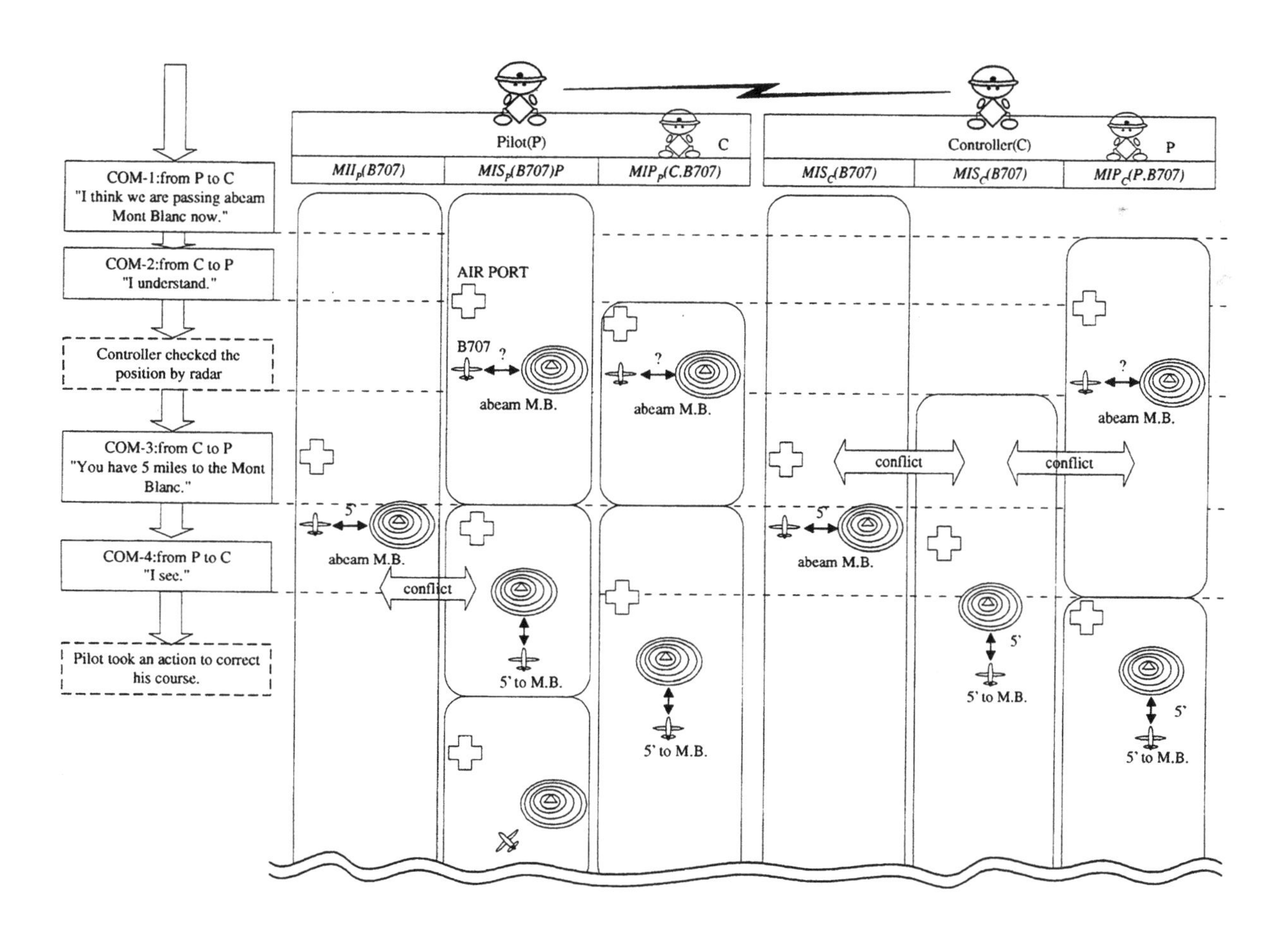

Fig.6 - (2) No Misunderstanding Case

$MIP_C(P, B707) = \{ ? \}$

Here, both $MII_P(B707)$ and $MII_C(B707)$ equal *{5 miles abeam M.B.}* since both the pilot and the controller beforehand had background knowledge of the normal and safety course to the airport.

The folowings are the process of their mental images.

(1) The controller recognized that the pilot was thinking his position located abeam M.B. by COM-1.
$MIP_C(P, B707) = \{?\} \rightarrow \{abeam\ M.B.\}$

(2) The pilot recognized that his report was correctly transferred to the controller by the answer back (COM-2).
$MIP_P(C, B707) = \{?\} \rightarrow \{abeam\ M.B.\}$

(3) Then the controller confirmed the plane's actual position by checking radar.
$MIS_C(B707) = \{?\} \rightarrow \{5'\ to\ M.B.\}$

And the discrepancy among the controller's mental images ($MIS_C(B707) <> MII_C(B707)$, ($MIS_C(B707) <> MIP_C(P, B707)$)) emerged. He felt necessity to correct the pilot's mind and sent the pilot the information.

(4) Though the pilot received the information (COM-3), he misunderstood it. He thought that the information was about the distance abeam M.B.
$MIS_P(B707) = \{abeam\ M.B.\} \rightarrow \{5'\ abeam\ M.B.\}$
$MIP_P(C, B707) = \{abeam\ M.B.\} \rightarrow \{5'\ abeam\ M.B\}$

(5) The controller believed firmly that he could change the pilot mind about the location only due to the answer back "I see." from him (COM-4).
$MIP_C(P, B707) = \{abeam\ M.B.\} \rightarrow \{5'\ to\ M.B.\}$

As a result all mental images of the pilot and controller began to accord each other from that time and no new action were taken until B707 collided to the mountain.

It is worthy of notice that there were chances to correct their misunderstanding, nevertheless, an important advice of the controller could never be transferred to the pilot correctly due to inadequate communication. Namely COM-3 had an ambiguous expression meaning "to" or "abeam". And furthermore the final answer back COM-4, "I see." , had no function to identify how the pilot understood the information. Thus, the change of mental images produced from the verbal communications between the pilot and the controller indicates a cause of the accident clearly.

Fig.6-(2) shows the assumptive case that if the pilot understood COM-3 information from the controller correctly. In this case, the discrepancy between $MII_P(B707)$ and $MIS_P(B707)$ of the pilot made him take a counter action soon after he received the COM-3.

8. CONCLUSIONS

This paper proposed the use of three kinds of mental images to analyze a teamwork operation. Especially the third one (*MIP*) is effective in representing mutual understanding of each member. The paper also showed that a deviation detected between any two mental images would initiate a transition from the monitoring mode to the problem-solving mode. From the viewpoint of human error analysis, the consideration of mental images mentioned above would become indispensable for the task analysis and the development of an operator model in teamwork.

REFERENCES

Foushee, H. C. and R. L. Helmreich (1988), Group Interaction and Flight Crew Performance, In: *Human Factor in Aviation*, pp.193-197, Academic Press

Itoh, K. (1997), Data Elicitation and Task Analysis for Cognitive Operations in Ship Navigation, In: *Proceeding of AIR&IHAS'97*, Wako

Kohda, T., T. Tanaka, Y. Nojiri and K. Inoue (1996), Reliability Analysis of Operating Procedure in Team Performance, *ESREL'96 - PSAM III* ,Greece

Lindo, M. (1993), Multilevel Flow Modeling, *AAAI93 Workshop on Reasoning about Function*, Washigton D.C.

Rasmussen, J. (1986), *Information Processing and Human-Machine Interaction*, North-Holland

Shepherd, A. (1989), Analysis and Training in Information Technology Tasks, In: *Task Analysis for Human-Computer Interaction* (D. Diaper (Ed)), Ellis Horwood

Vicente, J. and J. Rasmussen (1992), Ecological Interface Design: Theoretical Foundations, *IEEE Transactions on Systems, Man, and Cybernetics*, **Vol. 22**, No.4

APPLICATION OF MIND STATE ESTIMATION TO PLANT OPERATORS

Taketoshi Kurooka, Masafumi Kisa, Yuh Yamashita and Hirokazu Nishitani

*Graduate School of Information Science,
Nara Institute of Science and Technology, Japan
8916-5 Takayama, Ikoma, Nara 630-0101, JAPAN*

Abstract: We developed a linear regression model with Electroencephalograms(EEGs) to estimate the mind state of a plant operator. First, we defined three modes according to the typical thinking state of plant operators. The classification was validated by preparatory experiments with card games, mathematical problems, and puzzles. The estimation method was applied to a plant operator who operates a boiler plant simulator under abnormal situations. As a result, the method provided the plant operator's mind state, which is liable to be overlooked by analyzing with a simple interview and observation. *Copyright ©1998 IFAC*

Keywords: Human factors, Human-machine interface, Human supervisory control, Man/Machine interaction, Operators, Training

1. INTRODUCTION

Although a clear explanation has not been made about the generation mechanism of EEGs, EEGs provide useful information regarding cerebral functions. Recently, many researchers have been studying the mechanism, application to interface design and evaluation, simple measuring instruments, and so on (Saiwaki *et al.*, 1997; Funada *et al.*, 1997; Fuwamoto *et al.*, 1997; Tanaka *et al.*, 1997). We tried to estimate the plant operator's mind state with EEGs for the purpose of developing a useful support system for operators.

Musha et al. proposed a method called Emotion Spectrum Analysis Method, in which a linear regression model is used to estimate the emotional state of humans (Musha *et al.*, 1996; Musha *et al.*, 1997). They defined the human emotional state by four orthogonal components: anger/stress, sadness, joy and relaxation. In the method, EEGs are measured with 10 disk electrodes which are placed according to the international 10-20 method. Cross-correlation coefficients of 45 pairs of 10 electrodes are calculated in three types of bands: 5-8 Hz (theta rhythm), 8-13 Hz (alpha rhythm), and 13-20 Hz (beta rhythm). A set of 135 state variables of cross-correlation coefficients of EEGs are related to each element of the four-dimensional emotion state by a linear model. However, this definition of emotional state is not suitable to directly analyze plant operator's behavior because it is difficult to consider why such a sort of emotion was evoked. It is more useful to know the operator's thinking state rather than his/her emotional state in order to analyze the cognitive state of plant operators. We defined a three-dimensional vector of the thinking state. This report presents our experiment results of a plant operator's thinking state estimation with EEGs.

2. DEFINITION OF THE THINKING STATE

2.1 *Basic modes of the thinking state*

We classified the thinking state into 3 modes with emphasis on the operator's cognitive process of identifying the cause of plant failures. Each mode corresponds to an elementary thinking state.

Mode A; It is a state when the operator can easily identify the cause of the plant failure. An operator has confidence in his/her operational procedure, strategy or decision.

Mode B; It is a state when the operator has alternative inference about the cause of the failure, but the operator has no assurance of a conclusion. The operator is building hypotheses, and considering in a more complicated way compared to Mode A.

Mode C; It is a state when the operator has no idea what to do or the operator has too many things to deal with by him/herself. The operator is embarrassed or confused.

2.2 *Preparatory experiments*

The proposed classification was verified with preparatory experiments using card games, mathematical problems, and puzzles. The experiments were carried out as follows:

1) Pair matching card games, called Pelmanisim, were performed with 16 everted cards, and measured the EEGs of the subjects during the games. Each game contributes to evoking each mode. In order to evoke Mode A, the subject was given a reference sheet which illustrated a substance of everted cards, and the subject matched the pairs by referring to the sheets. In order to evoke Mode B, the cards were placed in order. The subject was informed of three types of rules which accounted for the arrangement of the cards. The subject matched the pairs keeping in consideration which rules corresponded to the present arrangement. In order to evoke Mode C, the cards were placed in random order. The subject then matched the pairs while being compelled to decipher what kind of arrangement the cards were following.
2) Linear regression models were made from the measured EEGs. A model was made for each subject.
3) The EEGs were measured while the subjects solved mathematical problems and puzzles, and estimated the thinking state from the model.
4) The estimation results were compared to observations and verified.

2.3 *Verification of the basic modes*

The preparatory experiments were carried out on 7 subjects following the above procedure. The estimation results corresponded well with observations, and the proposed classification was validated.

As an example, the verification result with a mathematical problem is given below. The solving process of a mathematical problem with referential time periods is shown in Fig. 1. The output transition of the thinking state model on the solving process according to passage of time is shown in Fig. 2. One step (scale unit) is equal to 5.12 seconds in abscissa. The each output levels of Mode A, Mode B and Mode C are shown from top in ordinates. The problem was shown to the subject at step 0, and the time limit was over at step 60. The subjects were informed before the experiment that the problem is based on mathematics for high school students and that the time limit is 5 minutes. It is possible to analyze the solving process in Fig. 1 with reference to distinctive features in Fig. 2 as follows:

1) On the whole period, the level of Mode B is higher in sections (a), (b), (c) (the first half) and the level of Mode C is higher in sections (d), (e), (f), (g) (the second half). It can be presumed that it shows the subject's confusion and irritation caused by his inability to find the solution and the approaching time limit. In section (g), there is some reduction in the level of Mode C. It can be estimated that the subject almost abandoned to solve the problem.
2) From section (a) to section (b), the level of Mode A, which is higher at step 0, decreases and the level of Mode B and Mode C increases. This aspect agrees with the subject's state which is estimated from Fig. 1, i.e., when the subject recognized that the problem is a factorization, and tried to recollect his memory but failed to immediately find a strategy for solving the problem.
3) The level of Mode B shows maximum peak in section (c) while the levels of Mode A and Mode C are low. As seen in Fig. 1, the subject resorted to trial and error during this time period. It can be estimated that the subject was getting accustomed to the equation and gradually recollecting memory about the factorization. Additionally, the level of Mode A is getting higher in the following section (d) compared to the other sections. It can be estimated that the subject succeeded in recollecting a solving technique of factorization and therefore he could try the technique again in the following time period.

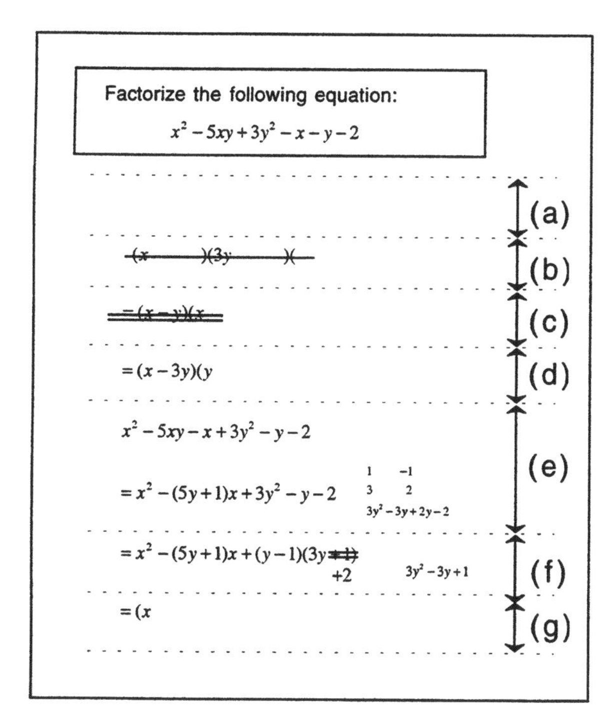

Fig. 1. Solving process of a mathematical problem

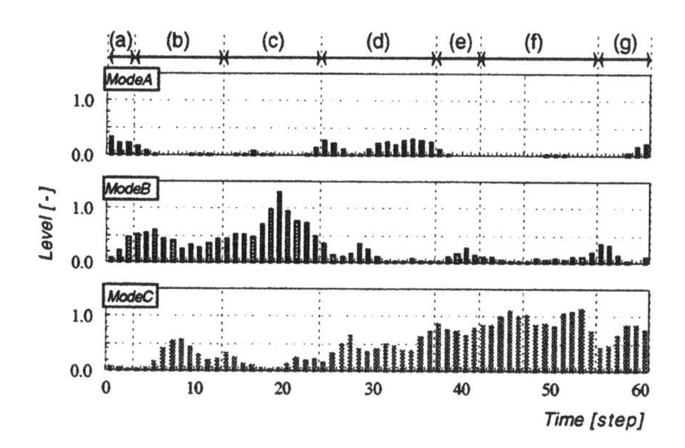

Fig. 2. Outputs of a thinking state model on solving process of a mathematical problem

Cognitive burdens can be much heavier in plant operation. However, logical thinking is required for the operator to investigate the cause of plant failure as well as it is to solve a mathematical problem. Therefore, the proposed classification based on typical thinking state is applicable to plant operation.

3. DEVELOPMENT OF OPERATOR'S THINKING STATE MODEL

3.1 *Experimental procedure*

An industrial training system of plant operation was used for the experiments. The system was composed of the plant operation control system called DCS and training simulators for plant operation. A boiler simulator was used with 10 kinds of malfunctions shown in Table 1. The experimental procedure is summarized as follows:

1) The subject was trained under 4 kinds of fundamental malfunctions to have the skill of boiler plant operation and fault diagnosis with DCS.
2) Experiments evoking the basic modes were performed to establish an operator's thinking state model.
3) Meaningful time periods when the basic modes were evoked were selected with reference to observation data, and an operator's thinking state model was established with the EEGs in these periods.

Table 1 Malfunctions used for experiments

Malfunctions	Subject Training	Model Building	Esti-mation
Failure of deaerator feed water valve	√		
Clogged high-pressure strainer of fuel system	√		
Breakage of drum safety valve (discharge of steam)	√	√ Mode A	
Misindication of boiler level meter (step change)		√ Mode B	
Misindication of steam flow rate (sinusoidal change)		√ Mode C	
Sticking of pressure control valve of heavy oil pump			√ Test1
Misindication of steam flow rate (step change)			√ Test2
Sticking of feed water valve			√ Test3
Breakage of fuel heater pipe	√		√ Test4
Misindication of boiler level meter (sinusoidal change)			√ Test5

3.2 *Experiments evoking basic modes*

Experiment evoking Mode A; The malfunction used in this experiment was the breakage of the drum safety valve (discharge of steam). This malfunction had been used for the training and thus the operator was accustomed to the malfunction. With consideration given to observation data, the operator's thinking process was summarized in Fig. 3. The operator recognized that the feed water flow rate was high by the alarm, conducted a wide range search of the cue to identify the cause of failure, and then identified the failure cause. During the identification, the operator built up two hypotheses, i.e., the leakage of discharge valve

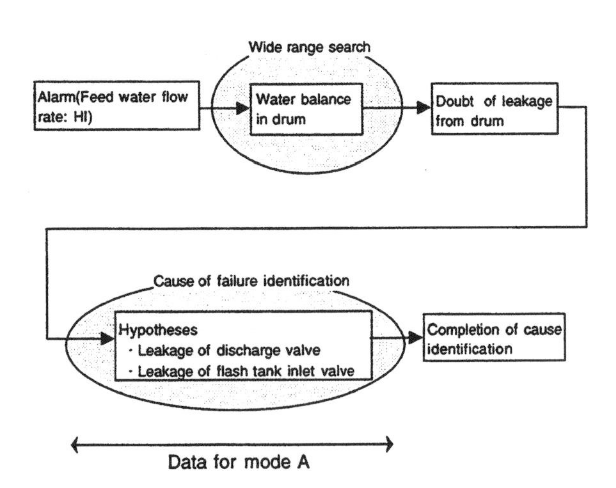

Fig. 3. Operator's thinking process in ModeA

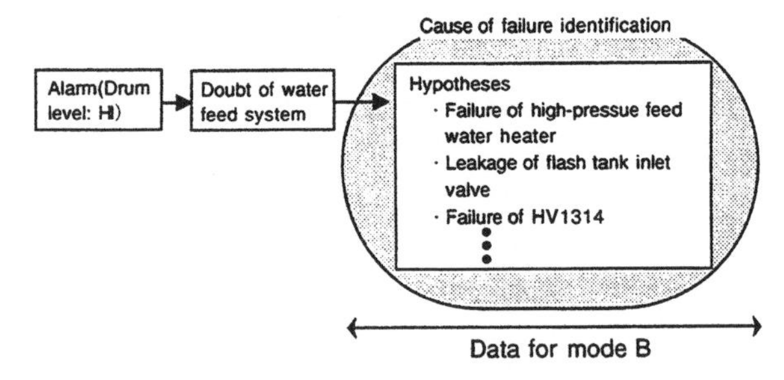

Fig. 4. Operator's thinking process in ModeB

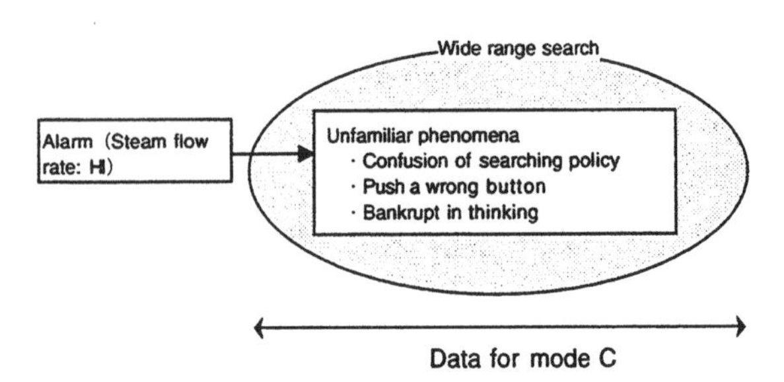

Fig. 5. Operator's thinking process in Mode C

and the leakage of flash tank inlet valve. This situation coincided with the definition of Mode A. Therefore, it was determined that the thinking state of Mode A was evoked in this period.

Experiment evoking Mode B; The malfunction used in this experiment was the misindication of a boiler level meter. The operator had never before experienced this malfunction. However, the process variables changed simply. Therefore, the operator easily succeeded in building a hypothesis for identifying the plant failure. With consideration given to the observation data, the operator's thinking process was summarized in Fig. 4. The operator recognized that the drum level was high by the alarm, and tried to identify the failure cause immediately. The operator built up at least 3 kinds of hypotheses and did not narrow the failure causes to one single point. This situation coincided with the definition of Mode B. Therefore, it was determined that the thinking state of Mode B was evoked in this period.

Experiment evoking Mode C; The malfunction used in this experiment was the misindication of a steam flow rate meter. The operator had never experienced this malfunction, and the plant condition changed in a complicated manner. With consideration given to the observation data, the operator's thinking process was summarized in Fig. 5. The operator tried to search the cue to identify the cause of failure but failed to build up a hypothesis. After a while, the operator told the experimenter that he came to a deadend trying to identify the reason for plant failure. This situation coincided with the definition of Mode C. Therefore, it was determined that the thinking state of Mode C was evoked in this period.

4. EXPERIMENTS FOR VERIFICATION

4.1 *Results*

Experiments of boiler plant operation were performed to verify the operator's thinking state model. The malfunctions used for the verification are also shown in Table 1. The verification procedure is summarized as follows:

1) The operator's thinking state was estimated under each malfunction. Operator's behavior was also represented with a flow graph from observation data.
2) The estimation results with EEGs were compared with the operator's behavior, and correspondence was examined in detail.

Cognitive state in plant operation was understood by noting the mode transition. As an example, the experiment #2 was examined in detail below.

The output transition of the thinking state model in experiment #2 is shown in Fig. 6. Right after the first alarm occurred, the level of Mode B denoted the first peak. As Mode B decreased, the level of Mode A got higher. Mode C appeared and gradually increased around step 20, where Mode A denoted the first peak and decreased. After a while, Mode C decreased, and Mode B increased in return. After that, Mode B's level gradually went lower and Mode A denoted a higher level again. Mode A continued indicating a higher level until the end of the operator's thinking process.

With consideration given to the observation data based on video image, audio recording and an

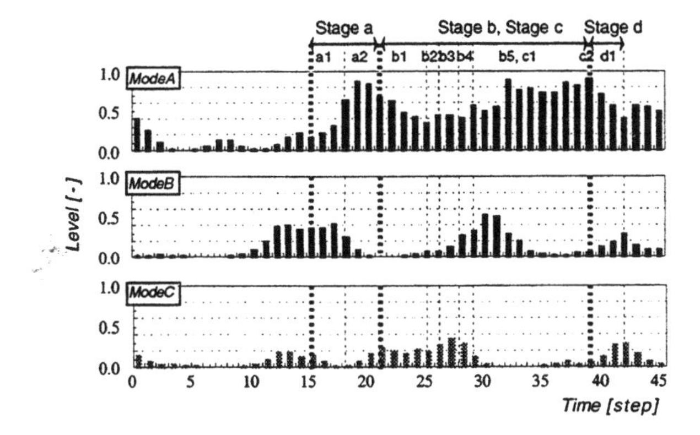

Fig. 6. Outputs of a thinking state model in experiment #2

interview, the operator's thinking process was summarized in Fig. 7. After the first alarm occurred, the operator checked the water balance in the drum twice. Consequently, the operator conducted a wide range search of the cue to identify the cause of failure. During this period, the operator doubted the feed water system, and steam system thereafter. The operator tried to search some cues in those systems but failed. After a while, the operator found an abnormality in the relationship between process variables by viewing a trend graph, and doubted a failure in the steam flow rate meter. The operator asked a fieldman to confirm the failure, and searched for other alternatives while he was waiting for an answer from the fieldman. After 30 seconds, the operator received the answer that the steam flow rate meter had in fact failed, and concluded his investigation.

4.2 *Discussions*

The operator's behavior was divided into 4 stages shown in Fig. 7, i.e., conventional tasks which had been learned during the training (Stage a), wide range search (Stage b), cause of failure identification (Stage c), and confirmation (Stage d). The operator's behavior based on the observation data and the estimation result were compared to validate the estimation model. The time period, such as ai, bi, ci and di in Fig. 7, was segmented according to the operator's task. Transition time from b5 to c1 was indistinct from the observation data. The correspondence between the transition of modes in Fig. 6 and the operator's behavior in Fig. 7 were summarized as follows:

1) In Stage a, the level of Mode A denoted the first peak. The operator checked the water balance of the drum as a conventional task, and easily found an abnormality. That is why Mode A was evoked.
2) In Stage b, the level of Mode C was higher from period b1 to period b4. In these periods the operator tried to find a cue to identify the cause of plant failure but failed. It is assumed that the operator was troubled. In particular, the level of Mode C was at the highest in period b3. The operator could not find any cue in the water system and, therefore, other possibility must be considered in period b3. That is why Mode C was strongly evoked. In period b5, the level of Mode C disappeared and Mode B increased. This corresponded with a fact that the operator found an abnormality in the relationship between process variables by viewing a trend graph and succeeded in building the hypotheses of the cause of failure.
3) In Stage c, the level of Mode A was higher. This is because the operator confidently doubted the steam flow meter failure.
4) In Step d, the level of Mode A slightly decreased, and the levels of Mode B and Mode C increased. The operator was waiting for the answer from the fieldman, and tried to find other alternatives of the cause of failure. Therefore, the operator might feel some anxiety and thus Mode B and Mode C were evoked. This explanation was supported by the fact that the levels of Mode B and Mode C decreased after the operator received the answer that his conclusion was correct.

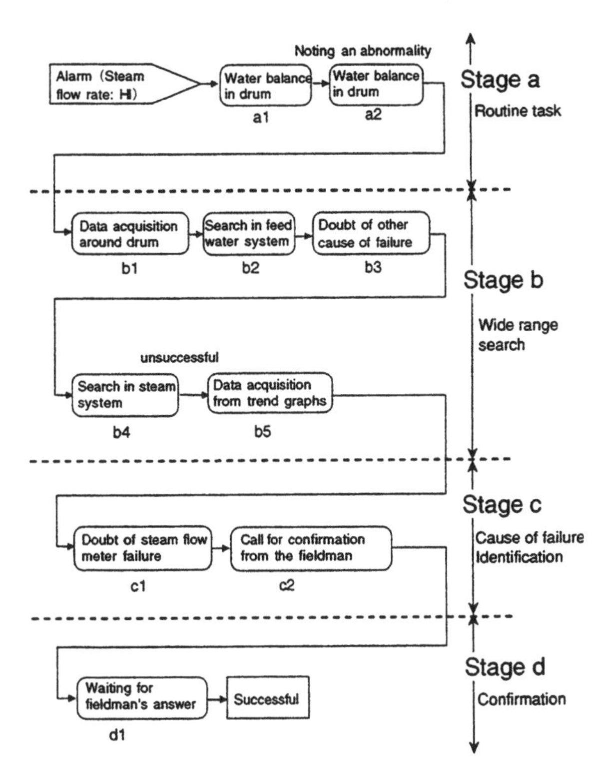

Fig. 7. Operator's behavior in experiment #2

As seen above, the estimation result and the observation data are in good agreement.

4.3 *Merits of plant operation analysis with EEG data*

The output of the thinking state estimation model provides useful information to examine the cognitive aspect of plant operation in detail, which is difficult with observation alone. For example, the plant operation in the experiment #2 can be analyzed based on the result in Fig. 6 as follows:

1) The operator checked the water balance in the drum twice in period a1 and a2. Mode B showed a higher level in period a1, while Mode B decreased and Mode A increased in period a2. From this information, it is inferred that the operator made his first check of the water balance as a routine task in period a1. He found some cue to identify the failure cause and carefully checked the water balance again in period a2.
2) It is not clear from the observation data when the operator doubted the steam flow meter failure in period c1. However, from the thinking state estimation with EEGs, it is assumed that it was at step 31, when Mode B changed to Mode A.

5. CONCLUSION

A plant operator's thinking state model was proposed. First, the human's logical thinking state was classified into three basic modes, and it was validated by mathematical problem solving that the proposed model is useful to explain the human thinking process. Secondly, the thinking state model was applied to plant operation. The observation data corresponded well with the outputs of the thinking state estimation model. The state is liable to be overlooked by analyzing with a simple interview and observation. This information is useful to know the cognitive state of the operator, who works for logical thinking jobs such as the cause of plant failure identification. This enables us to investigate human machine interaction in detail with referring to the human's thinking state.

ACKNOWLEDGMENTS

The authors gratefully acknowledge financial support from the Japan Society for the Promotion of Science under the research for the future program (JSPS-RFTF96R14301).

REFERENCES

Funada, M. F., Y. Yazu, K. Idogawa and S. P. Ninomija (1997). How efficient are combined tasks for effective HCI? — An objective evaluation through AR analysis of EEG —. In: *Advances in Human Factors/Ergonomics 21A; Design of Computing Systems: Cognitive Considerations* (G Salvendy M.J.Smith and R.J.Koubek, Eds.). Vol. 1. pp. 579–586. Elsevier. New York.

Fuwamoto, Y., E. Nishina, S. Nakamura, N. Kawai and T. Oohashi (1997). Evaluation of informational enviroment in the train cabin by the field-type EEG telemetry system. *Proceedings of the Thirteenth Symposium on Human Interface (Japan)* pp. 205–210.

Musha, T., Y. Terasaki, H. A. Haque and G. A. Ivanitsky (1997). Feature extraction from EEGs associated with emotions. *Artificial Life and Robotics* **1**, 15–19.

Musha, T., Y. Terasaki, T. Takahashi and H. A. Haque (1996). Numerical estimation of the state of mind. *Proceedings of the 4th International Conference on Soft Computing* **1**, 25–29.

Nishitani, H., T. Kurooka and T. Kitajima (1997). Evaluation method of human interface in CRT operation using protocol analysis. *Journal of the Society for Industrial Plant Human Factors of Japan* **2**, 16–25.

Saiwaki, N., K. Kato, T. Inoue and S. Inokuchi (1997). An approach to visualization of information flow of EEGs by direct coherence analysis. *Transactions of the Institute of Systems, Control and Information Engineering* **10**, 537–546.

Tanaka, H., T. Abe, Y. Nagashima, Y. Nishitani and H. Ide (1997). The realtime EEG analysis for the communication interface of intention. *Proceedings of the Thirteenth Symposium on Human Interface (Japan)* pp. 211–216.

CONTEXT IN MODELS OF HUMAN-MACHINE SYSTEMS

Todd J. Callantine

San Jose State University/NASA Ames Research Center

Abstract: All human-machine systems models represent context. This paper proposes a theory of context through which models may be usefully related and integrated for design. The paper presents examples of context representation in various models, describes an application to developing models for the Crew Activity Tracking System (CATS), and advances context as a foundation for integrated design of complex dynamic systems. *Copyright © 1998 IFAC*

Keywords: Models, Modeling, Man-Machine Systems, Design, States, Events, Constraints

1. INTRODUCTION

Models are powerful tools for the analysis, design, and evaluation of human-machine systems. However, their utility depends on the fidelity with which they represent interactions among humans, machines, tasks, and the environment, and the effort required to develop and apply them effectively. These factors have led to the development of a variety of models that address one or more facets of analysis, design, and evaluation of a particular human-machine system, or subsystem. Some models focus on machine behavior to better understand machine responses to human and environmental inputs and design displays, and ensure that a system will operate safely (e.g., Degani, and Heymann, 1998; Feary, *et al.*, 1997; Leveson *et al.*, 1997; Sherry, 1995). Others focus on operator activities to design and analyze interactions, and provide the foundation for analysis tools, training systems, and interface designs that incorporate knowledge-based displays and aiding functionality (e.g., Callantine, 1996; Callantine *et al.*, 1997, 1998; Funk and Lind, 1992; Mitchell, 1996, 1998).

These and other successful models prescribe and/or describe, with sufficient coverage and at appropriate levels of abstraction, the salient behaviors of relevant system elements in a form that supports both computational application(s) *and* communication between designers and practitioners in the domain of interest (cf. Heimdahl, *et al.*, 1997; Mitchell, 1996). As the complexity of systems and associated design efforts grows, it has become increasingly important to develop multi-purpose, computational models that meet these requirements and effectively support the design process. A new air traffic management system, for example, involves numerous designers and practitioners collaborating to concurrently develop new cockpit interfaces, air traffic control automation, and procedures—to name a few—all of which have extensive organizational impacts.

Integrated design on a large scale necessitates ways to effectively *combine* a variety of models with different focuses and forms, so that designers and stakeholders can use models suited to their areas of expertise to facilitate the design process and operation of the resulting system (cf. Vakil and Hansman, 1998). Models 'reverse-engineered' from available technical information about a new piece of technology can foster discrepancies in the resulting operator procedures and training (Smith and Moses, 1998). Instead, a collection of models should evolve together, creating a record of the design process that 'lives' in the design.

One possible solution lies in a feature common to all successful models, regardless of their focus or form: a means of representing context. Context is a central tenet of human-machine systems engineering (Mitchell, 1996). From the perspective of a given element in a human-machine system, context can be thought of as the relationship of the state of the given element to the dynamic collection of constraints imposed by other elements. Models of human interaction with complex systems, for example, typically represent operator activities together with conditions, or rules, that indicate the context in which the representation prescribes and/or describes operator task performance. Similarly, machine models represent the conditions under which an operator intervention results in a specific machine behavior. However, modeling efforts have attended more to the preferred units of analysis, coverage, and level(s) of abstraction for operator or machine behaviors than to the associated context. Context representations, while 'explicit,' are seldom hierarchical and lucid from various perspectives.

The thesis of this research is that, within a domain, models with different focuses and forms may be usefully related through context, and that context can provide the foundation from which different models can evolve along with a design while maintaining consistency. This paper proposes a theory of context to investigate these assertions. The theory posits classes of domain-specific context information can be combined to explicitly represent context in a general form, at multiple levels of abstraction. The theory arose from research on the Crew Activity Tracking System (ÇATS), which uses a model of correct task performance to predict operator activities and interpret operator actions in real-time (Callantine, 1996). A CATS model uses logical equations of Boolean-valued "context specifiers" to represent when an operator activity should be performed, as shown in Fig. 1. Similar expressions are used for other representing other conditions, such as when an activity is no longer applicable.

The paper gives examples of context representation in human-machine systems models, then describes the theory. Next, the paper presents a prototype modeling tool, called the CATS Modeler, that embodies the proposed theory. The CATS modeler enables domain-specific knowledge required by a CATS model to be specified and visualized graphically. The paper concludes with remarks on potential benefits of the theory.

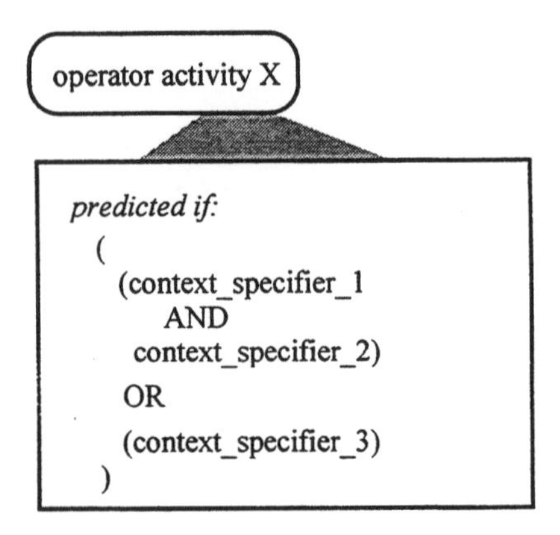

Fig. 1. Generic depiction of a logical equation of "context specifiers" as conditions for predicting an activity in a CATS model.

2. CONTEXT REPRESENTATION IN HUMAN-MACHINE SYSTEMS MODELS

Before describing the theory of context, the paper first provides examples of context representation in human-machine systems models. All human-machine systems models represent context; indeed, it is context that makes modeling human-machine interactions tractable (Mitchell, 1996). How context is represented, however, varies with the form, unit of analysis, level of abstraction, and intended application of a particular model. Context is typically represented at a level of abstraction congruent with the model hierarchy and intended application. Moreover, modelers commonly make inexplicit assumptions when specifying context.

First consider CATS models, which use operator activities as the unit of analysis. Table 1 lists examples of context specifiers defined to represent context in

Table 1. Examples of context specifiers that have been used as conditions in CATS models. Italics denote context specifiers expressed at an elemental level; others are expressed conceptually.

Name	Description	Name	Description
aircraft_heading_within_limits	Heading within +/- 0.5 degrees of cleared heading issued by ATC	*afs_engaged_roll_mode_LNAV*	Engaged autopilot roll mode is LNAV
mcp_altitude_outside_limits	MCP altitude does not match cleared altitude from ATC	aircraft_speed_outside_XR_limit	Aircraft cannot attain the speed required to meet the next waypoint crossing restriction
fms_target_speed_outside_limits	FMS-computed target speed more than 10 knots from speed issued by ATC or published procedure	cdu_descent_speed_built	A descent speed (*e.g.*, 320/) that matches the ATC-cleared descent speed has been "built" in the CDU scratchpad
fms_target_altitude_within_limits	FMS-computed target speed within +/- 100 feet of altitude directed by ATC or published procedure	cdu_descent_mach_entered	A descent mach (*e.g.*, .82/) built in the CDU scratchpad that matches the ATC-cleared descent mach has been selected to the appropriate line on the correct CDU page
afs_engaged_pitch_mode_FL_CH	Engaged autopilot pitch mode is Flight Level Change	cdu_execute_complete	FMS has computed a valid route after a short time delay, and the previous modified route is now the active route

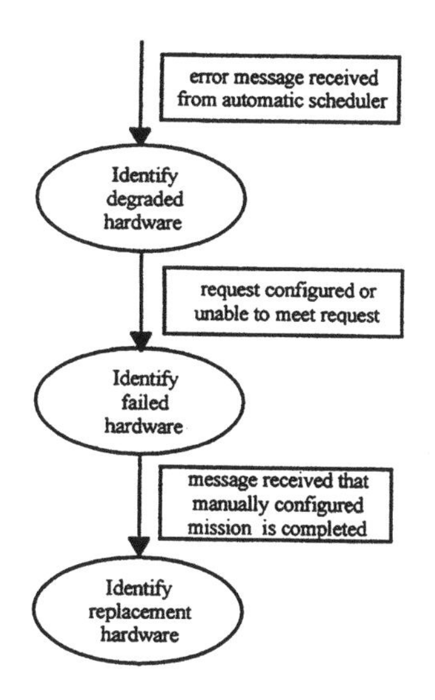

Fig. 2. Portion of an Operator Function Model for satellite ground controllers showing conceptual context information (Rubin, *et al.*, 1998).

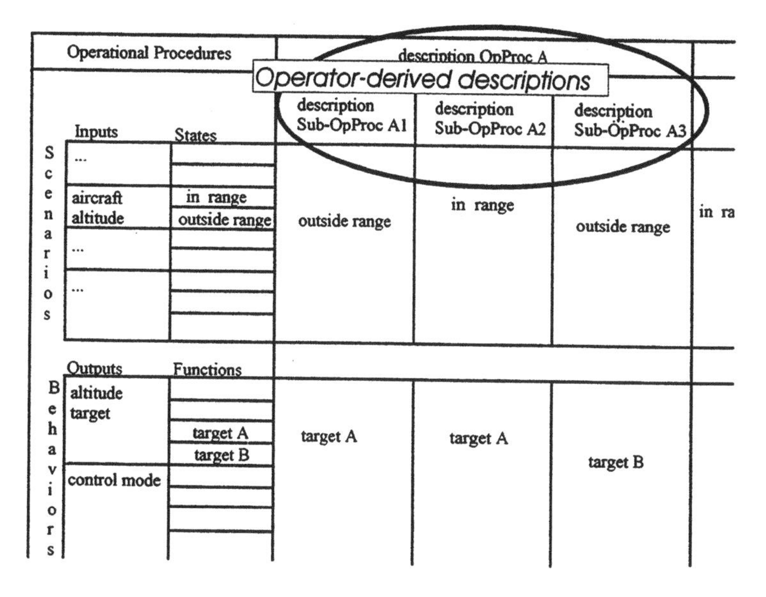

Fig. 3. Portion of an Operational Procedure (OpProc) table for flight management system design and training depicting some typical contextual information (e.g., aircraft altitude in range) and descriptions used by operators (Sherry, 1995; Feary et al., 1997).

CATS models (Callantine, 1996; Callantine, *et al.*, 1997, 1998), illustrating how they attempt to capture the manner in which supervisory controllers express when interventions with automated systems are needed. Some are expressed in terms of elemental state variables, while others are conceptual—they use concepts (e.g., target speed) that can have different meanings to different people or under different circumstances. CATS models are intended to be *memoryless*; evaluating the conditions in the model using the current state information should yield the currently preferred set of operator activities. A memoryless model can therefore readily answer "what if...?" queries.

However, memoryless context representations can be difficult to specify. When representing cognitive or perceptual activities along with overt operator actions, which is important (e.g., Mitchell, 1998), *exogenous* state information on which to base context may not be readily available; for example, it may hinge on the 'state' of verbal discourse between humans. In other cases, the information may be *endogenous*, for example, a constraint on available cognitive resources. For these reasons models often express context partly in terms of events, or as a script of behaviors. In Operator Function Models (OFMs), for example, operator-relevant concepts accompany elemental context information (Fig. 2) (Rubin, *et al.*, 1988). The proposed theory allows 'context models' to be used to explicate assumptions and support queries to the model in these cases, an idea that is further described below.

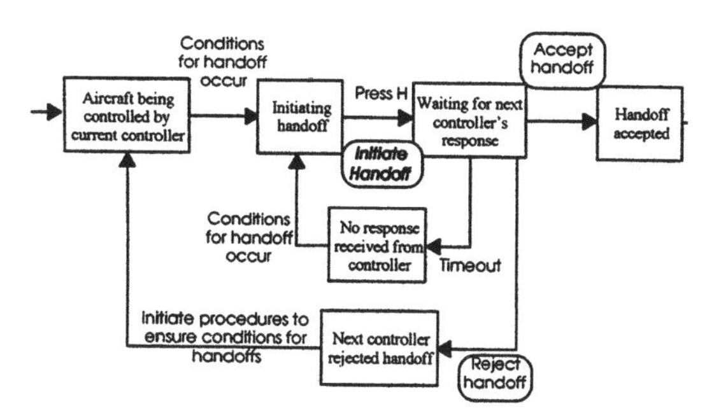

Fig 4. Portion of a SpecTRM-RL model of an air traffic controller procedure for "handing off" an aircraft to a different controller. Italics denote an activity performed by the 'handing off' controller (Leveson, *et al.*, 1997).

Models that represent machine behavior are typically state-based; Mitchell (1996, 1998) describes how operator activity models originated from state-based models. Sherry's (1995) Operational Procedure model (Fig. 3), Leveson *et al.*'s (1997) SpecTRM-RL (Fig. 4), and the Statechart models developed by Degani and Heymann (1998) (Fig. 5) all use *states* as the unit

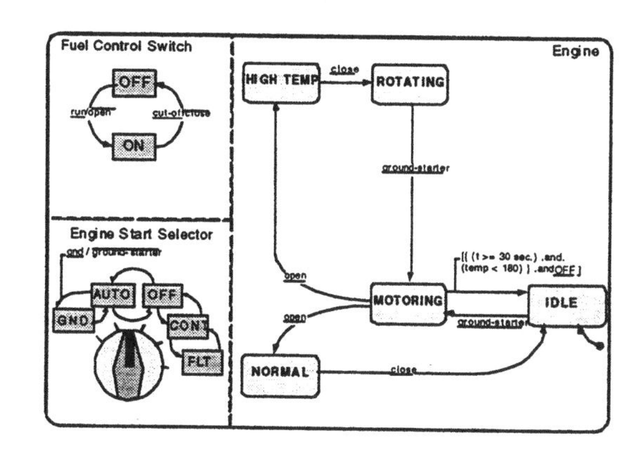

Fig. 5. Statechart model of a pilot procedure for "hot" engine starts uses both elemental and conceptual context information (Degani and Heymann, 1998).

of analysis, and include operator interventions as part of the contextual representation. State-based models are receptive to formal methods, making them advantageous for code generation (Sherry 1995), safety analyses (e.g., Leveson, *et al.*, 1997) and procedure design (e.g., Degani and Heymann, 1998).

Human-centered design efforts have included attempts to bridge the representational gap between state-based and event-based models. For example, Heimdahl *et al.* (1997) and Sherry (1995) both draw state-based models of machines toward event-based descriptions of operator activities, while Mitchell (1998) stresses the importance of representing system state in an OFM-based design methodology. Because research has attended more to extending models than relating them via context, some interesting crossover has occurred; for example, Fig. 5 depicts a state-based model developed to apply safety analyses to a new operator procedure. Because the procedure is part of a multi-agent system, this model also includes conceptual information about the actions of *another* air traffic controller, as well as the controller performing the procedure. This example suggests that even when context is expressed at a level likely to match that of an activity model, context information still needs to be decomposed to determine how the models may relate.

3. A THEORY OF CONTEXT

This section describes how context representations described above are theoretically related. *Context* is the relationship between the state of a system element—some partition of the system—and the dynamic collection of constraints imposed on it by other elements. A *constraint* is a set of bounds on a *state*—a property that describes some aspect of the system element's overall condition at a given time. Context may be represented by relating five classes of information: state variables, limit states, concepts, values, and modifiers. A *limit state* is a *state variable* that is part of a constraint. In aviation, for example, air traffic controllers can impose a constraint on aircraft heading; "cleared_heading" denotes a limit state based on "aircraft_heading," which is a state variable that represents an aspect of the aircraft's overall state (i.e., its heading).

Concepts are contextual information that is expressed at a level of abstraction higher than the elemental level at which a state variable represents an invariant with respect to the world. For example, in a context specifier such as "high_on_the_path," "path" is a concept used to represent a high-level constraint generated by an aircraft's Flight Management System. By contrast, "aircraft_ heading_within_limits" is a context specifier expressed in terms of an elemental state variable (i.e., "aircraft_heading") whose *meaning* is invariant—while there are different ways of expressing heading, they can be reliably converted. Concepts may also represent things that are temporally removed from the present. For example, predicted future states or *events* used to describe the attainment of some state are concepts. The remaining class of context infromation, *values*, may be Boolean, numeric, fuzzy, etc., as long as they have the same units as other context information they appear in a relation with.

The theory provides for the representation of concepts in one of two ways. First, they may be decomposed via logical equations of context specifiers representing lower-level concepts until the level of elemental state information is reached. Alternatively, a concept can be evaluated through the use of a "context model"—a model that takes state variables and limit states as input and produces output that matches the units of the relationship in which the concept appears. The model output might be a Boolean value, a predicted future state value, or a fuzzy-valued determination about progress toward meeting some constraint. Thus, through the notion of concepts, the theory affords both flexibility and the capability to represent context in terms salient to various designers and practitioners.

4. REPRESENTING CONTEXT USING THE CATS MODELER

A model's utility depends increasingly on a supporting framework that includes an application methodology and computer-based tools for developing and visualizing the model, so that it can be easily modified and understood by people with different areas of expertise. A model of a new procedure for example, may change frequently in response to decisions made by automation designers, and vice versa. The process of negotiating such changes requires conventions for communicating about them and thus benefits from a way to visualize the relevant models (e.g., Mitchell, 1998; Leveson, *et al.*, 1997). A prototype tool, called the CATS Modeler, is under development for the purpose of specifying and visualizing CATS models. The CATS Modeler is implemented in Java™, with an eye toward permitting collaborative model development via the World Wide Web.

CATS models are represented in computer-readable files that afford easy editing of the activity hierarchy and conditions (Callantine, 1996). In addition to making the specification process graphical, the CATS Modeler is designed to allow the CATS context specifiers to be defined graphically. Because CATS is designed to take system state information and constraints on operation as inputs for predicting operator activities according to the conditions in a model of operator activities, defining the data required and mechanism for evaluating a given context specifier at runtime is especially important.

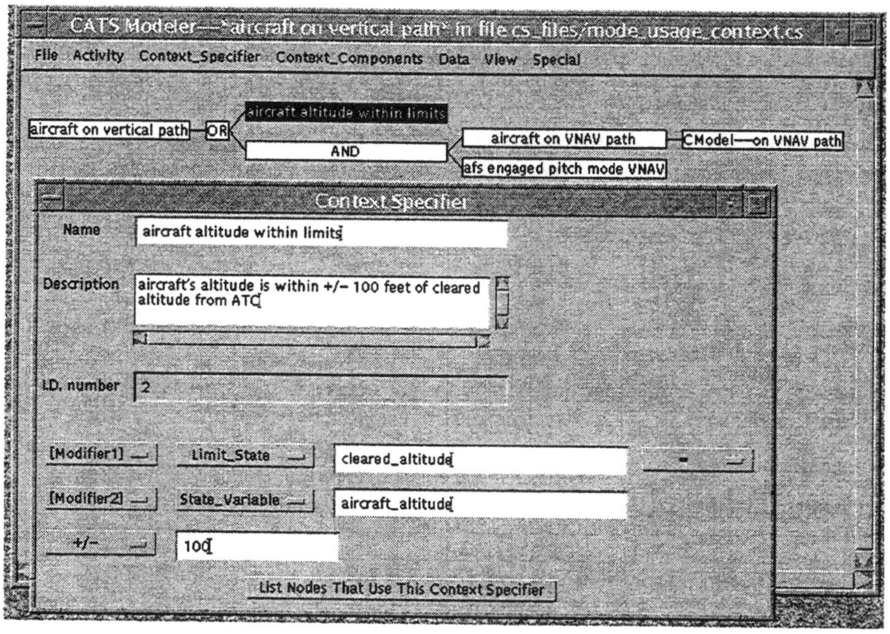

Fig. 6. Screen snapshot from the prototype CATS Modeler, showing the dialog used to define an elemental context specifier ("aircraft altitude within limits"), and a logical equation of context specifiers used to define a context specifier that is a *concept* ("aircraft on vertical path").

By implementing the proposed theory, the CATS Modeler offers a great deal of flexibility for specifying contextual information. Elemental context specifiers are defined using a simple dialog; context specifiers that are concepts can be defined as logical equations of other context specifiers. Fig. 6 shows an elemental context specifier ("aircraft altitude within limits"), and a context specifier that is a *concept* being defined for use in a CATS model. The AND/OR tree used to express the logical equation supports additions, deletions, and adjustments via graphical manipulation, while the elemental context specifier is defined by relating a state variable and limit state using a dialog to create a definition "sentence." Other choices offered in the dialog are enumerated in Fig. 7. One result of the context specification process is knowing that CATS must have access to a limit state called "cleared_altitude" and a state variable called "aircraft_altitude" in order to evaluate conditions, and concept-level context specifiers, that include the context specifier "aircraft altitude within limits." The second is that context specifier definitions created through the graphical specification process are explicit, hierarchical representations that eliminate the need to write code to evaluate context specifiers that appear in a CATS model—all of the knowledge that a CATS model represents can be specified and visualized graphically.

The 'context models' allowed for by the theory of context implemented in the CATS Modeler are exemplified by the "CModel—on VNAV path" node that appears in Fig 6. 'Context models' are a way of expressing, first, that it is too tedious to express the concept in question using logical equations of lower-level context specifiers—and that the model *is* making some assumptions here that have yet to be explicated. More importantly, however, they may serve as links through which different types of models can be integrated for various purposes. In this case, for example, a model certainly exists for how the Vertical Navigation (VNAV) mode of an aircraft 'knows' whether it is on the computed vertical path; the model could be 'attached' here to accurately provide this context information. Simulators or other sources are often not as accurate, or the information is distributed, making it difficult to be certain of its validity at any given time. In addition, context may depend on predicted future states that require a 'context model' for evaluation. The context specifier "aircraft_speed_outside_XR_ limit" in Table 1 exemplifies a concept that requires a predictive model.

Similarly, models attempting to predict what an operator will *actually* do in a given situation might use a 'context model' based on a theoretical model of cognition or perception to explicate the role of a particular decision making process or interface in determining context. Whether or not the models employed for this purpose are accurate, they nonetheless explicate assumptions about context that impact the utility of a human-machine systems model.

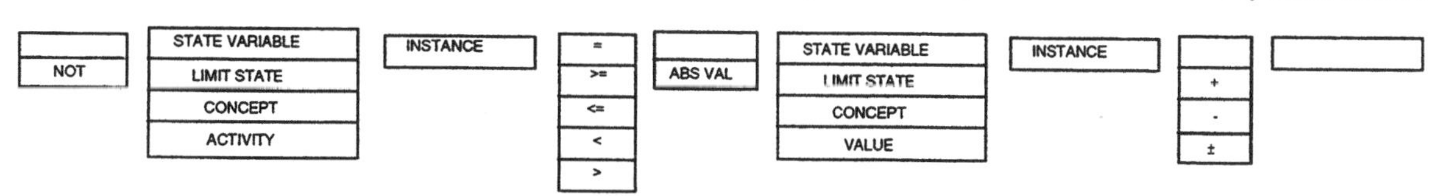

Fig. 7. Choices offered for creating a "sentence" that defines an elemental context specifier using the prototype CATS Modeler.

One additional noteworthy item is the appearance of the choice "ACTIVITY" in Fig. 7. Because CATS assigns 'statuses' to activities in a model as part of its processing methodology (Callantine, 1996), the CATS Modeler offers the opportunity to 'chain' contextual information as a modeling convenience, similar to scripting activities in other models. If the conditions under which an activity attains some status is well-defined, using the fact that the activity *has* that status at the current time is nearly as good as explicating the context. However, interactions between the predictive and interpretive portions of the CATS processing scheme (Callantine *et al.*, 1998) must be taken into account when exercising this option.

5. CONCLUDING REMARKS

The notion of 'context as constraints' is very powerful; in fact, any human-machine systems research or design endeavor can be cast in terms of context. For example, any interaction in a human-machine system can be viewed as conveying or describing progress toward some constraint. Designing complex systems essentially entails determining how to generate, communicate, assess, amend, and achieve dynamic constraints. Notions such as "cognitive complexity" and "mode awareness" describe situations when operators are tasked with converting constraints or assessing them in terms relevant to the machine, instead of the goals they are trying to achieve.

This paper proposed a theory of context as a means of relating human-machine systems models in one of two ways: first, decomposing context information into its elemental form, using state descriptors with invariant meaning or, second, using one model to support context representation in another. Application of the theory could make models more representative of the system element(s) they are abstractions of, easier to specify, and better suited for a particular application. The perspective taken establishes a starting point for integrated design, viz., the constraints that are the medium through which the system elements relate.

6. ACKNOWLEDGEMENTS

This work benefited from discussions with a number of insightful individuals, including Asaf Degani, Christine Mitchell, Everett Palmer, Nancy Leveson, Thomas Prevot, Micheal Shafto, Lance Sherry, and Charles Hynes.

7. REFERENCES

Callantine, T. J. (1996). Tracking operator activities in complex systems. Unpublished Doctoral Dissertation, Atlanta, GA: Georgia Institute of Technology.

Callantine, T. J., Palmer, E. A., and Smith, N. (1997). Model-based crew activity tracking for precision descent procedure refinement. *Proceedings of the 1997 IEEE Conference on Systems, Man, and Cybernetics*, Orlando, FL.

Callantine, T. J., Mitchell, C. M, and Palmer, E. A. (1998). GT-CATS: Tracking pilot mode usage activities in the glass cockpit. *International Journal of Aviation Psychology*, submitted for publication.

Degani, A. and Heymann, M. (1998). Formal aspects of procedures: The problem of sequential correctness, submitted for publication.

Feary, M., Alkin, M., Palmer, P., Sherry, L., McCrobie, D., & Polson, P. (1997). Behavior-based vs. system-based training and displays for automated vertical guidance. In *Proceedings of the Ninth International Symposium on Aviation Psychology*. Columbus, OH.

Funk, K. H., and Lind, J. H. (1992). Agent-based pilot-vehicle interfaces. *IEEE Transactions on Systems, Man, and Cybernetics, 22*(6), 1309-1322.

Heimdahl, M., Leveson, N. G., Reese, J. (1997). Intent specifications: An approach to building human-centered specifications. Everett, WA: Safeware Engineering Corporation.

Leveson, N. G., Sandys, S., Pinnel, D., Brown, M., Joslyn, S., Alfaro, L., Zabinsky, Z., and Shaw, A. (1997). Final report: Safety analysis of air traffic control upgrades. NASA Contractor Report: Ames Research Center.

Mitchell, C. M. (1996). Models for the design of human interaction with complex systems. In *Proceedings of the Conference on Cognitive Engineering Systems in Process Control*, Kyoto, Japan, November.

Mitchell, C. M. (1998). Model-based design of human interaction with complex systems. In W. B. Rouse and A. P. Sage (Eds.), *Systems Engineering & Management Handbook*. New York: Wiley (to appear).

Rubin, K.S., Jones P. M., and Mitchell, C. M. (1988). OFMspert: Inference of operator intentions in supervisory control using a blackboard architecture. *IEEE Transactions on Systems, Man, and Cybernetics, **18(4)***, 618-637.

Sherry, L. (1995). A formalism for the specification of operationally embedded reactive systems. In *Proceedings of the International Council of System Engineering*, St. Louis, MO.

Smith, N.and Moses, J. (1998). Personal Communication, NASA Ames Research Center.

Vakil, S. S. and Hansman, R. J. (1998). Functional models of flight automation systems to support design, certification, and operation. AIAA Report 98-1035. Reston, VA: American Institute of Aeronautics and Astronautics.

FULFILLING CUSTOMER REQUIREMENTS UNDER COMPLEXITY AND UNCERTAINTY - FORECASTING AND PLANNING VS FLEXIBILITY OF HUMAN-MACHINE SYSTEMS -

Thomas M. Buro

Airport Research Center (ARC), University of Technology (RWTH) D-52056 Aachen, Germany, buro@wtal.de

Abstract: This report deals with the concept of socio-technical systems (OSTO) and the ability to fulfil customer requirements in a flexible organisation. Examples developed by the Theory of Constraints (TOC) are presented and compared to traditional approaches of forecasting and planning. The methodology is then applied to the field of global transport systems and a new approach is suggested there. *Copyright © 1998 IFAC*

Keywords: Socio-Technical System Design, Constraints, Networks, Planning

1. INTRODUCTION

Today the analysis, design and evaluation of human-machine systems have to reflect the complexity of reality. Thus these systems need the ability to cope with a high level of complexity and uncertainty while fulfilling customer requirements.

The traditional approach to design such systems is to forecast the future and plan operations. This approach tries to reduce the complexity and ignores a high level of uncertainty. Human-machine systems designed for such an environment do not reflect the outside complexity. If forecast and reality do not match problems will occur.

A different approach is to understand and analyse an organisation as a socio-technical system and to accept the complexity of reality today. Such a system has to be designed in a different way with enough flexibility to fulfil its mission and reach its goal.

The basic concept of this approach is presented here. It is illustrated by examples of complex situations involving interdependency and statistical fluctuations. This approach is then used to analyse and develop a feasible solution for air transport systems.

2. A METHODOLOGY OF ANALYSING HUMAN-MACHINE SYSTEMS

As a method to analyse, redesign and monitor a socio-technical system, the OSTO approach has been suggested specifically by Hanna (1988), Rieckmann and Weissengruber (1992), Henning and Marks (1992). The abbreviation means "open, socio-technical-economic system", i.e. the open system contains social and technical as well as economic components. It is build on the Socio-Technical System Theory which was considerably influenced by members of the London Tavistock Institute of Human Relations.

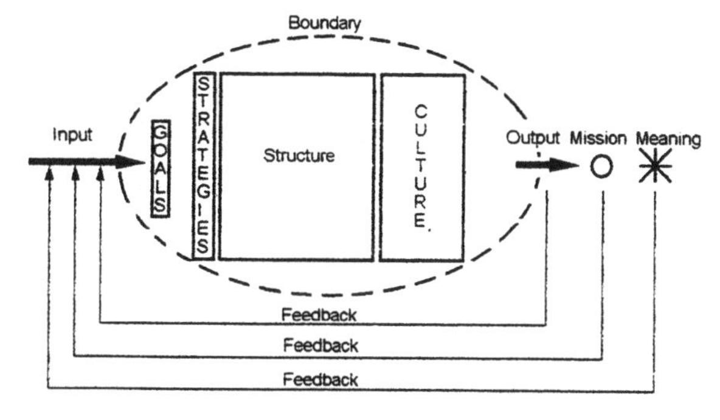

Fig. 1: OSTO approach, a recognised model of systemic management.

OSTO is focused on analysing work processes which are strongly inter-related. They are largely the result of a combination of a high level of technology and human communication. OSTO defines systems as "living systems" (open cybernetic systems). Such systems include humans with their processes of work and life. Feedback processes stabilise and renew this open living system (fig. 1.)

The system transforms input into *output* by a transformation process. Typical outputs are e.g. goods delivered, expenses, customers satisfaction, waste, staff attitudes etc. The *mission* is the reason for the existence of the organisation. It represents the unwritten contract between the system and its environment. In the long run, living systems cannot survive without defining its reason for existing. However, all processes within the system only become core processes, if they are orientated towards the mission.

Moreover, it is necessary for the survival of the system that the purpose and the *meaning* of the system is future-orientated. Only then, the members of the organisation maintain both the motivation for and the identification with the system. Thus, the system can maintain its acceptance in a wider social context.

Further elements of cybernetic systems are *feedback loops* which are either negative or positive (weakening or enforcing). They give to the system the qualities of stabilisation and renewal. In this context the organisation of feedback processes is an important managerial task with regard to the survival of the system.

System output and behaviour are regarded as consequences of the structure or design elements of the system (see fig. 2). The design elements of a system are linked with each other by dynamic and complex interrelations.

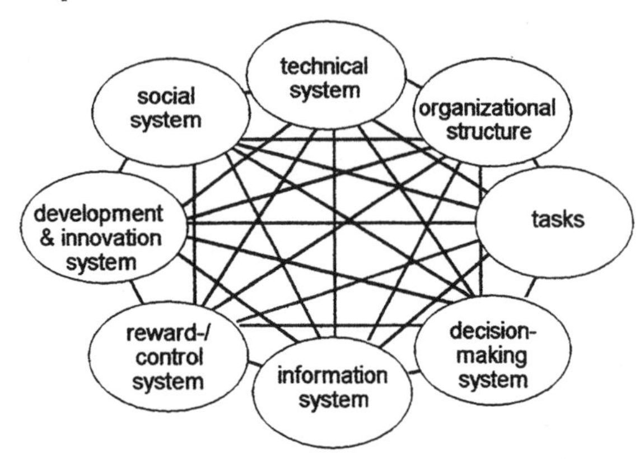

Fig. 2: Structure of an organisation

The main elements are the technical, social and organisational subsystems: technology, people and organisation. Modelling processes have to include these elements in order to be practice-oriented and "real".

A working model for complex human-machine systems is suggested by the Theory of Constraints (Goldratt et. al. 1984, 1990, 1994, Dettmer 1997). Systems are understood as a sequence of interdependent subsystems and human-machine systems. As the desired output of such systems is usually not infinite, the system must be constrained by a limited number of Constraints. The analogy of a chain is useful, as each chain has only one weakest link. These system Constraints or weakest links are the starting points for improvement. They can be part of the technical subsystem (e.g. the capacity of a machine) or part of any other design element of a system (e.g. a faulty policy).

The Theory of Constraints strives for the continuous improvement of the system as a whole and suggest a five-step process to deal with Constraints.

1. *Identify:* Identify the system constraint.
2. *Exploit:* Use every part of capacity as it exist now on the constraint as effective as possible
3. *Subordinate*: Change the design of the system (i.e. policies) in a way that the constraint is working most effective
4. *Elevate:* Change the system to eliminate the constraint. These changes are more severe, last longer and consume resources.
5: *Go Back, but don't allow INERTIA to form a system constraint:* As the constraint is broken actions and policies of steps 1-4 can be obsolete.

This methodology is an appropriate framework for modelling human-machine systems complex environments. Examples in production, supply chain management and transport networks are presented in the following paragraphs.

3. EXAMPLES FOR COMPLEX HUMAN-MACHINE SYSTEMS

A complex environment, which is traditionally managed by forecasting is distribution and supply chain management. The task of supply chains is to deliver products to the customer. The question is how many items of each product should be produced and shipped to different warehouses.

Usually marketing departments, regional warehouses etc. issue forecasts. These forecasts are aggregated and used for production and supplier schedules. Products are pushed according to the forecasts to the warehouses. The only problem is the customer who is not going to buy exactly what he should. This operation ends up with too much of one item or not enough of another item. Probably forecasting is inappropriate in complex and uncertain environments.

The numbers are the result of various guesses and judgements. These guesses end up in inventory pile ups, forecasts which are not in line with actual needs and production of items not longer needed (Newton, 1996).

A different approach is suggested by the Theory of Constraints. The main idea is to mirror the uncertain reality and use it to manage the flow of material in the system. The customers needs are not forecasted, but the actual rate of consumption is monitored carefully in a flexible distribution system.

In a supply chain the customer represents a constraint. The system can not sell more, than the customer buys and each item which is not available results in a lost sale. The system should have the ability to supply everything the customer wants to buy. The TOC application for this task is called drum-buffer-rope.

The rate of consumption is reported immediately to production to control the rate of finished goods released into the supply chain. This is called the drum. The amount of raw material, work-in-process inventory and finished goods in the pipeline represent the rope between production and the drum (constraint). A time buffer is used in the last tier of the supply chain in front of the customer. The buffer is necessary to cope with fluctuations on the supply chain (breakdowns etc.). The size of this buffer is determined by the rate of consumption, the ability to replenish the buffer and the fluctuations in the pipeline.

This distribution approach is not forecasting the consumption, but carefully monitoring the actual rate and reacting accordingly. The time buffer and the length of the rope adjusted according to the actual situation. Thus this mode of operation mirrors the complexity of reality instead of planning it.

Companies who have implemented the flexible Drum-Buffer-rope approach to their operations have improved their on-time performance with significantly less inventory (Deylen, 1998). The flexibility demanded by the customer and the TOC distribution system requires to mirror the complex and uncertain reality in production too.

The traditional mode of operation in production is too plan and schedule every task on every resource in advance. This has resulted in highly complex material planning systems (MRP II), reported as early as 1974 (Orlicky, 1974). The goal of these system is to maximise utilisation and reduce cost. The processing logic makes some assumptions which do not mirror the complexity of reality, which leads to serious distortions (Stein, 1996, Drexel et. al. 1995).

Such systems have to sub-optimise the following conflict: In order to maximise profit of the company, the cost of carrying inventory must be reduced and therefor small batches should be used. But on the other hand: in order to maximise profit the cost of set-ups must be reduced, thus big batch sizes should be used. This conflict is frequently compromised by formulas like Economic Order Quantity (EOQ) etc. These formulas are based on assumptions which do not mirror reality today (e.g. it assumes that the lot size remains constant, but a lot size might split in different process and transfer batches). This traditional production approach results in large batch sizes, long lead-times, high stock holding, high obsolescence and poor availability as the system is neither flexible nor is it an appropriate model.

The TOC process of continuous improvement starts with the identification of the constraint. This is probably the resource with the lowest capacity compared to demand. The next step is to exploit the constraint by building a schedule only for the constraint. The third step is to subordinate everything else with the drum-buffer-rope approach. The constraint determines the output of the whole system, thus it should be utilised maximal. A time buffer is built in front of the constraint to protect it from uncertainties in feeding operations. The constraint rate of production acts like a drum and controls the release of raw material into the production process. The amount of work-in-process inventory in front of the constraint represents the length of the rope.

Non constraint resources have by definition more capacity as needed. A part of this extra capacity is defined as protective capacity, which is necessary to recover from statistical fluctuations. Extra capacity which is not needed to cope with statistical fluctuations is defined excess capacity. Excess capacity on resources can be used to change set-ups more often, and handle small transfer batches.

The drum-buffer-rope approach is able to handle statistical fluctuations and interdependent activities in a flexible way. It keeps the human operator in full control and models his professional reality. The information system is focused on the constraining resources and gives the ability to take necessary actions, if the output of the constraint is in danger.

A methodology for analysing human-machine systems needs to mirror the complex professional reality. In this respect the model of complex, living systems and the Theory of Constraints is an appropriate framework for modelling transport networks, it is to be used for analysing "per se" systems, subsystems and hubs.

4. PRESENT GLOBAL TRANSPORT SYSTEMS

Global Air transport systems consist of several local or organisational subsystems with transitions between them. Transport networks which comprise of two or more different organisational subsystems are called combined systems. These organisational subsystems co-operate and form temporal transport chains.

A methodology of analysing this complex networked structure of transportation has to distinguish between two different approaches (see fig. 3): 1. analysing the system, subsystems and hubs "per se" and 2. analysing the specific system components which are relevant for transitions between the different local and organisational subsystems of transportation (Wollenweber, 1996). Such complex systems must constantly develop and redesign themselves to meet all requirements of their task.

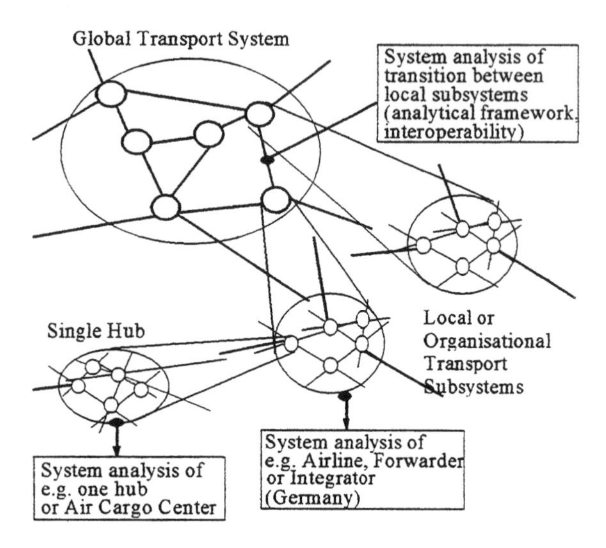

Fig. 3: Structure of system, subsystems and hubs.

The traditional system of air transport is formed by air cargo carriers (airlines) and freight forwarders with a network of agents around the world. Air cargo carriers provide global transport from airports to airports. The service is completed by freight forwarders between airport and customer. Thus a complete global transport system is only offered in co-operation of organisational subsystems. These networks are defined as combined systems

The financial goal of air cargo carriers is to make profits. The carrier is facing a conflict leading to several undesired effects, e.g. long lead times, low on-time performance, unused capacity, unnecessary rejected clients etc. This situation has not changed significantly over the last 25 years, e.g. the lead time for an average shipment from shipper to consignee is still 6 days (IATA, 1975, Bauer-Jones 1998).

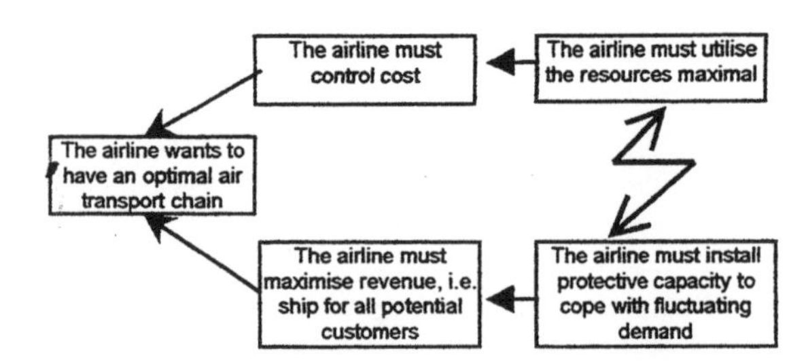

Fig. 4: Conflict Diagram for the airline situation.

The conflict is described in figure 4. The airline is interested in an optimal air transport chain to fulfil customer requirements and to sell the services it can provide. To built an optimal air transport chain the air cargo carrier has to control costs and utilise the resources. This is a common requirement in most organisations.

Furthermore it is necessary to fulfil customers needs on a high quality level. The airline has to maximise revenue, i.e. it is necessary to ship for all potential customers, but the demand for such services is fluctuating. One way to cope with uncertain demand is to install protective capacity (fig 5). These requirements result in a conflict as aircraft capacity is not flexible. There is a constant struggle between optimum utilisation of the resources and installing excess capacity to cope with uncertain peak demands.

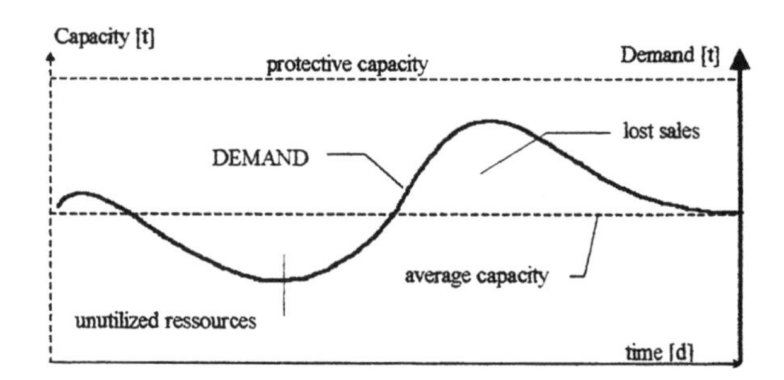

Fig. 5: Fluctuating demand and capacity.

Integrated transport systems (e.g. Federal Express, UPS, DHL etc.) are designed with sufficient protective capacity to cope with this fluctuating demand. The utilisation of aircraft based on tonne-kilometres is significantly lower compared to major air cargo carriers (IATA, 1996). This solution to the conflict is very customer oriented. Higher operating expenses are compensated by higher prices.

A common way of air cargo carriers to cope with the complex and uncertain professional reality is to book shipments to specific flights in advance, reserving space and weight on an aircraft. The procedure is similar to the booking of passenger flights and to the scheduling of each task on every resource in production. The actual weights and volumes of freight are uncertain. If the installed capacity can cope with the average demand there are lost sales in cases of higher demand; and there is free space in case of lower demand which forces price reductions (fig. 5).

Average capacity and actual demand will seldom match. The level of noise in the system is even higher for a passenger airlines which also carriers freight. The number of passengers and the amount of passenger luggage influences the available freight capacity too.

A lot of effort has been made to increase the accuracy of bookings and forecasting of the demand. Some suggestions include penalties for inaccurate bookings, selling options for air transport services to shippers and adjust the capacity due to the price level of these options etc.

In contrary an approach to global transport systems which takes into consideration the complex and uncertain environment is presented here.

5. A NEW APPROACH TO GLOBAL TRANSPORT SYSTEMS

A different approach has to solve the air cargo carriers conflict. The injection to break the conflict of the air cargo carrier is aimed at the policy Constraints as there are the traditional offer and booking procedures. The airline should sell transport lead times to customers instead of capacity on flights.

The customer is asked for the required delivery date by a price system which reflects the lead times of transportation. Thus capacity and actual demand can be uncoupled from specific flights. Less urgent shipments can be postponed more urgent shipments will be handled with priority. The human operator is supported by a human-machine system to take these decisions. Such a new transport system must be flexible to handle shipments on different routes and in different orders, which requires new handling processes and probably constructive changes in air cargo handling terminals.

The approach takes advantage of the statistic rule, that the standard deviation of a sum of independent fluctuating variables is the root of the sum of the standard deviations of the single variables (Bronstein, Semendjajew 1989). The demand and capacity in the transport environment can be considered as such variables.

Based on the statistical rule, the noise in the system levels out with the amount of shipments aggregated. Demand and capacity should be matched on the highest aggregation level as possible. The highest feasible level is probably the demand for shipments between two regions over a time period of several days compared to all capacity available between airports in these regions. Thus several flights from the same airport can be aggregated over a period of time and as well as the flights within a region of several airports.

In the present the air cargo carriers try to match the fluctuating demand with the fluctuating capacity for each aircraft, as each shipment is assigned to specific flights, at a low level of aggregation.

This new approach needs different human-machine systems compared to the ones used today, as schedules are not determined by booking. One system has to restrict the release of shipments into the system, i.e. it has to compare the actual load and urgency in the system (shipments under control of the air cargo carrier) with the available capacity for a given period. The other system has to manage the flow of shipments inside of the transport system by evaluating the urgency of shipments and assigning them to specific flights.

The new approach is similar to the Theory of Constraints drum-buffer-rope approach in production. The first step is to identify physical Constraints and define it as the drum, i.e. the transport connections between two regions. The rate of transport between these regions serves as a drum and is used to control the amount of shipments which can be accepted by the air cargo carrier and released into the system.

The constraint (drum) must be protected against uncertainty by a time-buffer to exploit it as much as possible. This is possible as the air cargo carrier offers different lead times and prices. Thus the airline accepts shipments with different lead times and it should be possible to have a time buffer of shipments in front of the constraint, e.g. shipments with long lead times waiting for transportation until there is free capacity.

Each customer should be treated like a constraint too and the shipments should be protected as well. There has to be a shipment buffer to protect the promised due date for each shipment.

The time buffer of a shipment t_b is the difference between the promised lead time $t_{l,p}$ and the minimal possible transport time to the destination $t_{min,0,}$:

$$t_b = t_{l,p} - t_{min.0}$$

The minimal transport time is a function of the distance to destination, the frequency of available flights and the possibility to expedite. The flow of actual shipments in the transport system can be managed using the concept of time buffer consumption (Goldratt 1990a, 1994).

The time buffer consumption β is the ratio between the shipment buffer gone so far compared to the original shipment buffer in the beginning. When the transport time so far $t_{l,x}$ and the actual minimal time to destination $t_{min,x}$ is known, then the time buffer consumption can be calculated as the ratio between the actual transport time (minus the change in minimal transport time) and the original time buffer:

$$\beta = (t_{l,x} - (t_{min,0} - t_{min,x})) / (t_{l,p} - t_{min,0})$$

The human-machine system should provide the time buffer consumption for all shipments. The operator is given the responsibility to decide on which route to send them. Urgent shipments should be handled on direct flights, less urgent shipments can be shipped also via deviations.

This approach to human-machine systems is more flexible, it accepts the uncertainty and interdependencies of the present professional reality. The human-machine systems have to be designed to give the human operators as much decision support as possible, focus his attention on the important issues and keep him in full control of the process.

6. SUMMARY

Global transport systems are highly complex networks in an environment of uncertainty and interdependencies. As a method to analyse a socio-technical system, the OSTO approach has been suggested.

As an application to uncertain and complex tasks the Theory of Constraints drum-buffer-rope process has been suggested and presented for environments (distribution, production) where forecasting and planning is the traditional approach. The design of human-machine systems today should mirror the complexity of the professional reality and should take into consideration statistical fluctuations and interdependencies. The systems should be designed as flexible as possible and support the human operator.

It is feasible to apply the approach to global transport systems. A different solution compared to the traditional mode of operation has been presented. And the required design of human-machine systems which support the human operator and keep him in full control of the process in an uncertain and complex environment has been described.

REFERENCES

Bauer-Jones, G.L. (1998). Changing a culture, Is anyone listening. IN: *Proceedings of the 19th International Air Cargo Forum*, May 4th, The International Air Cargo Association, Paris.

Bronstein, I.N. and Semendjajew K.A. (1989). *Taschenbuch der Mathematik*, 24th ed., Teubner, Leipzig.

Dettmer, H.W. (1997). Goldratt's *Theory of Constraints, A Systems Approach to Continuous Improvement*, ASQC Quality Press, Milwaukee, WI.

Deylen, L. (1998). The implementation of Drum-Buffer-Rope at Oregon Freeze Dry and it's role in the design and installation of an ERP System. IN: *Proceedings of the Theory of Constraints International Symposium*, 11-12 May 1998, A.G. Goldratt Institute, London.

Drexl, A., Haase, K., Kimms, A. (1995). Losgrößen- und Ablaufplanung in PPS-Systemen auf der Basis randomisierter Opportunitätskosten. *Zeitschrift für Betriebswirtschaft*, **65**, No. 3, pp. 267-285.

Goldratt, E.M., Cox, J. (1984). *The Goal, Excellence in Manufacturing*, North River Press, Croton-on-Hudson, NY.

Goldratt, E.M., Fox, R.E. (1986). *The Race*, North River Press, Croton-on-Hudson, NY.

Goldratt, E.M. (1990a). *The Haystack Syndrome, Sifting Information Out of The Data Ocean*, North River Press, Croton-on-Hudson, NY.

Goldratt, E.M. (1990b). *What is this thing called Theory of Constraints and how should it be implemented?*, North River Press, Croton-on-Hudson, NY.

Goldratt, E.M. (1994). E. M., *It's not luck*, North River Press, Great Barrington, MA

Hanna, D.P. (1988). *Designing organizations for high performance*. Addison-Wesley Publ., Reading, MA.

Henning, K. and Marks, S. (1992). *Kommunikations- und Organisationsentwicklung*. Augustinus, Aachen.

International Air Transport Association, IATA (1996). *CART Report*. IATA, London.

International Air Transport Association, IATA (1996). *World Air Transport Statistics*. IATA, London.

Newton, K.C. (1996). Playing the hot potato in the supply chain. *Baylor Business Review*, **14**, No. 1, p. 13.

Orlicky, J. (1974). *Material Requirements Planning*, McGraw-Hill, NY.

Rieckmann, H. and Weissengruber, P.H. (1990). Managing the Unmanagable? In: *Management Development im Wandel* (Kraus, H., Kailer, N., Sandner, K., Eds.) pp. 27-96. Mack, Wien.

Stein, R.E. (1996). *Re-engineering the manufacturing system: applying the Theory of Constraints*. Marcel Decker Publ., NY.

Wollenweber, D. (1996). *Systemic Concept of Transportation Chains*. PhD Thesis, Aachen.

POINTING DEVICES FOR NAVIGATION IN WINDOWS-BASED MACHINE-CONTROLLERS

D. Zühlke, L. Krauss

University Kaiserslautern, Institute for Production Automation, Germany

Abstract: Windows-orientated software systems, such as WINDOWS, OS/2, MAC OS have meanwhile become the state-of-the-art in the field of home and office computers. They considerably ease the handling of complex systems for beginners due to the utilisation of well known metaphors, such as paper basket, file, sand-glass as well as direct manipulation techniques (e.g. Drag-and-Drop). Hence it is not surprising, that these systems are also used more and more in the field of industrial applications. Unfortunately developers do not take into account several important differences between office and industrial applications. One important difference is surely the necessity for a mouse-replacement, as a standard office mouse is not applicable due to dirt and the absence of a rolling surface. This paper will describe the problem and evaluate alternative devices. *Copyright © 1998 IFAC*

Keywords: Human-machine interface, pointing systems, interaction mechanisms, manipulation tasks, evaluation, tests, industrial production systems, process automation.

1 INTRODUCTION

In the past machine controllers have been built using customised hard- and software. Today the era of open controllers based on international standards in hard- and software has begun. In order to develop cost-effective solutions those standards are more and more influenced by the PC-world.

However, problems are evident, the WINDOWS-operating system and other WINDOWS-like Systems are designed to use interactive communication techniques, e.g. drag-and-drop, virtual sliders and rulers, which require a mouse or something equivalent as a pointing and navigation device.

Based on the experience with the German open CNC controller project OSACA/HÜMNOS (Boll et al. 1997) tests were conducted with machine operators to determine the suitability of different pointing devices like mouse, joystick, trackball, touchscreen, touchpad in industrial environments.

2 STATE-OF-THE-ART IN INDUSTRIAL POINTING DEVICES

A characteristic feature of Windows-systems is the direct manipulation. It enables users to operate almost entirely by means of pointing actions. Thus users get the feeling to work with real objects, which can be moved (Drag-and-Drop), reduced or increased in size. The GUI is considered as the model of a real world and designed correspondingly (Zeidler and Zellner, 1994). The precondition is the availability of a suitable pointing device, e.g. a mouse. The user clicks on the desired object with the mouse key (selection) and certain commands are executed by certain mouse actions (function activation).

However, the mouse is not always the most appropriate device to interact with computers. In industrial applications the mouse can not very often be utilised due to dirt and due to the lack of a horizontal surface for moving it. An effective utilisation of WINDOWS-systems in this field requires the development of a suitable alternative. Several alternatives have already been introduced, however, each has its advantages and disadvantages.

Depending on the relation between action location and target location pointing devices for direct manipulation can be divided into direct-control and indirect-control pointing devices (Shneidermann 1997). Direct-control pointing devices enable the user to make inputs directly with the hand on the screen surface at locations where the process information is displayed. In case of indirect-control pointing devices action location and target location are separated from each other. Compared to direct-control pointing devices more cognitive processing and an increased hand-eye co-ordination is required in order to bring the onscreen cursor to the desired position. Numerous devices are available in both categories, which have specific advantages and disadvantages.

Direct-control pointing devices

The **lightpen** is a direct-control pointing device that was frequently utilised in the past. It enabled users to select an object directly on the screen and to perform a positioning or other task. However, users were disturbed by a cable, that was necessary to transfer the information of the selected point to the computer. Furthermore, the lightpen principle works only with CRT´s and not with LCD´s which are more and more used in industrial applications.

A technique similar to the lightpen is utilised for **touchscreens**. It allows an intuitive handling and control by the user (see-and-point). The touchscreen does not require an extra device, that must be picked up by the user, but it enables inputs directly with the finger on the screen. The disadvantages are smudging of the screen surface by finger prints especially in industrial applications (e.g. with cooling lubricant mist), and parts of the screen are obscured by the users' hand. The input resolution is rather low due to the finger size. Hence touchscreens should only be utilised for selecting large-surface objects. The high friction between finger and touchscreen results in a good attenuation against undesired minute movements on the one hand but impedes the drag-and-drop function on the other hand. Depending on the physical principle touchscreens may not react to gloves, which must be taken into account when selecting a certain field of application.

A further limitation is the absence of the mouse-buttons. Whereas clicking and double-clicking can be realized by short finger strokes on the touch-sensitive surface, dragging while ´holding down a button´ will mandatorily require a two hand operation.

Indirect-control pointing devices

The device which resembles the mouse the most is the **trackball**. The trackball is also utilised in many industrial applications and enables a very exact positioning on the one hand but reacts rather sensitively to vibrations on the other hand. Hence it should not be used in mobile work places, such as train or airplane cockpits. Experiments with various designs have shown, that a **trackball** should be attenuated by e.g. felt dampers in order to avoid uncontrolled rolling after an input and to attenuate light vibrations. In comparison to the mouse the trackball does not require a lot of space. Desk space or mouse pads are not necessary.

The **touchpads** are also touch-sensitive devices. The finger movement is detected by a pressure-sensitive sensor foil. In the meantime touchpads are used extensively in laptops which has resulted in improved reliability and lower costs. Precise positioning can be achieved and touchpads are rather insensitive to dirt due to the non-existence of any open or moving components. The relatively large-surface contact between finger and foil results in a high attenuation against undesired movements. Thus touchpads are very suitable for mobile work places. (For the first time in cockpit technology touchpads are used in the Boeing 777 as an inflight input device).

The **joystick** has proved itself as a pointing device for many years and can also be found as a robust industrial design. However, experiments have shown very clearly, that it is a rather unsuitable tool for navigating on screens. Its utilisation is only recommended for moving machine axes, when the movement directions of the axes correlate with the axes of the joystick.

The **trackpoint** or **mousestick** can be considered as a miniature version of the joystick. It is a small isometric joystick (often embedded in laptop keyboards). It has a rubber tip to facilitate the finger contact and to avoid slipping. With modest practice, it can be used quickly and accurately while keeping the fingers over the keyboard. In industrial applications where the input devices are predominantly installed vertically and full keyboards are not so often used, a trackpoint has no advantages.

The **mouse-button** mainly equals the trackpoint in terms of design, but the characteristics of a button have been incorporated. It is force-sensitive and can be moved in four directions with the finger. Tests revealed that users had difficulties in getting used to this device especially while handling it vertically at the eye-level, however it is cost-effective and can be easily embedded even in sealed panels. Today, the mouse-button is already used in several industrial controllers (e.g. Allen-Bradley, DASA).

Function keys still remain the most frequent pointing device. However, their utilisation results in renouncing major advantages of Windows-systems, such as drag-and-drop. Sliders, rulers, and other formatting elements can not be used either.

In order to enable a basic navigation the alphanumeric keys are supplemented by special navigation keys, such as cursor right, left, up, down, page down, up. The navigation is limited to larger steps, e.g. one symbol or one input field. However this kind of navigation is acceptable for many applications in the field of control technique.

Unfortunately mistakes are very often made with respect to the selection and array of these keys. Hence it is very important in terms of ergonomics to group all navigation keys in one block and array them in a clear lay-out according to the rules of natural mapping. In order to reduce costs in many cases the standard alphanumeric keyboard is used to input navigation commands by simultaneously pressing several keys, e.g. Shift + Up, which is rather confusing and user-unfriendly.

Three logic levels must be distinguished for reasons of operating logic. Four directions of movement can be attributed to each level.

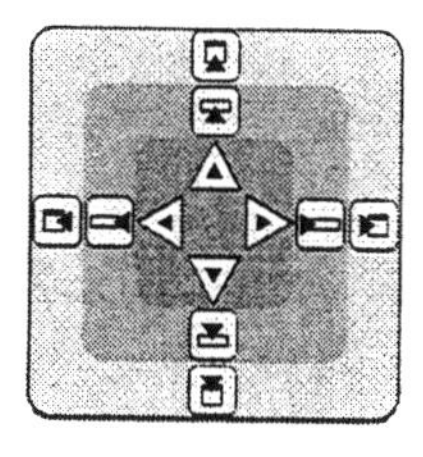

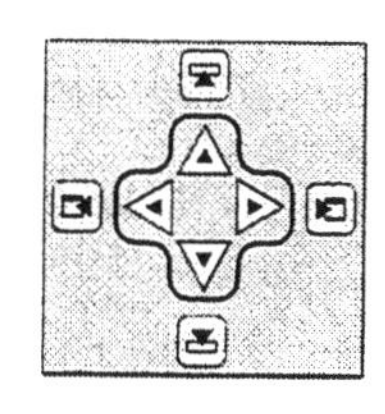

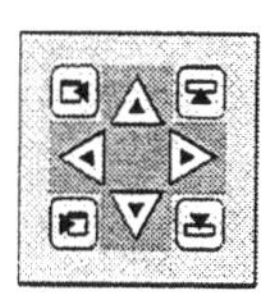

Figure 1 Array of cursor-movements

On the character-level the cursor is moved character by character, on the input field level from input field to input field and on the page or window level from page to page or from window to window. Since in many applications all four directions of movement are not required at all levels, it is possible to reduce the full arrangement shown in figure 1 left to the reduced versions displayed at the center and right.

Apart from the devices described in this study there are other devices, such as the Gyro-Mouse, the 6D-mouse, which will not be explained in detail, because their application in the field of process control has not become accepted yet.

3 VALIDATION AND COMPARISON

3.1 *Influence of operating position*

The operating position must always be taken into account before selecting a pointing device. Hence the mouse is a very suitable device while working in a sitting position at desks, however, it is absolutely inappropriate for working at vertically installed machine control panels.

Special emphasis must be put on the ergonomic requirements in the field of machine control, where panels are installed vertically and very often mounted in swivelling consoles.

The precise data input with pointing devices puts high demands on the users' fine-motoricity. These demands can only be met if on the one side the user's hand is supported (figure 2) and on the other side the pointing device is installed at an appropriate level in terms of ergonomics (mostly elbow-level).

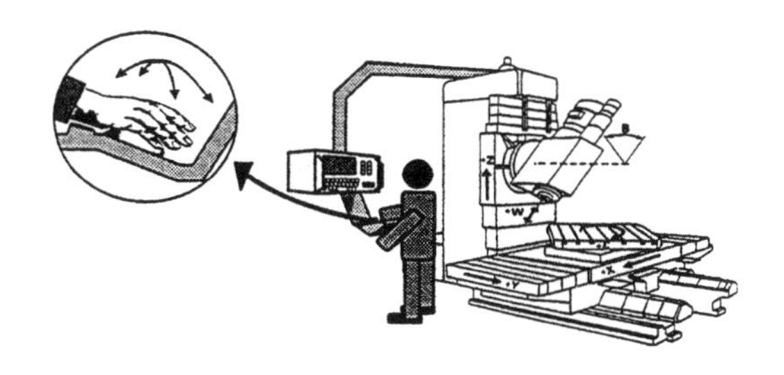

Figure 2 Ergonomic hand support at a machine controller

However, this requirement may conflict with the design of the control panel. If e.g. touchscreens are utilised an ergonomic operating position can only be achieved by installing the panel at elbow-level, but a good readability even from larger distances demands an installation at eye-level. Since both aims can not be accomplished at the same time, preference must be given to one or the other option. If the control panel is mounted in a swivelling console, any movement during operation must be impeded. Very precise pointing operations are made impossible by the slightest vibration. This especially applies to most force-sensitive industrial mouse-buttons. In order to perform very precise movements in two directions by using considerable power a wrist support and an absolutely rigid console construction are needed.

3.2 *Conducted Tests*

Problems with the input techniques for Windows-based software systems have only arisen recently. Therefore hardly any experience could be gained in this matter. Scientific results can only be obtained from realistic experiments with test persons. A first comparative investigation with 20 persons has been carried out at this institute and has provided some interesting results.

The test persons had to perform the following tasks:

1. Select a menu and pull-down to item

2. Click on several buttons

3. Select a large-surface window

4. Drag a rectangular to a given size and position

5. Select a text-string within a text-page

6. Track a given curve very precisely

A computer automatically recorded in a logfile the times needed by the test person to perform the tasks as well as the error rate (Ziegler and Ilg, 1993). Hence it can be assumed that the data is complete and not influenced by any disturbing factors. Since the logfile only shows times and does not give any information about the users intentions an additional questionnaire was handed to the users. It provides a subjective evaluation of each pointing device. Questions like „Do you have any difficulty in learning to operate the pointing device" have been asked among others. The suitability of the pointing devices for each test task was determined by a five-step ranking scale (Bortz and Döring, 1995). Impressions gained by observing personnel were recorded in an additional test protocol. The optimal operating position for each task was investigated by modifying the position of the pointing device horizontally or vertically.

As only one sample of each device was used for the first tests, the quantitatve results are not valid in generality. Comparisons with samples of different manufacturers unveiled significant differences especially for different touch technologies. Currently further tests are conducted with samples from different manufacturers to eliminate the technology influence. As most of these devices are not plug-compatible but instead require the installation of manufacturer specific drivers and a system-reboot, the tests turned out to be more time-consuming than previously expected. The latest and comprehensive quantitative values will be distributed during the conference.

4 RESULTS

Table 1 shows the comparison of measured times between the different pointing devices for each task in a desk-like workplace. The measured error rates representing the accuracy of operation mostly correspond to the time values. Only for Task 6 (precise curve tracking) the error rates of touchscreen and joystick are considerably higher

Table 1 Quantitative results for each task in relation to pointing devices (mean values, horizontal)

	Task-No.					
DEVICE	1	2	3	4	5	6
	Time [s]					
MOUSE	8,3	9,6	4,6	13,6	10,5	19,8
TOUCH-SCREEN	19,9	10,0	5,0	40,5	19,7	30,0
TRACK-BALL	9,9	11,6	5,8	14,0	12,3	28,0
TOUCH-PAD	18,7	17,5	11,1	37,9	19,5	45,0
JOY-STICK	20,2	20,3	11,1	58,1	30,3	35,8
KEYS	14,5	20,8	5,5	28,8	25,5	-----

Table 2 shows the qualitative results of the evaluation. The valuation criteria in the questionnaire were general valuations and explicit subjective criteria. For example:

Soil sensitiveness (A)

Attenuation (B)

Subjective valuation (C)

Optimal operating position (D). Some devices can be used in horizontal (h) or in vertical (v) position or both (h/v).

Table 2 Qualitative valuation criteria of pointing devices

	Criteria			
DEVICE	A	B	C	D
MOUSE	-	O	+	h
TOUCH-SCREEN	O	+	+	h/v
TRACK-BALL	O	O	+	h/v
TOUCH-PAD	+	+	O	h
JOY-STICK	+	O	-	h
KEYS	+	+	O	h/v

In case of an appropriate object display (object size adjusted to finger size) the touchscreen proves to be very suitable for quick pointing, due to an optimal hand-eye co-ordination (selection tasks). Its operation can be learned within a relatively short time. Major disadvantages become obvious if the touchscreen is utilised for movement tasks, i.e. if an object must be moved on the screen surface, a text must be highlighted or a line must be tracked. It is also unsuitable for exact positioning like on small standard Windows-elements (Radio-buttons, check-boxes).

As a pointing device which is very similar to the mouse the trackball achieved good results. It only shows disadvantages when it comes to exact tracking, which is reasoned by the finger position. A key must be pressed with the thumb or the index finger while the remaining fingers must move the ball at the same time. Most people have difficulties in performing the exact finger co-ordination required for this task.

The touchpad also achieved good results in simple pointing tasks. However, it shows a drawback at movement tasks, when cursors must be moved while pressing the Enter-key. The human finger co-ordination is unsuitable for the unfavourable construction. The index finger is utilised to select the desired object (e.g. painter or text marker), but the object can only be activated by pressing the ENTER-key with another finger. In most cases the thumb is used for this purpose.

Joysticks are a useful tool when e.g. a moving object must be tracked on the screen, but exact positioning and selection tasks can not be performed. Due to the usual way of operating a joystick (upright with loose fist) test persons considered it as very difficult to position the cursor in a reasonable time on a desired object.

The keyboard enables to select and activate single objects in a Window-surface but impedes spatial manipulation, i.e. movement of objects on the surface. However, if the system design has been adjusted especially to keyboard interaction as in the German Open CNC Controller Project OSACA/HÜMNOS (Boll et al. 1997), the keyboard proves to be a cost-effective and reliable alternative. Various input fields (e.g. text boxes) can be selected within a window by means of the group-change keys. A window-toggle key is used to jump from one window to another and the scroll keys to scroll within data fields (Fig. 3).

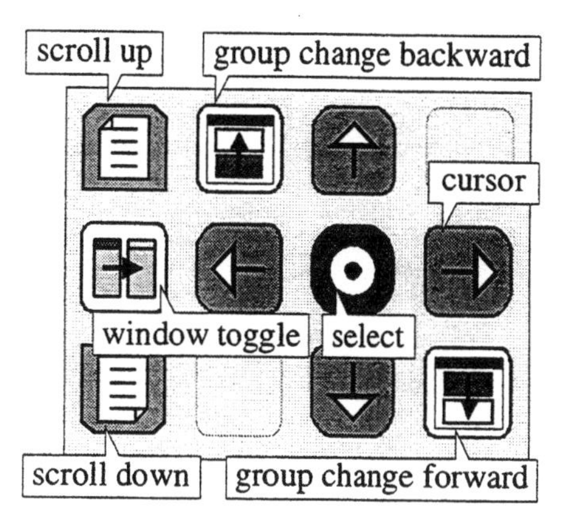

Figure 3 OSACA/HÜMNOS array of keys

5 CONCLUSION

If Windows-based systems are incorporated in the frame of the development of a new controller, a suitable navigation concept must be developed in the first place. The well-known techniques from office applications, such as mouse operation or drag-and-drop can not simply be adopted. The mere utilisation of a keyboard meets the industrial standard, but many advantages of modern software systems can not be made use of. New developments in the field of control technique will demand new solutions in the interaction with users. This study may facilitate to select an appropriate device. Further investigations need to follow resulting in concrete design guidelines for developers of control techniques. A final decision should only be taken after comparative tests with several test persons have been carried out under realistic process conditions.

REFERENCES

Shneidermann, B. (1997). *Designing the User Interface*. Addison-Wesley, Reading, Massachusetts.

Baumann, K., Lanz, H. (1997). *Mensch-Maschine-Schnittstellen elektronischer Geräte : Leitfaden für Design und Schaltungstechnik*. Springer, Berlin.

Zeidler, A., Zellner, R. (1994). *Software-Ergonomie : Techniken der Dialoggestaltung*. Oldenbourg, München.

Ziegler, J., Ilg, R. (1993). *Benutzergerechte Software-Gestaltung : Standards, Methoden und Werkzeuge*. Oldenbourg, München.

Bortz, J., Döring, N. (1995). Forschungsmethoden und Evaluation. Springer, Berlin.

Boll, J. et al. (1997). *OSACA / HÜMNOS Style Guide for Machine-Tool-GUI's : Ein Handbuch zur Gestaltung von Benutzungs-oberflächen für Werkzeugmaschinen*. Fraunhofer IRB, Stuttgart.

DESIGN AND EVALUATION OF LAPAROSCOPIC FORCEPS WITH ACCURATE FORCE FEEDBACK

J. L. Herder, K. T. den Boer, W. Sjoerdsma

Delft University of Technology, Faculty of Design, Engineering and Production, Section of Man-Machine Systems, Mekelweg 2, 2628 CD Delft, The Netherlands

Abstract: This paper presents newly designed laparoscopic forceps and describes the evaluation of the man-machine interface performance. Careful mechanical design has led to a purely mechanical instrument with excellent intrinsic feedback capacity, thus omitting the need for an electronic feedback system. This concept has relevance for the design of all manually operated instruments and tools. *Copyright © 1998 IFAC*

Keywords: backlash, friction, high-efficiency, inherent feedback, manipulators, medical applications, mechanical engineering, transmission characteristics.

1. INTRODUCTION

Laparoscopic surgery is an operating technique based on several small incisions in the abdominal wall instead of a single large one. In this procedure, workspace is created first, for instance by inflating the abdomen with carbondioxide gas. Next, an endoscopic camera and several long and slender instruments are inserted. The surgeon views the image of the endoscope on a monitor and carries out all manipulations with the instruments. The advantages for the patients (reduced pain, scarring, and hospitalization time) have led to widespread acceptance of this technique for several treatments (Cuschieri, 1995). Disadvantages of laparoscopy with respect to conventional, open surgery are revealed when the procedure of laparoscopic surgery is considered as a process of interaction between the surgeon and the patient (Grimbergen, 1997). Essentially, the 'man-machine' interface so to speak has been reduced to the handgrips of the instruments and the monitor screen only. As a result, surgeons have to rely on the two-dimensional image, on the sensory information passed through the instruments, and suffer from eye-hand coordination problems. A valuable and extensive elaboration on these difficulties is found in Breedveld (1998).

This study addresses the problem of the loss of force feedback. As opposed to the situation in open surgery, grasping instruments are the only source of tactile information for the laparoscopic surgeon. Therefore, the force transmission characteristics of these instruments are important features. Unfortunately, laparoscopic grasping instruments currently available possess much friction (obscuring the major part of the force feedback) and backlash (disturbing position feedback and hampering precise grasping). In addition, their force transmission ratios are strongly dependent on the opening angle of the gripper (Sjoerdsma *et al.,* 1997). This paper presents and evaluates the design of a purely mechanical grasping instrument, called the Wilmer® forceps, with negligible friction and a constant force transmission function.

2. METHODS

In principle there are three methods to solve the problem of friction: low friction mechanical design, compensation for friction, and teleoperation. These methods are briefly discussed next.

- *Mechanical solutions.* If friction could be eliminated from the mechanism itself this would clearly yield the most straightforward approach. However, conventional high-efficiency pivoting elements, such as ball bearings, cannot be applied due to space limitations and the fact that lubrication is not feasible.
- *Compensation for friction.* In the design phase, the amount of friction can be estimated. Thus, model-based feedforward could be applied: an actuator that appends the energy lost through friction can be connected to the operating lever. This method may work well when the mechanism is moving, but is not reliable in stationary situations, since then friction is indeterminate.
- *Teleoperation.* The problem of friction can be circumvented by the application of a master-slave servo system. In most master-slave systems, the grasping force is measured by a sensor in the jaws of the grasper and reproduced by an actuator on the operating handle (the master unit). Meanwhile, the position of the master is reproduced at the grasper (the slave unit) by a second actuator (*e.g.* Hill *et al.*, 1994; Howe *et al.*, 1995; Green *et al.*, 1995; Lazeroms *et al.*, 1997; Majima and Matsushima, 1991). This approach should allow easy adjustment of the transfer function, which may be advantageous for example in situations where operating forces are below the human sensory threshold, such as in microsurgery, and tremor can be filtered out before the master's movement is transmitted to the slave unit. However, master-slave system are complex and sensitive to disturbances and contact instability (Sheridan, 1996; Lazeroms *et al.*, 1997).

Worldwide, most research on laparoscopic instruments is directed at the application of master-slave systems. When the above three methods are inspected, it appears that teleoperation is the most promising indeed. If problems such as contact instability are overcome, the possibilities of haptic sensing (feedback of force distribution), operating at long distance (war scenes, space travel), and adjustment of transfer functions (force amplification, filtering of tremor) are attractive. This technical challenge appeals to surgeons as well as to engineers and researchers. However, one should not lose sight of reality. By far, most operations will take place with the patient and the surgeon in the same operating theatre. Therefore, the main effort should be directed to this situation where the surgeon stands (or sits) at the table the patient lies upon. Within this category, laparoscopy (endoscopic surgery of the abdomen) will remain a major operating area where long-distance operation or the adjustment of transfer functions have limited relevance. Moreover, the complexity, power requirement (electric wire), weight

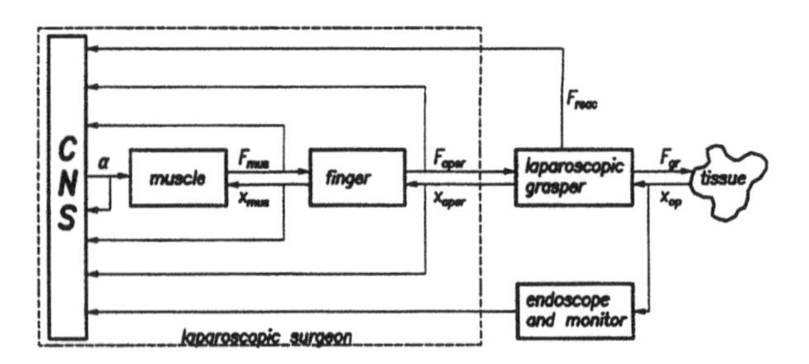

Figure 1. Block diagram of information paths in the control of laparoscopic grasping instruments. The upper part displays the available force feedback, whereas the lower part shows movement feedback. The operating force F_{oper} is the force on the movable part of the handgrip, whereas the reaction forces on the stationary part of the handgrip are represented by F_{reac}. Force feedback reaches the central nervous system (CNS) through the Golgi tendon organs (propriocepsis), and through tactile receptors in the skin of the hand.

and cost of master-slave systems justify a thorough study of the possibilities of improvements using the other methods. Since friction compensation has the same disadvantages as teleoperation without having its advantages, this method is not considered. The present study concentrates on the development of purely mechanical low-friction devices.

3. DESIGN

The grasping instrument described in this paper was designed by the WILMER section of the Man-Machine Systems group at Delft University of Technology, The Netherlands, who have specialized in mechanical design in medical and rehabilitation technology. The available theory on hand prosthesis design proved useful for manually operated instruments in general, including laparoscopic instruments. Figure 1 illustrates the concept of *extended physiological proprioception* (Simpson, 1974), originally derived for hand prostheses, applied to the situation of the laparoscopic surgeon. Movements (lower half of the figure) can be seen on the monitor and felt from the propriocepcis of the hand. Forces (upper half of the figure) can only be felt through the instruments: the surgeon is dependent on their force transmission characteristics, represented by the block labeled 'laparoscopic grasper'. As useful feedback is present only if there is a clear relationship between the grasping force (F_{gr}) and the operating force (F_{oper}), the design of the grasper places emphasis on its force transmission characteristics. Sjoerdsma *et al.* (1997) argue that a constant force transmission function with no energy dissipating phenomena such as friction or hysteresis due to viscoelasticity, and without backlash is favorable.

The design started with the selection of low-friction pivots. Due to space limitations and the unfeasibility of lubrication, ball bearings are not suitable.

Precision pivots used in measurement devices are not robust enough. Elastic hinges generate considerable counter moments having a negative influence on the force transmission characteristic. An alternative was found in rolling link mechanisms (Kuntz, 1995; Sieker, 1956). In this technique, elements roll directly on one another, and no specific bearing elements are needed. Since simplicity was taken as a point of departure, a single-sided movable jaw was decided upon, attached to a single roller. This roller was supported by the frame of the forceps, in such a way that minimal reaction forces were generated (Herder, 1998). The demand for a constant force transmission function was satisfied by the application of a symmetrical construction. Figure 2 presents a schematic view of the mechanism, with somewhat exaggerated dimensions to illustrate the working principle. Two small rollers with the same radius are placed opposite each other against two support areas, one at each end of the frame. These support areas are parts of an imaginary central cylinder, dashed in figure 2a, and they are fixed rigidly to the frame of the grasper. The rollers are connected by two rods. The pivots at the ends of these connecting rods also have rolling contact with the pins that are fixed to the rollers. This is achieved by fashioning the ends of the rods as small segments from two large rings, drawn with dashed lines in figure 2a. The cross in the middle of the imaginary cylinder is the center of symmetry during motion. When the grasper is being closed, both rollers roll clockwise over their support areas, from the drawn positions towards the dashed situation. Meanwhile, the rods' ends are rolling over the pins fixed to the rollers.

Thus, the complete mechanism contains only six rolling pivots, resulting in low overall friction. As a result of the symmetry, the rollers' angular velocity is equal and, in accordance with the principle of virtual work, their moments are transferred one on one from the grasper to the handle. Additionally, because of the symmetrical construction, the connecting rods remain at constant length and can therefore be made very stiff. As a consequence, the instrument has an outstanding internal stiffness. This is advantageous, as the stiffness of the tissue is now perceived separately instead of in series with the elasticity of the instrument: the apparent stiffness, as felt by the surgeon, corresponds with the stiffness of the tissue held in the grasper. Constructive details and photographs of the prototype can be found in Herder *et al.* (1997a).

4. EVALUATION

Laboratory measurements confirmed that friction in the Wilmer® forceps is low. The energy losses because of friction amount to only 4% yielding a mechanical efficiency of 96%, while the non-constancy of the force transmission function is only

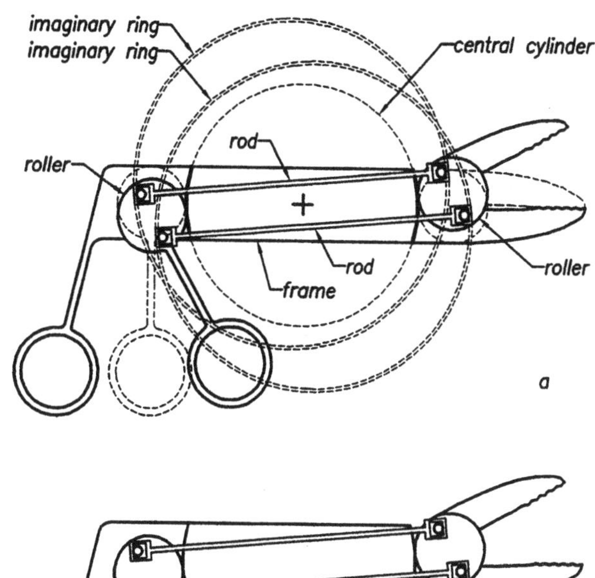

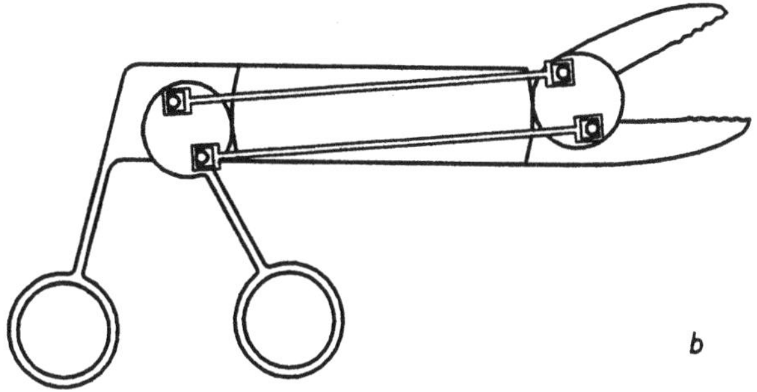

Figure 2. (a) Patented working principle of the Wilmer® forceps. The dimensions in this figure are not realistic, but chosen so as to elucidate the working principle. Two identical rollers can roll over small parts of an imaginary central cylinder. They are connected by two rods, being parts of two imaginary rings. The symmetrical construction has resulted in a constant force transmission function (see text for detailed explanation). Hence, the whole construction possesses only six rolling pivots, resulting in very low friction. The laparoscopic grasper design presented is currently under patent application. (b) The schematic representation of figure 3a without the dashed lines.

3%. For comparison, instruments currently available have mechanical efficiency values not exceeding 33%, and show deviations from a constant force transmission of up to a factor of six (Sjoerdsma *et al.*, 1997).

In addition to the mechanical measurements, the subjective sensitivity was assessed by means of a psychophysical experiment. It was decided to use the method of limits (Gescheider, 1976) to assess the absolute sensory threshold, which was taken as a measure of the subjective sensitivity. In addition to the evaluation of the Wilmer® forceps, the same experiment was also performed with bare fingers and with three laparoscopic forceps commercially available. These were a disposable instrument with a shaft diameter of 5 mm, and reusable instruments of 5 and 10 mm respectively.
An experimental set-up has been developed which meant to simulate a pulsating artery. By way of a computer controlled oscillator, pulses of varying

amplitude could be guided through a thin-walled silicon tube. The fluid pressure was measured near the section used for the grasping experiment. Subjects were asked to grasp the tube with their bare fingers, with the Wilmer® forceps, and with the other instruments. Using the psychophysical method of limits (Gescheider, 1976), sensitivity thresholds of the graspers were assessed and compared to the results obtained with bare fingers. Starting with a signal below the sensory threshold, the amplitude was gradually increased until the subject indicated to feel the pulse. Then it was increased further to be well above this transition point after which it was gradually decreased until the subject ceased to feel the pulse (second transition point).

Preliminary results are given next (see also Herder *et al.,* 1997b), whereas a much more extensive description of the experiment as well as many more and more accurate data are included in Den Boer *et al.* (1998). If the sensitivity threshold for bare fingers is normalized as unity, the threshold of the Wilmer® forceps amounts to 2, approximately, whereas the other instruments have threshold values of about 10 (some instruments much more). Furthermore, the scatter in the data from different subjects is much higher with the commercially available instruments than with the Wilmer® design and with bare fingers.

5. CONCLUSION

In minimally invasive surgery direct manual contact is lost. As a result, sensory feedback is considerably reduced. In principle, however, force feedback in laparoscopic forceps remains possible, provided these instruments are designed carefully. A man-machine-systems perspective reveals that laparoscopic instruments are a crucial link in a complex information transmission process: physical properties from tissue are to be perceived by the surgeon via the instruments.

In this study, the design and evaluation of the purely mechanical Wilmer® forceps is presented. Excellent force feedback quality has been achieved by using rolling pivots. Additionally, a constant force transmission function was attained by applying a symmetrical construction. Thus, a mechanical efficiency of 96% was achieved. In addition to the mechanical measurements, psychophysical measurements were performed to assess the subjective sensitivity. The sensitivity of the Wilmer® prototype was significantly higher compared to the other laparoscopic instruments tested. The Wilmer® prototype was only about two times less sensitive than bare fingers, whereas the other instruments were roughly ten times less sensitive. It is expected that this instrument will enable surgeons to operate more accurately, probably resulting in reduced risk of tissue damage, and decreased operating time.

ACKNOWLEDGMENT

This research is part of the Minimally Invasive Surgery and Interventional Techniques (MISIT) program of the Delft Interfaculty Research Center (DIOC) on Medical Engineering. The Authors wish to thank Michiel Hoogendoorn and Paul Heeman of the MTO (medical technology development-department) for manufacturing the test set-up respectively implementing the computer controlled test protocol, and Dr Dirk Meijer of the Test Center of Experimental Surgery, all at the Academic Medical Center (AMC) of the University of Amsterdam, The Netherlands. In addition, the authors wish to thank the participants for their cooperative support.

REFERENCES

Breedveld, P. (1998). Observation, manipulation and eye-hand coordination in minimally invasive surgery, overview of negative effects, experiments and supporting aids. Report N-510, Delft University of Technology, Delft.

Cushieri, A. (1995). Whither Minimal Access Surgery: Tribulations and Expectations. *The American Journal of Surgery,* **169**, 9-19.

Den Boer, K. T., J. L. Herder, W. Sjoerdsma, D. W. Meijer, D. J. Gouma, H. G. Stassen (1998). Sensitivity of laparoscopic dissectors, what can you feel? Submitted to *Surgical Endoscopy.*

Gescheider, G. A. (1976). *Psychophysics, method and theory.* John Whiley, NY.

Green, P. S., J. W. Hill, J. F. Jensen, A. Shah (1995). Telepresence Surgery. *IEEE Engineering in Medicine and Biology,* 324-329.

Grimbergen, C. A. (1997). Minimally invasive surgery: human-machine aspects and engineering approaches. In: *Perspectives on the Human Controller,* (Sheridan, T. B., A. Van Lunteren (Eds.)), pp. 223-231. Lawrence Erlbaum Associates, Mahwan, New Yersey.

Herder J. L. (1998). Force Directed Design of laparoscopic forceps. Accepted for publication at the Special Session on Force Synthesis, *25th Biennial Mechanisms Conference, ASME-DETC98*, Atlanta, Georgia.

Herder J. L., M. J. Horward, W. Sjoerdsma (1997a). A laparoscopic grasper with force perception. *Min Invas Ther & Allied Technol,* **6**, 279-286.

Herder J. L., K. den Boer, M. Hoogendoorn, W. Sjoerdsma, C. A. Grimbergen, H. G. Stassen (1997b). Evaluation of feedback in laparoscopic

forceps. *Min Invas Ther & Allied Technol,* **6**, Supp. 1, pp. 55.
Hill, J. W., P. S. Green, J. F. Jensen, Y. Gorfu, A. S. Shah (1994). Telepresence Surgery Demonstration System, *Proc 1994 IEEE Int. Conf. On Robotics & Automation,* San Diego, CA, pp. 2302-2307.
Howe, R. D., W. J. Peine, D. A. Kontarinis, J. S. Son (1995). Remote Palpation Technology, *IEEE Engineering in Medicine and Biology,* 318-323.
Majima S., K. Matsushima (1991). On a micro-manipulator for medical application, stability consideration of its bilateral controller. *Mechatronics,* **1**, 293-309.
Kuntz, J. P. (1995). *Rolling Link Mechanisms*, PhD Thesis, Delft University of Technology, Delft, The Netherlands.
Lazeroms M., W. Jongkind, G. Honderd. (1997). Telemanipulator Design for Minimally Invasive Surgery. *Proceedings of the American Control Conference,* Albuquerque, New Mexico, pp. 2982-2986.
Sieker, K. H. (1956). *Einfache Getriebe*, 2. Auflage, C.F. Winter'sche Verlagshandlung, Füssen.
Sheridan T. B. (1996). Human Factors in Telesurgery. In: *Computer integrated surgery, Technology and Clinical Applications,* (Taylor, R. H., S. Lavallée, G. C. Burdea, R. Mösges (Eds.)). MIT Press, Cambridge, Mass., pp. 223-229.
Simpson, D. C. (1974). The choice of control system for the multimovement prosthesis: extended physiological proprioception (e.p.p.). In: *The control of upper-extremity prostheses and orthoses.* (Herberts P., R. Kadefors, R. Magnusson, I. Peterson (Eds.)). C. C. Thomas, Springfield, Illinois, pp. 146-150.
Sjoerdsma, W., J. L. Herder, M. J. Horward, A. Jansen, J. J. G. Bannenberg, C. A. Grimbergen (1997). Force transmission of laparoscopic grasping instruments, *Min Invas Ther & Allied Technol,* **6**, 274-278.

DEVELOPMENT OF HEAD-ATTACHED INTERFACE DEVICE (HIDE) AND ITS FUNCTIONAL EVALUATION

H.SHIMODA*, N.HAYASHI**, Y.NIKAIDO*, N.UMEDA*
and H.YOSHIKAWA*

*Kyoto University, Graduate School of Energy Science, Gokasho, Uji, Kyoto, 611-0011, JAPAN
**Present; Sony Corporation, 5-9-12, Kitashinagawa, Shinagawa-ku, Tokyo, 141-0001, JAPAN

Abstract. Head-attached interface device (HIDE) has been developed as a new type of human interface device in an actual work environment. The HIDE has flexible integrated functions which consist of speech recognition and view direction detection as hand-free input channels, direct presentation of audio/visual information as output channels, and mobile system configuration. In this paper, the design concept of the HIDE was first introduced, and individual functions to configure the design concept were developed by their separate testing. An integrated prototype system HIDE was then produced, and its functional evaluation test was made to examine the individual functions as the whole system. It was confirmed that the HIDE would be feasible enough to introduce in practical use. *Copyright ©1998 IFAC*

Key Words. Human interface device; hands-free input; mobile system; see-through display

1. INTRODUCTION

With the recent progress of information technology, highly advanced computers have come into wide use. Especially, progress of information network represented by the Internet has been making it easy to pick up the desired information. However, almost all of the computers should be operated by keyboards and mouse in front of the desks on which the computers are placed. It is true that some light-weight mobile computers have been spreading recently, but they still require keyboards and pointing devices for their operation. For the purpose of supporting a task in an actual work environment with such computers, it will be rather inconvenient, if the worker should operate it by keyboard and mouse when conducting the task which needs both hands.

In order to solve this problem, Wearable computer has been developed as a new interface device by a research group of Carnegie Mellon university (Bass, *et al.*, 1995). It consists of one-eye head mounted display (the Private Eye) and specially designed computer in a waist-attached unit. When changing display information, the operator uses a dial and buttons on the waist-attached unit. It is true that the wearable computer made the great progress for machine maintenance task, but the following problems still remain.

- Computer operation by using the dial and buttons interrupts actual work which needs both hands.
- The head mounted display prevents from seeing outside view when displaying information.
- It uses special designed computer, so that development tools for application software are limited.

On the other hand, some interaction methods without using hands such as eye-movement detection (Park, *et al.*, 1996), speech recognition (Berkley, *et al.*, 1994; Gamm, *et al.*, 1996) and so on has been developed recently. These methods, however, are lack of flexibly integrated interaction functions, and portable system configuration. Actually, if the interface device which can provide those functional features is developed, the application field will be not only machine maintenance task but also expanded into the following areas:

- Navigation system for vehicles such as automobiles, trains and airplanes,
- Input support for physically handicapped and aged person,
- Reduction of the workload for stocktaking task in product warehouse, and so on.

In order to realize efficient support for above application fields, Head-attached interface device (HIDE) has been developed as a realization of the following design concepts.

1. Hands-free input methods: Detection of eye view direction and speech recognition.
2. Direct information presentations: See-through information display to user's left eye in order not to prevent from seeing outside view, and presentation of voice operating guidance through earphone.

3. Easy development for various application software: Application software can be developed on a DOS/V personal computer (PC) by using various development tools.
4. Mobile system: The device should be light and compact to be a mobile system.

From the next chapter, the functions, system configuration and its evaluation of the HIDE is described.

2. FUNCTIONS AND SYSTEM CONFIGURATION

In order to realize the above design concepts, the following functions will be needed.

1. Visual information presentation,
2. Audio information presentation,
3. Speech recognition, and
4. View direction detection.

If all of the above functions are attached on user's head, the device will become large and heavy, so that the devices mounted on the head should be limited to minimum input/output devices and the remaining devices are attached on user's waist. The PC which controls all function is not attached on user's body but placed apart from the user. It communicates to the device attached on the user's body with radio wave transmission. To sum up, the HIDE has visual and audio information presentation function as output channels to the user, while detection of eye view direction and speech recognition as input channels. The HIDE has been developed as a set of the three units: (a)head-attached unit, (b)waist-attached unit, and (c)information processing unit, in order to materialize all the functions mentioned above as the HIDE system in compact fashion. Figure 1 shows the overall configuration of the HIDE.

Head-attached Unit The head-attached unit presents visual and audio information, collect user's speech and detect his/her view direction. The mounted devices are a compact color LCD and optical parts for visual presentation, an earphone for audio presentation, a microphone for speech collection, and infrared optical electronic devices for view direction detection. These parts should be selected as light and compact as possible.

Waist-attached Unit The waist-attached unit consists of electronic circuit to drive the devices on the head-attached unit, radio transmitter/receiver to communicate to the information processing unit, and battery to supply required power to these devices. They should be also selected as light and compact as possible.

Information Processing Unit The information processing unit consists of radio transmitter and receiver to communicate to the units on user's body, and PC which recognizes user's speech and view direction, and generates visual/audio information.

3. CONFIGURATION OF FUNCTIONS

In this chapter, the configurations of the individual functions are described.

3.1. *Visual Information Presentation*

The visual information presentation is to present visual information generated by PC in the information processing unit to user's left eye. The function consists of visual signal transmission part and optical display part.

Visual signal transmission part The Visual signal transmission part presents images generated by PC on a compact display in the head-attached unit. The images are put out as VGA output of PC, and 1.8-inch color LCD is used as a compact display. The LCD is originally made for a portable TV, so that it has a TV tuner. Figure 2 shows the configuration of the visual signal transmission part. The RGB signal as VGA output from PC is first converted to NTSC signal by a video converter. And then, the converted NTSC signal is transferred to the head-attached unit via VHF by an AV transmitter. In the head-attached unit, the VHF signal is received by the TV tuner and displayed on the color LCD.

Optical display part The optical display part presents distant and enlarged virtual view of the displayed image on the LCD to the user's left eye. Figure 3 shows the configuration of the optical display part. The image displayed on the color LCD is first reflected on a surface evaporated mirror, and projected forward to the user through non-spherical lenses which has less distortion. The user can see the projected image which is reflected on a half mirror placed in front of the user's left eye. Since the half mirror is placed in front of the user, he/she can also see outside view at the same time.

3.2. *Audio Information Presentation*

The audio information presentation function presents audio information such as voice guidance generated by the PC to the user. Figure 4 shows the configuration of the audio presentation function. The voice guidance is first generated by the voice composition software in the PC, then modulated to sound signal of VHF and transmitted

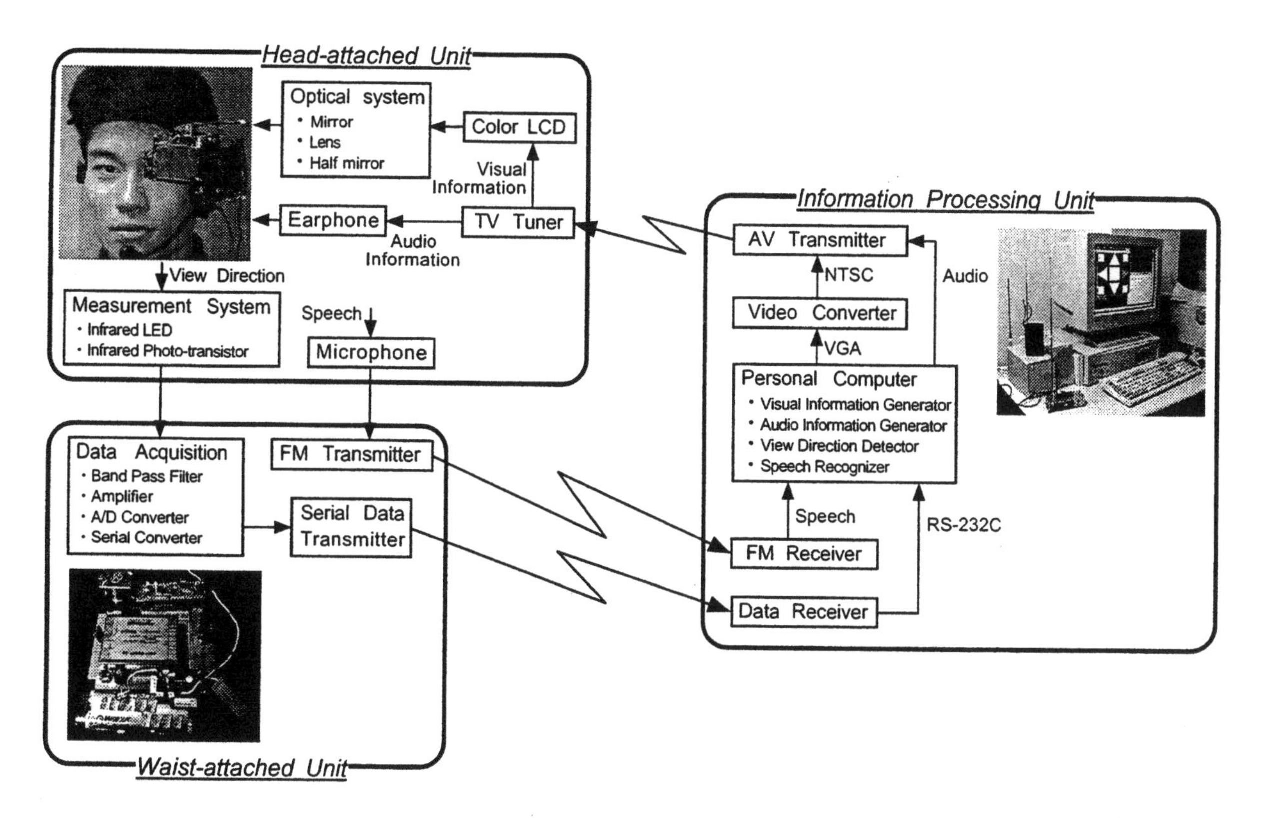

Fig. 1. Overall System Configuration of HIDE

with video signal to the head-attached unit by the AV transmitter. In the head-attached unit, the received signal is demodulated by the TV tuner and presented to the user's ear through the earphone.

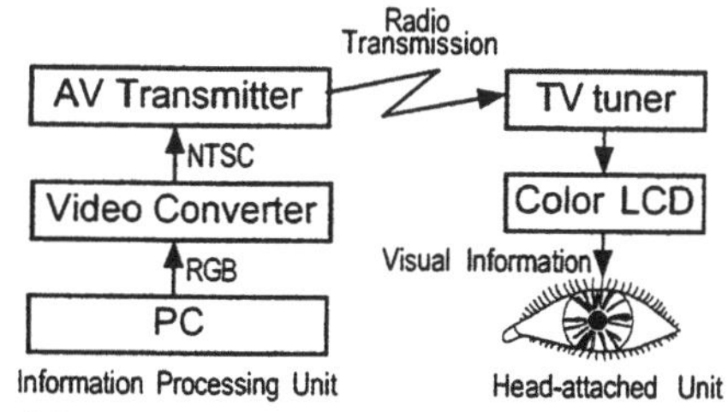

Fig. 2. Visual Signal Transmission Part

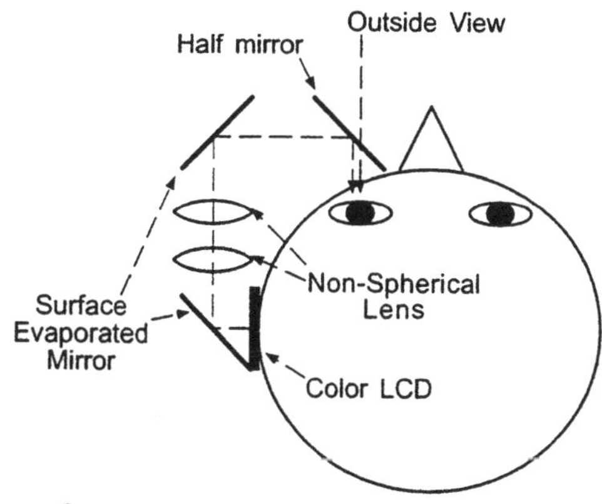

Fig. 3. Optical Display Part

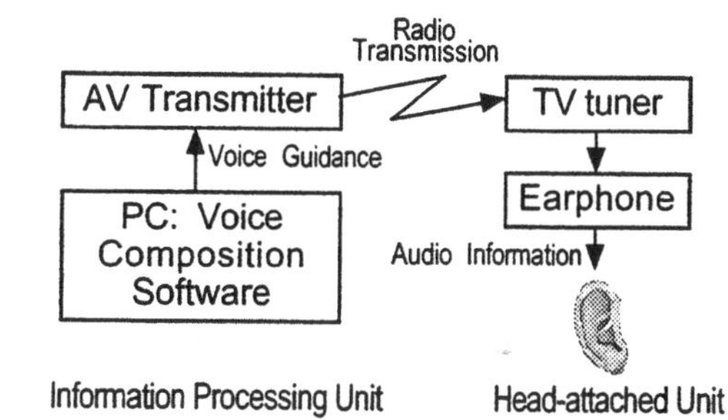

Fig. 4. Audio Information Presentation

3.3. *Speech Recognition*

The speech recognition function realizes voice control by recognizing the user's speech. Figure 5 shows the configuration of the speech recognition function. The user's speech is collected by a microphone in the head-attached unit, modulated and transmitted to the information processing unit by an FM transmitter. In the information processing unit, the received signal is demodulated by an FM receiver, then put into the PC and recognized by the PC-based speech recognition software.

3.4. *View Direction Detection*

The view direction detection recognizes user's view direction and eye blinking. Figure 6 shows the configuration of the view direction detection. The infrared ray modulated by 38kHz is irradiated from two infrared LED located right and left of the

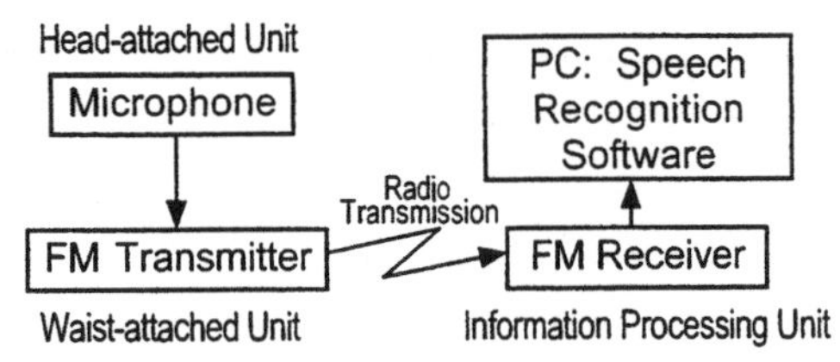

Fig. 5. Speech Recognition

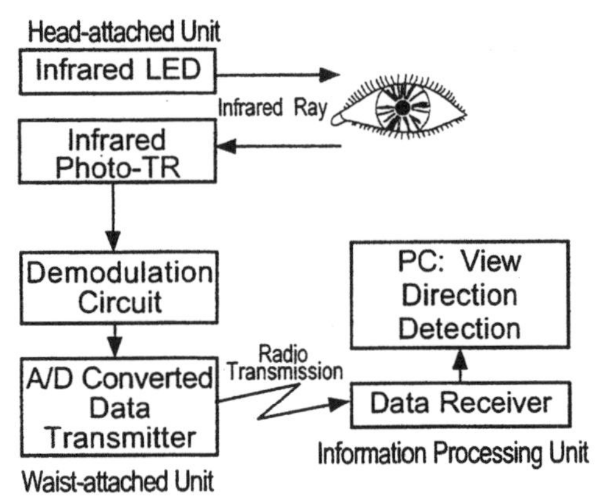

Fig. 6. View Direction Detection

user's eye to the sclera of left eye, then reflected ray is received by four photo-transistors located around the eye. Figure 7 shows the arrangement of the infrared LEDs and photo-transistors. The reason why infrared ray is modulated is to prevent from influence of outside infrared ray included in such as natural ray, white heat light and fluorescent light. The received signal is demodulated, amplified, A/D converted and transmitted to the information processing unit. In the unit, received data is put in the PC via serial data communication and detected the user's eye view direction and blinking.

The principle of view direction detection is as follows (see Figure 7). If the eyeball turns right, the infrared ray reflected on the sclera of right side increases, while that of left side decreases, so that the outputs of photo-transistors varies according to the eyeball rotation. On the other hand, when closing the eye, the reflected ray increases because the reflection ratio of skin is high. Thus, the view direction and eye-blinking can be detected from the difference of output of the four photo-transistors.

4. FUNCTIONAL EVALUATIONS

In this chapter, the evaluations of visual/audio information presentation, speech recognition and view direction detection are described.

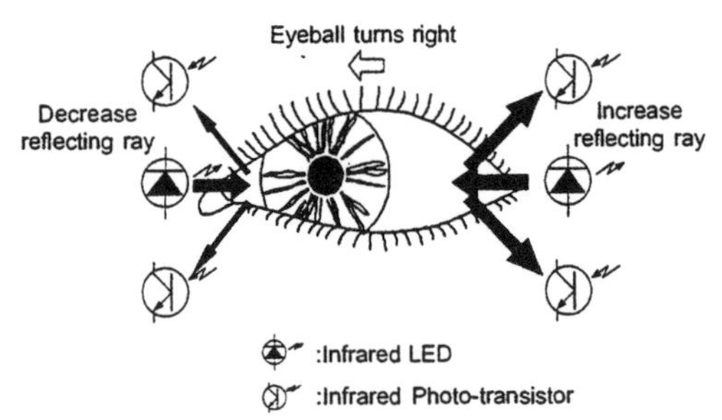

Fig. 7. Principle of View Direction Detection

4.1. *Evaluation of Visual Information Presentation*

The resolution of the image which is presented to the user was evaluated. In order to evaluate the function, vertical and horizontal stripe pattern images are used as sample images (see Figure 8). The stripe width of the displayed images was changed from 10 dots to 1 dot, and the subject who wears the HIDE was asked to answer the minimum stripe width which he/she can distinguish as stripes. 5 graduate school students, A, B, C, D and E, who have no ocular disease participated in the experiment as subjects. Table 1 shows the result of the experiment.

Table 1 Minimum Distinguishable Stripe Width

Subject	Horizontal Stripe	Vertical Stripe
A	4 dot	5 dot
B	5 dot	4 dot
C	4 dot	5 dot
D	4 dot	4 dot
E	3 dot	3 dot

According to the result, the resolution of the displayed image is 4 or 5 dot in both horizontal and vertical strips, while the displayed image size is 640x480(VGA).

4.2. *Evaluation of Speech Recognition*

The speech recognition rate was evaluated as follows. Subjects who wear the HIDE are asked to pronounce the four words 10 times as shown in Table 2. Since it takes a second for PC to recognize

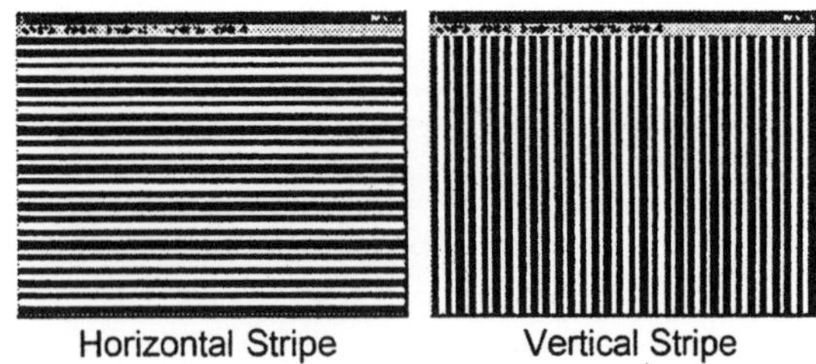

Fig. 8. Stripe Images

the subjects' speech, they make 2 second interval when pronouncing the words. The recognition results put out from the PC are one of (i)correct recognition, (ii)erroneous recognition and (iii)no recognition. 5 students, F, G, H, I and J participated in the experiments as subjects.

Table 3 shows the experimental result. There was no error recognition, and the average recognition rate was 95.5%, which is enough high for voice control function.

Table 2 Words for Speech Recognition

Word A	"Minimize"
Word B	"Voice Command List"
Word C	"Global Command"
Word D	"Start Button"

4.3. *Evaluation of View Direction Detection*

A glance input software which is an example of eye view control method is developed in order to evaluate the view direction detection. The user can input one of the 5 states, look-up, look-down, look-left, look-right and look-center by closing his/her eyes for 0.5 second after glancing one of the directions respectively. By using the software, the evaluation experiment was conducted as follows. Subjects who wear the HIDE are asked to close their eyes and then input the above 5 states 3 times, totally 15 times. The experiment was conducted in our laboratory computer room at night time, so that the outside noise of infrared ray came mainly from fluorescent lights. 5 students, K, L, M, N and O participated in the experiments as subjects. The results are correction rate of the inputs and time from opening their eyes to detection thier inputs. Table 4 shows the experimental result of glance input.

Table 4 Result of Glance Input Experiment

Subject	Correct/Total	Rate(%)	Time(sec)
K	15/15	100.0	1.89
L	15/15	100.0	3.08
M	12/15	80.0	2.07
N	15/15	100.0	1.98
O	14/15	93.3	1.74
Average		93.3	2.15

The average input rate was 93.3% and the average time is 2.15 sec, which is enough usable and reliable as an input method.

4.4. *Evaluation of Radio Transmission*

The radio transmission of the system is used for (1)visual information presentation, (2)audio information presentation, (3)speech recognition and (4)view direction detection. The maximum distance of the radio transmission of each function was evaluated. First, a subject who wears the HIDE stands next to the information processing unit, and checks whether each function is successful or not. Then he was asked to go apart from the information processing unit by 1m pitch with checking the functions. The thresholds to judge whether or not each function is available are set as follows;

1. Visual presentation
   The subject cannot distinguish 5 dot horizontal stripe image as shown in Figure 8.
2. Audio presentation
   The subject cannot hear the voice guidance.
3. Speech recognition
   The recognition rate becomes under 80%.
4. View direction detection
   The data receiver cannot receive the data correctly.

Table 5 shows the result of the maximum distance in which each function can work well.

Table 5 Maximum Distance of Successful Radio Transmission

Function	Distance
Visual Information Presentation	8 m
Audio Information Presentation	8 m
Speech Recognition	9 m
View Direction Detection	32 m

4.5. *Other Evaluations*

In this section, the operation time of the HIDE and its weight is measured. 2200mAh Ni-Cd rechargeable battery is used to supply power to the head-attached unit and waist-attached unit. The operation time was measured by using fully charged battery beforehand. 87 minutes after starting operation, the color LCD was out of operation. On the other hand, the weight of each unit which is attached user's body is measured as shown in Table 6. Both head-attached unit and waist attached unit are light enough for an actual work.

Table 7 shows the summary of above evaluation results. According to the table, control functions

Table 3 Result of Speech Recognition Experiment

Subject	Word A C.R.	Word A E.R.	Word A N.R.	Word B C.R.	Word B E.R.	Word B N.R.	Word C C.R.	Word C E.R.	Word C N.R.	Word D C.R.	Word D E.R.	Word D N.R.	Rate(%)
F	9	0	1	10	0	0	10	0	0	10	0	0	97.5
G	10	0	0	10	0	0	10	0	0	9	0	1	97.5
H	10	0	0	10	0	0	10	0	0	10	0	0	100.0
I	10	0	0	10	0	0	9	0	1	9	0	1	95.0
J	10	0	0	9	0	1	7	0	3	9	0	1	87.5
Average		98.0			98.0			92.0			94.0		98.5

C.R.:Correct Recognition E.R.:Erroneous Recognition N.R.:No Recognition

Table 6 Weight of Each Unit

Unit	Weight
Head-attached Unit	210 g
Waist-attached Unit	580 g
Total of body-attached units	790 g

such as speech recognition and view direction detection are reliable with little error. However, the maximum distance of radio transmission is only 8 m, and operation time is only for 87 min.

In order to realize effective support for actual work envoironment, it is important how two output channels and two input channels should be combined as an integrated interaction method by reflecting the results of the above evaluations. For example, it is appropriate that most of presented information is displayed as visual information and the audio information presentation is used for voice operation guidance, because the visual presentation can present much information in short time, while the audio presentation is apt to arrest the user's attention. As for the control method, the speech recognition is used for overall control for application software because there was a few "No Recongition" and no "Erroneous Recognition" in the conducted experiment as shown in Table 3. The view direction detection is used to change the displayed information because of its quick detection.

Table 7 Summary of Eveluation Result

Speech Recognition Rate	95.5%
View Direction Detection Rate	93.3%
View Direction Detection Time	2.15 sec
Resolution of Visual Information Presentation	4 or 5 dot
Distance of Radio Transmission	8 m
Operation Time	87 min

5. CONCLUSION

In this study, the prototype of head-attached interface device (HIDE) was produced as an on-site and on-demand information presentation interface device. The HIDE consists of see-through visual information presentation and audio information presentation as information output channels, and speech recognition and view direction detection as control channels. By using this prototype, its functions were evaluated. To sum up the conducted functional evaluation experiments, it was confirmed that the design concept of the HIDE was feasible enough as an effective and flexible interface device for an actual work environment. Since the HIDE is still under development, the following will be realized in the future.

- Improvement of displayed resolution,
- Extension of radio transmission distance,
- Lighter weight unit,
- Optimization of interface method, and
- Field study for actual work applications

6. REFERENCES

Bass,L., *et al.* (1995). *On Site Wearable Computer System*, CHI 95 Conference Companion, pp 83-84.

Berkley, D.A., *et al.* (1994). *A multimodal teleconferencing system using hands-free voice control*, Proc. of International Conference on Spoken Language Processing '94, **Vol.2**, pp.555-558.

Gamm,S., *et al.* (1996). *Finding with the design of a command-based speech interface for a voice mail system*, Proc. of 3rd IEEE Workshop on Interactive Voice Technology for Telecommunication Application, pp.93-96.

Park,K.S., and Lee, K.T. (1996). *Eye-controlled human/computer interface using the line of sight and intentional blink*, Comput. Ind. Eng. (UK), **Vol.30, No. 3**, pp.463-473.

SUPER OPERATOR - WEARABLE INFORMATION SYSTEM IN HIGHLY AUTOMATED MANUFACTURING

J. Stahre and A. Johansson

Chalmers University of Technology,
Dept. of Production Engineering,
S-412 96 Göteborg, SWEDEN
email: johan.stahre@pe.chalmers.se

Abstract: In advanced and highly automated manufacturing systems there is an increasing need for efficient distribution of accurate and instant information. A theoretical framework based on human supervisory control modeling, situation awareness, and the SRK taxonomy for human behavior has been combined with empirical studies to specify criteria for a wearable information system targeted at manufacturing system operators. A laboratory prototype of the "Super Operator" has been developed, providing multiple information channels, wireless connection to the internet, 3D-navigation in a virtual-reality model of the work-space, on-line machine manuals, error lists, decision support, etc. *Copyright©1998 IFAC*

Keywords: Supervisory control, user interfaces, manufacturing systems

1. INTRODUCTION

Industrial market demands require continuously increased manufacturing flexibility with regards to volume, variants, product mix, etc. To meet requirements, advanced manufacturing systems, e.g. car-body assembly plants, are subject to high degrees of automation, extensive use of industrial robots, and subsequent reductions in personnel. Ratios of 100 robots to 15 operators or less can frequently be seen. Although there are still several "simplistic" tasks (e.g. material loading) being done manually, the greatest responsibilities for the operators lies in maintenance and "nursing" of the complex machines and technical systems.

That paradox and "irony of automation", as forescen and described by Bainbridge (1982), creates for the operators not only physically harmful situations, but also mentally demanding worktasks. Future operators must be able to deal with an extensive number of new situations where instructions and previous knowledge will be insufficient. Competence requirements are high and their physical movement among the machines is restricted by mechanical, optical, and electronic safety barriers. Maintenance tasks, traditionally related to handling and repair of mechanical components, now to a great extent involve computer programming and use of abstract computer models of the manufacturing systems. Since the time required for resolving maintenance tasks is in direct relation to production down-time, it is of great economical concern for the company to minimize such tasks with regards to time and frequency. Further, the number of people in e.g. multi-competent operator teams is low and the persons are generally dispersed among a multitude of machines. Direct consequences are e.g. inability to have direct verbal communication due to physical distances, and also alienation within the group.

As described recently, by an operator in a Swedish car body assembly line:

Within the shift-group we hardly have the possibility to meet during the workday, not even to plan to have lunch together. Still they expect us to act as a well-trimmed team when the manufacturing system breaks down.

In summary, the work-situation for an "operator-of-the-future", in the context of a highly automated manufacturing plant, is approaching that of a soldier or control room operator.

Consequently, new ways of operator support must be explored to deal with the emerging set of problems. Support tools must, for example, provide instant and easy-to-use access to process and product knowledge, and provide transparency into the manufacturing system (hardware and software). Tools must also allow for peer-to-peer communication and manipulation of abstract system models. Finally, the support tools must be highly mobile and supply operators with multiple selections of input and output devices.

The purpose of this paper is to add a theoretical and empirical frame of reference to the primarily practice oriented requirements and criteria that can be found in advanced manufacturing systems of today. Further, to present a prototype implementation, in its early stages, providing a feasible solution to a selection of the problems described.

2. THEORETICAL FRAME OF REFERENCE

The selection of theoretical framework for this paper is based on the assumption that manufacturing system operators are in highly automated plants are approaching a situation similar to that of a pilot, control room operator, or soldier. An additional assumption made is that Bainbridge's paradox will continue to be valid. As a consequence, operators will have to maintain a variety of skills, ranging from manual handling to complex programming and planning. Given these assumptions, three important concepts emerge: Human supervisory control, Situation awareness, and skill-, rule, and knowledge-based behavior.

2.1 Human supervisory control

A fundamental reference for development of manufacturing operator support is Sheridan's classical Supervisory Control Model. The five basic roles of a system operator as suggested by Sheridan (1987), (i.e. Plan, Teach, Monitor, Intervene, and Learn) can all be identified in the manufacturing operator situation, as shown by Stahre (1995a). Support needs for tasks included in the roles can be identified through systematic empirical evaluation of operator tasks in a specific manufacturing system (Stahre,1995a) (Johansson, 1996).

2.2 Situation awareness

A complementary approach to support development is the concept of Situation Awareness (SA) suggested by Endsley (1995) and Wickens (1995), both primarily focused on the SA of aircraft pilots.
The concept of SA is abstract and deals with the human mental ability to handle situations, based on the combination of experiences in the past, perception of the present, and projections of the future.

As argued by Endsley, there are three levels of SA:

1. Perception of status, attributes, and dynamics of relevant elements in the environment
2. Comprehension of the current situation with respect to set goals
3. Projection of future status and actions of elements

Due to the dynamic and complex environment experienced by operators in highly automated environments, SA categorization would apply to a number of their tasks. Manufacturing system operators could therefore be supported on all three levels of SA.

2.3 S-R-K taxonomy for Human behavior

Finally, Rasmussen has provided a framework for analysis of tasks and human behavior, which apply also to manufacturing system operators, in the classical taxonomy of skill-, rule-, and knowledge-based behavior (Rasmussen, 1983).

2.4 Additional references

Stahre (1995a and 1995b) suggests that a matrix consisting of the SRK levels and the human supervisory control roles can be used to evaluate operator tasks in complex manufacturing systems. The matrix may be used to distinguish between operator tasks needing training, knowledge-, or rule-based decision support.

The significance of relations between Human supervisory control and the SRK-taxonomy have been discussed by Stassen et al. (1990). Their work was further elaborated on in a study on operator requirements for human-machine interfaces of CNC machine tools, as described by Schlick et al. (1996). In the study a mobile computer interface, in concept similar to the Super Operator project, was implemented for a machining cell.

Mobile information systems and "wearable computers" are emerging research and commercial areas. Few general conclusions from this area can be drawn presently. This is indicated by the fact that the first IEEE 1st International Symposium on Wearable Computers (ISWC '97) took place October 13-14, 1997 in Cambridge, Massachusetts, USA.
However, the applicability of and need for functionality promised by such systems is undisputable. For manufacturing system applications Wilson (1996) suggests the following criteria for two possible application areas of shopfloor information systems (i.e. operator maintenance information systems and communication for self-directed work teams):

- availability at the site of maintenance
- flexible information detail and complexity, according to user and task
- capability to provide graphical as well as alphanumeric information
- on-line access to parts stores, production records
- robustness, durability and, preferably, portability (mobility)
- systems must provide means for intra-team, inter-team, and inter-shift communication
- systems must be a resource to support team decision making or problem solving
- systems must support the team both as a production unit and in how it develops as a team
- systems must be a mechanism for collating "notes of good practice"

Wilson also discuss the use of emerging technologies such as miniature computers, head-mounted displays and "virtual reality" for applications on the shop-floor. His remarks are that: regardless technology selected, motivation to use new types of equipment is relying on the device's ability to "out-perform" old tools, e.g. paper, pencil, logbooks, charts, and dictaphones.

3. EMPIRICAL BACKGROUND AND DATA ACQUISITION

Empirical data has been acquired, by the authors, in operative CIM environments. By triangulating data acquisition methods (Johansson, 1996) several views of the manufacturing systems have been combined into a holistic picture of the operators' situation and support requirements. Examples of views mapped are work satisfaction, communication infrastructures, and competence levels.

Using the set of methods, several studies of advanced manufacturing systems have identified and analyzed large numbers of work tasks performed by operators (Stahre, 1995) (Johansson, 1996). When subjectively evaluating their complete range of tasks, operators tend to implicate maintenance and repair tasks as the most complicated, thus requiring decision support and increased competence. The same sample of operators further verifies Bainbridge's paradox by targeting the simple material handling tasks as being boring and posing the greatest risks for physical injury.

Further, available computer-based decision support systems are often not used to the extent to which they were developed. Problems are generally resolved through discussions among the operators, through telephone if physical distances makes it necessary.

Also, alienation is obvious, not only between the operators and the encapsulated machines, but between the widely dispersed operators in manufacturing systems with high degrees of automation.

4. THE SUPER OPERATOR

From the theoretical and empirical base a prototype of an operator-oriented decision support has been developed in the Manufacturing System Laboratory at Chalmers university of Technology.

The "Supervisory Control Operator", or "Super Operator" for short, is a modular WWW-application implemented in a wearable computer system.

4.1 Super Operator application

The Super Operator prototype (fig 1) is based on commercially available equipment and consists of a computer and battery-pack attached to a belt.

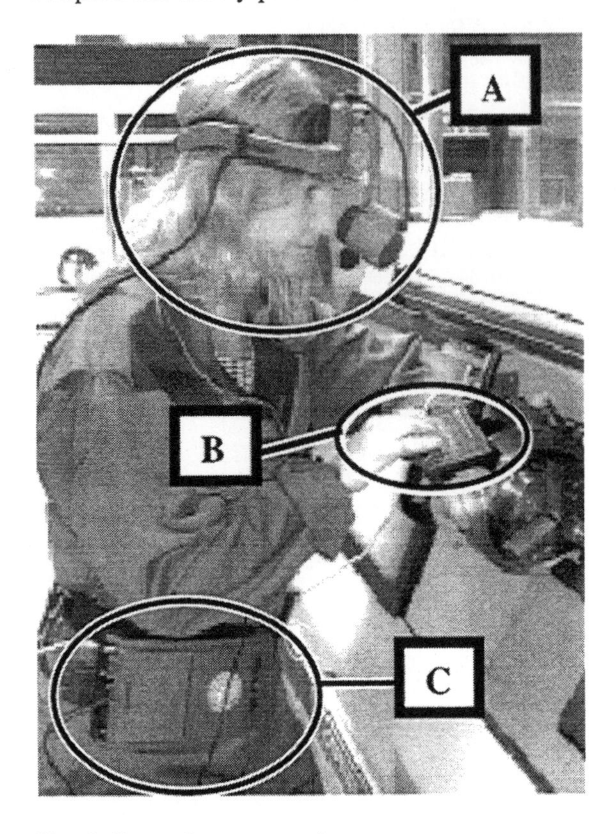

Fig. 1. Super Operator equipment.

Figure 1 shows the "Super Operator" prototype and letters indicate vital parts of the technical equipment:

A - Head-mounted display
B - Miniature keyboard
C - Wearable computer

The miniature keyboard is attached to the operator's palm. A head-up display (256 gray-scales) provides a miniature display for one eye which can be adjusted for either eye. A headset with earphone and microphone is attached to the display. Wireless communication is provided through a mobile/cellular telephone or via radio-LAN. System software is a standard Windows '95 operating system with a standard internet browser.

The main parts of the interface are indicated in fig. 2:

D - Navigable 3D virtual reality model
E - Links to on-line manuals
F - Links to Error reports
G - Links to Service reports.

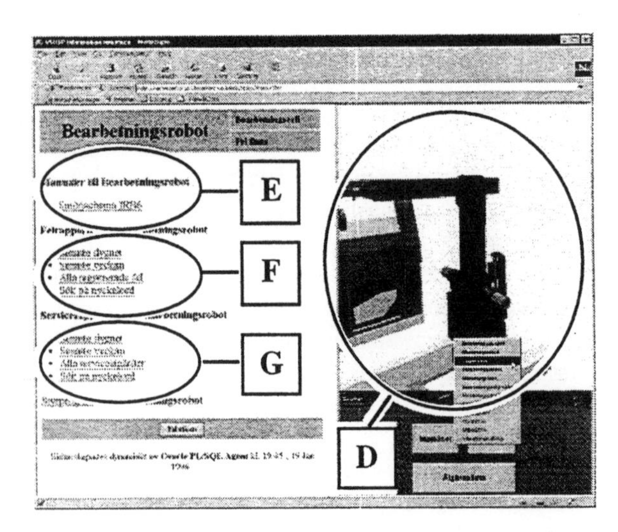

Fig. 2. Overview of the Super Operator WWW-based interface; with areas for 3D-navigation, comment input windows, and navigation buttons with texts in the operators' native language. (Bokesand and Melkstam, 1997).

The Super Operator implementation aims to use standard equipment and focuses development efforts on meeting the operators' requirements as identified above. Examples of functionality provided by the prototype system are:

- Internet/intranet connection
- Wireless communication, telephone capabilities
- Virtual manufacturing system transparency
- Access to mainframe relational database
- Access to (and ability to manipulate) 3-dimensional models of the manufacturing system and its machines
- On-line system status display
- System and machine error log
- Voice-controlled interface
- etc.

The Super Operator is in prototype and laboratory stage. Preparations are being made for full-scale tests for mobile maintenance personnel at a car-body assembly plant. Usability tests will provide data for further developments of the prototype.

5. DISCUSSION AND DELIMITATIONS

The Super Operator project aims at applying the requirements drawn from the theoretical base provided by the concepts of supervisory control, situation awareness, and the SRK behavior levels.
Further, to fulfill the criteria set by Wilson (1996). The resulting demonstration appears to be a possible platform for further work. Wearable computers are developing rapidly, mainly due to military demand.

It can be expected in the near future that wearable computing provides much greater functionality and price/performance ratio. Therefore, this paper does not go into details regarding insufficiencies in presently available equipment, e.g. battery consumption, weight, screen resolution, and lack of color displays. Focus is instead on content and background of information conveyed by the application in the specific context of highly automated manufacturing systems.

A limitation of the Super Operator project lies in repeatability of the studies since it relies on extensive studies of people and manufacturing systems in operation. Controlled laboratory experiments are virtually impossible in such environments due to minor but constant production-related changes of both equipment and personnel. By triangulation of data acquisition methods, general validity of results has been increased (Stahre and Johansson, 1997).

6. CONCLUSIONS

This paper suggests that the Super Operator concept provides a modern toolbox for mobile, supervisory control of advanced manufacturing systems. The implementation is based on requirements for operator situation awareness and provides support for human behavior on skill-, rule- and knowledge-based levels. Valid operator requirements have been systematically acquired, from several studies holistically targeting the situation of operators in highly automated, industrial manufacturing systems.

The resulting prototype application is based on the emerging technology of wearable computing. The "Super Operator" is a mobile information center implemented in a modular world-wide-web application. Although "wearables" is a very interesting technology as such, focus in the Super Operator project is mainly on identifying and fulfilling requirements of operators in their interaction with advanced manufacturing systems.

7. ACKNOWLEDGEMENTS

The authors would like to express their gratitude to A. Bokesand, D. Melkstam, and G. Stigler for the development of the technical Super Operator prototype. Thanks also to the personnel at Volvo's Car-body assembly plant, Göteborg. We gratefully acknowledge the National Swedish board for Technical Development for funding our research.

8. REFERENCES

Bainbridge, L. (1982). Ironies of automation. In: *Analysis, design, and evaluation of man—machine systems* (Johannsen, G., Rijnsdorp, J. E., (Eds.)), IFAC, Duesseldorf, 151-157.

Bokesand, A. and D. Melkstam (1997). *Virtual Reality — the modern toolbox for manufacturing systems.* Masters thesis, Dept. of Production Engineering, Chalmers Univ. of Technology, Göteborg.

Endsley, M. (1995). Toward a Theory of Situation Awareness in Dynamic Systems. *Human Factors,* **37**(1), Human Factors and Ergonomics Society.

Johansson, A. (1996). *An Initial Methodology for Operator Support tool Development in Advanced Manufacturing Systems.* Licentiate Thesis, Chalmers Univ. of Technology, Göteborg.

Rasmussen, J. (1983). Skills, rules, and knowledge; Signals, signs, and symbols, and other distinctions in human performance models. *IEEE Transactions on Systems, Man, and Cybernetics,* **13** (3). pp. 257-267.

Schlick, C., J. Springer and H. Luczak (1996). Support System design for man-machine interfaces of autonomous production cells. In: Proceedings of 13th Triennial World Congress (XXNN (Ed)), pp. 299-304, San Francisco, USA

Sheridan, T.B. (1987). Supervisory Control. In: *Handbook of Human Factors* (Salvendy, G. (Ed.)), John Wiley & Sons, New York.

Stahre, J. (1995a) *Towards Human Supervisory Control in Advanced Manufacturing Systems.* Doctoral thesis, Chalmers Univ. of Tech.,.

Stahre, J. (1995b). Evaluating Human/Machine Interaction Problems in Advanced Manufacturing. *Computer Integrated Manufacturing Systems* , **8** (2), Butterworth—Heinemann.

Stassen, H. G. Johannsen and N. Moray (1990). Internal representation, internal model, human performance model, and mental workload.", *Automatica* **26**: (4), pp. 811-820.

Stahre, J. and Johansson, A. (1997). Joint optimization of People and Technology — an Example from the Car Industry. In: *Proceedings of 7th Int. Conf, on Human—Computer Interaction* , San Francisco, USA.

Wilson, J.R. (1996). Information, Opportunity and Involvement: New Media for Local Control. In: *Manufacturing Agility and Hybrid Automation-I* (Koubek, R.J. and Karwowski, W (Eds)). IEA Press.

NETWORK SUPPORT FOR IMAGINATIVE COMMUNICATION IN TECHNOLOGICAL SYSTEMS AND ADVANCED SOCIETIES

Masami CHOUI, Gen KUWANA, Yu SHIBUYA, and Hiroshi TAMURA

Kyoto Institute of Technology, Matsugasaki, Sakyo-ku, 606-8585, Kyoto, JAPAN
Email: [choui, gen, shibuya, tamura]@hisol.dj.kit.ac.jp

Abstract: Experiments on network communication and the participants' behaviors were described. The metaphor of "Forest of Ideas" was developed to explain their behaviors, and to lead the creative communication among network participants. Based on this metaphor, a WWW-based discussion system was constructed and evaluated in technical courses experimentally. In order to bring up a beautiful forest, that is a good imaginative communication, proper supports of coordinators should be made timely depending on the situation. Furthermore, a lake as rest space is important to stimulate active communication among participants.

Keywords: network, communication, CSCW, WWW, Forest of Ideas, network conferencing

1. INTRODUCTION

Until recently we have learned from experiences that the communication among different work units are essentially important in developing a new product, in improving the quality of the product, the performance of the systems and maintenance quality or in providing timely service to customers. Those activities, often called KAIZEN (improvement), used the ways of communications, such as face to face meeting and sticking hand written papers on bulletin board, etc. Several trials have been done to use computer network (CSCW) or video network (TV conferencing) to enhance communication capability among cooperative units remotely located each others. In the applications such as telelearning and telework, of an essential importance is the development of network environments, which will support creative and imaginative communication, instead of mere transmission of data or knowledge.

In spite of many studies, most of the CSCW studies are dealing the task-oriented communication, in which people are apt to communicate what are necessary. They are necessary for the daily exchange of routine data, but are neither imaginative nor creative. In the more natural human communication, people are exchanging messages to find out the problems, or to catch the situations, or to catch the timing to communicate. These communications are necessary to enclose the situation awareness of the network participants.

In this paper some experiments on network communication and the participants behaviors are explained. The metaphor of "Forest of Ideas" is developed to explain the behaviors, and to lead the creative communication among network participants.

2. FOREST OF IDEAS

Forest is some area of land with many trees growing close together. In a forest, in addition to trees, weeds, flowers, climbing plants, and bushes are living together. Furthermore, various trees are different in shape, color, etc. and each tree consists of root, trunk, branches, and leaves. In order to bring up a forest, sunlight and water are necessary. In a beautiful forest, animals, insects and birds are also taking an important parts in ecology. A lake inside a forest might be important for animals to rest and to keep a forest beautiful.

Considering growth process of a forest, big trees never come out from the wasteland from the beginning. At first weeds of low height might come out. Then, bushes grow up among the weeds gradually. Finally, big trees start growing out among the bushes. However, there are still grasses under the big trees.

We expected that the above explanations of forests are also applicable to explain the discussions and interactions in the computer networked communication. In this paper, a concept named "Forest of Ideas" is introduced. For example, discussions in computer network systems might be explained by the growth of forests.

In Fig. 1, the concept of "Forest of Ideas" is illustrated. In "Forest of Ideas", participants (gardeners) bring themes (seeds or saplings) into discussion space (land), and bring up opinions (weeds, bushes, and trees). In usual forest, trees, bushes, and weeds are coexisting, and they are necessary to keep the forest growing up continuously. In "Forest of Ideas", it is also important that there exist various kinds of ideas simultaneously. In order to bring up ideas, coordinators (instructors of forests) prompt participants to write opinions (supply water), lead the discussion to desired direction (control sunlight), and give participants additional news (put fertilizer). The system manager (manager of forests) and coordinators should control the environment of "Forest of Ideas" and help participants to get ideas. Information management by coordinators becomes more important to improve the computer-networked discussion.

3. IMAGINATIVE COMMUNICATION IN FOREST OF IDEAS

In order to make an imaginative communication in Forest of Ideas, suitable supports should be offered timely. In Fig. 2, a progress of woods, that is a group of trees, is described in detail.

Before starting communication, a field has to be plowed to be suitable for the young plants to grow. Then some seeds, i.e. topics of communication, are sowed on the field.

After that, participants will visit the field and bring up trees and woods. If participants find an interesting seed or seedling, they will grow it. It is important that the users can find their interesting object easily. The managers or coordinators should support the participants by offering topics of communication in several ways. For example, multimedia presentation might be effective to stimulate participants' interests.

During growing up of woods, the managers or

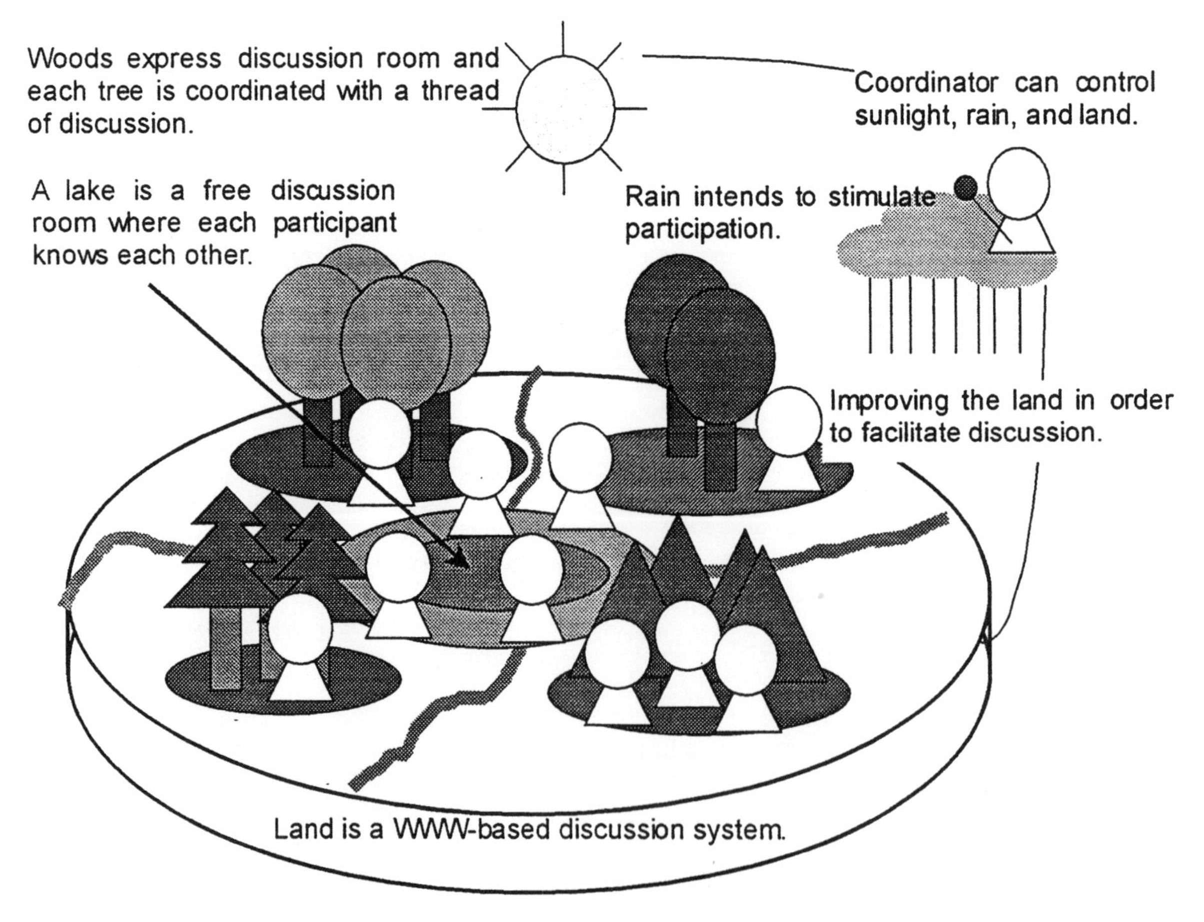

Fig. 1 Forest of Ideas

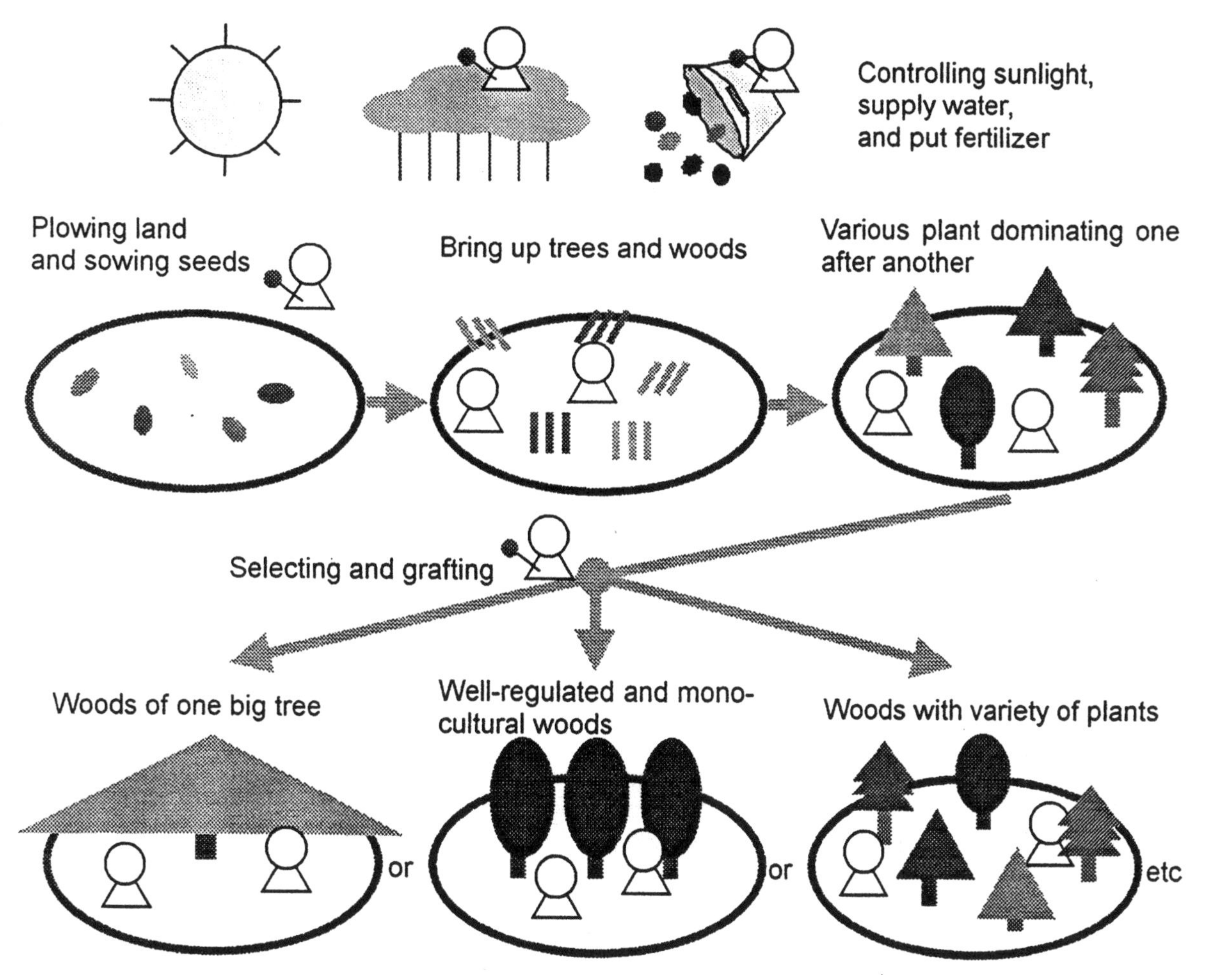

Fig. 2 Progress of woods

coordinators should support the users and lead proper direction some times. Managers should improve land to make participants walk in the forest easily. This means improving user interface of system. Coordinators should select and graft trees in order to keep forest beautiful. This means concluding opinions. If they allow trees to grow as they like, woods will be made of promiscuous trees. In this case, discussion will become diverse and no good communication will be expected.

Making beautiful woods or trees are one of several goals of bringing up them. That means that there are several kinds of woods or trees that will appear as results of conferencing. For example, making a consensus or conclusion of participants is similar to make only one tree in the woods. If a conference is expected to make not only one but also several number of good results, woods will consists of several kind of beautiful trees finally. Furthermore, if there is no beautiful tree but bush, that might be also good situation at the earlier stage of conferencing.

Furthermore, lake is very important as resting space in the forest. Participants know each other and get new information around the lake, and they become aware of their standpoints in the woods. This activates woods in the forest.

4. WWW-BASED DISCUSSION SYSTEM

Based on the concept of "Forest of Ideas", a discussion system was implemented in WWW networks and used for discussions among students of technical courses experimentally.

This system consists of a WWW/database server and clients that have WWW browsers. The OS of the server is Linux, which is compatible with UNIX. PostgreSQL is used for database server, CGI script called PHP/FI for interface between database, and WWW pages, and Apache for http server. Contents and details of opinions, data about participants, and logs of participants' behavior are stored in database and processed by language that looks like SQL. Using database, it becomes easy to store, search, and process information.

At the beginning of the class, a teacher offered students some topics and asked them to discuss about the topics on the discussion system. This kind of discussion might be regarded as a task-driven discussion. At that time, two topics were proposed. They were "mobile phones in the 21st century" and "diversification of functions of trains". Students were also allowed to discuss other topics. This kind of discussion might be a free discussion.

We prepared three kinds of discussion room according to each purpose; task-driven discussion room, free discussion room, and user supporting room. After a while we added concluding discussion room to conclude discussion.

Participants can enjoy informal communication in the free discussion room, and formal communication is required in the task-driven discussion room. We divided the two purposes between these discussion rooms. On the standpoint of user supporting, system manager answers participants' questions about this system in the user supporting room. In the concluding room, we tried to conclude discussion about "mobile phones in the 21st century".

5. BASIC TIME RESPONSES

The access log of the WWW server and database server were analyzed to characterize some of the basic time responses of network discussions.

5.1 Time Response of Network Discussion

The dynamic characteristics of network discussions are considered under ideal assumption. First of all, the network and the server are supposed to have sufficient performance even though a considerable number of participants are using at the same time. When one opinion is presented, to the network with N discussants, N discussant will open the opinion with some delay, which might show some distribution. If n opinions are present for the N discussants, N/n discussants in average are expected to access an opinion. However, discussants on line

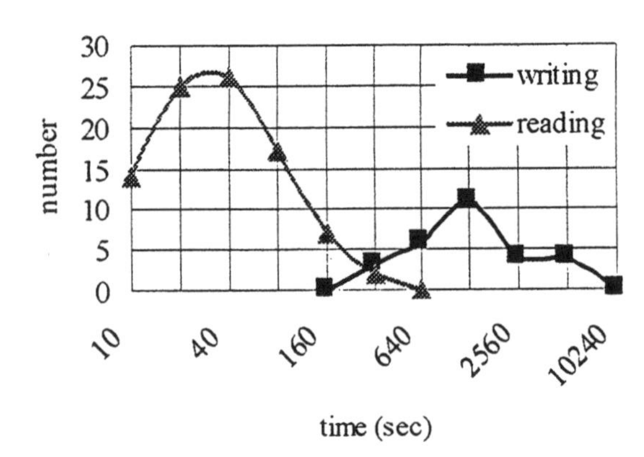

Fig. 3 Response after theme opinion was wrote

have the liberty to select one out of n opinions, based on the writer or title. If n is larger number, discussants spend more time in looking around the title list. A discussant, once opened an opinion might spend some time in reading the opinion and in deciding whether to respond or not. For these reasons, the number of the first reading to a new opinion increases with some delay, and then decreases again after a while. The first read and write responses of the discussants are shown in Fig. 3, where the abscissa is the time after an opinion is submitted to the network, in the logarithmic scale.

The discussants write their opinions less frequently with more delay than they read. The read responses might happen by different discussants after the first. A write response might be given by one of the readers. And then the opinion will drive new read activities. Fig.3 is derived by the analysis of

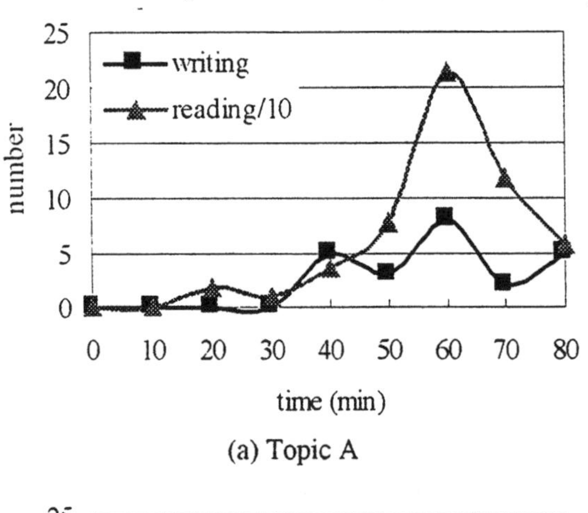

(a) Topic A

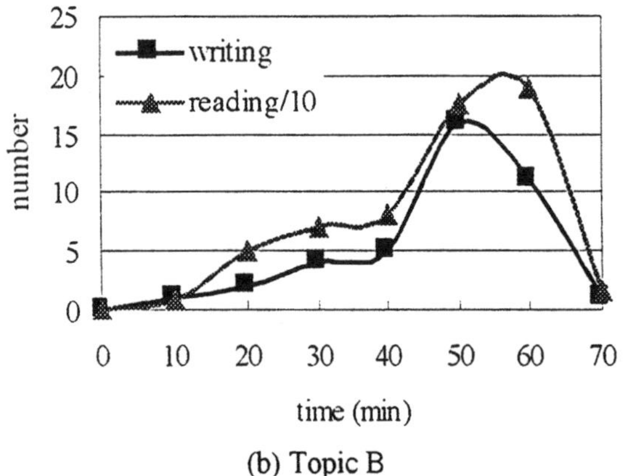

(b) Topic B

Fig. 4 Time Response of Two Topics

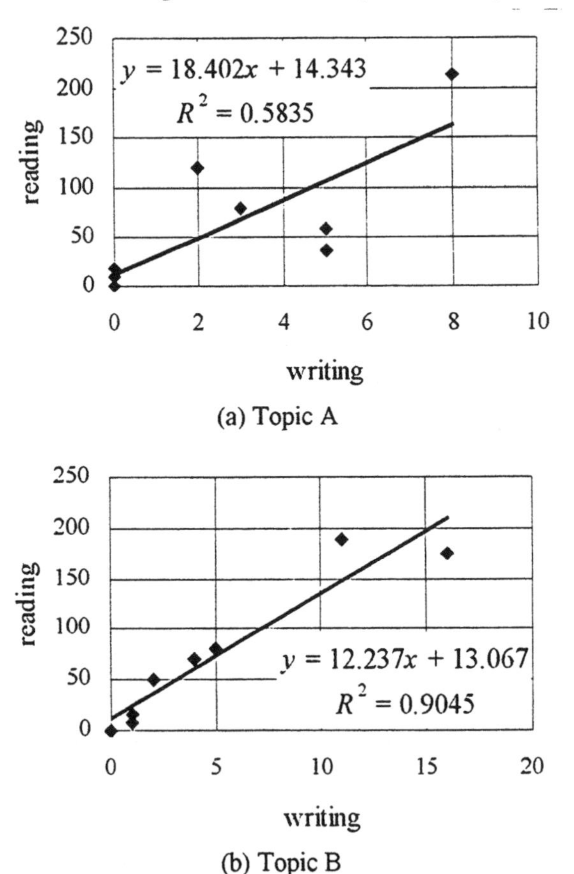

Fig. 5 Correlation between writing and reading

discussant's activities of about 100 minutes which looked most regular. The average time of first reading and writing comes 82sec., and 1227 sec. respectively after submission of an opinion.

Next the analysis is made on the discussions along a topic. Two topics are chosen for the explanation below. The Topic A was about 'the good manner of using mobile phones in a concert hall, etc', and the topic B was 'the technology to restrain the use of the mobile phones'. In Fig.4 the ordinate is the number of read in 1/10 scale and that of the write. The read responses are recorded about ten times more than the write. The increase in read is driven by the increase in write. The fluctuations in the number of read and write are due to the time necessary to write an opinion. The discussants spend more than 10 minutes to finish an opinion, while they normally stop reading. Following the submission of a valuable opinion the number of read will increase steeply. This might be explained as a step of growth in the forest.
The number of read and write actions shown in Fig. 4 are in close correlation, as shown in Fig. 5.

5.2 Comparisons between Discussion Rooms

The read and write characteristics of two discussion rooms are shown in Fig. 6. Fig.6(a) is the log of free discussion room, and (b) that of task-driven room. Basically, the number of read actions follows that of write. That means participants read the opinions as they are written. But the ratio of read and write is higher at the early stages of discussions. As the discussions mature, discussants spend more time in writing own opinions.

The discussions become active from the early stage of discussion in the free discussion room, while in the task-driven room, write actions increases gradually.

The write actions, however, increase quickly in the discussions of easy topics, such that are familiar to discussants or that can be answered simply by yes or no (Fig.7 (a)). The difficult topics (Fig.7 (b)) are such that discussants are required to create their own ideas and sentences. Discussants send responses for the easy topics at the early stage, and write opinions to the difficult topics at the later stages.

5.3 Role of Free Discussion Room

The number of opinions subjected to four discussion rooms is shown in Fig. 8. At the early stage of discussions (first week), free discussions were most active, and the task driven discussions increased gradually in the following weeks. We understand discussants are worrying about the other participants' behaviors and they want to know each other before submitting their proper opinions. In order to know each other, easy ways are talk about topics common to each others. After talking about such topics, they begin to write their opinion in task-driven discussion.

Free discussion rooms are similar to a lake in the forest where living creatures meet together and rest.

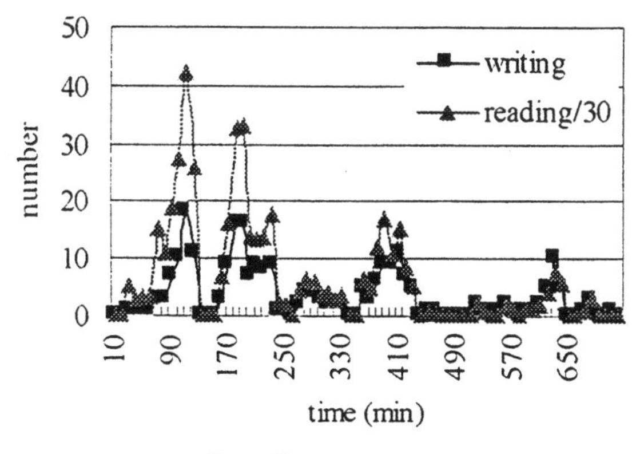

(a) Free discussion room

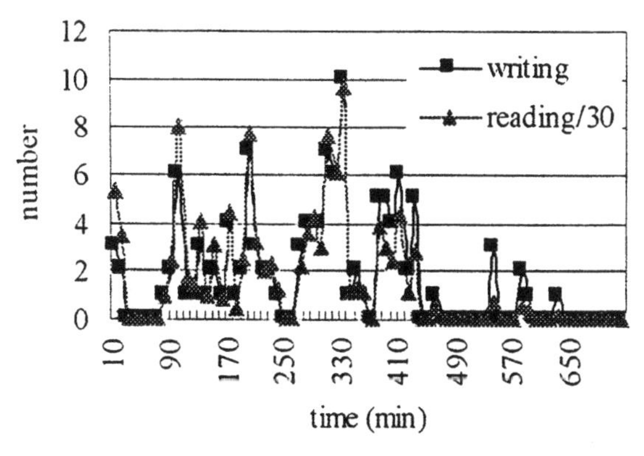

(b) Task-driven discussion room

Fig. 6 Difference of reaction between rooms

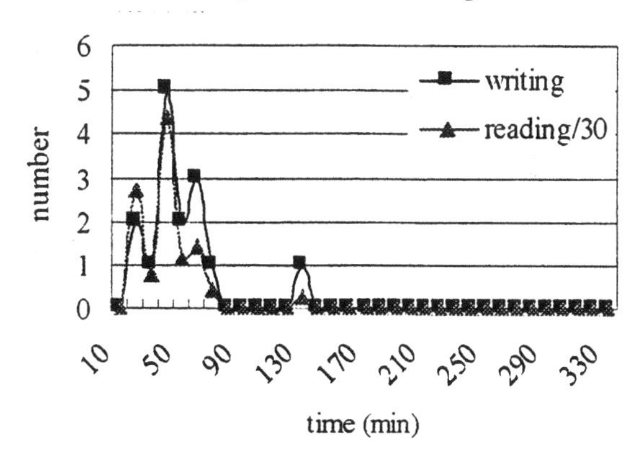

(a) Easy topic

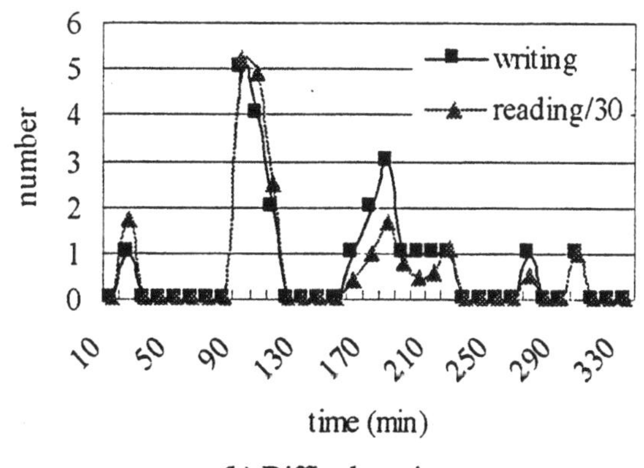

(b) Difficult topic

Fig. 7 Difference of reaction between topics

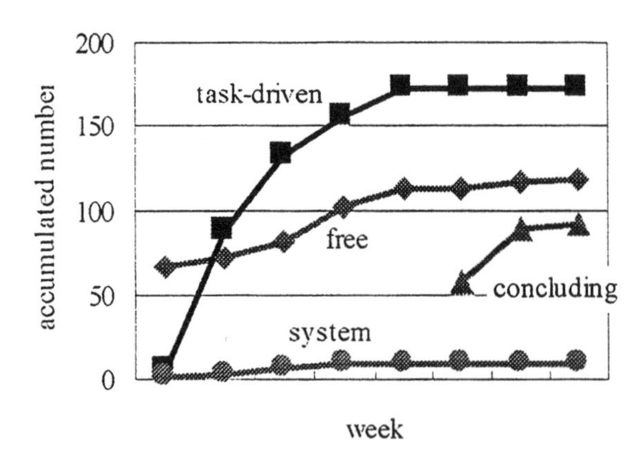

Fig. 8 Accumulated number of opinions

After recognizing an atmosphere on the network in free discussion room, the argument moves ahead to task-driven discussion room.

This system allows writing opinions by anonymous names if participants want to use. Fig. 9 shows the ratio of opinions submitted in anonymous names. The ratio is greater in the free discussion room than in the task-driven.

This result also shows participants use informal communication in order to recognize an atmosphere on the network. And they take responsibility for their opinions by writing opinions by less use of anonymous names in the formal communication.

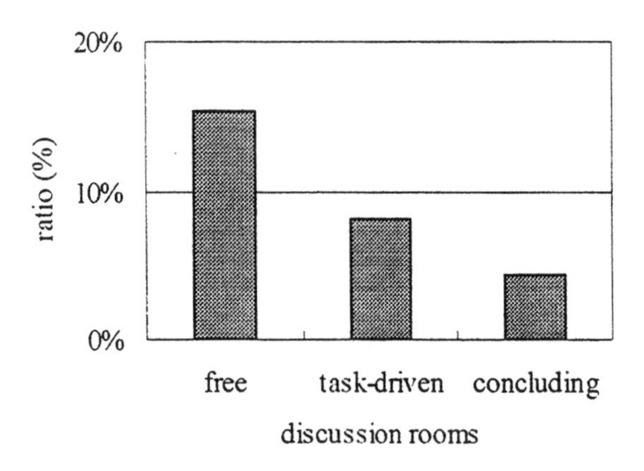

Fig. 9 Ratio of anonymous writers

5.4 The Importance of Coordinators

Furthermore, a discussion room that grows well has a leader, that is to say, a coordinator who controls whole of the room including participants' behavior. The coordinator writes opinions actively so that the number of his opinion is more than that of other participants. Two types of participants become coordinators. One is an intentional coordinator and another is a spontaneous coordinator. Anyway, both type of coordinator sometimes make some rules in the room, and classify opinions into some categories.

6. CONCLUSION

The experimental uses of network communication have been done with the students of technical courses on the imaginative topics, such as portable phones in the 21st century and mobile information terminals. The discussions are partially opened to the Internet for the public access.

At this moment we have to emphasis the importance in the forest, that is, free discussion, in order to activate and to improve the discussion. Good coordinators are necessary and have to play an important role in discussions. In WWW-based discussion systems, it is expected to have intentional or spontaneous coordinators.

As the problem in the future, we are interested in improving the ground of forest (in other words, improvement of the user interface of WWW-based discussion system). It is also important to put enough fertilizer (give participants useful news). Then, the more creative and active communication will be expected.

ACKNOWLEDGEMENT

This study is partly supported by Grand-in-Aid for Scientific Research from the Ministry of Education, Science, Sports, and Culture of Japan under Grand No. 09044029.

REFERENCES

Choui, M., Tamura, H., Kuwana, G., Shibuya, Y., (1997). Information Management in "Forest of Ideas", *Symposium on Human Interface*, **13**, pp.439-444.

Choui, M., Tamura, H., Shibuya, Y., (1998). Roles of Free Talking in Enhancing Network Conferencing, *Human Interface News and Report*, **13-1**, pp.19-26.

Dourish, P., and Bly, S.: Portholes Supporting Awareness in Distributed Work Group, Proc. CHI'92, pp.541-548, (1992).

Fish, R.S.,et al.:The Video window System in Informal Communications, Proc.CSCW'90,pp.1-12,(1990).

Human Interface Lectures, (1998), http://hilec.dj.kit.ac.jp/

Kuwana, G., Shibuya, Y., Tamura, H., Choui, M., (1997). Teaching Information Literacy in "Forest of Ideas", *Symposium on Human Interface*, **13**, pp.381-386.

Shibuya, Y., Kuwana, G., Tamura, H., (1997). Teaching Information Literacy in Technical Courses using WWW, *Advances in Human Factors/Ergonomics*, **21B**, pp.723-726.

Quantification of task related activity by statistical and analytical methods

M. Fjeld[1], S. Schluep[1] & M. Rauterberg[2]

[1]*Institute of Hygiene and Applied Physiology (IHA)*
Swiss Federal Institute of Technology (ETH)
Clausiusstr. 25, CH-8092 Zurich, Switzerland
{fjeld, schluep}@iha.bepr.ethz.ch

[2]*Center for Research on User-System Interaction (IPO)*
Technical University Eindhoven (TUE)
NL-5612 AZ Eindhoven, Netherlands
rauterberg@ipo.tue.nl

Abstract: Behaviour of expert and novice database users solving the same task was recorded. Several successful strategies were identified. Since there are more users than strategies, some users applied the same strategy. The aim was to develop methods grouping users with common strategy. Following three approaches (correlation, intersection, and exclusion), a metric among task solving behavioural sequences was defined. Measured data was organised in matrix systems relating all users. Statistical and analytical interpretation of matrices showed distinct groups. A common denominator for a group can indicate a strategy. *Copyright © 1998 IFAC*

Keywords: automatic recognition, model based recognition, behavioural science, user interface, problem solvers, multidimensional systems, statistical analysis.

1. INTRODUCTION

The aim of this paper is to develop automatic methods finding task solving strategies. Such knowledge is of interest to understand how users behave in a newly designed system, and to thereby giving them better support. Also, it may also help understanding how expert users behave in highly complex systems.

Under certain conditions, strategies may also be obtained by protocol analysis (Ericsson and Simon, 1984). Protocol analysis implies manual inspection of video and verbal utterances in addition to logfiles. With simple tasks, this work can be overcome. For more complex tasks, protocol analysis has proved cumbersome. Semi-automatic generation of process models was studied by Ritter and Larkin (1994). Guided by their work, further principles for automatic recognition of user strategies and plans will be suggested.

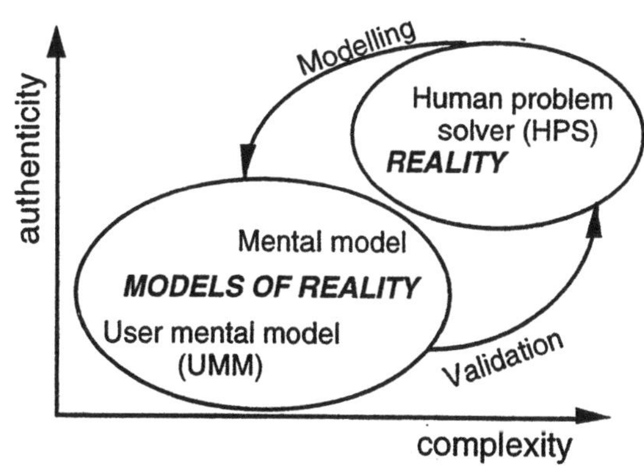

Fig. 1. A scheme showing the relation between models of reality and real humans (HPSs). Models are meant to represent objects and processes existing in reality.

This paper treats computer mediated, everyday task solving. A special case of mental models, called user mental model (Tauber, 1985) (UMM, Fig. 1) is introduced. UMMs can bring understanding about strategies people use when solving specific problems. UMMs can be represented in many ways, using plain text, Petri nets or state-transition vectors. In this paper, the latter representation was chosen to elaborate UMMs based on observable task solving strategy.

In general, a lot of task solving behaviour that is not strictly *task related* can be observed. If one single HPS is studied, it is hardly possible to single out the successful *strategy* from the *remaining behaviour*. One approach may be to study many users solving the same task. Since they all solve the same problem, it is likely that their common behaviour is what was required to solve the task.

A *strategy* is defined to be one (of many), possibly error free, successful task solving behavioural sequences for the current system and task. As soon as a complete strategy is accomplished, task solving is over. If users follow different strategies, a group of users may have one strategy in common, an other group a second one. Modelling means finding a measure for relation among users and thereby grouping users with common strategy. The common denominator for each group indicates a strategy.

Which strategy a user prefers, as well as other behaviour can tell us something about the particular HPS; for instance how a strategy was acquired. Given a behavioural task solving sequence, it is of interest to separate the *strategy* (which is more related to the task-system combination) from the *remaining behaviour* (which is more related to the HPS). The remaining behaviour may consist of partial strategies or strategies that would be successful within an other system and/or task.

In this paper, human perception and verbalisation will not be considered as part of the problem solving. Purely based on observable task solving behaviour, the aim is to develop automatic methods, applicable with simple as well as with complex tasks. Protocol analysis will only be used to validate the elaborated automatic methods.

2. SYSTEM DESCRIPTION

The system studied is a relational database program with 153 different dialogue states. A transition consists of a dialogue state and one (of many) possible user actions in that state. All possible transitions of the system are represented by a state-transition vector space. A state-transition-vector (STV, Formula 1) summarises a subject's task solving behaviour.

$$\{e_i^p\} \qquad (1)$$

$e_i^p \in N_0$ — STV component

$p \in \{N_1, ..., N_6, E_1, ..., E_6\}$ — user index

$i = 1, ..., n$ — transition index

$n = 978$ — number of transitions

The total number of possible transitions for the complete database program is given by n. An STV component's value tells how many times a certain transition was activated to solve the task. Talking about a *user* without any further detail, refers to the corresponding *STV* of that user. In both cases, it means the observed task solving behaviour of the user, expressed by an STV.

Since the order of activated transitions is not contained in the STV, order of user behaviour is only partly conserved. It is stored in an implicit form, given by the system dialogue structure and embedded in the STV structure.

3. TASK DOMAIN

An empirical investigation was carried out by Rauterberg (1992) to compare different types of expertise. For the reconstruction of UMMs, logfiles of six novice {N1,...,N6} and six expert {E1,...,E6} users were used, all solving the same task. The task was to find out how many data records there are in a given database consisting of three files. As soon as the required results were found, task solving activity was finished. An example UMM of a task solving process, based on E4, is presented in Rauterberg et al. (1997). For E4, 15 *different* transitions (number of positive STV components) were activated to solve the task. (Over all the users, *different* transitions was between 14 and 42). Some transitions were activated repeatedly, so the *total* number of activated transitions (sum of STV components) is 25. (Over all the users, *total* number of activated transitions was between 25 and 175).

4. BASIC QUESTIONS AND METHODOLOGY

Studying an STV of one user can tell us which system states the user passed by, which transitions that were activated in those states and how many times that occurred. Different users working with the same system are directly comparable, since their behavioural sequences only differ by the value of the vector components. In order to group users according to their behaviour, there is first a need to relate behavioural sequences.

Table 1: Relation among task solving behaviour has various *qualitative* aspects with corresponding *quantities*.

Quality	Quantity
Interaction	Correlation
Similarity	Intersection
Difference	Exclusion

Table 1 shows a few qualities of relationship and corresponding quantitative methods. A classical method is that of correlation. An alternative is to look for analytical methods. For instance, STVs can be considered as sets and represented by ellipses, as in Fig. 2, or by rectangles as in Fig. 4. The area of an ellipse or rectangle corresponds to the sum of the STV components values. Intersection area can be understood as (symmetric) similarity between two STVs. Exclusion areas can be understood as the (asymmetric) differences between two STVs.

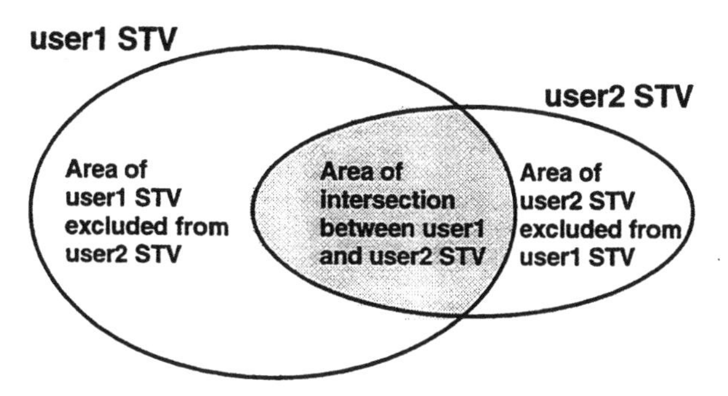

Fig. 2. Intersection area and exclusion areas between user1 and user2 STV. .

Based on such considerations, the following questions are raised and corresponding methods are suggested:

1. What is the interaction among two behavioural sequences? Method: *correlation.*
2. What do two behavioural sequences have in common (similarity)? Method: *intersection* (Fig. 2).
3. What do two behavioural sequences not have in common (difference)? Method: *exclusion* (Fig. 2).

Table 2: The three suggested methods, metrics, metric nature and grouping algorithms. CORR means a standard correlation method, the other metrics are defined by Formula 2, 3 and 4.

Method	Metric	Metric nature	Grouping algorithm
Correlation	CORR	Statistical	Statistical
Intersection	$M^{IS}_{p,q}$, $M^{BIS}_{p,q}$	Analytical	Statistical
Exclusion	$M^{EX}_{p,q}$	Analytical	Analytical

For each method, a metric (Table 2) is elaborated. The order of the metric may be symmetrical (the metric applied from user1 STV to user2 STV is the same as the metric applied from user2 STV to user1 STV; correlation and intersection) or asymmetric (the metric applied from user1 STV to user2 STV is *not* the same as the metric applied from user2 STV to user1 STV; exclusion). Each of these metrics can be applied between all possible pairs of STVs, giving a matrix system. A grouping algorithm is applied on this matrix, indicating groups with related behaviour.

For each group suggested by the grouping algorithm, a strategy may be approximated. The procedure is to create an STV with maximum number of non-zero components smaller or equal to the STVs components of that group. This step will not be explored any further in this paper.

The following presentation will proceed from more statistically based to more analytically (non-statistically) based methods.

4.1 Correlation method

In this method correlation applied among pairs of STVs measures degree of interaction among users. The matrix is then analysed by multi-dimensional-scaling (MDS, Systat, 1989) to indicate groups of users.

Metric

Pearson correlation is applied to measure interaction between two STVs. This procedure gives a diagonal dominant, symmetrical matrix with possible values between minus one (opposite interaction), via zero (no interaction) and one (complete interaction). For Fig. 3 the observed values are between -0.003 and 0.948 (without considering the diagonal elements).

Grouping algorithm

The correlation matrix is interpreted by two dimensional MDS, giving the plot of Fig. 3.

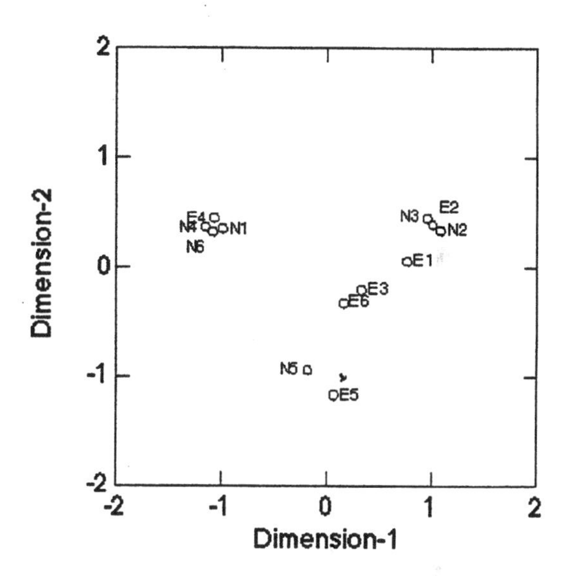

Fig. 3. MDS (r=1, Kruskal, Mono) plot with a Pearson correlation matrix gives RSQ=0.870.

Outcome

Fig. 3 shows that users may be grouped: {N1, N4, N6, E4}, {N2, N3, E1, E2, E3, E6} and {N5, E5}. Some of the users, like N5 and E5, may be interpreted as partly members of other groups, indicating parts of or combinations of strategies. According to the proportion of variance (RSQ=0.870), MDS explains some of the user data variance, but 13% remains unexplained.

4.2 Intersection method

This method is based on the observation that if two users follow the same strategy, that strategy will belong to the intersection of the two users STVs. However, this is no condition for this method to work. Two STVs may have a significant intersection, without having solved the problem with the same strategy.

Two STVs common part is the same, so the interaction matrix is symmetric and can be analysed by MDS.

Metric

Similar behaviour is measured by summing up the smaller STV components of the two STVs, thus considering the number of activated transitions common to both users.

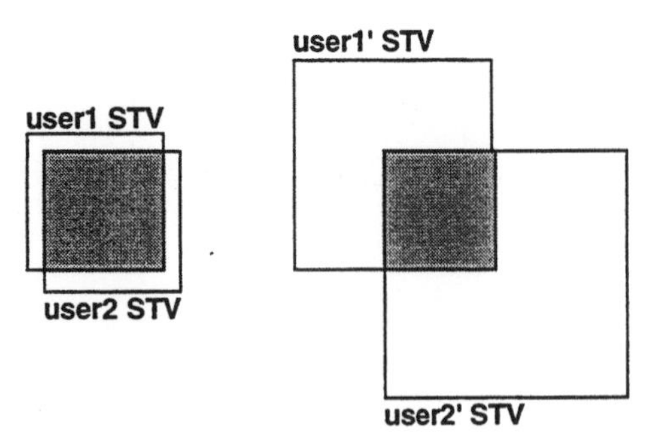

Fig. 4. User1 and user2 STVs represented by rectangles. In both situations, the intersection is the same, but the similarity between STVs are not. This indicates that a *normalisation* is required.

User1 and user2 STV can also be seen as sets, represented by rectangles (Fig. 4). For the intersection area (measured quantity) to be a valid measure of similarity (desired quality), a *normalisation* is required. It is possible to scale degree of intersection by the larger (max.), the average (mean) or the smaller (min.) sum of the intersection areas. Scaling by the smaller of the areas corresponds to scaling by the maximum possible intersection. Expressed in the state-transition vector space gives the intersection metric of Formula 2.

$$M^{IS}_{p,q} = \frac{\sum_{i=1}^{n} \min\left(e_i^p, e_i^q\right)}{\min\left(\sum_{i=1}^{n} e_i^p, \sum_{i=1}^{n} e_i^q\right)} \qquad (2)$$

$p,q \in \{N_1,\ldots,N_6,E_1,\ldots,E_6\}, p \neq q$ user indices

Ignoring repetitive behaviour is a mean to reduce complexity. Replacing each STV component >1, by 1, results in a *binary* state-transition-vector. An intersection metric based on this vector is given by Formula 3.

$$M^{BIS}_{p,q} = \frac{\sum_{i=1}^{n} \min\left(e_i^p \cdot e_i^q, 1\right)}{\min\left(\sum_{i=1}^{n} \min\left(e_i^p, 1\right), \sum_{i=1}^{n} \min\left(e_i^q, 1\right)\right)} \qquad (3)$$

$p,q \in \{N_1,\ldots,N_6,E_1,\ldots,E_6\}, p \neq q$ user indices

Formula 2 and 3 both give a symmetrical matrix where the elements take possible values between zero (no similarity) and one (equality). For Fig. 5 (Formula 2) the observed values are between 0.078 and 0.929 (without considering the diagonal elements). For Fig. 6 (Formula 3) the comparable values are between 0.182 and 0.882.

Grouping algorithm

The symmetrical exclusion matrix is interpreted by MDS, giving the plots in Figs. 5 and 6. In both cases, the users can be divided into three groups, {N1, N4, N5, N6, E4}, {N2, N3, E1, E2, E5} and {E3, E6}. Again, some users, like E3 and E6, may be interpreted as members of other groups, indicating a combination of strategies.

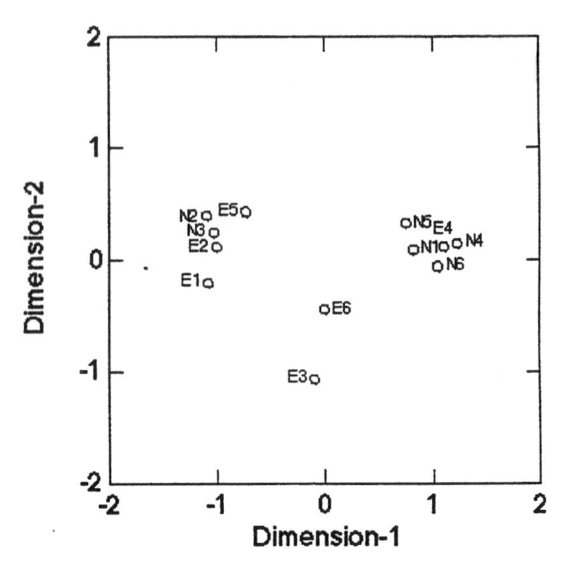

Fig. 5. MDS (r=1, Kruskal, Mono) plot with a normalised intersection matrix gives RSQ=0.975.

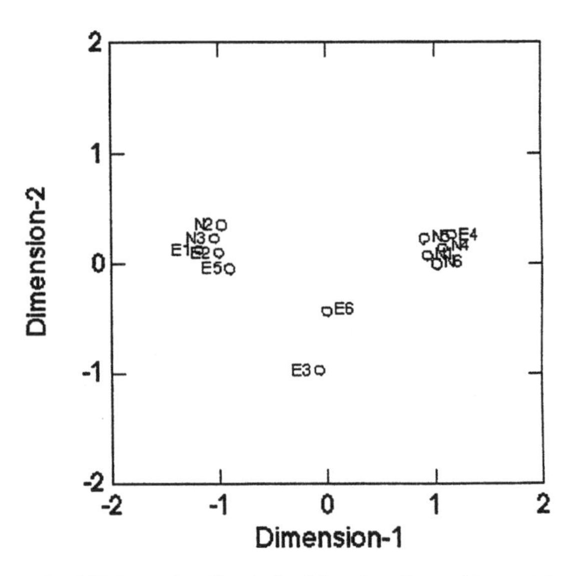

Fig. 6. MDS (r=1, Kruskal, Mono) plot with a binary normalised intersection matrix gives RSQ=0.995.

Outcome

According to the RSQ of Fig. 5 (RSQ=0.975) and of Fig. 6 (RSQ=0.995), most of the variance in the user data can be explained. However, the binary based plot of Fig. 6 is slightly better than that of Fig. 5. This is surprising, since the binary metric ignores information about repetitive behaviour. Maybe information about repetition is redundant in the context of this method.

4.3 Exclusion method

This method is based on the exclusion as a metric of difference. Exclusion among two users can be seen as two disjoint areas (Fig. 2), unless there is equality. The area of user1 STV (Fig. 2) excluded from user2 STV (Fig. 2), is not the same as the area of user2 STV excluded from user1 STV. Since the two exclusion areas are different, the resulting matrix is asymmetric and the method does not allow for MDS as grouping algorithm.

Metric

This method applies the exclusion metric of Formula 4 between all users p and q, giving an asymmetrical matrix (Table 3). The metric gives the difference between two STVs by estimating how much of user p STV (column, Table 2) is excluded from user q STV (row).

$$M_{p,q}^{EX} = \sum_{i=1}^{n} \left| \min\left(e_i^p - e_i^q, 0\right) \right| \tag{4}$$

$p, q \in \{N_1, ..., N_6, E_1, ..., E_6\}$ user indices

Table 3: Matrix elements, given by Formula 4, quantify exclusion of user q (column) STV from user p (row) STV.

	N1	N2	N3	N4	N5	N6	E1	E2	E3	E4	E5	E6
E6	6	43	47	51	70	50	47	35	73	5	171	0
E5	17	15	14	69	48	67	23	7	64	21	0	24
E4	9	44	47	56	70	55	47	35	73	0	171	8
E3	17	41	41	68	77	62	28	28	0	21	162	24
E2	17	15	16	68	81	67	19	0	66	21	143	24
E1	20	16	19	73	85	72	0	7	54	21	147	24
N6	3	41	44	28	48	0	47	30	63	4	166	2
N5	3	39	42	41	0	33	45	29	63	4	132	7
N4	2	41	42	0	55	27	47	30	68	4	167	2
N3	16	11	0	68	82	69	19	4	67	21	138	24
N2	18	0	15	71	83	70	20	7	71	22	143	24
N1	0	41	43	55	70	55	47	32	70	10	168	10

Grouping algorithm

Grayscale representation (Fig. 7) is based on the exclusion matrix (Table 3) and generated by Mathematica (Wolfram, 1991) *ListDensityPlot* with the inverted exclusion matrix as input. The matrix is inverted to achieve a consistent plot. Fig. 7 is only meant as a visualisation of Table 3, and is not an exact mapping. Since division by zero is not defined, the diagonal elements of Table 3 were directly mapped to the darkest graytone.

A group is defined as users with few differences. Hence, the inverted quantity of exclusion is significant; *inclusion*. Table 3 and Fig. 7 will be now be interpreted as an indicators for *inclusion*. Lower values, respectively darker matrix elements correspond to higher degree of inclusion.

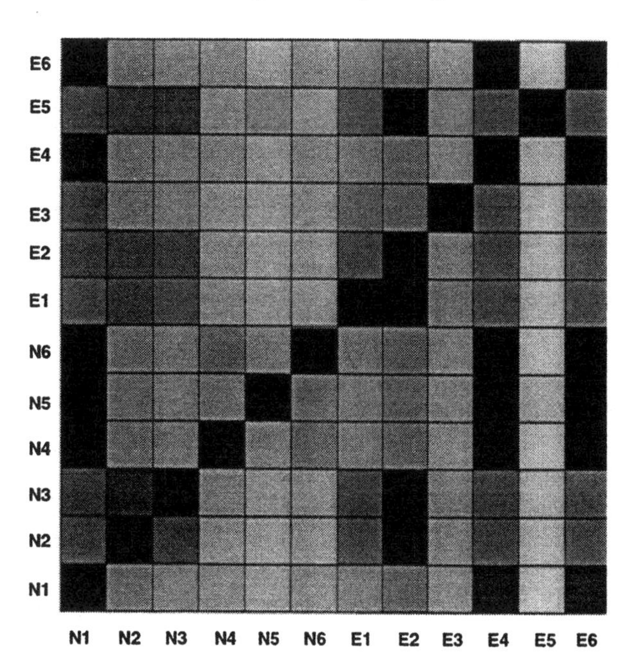

Fig. 7 Grayscale exclusion matrix. Darker elements mean lower exclusion, i.e. higher inclusion of user q (column) in user p (row) STV.

An iterative *predictor-corrector* algorithm is suggested to interpret Table 3. The *corrector* is an estimator for a threshold value so that considering elements between zero and corrector will give predicted number (*predictor*) of user groups. The stop criterion for the iteration method is that number of user groups given by the corrector, equals the value of the predictor. Research on converge is part of future work, so it is simply assumed. For each iteration the corrector is modified in order to meet the stop criterion, according to the following rules: If too few inclusion relations are considered (i.e. the corrector is too close to zero), the number of groups will be higher than the predictor. If too many inclusion relations are considered (i.e. the corrector is too far from zero), many or all of the users will be related by inclusion statements, and the number of groups will be lower than the predictor. The predictor is given the value three. By visual inspection of Fig. 7 it appears reasonable to consider the darkest matrix elements only. Since these elements have numerical values equal to or below 8 (Table 3), the initial value of the corrector is chosen to be 8. Each STV is similar to itself, so diagonal elements are ignored, giving Table 4.

Table 4: These four inclusion relations can be derived.

Relation no.	User q included in user p STV
1	q = N1, p ∈ {N4, N5, N6, E6}
2	q = E4, p ∈ {N4, N5, N6, E6}
3	q = E6, p ∈ {N4, N5, N6, E4}
4	q = E2, p ∈ {N2, N3, E1, E5}

All users that are related by an inclusion relation are defined to belong to one group. Since the three first inclusion relations are interrelated, they give one group. The remaining, fourth similarity relation gives a second group. Users not appearing in any similarity relation each define a separate group.

Outcome

Hence, Table 4 gives these groups: {N1, N4, N5, N6, E4, E6}, {N2, N3, E1, E2, E5} and {E3}. The number of groups was assumed to be three, so the stop criterion has been met. If it had not been met, it would have been necessary to try with a higher or lower corrector (according to the above mentioned rules) and go back to the start of the predictor-corrector algorithm. The algorithm is repeated until the stop criterion is met (convergence).

4.4 Validation method

To validate the outcome of the three preceding methods, protocol analysis (Ericsson and Simon, 1984) of the task was performed. This is manual work, based on analysis of video and verbal utterances in addition to logfiles. It is mostly feasible for simple tasks, where users follow one or a few strategies. The analysis showed that there are three distinct strategies solving the task; S1, S2 and S3. Table 5 shows the users according to their strategy.

Table 5: Validation data show three distinct strategies and group the users according to the strategy they applied.

Strategy	Users succeeding by strategy
S1	N1, N4, N5, N6, E4, E6
S2	N2, N3, E1, E2, E5
S3	E3

5. DISCUSSION

The correlation and intersection methods do not correspond fully with the validation data (Table 5), whereas the exclusion method gives the same results. So, with the current combination of system, task and users behaviour, the exclusion method is the best one. This means, for this case, that the purely analytical method was the best one. Measured by the RSQ, intersection is better than correlation method, which is purely statistical. So, in the context of this paper, statistical methods offer less explaining power than the analytical methods.

However, the exclusion method does not say anything about possible combinations or parts of strategies applied. For such questions, the statistically based methods seem more relevant. This is confirmed by Hanson et. al. (1991), treating class (or: group) assignment with Bayesian methods: "Such classes are also 'fuzzy'; instead of each case being assigned to a class, a case has a probability of being a member of each of the different classes".

6. CONCLUSION

For the present combination of system, task and user behaviour, it was possible to develop methods grouping users according to their task solving strategy.

7. FUTURE PERSPECTIVES

Results for one task only were acquired. To make the methods more reliable, it is necessary to evaluate several tasks. For each task the methods will be validated by manual protocol analysis. It is of particular interest to find out the exclusion method performs with other, more complex tasks.

It is also planned to study learning experiments, in order to recognise the acquisition process of strategies.

ACKNOWLEDGEMENT

Morten Fjeld is a PhD fellow of the Research Council of Norway.

REFERENCES

Ericsson, K. A. and Simon H. A. (1984). *Protocol analysis, verbal reports as data*. The MIT Press.

Hanson, R., Stutz, J. and Cheeseman, P. (1991). *Bayesian Classification Theory*. Technical Report FIA-90-12-7-01, NASA, Ames Research Center, AI Branch.

Rauterberg, M. (1992). An empirical comparison of menu selection (CUI) and desktop (GUI) computer programs carried out by beginners and experts. *Behaviour and Information Technology*, **11**, pp. 227-236.

Rauterberg, M. (1996). A Petri net based analyzing and modelling tool kit for logfiles in human computer interaction. *Proceedings Cognitive Systems Engineering in Process Control SCEPC'96* (Yoshikawa, H. and Hollnagel. E., (Ed.)), Kyoto University, pp. 268-275.

Rauterberg, M., Fjeld, M. and Schluep, S. (1997). Parallel or event driven goal setting mechanism in Petri net based models of expert decision behavior. *Proceedings of CSPAC'97* (Bagnara, S., & Hollnagel, E., Mariani, M. and Norros, L. (Ed.)), Roma, CNR, pp. 98-102.

Ritter, F. E. and Larkin, J. H. (1994). Developing Process Models as Summaries of HCI Action Sequences. *Human Computer Interaction*, **9**, pp. 345-383.

SYSTAT Inc. (1989). SYSTAT®: The system for statistics. pp 93-166. SYSTAT PC program, version 7.0.1.

Tauber, M. J. (1985). Top down design of human-computer systems from the demands of human cognition to the virtual machine - an interdisciplinary approach to model interfaces in human-computer interaction. *Proceedings of the IEEE workshop on languages for automation*, Palma De Mallorca (E), pp. 132-140.

Wolfram, S. (1991). Mathematica®, A system for Doing Mathematics by Computer, 2nd Ed. Mathematica program version X3.0.1.1 for Silicon Graphics IRIX. AddisonWesley, pp. 164, 395, 819.

SUPPORT COOPERATION IN SMALL GROUP CONTEXT USING WWW APPLICATIONS

Cheng ZHANG **Kay Chuan TAN**

Department of Industrial and systems Engineering
National University of Singapore
10 Kent Ridge Crescent, Singapore 119260
Email: engp7582@nus.edu.sg isetankc@nus.edu.sg

Abstract: This research looks into the use of WWW applications for cooperation and support within a small group context. Formal and informal cooperative work between individuals in a small academic department at a local university were identified and analyzed. Useful applications that were widely available on the WWW were modified to result in a prototype system used to support both formal and informal cooperative work within the Department. This study looked into the ways that small groups cooperate without CSCW. CSCW was then introduced, taking into account the contextual capabilities and limitations. The results of the study should benefit CSCW design for small groups. *Copyright © 1998 IFAC*

Keywords: CSCW, WWW applications, formal and informal cooperative work, small group

1 INTRODUCTION

The study of CSCW (Computer Supported Cooperative Work) attracts scientists from a variety of disciplines including sociology, anthropology, psychology, computer science, office automation, human factors, management science and organization design. It emerged in response to the increasing research and development activities associated with groups of people using computers. An understanding of the cooperative nature of the R&D activities is crucial if one is to engage in progressive CSCW design (Schmidt and Bannon, 1991).

In order to build effective CSCW applications, there is a need to understand how cooperation occurs and how conflict arises. This understanding is required at both a general and at a specific level. At a general level, the need is to understand cooperative principles that apply widely across different contexts such as varying group sizes and different workflow patterns. Thereafter, certain concepts and procedures are selected to apply at a specific level (e.g., a particular group size, or a particular workflow pattern) in order to understand CSCW in that particular context.

Given the present state of technology, the range of CSCW systems that can be developed and directly deployed is limited (Bentley, *et al.*, 1997). The development of World Wide Web (WWW) applications and Internet tools provide good opportunities for CSCW systems design. Although cooperation can be supported by non-computer-based tools such as telephone and video conference technologies, computer systems provide easier access to shared information for distributed working group by using minimal technical infrastructure (Gorton, *et al.*, 1996). The Web can be adopted as a globally accessible, platform independent infrastructure for CSCW systems, which can provide richer support to cooperative work and are more convenient to distributed users.

In this research, the study of cooperation and conflict in small group was conducted in Industrial and Systems Engineering (ISE) Department of National University of Singapore (NUS). The objective here was to analyze the current cooperative activities in order to get a clear understanding of how cooperation and conflict occur. A prototype CSCW system was developed that uses Web applications which provide support to cooperative activities.

2 COOPERATION AND CONFLICT

Cooperation and conflict are common phenomenons that occur in interactions between individuals within a group. As CSCW is concerned with the design of systems which enable people to work together, an examination of cooperation and conflict, and also the proper ways of dealing them, would be of great benefit.

There have been many attempts to define the concepts of cooperation and conflict. As early as 1867, Karl Marx (1957) considered that the cooperation referred to "multiple individuals working together in a planned way in the same production process or in different but connected production processes," and that "conflict referred that the ruling class, using the instruments of the state, imposes order and suppresses dissent". More recently, Argyris (1991) defined cooperation as "acting together, in a coordinated way at work, leisure or social relationships, in the pursuit of shared goals, the enjoyment of social activity, or simply further the relationship". Putnam & Poole (1987) gave the following definition of conflict: " the interaction of interdependent people who perceive opposition of goals, aims, and values, and who see the other party as potentially interfering with the realization of these goals".

The cooperation mechanisms used by organizations or small groups differ in their degrees of formality – that is, in their degree of perspecification, conventionality, and rule boundedness (Robert, *et al.*, 1990). In the formal sense, cooperation is accomplished by adherence to common rules, regulations, and standard operating procedures; through preestablished plans, schedules, and forecasts; and through memos, management information reports, and other standardized communications. Informal cooperation, on the other hand, is that which remains when rules and hierarchies are eliminated from cooperative activities.

3 WWW AS AN ENABLING TECHNOLOGY FOR CSCW

World Wide Web (WWW) is conducive to cooperative technology because it allows people to share information. Most CSCW systems now developed are based on particular platforms and are only usable within the particular organizations for which they are developed. The diversity of machines, operation systems and application software in the work domain may make it difficulty for CSCW system to be deployed in other environment. In contrast, the WWW offers a globally accessible, platform independent infrastructure. It can be seen as an enabling technology for CSCW in several ways (Bentley, *et al.*, 1997):

- WWW provides access to information in a platform independent manner because its client programs (browsers) are available for all popular computing platforms and operation systems;
- Browsers offer a simple user interface and consistent information presentation across these platforms;
- Browsers are already part of the computing environment in an increasing number of organizations, requiring no additional installation or maintenance of software for users to cooperate using the WWW;
- Many organizations have installed their own web severs as part of an Internet presence or a corporate Intranet and have familiarity with sever maintenance and sever extension through programming the sever API.

4 CASE STUDY IN ISE DEPARTMENT

ISE department was selected as research site because this department is rather small comparing with other department, but at the same time, has the same mechanism of cooperation and conflict as others.

4.1 The department overview

The ISE department at NUS consists of office clerks, teaching staff, and research students. This small group can be divided into smaller clusters, which include the staff and research students who have the same or similar research interests. The office clerks often serve as a bridge connecting the Department with outside contacts. They also communicate with the other departments and administrative office of university. Figure 1 portrays the inter-activity flow of the Department.

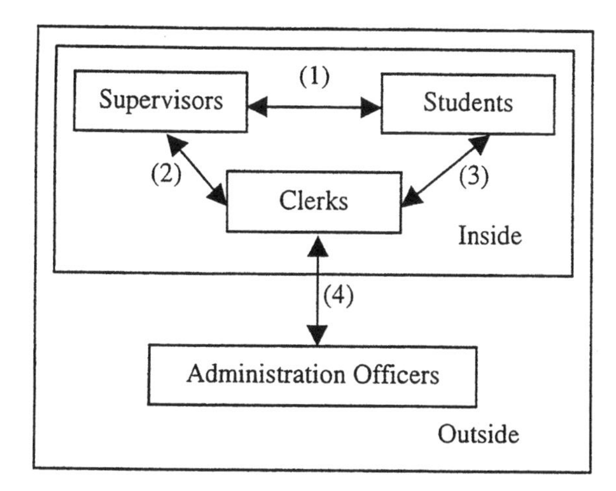

Fig. 1. Inter-activities of the ISE Department

From the figure above, four kinds of interactions are defined:

Interactions between supervisors and students (1). This type of interactions happens mostly in formal ways. Formal cooperative work occurs in two ways: weekly report and group discussion. Students hand in their weekly reports to supervisors and group discussion, therefore, can be set up for teaching and learning.

Interactions between supervisors/students and clerks (2&3). For the main administration affairs inside the Department, some work needs to be done under the cooperation of supervisors/students and clerks, e.g. the selection of the courses and the arrangement of seminars.

Interactions between clerks and administrative officers (4). Clerks always serve as a major connection between the Department and the university. They collect the opinions inside of the Department and then transmit them to school officers, and vise versa.

The formal cooperation of work activities in this department is achieved mainly through documentation and preset procedures. The principle document used in department is the "Progress Card for Research Students"(PCRS). The seminars are a formal way by which people in the Department interact with each other. Generally, seminars are conducted twice a month by supervisors or students in their related research fields.

4.2 Analysis and Discussion

The analysis of this case study focuses on the deviations and errors happened in the formal cooperation. The deviations and the ways of coping with such situations vary greatly on a case by case basis. The main deviations from the formal cooperation procedure are listed as following.

(1) Incomplete PCRS. The students need to hand in their progress reports to their supervisors every six months. Since the students can enroll in the university anytime during the year, it is not easy for supervisor to remember the exact time for students to hand in their reports. Thus, some incomplete reports may appear in the PCRS.

(2) Lost of PCRS. There are no specific persons to charge of keeping the PCRS. Sometimes students keep them, at other times; supervisors keep them; and sometimes department officers keep them. The losing of PCRS occurs at times.

(3) Informal Practice. The main channels of information interchange are through electronic mail and conversations. These two methods are self-absorbed, uninhibited, and non-conformist. The massages sent focus more on the interests of sender's than the receiver's interests. In addition, about half of the massages were not work related. Therefore, in some extent, e-mail has brought some amount of time wastage at least on the part of the receivers.

(4) Absence to seminars. Regular seminars are conducted twice a month. However, absences due to variety of reasons are common. Instances occur where individuals interested in a particular seminar are not able to make it at the given date and time. These seminars, once missed, can not be made up.

In summary, the cooperation activities within the Department are conducted in the atmosphere of shared information. Staffs and students get in touched, or communicate with each other not only by the means of formal procedures (e.g., PCRS and seminars), but also through informal communications. In this case, members of the Department can not achieve all the information by the means of PCRS and seminars alone. These formal channels of communication are supplemented by a great deal of informal communication channels such as everyday chatting, telephone call, email, etc. Informal communication can, in some degree, lead to more successful cooperation.

4.3 Implications for CSCW Design

The objective of this case study is to consider contextual factors into CSCW design. Software designed to support standard procedures is easy to breakdown since the work often is consisted of formal procedure and informal procedure. In this case, we give some implications to CSCW design as following:

(1) Establish a soft-copy of PCRS in computer systems so that some basic information can be retrieved directly from other systems such as student personal information, student examination scores, leave record, etc.

(2) Have access to the PCRS restricted to certain individuals. For example, the students can input some information under their rights, and the supervisors hereby can view the records of their students and sign for their progress. Also, the administrator can collect all the information from it and print out a hard copy.

(3) Informal communication within the Department can be combined with formal communication by building some chatting and massaging functions into the design. A Bulletin Board can be used for research area and topics study.

(4) For informal interaction to occur, people need an environmental mechanism that brings them together in the same place and at the same time. In CSCW design, a place can be set in the environment to support the real time chatting like telephone. The same principle could apply to formal communication. For example, a virtual meeting room can be designed to support real time meeting through visual channels.

5 PROTOTYPE OF THE SYSTEM BY USING WWW TOOLS

Through use of technology and software currently available, a prototype of the CSCW system was established in this small group. The functions of this system include creating and maintaining the soft-copy of the PCRS in the network, finding a person by using the "I Seek You" (ICQ) software, and conducting net meetings through the NetMeeting software.

5.1 Creating Soft-copy of Student Progress Card

A soft-copy of the PCRS was built into the information system. The security of the system is ensured via the use of password access. People having proper authority can access the PCRS to view and modify it. All the modifications are recorded in a file. It prototype is able to automatically send messages to students and supervisors when the deadline for filling in the form comes. It also sends the PCRS to the University's main information database when the filling is completed.

5.2 Finding a Person on the Net

The "I Seek You" (ICQ) software provides useful functions for checking if a person is online. By the use of this tool, supervisors need not call or go by themselves to check if the students are online. They can send message to the students to give them some instructions or arrange a conference rather than using e-mail. This tool also can be used to send files or Internet pages.

5.3 The Procedure of Cooperation

Net conference. Conferences and seminars can be held at distributed sites as long as the sites have computers connected to the Internet. The procedure for organizing a conference is as follows. First, the conference organizer sends the conference notice (including the time, topic and the speaker) to the participants using an ICQ message. When the time comes, he checks if all the participants are online by using ICQ software. He then organizes the conference using the NetMeeting software. All participants can join in the conference. During the time when conference is held, participants can talk and chat freely. The results and all information transacted during the conference can be saved in the computer. The organizer is able to reorganize the saved files of this conference and send them to all the participators using ICQ.

Cooperative work. The cooperative work is supported by computers connected to the Internet. The supervisor can check if the students are online. Then he can ask students to work on the topics in different working sites. By the use of the NetMeeting software, they can simultaneously work on the same piece of document. Finally, they can save their cooperative work in the computer for future usage.

5.4 Group discussion.

A bulletin board in the WWW was set up to support discussion on any topic of interest be it research or coursework related. Supervisors and students can raise a topic or question to be discussed. Issues concerned with the seminars can also be discussed.

6 THE EFFECTS OF PROTOTYPE SYSTEM

The prototype system integrated traditional data and text processing with teleconferencing technology. This turns out a satisfactory result. During a meeting, computer-generated graphics can be used to present the results to the conference participants. The reports of whole meeting can be saved and sent to the individual participants. The cooperative work such as working on the same text report can be accessed, updated, and transmitted as well. The prototype system provides effective support for cooperation in the ISE Department. In general, it contains some advantages and disadvantages described as follows:

Easy to build and use. The system is built by using available technologies and software from the Internet. It provides an efficient way to support cooperative work. For example, by using this system, it becomes easier to gather the right people at meetings, to have better preparation before and during the meetings, and allow better coordination among decentralized sites.

Opportunity enhancement. This small system was established according to the analysis of informal and formal communication that takes place in the Department. It can increase flexibility in frequency and timing of communication between individuals. Some informal communication can also be implemented through this system. In addition, people may get some research work information by informal communication.

The prototype is able to save useful information for research purposes. It also can enhance the opportunities for the students benefiting from working together. Under this system, the displayed information can be dynamically edited and processed, permanent records can be saved, and new information relevant to the discussions can be retrieved for display at any time. Moreover, it provides a private space for individual participants to view relevant private data and to compute the information before it is discussed by the group.

High cost and hardware problem. This small system does not including video transfer function because of the cost problem and transmission speed. The video transmission can add an illusion of physical presence to participants. However, it can also decrease the speed of data transmission. Video is valuable when nonverbal communication, in the form of gestures and facial expressions, is an important part of the discussion and negotiation that takes place (Greif, 1988). Video communication may be an enhancement to the prototype system, but it is less critical than voice communication and the ability to work in the same context.

Another problem is that of hardware. The software used in this system is available through the Internet. However, the software provider does not provide full usability for all versions on different computing platforms. In this respect, the are some limits to the prototype system. It only works in Windows 95/NT operation system. For the time being, the versions for UNIX and Macintosh are not available.

Some negative effects. Some easily arranged meeting can be implemented through telephone or direct face-to-face meeting other than using a lot of unproductive time spent for preparations (Robert and Christine 1988). Even more, since people do not meet face-to-face, lower morale may appear compared with direct people contact. Another problem is that this system uses the technology of remote communication, which means that the freedom of controlling a remote site is deceased.

7 CONCLUSION AND FUTURE PLANS

This case study gives an applicable solution to CSCW designers through analysis of a small work group. The method adopted here can be used for the other small groups in order to provide better support for cooperative work.

For the designers, the informal interaction between individuals should be paid much more attention in order to achieve a pleasant solution to support cooperative work and avoid conflict. Informal communication is difficult to predict and may also break down. Thus, an analysis of human behavior will greatly benefit the design of a CSCW system.

In this study, an application prototype using the WWW to support cooperative activities is provided. Since the has Internet become more and more common, it is very useful to take good use of Net tools to build applications, especially considering that they are comparably easy to extend widely. For the time being, the prototype turns out good results to solve the conflicts and to support cooperation. With time, increasingly useful Internet software will appear to support interactive activities. It will be rather easy to adopt some to make up the functions not provided now.

This prototype provides a good opportunity for studying cooperative work and making good use of Internet tools available. We foresee a large number of Internet applications will appear to give people more support. Two major issues need further research (Dix, 1997). They are: i) which parts of the Web infrastructure need to be modified or extended; and ii) what parts of the Web infrastructure are used by an application. This method of adopting WWW tools into small groups can be used to support large group cooperation. The successful applications, though, require matured communication technologies, hardware and software supporting.

REFERENCES

Argyris, M. (1991). *Cooperation: the Basis of Sociability*, London: Routledge.

Bentley, R., T. Horstmann, and J. Trevor(1997). The World Wide Web as Enabling Technology for CSCW: The Case of BSCW. In *Computer Supported Cooperative Work: The Journal of Cooperative Computing,* **Volume 6** pp. 111-134, Netherlands: Kluwer Academic Publishers.

Dix, A. (1997). Challenges for Cooerative Work on the Web: An Analytical Approach. In *Computer Supported Cooperative Work: The Journal of Cooperative Computing,* **Volume 6** pp. 135-156, Netherlands: Kluwer Academic Publishers.

Gorton, I., I Hawryszkiewyca, and L. Fung (1996). Enabling Software Shift Work with Groupware: A Case Study. In *Proceedings of 19th Hawaii International Conference on System Sciences,* **Volume 3** pp. 72-81, Los Alanitos, CA: IEEE Computer Society Press.

Greif, I. (1988). Computer-based Real-time Conferencing Systems. In *Computer-Supported Cooperative Work: A Book of Readings*, (I. Greif (Ed)), pp. 397-420, San Mateo: Morgan Kaufmann.

Marx, K.(1957). *Capital* (Translated from the 4th German ed. by Eden and Cedar Paul. Introduction by G. D. H. Cole), London : J. M. Dent.

Putnam, L. L. and M. S. Poole (1987). Conflict and Negotiation. In *Handbook of Organizational Communication: An Interdisciplinary Perspective*, (L.W. Porter (Ed)), pp. 549-599, Newbury Park: Sage.

Robert, E. K., S.F. Robert, W. R. Robert, and L. C. Barbara (1990). Informal Communication in Organizations: Form, Function, and Technology. In *People's Reactions to Technology in Factories, Office, and Aerospace*, (S. Oskamp and S. Spacapan (Ed)), pp. 145-199, Sage.

Robert, J., and B. Christine (1988). Thinking ahead: What to Expect from Teleconferencing. In *Computer-Supported Cooperative Work: A Book of Readings*, (I. Greif (Ed)), pp. 185-198, San Mateo: Morgan Kaufmann.

Schmidt, K. and L. Bannon (1991). CSCW: Four Characters in Search of a Context. In *Studies in Computer Supported Cooperative Work*, (J. Bowers and S. Benford (Ed)), pp. 3-16. Amsterdam: North Holland/Elsevier Science Publishers.

AUTHOR INDEX

Printed and bound by CPI Group (UK) Ltd, Croydon, CR0 4YY
10/05/2025
01866278-0001